Grundlehren der mathematischen Wissenschaften 337

A Series of Comprehensive Studies in Mathematics

Vladimir G. Maz'ya
Tatyana O. Shaposhnikova

Theory of Sobolev Multipliers

With Applications to Differential and Integral Operators

Springer

Vladimir Maz'ya
Department of Mathematical Sciences
M&O Building
University of Liverpool
Liverpool L69 3BX
UK

and

Department of Mathematics
Linköping University
SE-581 83 Linköping
Sweden
vlmaz@mai.liu.se

Tatyana Shaposhnikova
Department of Mathematics
Linköping University
SE-581 83 Linköping
Sweden
tasha@mai.liu.se

ISBN: 978-3-540-69490-8 e-ISBN: 978-3-540-69492-2

Library of Congress Control Number: 2008932182

Mathematics Subject Classification Numbers (2000): 26D10, 46E25, 42B25

Cover design: WMX Design GmbH, Heidelberg

Printed on acid-free paper

springer.com

Contents

Introduction

'I never heard of "Uglification," Alice ventured to say. 'What is it?''

Lewis Carroll,
"Alice in Wonderland"

Subject and motivation. The present book is devoted to a theory of multipliers in spaces of differentiable functions and its applications to analysis, partial differential and integral equations. By a multiplier acting from one function space S_1 into another S_2, we mean a function which defines a bounded linear mapping of S_1 into S_2 by pointwise multiplication. Thus with any pair of spaces S_1, S_2 we associate a third one, the space of multipliers $M(S_1 \to S_2)$ endowed with the norm of the operator of multiplication. In what follows, the role of the spaces S_1 and S_2 is played by Sobolev spaces, Bessel potential spaces, Besov spaces, and the like.

The Fourier multipliers are not dealt with in this book. In order to emphasize the difference between them and the multipliers under consideration, we attach Sobolev's name to the latter. By coining the term Sobolev multipliers we just hint at various spaces of differentiable functions of Sobolev's type, being fully aware that Sobolev never worked on multipliers. After all, Fourier never did either.

Sobolev multipliers arise in many problems of analysis and theories of partial differential and integral equations. Coefficients of differential operators can be naturally considered as multipliers. The same is true for symbols of more general pseudo-differential operators. Multipliers also appear in the theory of differentiable mappings preserving Sobolev spaces. Solutions of boundary value problems can be sought in classes of multipliers. Because of their algebraic properties, multipliers are suitable objects for generalizations of the basic facts of calculus (theorems on implicit functions, traces and extensions, point mappings and their compositions etc.) Moreover, some basic operators

V.G. Maz'ya, T.O. Shaposhnikova, *Theory of Sobolev Multipliers*,
Grundlehren der mathematischen Wissenschaften 337,

of harmonic analysis, like the classical maximal and singular integral operators, act in certain classes of multipliers.

We believe that the calculus of Sobolev multipliers provides an adequate language for future work in the theory of linear and nonlinear differential and pseudodifferential equations under minimal restrictions on the coefficients, domains, and other data.

Before the 1970s, the word *multiplier* was usually associated with the name of Fourier, and a deep theory of L_p-Fourier multipliers created by Marcinkiewicz, Mikhlin, Hörmander *et al* was quite popular. As for the multipliers preserving a space of differentiable functions, only a few isolated results were known (Devinatz and Hirschman [DH], Hirschman [Hi1], [Hi2], Strichartz [Str], Polking [Pol1], Peetre [Pe2]), while the multipliers in pairs of such spaces were not considered at all.

The first (and the only one for the time being) attempt to work out a more or less comprehensive theory of multipliers acting either in one or in a pair of spaces of Sobolev type was undertaken by the authors in the late 1970s and early 1980s [Maz10], [Maz12], [MSh1]–[MSh16]. Results of that theory were collected in our monograph "Theory of Multipliers in Spaces of Differentiable Functions" (Pitman, 1985) [MSh16]. During the last two decades, we continued to work in the area, adding new results and developing further applications [Sh2]–[Sh14], [MSh17]–[MSh23]. We wish to reflect the present state of our theory in this book. An essential part of the aforementioned monograph is also included here.

No results concerning multipliers in spaces of analytic functions are mentioned in what follows, in contrast to [MSh16]. To describe progress in this area achieved during the last twenty five years would require a disproportionate growth of the book.

Structure of the book. The book consists of two parts. Part I is devoted to the theory of multipliers and covers the following topics:

- Trace inequalities
- Analytic characterization of multipliers
- Relations between spaces of Sobolev multipliers and other function spaces
- Maximal subalgebras of multiplier spaces
- Traces and extensions of multipliers
- Essential norm and compactness of multipliers
- Miscellaneous properties of multipliers (spectrum, composition and implicit function theorems, point mappings preserving Sobolev spaces, etc.)

In Part II we dwell upon several applications of this theory. Their list is as follows:

- Continuity and compactness of differential operators in pairs of Sobolev spaces
- Multipliers as solutions to linear and quasilinear elliptic equations

- Higher regularity in the single and double layer potential theory for Lipschitz domains
- Regularity of the boundary in L_p-theory of elliptic boundary value problems
- Singular integral operators in Sobolev spaces

Each chapter starts with a short introductory outline of the included material.

Readership. The volume is addressed to mathematicians working in functional analysis and in the theories of partial differential, integral, and pseudodifferential operators. Prerequisites for reading this book are undergraduate courses in these subjects.

Acknowledgments. V. Maz'ya was partially supported by the National Science Foundation (Grant DMS-0500029, USA) and EPSRC (Grant EP/F005563/1, UK). T. Shaposhnikova gratefully acknowledges support from the Swedish Natural Science Research Council (VR).

Part I

Description and Properties of Multipliers

1

Trace Inequalities for Functions in Sobolev Spaces

In this chapter we characterize the best constant C in the so-called trace inequality

$$\Big(\int |\nabla_l u|^p d\mu\Big)^{1/p} \le C\, \|u\|_{W_p^m}, \qquad u \in C_0^\infty, \tag{1.0.1}$$

with an arbitrary measure μ on the left-hand side. When the domain is not indicated in the notation of a space or a norm, then it is assumed to be $\mathbb{R}^n$. Another variant of (1.0.1) will be with W_p^m replaced by w_p^m. Here W_p^k and w_p^k are completions of the space C_0^∞ with respect to the norms $\|\nabla_k u\|_{L_p} + \|u\|_{L_p}$ and $\|\nabla_k u\|_{L_p}$, $\nabla_k = \{\partial^k/\partial x_1^{\alpha_1} \cdots \partial x_n^{\alpha_n}\}$. Two-sided estimates for C are given in different terms for $p = 1$ (Sect. 1.1) and for $p \in (1, \infty)$ (Sect. 1.2). The last Sect. 1.3 concerns the case of p on the left-hand side of (1.0.1) replaced by $q \ne p$.

In what follows, we denote by c, c_1, c_2 various positive constants which depend only on m, l, p, n and similar parameters. The values a and b are called equivalent ($a \sim b$) if

$$c_1\, a \le b \le c_2\, a.$$

Here and henceforth Ω is an open set in $\mathbb{R}^n$ and $C_0^\infty(\Omega)$ is the space of infinitely differentiable functions with compact supports in Ω, and $\mathcal{B}_r(x) = \{y \in \mathbb{R}^n : |y - x| < r\}$, $\mathcal{B}_r = \mathcal{B}_r(0)$.

1.1 Trace Inequalities for Functions in w_1^m and W_1^m

We start with the results concerning $p = 1$ obtained in [Maz11], see also [Maz14], Sect. 1.4.

1.1.1 The Case $m = 1$

The following lemma gives a representation of the n-dimensional variation of a function as an integral of the area of a level surface.

Lemma 1.1.1. *Let $u \in C^\infty(\Omega)$ and let*

$$N_t = \{x \in \Omega : |u(x)| \geq t\}.$$

Then

$$\int_\Omega |\nabla u(x)| dx = \int_0^\infty s(\Omega \cap \partial N_t) dt, \tag{1.1.1}$$

where s is the $(n-1)$-dimensional area (It is well known that ∂N_t is a smooth $(n-1)$-dimensional manifold for almost all $t > 0$, see [Mo].)

For more general classes of functions, (1.1.1) was proved in [Kr] for $n = 2$ and in [Fe1].

We give a simple proof of (1.1.1) for $u \in C^\infty(\Omega)$.

Proof. Let $w = (w_1, \cdots, w_n), w_j \in C_0^\infty(\Omega)$. Integrating by parts, we obtain

$$\int_\Omega w \nabla u \, dx = \int_\Omega u \text{ div } w \, dx = -\int_{u>0} u \text{ div } w \, dx - \int_{u<0} u \text{ div } w \, dx.$$

From the definition of the Lebesgue integral we get

$$\int_{u>0} u \text{ div } w \, dx = \int_0^\infty dt \int_{u>t} \text{div } w \, dx.$$

For almost all $t > 0$

$$\int_{u>t} \text{div } w \, dx = -\int_{u=t} w\nu \, ds = -\int_{u=t} \frac{w \nabla u}{|\nabla u|} ds,$$

where ν is the inward normal to $\{x : u(x) \geq t\}$. Therefore

$$\int_{u>0} u \text{ div } w \, dx = -\int_0^\infty \int_{u=t} \frac{w \nabla u}{|\nabla u|} ds \, dt.$$

The transformation of the integral

$$\int_{u<0} u \text{ div } w \, dx$$

is quite similar. Thus,

$$\int_\Omega w \nabla u \, dx = \int_0^\infty dt \int_{\Omega \cap \partial N_t} \frac{w \nabla u}{|\nabla u|} ds.$$

In this identity, we set

$$w = \varphi_j \frac{\nabla u}{(|\nabla u|^2 + j^{-1})^{1/2}},$$

where $j = 1, 2, \cdots$ and $\{\varphi_j\}$ is a non-decreasing sequence of nonnegative functions in $C_0^\infty(\Omega)$, convergent to one in Ω. Then

$$\int_\Omega \varphi_j \frac{(\nabla u)^2 dx}{(|\nabla u|^2 + j^{-1})^{1/2}} = \int_0^\infty dt \int_{\Omega \cap \partial N_t} \frac{\varphi_j \, |\nabla u| ds}{(|\nabla u|^2 + j^{-1})^{1/2}}.$$

Passing to the limit as $j \to \infty$ and using the monotone convergence theorem, we obtain (1.1.1). The proof is complete. □

Corollary 1.1.1. *Let $u \in C^\infty(\Omega)$ and let Φ be a nonnegative lower semi-continuous function on Ω. Then*

$$\int_\Omega \Phi(x) |\nabla u(x)| \, dx = \int_0^\infty dt \int_{\Omega \cap \partial N_t} \Phi(x) ds_x.$$

Proof. The result follows from the chain of identities:

$$\begin{aligned} \int_\Omega \Phi(x)|\nabla u(x)| dx &= \int_0^\infty \int_{\Phi(x) > \rho} |\nabla u(x)| \, dx \, d\rho \\ &= \int_0^\infty \int_0^\infty s(\{x \in \partial N_t : \Phi(x) > \rho\}) d\rho \, dt \\ &= \int_0^\infty dt \int_{\Omega \cap \partial N_t} \Phi(x) ds_x. \end{aligned}$$

□

Formula (1.1.1) leads to a relation between the estimate

$$\int |u| \, d\mu \leq C \|\nabla u\|_{L_1}, \quad u \in C_0^\infty, \tag{1.1.2}$$

and an isoperimetric inequality. Namely, we have the following assertion.

Lemma 1.1.2. *The exact constant C in (1.1.2) is equal to*

$$\sup_g \frac{\mu(g)}{s(\partial g)}, \tag{1.1.3}$$

where g is any open set in $\mathbb{R}^n$ with compact closure and smooth boundary.

Proof. We have

$$\int |u| \, d\mu = \int_0^\infty \mu(N_t) \, dt \leq \sup_g \frac{\mu(g)}{s(\partial g)} \int_0^\infty s(\partial N_t) \, dt,$$

which, together with Lemma 1.1.1, gives the upper bound for C.

Let $\delta(x) = \operatorname{dist}(x, g)$ and $g_t = \{x : \delta(x) < t\}$. It is well known that there exists a small $\epsilon > 0$ such that the surface ∂g_t is smooth for $t \leq \epsilon$. We substitute the function $u_\epsilon(x) = \alpha[\delta(x)]$, where $\alpha \in C^\infty([0,\infty))$ and $\alpha(0) = 1$, $\alpha(t) = 0$ for $t > \epsilon$, into (1.1.2). According to Corollary 1.1.1,

$$\int |\nabla u_\epsilon|\,dx = \int_0^\epsilon \alpha'(t) s(\partial g_t)\,dt.$$

Since $s(\partial g_t) \to s(\partial g)$ as $t \to +0$, it follows that

$$\int |\nabla u_\epsilon|\,dx \to s(\partial g). \tag{1.1.4}$$

Also,

$$\int |u_\epsilon|\,d\mu \geq \mu(g). \tag{1.1.5}$$

Combining (1.1.4), (1.1.5) and (1.1.2), we obtain $\mu(g) \leq C s(\partial g)$.

The following more general assertion, which will be used in Sect. 5.1.1, is proved in the same way.

Proposition 1.1.1. *The best constant C in*

$$\int |u|\,d\mu \leq C\|\Phi\,\nabla u\|_{L_1},$$

where $\Phi \in C(\mathbb{R}^n)$ and u is an arbitrary function in $C_0^\infty(\mathbb{R}^n)$, is equal to

$$\sup_g \frac{\mu(g)}{\displaystyle\int_{\partial g} \Phi(x) ds_x}.$$

Here g is any open set in $\mathbb{R}^n$ with compact closure, bounded by a smooth surface, as in Lemma 1.1.2.

Further, we prove that

$$\sup_g \frac{\mu(g)}{s(\partial g)} \sim \sup_{x\in\mathbb{R}^n, r>0} r^{1-n}\mu(\mathcal{B}_r(x)). \tag{1.1.6}$$

With this aim in view, we present certain known auxiliary assertions. We start with the formulation of the classical Besicovitch covering theorem (see [Guz]).

Lemma 1.1.3. *Let E be a bounded set in $\mathbb{R}^n$ and let $\mathcal{B}_{r(x)}(x)$ be a ball with $r(x) > 0$ and $x \in E$. By L we denote the totality of these balls. Then one can choose a sequence of balls $\{\mathcal{B}^{(m)}\}$ from L such that*

(i) $E \subset \bigcup_m \mathcal{B}^{(m)}$;

(ii) there exists a number N, depending only on the dimension of the space, such that every point of the space belongs to at most N balls from $\{\mathcal{B}^{(m)}\}$;

(iii) the balls $(1/3)\mathcal{B}^{(m)}$ are disjoint.

We present one more well-known geometric lemma (see [Fe2]).

Lemma 1.1.4. *Let g be an open subset of $\mathbb{R}^n$ with a smooth boundary such that*

$$2\,\mathrm{mes}_n(\mathcal{B}_r \cap g) = \mathrm{mes}_n(\mathcal{B}_r).$$

Then

$$s(\mathcal{B}_r \cap \partial g) \geq c_n r^{n-1},$$

where c_n is a positive constant depending only on n.

Proof. Let χ and ψ be characteristic functions of the sets $g \cap \mathcal{B}_r$ and $\mathcal{B}_r \backslash g$. For any vector $z \neq 0$ we introduce the projection mapping p_z onto a $(n-1)$-dimensional subspace orthogonal to z. By Fubini's theorem,

$$\begin{aligned}(1/4)\big(\mathrm{mes}_n(\mathcal{B}_1)r^n\big)^2 &= \mathrm{mes}_n(g \cap \mathcal{B}_r)\,\mathrm{mes}_n(\mathcal{B}_r \backslash g)\\ &= \int\!\!\int \chi(x)\psi(y)dxdy = \int\!\!\int \chi(x)\psi(x+z)dzdx\\ &= \int_{|z|\leq 2r} \mathrm{mes}_n(\{x : x \in \mathcal{B}_r \cap g, (x+z) \in \mathcal{B}_r \backslash g\})dz.\end{aligned}$$

Since any segment which joins $x \in g \cap \mathcal{B}_r$ with $(x+z) \in \mathcal{B}_r \backslash g$ intersects $\mathcal{B}_r \cap \partial g$, the last integral does not exceed

$$2r \int_{|z|\leq 2r} \mathrm{mes}_{n-1}[p_z(\mathcal{B}_r \cap \partial g)]dz \leq (2r)^{n+1}\mathrm{mes}_n(\mathcal{B}_1)\, s(\mathcal{B}_r \cap \partial g).$$

The result follows. □

The following covering lemma is due to Gustin [Gus]. We give here the proof found by Federer [Fe2].

Lemma 1.1.5. *Let g be a bounded open subset of $\mathbb{R}^n$ with smooth boundary. There exists a covering of g by a sequence of balls with radii ρ_i, $i = 1, 2, \cdots$, such that*

$$\sum_j \rho_j^{n-1} \leq c\, s(\partial g), \tag{1.1.7}$$

where c is a constant which depends only on n.

Proof. Each point $x \in g$ is the center of a ball $\mathcal{B}_r(x)$ for which

$$\frac{\mathrm{mes}_n(\mathcal{B}_r(x) \cap g)}{\mathrm{mes}_n(\mathcal{B}_r(x))} = \frac{1}{2}. \tag{1.1.8}$$

(This ratio is a continuous function of r equal to 1 for small r and tending to zero as $r \to \infty$.) By Lemma 1.1.3, there exists a sequence of disjoint balls $\mathcal{B}_{r_j}(x_j)$ for which

$$g \subset \bigcup_{j=1}^{\infty} \mathcal{B}_{3r_j}(x_j).$$

From Lemma 1.1.4 and (1.1.8) we get

$$s(\mathcal{B}_{r_j}(x_j) \cap \partial g) \geq c_n r_j^{n-1}.$$

Therefore

$$s(\partial g) \geq \sum_j s(\mathcal{B}_{r_j}(x_j) \cap \partial g) \geq 3^{1-n} c_n \sum_j (3r_j)^{n-1}.$$

Thus, $\{\mathcal{B}_{r_j}(x_j)\}$ is the required covering. □

Corollary 1.1.2. *The best constant in* (1.1.2) *is equivalent to*

$$K = \sup_{x \in \mathbb{R}^n, r>0} r^{1-n} \mu(\mathcal{B}_r(x)).$$

Proof. By Lemma 1.1.2, it is sufficient to show that $\mu(g) \leq c\, K s(\partial g)$ for any admissible set g. Let $\{\mathcal{B}_{\rho_j}(x_j)\}$ be a covering of g constructed in Lemma 1.1.5. It is clear that

$$\mu(g) \leq \sum_j \mu(\mathcal{B}_{\rho_j}(x_j)) \leq K \sum_j \rho_j^{n-1} \leq c\, K\, s(\partial g).$$

The proof is complete. □

1.1.2 The Case $m \geq 1$

Theorem 1.1.1. *Let m and l be integers with $m \geq l \geq 0$. Then the best constant in*

$$\int |\nabla_l u|\, d\mu \leq C \|u\|_{w_1^m}, \quad u \in C_0^\infty, \tag{1.1.9}$$

is equivalent to

$$K = \sup_{x \in \mathbb{R}^n, r>0} r^{m-l-n} \mu(\mathcal{B}_r(x)). \tag{1.1.10}$$

Proof. (*i*) We start with the estimate $C \geq c\, K$, setting

$$u(\xi) = (x_1 - \xi_1)^l \, \varphi(r^{-1}(x - \xi))$$

in (1.1.9), where $\varphi \in C_0^\infty(\mathcal{B}_2)$ and $\varphi = 1$ on $\mathcal{B}_1$. Since

$$\int |\nabla_l u| d\mu \geq l!\, \mu(\mathcal{B}_r(x)), \quad \|\nabla_m u\|_{L_1} = c\, r^{n-m+l},$$

it follows that $C \geq c\, K$.

(*ii*) Now we establish the estimate $C \leq c\, K$. Let us start with the case $l = 0$. We have

$$\int |u|\,d\mu(x) = c\int \left|\int \frac{(\xi - x)\nabla_\xi u(\xi)}{|\xi - x|^n} d\xi\right| d\mu(x) \le c\int |\nabla u|\,g\,dx, \qquad (1.1.11)$$

where

$$g(x) = \int \frac{d\mu(y)}{|x-y|^{n-1}}.$$

We argue by induction on m. For $m = 1$ the result is contained in Corollary 1.1.2. The last integral in (1.1.11) does not exceed

$$c \sup_{x\in\mathbb{R}^n, r>0} \Big(r^{m-n-1}\int_{\mathcal{B}_r(x)} g(\xi)d\xi\Big) \|\nabla u\|_{w_1^{m-1}}.$$

Clearly,

$$\int_{\mathcal{B}_r(x)} g(\xi)d\xi = \int_{\mathcal{B}_r(x)} d\xi \int_{\mathcal{B}_{2r}(x)} |\xi-\sigma|^{1-n} d\mu(\sigma)$$
$$+ \int_{\mathcal{B}_r(x)} d\xi \int_{\mathbb{R}^n\setminus\mathcal{B}_{2r}(x)} |\xi-\sigma|^{1-n} d\mu(\sigma).$$

The first integral on the right-hand side is majorized by $cr\mu(\mathcal{B}_{2r}(x))$ and the second one is not greater than

$$cr^n \int_{\mathbb{R}^n\setminus\mathcal{B}_{2r}(x)} |x-\sigma|^{1-n} d\mu(\sigma) = c\,(n-1)r^n \int_{2r}^{\infty} \mu\{\sigma : 2r \le |x-\sigma| < t\} t^{-n} dt.$$

Thus

$$r^{m-n-1}\int_{\mathcal{B}_r(x)} g(\xi)\,d\xi \le c \sup_{x\in\mathbb{R}^n, r>0} r^{m-n}\mu(\mathcal{B}_r(x)).$$

For $l \ge 1$ the result follows by induction. □

Remark 1.1.1. It is clear that for $m - l > n$ the finiteness of (1.1.10) means that $\mu = 0$. In the case $m - l = n$, the value (1.1.10) is equal to $\mu(\mathbb{R}^n)$.

We give an analogue of Theorem 1.1.1 for the space W_1^m.

Theorem 1.1.2. *Let m and l be integers, $m \ge l \ge 0$. Then the best constant in*

$$\int |\nabla_l u|\,d\mu \le C\|u\|_{W_1^m}, \quad u \in C_0^\infty, \qquad (1.1.12)$$

is equivalent to

$$K = \sup_{x\in\mathbb{R}^n, r\in(0,1]} r^{m-l-n}\mu(\mathcal{B}_r(x)). \qquad (1.1.13)$$

Proof. The estimate $C \ge c\,K$ is obtained in precisely the same way as the analogous one in Theorem 1.1.1. To prove the converse inequality, we introduce a partition of unity $\{\varphi_j\}_{i\ge1}$ subordinate to the covering of $\mathbb{R}^n$ by unit balls with centers in nodes of a sufficiently small coordinate grid. We apply Theorem 1.1.1 to the integral

$$\int |\nabla_l(\varphi_j u)| d\mu_j,$$

where μ_j is the restriction of μ to the support of the function φ_j. Then

$$\int |\nabla_l u| d\mu \leq \sum_j \int |\nabla_l(\varphi_j u)| d\mu_j \leq c\, K \sum_j \|\varphi_j u\|_{w_1^m}$$
$$\leq c\, K \|u\|_{W_1^m}.$$

□

Remark 1.1.2. Obviously, in the case $m - l \geq n$ the value K defined in (1.1.13) is equal to

$$\sup_{x \in \mathbb{R}^n} \mu(\mathcal{B}_1(x)).$$

1.2 Trace Inequalities for Functions in w_p^m and W_p^m, $p > 1$

1.2.1 Preliminaries

In this and subsequent chapters we often use operators of the form

$$k(D) = F^{-1} k(\xi) F,$$

where F is the Fourier transform in $\mathbb{R}^n$ and k is a function or a vector-valued function which is called the symbol. In particular, $D = -i\nabla$ and $D^\alpha = (-i)^{|\alpha|} \partial x_1^{\alpha_1} \cdots \partial x_n^{\alpha_n}$.

The following assertion is a variant of the Mikhlin theorem on Fourier multipliers (see [Liz]).

Lemma 1.2.1. *Let the function k and its derivatives $\partial^m k / \partial\xi_{j_1} \cdots \partial\xi_{j_m}$, where $0 \leq j_1 + \cdots + j_m = m \leq n$ and $j_1, \cdots, j_m$ are distinct, be continuous on the set $\{\xi \in \mathbb{R}^n : \xi_1 \cdots \xi_n \neq 0\}$ and let*

$$\left| \xi_{j_1} \cdots \xi_{j_m} \frac{\partial^m k}{\partial \xi_{j_1} \cdots \partial \xi_{j_m}} \right| \leq \text{const.} \tag{1.2.1}$$

Then the operator $k(D)$ is continuous in L_p, $p \in (1, \infty)$.

In particular, the singular integral operator with a symbol $k \in C^n(\mathbb{R}^n \backslash 0)$ is continuous in L_p.

In what follows, the operators $(-\Delta)^{r/2}$ and $(1 - \Delta)^{s/2}$ with the symbols $|\xi|^r$ and $(1 + |\xi|^2)^{s/2}$, $r > -n$, $s \in \mathbb{R}^1$, play an important role.

Lemma 1.2.2. *Let* $1 = 1, 2, \cdots$, $p \in (1, \infty)$. *There exists a constant* $c > 1$, *depending only on* n, p, l, *such that*

$$c^{-1}\|(-\Delta)^{l/2}u\|_{L_p} \le \|\nabla_l u\|_{L_p} \le c\,\|(-\Delta)^{l/2}u\|_{L_p} \qquad (1.2.2)$$

for all $u \in C_0^\infty$.

Proof. Let α be a multi-index of order l. Then

$$F^{-1}\xi^\alpha F u = F^{-1}\xi^\alpha|\xi|^{-l}|\xi|^l F u.$$

The function $\xi^\alpha|\xi|^{-l}$ satisfies the condition of Lemma 1.2.1 which implies the right inequality in (1.2.2). Also,

$$|\xi|^l = |\xi|^{2l}|\xi|^{-l} = \Big(\sum_{|\alpha|=l} c_\alpha \xi^\alpha \xi^\alpha\Big)|\xi|^{-l},$$

where $c_\alpha = l!/\alpha!$. Therefore,

$$F^{-1}|\xi|^l F u = \sum_{|\alpha|=l} c_\alpha F^{-1}\frac{\xi^\alpha}{|\xi|^l}\xi^\alpha F u.$$

Applying Lemma 1.2.1 once again, we obtain the left part of (1.2.2). □

The following assertion has a similar proof.

Lemma 1.2.3. *Let* $l = 1, 2, \cdots, p \in (1, \infty)$. *There exists a constant* $c > 1$, *depending only on* n, p, l, *such that*

$$c^{-1}\|(1-\Delta)^{l/2}u\|_{L_p} \le \|u\|_{W_p^l} \le c\,\|(1-\Delta)^{l/2}u\|_{L_p} \qquad (1.2.3)$$

for all $u \in C_0^\infty$.

The operator $I_l := (-\Delta)^{-l/2}$ is the integral convolution operator with the kernel $c\,|x|^{l-n}$, $c =$ const., for $l \in (0, n)$. It is usually called the Riesz potential of order l.

For $l > 0$ the operator $(1-\Delta)^{-l/2}$ has the representation

$$(1-\Delta)^{-l/2} f = G_l * f,$$

where G_l is the function with the Fourier transform $(1+|\xi|^2)^{-l/2}$.

The function G_l can be written in the form

$$G_l(x) = c\int_0^\infty e^{-t-x^2/4t}\, t^{-n/2+l/2-1}dt$$

or in the form

$$G_l(x) = c\,K_{(n-l)/2}(|x|)|x|^{(l-n)/2},$$

where K_γ is the modified Bessel function of the third kind.

The function G_l is positive and decreases with the growth of $|x|$. It satisfies the following asymptotic estimates. For $|x| \to 0$,

$$G_l(x) \sim \begin{cases} |x|^{l-n}, & 0 < l < n, \\ \log |x|^{-1}, & l = n, \\ 1, & l > n. \end{cases} \tag{1.2.4}$$

For $|x| \to \infty$ the following relation holds:

$$G_l(x) \sim |x|^{(l-n-1)/2} \, e^{-|x|}. \tag{1.2.5}$$

The integral operator

$$F \xrightarrow{J_l} G_l * f$$

is called the Bessel potential of order l.

For properties of Riesz and Bessel potentials see [AMS], [St2], [Str].

We introduce the maximal Hardy–Littlewood operator $\mathcal{M}$ defined by

$$(\mathcal{M}f)(x) = \sup_{r>0} \frac{1}{\mathrm{mes}_n \mathcal{B}_r} \int_{\mathcal{B}_r(x)} |f(y)| dy.$$

By the Hardy–Littlewood theorem (see [St2]), the operator $\mathcal{M}$ is bounded in L_p, $p \in (1, \infty)$.

1.2.2 The (p, m)-Capacity

We define the (p, m)-capacity of a compact set $e \subset \mathbb{R}^n$ by

$$C_{p,m}(e) = \inf\{\|f\|^p_{L_p} : f \in L_p, \ f \geq 0, \ J_m f \geq 1 \text{ on } e\}, \tag{1.2.6}$$

where J_m is the Bessel potential of order m. This capacity satisfies

$$C_{p,m}(e) = \inf\{\|(1 - \varDelta)^{m/2} u\|^p_{L_p} : u \in C_0^\infty, \ u \geq 1 \text{ on } e\} \tag{1.2.7}$$

(see Meyers [Me]). In view of the boundedness of the singular integral operator in L_p,

$$C_{p,m}(e) \sim \inf\{\|u\|^p_{W_p^m} : u \in C_0^\infty, u \geq 1 \text{ on } e\}. \tag{1.2.8}$$

Replacing the Bessel potential J_m in (1.2.6) by the Riesz potential I_m, we obtain the definition of the capacity $c_{p,m}(e)$. These and analogous set functions used in this book have been the subject of active study (see Maz'ya [Maz7], Maz'ya, Havin [MH1], [MH2], Meyers [Me], Hedberg [Hed2], Adams, Meyers [AM], Sjödin [Sj], Adams, Hedberg [AH]).

We describe certain simple properties of the capacities $C_{p,m}(e)$ and $c_{p,m}(e)$ which will be used in this chapter.

Proposition 1.2.1. *The capacities $C_{p,m}(e)$ and $c_{p,m}(e)$ are non-decreasing functions of the set e.*

The proof is obvious.

Proposition 1.2.2. *If $mp > n$, then*

$$C_{p,m}(e) \sim 1$$

for all compact sets $e \neq \emptyset$ with diameter less than one.

Proof. Obviously, $C_{p,m}(e) \leq C_{p,m}(\mathcal{B}_1)$. On the other hand by Sobolev's theorem on the imbedding $W_p^m \subset L_\infty$, we have

$$c \, \|u\|_{W_p^m} \geq \|u\|_{L_\infty} \geq 1$$

for any function $u \in C_0^\infty$ which exceeds one on e. Consequently,

$$C_{p,m}(e) \geq c^{-p}.$$

□

Proposition 1.2.3. *If $mp < n, p \in (1, \infty)$, then*

$$c_{p,m}(e) \geq c \, (\text{mes}_n e)^{(n-mp)/n}. \tag{1.2.9}$$

The proof follows from the definition of the capacity and from Sobolev's inequality $\|u\|_{L_q} \leq c \, \|u\|_{w_p^m}$, where $q = pn/(n - mp)$, $u \in C_0^\infty$. □

To prove an estimate similar to (1.2.9) in the case $mp = n$ we need the following known assertion (see Yudovič [Yu], Pohozhaev [Poh1], Trudinger [Tru]) which is given here with the proof for the reader's convenience.

Lemma 1.2.4. *If $mp = n$ and $p \in (1, \infty)$, then*

$$\int \Phi \left(c \frac{|u|^{p'}}{\|u\|_{W_p^m}^{p'}} \right) dx \leq 1 \tag{1.2.10}$$

for all $u \in C_0^\infty$, where c is a constant independent of u, $p + p' = pp'$, and

$$\Phi(t) = e^t - \sum_{j=0}^{[p]} t^j / j!.$$

Proof. Let $u = J_m f = G_m * f$. By Lemma 1.2.3, it suffices to give the proof under the assumption $\|f\|_{L_p} = 1$. Obviously,

$$\int \Phi(c \, |u|^{p'}) dx = \sum_{j=[p]+1}^{\infty} \frac{c^j}{j!} \|u\|_{L_{p'j}}^{p'j}. \tag{1.2.11}$$

By Young's inequality for $q \geq p$,

$$\|u\|_{L_q} \leq \|G_m\|_{L_s} \|f\|_{L_p}, \quad s = \frac{qp'}{q + p'}, \quad p + p' = pp'. \tag{1.2.12}$$

Using the estimates (1.2.4) and (1.2.5) for the function G_m, one can show that

$$\|G_m\|_{L_s}^s \leq c_0 q, \tag{1.2.13}$$

where $c_0 = c_0(p, n)$. From (1.2.12), (1.2.13), where $q = p'j$ and $s = p'j/(j+1)$, it follows that the right-hand side of (1.2.11) does not exceed

$$\sum_{j=[p]+1}^{\infty} c^j (c_0 p' j)^{j+1}/j!. \tag{1.2.14}$$

This series converges if $c\,c_0\,p'e < 1$. Diminishing c, one can make the sum (1.2.14) arbitrarily small. □

Proposition 1.2.4. *If $mp = n, p \in (1, \infty)$ and $d(e) \leq 1$, then*

$$C_{p,m}(e) \geq c \left(\log \frac{2^n}{\text{mes}_n e} \right)^{1-p}. \tag{1.2.15}$$

Proof. Let $u \in C_0^\infty$, $u \geq 1$ on e. It follows from (1.2.10) that

$$\varPhi\big(c\,\|u\|_{W_p^m}^{-p'}\big)\,\text{mes}_n e \leq 1.$$

Hence

$$\varPhi(c\,[C_{p,m}(e)]^{1/(1-p)}) \leq (\text{mes}_n e)^{-1}. \tag{1.2.16}$$

Since the argument of the function $\varPhi$ in (1.2.16) is bounded away from zero, we have

$$\exp(c\,[C_{p,m}(e)]^{1/(1-p)}) \leq c_0 (\text{mes}_n e)^{-1}.$$

□

Proposition 1.2.5. *If $mp < n$, then*

$$c_{p,m}(\mathcal{B}_r) = c\,r^{n-mp}.$$

Proof. Using the dilation, we obtain

$$c_{p,m}(\mathcal{B}_r) = r^{n-mp}\,c_{p,m}(\mathcal{B}_1).$$

□

Proposition 1.2.6. *If $mp < n$ and $0 < r \leq 1$, then*

$$C_{p,m}(\mathcal{B}_r) \sim r^{n-mp}.$$

Proof. The lower bound for the capacity follows from (1.2.9). The upper one is obtained after the substitution of the function $x \to \eta(x/r)$ into the norm $\|u\|_{W_p^m}$, where $\eta \in C_0^\infty$ and $\eta = 1$ on the ball $\mathcal{B}_1$. □

Proposition 1.2.7. *If* $mp = n$, $p \in (1, \infty)$, *and* $0 < r \leq 1$, *then*

$$C_{p,m}(\mathcal{B}_r) \sim (\log 2/r)^{1-p}.$$

Proof. The lower bound for the capacity follows from (1.2.15). Let us justify the upper bound. We introduce the function

$$v(x) = (\log 2/r)^{-1} \log 2/|x|$$

and by α denote a function in the space $C^\infty(\mathbb{R}^1)$ such that $\alpha(t) = 0$ for $t < 0$, $\alpha(t) = 1$ for $t > 1$. Further, let $u(x) = \alpha[v(x)]$. Clearly, $u \in C_0^\infty(\mathcal{B}_2)$ and $u = 1$ on $\mathcal{B}_r$. Moreover, one can check that

$$|\nabla_m u(x)| \leq c\,(\log 2/r)^{-1}|x|^{-m}$$

on $\mathcal{B}_2 \backslash \mathcal{B}_r$. This implies that

$$\begin{aligned} C_{p,m}(\mathcal{B}_r) &\leq c\,\|\nabla_m u; \mathcal{B}_2\|_{L_p}^p \\ &\leq c\,(\log 2/r)^{-p} \int_{\mathcal{B}_2 \backslash \mathcal{B}_r} |x|^{-mp} dx = c\,(\log 2/r)^{1-p}. \end{aligned}$$

1.2.3 Estimate for the Integral of Capacity of a Set Bounded by a Level Surface

The following assertion is proved in [Hed3].

Lemma 1.2.5. *Let* $0 < \theta < 1$, $0 < r < n$ *and let* $I_r f$ *be the Riesz potential of order* r *with a nonnegative density* f. *Then*

$$(I_{r\theta} f)(x) \leq c\,\big((I_r f)(x)\big)^\theta \big((\mathcal{M} f)(x)\big)^{1-\theta}, \tag{1.2.17}$$

where $\mathcal{M}$ *is the Hardy–Littlewood maximal operator.*

Proof. Let t be an arbitrary positive number to be chosen later. We use the equality

$$\begin{aligned} \int_{\mathcal{B}_t(x)} \frac{f(y)dy}{|x-y|^{n-r\theta}} &= (n - r\theta) \int_0^t \int_{\mathcal{B}_s(x)} f(y)\,dy \frac{ds}{s^{n-r\theta+1}} \\ &\quad + t^{r\theta - n} \int_{\mathcal{B}_t(x)} f(y)\,dy \end{aligned} \tag{1.2.18}$$

which is checked by changing the order of integration on the right-hand side. Hence

$$\int_{\mathcal{B}_t(x)} \frac{f(y)dy}{|x-y|^{n-r\theta}} \leq c\,t^{r\theta} (\mathcal{M} f)(x). \tag{1.2.19}$$

Clearly we have

$$\int_{\mathbb{R}^n\setminus\mathcal{B}_t(x)} \frac{f(y)dy}{|x-y|^{n-r\theta}} \leq t^{r(\theta-1)} \int_{\mathbb{R}^n\setminus\mathcal{B}_t(x)} \frac{f(y)dy}{|x-y|^{n-r}}$$

$$\leq t^{r(\theta-1)}(I_r f)(x). \tag{1.2.20}$$

Adding this inequality to (1.2.19), we obtain

$$(I_{r\theta} f)(x) \leq c\, t^{r\theta}(\mathcal{M}f)(x) + t^{r(\theta-1)}(I_r f)(x).$$

Minimization of the right-hand side in t completes the proof. □

Corollary 1.2.1. *Let m be an integer, $0 < m < n$, $I_m f = |x|^{m-n} * f$ with $f \geq 0$ and let F be a function in $C^m(0,\infty)$ such that*

$$t^{k-1}|F^{(k)}(t)| \leq Q, \qquad k = 0, 1, \cdots, m, \quad Q = \text{const}.$$

Then

$$|\nabla_m F(I_m f)| \leq c Q(\mathcal{M}f + |\nabla_m I_m f|) \tag{1.2.21}$$

almost everywhere in $\mathbb{R}^n$.

Proof. Let $u = I_m f$. One can verify by induction that

$$|\nabla_m F(u)| \leq c \sum_{k=1}^{m} |F^{(k)}(u)| \sum_{j_1+\cdots+j_k=m} |\nabla_{j_1} u| \cdots |\nabla_{j_k} u|. \tag{1.2.22}$$

Consequently,

$$|\nabla_m F(u)| \leq c Q \sum_{k=1}^{m} \sum_{j_1+\cdots+j_k=m} \frac{|\nabla_{j_1} u|}{u^{1-j_1/m}} \cdots \frac{|\nabla_{j_k} u|}{u^{1-j_k/m}}. \tag{1.2.23}$$

Since $|\nabla_s u| \leq c\, I_{m-s} f$, it follows from (1.2.23) that

$$|\nabla_m F(u)| \leq c Q\Big(|\nabla_m I_m f| + \sum_{k=1}^{m} \sum_{j_1+\cdots+j_k=m}{}' \frac{I_{m-j_1} f \cdots I_{m-j_k} f}{(I_m f)^{1-j_1/m} \cdots (I_m f)^{1-j_k/m}}\Big),$$

where the sum $\sum'$ is taken over the collections of numbers $j_1, \cdots, j_k$ each less than m. Applying Lemma 1.2.5, we complete the proof. □

Our aim is the following assertion.

Theorem 1.2.1. *Let $p \in (1,\infty), m = 1, 2, \cdots$ and $mp < n$. Then, for any function $u \in C_0^\infty$,*

$$\int_0^\infty c_{p,m}(N_t)\, t^{p-1} dt \leq c\, \|u\|^p_{w_p^m}, \tag{1.2.24}$$

where $N_t = \{x : |u(x)| \geq t\}$ and c is a constant depending only on n, p, m.

Proof. Let $u = I_m f$ and $v = I_m|f|$. It is easily seen that $v \in C^m(\mathbb{R}^n)$ and $v(x) = \mathcal{O}(|x|^{m-n})$ as $|x| \to \infty$. Thus the set $\{x : v(x) \geq t\}$ is compact for any $t > 0$. Putting $t_j = 2^j$, $j = 0, \pm 1, \cdots$, and using the inequality $v(x) \geq |u(x)|$, we obtain

$$\int_0^\infty c_{p,m}(N_t)\, t^{p-1} dt$$

$$\leq c \sum_{j=-\infty}^{+\infty} (t_{j+1} - t_j)^p c_{p,m}(\{x : v(x) \geq t_j\}). \tag{1.2.25}$$

Let $\gamma \in C^\infty(\mathbb{R}^1)$, $\gamma(\tau) = 0$ for $\tau < \epsilon, \gamma(\tau) = 1$ for $\tau > 1$, where $\epsilon > 0$. We introduce the function $v \to F \in C^\infty(0, \infty)$ equal to

$$F_j(v) = t_j + (t_{j+1} - t_j)\gamma((v - t_j)(t_{j+1} - t_j)^{-1})$$

on the segment $[t_j, t_{j+1}]$.

According to the definition of capacity $c_{p,m}$, the sum on the right-hand side of (1.2.25) does not exceed

$$\sum_{j=-\infty}^{\infty} \|F_j(v)\|^p_{w_p^m} = \|F(v)\|^p_{w_p^m}.$$

By Corollary 1.2.1, the last norm is majorized by

$$c\,(\|\mathcal{M}|f|\|_{L_p} + \|\nabla_m I_m |f|\|_{L_p}). \tag{1.2.26}$$

Since the operator $\mathcal{M}$ and the singular integral operator $\nabla_m I_m$ are continuous in L_p, the sum (1.2.26) does not exceed $c\,\|f\|_{L_p}$. By Lemma 1.2.2,

$$\|f\|_{L_p} \sim \|u\|_{w_p^m}.$$

The theorem is proved. □

Together with (1.2.24), in this chapter we use the inequality

$$\int_0^\infty C_{p,m}(N_t)\, t^{p-1} dt \leq c\,\|u\|^p_{W_p^m}, \tag{1.2.27}$$

where $p \in (1, \infty)$ and $m = 1, 2, \cdots$. The proof of (1.2.27) is similar to that of (1.2.24), the role of (1.2.17) being played by

$$(J_{r\theta} f)(x) \leq c\,\big((J_r f)(x)\big)^\theta \big((\mathcal{M} f)(x)\big)^{1-\theta}, \tag{1.2.28}$$

where $0 < \theta < 1, r > 0$ and $J_r f$ is the Bessel potential of order r with a nonnegative density f. We do not dwell on a similar though more cumbersome proof of (1.2.28). A more general inequality will be proved in Lemma 4.2.3.

The corollary and its proof remain valid if I_m is replaced by J_m. To obtain (1.2.27), it is sufficient to use the following chain of inequalities:

$$\int_0^\infty C_{p,m}(N_t)\, t^{p-1}dt \le c \sum_{j=-\infty}^{\infty} (t_{j+1}-t_j)^p C_{p,m}(\{x : |v(x)| \ge t_j\})$$

$$\le c \sum_{j=-\infty}^{\infty} \|F_j(v)-t_j\|^p_{W_p^m} \le c\left(\|\nabla_m(F(v))\|^p_{L_p} + \|v\|^p_{L_p}\right)$$

and to duplicate the end of the proof of Theorem 1.2.1.

Remark 1.2.1. The existence of inequalities of the type (1.2.24) was demonstrated in [Maz3], where (1.2.24) (and for $m=1$ even a stronger inequality, in which the capacity of the condenser $N_t \backslash N_{2t}$ plays the role of the capacity of the set N_t) was obtained only for $m=1$ and $m=2$. In the more difficult case $m=2$ the proof was based on the 'smooth truncation' of a potential near equipotential surfaces. Unifying this procedure with Hedberg's inequality (1.2.17), Adams [Ad3] obtained the above proof for all integers m. Further references can be found in [Maz18].

1.2.4 Estimates for Constants in Trace Inequalities

A simple though important corollary of inequalities (1.2.27) and (1.2.24) is:

Theorem 1.2.2. *Let $p \in (1,\infty), m = 1, 2, \cdots$ and let μ be a measure in $\mathbb{R}^n$.*

(i) The best constant C in

$$\int |u|^p d\mu \le C\, \|u\|^p_{W_p^m}, \quad u \in C_0^\infty, \tag{1.2.29}$$

is equivalent to

$$\sup_e \frac{\mu(e)}{C_{p,m}(e)}, \tag{1.2.30}$$

where e is an arbitrary compact set in $\mathbb{R}^n$ of positive capacity $C_{p,m}(e)$.

(Expressions similar to (1.2.30) often occur in this book. In what follows we do not mention the positivity of capacities in denominators.)

(ii) If $mp < n$, then the best constant C in

$$\int |u|^p d\mu \le C\, \|u\|^p_{w_p^m}, \quad u \in C_0^\infty, \tag{1.2.31}$$

is equivalent to

$$\sup_e \frac{\mu(e)}{c_{p,m}(e)}. \tag{1.2.32}$$

Proof. (*i*) From the definition of Lebesgue integral we obtain

$$\int |u|^p d\mu = \int_0^\infty \mu(N_t)\, d(t^p).$$

Therefore

$$\int |u|^p d\mu \le \sup_e \frac{\mu(e)}{C_{p,m}(e)} \int_0^\infty C_{p,m}(N_t)\, d(t^p).$$

Now (1.2.29) follows from (1.2.27).

Minimizing the right-hand side of (1.2.29) over the set

$$\{u \in C_0^\infty : u \ge 1 \text{ on } e\},$$

we get

$$C \ge \sup_e \frac{\mu(e)}{C_{p,m}(e)}.$$

Case (*ii*) is treated in the same way.

Lemma 1.2.6. *The best constants C_0 and C in the inequalities*

$$\int |\nabla_l u|^p d\mu \le C_0 \|u\|^p_{w_p^m}, \tag{1.2.33}$$

$$\int |u|^p d\mu \le C \|u\|^p_{w_p^{m-l}}, \tag{1.2.34}$$

where $m > l$ and $u \in C_0^\infty$, are equivalent.

Proof. The estimate $C_0 \le c\, C$ is obvious. We show that $C_0 \ge c\, C$. It is clear that

$$u = \sum_{|\alpha|=l} (l!/\alpha!) D^{2\alpha}(-\Delta)^{-l} u.$$

From (1.2.33) and Lemma 1.2.2 we get

$$\int |D^{2\alpha}(-\Delta)^{-l} u|^p d\mu \le C_0 \|D^\alpha(-\Delta)^{-l} u\|^p_{w_p^m} \le c\, C_0 \|u\|^p_{w_p^{m-l}}.$$

Hence (1.2.34) holds with $C \le c\, C_0$. □

Lemma 1.2.7. *The best constants C_0 and C in the inequalities*

$$\int (|\nabla_l u|^p + |u|^p) d\mu \le C_0 \|u\|^p_{W_p^m}, \tag{1.2.35}$$

$$\int |u|^p d\mu \le C \|u\|^p_{W_p^{m-l}}, \tag{1.2.36}$$

where $m > l$ and $u \in C_0^\infty$, are equivalent.

Proof. The estimate $C_0 \le c\, C$ is obvious. We prove the converse. Let $x \to \sigma$ be a smooth positive function on $[0,\infty)$, equal to x for $x > 1$. For any $u \in C_0^\infty$ we have the representation

$$u = (-\Delta)^l [\sigma(-\Delta)]^{-l} u + T(-\Delta),$$

where T is a function from $C_0^\infty([0,\infty))$. Since

$$(-\Delta)^l = (-1)^l \sum_{|\alpha|=l} (l!/\alpha!) D^{2\alpha},$$

it follows from (1.2.35) and Lemma 1.2.3 that

$$\int |u|^p d\mu \le c\, C_0 \big(\|\nabla_l [\sigma(-\Delta)]^{-l} u\|^p_{W_p^m} + \|Tu\|^p_{W_p^m} \big) \le c_1 C_0 \|u\|^p_{W_p^{m-l}}.$$

The proof is complete. □

We give one more expression equivalent to the best constant C in (1.2.29).

Corollary 1.2.2. *The exact constant C in (1.2.29) is equivalent to*

$$\sup_{\{e: d(e) \le 1\}} \frac{\mu(e)}{C_{p,m}(e)},$$

where $d(e)$ is the diameter of e.

Proof. The lower bound for C follows from Theorem 1.2.2. We prove the upper bound. Let κ be an arbitrary compact set in $\mathbb{R}^n$. Further, let closed cubes Q_j form the coordinate grid with step $n^{-1/2}$ and let $2Q_j$ be homothetic open cubes with double edge length. By u we denote a function in C_0^∞ such that $u \ge 1$ on κ. Let η_j be a function in $C_0^\infty(2Q_j)$, equal to one on Q_j. Since the multiplicity of intersection of $2Q_j$ is finite and depends only on n, we have

$$\begin{aligned}\sum_j C_{p,m}(\kappa \cap Q_j) &\le c_1 \sum_j \|\eta_j u; 2Q_j\|^p_{W_p^m} \\ &\le c_2 \sum_j \|u; 2Q_j\|^p_{W_p^m} \le c_3 \|u\|^p_{W_p^m}.\end{aligned}$$

Minimizing the last norm, we get

$$C_{p,m}(\kappa) \ge c \sum_j C_{p,m}(\kappa \cap Q_j). \tag{1.2.37}$$

Clearly,

$$\mu(\kappa \cap Q_j) \le \sup_{\{e: d(e) \le 1\}} \frac{\mu(e)}{C_{p,m}(e)}\, C_{p,m}(\kappa \cap Q_j).$$

Summing over j and using (1.2.37), we arrive at the inequality

$$\mu(\kappa) \le c \sup_{\{e:d(e)\le 1\}} \frac{\mu(e)}{C_{p,m}(e)} C_{p,m}(\kappa).$$

The result follows. □

Corollary 1.2.3. *If $mp > n$, then the best constant in* (1.2.29) *is equivalent to*

$$\sup_{x\in\mathbb{R}^n} \mu(\mathcal{B}_1(x)).$$

Proof. By Proposition 1.2.2, $C_{p,m}(e) \sim 1$ for any non-empty compact set e with $d(e) \le 1$. It remains to refer to Corollary 1.2.2. □

Using the estimates for capacity by the Lebesgue measure obtained in Propositions 1.2.3 and 1.2.4, we immediately obtain:

Proposition 1.2.8. *The following inequalities hold*

$$\sup_{\{e:d(e)\le 1\}} \frac{\mu(e)}{C_{p,m}(e)} \le \begin{cases} c \sup\limits_{\{e:d(e)\le 1\}} \dfrac{\mu(e)}{(\mathrm{mes}_n e)^{(n-pm)/n}} & \text{for } mp < n, \\ c \sup\limits_{\{e:d(e)\le 1\}} \left(\log \dfrac{4^n}{\mathrm{mes}_n e}\right)^{p-1} \mu(e) & \text{for } mp = n; \end{cases}$$

$$\sup_e \frac{\mu(e)}{c_{p,m}(e)} \le c \sup_e \frac{\mu(e)}{(\mathrm{mes}_n e)^{(n-pm)/n}} \quad \text{for } mp < n.$$

A direct corollary of Propositions 1.2.5 - 1.2.7 is

Proposition 1.2.9. *The following inequalities hold*

$$\sup_{\{e:d(e)\le 1\}} \frac{\mu(e)}{C_{p,m}(e)} \ge \begin{cases} c \sup\limits_{x\in\mathbb{R}^n, \rho\in(0,1)} \rho^{mp-n}\mu(\mathcal{B}_\rho(x)) & \text{for } mp < n, \\ c \sup\limits_{x\in\mathbb{R}^n, \rho\in(0,1)} \left(\log \dfrac{2}{\rho}\right)^{p-1} \mu(\mathcal{B}_\rho(x)) & \text{for } mp = n; \end{cases}$$

$$\sup_e \frac{\mu(e)}{c_{p,m}(e)} \ge c \sup_{x\in\mathbb{R}^n, \rho>0} \rho^{mp-n}\mu(\mathcal{B}_\rho(x)) \quad \text{for } mp < n.$$

1.2.5 Other Criteria for the Trace Inequality (1.2.29) with $p > 1$

Now we overview several other conditions which are necessary and sufficient for (1.2.29)

We start with a remark due to D. R. Adams [Ad3] stating that

$$\sup_e \frac{\mu(e)}{C_{p,m}(e)} \sim \sup_e \left[\frac{\displaystyle\int (J_m \mu_e(x))^{p'} dx}{\mu(e)} \right]^{p-1} \tag{1.2.38}$$

where the suprema are taken either over arbitrary compact sets $e \subset \mathbb{R}^n$ or over compact sets whose diameters do not exceed one, and μ_e stands for the restriction of μ to e.

In fact, let the left and right-hand sides of (1.2.38) be denoted by A and B, respectively. Further, let u be an arbitrary function in C_0^∞ with $u \geq 1$ on e. We have

$$\mu(e) \leq \int u(x)\, d\mu_e(x) \leq \|(1-\Delta)^{-m/2}\mu_e\|_{L_{p'}} \|(1-\Delta)^{m/2} u\|_{L_p}$$

which can be rewritten as

$$\mu(e) \leq c\, \|J_m \mu_e\|_{L_{p'}} \|u\|_{W_p^m}.$$

Minimizing the right-hand side over all functions u, we obtain

$$\mu(e) \leq c B^{1/p} \mu(e)^{1/p'} [C_{p,m}(e)]^{1/p},$$

i.e. $A \leq cB$. Now we check the converse estimate. According to part (i) of Theorem 1.2.2,

$$\int |u|^p d\mu \leq c\; A \|u\|^p_{W_p^m}$$

for all $u \in C_0^\infty$. Consequently,

$$\left| \int u\, d\mu_e \right|^p \leq c\; A\, \mu(e)^{p-1} \|(1-\Delta)^{m/2} u\|^p_{L_p},$$

and therefore

$$\|J_m \mu_e\|_{L_{p'}} \leq c\, A^{1/p} \mu(e)^{1/p'}.$$

Thus $B \leq c\, A$. □

The relation

$$\sup_e \frac{\mu(e)}{c_{p,m}(e)} \sim \sup_e \left[\frac{\displaystyle\int (I_m \mu_e(x))^{p'} dx}{\mu(e)} \right]^{p-1}, \tag{1.2.39}$$

where e is an arbitrary compact set in $\mathbb{R}^n$, $mp < n$, $p \in (1, \infty)$, can be established in precisely the same manner. □

By (1.2.38), the upper estimate

$$\sup_e \frac{\mu(e)}{C_{p,m}(e)} \leq c \sup_{\mathbb{R}^n} J_m(J_m\mu)^{1/(p-1)} \tag{1.2.40}$$

holds if $mp \leq n$. In particular, for $p = 2$ it becomes

$$\sup_e \frac{\mu(e)}{C_{2,m}(e)} \leq c \sup_{\mathbb{R}^n} J_{2m}\mu. \tag{1.2.41}$$

Upper estimates similar to (1.2.40) and (1.2.41), with $c_{p,m}$ and I instead of $C_{p,m}$ and J, stem from (1.2.39) in the case $mp < n$. □

One cannot allow arbitrary sets e in the capacitary upper bounds on the left-hand sides of (1.2.38) and (1.2.39). This is in contrast with the suprema on the right-hand sides of (1.2.38) and (1.2.39). In fact, it turned out, as shown by Kerman and Sawyer in [KeS], that the role of e on the right-hand side of (1.2.39) can be played by an arbitrary cube Q, i.e.

$$\sup_e \frac{\mu(e)}{c_{p,m}(e)} \sim \sup_Q \left[\frac{\int (I_m\mu_Q(x))^{p'} dx}{\mu(Q)} \right]^{p-1}, \tag{1.2.42}$$

With minor technical changes in the proof given in [KeS], one verifies that the set $\{e\}$ on the right-hand side of (1.2.38) can be reduced to a set of cubes Q, i.e.

$$\sup_e \frac{\mu(e)}{C_{p,m}(e)} \sim \sup_{\{Q:d(Q)\leq 1\}} \left[\frac{\int (J_m\mu_Q(x))^{p'} dx}{\mu(Q)} \right]^{p-1}, \tag{1.2.43}$$

where $d(Q)$ is the diameter of Q.

Other conditions, necessary and sufficient for (1.2.29) and (1.2.31), which do not involve arbitrary sets and even cubes and which are of a purely pointwise nature, were found in [MV1]. It is shown in [MV1] that

$$\sup_e \frac{\mu(e)}{C_{p,m}(e)} \sim \sup_{x\in\mathbb{R}^n} \left[\frac{J_m(J_m\mu)^{p'}(x)}{J_m\mu(x)} \right]^{p-1}, \tag{1.2.44}$$

where $mp \leq n$, and

$$\sup_e \frac{\mu(e)}{c_{p,m}(e)} \sim \sup_{x\in\mathbb{R}^n} \left[\frac{I_m(I_m\mu)^{p'}(x)}{I_m\mu(x)} \right]^{p-1}, \tag{1.2.45}$$

where $mp < n$. □

We note that the following three criteria for (1.2.31) which result from (1.2.32), (1.2.42), and (1.2.45), respectively,

$$\mu(e) \le C\, c_{p,m}(e), \tag{1.2.46}$$

$$\int (I_m \mu_Q(x))^{p'}\, dx \le C\, \mu(Q), \tag{1.2.47}$$

$$I_m (I_m \mu)^{p'}(x) \le C\, I_m \mu(x) \tag{1.2.48}$$

have been obtained independently of each other.

These criteria lead to different simpler conditions, either necessary or sufficient, for (1.2.31), and each criterion (1.2.46)–(1.2.48) has its own range of applications. For example, the sufficient condition

$$\mu(e) \le C(\text{mes}_n e)^{(n-pm)/n}$$

follows readily from (1.2.46) (see Proposition 1.2.8) but its direct derivation from either (1.2.47) or (1.2.48) has not been obtained so far.

We finish this subsection with one more condition, necessary and sufficient for the trace inequality (1.2.31) to hold, which was obtained by Verbitsky [Ver1]: for every dyadic cube P_0 in $\mathbb{R}^n$

$$\sum_{P \subseteq P_0} \left[\frac{\mu(P)}{(\text{mes}_n P)^{1-m/n}} \right]^{p'} \text{mes}_n P \le C\, \mu(P_0),$$

where the sum is taken over all dyadic cubes P contained in P_0 and the constant does not depend on P_0.

Adding the restriction that the side length of P_0 does not exceed 1, we have a necessary and sufficient condition for the trace inequality (1.2.29).

1.2.6 The Fefferman and Phong Sufficient Condition

It was shown by Fefferman and Phong [F2] that the trace inequality (1.2.31) is true for $p > 1$, $mp < n$ and for the measure μ, absolutely continuous with respect to the Lebesgue measure:

$$d\mu(x) = g(x)\, dx, \tag{1.2.49}$$

if there exists $t > 1$ such that

$$\int_{\mathcal{B}_r(x)} [g(y)]^t dy \le c\, r^{n-mpt}. \tag{1.2.50}$$

In order to prove this we make use of the inequality

$$\int_{\mathbb{R}^n} ((\mathcal{M}_{mt} f)(y))^p d\nu(y) \le c \sup_{x \in \mathbb{R}^n,\, r>0} \frac{\nu(\mathcal{B}_r(x))}{r^{n-mpt}} \int_{\mathbb{R}^n} |f(y)|^p dy \tag{1.2.51}$$

(see [SW]), where $\mathcal{M}_l f$ is the fractional maximal function defined by

$$(\mathcal{M}_l f)(x) = \sup_{r>0} r^{l-n} \int_{\mathcal{B}_r(x)} |f(y)| dy.$$

Obviously, for any $\delta > 0$

$$I_m f(x) = (n-m) \int_0^\infty r^{m-n-1} \int_{\mathcal{B}_r(x)} f(y) dy\, dr$$
$$\leq c\Big(\delta^m (\mathcal{M}f)(x) + \delta^{m(1-t)} (\mathcal{M}_{mt} f)(x)\Big),$$

where $\mathcal{M}$ is the Hardy–Littlewood maximal operator. Minimizing the right-hand side in δ, we arrive at the inequality

$$I_m f(x) \leq c\big((\mathcal{M}_{mt} f)(x)\big)^{1/t} \big((\mathcal{M}f)(x)\big)^{1-1/t}$$

(see [AH], Sect. 3.1). Hence

$$\|I_m f\|_{L_p(g dx)} \leq c \,\|\mathcal{M}_{mt} f\|^{1/t}_{L_p(g^t dx)} \|\mathcal{M} f\|^{1-1/t}_{L_p}.$$

Therefore, by (1.2.51) and the boundedness of $\mathcal{M}$ in L_p, we find that

$$\|I_m f\|_{L_p(g dx)} \leq c \sup_{\substack{x\in\mathbb{R}^n,\\ r>0}} \Big(\frac{1}{r^{n-mpt}} \int_{\mathcal{B}_r(x)} (g(y))^t dy\Big)^{1/pt} \|f\|_{L_p}$$

which implies the above mentioned result in [F2]. □

In the case $p = 2$ condition (1.2.50) was improved in [ChWW], where it is shown that, if φ is an increasing function: $[0,\infty) \to [1,\infty)$ subject to

$$\int_1^\infty (\tau\varphi(\tau))^{-1} d\tau < \infty, \tag{1.2.52}$$

then the condition

$$\sup_{\mathcal{B}} \frac{\int_{\mathcal{B}} g(x)\varphi\big(g(x)(\text{diam}\mathcal{B})^2\big) dx}{(\text{mes}_n \mathcal{B})^{1-2/n}} < \infty$$

is sufficient for (1.2.31) to hold with μ given by (1.2.49). The assumption (1.2.52) is sharp.

1.3 Estimate for the L_q-Norm with respect to an Arbitrary Measure

In this section we collect various assertions on the best constant in the inequality

$$\Big(\int |u|^q \, d\mu\Big)^{1/q} \leq C \,\|u\|_{W_p^m}, \tag{1.3.1}$$

where $u \in C_0^\infty(\mathbb{R}^n)$. Since in this book our concern is the case $p = q$, we restrict ourselves to a short survey of known results for $p \neq q$.

1.3.1 The case $1 \le p < q$

As one can see below, for $p < q$ all results on the best constant in (1.3.1) can be given in non-capacitary terms and with balls rather than arbitrary compact sets.

We start with the following criterion (see [Ad1] for $p > 1$ and [Maz11] for $p = 1$).

Lemma 1.3.1. *Let $1 \le p < q$ and let $mp < n$. Then the best constant in (1.3.1) is equivalent to*

$$\sup_{x\in\mathbb{R}^n,\rho\in(0,1)} \rho^{m-n/p}\big(\mu(\mathcal{B}_\rho(x))\big)^{1/q}. \tag{1.3.2}$$

The case $mp = n$ considered in the next lemma was treated in [MP], see also [Maz14], Sect. 8.6.

Lemma 1.3.2. *Let $1 < p < q$ and let $mp = n$. Then the best constant in (1.3.1) is equivalent to*

$$\sup_{x\in\mathbb{R}^n,\rho\in(0,1)} (\log 2/\rho)^{(p-1)/p}\big(\mu(\mathcal{B}_\rho(x))\big)^{1/q}. \tag{1.3.3}$$

The next lemma is an obvious corollary of Sobolev's theorem on the imbedding $W_p^m \subset L_\infty$ which holds for $mp > n$.

Lemma 1.3.3. *Let $1 < p < q$ and let $mp > n$ or $1 = p < q$, $m \ge n$. Then the best constant in (1.3.1) is equivalent to*

$$\sup_{x\in\mathbb{R}^n} \big(\mu(\mathcal{B}_1(x))\big)^{1/q}. \tag{1.3.4}$$

1.3.2 The case $q < p \le n/m$

To state the next assertion, proved in [MN], we use the following function of one variable

$$(0,\infty) \ni s \to \nu_{p,m}(\mu; s) = \inf_{\{e:\mu(e)>s\}} C_{p,m}(e). \tag{1.3.5}$$

Lemma 1.3.4. *Let $1 < p < \infty$ and let $0 < q < p$. Then the best constant in (1.3.1) is equivalent to*

$$\int_0^\infty \Big(\frac{s^p}{\nu_{p,m}(\mu; s)^q}\Big)^{\frac{1}{p-q}} \frac{ds}{s}. \tag{1.3.6}$$

Another completely different characterization of the trace inequality (1.3.1) can be found in [COV1] for $q > 1$ and [COV2] for $q > 0$.

Lemma 1.3.5. *Let* $1 < p < \infty$ *and let* $0 < q < p$. *Then the best constant in* (1.3.1) *is equivalent to*

$$\int_{\mathbb{R}^n} (W_{m,p}\, \mu(x))^{\frac{q(p-1)}{p-q}}\, dx, \tag{1.3.7}$$

where

$$(W_{m,p}\, \mu)(x) = \int_0^1 \left(\frac{\mu(\mathcal{B}_r(x))}{r^{n-mp}} \right)^{\frac{1}{p-1}} \frac{dr}{r} \tag{1.3.8}$$

is the so-called nonlinear Wolff potential.

2

Multipliers in Pairs of Sobolev Spaces

2.1 Introduction

In the present chapter we study multipliers acting in pairs of spaces W_p^k and w_p^k, where k is a nonnegative integer.

The concepts of this chapter prove to be prototypes for the subsequent study of multipliers in other pairs of spaces. Using the result of Sects. 1.2 and 1.1, we derive necessary and sufficient conditions for a function to belong to the space of multipliers $M(W_p^m \to W_p^l)$ and $M(w_p^m \to w_p^l)$, where $m \geq l \geq 0$ and $p \in [1, \infty)$ (Sects. 2.2, 2.3, and 2.8). The case of the half-space $\mathbb{R}_+^n$ is treated in Sect. 2.4. Section 2.5 contains conditions for the inclusion $\gamma \in M(W_p^m \to W_p^{-k}), k > 0$. In Sect. 2.6 we present a brief description of the space $M(W_p^m \to W_q^l)$. Section 2.7 deals with certain properties of multipliers. In the concluding Sect. 2.9 we give a description of multipliers preserving spaces of functions with bounded variation. As usual, we omit $\mathbb{R}^n$ in notations of spaces, norms, and integrals.

Let $\gamma \in M(W_p^m \to W_p^l)$, $u_n \to u$ in W_p^m and $\gamma u_n \to v$ in W_p^l. Then there exists a sequence $\{n_k\}_{k \geq 1}$, such that for almost all x

$$u_{n_k}(x) \to u(x), \quad \gamma(x) u_{n_k}(x) \to v(x).$$

Hence $v = \gamma u$ almost everywhere in $\mathbb{R}^n$, and therefore the operator

$$W_p^m \ni u \to \gamma u \in W_p^l$$

is closed. Since this operator is defined on the whole of W_p^m, it is bounded by the Banach theorem.

The norm in $M(W_p^m \to W_p^l)$ is defined as the norm of the operator of multiplication

$$\|\gamma\|_{M(W_p^m \to W_p^l)} = \sup\{\|\gamma u\|_{W_p^l} : \|u\|_{W_p^m} \leq 1\}.$$

We use the notation MW_p^l instead of $M(W_p^l \to W_p^l)$.

V.G. Maz'ya, T.O. Shaposhnikova, *Theory of Sobolev Multipliers*, Grundlehren der mathematischen Wissenschaften 337,

It is worth noting that we can always assume that $m \geq l$, since in the opposite case $M(W_p^m \to W_p^l) = \{0\}$.[1] In fact, let $m < l$ and let $\gamma \in M(W_p^m \to W_p^l)$. We have

$$\|\gamma\|_{M(W_p^m \to W_p^l)} \geq \frac{\|\gamma\, e^{itx_1}\, \eta\|_{W_p^l}}{\|e^{itx_1}\, \eta\|_{W_p^m}},$$

where $t > 0$ and η is an arbitrary non-zero function in C_0^∞. Therefore,

$$\|\gamma\|_{M(W_p^m \to W_p^l)} \geq t^{l-m}\Big(\frac{\|\gamma\eta\|_{L_p}}{\|\eta\|_{L_p}} + o(1)\Big)$$

as $t \to \infty$ and this is possible only if $\gamma = 0$.

By $W_{p,\mathrm{unif}}^l$ we denote the space endowed with the norm

$$\|u\|_{W_{p,\mathrm{unif}}^l} = \sup_{z \in \mathbb{R}^n} \|\eta_z u\|_{W_p^l},$$

where $\eta_z(x) = \eta(x - z)$, $\eta \in C_0^\infty$, and $\eta = 1$ on $\mathcal{B}_1$.

We also need the space

$$W_{p,\mathrm{loc}}^l = \{u :\ \eta u \in W_p^l \ \text{ for all } \eta \in C_0^\infty\}.$$

Throughout this book similar notations S_{unif} and S_{loc} will be used for other Banach spaces S of functions defined on $\mathbb{R}^n$.

The following is the main result of this chapter.

Theorem 2.1.1. *(i) Let $p > 1$, $mp > n$ or $p = 1$, $m \geq n$. Then*

$$\|\gamma\|_{M(W_p^m \to W_p^l)} \sim \sup_{x \in \mathbb{R}^n} \|\gamma; \mathcal{B}_1(x)\|_{W_p^l}.$$

(ii) Let $p = 1$, $m < n$. Then

$$\|\gamma\|_{M(W_1^m \to W_1^l)} \sim \sup_{\substack{x \in \mathbb{R}^n \\ r \in (0,1)}} r^{m-n} \|\nabla_l \gamma; \mathcal{B}_r(x)\|_{L_1}$$

$$+ \begin{cases} \sup\limits_{x \in \mathbb{R}^n} \|\gamma; \mathcal{B}_1(x)\|_{L_1}, & m > l, \\ \|\gamma; \mathbb{R}^n\|_{L_\infty}, & m = l. \end{cases}$$

(iii) Let $mp \leq n$, $p > 1$. Then

$$\|\gamma\|_{M(W_p^m \to W_p^l)} \sim \sup_{\substack{e \subset \mathbb{R}^n \\ d(e) \leq 1}} \frac{\|\nabla_l \gamma; e\|_{L_p}}{(C_{p,m}(e))^{1/p}}$$

$$+ \begin{cases} \sup\limits_{x \in \mathbb{R}^n} \|\gamma; \mathcal{B}_1(x)\|_{L_1}, & m > l, \\ \|\gamma\|_{L_\infty}, & m = l, \end{cases}$$

where $d(e)$ is the diameter of e.

---

[1] In other words, the multipliers, acting between two different spaces, *uglify* their domain, in full correspondence with Mock Turtle's terminology in the epigraph to the present book.

2.2 Characterization of the Space $M(W_1^m \to W_1^l)$

Theorem 2.2.1. *For any* $m > 0$

$$\|\gamma\|_{M(W_1^m \to L_1)} \sim \sup_{\substack{x\in\mathbb{R}^n \\ r\in(0,1)}} r^{m-n}\|\gamma; \mathcal{B}_r(x)\|_{L_1}. \tag{2.2.1}$$

Proof. The result follows by Theorem 1.1.2.

Theorem 2.2.2. *For any* $0 < l \le m$ *the relation*

$$\|\gamma\|_{M(W_1^m \to W_1^l)} \sim \sup_{\substack{x\in\mathbb{R}^n \\ r\in(0,1)}} r^{m-n}\big(\|\nabla_l\gamma; \mathcal{B}_r(x)\|_{L_1} + r^{-l}\|\gamma; \mathcal{B}_r(x)\|_{L_1}\big) \tag{2.2.2}$$

holds.

Proof. First we note that

$$r^{j-l}\|\nabla_j\gamma; \mathcal{B}_r(x)\|_{L_1} \le c\,\big(\|\nabla_l\gamma; \mathcal{B}_r(x)\|_{L_1} + r^{-l}\|\gamma; \mathcal{B}_r(x)\|_{L_1}\big).$$

Hence the equivalence relation (2.2.2) can be written as

$$\|\gamma\|_{M(W_1^m \to W_1^l)} \sim \sup_{x\in\mathbb{R}^n, r\in(0,1)} r^{m-n}\sum_{j=0}^{l} r^{j-l}\|\nabla_j\gamma; \mathcal{B}_r(x)\|_{L_1}. \tag{2.2.3}$$

We start with the lower estimate for the multiplier norm of γ. Let $u(y) = \varphi((y-x)/r)$, where $r \in (0,1)$ for $m < n$ and $r = 1$ for $m \ge n$, $\varphi \in C_0^\infty(\mathcal{B}_2)$, $\varphi = 1$ on $\mathcal{B}_1$. We set this u into

$$\|\gamma u\|_{W_1^m} \le \|\gamma\|_{M(W_1^m \to W_1^l)}\|u\|_{W_1^m}$$

and use the inequality

$$r^{j-l}\|\nabla_j(\gamma u); \mathcal{B}_{2r}(x)\|_{L_1} \le c\,\|\gamma u\|_{W_1^l}, \quad j = 0, 1, \ldots, l,$$

valid because $\operatorname{supp}\gamma u \subset \mathcal{B}_{2r}(x)$. Hence

$$r^{j-l}\|\nabla_j\gamma; \mathcal{B}_r(x)\|_{L_1} \le c\,\|\gamma\|_{M(W_1^m \to W_1^l)} r^{n-m}$$

which gives the lower estimate for $\|\gamma\|_{M(W_1^m \to W_1^l)}$.

To obtain the upper estimate, we combine the obvious inequality

$$\|\nabla_l(\gamma u)\|_{L_1} \le c\sum_{j=0}^{l} \|\,|\nabla_j\gamma|\,|\nabla_{l-j}u|\,\|_{L_1}$$

with the estimate

$$\|\,|\nabla_j\gamma|\,|\nabla_{l-j}u|\,\|_{L_1} \le c \sup_{x\in\mathbb{R}^n, r\in(0,1)} r^{m-l+j-n}\|\nabla_j\gamma; \mathcal{B}_r(x)\|_{L_1}\|\nabla_{l-j}u\|_{W_1^{m-l+j}}$$

which holds by Theorem 1.1.2. Hence

$$\|\nabla_l(\gamma u)\|_{L_1} \leq c \sup_{x\in\mathbb{R}^n, r\in(0,1)} r^{m-n} \sum_{j=0}^{l} r^{j-l} \|\nabla_j\gamma; \mathcal{B}_r(x)\|_{L_1} \|u\|_{W_1^m}.$$

Also, by the same Theorem 1.1.2,

$$\|\gamma u\|_{L_1} \leq c \sup_{x\in\mathbb{R}^n, r\in(0,1)} r^{m-l-n} \|\gamma; \mathcal{B}_r(x)\|_{L_1} \|u\|_{W_1^{m-l}}.$$

Adding together the last two inequalities and noting (2.2.3), we complete the proof. □

Corollary 2.2.1. *Let* $0 \leq l < m$. *Then*

$$\|\gamma\|_{M(W_1^{m-l}\to L_1)} \leq c\,\|\gamma\|_{M(W_1^m\to W_1^l)}. \tag{2.2.4}$$

Proof. The result follows directly from Theorems 2.2.2 and 2.2.1.

The equivalence relation (2.2.2) is modified in the following assertion.

Theorem 2.2.3. (*i*) *If* $m \geq n$, *and* $m \geq l$, *then*

$$\|\gamma\|_{M(W_1^m\to W_1^l)} \sim \sup_{x\in\mathbb{R}^n} \|\gamma; \mathcal{B}_1(x)\|_{W_1^l}. \tag{2.2.5}$$

(*ii*) *If* $l < n$, *then*

$$\|\gamma\|_{MW_1^l} \sim \sup_{x\in\mathbb{R}^n, r\in(0,1)} r^{l-n} \|\nabla_l\gamma; \mathcal{B}_r(x)\|_{L_1} + \|\gamma\|_{L_\infty}. \tag{2.2.6}$$

(*iii*) *If* $l < m < n$, *then*

$$\|\gamma\|_{M(W_1^m\to W_1^l)} \sim \sup_{x\in\mathbb{R}^n, r\in(0,1)} r^{m-n} \|\nabla_l\gamma; \mathcal{B}_r(x)\|_{L_1} + \sup_{x\in\mathbb{R}^n} \|\gamma; \mathcal{B}_1(x)\|_{L_1}. \tag{2.2.7}$$

Proof. Relations (2.2.5) and (2.2.6) follow from Theorem 2.2.2.

Let $l < m < n$. The lower estimate is a direct consequence of Theorem 2.2.2. To derive the upper estimate we show that

$$r^{m-n-l} \|\gamma; \mathcal{B}_r(x)\|_{L_1} \leq c \left(\|\gamma; \mathcal{B}_1(x)\|_{L_1} + \sup_{\rho\in(0,1)} \rho^{m-n} \|\nabla_l\gamma; \mathcal{B}_\rho(x)\|_{L_1} \right) \tag{2.2.8}$$

for any $x \in \mathbb{R}^n$ and $r \in (0,1)$. By the Sobolev integral representation (see, for instance, [Maz14], Subsect. 1.1.10),

$$|\gamma(z)| \leq c\Big(\|\gamma; \mathcal{B}(x)\|_{L_1} + \int_{\mathcal{B}(x)} \frac{|\nabla_l\gamma(y)|}{|z-y|^{n-l}} dy \Big). \tag{2.2.9}$$

Hence

$$\|\gamma; \mathcal{B}_r(x)\|_{L_1} \le c \left(r^n \|\gamma; \mathcal{B}_1(x)\|_{L_1} + \int_{\mathcal{B}_r(x)} \int_{\mathcal{B}_1(x)} \frac{|\nabla_l \gamma(y)|}{|z-y|^{n-l}} dydz \right)$$

$$\le c\Big(r^n \|\gamma; \mathcal{B}_1(x)\|_{L_1} + r^l \|\nabla_l \gamma; \mathcal{B}_{2r}(x)\|_{L_1} + \int_{\mathcal{B}_r(x)} \int_{\mathcal{B}_2(x)\backslash \mathcal{B}_{2r}(x)} |\nabla_l \gamma(y)| \frac{dydz}{|y|^{n-l}} \Big).$$

Combining this fact with the obvious inequality

$$\int_{\mathcal{B}_r(x)} \int_{\mathcal{B}_1(x)\backslash \mathcal{B}_{2r}(x)} |\nabla_l \gamma(y)| \frac{dydz}{|y|^{n-l}} \le c\, r^{l-m+n} \sup_{\rho \in (0,1)} \rho^{m-n} \|\nabla_l \gamma; \mathcal{B}_\rho(x)\|_{L_1},$$

we complete the proof. □

Theorem 2.2.3 implies an interpolation inequality for elements of the space $M(W_1^m \to W_1^l)$.

Corollary 2.2.2. *Let $\gamma \in M(W_1^m \to W_1^l)$. Then*

$$\|\gamma\|_{M(W_1^{m-j} \to W_1^{l-j})} \le c\, \|\gamma\|^{1-j/l}_{M(W_1^m \to W_1^l)} \|\gamma\|^{j/l}_{M(W_1^{m-l} \to L_1)}, \tag{2.2.10}$$

where $j = 0, 1, \ldots, l$.

Proof. Making the dilation in the well-known inequality

$$\|\nabla_{l-j} \gamma; \mathcal{B}_1\|_{L_1} \le c\, \|\gamma; \mathcal{B}_1\|^{1-j/l}_{W_1^l} \|\gamma; \mathcal{B}_1\|^{j/l}_{L_1},$$

we obtain

$$r^{l-j} \|\nabla_{l-j} \gamma; \mathcal{B}_r(x)\|_{L_1} \le c\, \big(r^l \|\nabla_l \gamma; \mathcal{B}_r(x)\|_{L_1} + \|\gamma; \mathcal{B}_r(x)\|_{L_1} \big)^{1-j/l} \|\gamma; \mathcal{B}_r(x)\|^{j/l}_{L_1}.$$

Hence

$$r^{m-j-n} \|\nabla_{l-j} \gamma; \mathcal{B}_r(x)\|_{L_1} + r^{m-n-l} \|\gamma; \mathcal{B}_r(x)\|_{L_1}$$

$$\le c \big(r^{m-n} \|\nabla_l \gamma; \mathcal{B}_r(x)\|_{L_1} + r^{m-n-l} \|\gamma; \mathcal{B}_r(x)\|_{L_1} \big)^{1-j/l} \big(r^{m-n-l} \|\gamma; \mathcal{B}_r(x)\|_{L_1} \big)^{j/l}.$$

Reference to Theorem 2.2.2 completes the proof. □

Corollary 2.2.3. *Let $0 < l < m$. Then*

$$\|\gamma\|_{M(W_1^m \to W_1^l)} \sim \sum_{j=0}^{l} \|\nabla_{l-j} \gamma\|_{M(W_1^{m-j} \to L_1)} \tag{2.2.11}$$

and

$$\|\gamma\|_{M(W_1^m \to W_1^l)} \sim \|\nabla_l \gamma\|_{M(W_1^m \to L_1)} + \|\gamma\|_{L_1,\mathrm{unif}}. \tag{2.2.12}$$

For $m = l$ the norm $\|\gamma\|_{L_1,\mathrm{unif}}$ should be replaced by $\|\gamma\|_{L_\infty}$.

Proof. Estimate 2.2.12 results from Theorems 2.2.3 and 2.2.1.

By Theorems 2.2.1 and 2.2.3

$$\|\nabla_{l-j}\gamma\|_{M(W_1^{m-j}\to L_1)} \le c \sup_{x\in\mathbb{R}^n, r\in(0,1)} r^{m-j-n}\|\nabla_{l-j}\gamma;\mathcal{B}_r(x)\|_{L_1}$$

$$\le c\,\|\gamma\|_{M(W_1^{m-j}\to W_1^{l-j})}$$

which is dominated by $\|\gamma\|_{M(W_1^m\to W_1^l)}$ in view of Corollary 2.2.2. The lower estimate 2.2.11 follows. □

2.3 Characterization of the Space $M(W_p^m \to W_p^l)$ for $p > 1$

Here we derive necessary and sufficient conditions for a function to belong to the space $M(W_p^m \to W_p^l)$ for $p > 1$.

The next assertion contains an inequality between multipliers and their mollifiers.

Lemma 2.3.1. *Let γ_ρ denote a mollifier of a function γ which is defined as*

$$\gamma_\rho(x) = \rho^{-n}\int K(\rho^{-1}(x-\xi))\gamma(\xi)d\xi,$$

where $K \in C_0^\infty(\mathcal{B}_1)$, $K \ge 0$, and $\|K\|_{L_1} = 1$. Then

$$\|\gamma_\rho\|_{M(W_p^m\to W_p^l)} \le \|\gamma\|_{M(W_p^m\to W_p^l)} \le \liminf_{\rho\to 0}\|\gamma_\rho\|_{M(W_p^m\to W_p^l)}. \tag{2.3.1}$$

Proof. Let $u \in C_0^\infty$. By Minkowski's inequality

$$\Big(\int\Big|\nabla_{j,x}\int\rho^{-n}K(\xi/\rho)\gamma(x-\xi)u(x)d\xi\Big|^p dx\Big)^{1/p}$$

$$\le \int\rho^{-n}K(\xi/\rho)\Big(\int|\nabla_{j,y}(\gamma(y)u(y+\xi))|^p dy\Big)^{1/p}d\xi,$$

where $j = 0, l$. Therefore,

$$\|\gamma_\rho u\|_{W_p^l} \le \|\gamma\|_{M(W_p^m\to W_p^l)}\int\rho^{-n}K(\xi/\rho)\Big(\Big(\int|\nabla_{m,y}u(y+\xi)|^p dy\Big)^{1/p}$$

$$+\Big(\int|u(y+\xi)|^p dy\Big)^{1/p}\Big)d\xi \le \|\gamma\|_{M(W_p^m\to W_p^l)}\|u\|_{W_p^m}.$$

This gives the left inequality (2.3.1). The right inequality (2.3.1) follows from

$$\|\gamma u\|_{W_p^l} = \liminf_{\rho\to 0}\|\gamma_\rho u\|_{W_p^l} \le \liminf_{\rho\to 0}\|\gamma_\rho\|_{M(W_p^m\to W_p^l)}\|u\|_{W_p^m}.$$

The proof is complete. □

The following assertion is a particular case of Lemma 1.2.7.

Lemma 2.3.2. *Let $\gamma \in L_{p,\mathrm{loc}}$, $p \in (1, \infty)$, and let u be an arbitrary function in C_0^∞. The best constant in the inequality*

$$\|\gamma \nabla_l u\|_{L_p} + \|\gamma u\|_{L_p} \le C \, \|u\|_{W_p^m}$$

is equivalent to the norm $\|\gamma\|_{M(W_p^{m-l} \to L_p)}$.

The next lemma concerns derivatives of multipliers.

Lemma 2.3.3. *Suppose that*

$$\gamma \in M(W_p^m \to W_p^l) \cap M(W_p^{m-l} \to L_p), \quad p \in (1, \infty).$$

Then, for any multi-index α of order $|\alpha| \le l$,

$$D^\alpha \gamma \in M(W_p^m \to W_p^{l-|\alpha|})$$

and

$$\|D^\alpha \gamma\|_{M(W_p^m \to W_p^{l-|\alpha|})} \le \varepsilon \, \|\gamma\|_{M(W_p^{m-l} \to L_p)} + c(\varepsilon) \, \|\gamma\|_{M(W_p^m \to W_p^l)}, \tag{2.3.2}$$

where ε is an arbitrary positive number.

Proof. Let $u \in W_p^l$ and let φ be an arbitrary function in C_0^∞. Applying the Leibniz formula

$$D^\alpha(\varphi u) = \sum_{\{\beta : \alpha \ge \beta \ge 0\}} \frac{\alpha!}{\beta!(\alpha - \beta)!} D^\beta \varphi \, D^{\alpha-\beta} u,$$

we obtain

$$\int \varphi u (-D)^\alpha \gamma dx = \int \gamma D^\alpha(\varphi u) dx = \sum_{\{\beta : \alpha \ge \beta \ge 0\}} \frac{\alpha!}{\beta!(\alpha - \beta)!} \gamma D^\beta \varphi \, D^{\alpha-\beta} u \, dx$$

$$= \int \varphi \sum_{\{\beta : \alpha \ge \beta \ge 0\}} \frac{\alpha!}{\beta!(\alpha - \beta)!} (-D)^\beta (\gamma D^{\alpha-\beta} u) dx.$$

Therefore,

$$u D^\alpha \gamma = \sum_{\{\beta : \alpha \ge \beta \ge 0\}} \frac{\alpha!}{\beta!(\alpha - \beta)!} D^\beta \big(\gamma (-D)^{\alpha-\beta} u\big),$$

which implies the estimate

$$\|u D^\alpha \gamma\|_{W_p^{l-|\alpha|}} \le c \sum_{\{\beta : \alpha \ge \beta \ge 0\}} \|\gamma D^{\alpha-\beta} u\|_{W_p^{l-|\alpha|+|\beta|}}.$$

Hence, it suffices to prove (2.3.2) for $|\alpha| = 1$, $l \geq 1$. We have

$$\|u\nabla\gamma\|_{W_p^{l-1}} \leq \|u\gamma\|_{W_p^l} + \|\gamma\nabla u\|_{W_p^{l-1}}$$
$$\leq \big(\|\gamma\|_{M(W_p^m\to W_p^l)} + \|\gamma\|_{M(W_p^{m-1}\to W_p^{l-1})}\big)\|u\|_{W_p^m}.$$

Estimating the norm $\|\gamma\|_{M(W_p^{m-1}\to W_p^{l-1})}$ by (2.3.8), we arrive at (2.3.2). □

The equivalent representation for the norm in $M(W_p^m \to L_p)$ is a direct consequence of Theorem 1.2.2. Namely,

$$\|\gamma\|_{M(W_p^m\to L_p)} \sim \sup_e \frac{\|\gamma;e\|_{L_p}}{(C_{p,m}(e))^{1/p}}. \tag{2.3.3}$$

By Corollary 1.2.2 this can also be written in the form

$$\|\gamma\|_{M(W_p^m\to L_p)} \sim \sup_{e:d(e)\leq 1} \frac{\|\gamma;e\|_{L_p}}{(C_{p,m}(e))^{1/p}}, \tag{2.3.4}$$

where $d(e)$ is the diameter of e.

Now we pass to two-sided estimates for the norms in $M(W_p^m \to W_p^l)$, $p \in (1,\infty)$, given in terms of the spaces $M(W_p^k \to L_p)$. We start with lower estimates.

Lemma 2.3.4. *Let $\gamma \in M(W_p^m \to W_p^l)$. Then*

$$\|\nabla_l\gamma\|_{M(W_p^m\to L_p)} + \|\gamma\|_{M(W_p^{m-l}\to L_p)} \leq c\,\|\gamma\|_{M(W_p^m\to W_p^l)}. \tag{2.3.5}$$

Proof. Suppose first that $\gamma \in M(W_p^{m-l} \to L_p)$. We have

$$\|\gamma\nabla_l u\|_{L_p} \leq \|\gamma\|_{M(W_p^m\to W_p^l)}\|u\|_{W_p^m} + c \sum_{\substack{|\alpha|+|\beta|=l,\\ \beta\neq 0}} \|D^\alpha u D^\beta\gamma\|_{L_p}$$

$$\leq \Big(\|\gamma\|_{M(W_p^m\to W_p^l)} + c\sum_{j=1}^{l} \|\nabla_j\gamma\|_{M(W_p^{m-l+j}\to L_p)}\Big)\|u\|_{W_p^m}. \tag{2.3.6}$$

Lemma 2.3.3 implies

$$\|\nabla_j\gamma\|_{M(W_p^{m-l+j}\to L_p)}$$
$$\leq \varepsilon\,\|\gamma\|_{M(W_p^{m-l}\to L_p)} + c(\varepsilon)\,\|\gamma\|_{M(W_p^{m-l+j}\to W_p^j)}. \tag{2.3.7}$$

By an interpolation property of Sobolev spaces (see [Tr4]) we have

$$\|\gamma\|_{M(W_p^{m-j}\to W_p^{l-j})} \leq c\,\|\gamma\|^{(l-j)/l}_{M(W_p^m\to W_p^l)}\|\gamma\|^{j/l}_{M(W_p^{m-l}\to L_p)}, \tag{2.3.8}$$

where $0 \le j \le l$. Estimating the last norm in (2.3.7) by (2.3.8), we obtain

$$\|\nabla_j \gamma\|_{M(W_p^{m-l+j} \to L_p)} \le \varepsilon \, \|\gamma\|_{M(W_p^{m-l} \to L_p)} + c(\varepsilon) \, \|\gamma\|_{M(W_p^m \to W_p^l)}.$$

Substitution of this inequality into (2.3.6) gives

$$\|\gamma \nabla_l\|_{L_p} \le \Big(\varepsilon \, \|\gamma\|_{M(W_p^{m-l} \to L_p)} + c(\varepsilon) \, \|\gamma\|_{M(W_p^m \to W_p^l)}\Big) \|u\|_{W_p^m}. \tag{2.3.9}$$

Also,

$$\|\gamma u\|_{L_p} \le \|\gamma\|_{M(W_p^m \to W_p^l)} \|u\|_{W_p^m}. \tag{2.3.10}$$

Combining the last two estimates and applying Lemma 2.3.2, we arrive at

$$\|\gamma\|_{M(W_p^{m-l} \to L_p)} \le \varepsilon \, \|\gamma\|_{M(W_p^{m-l} \to L_p)} + c(\varepsilon) \, \|\gamma\|_{M(W_p^m \to W_p^l)}.$$

Hence,

$$\|\gamma\|_{M(W_p^{m-l} \to L_p)} \le c \, \|\gamma\|_{M(W_p^m \to W_p^l)}. \tag{2.3.11}$$

Next we remove the assumption $\gamma \in M(W_p^{m-l} \to L_p)$. Since $\gamma \in M(W_p^m \to W_p^l)$, then

$$\|\gamma \eta\|_{L_p} \le c \, \|\eta\|_{W_p^m},$$

where $\eta \in C_0^\infty(\mathcal{B}_2(x))$, $\eta = 1$ on $\mathcal{B}_1(x)$, and x is an arbitrary point in $\mathbb{R}^n$. Hence

$$\sup_x \|\gamma; \mathcal{B}_1(x)\|_{L_p} < \infty$$

and for any $k = 0, 1, \ldots$ there exists a constant c_ρ such that $|\nabla_k \gamma_\rho| \le c_\rho$. Since the function γ_ρ and all its derivatives are bounded, it follows that γ_ρ is a multiplier in W_p^k for any $k = 1, 2, \ldots$, and thus $\gamma_\rho \in M(W_p^{m-l} \to L_p)$. By (2.3.11)

$$\|\gamma_\rho\|_{M(W_p^{m-l} \to L_p)} \le c \, \|\gamma_\rho\|_{M(W_p^m \to W_p^l)}.$$

Letting $\rho \to 0$ and using Lemma 2.3.1 we arrive at (2.3.11) for all $\gamma \in M(W_p^m \to W_p^l)$.

To estimate the first term on the right-hand side of (2.3.5) we combine (2.3.11) with (2.3.7) for $j = l$. □

The estimate inverse to (2.3.5) is contained in the following lemma.

Lemma 2.3.5. *Let* $\gamma \in M(W_p^{m-l} \to L_p)$ *and let* $\nabla_l \gamma \in M(W_p^m \to L_p)$. *Then* $\gamma \in M(W_p^m \to W_p^l)$ *and the estimate*

$$\|\gamma\|_{M(W_p^m \to W_p^l)} \le c\Big(\|\nabla_l \gamma\|_{M(W_p^m \to L_p)} + \|\gamma\|_{M(W_p^{m-l} \to L_p)}\Big) \tag{2.3.12}$$

holds.

Proof. Inequality (2.3.8) along with Lemma 2.3.3 gives

$$\|\nabla_j \gamma\|_{M(W_p^{m-l+j}\to L_p)} \leq c\, \|\gamma\|^{j/l}_{M(W_p^m\to W_p^l)} \|\gamma\|^{1-j/l}_{M(W_p^{m-l}\to L_p)}, \tag{2.3.13}$$

where $j = 1, \dots, l-1$. For any $u \in C_0^\infty$,

$$\|\nabla_l(\gamma u)\|_{L_p} \leq c \sum_{j=0}^{l} \|\, |\nabla_j \gamma|\, |\nabla_{l-j} u|\, \|_{L_p} \leq c\Big(\|\nabla_l \gamma\|_{M(W_p^m\to L_p)}$$

$$+ \|\gamma\|_{M(W_p^{m-l}\to L_p)} + \sum_{j=1}^{l-1} \|\nabla_j\gamma\|_{M(W_p^{m-l+j}\to L_p)}\Big) \|u\|_{W_p^m}.$$

Then it follows from (2.3.13) that

$$\|\nabla_l(\gamma u)\|_{L_p} \leq c\, \Big(\|\nabla_l \gamma\|_{M(W_p^m\to L_p)} + \|\gamma\|_{M(W_p^{m-l}\to L_p)}\Big) \|u\|_{W_p^m}.$$

It remains to note that

$$\|\gamma u\|_{L_p} \leq \|\gamma\|_{M(W_p^{m-l}\to L_p)} \|u\|_{W_p^{m-l}}.$$

□

Unifying Lemmas 2.3.4 and 2.3.5, we arrive at the following assertion.

Theorem 2.3.1. *Let m and l be integers, and let $p \in (1, \infty)$. A function γ belongs to the space $M(W_p^m \to W_p^l)$ if and only if $\gamma \in W^l_{p,\mathrm{loc}}$, $\gamma \in M(W_p^{m-l} \to L_p)$, and $\nabla_l \gamma \in M(W_p^m \to L_p)$. Moreover,*

$$\|\gamma\|_{M(W_p^m\to W_p^l)} \sim \|\nabla_l \gamma\|_{M(W_p^m\to L_p)} + \|\gamma\|_{M(W_p^{m-l}\to L_p)}.$$

Relation (2.3.3) leads to the following reformulation of Theorem 2.3.1.

Theorem 2.3.2. *Let m and l be integers, and let $p \in (1, \infty)$. A function γ belongs to the space $M(W_p^m \to W_p^l)$ if and only if $\gamma \in W^l_{p,\mathrm{loc}}$ and, for any compact set $e \subset \mathbb{R}^n$,*

$$\|\nabla_l \gamma; e\|^p_{L_p} \leq c\, C_{p,m}(e)$$

and

$$\|\gamma; e\|^p_{L_p} \leq c\, C_{p,m-l}(e).$$

Moreover,

$$\|\gamma\|_{M(W_p^m\to W_p^l)} \sim \sup_e \Big(\frac{\|\nabla_l \gamma; e\|_{L_p}}{(C_{p,m}(e))^{1/p}} + \frac{\|\gamma; e\|_{L_p}}{(C_{p,m-l}(e))^{1/p}} \Big). \tag{2.3.14}$$

An important particular case of Theorem 2.3.2 is $m = l$.

Corollary 2.3.1. *Let l be an integer and let $p \in (1, \infty)$. A function γ belongs to the space MW_p^l if and only if $\gamma \in W_{p,\mathrm{loc}}^l$ and, for any compact set $e \subset \mathbb{R}^n$,*

$$\|\nabla_l \gamma; e\|_{L_p}^p \le c\, C_{p,l}(e).$$

Moreover,

$$\|\gamma\|_{MW_p^l} \sim \sup_e \frac{\|\nabla_l \gamma; e\|_{L_p}}{(C_{p,l}(e))^{1/p}} + \|\gamma\|_{L_\infty}. \tag{2.3.15}$$

By (2.3.4), one can use only compact sets with diameters not exceeding 1 in Theorem 2.3.2 and Corollary 2.3.1 and rewrite (2.3.14) and (2.3.15) as follows:

$$\|\gamma\|_{M(W_p^m \to W_p^l)} \sim \sup_{e: d(e) \le 1} \left(\frac{\|\nabla_l \gamma; e\|_{L_p}}{(C_{p,m}(e))^{1/p}} + \frac{\|\gamma; e\|_{L_p}}{(C_{p,m-l}(e))^{1/p}} \right), \tag{2.3.16}$$

$$\|\gamma\|_{MW_p^l} \sim \sup_{e: d(e) \le 1} \frac{\|\nabla_l \gamma; e\|_{L_p}}{(C_{p,l}(e))^{1/p}} + \|\gamma\|_{L_\infty}. \tag{2.3.17}$$

2.3.1 Another Characterization of the Space $M(W_p^m \to W_p^l)$ for $0 < l < m$, $pm \le n$, $p > 1$

Lemma 2.3.6. *Let $p \in (1, \infty)$, $0 < \nu < \mu$, and let φ be a nonnegative function in $L_{p\mu,\mathrm{loc}}$. Then, for any compact set e of positive measure,*

$$\sup_e \left(\frac{\int_e \varphi^{\nu p} dx}{C_{p,\nu}(e)} \right)^{1/\nu} \le \sup_e \left(\frac{\int_e \varphi^{\mu p} dx}{C_{p,\mu}(e)} \right)^{1/\mu}. \tag{2.3.18}$$

The same assertion holds for compact sets e of positive measure with diameter not exceeding 1.

Proof. Let $u \in C_0^\infty$ and let $f = J_\nu u$. By (1.2.17) and Hölder's inequality

$$\int \varphi^{\nu p} |u|^p dx \le c \int \varphi^{\nu p} (J_\mu |f|)^{\nu p/\mu} (\mathcal{M} f)^{(\mu-\nu)p/\mu} dx$$

$$\le c\Big(\int \varphi^{\mu p} (J_\mu |f|)^p dx \Big)^{\nu/\mu} \Big(\int (\mathcal{M} f)^p dx \Big)^{(\mu-\nu)/\mu}.$$

Using the continuity of the Hardy–Littlewood operator $\mathcal{M}$ in L_p, we find

$$\int \varphi^{\nu p} |u|^p dx \le c \sup_e \left(\frac{\int_e \varphi^{\nu p} dx}{C_{p,\mu}(e)} \right)^{\nu/\mu} \|J_\mu |f|\, \|_{W_p^\mu}^{\nu p/\mu} \|f\|_{L_p}^{(\mu-\nu)p/\mu}.$$

Since $\| J_\mu |f| \, \|_{W_p^\mu} \le c \, \|f\|_{L_p}$, it follows that

$$\int \varphi^{\nu p} |u|^p dx \le c \, \sup_e \left(\frac{\int_e \varphi^{\nu p} dx}{C_{p,\mu}(e)} \right)^{\nu/\mu} \|u\|^p_{W_p^\mu}.$$

The result follows by Theorem 1.2.2. □

Corollary 2.3.2. *For any $m > l \ge 0$ and $p \in (1, \infty)$,*

$$\sup_e \frac{\|\gamma; e\|_{L_p}}{(C_{p,m-l}(e))^{1/p}} \le c \, \sup_e \frac{\|\gamma; e\|_{L_{pm/(m-l)}}}{(C_{p,m}(e))^{(m-l)/pm}}. \tag{2.3.19}$$

The same assertion holds for compact sets e of positive measure with diameter not exceeding 1.

Proof. The result follows by Lemma 2.3.6 with $\varphi = |\gamma|^{1/(m-l)}$, $\nu = m - l$, and $\mu = m$. □

The following assertion was obtained in [Ad2].

Lemma 2.3.7. *Let $0 < l < m \le n/p$, $p \in (1, \infty)$, and let $I_l f$ be the Riesz potential with a nonnegative density $f \in L_{p,\mathrm{loc}}$. Then*

$$(I_l f)(x) \le c \left(\sup_{r>0} r^{m-n/p} \|f; \mathcal{B}_r(x)\|_{L_p} \right)^{l/m} \left((\mathcal{M} f)(x) \right)^{(m-l)/m}.$$

Proof. It is enough to put $x = 0$. For any $\delta > 0$ we have

$$(I_l f)(0) = \int_{\mathcal{B}_\delta} \frac{f(z)dz}{|z|^{n-l}} + \int_{\mathbb{R}^n \setminus \mathcal{B}_\delta} \frac{f(z)dz}{|z|^{n-l}}. \tag{2.3.20}$$

Clearly,

$$\int_{\mathcal{B}_\delta} \frac{f(z)dz}{|z|^{n-l}} \le c \, \delta^l (\mathcal{M} f)(0). \tag{2.3.21}$$

The second integral in (2.3.20) can be written as

$$\int_{\mathbb{R}^n \setminus \mathcal{B}_\delta} \frac{f(z)dz}{|z|^{n-l}} = (n-l) \int_\delta^\infty \int_{\mathcal{B}_r} f(\xi) d\xi \frac{dr}{r^{n-l+1}} - \delta^{l-n} \int_{\mathcal{B}_\delta} f(\xi) d\xi.$$

By Hölder's inequality the right-hand side does not exceed

$$c \Big(\int_\delta^\infty \Big(\int_{\mathcal{B}_r} f(\xi)^p d\xi \Big)^{1/p} \frac{dr}{r^{1-l+n/p}} + \delta^{l-n/p} \Big(\int_{\mathcal{B}_\delta} f(\xi)^p d\xi \Big)^{1/p} \Big).$$

Hence

$$\int_{\mathbb{R}^n \setminus \mathcal{B}_\delta} \frac{f(z)dz}{|z|^{n-l}} \leq c\, \delta^{l-m} \sup_{r>0} r^{m-n/p} \|f; \mathcal{B}_r(x)\|_{L_p}$$

which together with (2.3.21) implies that

$$(I_l f)(0) \leq c\, \big(\delta^l (\mathcal{M} f)(0) + \delta^{l-m} \sup_{r>0} r^{m-n/p} \|f; \mathcal{B}_r(x)\|_{L_p}\big).$$

The result follows by minimizing the right-hand side in δ. □

The next lemma is due to Verbitsky. For its proof see [MSh16], Sect. 2.6 and [MV1], Sect. 3.

Lemma 2.3.8. *Let $p \in (1, \infty)$ and let $0 < m \leq n/p$. Then*

$$\sup_e \frac{\|\mathcal{M} f; e\|_{L_p}}{(C_{p,m}(e))^{1/p}} \leq c \sup_e \frac{\|f; e\|_{L_p}}{(C_{p,m}(e))^{1/p}}. \tag{2.3.22}$$

This inequality plays a crucial role in the next assertion, also proved by Verbitsky, see [MSh16], Sect. 2.6.

Lemma 2.3.9. *Let m and l be integers, $0 < l < m \leq n/p$, and let $p \in (1, \infty)$. Then*

$$\sup_e \frac{\|\gamma; e\|_{L_{pm/(m-l)}}}{(C_{p,m}(e))^{(m-l)/pm}}$$

$$\leq c \Big(\sup_e \frac{\|\nabla_l \gamma; e\|_{L_p}}{(C_{p,m}(e))^{1/p}} + \sup_{x \in \mathbb{R}^n} \|\gamma; \mathcal{B}_1(x)\|_{L_1}\Big). \tag{2.3.23}$$

Proof. By the Sobolev integral representation [Sob] (see, for instance, [Maz14], Subsect. 1.1.10 and [Bur], Sect. 3.4)

$$|\gamma(x)| \leq c \Big(\int_{\mathcal{B}_2} \frac{|\nabla_l \gamma(x+z)|}{|z|^{n-l}} dz + \sup_{x \in \mathbb{R}^n} \|\gamma; \mathcal{B}_1(x)\|_{L_1}\Big).$$

Combining this inequality with Lemma 2.3.7, we obtain

$$|\gamma(x)| \leq c \big((\sup_{z \in \mathbb{R}^n, r \in (0,1)} r^{m-n/p} \|\nabla_l \gamma; \mathcal{B}_r(z)\|_{L_p})^{l/m} ((\mathcal{M} \nabla_l \gamma)(x))^{1-l/m}$$

$$+ \sup_{x \in \mathbb{R}^n} \|\gamma; \mathcal{B}_1(x)\|_{L_1}\big) \tag{2.3.24}$$

for almost all $x \in \mathbb{R}^n$.

Adopting the notation

$$\mathcal{K} = \sup_{z \in \mathbb{R}^n, r \in (0,1)} r^{m-n/p} \|\nabla_l \gamma; \mathcal{B}_r(z)\|_{L_p},$$

we deduce from (2.3.24) that

$$\int_e |\gamma|^{pm/(m-l)} dx$$
$$\leq c\Big(\mathcal{K}^{pl/(m-l)} \int_e |(\mathcal{M}\nabla_l \gamma)(x)|^p dx + (\sup_{x\in\mathbb{R}^n} \|\gamma; \mathcal{B}_1(x)\|_{L_1})^{pm/(m-l)} \text{mes}_n e\Big).$$

Using here the obvious estimate $\text{mes}_n e \leq C_{p,m}(e)$, we find that

$$\left(\frac{\int_e |\gamma(x)|^{pm/(m-l)} dx}{C_{p,m}(e)}\right)^{(m-l)/pm}$$
$$\leq c\Big(\mathcal{K}^{l/m}\Big(\sup_e \frac{\|\mathcal{M}\nabla_l\gamma; e\|_{L_p}}{(C_{p,m}(e))^{1/p}}\Big)^{1-l/m} + \sup_{x\in\mathbb{R}^n} \|\gamma; \mathcal{B}_1(x)\|_{L_1}\Big).$$

Reference to Lemma 2.3.8 and the inequality

$$\mathcal{K} \leq \sup_e \frac{\|\nabla_l\gamma; e\|_{L_p}}{(C_{p,m}(e))^{1/p}},$$

completes the proof. □

The following assertion gives one more representation of a norm in $M(W_p^m \to W_p^l)$.

Theorem 2.3.3. *Let m and l be integers, $0 < l < m \leq n/p$, and let $p \in (1,\infty)$. A function γ belongs to the space $M(W_p^m \to W_p^l)$ if and only if $\gamma \in W^l_{p,\text{loc}}$ and for any compact set $e \subset \mathbb{R}^n$*

$$\|\nabla_l\gamma; e\|^p_{L_p} \leq c\, C_{p,m}(e).$$

Moreover,

$$\|\gamma\|_{M(W_p^m\to W_p^l)} \sim \sup_e \frac{\|\nabla_l\gamma; e\|_{L_p}}{(C_{p,m}(e))^{1/p}} + \sup_{x\in\mathbb{R}^n} \|\gamma; \mathcal{B}_1(x)\|_{L_1}. \tag{2.3.25}$$

Here one can use only compact sets with diameter not exceeding 1.

Proof. We start with the lower estimate. By Lemma 2.3.4,

$$\|\gamma\eta\|_{L_p} \leq \|\gamma\|_{M(W_p^{m-l}\to L_p)} \|\eta\|_{W_p^{m-l}}$$

for any $\eta \in C_0^\infty(\mathcal{B}_2(x))$ with $\eta = 1$ on $\mathcal{B}_1(x)$, where x is an arbitrary point in $\mathbb{R}^n$. Therefore,

$$\sup_{x\in\mathbb{R}^n} \|\gamma; \mathcal{B}_1(x)\|_{L_p} \leq c\, \|\gamma\|_{M(W_p^{m-l}\to L_p)}.$$

The upper estimate is a direct corollary of (2.3.14), Corollary 2.3.2, and Verbitsky's Lemma 2.3.9. □

Corollary 2.3.3. *Let m and l be integers, $0 < l < m$ and let $p \in (1, \infty)$. Then*

$$\|\gamma\|_{M(W_p^m \to W_p^l)} \sim \sum_{j=0}^{l} \|\nabla_{l-j}\gamma\|_{M(W_p^{m-j} \to L_p)} \tag{2.3.26}$$

and

$$\|\gamma\|_{M(W_p^m \to W_p^l)} \sim \|\nabla_l \gamma\|_{M(W_p^m \to L_p)} + \|\gamma\|_{L_{1,\mathrm{unif}}}. \tag{2.3.27}$$

For $m = l$ the norm $\|\gamma\|_{L_{1,\mathrm{unif}}}$ should be replaced by $\|\gamma\|_{L_\infty}$.

Proof. The upper estimates follow from (2.3.25) and (2.3.3). The lower estimate in (2.3.26) results from

$$\|\nabla_{l-j}\gamma\|_{M(W_p^{m-j} \to L_p)} \le c\,\|\gamma\|_{M(W_p^{m-j} \to W_p^{l-j})} \le c\,\|\gamma\|_{M(W_p^m \to W_p^l)}.$$

□

We finish this subsection with one more two-sided estimate for the norm in $M(W_p^m \to W_p^l)$.

Corollary 2.3.4. *Let m and l be integers, $0 < l < m$ and let $p \in (1, \infty)$. Then*

$$c \sum_{j=0}^{l} \sup_e \frac{\|\nabla_j \gamma; e\|_{L_p}}{[C_{p,m-l+j}(e)]^{1/p}} \le \|\gamma\|_{M(W_p^m \to W_p^l)}$$

$$\le c\Big(\sup_e \frac{\|\nabla_l \gamma; e\|_{L_p}}{[C_{p,m}(e)]^{1/p}} + \|\gamma\|_{L_{1,\mathrm{unif}}}\Big). \tag{2.3.28}$$

For $m = l$ the norm $\|\gamma\|_{L_{1,\mathrm{unif}}}$ should be replaced by $\|\gamma\|_{L_\infty}$.

Proof. The lower estimate follows from (2.3.26) and (2.3.3). The upper one is contained in (2.3.25). □

2.3.2 Characterization of the Space $M(W_p^m \to W_p^l)$ for $pm > n$, $p > 1$

For $pm > n$ the space $M(W_p^m \to W_p^l)$ has a simple description which is contained in the next assertion.

Theorem 2.3.4. *If $pm > n$, $p \in (1, \infty)$, then*

$$\|\gamma\|_{M(W_p^m \to W_p^l)} \sim \sup_{x \in \mathbb{R}^n} \|\gamma; \mathcal{B}_1(x)\|_{W_p^l}. \tag{2.3.29}$$

Proof. Since for compact sets e with diameter less than 1 the equivalence relation

$$C_{p,m}(e) \sim 1$$

holds, the result follows from (2.3.25). □

Remark 2.3.1. For $pm > n$, $p \in (1, \infty)$, the relation (2.3.29) can be written as

$$\|\gamma\|_{M(W_p^m \to W_p^l)} \sim \|\gamma\|_{W^l_{p,\mathrm{unif}}}.$$

2.3.3 One-Sided Estimates for Norms of Multipliers in the Case $pm \le n$

For $mp \le n$ we can give different upper and lower bounds for norms in $M(W_p^m \to W_p^l)$ which do not involve capacity. In other words, we obtain separate non-capacitary necessary and sufficient conditions for a function to belong to this space of multipliers.

Using the estimates of the capacity of a ball given in Proposition 1.2.9 and Theorem 2.3.3 we immediately arrive at the following lower estimates for norms of multipliers.

Proposition 2.3.1. *(i) If $pm < n$, $p \in (1, \infty)$, then*

$$\|\gamma\|_{M(W_p^m \to W_p^l)} \ge c\Big(\sup_{x \in \mathbb{R}^n, r \in (0,1)} r^{m-n/p} \|\nabla_l \gamma; \mathcal{B}_r(x)\|_{L_p} + \sup_{x \in \mathbb{R}^n} \|\gamma; \mathcal{B}_1(x)\|_{L_1}\Big). \tag{2.3.30}$$

(ii) If $pm = n$, $p \in (1, \infty)$, then

$$\|\gamma\|_{M(W_p^m \to W_p^l)} \ge c\Big(\sup_{x \in \mathbb{R}^n, r \in (0,1)} ((\log(2/r))^{1-1/p} \|\nabla_l \gamma; \mathcal{B}_r(x)\|_{L_p} + \sup_{x \in \mathbb{R}^n} \|\gamma; \mathcal{B}_1(x)\|_{L_1}\Big). \tag{2.3.31}$$

Next, we give upper estimates for the norm in $M(W_p^m \to W_p^l)$ which result directly from Theorem 2.3.3 and Proposition 1.2.8.

Proposition 2.3.2. *(i) If $pm < n$, $p \in (1, \infty)$, then*

$$\|\gamma\|_{M(W_p^m \to W_p^l)} \le c\Big(\sup_{e: d(e) \le 1} \frac{\|\nabla_l \gamma; e\|_{L_p}}{(\text{mes}_n e)^{1/p - m/n}} + \sup_{x \in \mathbb{R}^n} \|\gamma; \mathcal{B}_1(x)\|_{L_1}\Big), \tag{2.3.32}$$

where $d(e)$ is the diameter of e.

(ii) If $pm = n$, $p \in (1, \infty)$, then

$$\|\gamma\|_{M(W_p^m \to W_p^l)} \le c\Big(\sup_{e: d(e) \le 1} (\log(2^n/\text{mes}_n e))^{1-1/p} \|\nabla_l \gamma; e\|_{L_p} + \sup_{x \in \mathbb{R}^n} \|\gamma; \mathcal{B}_1(x)\|_{L_1}\Big). \tag{2.3.33}$$

For $m = l$ one should replace $\sup_{x \in \mathbb{R}^n} \|\gamma; \mathcal{B}_1(x)\|_{L_1}$ by $\|\gamma\|_{L_\infty}$.

The next assertion follows immediately from Proposition 2.3.2.

Corollary 2.3.5. *(i) If $pm < n$, $l < m$, and $p \in (1, \infty)$, then*

$$\|\gamma\|_{M(W_p^m \to W_p^l)} \le c\, \|\gamma\|_{W^l_{n/m,\text{unif}}}.$$

(ii) If $lp < n$, $p \in (1, \infty)$, then

$$\|\gamma\|_{MW_p^l} \le c\Big(\sup_{x \in \mathbb{R}^n} \|\nabla_l \gamma; \mathcal{B}_1(x)\|_{L_{n/l}} + \|\gamma\|_{L_\infty}\Big).$$

2.3.4 Examples of Multipliers

Propositions 2.3.1, 2.3.1 and Theorem 2.3.4 enable one to verify conditions for inclusion of individual functions into spaces of multipliers. We give three examples of this kind.

Example 2.3.1. Let

$$\gamma(x) = \eta(x)|x|^{\alpha+i\beta},$$

where $\eta \in C_0^\infty$, $\eta(0) = 1$, $\alpha \in \mathbb{R}$, and $\beta \in \mathbb{R}\backslash\{0\}$.

Let $mp > n$. Clearly, $\gamma \in W^l_{p,\mathrm{unif}} = M(W_p^m \to W_p^l)$ if and only if $\alpha > l - n/p$.

Suppose that $mp \le n$. If $\alpha \le l - n/p$, then $\gamma \notin W^l_{p,\mathrm{loc}}$ and hence $\gamma \notin M(W_p^m \to W_p^l)$ because of (2.3.30) and (2.3.31). Consider the case $mp \le n$, $\alpha > l - n/p$. We obtain

$$\|\nabla_l \gamma; e\|_{L_p} \sim \||x|^{\alpha-l}; e\|_{L_p} \le c\,(\mathrm{mes}_n e)^{\alpha-l+n/p}$$

for all compact sets e with $d(e) \le 1$. Using (2.3.32) and (2.3.33), we conclude that $\gamma \in M(W_p^m \to W_p^l)$. Summarizing, we have

$$\gamma \in M(W_p^m \to W_p^l) \Longleftrightarrow \alpha > l - n/p.$$

Example 2.3.2. Let $\mu > 0$ and let

$$\gamma(x) = \eta(x)\exp(i|x|^{-\mu}),$$

where $\eta \in C_0^\infty$, $\eta(0) = 1$. Clearly,

$$|\nabla_l \gamma(x)| \sim |x|^{-l(\mu+1)} \quad \text{as } x \to 0.$$

Therefore, $\gamma \in W_p^l(\mathbb{R}^n)$ is equivalent to $n > pl(\mu+1)$. Let us find a criterion for $\gamma \in M(W_p^m \to W_p^l)$.

In view of Theorem 2.3.4, the same inequality $n > pl(\mu+1)$ is necessary and sufficient for γ to belong to $M(W_p^m(\mathbb{R}^n) \to W_p^l(\mathbb{R}^n))$ for $mp > n$.

Suppose that $mp < n$. We have

$$\|\nabla_l \gamma; \mathcal{B}_r\|_{L_\infty} \sim \||x|^{-l(\mu+1)}; \mathcal{B}_r\|_{L_p}$$

and

$$\lim_{r\to 0} r^{m-n/p}\|\nabla_l \gamma; \mathcal{B}_r\|_{L_p} = \infty$$

for $m < l(\mu+1)$. According to Proposition 2.3.1, this means that $\gamma \notin M(W_p^m \to W_p^l)$ for $m < l(\mu+1)$. If $m \ge l(\mu+1)$, then

$$\|\nabla_l \gamma; e\|_{L_p} \le c\||x|^{-l(\mu+1)}; e\|_{L_p} \le c\,(\mathrm{mes}_n e)^{-l(\mu+1)/n+1/p}$$

for any compact set e with diameter $d(e) \le 1$. This, together with Proposition 2.3.2, implies that $\gamma \in M(W_p^m \to W_p^l)$. Thus, for $mp < n$,

$$\gamma \in M(W_p^m \to W_p^l) \Longleftrightarrow m \geq l(\mu + 1).$$

In the same way we verify that

$$\gamma \in M(W_p^m \to W_p^l) \Longleftrightarrow m > l(\mu + 1)$$

for $mp = n$.

Example 2.3.3. Let $\mu, \nu > 0$, $\eta \in C_0^\infty(\mathcal{B}_1(0))$, $\eta(0) = 1$ and

$$\gamma(x) = \eta(x)(\log |x|^{-1})^{-\nu} \exp(i\,(\log |x|^{-1})^{\mu}).$$

Clearly,

$$|\nabla_l \gamma(x)| \sim c\,|x|^{-l}(\log |x|^{-1})^{l(\mu-1)-\nu}.$$

Using the same arguments as in Example 2.3.2, from this relation and Propositions 2.3.1 and 2.3.2 we obtain

$$\lambda \in W_p^l(\mathbb{R}^n) \Longleftrightarrow l(\mu - 1) < \nu - 1/p,$$
$$\lambda \in MW_p^l(\mathbb{R}^n) \Longleftrightarrow l(\mu - 1) \leq \nu - 1$$

for $lp = n$. □

2.4 The Space $M(W_p^m(\mathbb{R}_+^n) \to W_p^l(\mathbb{R}_+^n))$

2.4.1 Extension from a Half-Space

Let $\mathbb{R}_+^n = \{z = (x, x_n) : x \in \mathbb{R}^{n-1},\ x_n > 0\}$. The classical extension operator π is defined for functions given on $\mathbb{R}_+^n$ by

$$\pi(v)(z) = \begin{cases} v(z) & \text{for } x_n > 0, \\ \displaystyle\sum_{j=1}^{l} \alpha_j\, v(x, -jx_n) & \text{for } x_n < 0, \end{cases}$$

where α_j satisfy the conditions

$$\sum_{j=1}^{l} (-1)^k j^k \alpha_j = 1, \qquad 0 \leq k \leq l - 1.$$

Lemma 2.4.1. *Suppose that* $\gamma \in M(W_p^m(\mathbb{R}_+^n) \to W_p^l(\mathbb{R}_+^n))$, *where* $0 \leq l \leq m$ *and* $p \in [1, \infty)$. *Then*

$$\pi(\gamma) \in M(W_p^m(\mathbb{R}^n) \to W_p^m(\mathbb{R}^n))$$

and

$$\|\pi(\gamma); \mathbb{R}^n\|_{M(W_p^m \to W_p^l)} \leq c\,\|\gamma; \mathbb{R}_+^n\|_{M(W_p^m \to W_p^l)}. \tag{2.4.1}$$

Proof. Since $\gamma \in W_{p,\mathrm{loc}}^l(\overline{\mathbb{R}_+^n})$, it follows by the well-known property of the operator π that $\pi(\gamma) \in W_{p,\mathrm{loc}}^l(\mathbb{R}^n)$. Hence $\pi(\gamma)u \in W_p^l(\mathbb{R}^n)$ for any $u \in C_0^\infty(\mathbb{R}^n)$. We have

$$\|\pi(\gamma);\mathbb{R}^n\|_{W_p^l}^p = \|\gamma u;\mathbb{R}_+^n\|_{W_p^l} + \|\pi(\gamma)u;\mathbb{R}_-^n\|_{W_p^l}$$

$$\leq \|\gamma u;\mathbb{R}_+^n\|_{W_p^l} + c\sum_{j=1}^{l} \|\gamma\, u_j;\mathbb{R}_+^n\|_{W_p^l}^p,$$

where

$$u_j(x,x_n) = u(x,-x_n/j)$$

and $\mathbb{R}_-^n = \{z=(x,x_n) : x \in \mathbb{R}^{n-1},\ x_n < 0\}$. Therefore,

$$\|\pi(\gamma);\mathbb{R}^n\|_{W_p^l}^p \leq c\,\|\gamma;\mathbb{R}_+^n\|_{M(W_p^m\to W_p^l)}\big(\|u;\mathbb{R}_+^n\|_{W_p^m} + \sum_{j=1}^{l}\|u_j;\mathbb{R}_+^n\|_{W_p^m}\big)$$

and, since

$$\|u_j;\mathbb{R}_+^n\|_{W_p^m} \leq c\,\|u;\mathbb{R}_-^n\|_{W_p^m},$$

it follows that

$$\|\pi(\gamma);\mathbb{R}^n\|_{W_p^l}^p \leq c\,\|\gamma;\mathbb{R}_+^n\|_{M(W_p^m\to W_p^l)}\|u;\mathbb{R}^n\|_{W_p^m}.$$

The lemma is proved. □

2.4.2 The Case $p > 1$

In this subsection and elsewhere we use the notation

$$\mathcal{B}_r^\pm(Y) = \mathcal{B}_r(Y) \cap \overline{\mathbb{R}_\pm^n}.$$

Theorem 2.4.1. *A function γ belongs to $M(W_p^m(\mathbb{R}_+^n) \to W_p^m(\mathbb{R}_+^n))$ with $p \in (1,\infty)$ if and only if $\gamma \in W_{p,\mathrm{loc}}^l(\mathbb{R}_+^n)$ and, for any compact set $e \subset \overline{\mathbb{R}_+^n}$,*

$$\|\nabla_l\gamma;e\|_{L_p}^p \leq c\,C_{p,m}(e). \tag{2.4.2}$$

Moreover,

$$c_1 \sup_e \sum_{k=0}^{l} \frac{\|\nabla_k\gamma;e\|_{L_p}}{(C_{p,m-l+k}(e))^{1/p}} \leq \|\gamma;\mathbb{R}_+^n\|_{M(W_p^m\to W_p^l)}$$

$$\leq c_2\Big(\sup_e \frac{\|\nabla_l\gamma;e\|_{L_p}}{(C_{p,m}(e))^{1/p}} + \sup_{X\in\mathbb{R}_+^n}\|\gamma;\mathcal{B}_1^+(X)\|_{L_1}\Big). \tag{2.4.3}$$

Proof. By Theorem 2.3.3

$$\sup_{e\subset\overline{\mathbb{R}^n_+}} \frac{\|\nabla_k\gamma; e\|_{L_p}}{(C_{p,m-l+k}(e))^{1/p}} \leq \sup_{e\subset\mathbb{R}^n} \frac{\|\nabla_k\pi(\gamma); e\|_{L_p}}{(C_{p,m-l+k}(e))^{1/p}} \leq c\,\|\pi(\gamma); \mathbb{R}^n\|_{M(W^m_p\to W^l_p)}.$$

Reference to Lemma 2.4.1 implies the lower estimate for $\|\gamma; \mathbb{R}^n_+\|_{M(W^m_p\to W^l_p)}$.

We turn to the proof of the upper estimate in (2.4.3). Let $u \in C_0^\infty(\overline{\mathbb{R}^n_+})$ and let

$$\gamma \in M(W^m_p(\mathbb{R}^n_+) \to W^l_p(\mathbb{R}^n_+)).$$

We have

$$\|\gamma u; \mathbb{R}^n_+\|_{W^l_p} \leq \|\pi(\gamma)\,\pi(u); \mathbb{R}^n\|_{W^l_p}. \tag{2.4.4}$$

By Theorem 2.3.3 the right-hand side does not exceed

$$c\Big(\sup_{e\subset\mathbb{R}^n} \frac{\|\nabla_l\pi(\gamma); e\|_{L_p}}{(C_{p,m}(e))^{1/p}} + \sup_{z\in\mathbb{R}^n} \|\pi(\gamma); \mathcal{B}_1(z)\|_{L_1}\Big)\|\pi(u); \mathbb{R}^n\|_{W^m_p}.$$

Since

$$\|\pi(u); \mathbb{R}^n\|_{W^m_p} \leq c\,\|u; \mathbb{R}^n_+\|_{W^m_p} \tag{2.4.5}$$

and

$$\sup_{z\in\mathbb{R}^n} \|\pi(\gamma); \mathcal{B}_1(z)\|_{L_1} \leq c \sup_{X\in\mathbb{R}^n_+} \|\gamma; \mathcal{B}^+_1(X)\|_{L_1}, \tag{2.4.6}$$

it remains to prove the inequality

$$\sup_{e\subset\mathbb{R}^n} \frac{\|\nabla_l\pi(\gamma); e\|_{L_p}}{(C_{p,m}(e))^{1/p}} \leq c \sup_{e\subset\overline{\mathbb{R}^n_+}} \frac{\|\nabla_l\gamma; e\|_{L_p}}{(C_{p,m}(e))^{1/p}}. \tag{2.4.7}$$

Let $e_\pm = e\cap\overline{\mathbb{R}^n_\pm}$. The quotient on the left-hand side of (2.4.7) does not exceed the sum

$$\frac{\|\nabla_l\gamma; e_+\|_{L_p}}{(C_{p,m}(e_+))^{1/p}} + \frac{\|\nabla_l\pi(\gamma); e_-\|_{L_p}}{(C_{p,m}(e_-))^{1/p}}. \tag{2.4.8}$$

We put

$$e_j = \{z : (x, -x_n/j) \in e_-\}, \qquad j = 1, \ldots, l.$$

Using the inequality

$$\|\nabla_l\pi(\gamma); e_-\|_{L_p} \leq c\,\|\nabla_l\gamma; e_j\|_{L_p}$$

along with the equivalence relation

$$C_{p,m}(e_-) \sim C_{p,m}(e_j),$$

we conclude that the second term in (2.4.8) is majorized by

$$c \sup_{e\subset\overline{\mathbb{R}^n_+}} \frac{\|\nabla_l\gamma; e\|_{L_p}}{(C_{p,m}(e))^{1/p}}.$$

The proof is complete. □

The following assertion follows directly from the last theorem.

Corollary 2.4.1. *If $\gamma \in M(W_p^m(\mathbb{R}_+^n) \to W_p^l(\mathbb{R}_+^n))$, then*

$$D^\alpha \gamma \in M(W_p^m(\mathbb{R}_+^n) \to W_p^{l-|\alpha|}(\mathbb{R}_+^n))$$

for any multi-index α of order $|\alpha| \le l$. The estimate

$$\|D^\alpha \gamma; \mathbb{R}_+^n\|_{M(W_p^m \to W_p^{l-|\alpha|})} \le c\,\|\gamma; \mathbb{R}_+^n\|_{M(W_p^m \to W_p^l)}$$

holds.

2.4.3 The Case $p = 1$

Theorem 2.4.2. *A function γ belongs to $M(W_1^m(\mathbb{R}_+^n) \to W_1^l(\mathbb{R}_+^n))$ if and only if $\gamma \in W_{1,\mathrm{loc}}^l(\mathbb{R}_+^n)$ and*

$$\|\nabla_l \gamma; \mathcal{B}_r^+(X)\|_{L_1} \le c\, r^{n-m} \tag{2.4.9}$$

for any $X \in \mathbb{R}_+^n$ and $r \in (0,1)$. Moreover,

$$\|\gamma; \mathbb{R}_+^n\|_{M(W_1^m \to W_1^l)} \sim$$

$$\sup_{\substack{X \in \mathbb{R}_+^n \\ r \in (0,1)}} r^{m-n} \|\nabla_l \gamma; \mathcal{B}_r^+(X)\|_{L_1} + \sup_{X \in \mathbb{R}_+^n} \|\gamma; \mathcal{B}_1^+(X)\|_{L_1}. \tag{2.4.10}$$

Proof. By Theorem 2.2.3

$$\sup_{\substack{X \in \mathbb{R}_+^n \\ r \in (0,1)}} r^{m-n} \|\nabla_l \gamma; \mathcal{B}_r^+(X)\|_{L_1} \le \sup_{\substack{Y \in \mathbb{R}_+^n \\ r \in (0,1)}} r^{m-n} \|\nabla_l \pi(\gamma); \mathcal{B}_r(Y)\|_{L_1}$$

$$\le c\,\|\pi(\gamma); \mathbb{R}^n\|_{M(W_1^m \to W_1^l)}.$$

Reference to Lemma 2.4.1 implies the lower estimate for $\|\gamma; \mathbb{R}_+^n\|_{M(W_1^m \to W_1^l)}$.

To obtain the upper estimate, we take $U \in C_0^\infty(\overline{\mathbb{R}_+^n})$ and $\gamma \in M(W_1^m(\mathbb{R}_+^n) \to W_1^l(\mathbb{R}_+^n))$. In view of (2.4.4), which is also valid for $p = 1$, the norm $\|\gamma u; \mathbb{R}_+^n\|_{W_1^l}$ is dominated by

$$c\Big(\sup_{\substack{Y \in \mathbb{R}^n \\ r \in (0,1)}} r^{m-n} \|\nabla_l \pi(\gamma); \mathcal{B}_r(Y)\|_{L_1} + \sup_{Y \in \mathbb{R}^n} \|\pi(\gamma); \mathcal{B}_1(Y)\|_{L_1}\Big) \|\pi(U); \mathbb{R}^n\|_{W_1^m}.$$

By (2.4.5), which holds also for $p = 1$, and (2.4.4), it remains to obtain the majorant for $\|\nabla_l \pi(\gamma); \mathcal{B}_r(Y)\|_{L_1}$. We have

$$\|\nabla_l \pi(\gamma); \mathcal{B}_r(Y)\|_{L_1} \le \|\nabla_l \pi(\gamma); \mathcal{B}_r^+(Y)\|_{L_1} + \|\nabla_l \pi(\gamma); \mathcal{B}_r^-(Y)\|_{L_1}.$$

Putting

$$\mathcal{B}_{r,j} = \{z : (x, -x_n/j) \in \mathcal{B}_r^-(Y)\}$$

and using the inequalities

$$\|\nabla_l \pi(\gamma); \mathcal{B}_r^-(Y)\|_{L_1} \le c\,\|\nabla_l \pi(\gamma); \mathcal{B}_{r,j}\|_{L_1} \le c \sup_{X \in \mathbb{R}_+^n} \|\nabla \gamma; \mathcal{B}_r^+(X)\|_{L_1},$$

we complete the proof. □

2.5 The Space $M(W_p^m \to W_p^{-k})$

Let m and k be positive integers, and let W_p^{-k} stand for the dual space $(W_{p'}^k)'$, where $p + p' = pp'$. The following assertion contains a sufficient condition for inclusion into the distribution space $M(W_p^m \to W_p^{-k})$.

Theorem 2.5.1. *(i) Let $p \in (1, \infty)$, $0 < m \le k$. If*

$$\gamma = \sum_{|\alpha| \le k} D^\alpha \gamma_\alpha \tag{2.5.1}$$

with

$$\gamma_\alpha \in M(W_{p'}^k \to W_{p'}^{k-m}) \cap M(W_p^m \to L_p), \tag{2.5.2}$$

then $\gamma \in M(W_p^m \to W_p^{-k})$.

(ii) Let $p \in (1, \infty)$, $m \ge k > 0$. If

$$\gamma = \sum_{|\alpha| \le m} D^\alpha \gamma_\alpha$$

with

$$\gamma_\alpha \in M(W_p^m \to W_p^{m-k}) \cap M(W_{p'}^k \to L_{p'}),$$

then $\gamma \in M(W_p^m \to W_p^{-k})$.

Proof. It suffices to prove only (i), since (ii) follows from (i) by duality.

Let $u \in W_p^m$, $m \le k$. Since

$$uD^\alpha \gamma_\alpha = \sum_{\lambda \le \alpha} c_{\lambda\alpha} D^\lambda (\gamma_\alpha D^{\alpha-\lambda} u), \quad c_{\lambda\alpha} = \text{const},$$

we have

$$\|\gamma u\|_{W_p^{-k}} \le c \sum_{|\lambda| \le |\alpha| \le k} \|\gamma_\alpha D^{\alpha-\lambda} u\|_{W_p^{|\lambda|-k}}$$

$$\le c \sum_{|\lambda| \le |\alpha| \le k} \|\gamma_\alpha\|_{M(W_p^{m-k+|\lambda|} \to W_p^{|\lambda|-k})} \|u\|_{W_p^{m+|\alpha|-k}}. \tag{2.5.3}$$

Applying the interpolation inequality

$$\|\gamma_\alpha\|_{M(W_p^{m-k+|\lambda|} \to W_p^{|\lambda|-k})} \le c \|\gamma_\alpha\|^{(k-|\lambda|)/k}_{M(W_p^{m-k} \to W_p^{-k})} \|\gamma_\alpha\|^{|\lambda|/k}_{M(W_p^m \to L_p)},$$

which results from the interpolation property of Sobolev spaces (see [Tr4], Sect. 2.4), we obtain from (2.5.3)

$$\|\gamma u\|_{W_p^{-k}} \le c \, (\|\gamma_\alpha\|_{M(W_p^{m-k} \to W_p^{-k})} + \|\gamma_\alpha\|_{M(W_p^m \to L_p)}) \|u\|_{W_p^m}.$$

It remains to note that

$$\|\gamma_\alpha\|_{M(W_p^{m-k} \to W_p^{-k})} = \|\gamma_\alpha\|_{M(W_{p'}^k \to W_{p'}^{k-m})}.$$

The proof is complete. □

The next assertion shows that Theorem 2.5.1 provides a complete characterization of $M(W_p^m \to W_p^{-k})$ which holds under some conditions involving k, m, p, and n.

Theorem 2.5.2. *Let k and m be positive integers and let either $k \geq m > 0$ and $k > n/p'$ or $m \geq k > 0$ and $m > n/p$. Then $\gamma \in M(W_p^m \to W_p^{-k})$ if and only if*

$$\gamma \in W_{p,\mathrm{unif}}^{-k} \cap W_{p',\mathrm{unif}}^{-m}. \tag{2.5.4}$$

In particular, if $\max\{k, m\} > n/2$ *then $M(W_2^m \to W_2^{-k})$ is isomorphic to* $W_2^{-\min\{m,k\}}$.

Proof. It suffices to consider the case $k \geq m > 0$, $k > n/p'$, because the case $m \geq k > 0$, $m > n/p$ results by duality.

Necessity. It follows from the inclusion $\gamma \in M(W_p^m \to W_p^{-k})$ that $\gamma \in W_{p,\mathrm{unif}}^{-k}$. Since $M(W_p^m \to W_p^{-k})$ is isomorphic to $M(W_{p'}^k \to W_{p'}^{-m})$, we have $\gamma \in W_{p',\mathrm{unif}}^{-m}$ as well.

Sufficiency. It is standard and easily proved that

$$\gamma \in W_{p,\mathrm{unif}}^{-k} \cap W_{p',\mathrm{unif}}^{-m}$$

if and only if (2.5.1) holds with

$$\gamma_\alpha \in L_{p,\mathrm{unif}} \cap W_{p',\mathrm{unif}}^{k-m}.$$

Since $M(W_{p'}^k \to W_{p'}^{k-m})$ is isomorphic to $W_{p',\mathrm{unif}}^{k-m}$ for $p'k > n$, it follows that $\gamma_\alpha \in M(W_{p'}^k \to W_{p'}^{k-m})$.

It remains to show that $\gamma_\alpha \in M(W_p^m \to L_p)$. We choose q and r to satisfy

$$1/q > \max\{0, 1/p - m/n\} > -\varepsilon + 1/q,$$

$$1/r > \max\{0, 1/p' - (k-m)/n\} > -\varepsilon + 1/r$$

with a sufficiently small ε. Since $1/p > 1 - k/n$, we have $1/p > 1/q + 1/r$. By Hölder's inequality

$$\|\gamma_\alpha u\|_{L_{p,\mathrm{unif}}} \leq c\, \|\gamma_\alpha\|_{L_{r,\mathrm{unif}}} \|u\|_{L_{q,\mathrm{unif}}}$$

and by Sobolev's imbedding theorem

$$\|\gamma_\alpha u\|_{L_{p,\mathrm{unif}}} \leq c\, \|\gamma_\alpha\|_{W_{p',\mathrm{unif}}^{k-m}} \|u\|_{W_{p,\mathrm{unif}}^m}.$$

This means that $\gamma_\alpha \in M(W_p^m \to L_p)$. The proof is completed by reference to assertion (i) of Theorem 2.5.1. □

Remark 2.5.1. Note that by Sobolev's imbedding theorem

$$W^{-m}_{p',\mathrm{unif}} \subset W^{-k}_{p,\mathrm{unif}}, \qquad k \geq m,$$

if and only if either $n \leq (k-m)p$ or

$$n > (k-m)p, \quad \frac{k-m}{n} \geq \frac{2-p}{p}.$$

Under these conditions, $M(W^m_p \to W^{-k}_p)$ is isomorphic to $W^{-m}_{p',\mathrm{unif}}$ if $kp' > n$. Analogously, if $m \geq k$, $mp > n$ and either $n \leq (m-k)p'$ or

$$n > (m-k)p', \quad \frac{m-k}{n} \geq \frac{p-2}{p},$$

then $M(W^m_p \to W^{-k}_p)$ is isomorphic to $W^{-k}_{p,\mathrm{unif}}$.

We now state a direct application of Theorem 2.5.2 to the theory of differential operators.

Corollary 2.5.1. *Let k and m be integers and let $\mathcal{L}(D)$ denote a differential operator of order $m+k$ with constant coefficients. If either $k \geq m$ and $kp' > n$, or $m \geq k > 0$ and $mp > n$, then the operator*

$$W^m_p \ni u \to \mathcal{L}(D)u + \gamma(x)u \in W^{-k}_p$$

is continuous if and only if

$$\gamma \in W^{-k}_{p,\mathrm{unif}} \cap W^{-m}_{p',\mathrm{unif}}.$$

We conclude this subsection with a simple description of nonnegative elements of the space $M(W^m_2 \to W^{-m}_2)$.

Theorem 2.5.3. *Let $\gamma \geq 0$. Then $\gamma \in M(W^m_2 \to W^{-m}_2)$ if and only if $\gamma^{1/2} \in M(W^m_2 \to L_2)$. Moreover,*

$$\|\gamma\|_{M(W^m_2 \to W^{-m}_2)} = \|\gamma^{1/2}\|^2_{M(W^m_2 \to L_2)}.$$

Proof. Let $u \in W^m_2$, $v \in W^m_2$. We have

$$\left|\int \gamma u \bar{v}\, dx\right| \leq \|\gamma^{1/2}u\|_{L_2}\|\gamma^{1/2}v\|_{L_2} \leq \|\gamma^{1/2}\|^2_{M(W^m_2 \to L_2)}\|u\|_{W^m_2}\|v\|_{W^m_2}.$$

Hence

$$\|\gamma\|_{M(W^m_2 \to W^{-m}_2)} \leq \|\gamma^{1/2}\|^2_{M(W^m_2 \to L_2)}.$$

To obtain the converse inequality, we first note that

$$\left|\int \gamma u \bar{v}\, dx\right| \leq \|\gamma\|_{M(W^m_2 \to W^{-m}_2)}\|u\|_{W^m_2}\|v\|_{W^m_2}.$$

Putting here $u = v$, we obtain

$$\int |\gamma^{1/2} u|^2 \, dx \leq \|\gamma\|_{M(W_2^m \to W_2^{-m})} \|u\|^2_{W_2^m}.$$

Thus,

$$\|\gamma\|_{M(W_2^m \to W_2^{-m})} \geq \|\gamma^{1/2}\|^2_{M(W_2^m \to L_2)}.$$

□

2.6 The Space $M(W_p^m \to W_q^l)$

In this book consideration is limited to classes of the type $M(S_p^m \to S_p^l)$, i.e. to multipliers acting in one scale of spaces preserving the integrability degree p. Some generalizations to the pairs (S_p^m, S_q^l) are known. This is true, in particular, for the class $M(W_p^m \to W_q^l)$ with nonnegative and integer m and l, which is briefly described in the present subsection.

Using the same arguments as in the proof of Theorem 2.3.1, we obtain

$$\|\gamma\|_{M(W_p^m \to W_q^l)} \sim \|\nabla_l \gamma\|_{M(W_p^m \to L_q)} + \|\gamma\|_{M(W_p^{m-l} \to L_q)}, \tag{2.6.1}$$

where $p, q \in (1, \infty)$. Thus, the problem is reduced to a description of the class $M(W_p^m \to L_q)$. This can be obtained from Lemmas 1.3.1 – 1.3.5. In particular, by (2.6.1) and the same lemmas we have the following assertions concerning the norm in $M(W_p^m \to W_q^l)$ with $p < q$.

Theorem 2.6.1. *If* $1 < p < q$, *then*

$$\|\gamma\|_{M(W_p^m \to W_q^l)} \sim \begin{cases} \sup\limits_{x \in \mathbb{R}^n, r \in (0,1)} r^{m-n/p} \|\nabla_l \gamma; \mathcal{B}_r(x)\|_{L_q} + \sup\limits_{x \in \mathbb{R}^n} \|\gamma; \mathcal{B}_1(x)\|_{L_q} \\ \text{for } mp < n, \\ \sup\limits_{x \in \mathbb{R}^n, r \in (0,1)} (\log 2/r)^{1/p'} \|\nabla_l \gamma; \mathcal{B}_r(x)\|_{L_q} + \sup\limits_{x \in \mathbb{R}^n} \|\gamma; \mathcal{B}_1(x)\|_{L_q} \\ \text{for } mp = n, \\ \sup\limits_{x \in \mathbb{R}^n} \|\gamma; \mathcal{B}_1(x)\|_{W_q^l} \\ \text{for } mp > n. \end{cases}$$

Applying the same arguments as in Sect. 2.2, one can derive the following result from relations (1.3.2) and (1.3.4) with $p = 1$.

Theorem 2.6.2. *If* $q > 1$, *then*

$$\|\gamma\|_{M(W_1^m \to W_q^l)} \sim \begin{cases} \sup\limits_{x \in \mathbb{R}^n, r \in (0,1)} r^{m-n} \|\nabla_l \gamma; \mathcal{B}_r(x)\|_{L_q} + \sup\limits_{x \in \mathbb{R}^n} \|\gamma; \mathcal{B}_1(x)\|_{L_q} \\ \text{for } m < n, \\ \sup\limits_{x \in \mathbb{R}^n} \|\gamma; \mathcal{B}_1(x)\|_{W_q^l} \\ \text{for } m \geq n. \end{cases}$$

Choosing

$$\mu(e) = \int_e |\nabla_l \gamma(x)|^q dx$$

in Lemmas 1.3.4, 1.3.5 and using again the equivalence relation (2.6.1), we can obtain descriptions of $M(W_p^m \to W_q^l)$ with $p > q$, $p > 1$.

2.7 Certain Properties of Multipliers

In this section we study some simple properties of elements of the space $M(W_p^m \to W_p^l)$ with $p \in [1, \infty)$.

Proposition 2.7.1. *The space $M(W_p^m \to W_p^l)$ is contained in $M(W_p^{m-j} \to W_p^{l-j})$, $j = 1, \dots, l$, and*

$$\|\gamma\|_{M(W_p^{m-j} \to W_p^{l-j})} \leq c \, \|\gamma\|_{M(W_p^m \to W_p^l)}.$$

Proof. The inequality

$$\|\gamma\|_{M(W_p^{m-l} \to L_p)} \leq c \, \|\gamma\|_{M(W_p^m \to W_p^l)}$$

is proved in Lemma 2.2.2 for $p = 1$ and in Lemma 2.3.4 for $p > 1$. It remains to use interpolation inequalities (2.3.8) and (2.2.10). □

Proposition 2.7.2. *If a function γ depends only on variables $x_1, \dots, x_s$, $s < n$, and*

$$\gamma \in M(W_p^m(\mathbb{R}^s) \to W_p^l(\mathbb{R}^s)),$$

then

$$\gamma \in M(W_p^m(\mathbb{R}^n) \to W_p^l(\mathbb{R}^n))$$

and

$$\|\gamma; \mathbb{R}^n\|_{M(W_p^m \to W_p^l)} \leq c \, \|\gamma; \mathbb{R}^s\|_{M(W_p^m \to W_p^l)}.$$

The proof is obvious.

Proposition 2.7.3. *If $\gamma \in M(W_p^m \to W_p^l)$ and k is a positive integer satisfying $k \leq m/(m-l)$, then*

$$\gamma^k \in M(W_p^m \to W_p^{m-k(m-l)})$$

and

$$\|\gamma^k\|_{M(W_p^m \to W_p^{m-k(m-l)})} \leq c \, \|\gamma\|^k_{M(W_p^m \to W_p^l)}.$$

The proof is obvious.

Proposition 2.7.4. *The estimate*

$$\|\gamma\|_{L_\infty} \le \|\gamma\|_{MW_p^l} \tag{2.7.1}$$

holds.

Proof. For any $N = 1, 2, \ldots$, and arbitrary $u \in C_0^\infty$ we have

$$\|\gamma^N u\|_{L_p}^{1/N} \le \|\gamma^N u\|_{W_p^l}^{1/N} \le \|\gamma\|_{MW_p^l} \|u\|_{W_p^l}^{1/N}.$$

Passing to the limit as $N \to \infty$, we obtain (2.7.1). □

Proposition 2.7.5. *Let $\gamma \in MW_p^l$ and let σ be a segment on the real axis such that $\gamma(x) \in \sigma$ for almost all $x \in \mathbb{R}^n$. Further let $f \in C^{l-1,1}(\sigma)$. Then $f(\gamma) \in MW_p^l$ and*

$$\|f(\gamma)\|_{MW_p^l} \le c \sum_{j=0}^{l} \|f^{(j)}; \sigma\|_{L_\infty} \|\gamma\|_{MW_p^l}^j.$$

Proof. The assertion is obvious for $l = 1$. Suppose it is true for $l - 1$. For all $u \in C_0^\infty$ we have

$$\|u\, f(\gamma)\|_{W_p^l} \le \|f(\gamma)\nabla u\|_{W_p^{l-1}} + \|u\, f'(\gamma)\nabla\gamma\|_{W_p^{l-1}} + \|u\, f(\gamma)\|_{L_p}. \tag{2.7.2}$$

By the induction assumption, the first term on the right-hand side of (2.7.2) does not exceed

$$c\, \|\nabla u\|_{W_p^{l-1}} \sum_{j=0}^{l-1} \|f^{(j)}; \sigma\|_{L_\infty} \|\gamma\|_{MW_p^{l-1}}^j.$$

For the same reason, the second term on the right-hand side of (2.7.2) is dominated by

$$c\, \|u\nabla\gamma\|_{W_p^{l-1}} \sum_{j=0}^{l} \|f^{(j+1)}; \sigma\|_{L_\infty} \|\gamma\|_{MW_p^{l-1}}^j.$$

From (2.3.2) and Proposition 2.7.1 it follows that

$$\|\nabla\gamma\|_{M(W_p^l \to W_p^{l-1})} \le c\, \|\gamma\|_{MW_p^l}, \qquad \|\gamma\|_{MW_p^{l-1}} \le c\, \|\gamma\|_{MW_p^l}.$$

Hence the right-hand side of (2.7.2) is dominated by

$$c\, \|u\|_{W_p^l} \sum_{j=0}^{l} \|f^{(j)}; \sigma\|_{L_\infty} \|\gamma\|_{MW_p^l}^j.$$

The proof is complete. □

Corollary 2.7.1. *If $\gamma \in MW_p^l$ and $\|\gamma^{-1}\|_{L_\infty} < \infty$, then $\gamma^{-1} \in MW_p^l$ and*

$$\|\gamma^{-1}\|_{MW_p^l} \leq c\,\|\gamma^{-1}\|_{L_\infty}^{l+1}\,\|\gamma\|_{MW_p^l}^l.$$

Proof. The result follows from Proposition 2.7.5 with $f(\gamma) = \gamma^{-1}$ and from the inequality

$$\|\gamma^{-1}\|_{L_\infty}\,\|\gamma\|_{MW_p^l} \geq 1$$

which is a consequence of Proposition 2.7.4. □

Remark 2.7.1. All the assertions of the present section can be reformulated for the space $M(w_p^m \to w_p^l)$.

2.8 The Space $M(w_p^m \to w_p^l)$

In this section we assume that $mp < n$, $p \in (1, \infty)$ or $m \leq n$, $p = 1$.

Lemma 2.8.1. (*i*) *The inequality*

$$\|\gamma\|_{M(W_p^m \to L_p)} \leq \|\gamma\|_{M(w_p^m \to L_p)} \tag{2.8.1}$$

holds.

(*ii*) *Let $\rho > 0$ and let $\gamma \in M(w_p^m \to L_p)$. Then*

$$\lim_{\rho \to 0} \|\rho^{-m}\,\gamma(\cdot/\rho)\|_{M(W_p^m \to L_p)} = \|\gamma\|_{M(w_p^m \to L_p)}. \tag{2.8.2}$$

(*iii*) *The function γ satisfies*

$$\|\gamma\|_{M(w_p^m \to L_p)} \geq c \sup_{\substack{x \in \mathbb{R}^n \\ r > 0}} r^{m-n/p} \|\gamma; \mathcal{B}_r(x)\|_{L_p}. \tag{2.8.3}$$

Proof. (*i*) We have

$$\|\gamma\|_{M(W_p^m \to L_p)} = \sup_{u \in C_0^\infty(\mathbb{R}^n)} \frac{\|\gamma u\|_{L_p}}{\|u\|_{W_p^m}} \tag{2.8.4}$$

and (2.8.1) follows from the inequality

$$\|u\|_{W_p^m} \geq \|u\|_{w_p^m}.$$

(*ii*) Clearly,

$$\|\rho^{-m}\,\gamma(\cdot/\rho)\|_{M(W_p^m \to L_p)} = \sup_{U \in C_0^\infty(\mathbb{R}^n)} \frac{\|\rho^{-m}\,\gamma(\cdot/\rho)U\|_{L_p}}{\|U\|_{W_p^m}}$$

$$= \sup_{u \in C_0^\infty(\mathbb{R}^n)} \frac{\|\rho^{-m}\,\gamma(\cdot/\rho)u(\cdot/\rho)\|_{L_p}}{(\|\nabla_m u(\cdot/\rho)\|_{L_p}^p + \|u(\cdot/\rho)\|_{L_p}^p)^{1/p}}.$$

The right-hand side majorizes

$$\frac{\|\gamma u\|_{L_p}}{(\|\nabla_m u\|^p_{L_p} + \rho^{mp}\|u\|^p_{L_p})^{1/p}}.$$

Hence

$$\liminf_{\rho\to 0} \|\rho^{-m}\,\gamma(\cdot/\rho)\|_{M(W_p^m \to L_p)} \geq \|\gamma\|_{M(w_p^m \to L_p)}. \tag{2.8.5}$$

Noting that

$$\|\rho^{-m}\,\gamma(\cdot/\rho)\|_{M(w_p^m \to L_p)} = \|\gamma\|_{M(w_p^m \to L_p)}$$

and using (2.8.1), we obtain

$$\limsup_{\rho\to 0} \|\rho^{-m}\,\gamma(\cdot/\rho)\|_{M(W_p^m \to L_p)} \leq \|\gamma\|_{M(w_p^m \to L_p)}. \tag{2.8.6}$$

The result follows by combining (2.8.5) and (2.8.6).

(*iii*) Let $\eta \in C_0^\infty(\mathcal{B}_2)$, $\eta = 1$ on $\mathcal{B}_1$. The estimate (2.8.3) follows from (2.8.4) by choosing the test function $u(\xi) = \eta((\xi - x)/r)$. □

Lemma 2.8.2. (*i*) *Let $m \geq l$ and let $\gamma \in M(w_p^m \to w_p^l)$. Then*

$$\|\gamma\|_{M(W_p^m \to W_p^l)} \leq c\,\|\gamma\|_{M(w_p^m \to w_p^l)}. \tag{2.8.7}$$

(*ii*) *The inequality*

$$\liminf_{\rho\to 0} \|\rho^{l-m}\,\gamma(\cdot/\rho)\|_{M(W_p^m \to W_p^l)} \geq \|\gamma\|_{M(w_p^m \to w_p^l)} \tag{2.8.8}$$

holds.

Proof. (*i*) Let η be the same as in the proof of Lemma 2.8.1 (iii), and let $\eta_x(\xi) = \eta(x - \xi)$. We use the inequality

$$\|\gamma\|_{M(W_p^m \to W_p^l)} \leq c \sup_{x\in\mathbb{R}^n} \|\eta_x \gamma\|_{M(W_p^m \to W_p^l)},$$

and observe that the norm on the right-hand side is equal to

$$\sup_{u\in C_0^\infty} \frac{\|\eta_x \gamma u\|_{W_p^l}}{\|u\|_{W_p^m}}.$$

In view of the imbedding $w_p^l(\mathbb{R}^n) \subset L_{pn/(n-lp)}(\mathbb{R}^n)$, the norm in the numerator is equivalent to the norm $\|\eta_x \gamma u\|_{w_p^l}$. Hence

$$\|\gamma\|_{M(W_p^m \to W_p^l)} \leq c \sup_{x,u} \frac{\|\eta_x \gamma u\|_{w_p^l}}{\|u\|_{W_p^m}}$$

$$\leq c\,\|\gamma\|_{M(w_p^m \to w_p^l)} \sup_{x,u} \frac{\|\eta_x u\|_{w_p^m}}{\|u\|_{W_p^m}}.$$

The result follows.

(*ii*) Obviously, for any $u \in C_0^\infty(\mathbb{R}^n)$,

$$\|\rho^{l-m}\,\gamma(\cdot/\rho)\|_{M(W_p^m \to W_p^l)} \geq \frac{\|\rho^{l-m}\,\gamma(\cdot/\rho)u(\cdot/\rho)\|_{W_p^l}}{\|u(\cdot/\rho)\|_{W_p^m}}$$

$$= \Big(\frac{\|\nabla_l(\gamma u)\|^p_{L_p} + \rho^{pl}\|\gamma u\|^p_{L_p}}{\|\nabla_m u\|^p_{L_p} + \rho^{pm}\|u\|^p_{L_p}}\Big)^{1/p}.$$

Passing to the limit as $\rho \to 0$ we complete the proof. □

Now we give a description of the space $M(w_p^m \to w_p^l)$ which follows essentially from Corollary 2.3.3.

Theorem 2.8.1. *Let $mp < n$, $m \geq l$, $p \in (1, \infty)$ or $m \leq n$, $p = 1$. Then $\gamma \in M(w_p^m \to w_p^l)$ if and only if $\gamma \in w^l_{p,loc}$,*

$$\nabla_l \gamma \in M(w_p^m \to L_p),$$

and

$$\begin{aligned} &\gamma \in L_\infty(\mathbb{R}^n) && \text{for } m = l, \\ &\lim_{r\to\infty} r^{-n}\|\gamma; \mathcal{B}_r\|_{L_1} = 0 && \text{for } m > l. \end{aligned} \tag{2.8.9}$$

The norm in the space $M(w_p^m \to w_p^l)$, $m > l$, satisfies the equivalence relation

$$\|\gamma\|_{M(w_p^m \to w_p^l)} \sim \|\nabla_l \gamma\|_{M(w_p^m \to L_p)}. \tag{2.8.10}$$

For $m = l$ the norm $\|\gamma\|_{L_\infty}$ should be added to the right-hand side of this relation.

The equivalence relation

$$\|\gamma\|_{M(w_p^m \to w_p^l)} \sim \sum_{j=0}^{l} \|\nabla_{l-j}\gamma\|_{M(w_p^{m-j} \to L_p)} \tag{2.8.11}$$

holds.

Proof. We replace γ by $\rho^{l-m}\gamma(\cdot/\rho)$ in the equivalence relation (2.3.26). Then (2.8.11) follows from Lemmas 2.8.1 and 2.8.2 as $\rho \to 0$.

We put $\rho^{l-m}\gamma(\cdot/\rho)$ as γ in (2.3.27) to obtain

$$\|\rho^{l-m}\,\gamma(\cdot/\rho)\|_{M(W_p^m \to W_p^l)}$$

$$\leq c\big(\|\nabla_l(\rho^{l-m}\,\gamma(\cdot/\rho))\|_{M(W_p^m \to L_p)} + \sup_{\substack{x\in\mathbb{R}^n,\\ R>0}} R^{m-l-n}\|\rho^{l-m}\gamma(\cdot/\rho)\mathcal{B}_R(x)\|_{L_1}\big).$$

Since the second term on the right is equal to

$$\sup_{x,r} r^{m-l-n} \|\gamma; \mathcal{B}_r(x)\|_{L_1},$$

and the first term tends to $\|\nabla_l \gamma\|_{M(w_p^m \to L_p)}$ as $\rho \to 0$ by (2.8.2), the reference to (2.8.8) gives

$$\|\gamma\|_{M(w_p^m \to w_p^l)} \le c\big(\|\nabla_l \gamma\|_{M(w_p^m \to L_p)} + \sup_{\substack{x \in \mathbb{R}^n,\\ r>0}} r^{m-l-n} \|\gamma; \mathcal{B}_r(x)\|_{L_1}\big). \quad (2.8.12)$$

It remains to remove the second term on the right-hand side in the case $m > l$.

Consider the case $p \in (1, \infty)$. We use the inequality

$$|\gamma(\xi)| \le c\,\big((\mathcal{M}\nabla_l \gamma)(\xi)\big)^{1-l/m} \big(\sup_{\substack{x \in \mathbb{R}^n,\\ r>0}} r^{m-n/p} \|\nabla_l \gamma; \mathcal{B}_r(x)\|_{L_p}\big)^{l/m} \quad (2.8.13)$$

which follows from (2.3.24) with the term $\|\gamma\|_{L_{1,\mathrm{unif}}}$ on the right-hand side omitted due to condition (2.8.9). Integrating (2.8.13) over an arbitrary ball $\mathcal{B}_r(x)$, we arrive at

$$\|\gamma; \mathcal{B}_r(x)\|_{L_1} \le c\big(\sup_{\substack{x \in \mathbb{R}^n\\ r>0}} r^{m-n/p} \|\nabla_l \gamma; \mathcal{B}_r(x)\|_{L_p}\big)^{l/m} \|(\mathcal{M}\nabla_l \gamma)^{1-l/m}; \mathcal{B}_r(x)\|_{L_1}.$$

By Hölder's inequality

$$r^{m-l-n} \|\gamma; \mathcal{B}_r(x)\|_{L_1}$$
$$\le c\big(\sup_{\substack{x \in \mathbb{R}^n\\ r>0}} r^{m-\frac{n}{p}} \|\nabla_l \gamma; \mathcal{B}_r(x)\|_{L_p}\big)^{\frac{l}{m}} \big(r^{m-\frac{n}{p}} \|\mathcal{M}\nabla_l \gamma; \mathcal{B}_r(x)\|_{L_p}\big)^{1-\frac{l}{m}}. \quad (2.8.14)$$

In view of Lemma 2.3.8,

$$r^{m-n/p} \|\mathcal{M}\nabla_l \gamma; \mathcal{B}_r(x)\|_{L_p} \le c \sup_e \frac{\|\mathcal{M}\nabla_l \gamma; e\|_{L_p}}{(C_{p,m}(e))^{1/p}} \le c\,\|\nabla_l \gamma\|_{M(w_p^m \to L_p)},$$

which along with (2.8.14) leads to

$$\sup_{\substack{x \in \mathbb{R}^n,\\ r>0}} r^{m-l-n} \|\gamma; \mathcal{B}_r(x)\|_{L_1} \le c\,\|\nabla_l \gamma\|_{M(w_p^m \to L_p)}.$$

The result follows by (2.8.12).

For $p = 1$ the second term on the right-hand side of (2.8.12) is dominated by $c\,\|\nabla_l \gamma\|_{M(w_1^m \to L_1)}$ by Theorem 1.1.1. □

2.9 Multipliers in Spaces of Functions with Bounded Variation

In the 1960s, the family of differentiable functions was complemented by the space *bv* of functions with bounded variation which turned to be useful in geometric measure theory, the calculus of variations and the theory of quasi-linear partial differential equations.

A function u locally integrable on $\mathbb{R}^n$ has bounded variation if its gradient, understood in the sense of generalized functions, is a vector charge. In other words, the functional

$$(C_0^\infty)^n \ni \mathbf{g} \to (u, \operatorname{div} \mathbf{g})$$

satisfies the estimate

$$|(u, \operatorname{div} \mathbf{g})| \le C \max |\mathbf{g}|,$$

where C is a constant independent of $\mathbf{g}$.

Let $0 < \rho < R$ and let $\bar{u}(r)$ stand for the mean value of u on the sphere $\partial\mathcal{B}_r$. We introduce

$$\mathbf{g}_\varepsilon(x) = x\,|x|^{-n}\eta_\varepsilon(x),$$

where η_ε is a smooth function equal to one on $\mathcal{B}_{R-\varepsilon}\backslash\mathcal{B}_{\rho+\varepsilon}$ and to zero on $\mathcal{B}_R\backslash\mathcal{B}_\rho$, $0 \le \eta_\varepsilon \le 1$. We have

$$|(u, \operatorname{div} \mathbf{g}_\varepsilon)| \le (\operatorname{var} \nabla u)(\mathcal{B}_R\backslash\mathcal{B}_\rho).$$

Making simple calculations and passing to the limit as $\varepsilon \to 0$, we find that

$$|\bar{u}(R) - \bar{u}(\rho)| \le c\,(\operatorname{var} \nabla u)(\mathcal{B}_R\backslash\mathcal{B}_\rho)$$

for almost all ρ and R with $\rho < R$. Therefore, any function with finite variation has the limit

$$u_\infty = \lim_{r\to\infty} \bar{u}(r).$$

The set of functions with finite variation for which $u_\infty = 0$ is called the space bv. The norm in bv is defined as the variation of the charge ∇u. Endowed with this norm, bv becomes a Banach space.

Not every function in bv can be approximated by functions in C_0^∞ in the semi-norm $\|u\|_{bv}$, because the completion of C_0^∞ in this semi-norm is the Sobolev space w_1^1. However, functions in bv can be approximated by functions in C_0^∞ in the following weak sense. If $u \in bv$, then there exists a sequence $\{u_m\}_{m\ge1}$ of functions in C_0^∞ such that $u_m \to u$ in $L_{1,\mathrm{loc}}$ and

$$\lim_{m\to\infty} \int |\nabla u_m|\,dx = \|u\|_{bv}. \tag{2.9.1}$$

The Banach space $BV = bv \cap L_1$, endowered with the norm

$$\|u\|_{BV} = \|u\|_{bv} + \|u\|_{L_1},$$

possesses a similar property.

The existence of the sequence $\{u_m\}$ and the classical isoperimetric inequality

$$\big(\operatorname{mes}_n(g)\big)^{(n-1)/n} \le n^{-1}\big(\operatorname{mes}_n(\mathcal{B}_1)\big)^{-1/n} s(\partial g), \tag{2.9.2}$$

where g is an arbitrary open subset of $\mathbb{R}^n$ with compact closure and smooth boundary, imply the inequality

$$\|u\|_{L_{n/(n-1)}} \leq n^{-1}\big(\mathrm{mes}_n(\mathcal{B}_1)\big)^{-1/n}\|u\|_{bv}, \qquad u \in bv. \tag{2.9.3}$$

By (2.9.1), we may assume that $u \in C_0^\infty$. We have

$$\|u\|^{n/(n-1)}_{L_{n/(n-1)}} = \int_0^\infty \mathrm{mes}_n(N_t)\, d(t^q), \tag{2.9.4}$$

where $q = n/(n-1)$ and $N_t = \{x : |u(x)| \geq t\}$. The last integral has the estimate

$$q\int_0^\infty \mathrm{mes}_n(N_t)t^{q-1}dt \leq q\int_0^\infty \Big(\int_0^t \mathrm{mes}_n(N_\tau)^{1/q}d\tau\Big)^{q-1}\mathrm{mes}_n(N_t)^{1/q}dt$$

$$= \Big(\int_0^\infty \big(\mathrm{mes}_n(N_t)\big)^{1/q}dt\Big)^q.$$

This fact and (2.9.2) imply the estimate

$$\|u\|_{L_{n/(n-1)}} \leq n^{-1}\big(\mathrm{mes}_n(\mathcal{B}_1)\big)^{-1/n}\int_0^\infty s(\partial N_t)dt,$$

which is equivalent to (2.9.3) in view of (1.1.1).

Setting the characteristic function of a ball into (2.9.3), we conclude that the constant in (2.9.3) is sharp. This inequality with a non-sharp constant was proved first in [Gag1] by a different method.

The space bv is closely related to the notion of the perimeter of a set which, to a large extent, is the reason for its importance in analysis. The perimeter $P(E)$, in the sense of Cacciopoli and De Giorgi, of a Lebesgue measurable set $E \subset \mathbb{R}^n$ is defined by

$$P(E) = \inf_{\Pi_m} \liminf_{m\to\infty} s(\partial\Pi_m),$$

where $\{\Pi_m\}$ is a sequence of polyhedra converging in volume to E. This means that the volume of the symmetric difference

$$(\Pi_m\backslash E) \cup (E\backslash\Pi_m)$$

tends to zero.

This definition, combined with (2.9.2) implies the isoperimetric inequality

$$\min\big\{\big(\mathrm{mes}_n(E)\big)^{(n-1)/n},\ \big(\mathrm{mes}_n(\mathbb{R}^n\backslash E)\big)^{(n-1)/n}\big\}$$

$$\leq n^{-1}\big(\mathrm{mes}_n(\mathcal{B}_1)\big)^{-1/n}P(E).$$

The characteristic function χ_E of a set E belongs to bv if and only if $P(E) < \infty$. Moreover,

$$P(E) = \|\chi_E\|_{bv}$$

(see [Fe3]).

The perimeter $P(E)$ does not exceed the Hausdorff measure $s(\partial E)$; in particular, the inequality $P(E) < s(\partial E)$ is not excluded. The following generalization of the notion of the normal to a smooth surface is useful for revealing a deeper relation between the perimeter and the measure s.

A unit vector ν is called the exterior normal to a set E at a point x in the sense of Federer if

$$\lim_{\rho\to 0} \rho^{-n} \mathrm{mes}_n \big\{ y : y \in E \cap \mathcal{B}_\rho(x),\ (y-x)\nu > 0 \big\} = 0,$$

$$\lim_{\rho\to 0} \rho^{-n} \mathrm{mes}_n \big\{ y : y \in \mathcal{B}_\rho(x) \backslash E,\ (y-x)\nu < 0 \big\} = 0.$$

The set of all points $x \in \partial E$ for which the normal to E exists is called the reduced boundary of the set E and is denoted by $\partial^* E$. A set E is said to be a set with a local finite perimeter if

$$P(E \cap \mathcal{B}_r(x)) < \infty$$

for all balls $\mathcal{B}_r(x)$.

The following assertion is a crucial result in the theory of perimeter (see [Fe3]).

Theorem 2.9.1. *If E is a set with a locally finite perimeter, then its reduced boundary $\partial^* E$ is measurable with respect to s and* var $\nabla \chi_E$. *Moreover,*

$$\mathrm{var}\, \nabla \chi_E (\mathbb{R}^n \backslash \partial^* E) = 0,$$

and for any set $\mathcal{L} \subset \partial^* E$

$$(\nabla \chi_E)(\mathcal{L}) = - \int_{\mathcal{L}} \nu(x)\, s(dx).$$

This implies that $P(E) = s(\partial^* E)$.

Note that for any $u \in bv$ the generalization of (1.1.1) holds:

$$\|u\|_{bv} = \int_{-\infty}^{\infty} P(M_t)\, dt, \tag{2.9.5}$$

where

$$M_t = \{x : u(t) > t\}$$

(see [FR]).

2.9.1 The Spaces Mbv and MBV

The next assertion contains a description of the space Mbv.

Theorem 2.9.2. *A function γ belongs to the space Mbv if and only if $\gamma \in bv_{\rm loc} \cap L_\infty$ and for any ball $\mathcal{B}_r(x)$*

$$\operatorname{var} \nabla\gamma(\mathcal{B}_r(x)) \le c\, r^{n-1}. \tag{2.9.6}$$

The relation

$$\|\gamma\|_{Mbv} \sim \sup_{x \in \mathbb{R}^n,\, r>0} r^{1-n} \operatorname{var} \nabla\gamma(\mathcal{B}_r(x))$$

holds.

Proof. Sufficiency. Let $u \in bv$ and let γ be a function in $bv_{\rm loc} \cap L_\infty$ subject to (2.9.6). By $\{u_m\}$ we denote a sequence of functions in C_0^∞ convergent in $L_{1,{\rm loc}}$ for which (2.9.1) holds. Then, for any $\varphi \in C_0^\infty$,

$$|(\nabla\varphi, \gamma u)| = |\lim_{m\to\infty} (\nabla\varphi, \gamma u_m)|$$

$$\le |\limsup_{m\to\infty} (\varphi, u_m \nabla\gamma)| + |\limsup_{m\to\infty} (\varphi, \gamma\nabla u_m)|.$$

Hence

$$|(\nabla\varphi, \gamma u)| \le \|\varphi\|_{L_\infty} \limsup_{m\to\infty} \int |u_m| \operatorname{var} \nabla\gamma(dx)$$

$$+ \|\varphi\|_{L_\infty} \|\gamma\|_{L_\infty} \limsup_{m\to\infty} \int |\nabla u_m| dx.$$

Applying Corollary 1.1.2 to the first integral, we find that

$$|(\nabla\varphi, \gamma u)| \le \|\varphi\|_{L_\infty} \big(c \sup_{x \in \mathbb{R}^n,\, r>0} r^{1-n} \operatorname{var} \nabla\gamma(\mathcal{B}_r(x)) + \|\gamma\|_{L_\infty}\big) \|u\|_{bv}.$$

Thus the sufficiency and the upper estimate for the norm $\|\gamma\|_{Mbv}$ are proved.

Necessity. If $\gamma \in Mbv$, then for any $N = 1, 2, \ldots$

$$\|\gamma^N v\|_{bv} \le \|\gamma\|^N_{Mbv} \|v\|_{bv}.$$

Applying inequality (2.9.3), we find that

$$\|\gamma^N v\|^{1/N}_{L_{n/(n-1)}} \le \|\gamma\|_{Mbv} \|v\|^{1/N}_{bv}.$$

Passing to the limit as $N \to \infty$, we conclude that

$$\|\gamma\|_{L_\infty} \le \|\gamma\|_{Mbv}.$$

Note that a mollification v_h of the characteristic function of the ball $\mathcal{B}_{(1+\varepsilon)r}(y)$ obeys the inequality

$$\int |v_h| \operatorname{var} \nabla\gamma(dx) \le \|\gamma v_h\|_{bv} + \int |\gamma|\, |\nabla v_h|\, dx.$$

Passing to the limit as $\varepsilon \to 0$, we obtain

$$\mathrm{var}\,\nabla\gamma(\mathcal{B}_r(y)) \le (\|\gamma\|_{Mbv} + \|\gamma\|_{L_\infty}) r^{n-1}.$$

Thus,

$$r^{1-n}\mathrm{var}\,\nabla\gamma(\mathcal{B}_r(y)) \le 2\,\|\gamma\|_{Mbv}.$$

The proof is complete. □

We formulate a similar result for the space MBV which is proved in the same way.

Theorem 2.9.3. *A function γ belongs to the space MBV if and only if $\gamma \in bv_{\mathrm{loc}} \cap L_\infty$ and, for any ball $\mathcal{B}_r$ with $r \in (0,1)$, (2.9.6) holds. Moreover,*

$$\|\gamma\|_{MBV} \sim \sup_{\substack{x\in\mathbb{R}^n,\\ r\in(0,1)}} r^{1-n}\mathrm{var}\,\nabla\gamma(\mathcal{B}_r(x)).$$

Theorems 2.9.2 and 2.9.3 imply the following necessary and sufficient conditions for inclusion of the characteristic function χ_E of a set $E \subset \mathbb{R}^n$ into Mbv and MBV.

Corollary 2.9.1. (*i*) *The function χ_E belongs to the space Mbv if and only if*

$$s(\mathcal{B}_r(x) \cap \partial^* E) \le c\, r^{n-1} \tag{2.9.7}$$

for any set E with a locally finite perimeter and any ball $\mathcal{B}_r(x)$.

(*ii*) *Adding to this statement the condition $r \in (0,1)$, one gets a necessary and sufficient condition for the inclusion $\chi_E \in MBV$.*

Proof. It is enough to refer to Theorems 2.9.2 and 2.9.3, noting additionally that (2.9.7) is equivalent to

$$\mathrm{var}\,\nabla\chi_E(\mathcal{B}_r(x)) \le c\, r^{n-1}.$$

Indeed, by Theorem 2.9.1

$$s(\mathcal{B}_r(x) \cap \partial^* E) = \mathrm{var}\,\nabla\chi_E(\mathcal{B}_r(x) \cap \partial^* E) = \mathrm{var}\,\nabla\chi_E(\mathcal{B}_r(x)).$$

The proof is complete. □

3

Multipliers in Pairs of Potential Spaces

In this chapter we study the space of multipliers $M(H_p^m \to H_p^l)$ and $M(h_p^m \to h_p^l)$, $m \geq l \geq 0$, where H_p^s and h_p^s are the space of Bessel and Riesz potentials of order s with densities in L_p. (The case $m < l$ is not interesting, since then $M(H_p^m \to H_p^l) = \{0\}$, as can be shown by the argument used for Sobolev spaces in Sect. 2.1.)

The introductory Sect. 3.1 gives information on Bessel linear and nonlinear potentials, on capacity and on imbedding theorems, which is used in subsequent sections. A characterization of the spaces $M(H_p^m \to H_p^l)$ and $M(h_p^m \to h_p^l)$ is given in Sects. 3.2 and 3.7. In Sects. 3.3 and 3.4 we obtain either necessary or sufficient conditions for a function to belong to $M(H_p^m \to H_p^l)$, formulated in terms of different classes of functions. In Sect. 3.5 certain properties of elements of $M(H_p^m \to H_p^l)$ are studied. In particular, we consider the imbedding of $M(H_p^m \to H_p^l)$ into $M(H_q^{m-j} \to H_q^{l-j})$. Descriptions of the point, residual, and continuous spectra of multipliers in H_p^l and $H_{p'}^{-l}$ are given in 3.6. Finally, Sect. 3.8 contains a characterization of positive homogeneous elements of the spaces $M(H_p^m \to H_p^l)$ and $M(h_p^m \to h_p^l)$.

3.1 Trace Inequality for Bessel and Riesz Potential Spaces

Let μ be a Radon measure in $\mathbb{R}^n$ and let S_p^l be a certain space of Sobolev type with p and l being the integrability and smoothness parameters, $p \geq 1$, $l > 0$. As in Chap. 2, characterizations of the space $M(S_p^m \to S_p^l)$, $m \geq l$, to be obtained in the sequel are based on necessary and sufficient conditions ensuring the trace inequality

$$\Big(\int_{\mathbb{R}^n} |u|^p \, d\mu \Big)^{1/p} \leq c \, \|u\|_{S_p^l},$$

V.G. Maz'ya, T.O. Shaposhnikova, *Theory of Sobolev Multipliers*,
Grundlehren der mathematischen Wissenschaften 337,

where u is an arbitrary function in $C_0^\infty(\mathbb{R}^n)$. In this section we present such conditions for the spaces of Bessel and Riesz potentials.

3.1.1 Properties of Bessel Potential Spaces

Here we survey some known facts on Bessel potential spaces to be used in the sequel.

Given any real μ, we put

$$\Lambda^\mu = (-\Delta + 1)^{\mu/2} = F^{-1}(1 + |\xi|^2)^{\mu/2} F \, ,$$

where F is the Fourier transform in $\mathbb{R}^n$.

Let $1 < p < \infty$, $m \geq 0$. We introduce the space H_p^m of Bessel potentials as the completion of C_0^∞ with respect to the norm

$$\|u\|_{H_p^m} = \|\Lambda^m u\|_{L_p}.$$

If m is integer then, according to Lemma 1.2.3, the spaces W_p^m and H_p^m are isomorphic.

It follows from the definition of H_p^m that the function u belongs to H_p^m if and only if $u = \Lambda^{-m} f$, where $f \in L_p$.

Replacing Λ^m in the definition of H_p^m by the operator $(-\Delta)^{m/2}$, we arrive at the definition of the space h_p^m. It is known that the space h_p^m with $mp < n$ is isomorphic to the space of Riesz potentials of order m with density in L_p (see [MH2]).

Definition 3.1.1. *We define* $(S_m u)(x) = |\nabla_m u(x)|$ *for integer* $m \geq 0$ *and*

$$(S_m u)(x) = \Big(\int_0^\infty \Big[\int_{\mathcal{B}_1} |\nabla_{[m]} u(x+\theta y) - \nabla_{[m]} u(x)| d\theta \Big]^2 y^{-1-2\{m\}} dy \Big)^{1/2} \quad (3.1.1)$$

for noninteger $m > 0$.

We present without proof a normalization of potential spaces due to Strichartz [Str].

Theorem 3.1.1. *The equivalence relations hold:*

$$\|u\|_{h_p^m} \sim \|S_m u\|_{L_p} \, , \quad (3.1.2)$$

$$\|u\|_{H_p^m} \sim \|S_m u\|_{L_p} + \|u\|_{L_p} \, . \quad (3.1.3)$$

The last formula implies the following uniform localization property for the space H_p^m (see [Str]).

Theorem 3.1.2. *Let* $\{\mathcal{B}^{(j)}\}_{j \geq 0}$ *be a covering of* $\mathbb{R}^n$ *by balls with unit diameter. Let this covering have a finite multiplicity, depending only on* n. *Further, let* $O^{(j)}$ *be the centre of* $\mathcal{B}^{(j)}$, $O^{(0)} = 0$ *and* $\eta_j(x) = \eta(x - O^{(j)})$, *where* $\eta \in C_0^\infty(2\mathcal{B}^{(0)})$ *and* $\eta = 1$ *on* $\mathcal{B}^{(0)}$. *Then*

$$\|u\|_{H_p^m} \sim \Big(\sum_{j \geq 0} \|u \eta_j\|_{H_p^m}^p \Big)^{1/p} . \quad (3.1.4)$$

We formulate the Sobolev imbedding theorem for the space H_p^m.

Theorem 3.1.3. *(i) If $mp < n$, $p \le q \le np/(n-mp)$ or $mp = n$, $p \le q < \infty$, then for all $u \in H_p^m$*

$$\|u\|_{L_q} \le c\,\|u\|_{H_p^m}.$$

(ii) If $mp > n$, then for all $u \in H_p^m$

$$\|u\|_{L_\infty} \le c\,\|u\|_{H_p^m}.$$

The following generalization of the function $S_l u$ is the nonlinear operator of fractional differentiation defined by Polking [Pol1]:

$$(S_{q,\theta}^l u)(x) = \Big(\int_0^\infty \Big(\int_{\mathcal{B}_1} |u(x+\rho y) - u(x)|^q dy\Big)^{\theta/q} \frac{d\rho}{\rho^{1+l\theta}}\Big)^{1/\theta}.$$

Lemma 3.1.1. [Pol1] *Let $0 < l < 1$, $1 \le q \le \theta$, $2 \le \theta < \infty$. If $1 < p < \infty$ and $p > nq/(n+lq)$, then there exists a constant c depending only on l, q, p, θ, n such that*

$$\|S_{q,\theta}^l u\| \le c\,\|u\|_{H_p^l}$$

for all $u \in H_p^l$.

3.1.2 Properties of the (p, m)-Capacity

For positive noninteger m the (p, m)-capacity is defined by (1.2.6), i.e.

$$C_{p,m}(e) = \inf\{\|f\|_{L_p}^p : f \in L_p,\ f \ge 0,\ J_m f \ge 1 \text{ on } e\}.$$

Similarly to (1.2.7), one has

$$C_{p,m}(e) \sim \inf\{\|u\|_{H_p^m}^p : u \in C_0^\infty,\ u \ge 1 \text{ on } e\}.$$

Another capacity, introduced for any positive and noninteger m, is

$$c_{p,m}(e) = \inf\{\|f\|_{L_p}^p : f \in L_p,\ f \ge 0,\ I_m f \ge 1 \text{ on } e\}.$$

We list certain 'metric' properties of these capacities which will be used later. For integer $m > 0$ they were proved in Propositions 1.2.2, 1.2.4, and 1.2.6.

Proposition 3.1.1. *If $mp > n$, then for all compact sets $e \ne \varnothing$ with $d(e) \le 1$ the relation $C_{p,m}(e) \sim 1$ holds.*

The proof follows from part *(ii)* of Theorem 3.1.3.

Proposition 3.1.2. *If $mp < n$, then*

$$C_{p,m}(e) \ge c\,(\text{mes}_n e)^{(n-mp)/n}\,. \tag{3.1.5}$$

The proof follows from part *(i)* of Theorem 3.1.3.

Proposition 3.1.3. *If $mp = n$ and $d(e) \le 1$, then*

$$C_{p,m}(e) \ge c\Big(\log \frac{2^n}{\mathrm{mes}_n e}\Big)^{1-p}. \tag{3.1.6}$$

Proof. See Proposition 1.2.4.

Proposition 3.1.4. (*i*) *If $mp < n$ and $0 < r \le 1$, then*

$$C_{p,m}(\mathcal{B}_r) \sim r^{n-mp}.$$

(*ii*) *If $mp = n$, $0 < r \le 1$, then*

$$C_{p,m}(\mathcal{B}_r) \sim (\log 2/r)^{1-p}.$$

(*iii*) *If $r > 1$, then*

$$C_{p,m}(\mathcal{B}_r) \sim r^n.$$

For the proof of these relations see [AH], Sect. 5.1.

In the next proposition $\{\mathcal{B}^{(j)}\}$ is the same covering as in Theorem 3.1.2.

Proposition 3.1.5. *For any compact e*

$$C_{p,m}(e) \sim \sum_{j \ge 0} C_{p,m}(e \cap \mathcal{B}^{(j)}).$$

Proof. Let $u \in C_0^\infty$, $u \ge 1$ on e. From the definition of $C_{p,m}$ it follows that

$$\sum_{j\ge 0} C_{p,m}(e \cap \mathcal{B}^{(j)}) \le \sum_{j \ge 0} \|u\eta_j\|^p_{H^m_p},$$

where $\{\eta_j\}_{j\ge 0}$ is the sequence defined in Theorem 3.1.2. By (3.1.4), the right-hand side is dominated by $c\,\|u\|^p_{H^m_p}$. Minimizing this value, we obtain the lower bound for $C_{p,m}(e)$. The required upper bound is a direct corollary of the semi-additivity of the capacity.

Proposition 3.1.6. *Let $mp < n$ and let e be a compact subset of $\mathbb{R}^n$ of diameter $d(e) \le 1$. Then*

$$C_{p,m}(e) \sim c_{p,m}(e). \tag{3.1.7}$$

Proof. The estimate $C_{p,m}(e) \ge c_{p,m}(e)$ is obvious. We prove that

$$C_{p,m}(e) \le c\, c_{p,m}(e).$$

By definition of the capacity $c_{p,m}(e)$, for any $\varepsilon > 0$ there exists a function $u \in C_0^\infty$ such that $u \ge 1$ on e and

$$\|S_m u\|^p_{L_p} \le c_{p,m}(e) + \varepsilon. \tag{3.1.8}$$

We introduce a function $\eta \in C_0^\infty(\mathcal{B}_2)$ such that $\eta \geq 1$ on e. Then

$$\|S_m(\eta u)\|_{L_p}^p + \|\eta u\|_{L_p}^p \leq c\big(\|S_m u\|_{L_p}^p + \sum_{k=0}^{[m]} \|\,|x|^{k-m}\nabla_k u\|_{L_p}^p\big)$$

which does not exceed $c\,\|S_m u\|_{L_p}^p$ by Hardy's inequality. Reference to estimate (3.1.8) completes the proof. □

Corollary 3.1.1. *Let $mp < n$ and let e be a compact subset of $\mathbb{R}^n$ of diameter $d(e) \leq 1$. Then, for any $\delta \in (0,1)$,*

$$C_{p,m}(e) \sim \delta^{n-mp} C_{p,m}(\delta\, e). \tag{3.1.9}$$

Proof. On the one hand,

$$C_{p,m}(e) \leq \delta^{n-mp}\, C_{p,m}(\delta\, e)$$

by dilation. On the other hand,

$$C_{p,m}(e) \geq c_{p,m}(e) = \delta^{n-mp} c_{p,m}(\delta\, e).$$

Reference to Proposition 3.1.6 completes the proof. □

We give an estimate for the integral of the capacity of a set bounded by a level surface which contains (1.2.27) as a particular case. For the history and proof see [AH], Ch. 7.

Proposition 3.1.7. (*i*) *Let $u \in H_p^m$, $p \in (1,\infty)$, $m > 0$, and let*

$$N_t = \{x : |u(x)| \geq t\}.$$

Then

$$\int_0^\infty C_{p,m}(N_t) t^{p-1}\, dt \leq c\,\|u\|_{H_p^m}^p\,. \tag{3.1.10}$$

(*ii*) *If $u \in h_p^m$, $p \in (1, n/m)$, $m > 0$, then*

$$\int_0^\infty c_{p,m}(N_t) t^{p-1}\, dt \leq c\,\|u\|_{h_p^m}^p\,. \tag{3.1.11}$$

3.1.3 Main Result

Here we present a generalization to H_p^m of Theorem 1.2.2 on integrability of functions in W_p^m with respect to a measure μ.

Using (3.1.10) and (3.1.11), we can prove the following theorem in the same way as Theorem 1.2.2.

Theorem 3.1.4. *(i) The best constant C in*

$$\int |u|^p \, d\mu \le C \, \|u\|^p_{H^m_p}, \quad u \in C_0^\infty, \tag{3.1.12}$$

is equivalent to

$$\sup_e \frac{\mu(e)}{C_{p,m}(e)},$$

where e is an arbitrary compact set of positive capacity $C_{p,m}(e)$.

(ii) Let $p > 1$ and $mp < n$. The best constant C in

$$\int |u|^p \, d\mu \le C \, \|u\|^p_{h^m_p}, \quad u \in C_0^\infty, \tag{3.1.13}$$

is equivalent to

$$\sup_e \frac{\mu(e)}{c_{p,m}(e)},$$

where e is an arbitrary compact set of positive capacity $c_{p,m}(e)$.

Remark 3.1.1. From Proposition 3.1.5 it follows that

$$\sup_e \frac{\mu(e)}{C_{p,m}(e)} \sim \sup_{\{e: d(e) \le 1\}} \frac{\mu(e)}{C_{p,m}(e)}. \tag{3.1.14}$$

Remark 3.1.2. Obviously, the constant C in Theorem 3.1.4 satisfies the inequality

$$C \ge \sup_{x \in \mathbb{R}^n, \rho \in (0,1/2)} \frac{\mu(\mathcal{B}_\rho(x))}{C_{p,m}(\mathcal{B}_\rho(x))}. \tag{3.1.15}$$

If the converse estimate (up to a factor $c = c(n, p, l)$) were valid we could avoid the notion of capacity in this book (see Proposition 3.1.4/4). According to Proposition 3.1.1, this is really the case for $mp > n$ when the right-hand and left-hand sides of (3.1.15) are equivalent to

$$\sup\{\mu(\mathcal{B}_1(x)) : x \in \mathbb{R}^n\}.$$

However, as was noted by Adams [Ad3], the finiteness of the right-hand side of (3.1.15) for $n \ge mp$ does not imply $C < \infty$.

Before we prove the last assertion, we recall the definition of the Hausdorff φ-measure of a set $E \subset \mathbb{R}^n$, where φ is a non-decreasing positive function on $[0, 1]$: namely,

$$H(E, \varphi) = \lim_{\varepsilon \to +0} \inf_{\{\mathcal{B}^{(i)}\}} \sum_i \varphi(r_i).$$

Here $\{\mathcal{B}^{(i)}\}$ is any covering of E by open balls $\mathcal{B}^{(i)}$ with radii $r_i < \varepsilon$. We put $\varphi(t) = t^{n-mp}$ for $n > mp$ and $\varphi(t) = |\log t|^{1-p}$ for $n = mp$. Let E be a Borel set in $\mathbb{R}^n$ such that its diameter $d(E)$ satisfies $d(E) < 1$ and $0 < H(E, \varphi) < \infty$.

We may assume E to be closed and bounded, since any Borel set of positive Hausdorff measure contains a closed subset with the same property.

By Frostman's theorem (see Carleson [Car], Theorem 1, Ch. 2) there exists a non-zero measure μ with support in E such that

$$\mu(\mathcal{B}_\varrho(x)) \le c\,\varphi(\rho)$$

with a constant c independent of x and ρ. By virtue of Proposition 3.1.4, this means that the right-hand side of (3.1.15) is finite.

On the other hand, by the theorem of Meyers [Me] and Havin and Maz'ya [MH1], [MH2], the finiteness of the measure $H(E,\varphi)$ implies $C_{p,m}(E) = 0$. So $C = \infty$ (see Theorem 3.1.4), although the right-hand side of (3.1.15) is finite.

Remark 3.1.3. According to [KeS], [MV1], and [Ver1], the criteria for (1.2.31) formulated at the end of Sect. 1.2.5 for integer m hold for all positive m. The same concerns the Fefferman-Phong sufficient condition dealt with in Sect. 1.2.6. For surveys of these and related results see [Ver2] and [Ver3].

3.2 Description of $M(H_p^m \to H_p^l)$

3.2.1 Auxiliary Assertions

We formulate the Calderon interpolation theorem [Cal] for spaces of Bessel potentials, of which a particular case is the inequality (2.3.8) used in the study of the space $M(W_p^m \to W_p^l)$.

Proposition 3.2.1. *Let $p_0, p_1 \in (1,\infty)$, $\theta \in (0,1)$, $\mu \in \mathbb{R}^1$ and*

$$\frac{1}{p} = \frac{\theta}{p_0} + \frac{1-\theta}{p_1}, \qquad l = \theta l_0 + (1-\theta) l_1 .$$

Further, let L be a linear operator mapping

$$H_{p_0}^{l_0+\mu} \cap H_{p_1}^{l_1+\mu} \quad \text{into} \quad H_{p_0}^{l_0} \cap H_{p_1}^{l_1}$$

and admitting an extension to continuous operators:

$$H_{p_0}^{l_0+\mu} \to H_{p_0}^{l_0} \quad \text{and} \quad H_{p_1}^{l_1+\mu} \to H_{p_1}^{l_1}.$$

Then L can be extended to a continuous operator: $H_p^{l+\mu} \to H_p^l$, and the interpolation inequality

$$\|L\|_{H_p^{l+\mu} \to H_p^l} \le c\, \|L\|^{\theta}_{H_{p_0}^{l_0+\mu} \to H_{p_0}^{l_0}} \|L\|^{1-\theta}_{H_{p_1}^{l_1+\mu} \to H_{p_1}^{l_1}}$$

holds.

Setting $L=\gamma$ in this proposition, we obtain

$$\|\gamma\|_{M(H_p^{l+\mu}\to H_p^l)} \le c\,\|\gamma\|^{\theta}_{M(H_{p0}^{l_0+\mu}\to H_{p0}^{l_0})}\|\gamma\|^{1-\theta}_{M(H_{p1}^{l_1+\mu}\to H_{p1}^{l_1})}\,. \tag{3.2.1}$$

Lemma 3.2.1. *Let γ_ρ be a mollification of $\gamma \in H^l_{p,\mathrm{loc}}(\mathbb{R}^n)$, $1<p<\infty$, with kernel $K \ge 0$ and radius ρ. Then*

$$\|\gamma_\rho\|_{M(H_p^m\to H_p^l)} \le \|\gamma\|^{\theta}_{M(H_p^m\to H_p^l)} \le \liminf_{\rho\to 0}\|\gamma_\rho\|_{M(H_p^m\to H_p^l)}\,. \tag{3.2.2}$$

Proof. Let $u \in C_0^\infty$ and $\{l\}>0$. By (3.1.3)

$$\|\gamma_\rho u\|_{H_p^l} \le c\Big\{\int\Big(\int_0^\infty\Big[\int_{\mathcal{B}_1}\Big|\int \rho^{-n}K(\xi/\rho)$$
$$\times\nabla_{[l],x}(Q(x+\theta y,\xi)-Q(x,\xi))\,d\xi\Big|\,d\theta\Big]^2 y^{-1-2\{l\}}\,dy\Big)^{p/2}dx\Big\}^{1/p}$$
$$+c\Big\{\int\Big|\int\rho^{-n}K(\xi/\rho)Q(x,\xi)\,d\xi\Big|^p dx\Big\}^{1/p},$$

where $Q(x,\xi)=\gamma(x-\xi)u(x)$. By the Minkowski inequality

$$\|\gamma_\rho u\|_{H_p^l} \le \int \rho^{-n}K(\xi/\rho)\|S_l Q(\cdot,\xi)\|_{L_p}\,d\xi + \int\rho^{-n}K(\xi/\rho)\|Q(\cdot,\xi)\|_{L_p}\,d\xi\,.$$

Since

$$\|Q(\cdot,\xi)\|_{H_p^l} \le \|\gamma\|_{M(H_p^m\to H_p^l)}\|u\|_{H_p^m},$$

the left estimate (3.2.2) follows. The right inequality (3.2.2) results from

$$\|\gamma u\|_{H_p^l} = \lim_{\rho\to 0}\|\gamma_\rho u\|_{H_p^l} \le \liminf_{\rho\to 0}\|\gamma_\rho\|_{M(H_p^m\to H_p^l)}\|u\|_{H_p^m}\,.$$

For the case $\{l\}=0$ see Lemma 2.3.1. □

3.2.2 Imbedding of $M(H_p^m \to H_p^l)$ into $M(H_p^{m-l} \to L_p)$

In Theorem 3.1.4 and Remark 3.1.1 the following description of the space $M(H_p^k \to L_p)$ is contained:

Lemma 3.2.2. *The relations*

$$\|\gamma\|_{M(H_p^k\to L_p)} \sim \sup_e \frac{\|\gamma; e\|_{L_p}}{[C_{p,k}(e)]^{1/p}}$$

and

$$\|\gamma\|_{M(H_p^k\to L_p)} \sim \sup_{\{e:d(e)\le 1\}} \frac{\|\gamma; e\|_{L_p}}{[C_{p,k}(e)]^{1/p}}$$

hold, where $d(e)$ is the diameter of e.

Lemma 3.2.3. *The inequality*

$$\|\gamma\|_{M(H_p^{m-l}\to L_p)} \le c\,\|\gamma\|_{M(H_p^m\to H_p^l)} \tag{3.2.3}$$

holds.

Proof. Let $\gamma \in M(H_p^m \to H_p^l)$ and let γ_ρ be a mollification of γ with radius ρ. Since $M(H_p^m \to H_p^l) \subset M(H_p^m \to L_p)$, then $\gamma \in L_{p,\mathrm{unif}}$. Therefore $\gamma_\rho \in L_\infty$ and consequently $\gamma_\rho \in M(H_p^{m-l} \to L_p)$. This property of mollifications will be used in what follows.

1. The case $m \ge 2l$. Let $u = J_{m-l}f$, $f \in L_p$. By Lemma 1.2.5,

$$|u| \le c\,(J_m|f|)^{1-l/m}(\mathcal{M}f)^{l/m}\,.$$

Hence

$$\|\gamma_\rho u\|_{L_p} \le c\,\|f\|_{L_p}^{l/m}\,\|\gamma_\rho^{l/(l-m)}\gamma_\rho J_m|f|\|_{L_p}^{1-l/m}\,.$$

This inequality and Lemma 3.2.2 imply that

$$\|\gamma_\rho u\|_{L_p} \le c\,\|f\|_{L_p}^{l/m}\,\|\gamma_\rho J_m|f|\|_{H_p^l}^{1-l/m} \sup_e \left(\frac{\int_e |\gamma_\rho|^{pl/(m-l)}\,dx}{C_{p,l}(e)}\right)^{(m-l)/pm}.$$

We use Lemma 2.3.6, which is valid for all ν and μ, $0<\nu<\mu$, with $\varphi = |\gamma_\rho|^{1/(m-l)}$, $\nu = l$, $\mu = m-l$. Then the last supremum does not exceed

$$c\left(\sup_e \frac{\|\gamma_\rho; e\|_{L_p}}{[C_{p,m-l}(e)]^{1/p}}\right)^{l/m}$$

which, together with Lemma 3.2.2, gives

$$\begin{aligned}\|\gamma_\rho u\|_{L_p} &\le c\,\|\gamma_\rho\|_{M(H_p^{m-l}\to L_p)}^{l/m}\,\|\gamma_\rho\|_{M(H_p^m\to H_p^l)}^{1-l/m}\,\|f\|_{L_p}^{l/m}\,\|J_m|f|\|_{H_p^m}^{1-l/m}\\ &\le c\,\|\gamma_\rho\|_{M(H_p^{m-l}\to L_p)}^{l/m}\,\|\gamma_\rho\|_{M(H_p^m\to H_p^l)}^{1-l/m}\,\|u\|_{H_p^{m-l}}\,.\end{aligned}$$

Then reference to Lemma 3.2.1 yields (3.2.3) for $m \ge 2l$.

2. Suppose that $m = l$. For any positive integer N we have

$$\|\gamma^N u\|_{L_p}^{1/N} \le \|\gamma^N u\|_{H_p^l}^{1/N} \le \|\gamma\|_{MH_p^l}\|u\|_{H_p^l}^{1/N}\,.$$

Consequently,

$$\|\gamma\|_{ML_p} = \|\gamma\|_{L_\infty} \le \|\gamma\|_{MH_p^l}.$$

3. Now let $2l > m > l$. By ε we denote a positive number such that $\varepsilon < m-l$. Since $m-l+\varepsilon > 2\varepsilon$, it follows from the first part of the proof that

$$\|\gamma_\rho\|_{M(H_p^{m-l}\to L_p)} \le c\,\|\gamma_\rho\|_{M(H_p^{m-l+\varepsilon}\to H_p^\varepsilon)}\,.$$

By (3.2.1) we have

$$\|\gamma_\rho\|_{M(H_p^{m-l+\varepsilon}\to H_p^\varepsilon)} \le c\,\|\gamma_\rho\|^{1-\varepsilon/l}_{M(H_p^{m-l}\to L_p)}\|\gamma_\rho\|^{\varepsilon/l}_{M(H_p^m\to H_p^l)}$$

which, together with the preceding estimate and Lemma 3.2.1, implies (3.2.3). □

3.2.3 Estimates for Derivatives of a Multiplier

Lemma 3.2.4. *If $\gamma \in M(H_p^m \to H_p^l)$, then $D^\alpha\gamma \in M(H_p^m \to H_p^{l-|\alpha|})$ for any multi-index α with $|\alpha| \le l$. The estimate*

$$\|D^\alpha\gamma\|_{M(H_p^m\to H_p^{l-|\alpha|})} \le c\,\|\gamma\|_{M(H_p^m\to H_p^l)} \tag{3.2.4}$$

holds.

Proof. It suffices to consider the case $|\alpha| = 1$, $l \ge 1$. Obviously,

$$\begin{aligned}\|u\nabla\gamma\|_{H_p^{l-1}} &\le \|u\gamma\|_{H_p^l} + \|\gamma\nabla u\|_{H_p^{l-1}}\\ &\le \big(\|\gamma\|_{M(H_p^m\to H_p^l)} + \|\gamma\|_{M(H_p^{m-1}\to H_p^{l-1})}\big)\|u\|_{H_p^m}\,.\end{aligned}$$

Using (3.2.1) and (3.2.3), we obtain

$$\begin{aligned}\|\gamma\|_{M(H_p^{m-l}\to H_p^{l-1})} &\le c_1\|\gamma\|^{1-1/l}_{M(H_p^{m-l}\to L_p)}\|\gamma\|^{1/l}_{M(H_p^m\to H_p^l)}\\ &\le c_2\|\gamma\|_{M(H_p^m\to H_p^l)}\,.\end{aligned}$$

Consequently,

$$\|u\nabla\gamma\|_{H_p^{l-1}} \le c\,\|\gamma\|_{M(H_p^m\to H_p^l)}\|u\|_{H_p^m}\,.$$

□

This result and Lemma 3.2.3 give:

Corollary 3.2.1. *If $\gamma \in M(H_p^m \to H_p^l)$, then $D^\alpha\gamma \in M(H_p^{m-l+|\alpha|} \to L_p)$ for any multi-index α of order $|\alpha| \le l$. The inequality*

$$\|D^\alpha\gamma\|_{M(H_p^{m-l+|\alpha|}\to L_p)} \le c\,\|\gamma\|_{M(H_p^m\to H_p^l)}$$

holds.

3.2.4 Multiplicative Inequality for the Strichartz Function

Lemma 3.2.5. *Let* $m \geq l$, $0 < \delta < l < 1$. *Then*

$$S_{l-\delta}\gamma \leq c\left(S_l\gamma + \|\gamma\|_{M(H_p^{m-l}\to L_p)}\right)^{1-\delta/m} \|\gamma\|^{\delta/m}_{M(H_p^{m-l}\to L_p)} \,. \tag{3.2.5}$$

Proof. For any $R > 0$

$$(S_{l-\delta}\gamma)(x) \leq c\Big(R^\delta\Big[\int_0^R \Big(\int_{\mathcal{B}_1} |\gamma(x+\theta y) - \gamma(x)|\, d\theta\Big)^2 y^{-1-2l}\, dy\Big]^{1/2}$$
$$+\Big[\int_R^\infty \Big(\int_{\mathcal{B}_1} |\gamma(x+\theta y)|\, d\theta\Big)^2 y^{-1-2(l-\delta)}\, dy\Big]^{1/2} + |\gamma(x)| R^{\delta - l}\Big)\,.$$

Since

$$c\,|\gamma(x)| \leq \int_{\mathcal{B}_1} |\gamma(x+\theta y) - \gamma(x)|\, d\theta + \int_{\mathcal{B}_1} |\gamma(x+\theta y)|\, d\theta\,,$$

it follows that

$$c\,R^{-l}|\gamma(x)| \leq \Big(\int_R^\infty \Big(\int_{\mathcal{B}_1} |\gamma(x+\theta y) - \gamma(x)|\, d\theta\Big)^2 y^{-1-2l}\, dy\Big)^{1/2}$$
$$+\Big(\int_R^\infty \Big(\int_{\mathcal{B}_1} |\gamma(x+\theta y)|\, d\theta\Big)^2 y^{-1-2l}\, dy\Big)^{1/2}\,.$$

Henceforth we assume that $R \leq 1$. We have

$$(S_{l-\delta}\gamma)(x) \leq c\Big[R^\delta (S_l\gamma)(x) + \Big(\int_R^1 \Big(\int_{\mathcal{B}_1} |\gamma(x+\theta y)|\, d\theta\Big)^2 y^{-1-2(l-\delta)}\, dy\Big)^{1/2}$$
$$+\Big(\int_1^\infty \Big(y^{-n}\int_{\mathcal{B}_y} |\gamma(x+s)|\, ds\Big)^2 y^{-1-2(l-\delta)}\, dy\Big)^{1/2}\Big]$$
$$\leq c\Big[R^\delta (S_l\gamma)(x) + R^{\delta-m} \sup_{x\in\mathbb{R}^n, \rho\in(0,1)} \frac{\|\gamma; \mathcal{B}_\rho(x)\|_{L_p}}{\rho^{n/p-m+l}}$$
$$+ \sup_{x\in\mathbb{R}^n} \|\gamma; \mathcal{B}_1(x)\|_{L_p}\Big]\,.$$

It is clear that the last term can be thrown away by changing the constant c. By Lemma 3.2.2,

$$\sup_{x\in\mathbb{R}^n, \rho\in(0,1)} \frac{\|\gamma; \mathcal{B}_\rho(x)\|_{L_p}}{\rho^{n/p-m+l}} \leq c\,\|\gamma\|_{M(H_p^{m-l}\to L_p)}\,.$$

Thus, for all $R \in (0,1]$

$$(S_{l-\delta}\gamma)(x) \leq c\,R^\delta\left((S_l\gamma)(x) + R^{-m}\|\gamma\|_{M(H_p^{m-l}\to L_p)}\right)\,. \tag{3.2.6}$$

If

$$(S_{l-\delta})\gamma(x) \leq \|\gamma\|_{M(H_p^{m-l}\to L_p)},$$

we arrive at (3.2.5) by putting $R = 1$ in (3.2.6). In the opposite case, (3.2.5) follows from (3.2.6), with

$$R^m = \frac{\|\gamma\|_{M(H_p^{m-l}\to L_p)}}{(S_l\gamma)(x)}.$$

The lemma is proved. □

3.2.5 Auxiliary Properties of the Bessel Kernel G_l

We formulate some asymptotic properties of the kernel G_l of the Bessel potential which will be used later (see [AMS]):

(*i*) For $|x| \to 0$,

$$G_l(x) \sim |x|^{l-n} \quad \text{if} \quad 0 < l < n\,, \tag{3.2.7}$$

$$|G_l(x) - c_1 \log|x|^{-1}| \leq c_2 \quad \text{if} \quad l = n\,, \tag{3.2.8}$$

$$|G_l(x) - c| \leq c\,|x|^{\min(l-n,1)} \quad \text{if} \quad l > n\,, \tag{3.2.9}$$

$$|\nabla G_l(x)| \leq \begin{cases} c\,|x|^{l-n-1} & \text{if} \quad l \leq n+1\,, \\ c & \text{if} \quad l \geq n+1\,. \end{cases} \tag{3.2.10}$$

(*ii*) For $|x| \to \infty$,

$$G_l(x) \sim |x|^{(l-n-1)/2} e^{-|x|}\,, \tag{3.2.11}$$

$$|\nabla G_l(x)| \sim |x|^{(l-n-1)/2} e^{-|x|}\,. \tag{3.2.12}$$

Using these relations we prove the following:

Lemma 3.2.6. *The estimate*

$$|G_l(x) - G_l(y)| \leq c\,|x-y|^\delta \big(G_{l-\delta}(x/4) + G_{l-\delta}(y/4)\big) \tag{3.2.13}$$

holds with $\delta \in (0,1]$ and $l > \delta$.

Proof. It suffices to consider the case $|y| > |x|$. If $2|x-y| < |x|$, then by (3.2.10), for $|x| < 2$,

$$|G_l(x) - G_l(y)| \leq \begin{cases} c\,|x-y||x|^{l-n-1}\,, & l \leq n+1\,, \\ c\,|x-y|\,, & l \geq n+1\,, \end{cases}$$

and for $|x| \geq 2$ we deduce from (3.2.11) that

$$|G_l(x) - G_l(y)| \leq c\,|x-y||x|^{(l-n-1)/2} e^{-|x|/2}\,.$$

These estimates and (3.2.12) yield

$$|G_l(x) - G_l(y)| \leq c\,|x-y|^\delta G_{l-\delta}(x/4)$$

for $2|x-y| < |x|$.

Now let $2|x-y| > |x|$. If $l > n$ and $|y| < 2$, then by (3.2.9) we have

$$|G_l(x) - G_l(y)| \le c\,|y|^{\min(l-n,1)}$$

which, together with $|y| \le 3\,|x-y|$, gives

$$|G_l(x) - G_l(y)| \le c\,|x-y|^{\delta}|y|^{\min(l-n,1)-\delta}\,.$$

Combining this estimate with (3.2.7)–(3.2.9), we obtain (3.2.13) in the case $l > n$.

For the same values of x and y we have

$$|G_l(x) - G_l(y)| \le c_1|\log(|x|/|y|)| + c_2 \le c\,(|y|/|x|)^{\delta} \le c\,|x-y|^{\delta}|x|^{-\delta}$$

if $l = n$, and

$$|G_l(x) - G_l(y)| \le c\,(|x|^{l-n} + |y|^{l-n}) \le c\,|x-y|^{\delta}(|x|^{l-n-\delta} + |y|^{l-n-\delta})$$

if $l < n$. Using (3.2.8) again, we arrive at (3.2.13).

It remains to deal with the case

$$2|x-y| \ge |x|, \quad |y| \ge 2.$$

By (3.2.7)–(3.2.9) and (3.2.11) we obtain

$$|G_l(x) - G_l(y)| \le c\,G_l(x) \le c\,|x-y|^{\delta}|x|^{-\delta}G_l(x) \le c\,|x-y|^{\delta}G_{l-\delta}(x/4)$$

for $|x| > 1$. If $|x| < 1$, then $|x-y| \ge 1$ and therefore

$$|G_l(x) - G_l(y)| \le c\,G_l(x) \le c\,|x-y|^{\delta}G_l(x) \le c\,|x-y|^{\delta}G_{l-\delta}(x/4)\,.$$

The proof is complete. □

3.2.6 Upper Bound for the Norm of a Multiplier

We obtain a sufficient condition for a function to belong to the space $M(H_p^m \to H_p^l)$.

Lemma 3.2.7. *Let $\gamma \in H^l_{p,\mathrm{loc}}$, $S_l\gamma \in M(H_p^m \to L_p)$, and $\gamma \in M(H_p^{m-l} \to L_p)$. Then $\gamma \in M(H_p^m \to H_p^l)$ and*

$$\|\gamma\|_{M(H_p^m \to H_p^l)} \le c\,(\|S_l\gamma\|_{M(H_p^m \to L_p)} + \|\gamma\|_{M(H_p^{m-l} \to L_p)})\,. \tag{3.2.14}$$

(The function S_l was introduced in Definition 3.1.1.)

Proof. By Theorem 3.1.1 one can easily verify that a function in L_∞ with uniformly bounded derivatives of any order belongs to the space $M(H_p^m \to H_p^l)$.

Let γ_ρ be a mollification of γ with nonnegative kernel and radius ρ. Since $M(H_p^m \to H_p^l) \subset M(H_p^m \to L_p)$, it follows that $\gamma \in L_{p,\mathrm{unif}}$. Therefore $\nabla_j \gamma_\rho \in L_\infty$, $j = 0, 1, \ldots$, and $\gamma_\rho \in M(H_p^m \to H_p^l)$.

The assertion was proved for the case $\{l\} = 0$ in Lemma 2.3.5.

Suppose that $\{l\} > 0$. For any $u \in C_0^\infty$,

$$\|\gamma_\rho u\|_{L_p} \le \|\gamma_\rho\|_{M(H_p^{m-l} \to L_p)} \|u\|_{H_p^{m-l}} \,. \tag{3.2.15}$$

Clearly,

$$\|S_l(\gamma_\rho u)\|_{L_p} \le c \sum_{\substack{0 \le \beta \le \alpha \\ |\alpha| = [l]}} \|S_{\{l\}}(D^\beta \gamma_\rho D^{\alpha-\beta} u)\|_{L_p} \,.$$

First consider the terms corresponding to multi-indices of order $|\beta| = j < [l]$. By Lemma 3.2.4 and interpolation inequality (3.2.1), for any $k \in (0, l-j)$ we have

$$\begin{aligned} \|\nabla_j \gamma_\rho\|_{M(H_p^{m-k} \to H_p^{l-k-j})} &\le c\, \|\gamma_\rho\|_{M(H_p^{m-k} \to H_p^{l-k})} \\ &\le c\, \|\gamma_\rho\|^{1-k/l}_{M(H_p^m \to H_p^l)} \|\gamma_\rho\|^{k/l}_{M(H_p^{m-l} \to L_p)} . \end{aligned} \tag{3.2.16}$$

Consequently, for $|\beta| = j < [l]$,

$$\begin{aligned} &\|S_{\{l\}}(D^\beta \gamma_\rho D^{\alpha-\beta} u)\|_{L_p} \\ &\le c\, \|\gamma_\rho\|^{(\{l\}+j)/l}_{M(H_p^m \to H_p^l)} \|\gamma_\rho\|^{([l]-j)/l}_{M(H_p^{m-l} \to L_p)} \|u\|_{H_p^m} \,. \end{aligned} \tag{3.2.17}$$

This inequality together with (3.2.15) implies that

$$\begin{aligned} \|\gamma_\rho u\|_{H_p^l} \le (\varepsilon \|\gamma_\rho\|_{M(H_p^m \to H_p^l)} + c(\varepsilon) \|\gamma_\rho\|_{M(H_p^{m-l} \to L_p)}) \|u\|_{H_p^m} \\ + \|S_{\{l\}}(u \nabla_{[l]} \gamma_\rho)\|_{L_p} \end{aligned} \tag{3.2.18}$$

for any $\varepsilon > 0$. It remains to estimate the last term on the right-hand side. We have

$$\begin{aligned} \|S_{\{l\}}(u \nabla_{[l]} \gamma_\rho)\|_{L_p} \le \|u S_l \gamma_\rho\|_{L_p} + \| |\nabla_{[l]} \gamma_\rho| S_{\{l\}} u \|_{L_p} + \\ \Big\{ \int \Big(\int_0^\infty \Big[\int_{\mathcal{B}_1} |\Delta_{y\theta} u(x)| |\Delta_{y\theta} (\nabla_{[l]} \gamma_\rho)(x)|\, d\theta \Big]^2 y^{-1-2\{l\}} dy \Big)^{p/2} dx \Big\}^{1/p}, \end{aligned} \tag{3.2.19}$$

where $\Delta_z u(x) = u(x+z) - u(x)$. Obviously,

$$\|u S_l \gamma_\rho\|_{L_p} \le c\, \|S_l \gamma_\rho\|_{M(H_p^m \to L_p)} \|u\|_{H_p^m} \,. \tag{3.2.20}$$

Consider the second norm on the right-hand side of (3.2.19). Applying Minkow- ski's inequality, we obtain

$$S_{\{l\}}u \le \Lambda^{-[l]} S_{\{l\}} \Lambda^{[l]} u\,.$$

This and (3.2.16) yield

$$\| |\nabla_{[l]}\gamma_\rho| S_{\{l\}}u\|_{L_p} \le \|\nabla_{[l]}\gamma_\rho\|_{M(H_p^{m-\{l\}} \to L_p)} \|\Lambda^{\{l\}-m} S_{\{l\}} \Lambda^{m-\{l\}} u\|_{H_p^{m-\{l\}}}$$

$$\le c\, \|\gamma_\rho\|^{[l]/l}_{M(H_p^m \to H_p^l)} \|\gamma_\rho\|^{\{l\}/l}_{M(H_p^{m-l} \to L_p)} \|u\|_{H_p^m}\,. \tag{3.2.21}$$

Now we estimate the third term on the right-hand side of (3.2.19). Let $u = \Lambda^{-m} f$ and let δ be a sufficiently small positive number. We notice that

$$|u(x+y\theta) - u(x)| \le \int |G_m(x-\xi+y\theta) - G_m(x-\xi)|\, |f(\xi)|\, d\xi$$

and use Lemma 3.2.6. Then

$$|u(x+y\theta) - u(x)| \le c\, y^\delta \left[(\Lambda^{\delta-m}|f|)\Big(\frac{x+y\theta}{4}\Big) + (\Lambda^{\delta-m}|f|)\Big(\frac{x}{4}\Big) \right].$$

Thus, the third term in (3.2.19) does not exceed

$$c\Big\{ \int \Big(\int_0^\infty \Big[\int_{B_1} (\Lambda^{\delta-m}|f|)\Big(\frac{x+y\theta}{4}\Big) |\Delta_{y\theta}(\nabla_{[l]}\gamma_\rho)(x)| d\theta \Big]^2 y^{-1-2(\{l\}-\delta)} dy \Big)^{\frac{p}{2}} dx \Big\}^{\frac{1}{p}}$$

$$+c\Big\{ \int \Big[(\Lambda^{\delta-m}|f|)\Big(\frac{x}{4}\Big)\Big]^p (S_{l-\delta}\gamma_\rho)^p\, dx \Big\}^{\frac{1}{p}}.$$

After simple calculations we obtain that this sum is majorized by

$$c\ \Big(\Big\| (\Lambda^{\delta-m}|f|)\Big(\frac{\cdot}{4}\Big) \nabla_{[l]}\gamma_\rho \Big\|_{H_p^{\{l\}-\delta}} + \Big\| |\nabla_{[l]}\gamma_\rho| S_{\{l\}-\delta} \Big[(\Lambda^{\delta-m}|f|)\Big(\frac{\cdot}{4}\Big) \Big] \Big\|_{L_p}$$
$$+ \Big\| (\Lambda^{\delta-m}|f|)\Big(\frac{\cdot}{4}\Big) S_{\{l\}-\delta}(\nabla_{[l]}\gamma_\rho) \Big\|_{L_p} \Big).$$

Let the last norms be denoted by N_1, N_2 and N_3 respectively. Using (3.2.16), we get

$$N_1 \le \|\nabla_{[l]}\gamma_\rho\|_{M(H_p^{m-\delta} \to H_p^{\{l\}-\delta})} \Big\| (\Lambda^{\delta-m}|f|)\Big(\frac{\cdot}{4}\Big) \Big\|_{H_p^{m-\delta}}$$
$$\le c\, \|\gamma_\rho\|^{1-\delta/l}_{M(H_p^m \to H_p^l)} \|\gamma_\rho\|^{\delta/l}_{M(H_p^{m-l} \to L_p)} \|f\|_{L_p}\,. \tag{3.2.22}$$

Similarly,

$$N_2 \le \|\nabla_{[l]}\gamma_\rho\|_{M(H_p^{m-\{l\}} \to L_p)} \Big\| S_{\{l\}-\delta} \Big[(\Lambda^{\delta-m}|f|)\Big(\frac{\cdot}{4}\Big) \Big] \Big\|_{H_p^{m-\{l\}}}$$
$$\le c\, \|\nabla_{[l]}\gamma_\rho\|_{M(H_p^{m-\{l\}} \to L_p)} \Big\| (\Lambda^{\delta-m}|f|)\Big(\frac{\cdot}{4}\Big) \Big\|_{H_p^{m-\delta}}$$
$$\le c\, \|\gamma_\rho\|^{[l]/l}_{M(H_p^m \to H_p^l)} \|\gamma_\rho\|^{\{l\}/l}_{M(H_p^{m-l} \to L_p)} \|f\|_{L_p}\,. \tag{3.2.23}$$

Now we estimate the norm N_3. According to Lemma 3.2.5,

$$S_{\{l\}-\delta}(\nabla_{[l]}\gamma_\rho) \le c\,(S_l\gamma_\rho)^{1-\delta/m}\|\nabla_{[l]}\gamma_\rho\|^{\delta/m}_{M(H_p^{m-\{l\}}\to L_p)}$$
$$+c\,\|\nabla_{[l]}\gamma_\rho\|_{M(H_p^{m-\{l\}}\to L_p)}\,.$$

Further we notice that

$$\Lambda^{\delta-m}|f| \le c\,(\Lambda^{-m}|f|)^{1-\delta/m}(\mathcal{M}f)^{\delta/m}$$

(see Lemma 1.2.5). Thus we have proved the estimate

$$N_3 \le c\,\|\nabla_{[l]}\gamma_\rho\|^{\delta/m}_{M(H_p^{m-\{l\}}\to L_p)}\Big\|\Big[(\Lambda^{-m}|f|)\Big(\frac{\cdot}{4}\Big)S_l\gamma_\rho\Big]^{1-\delta/m}\Big[(\mathcal{M}f)\Big(\frac{\cdot}{4}\Big)\Big]^{\delta/m}\Big\|_{L_p}$$
$$+c\,\|\nabla_{[l]}\gamma_\rho\|_{M(H_p^{m-\{l\}}\to L_p)}\|\Lambda^{\delta-m}|f|\|_{L_p}\,.$$

Consequently,

$$N_3 \le c\,\|\nabla_{[l]}\gamma_\rho\|^{\delta/m}_{M(H_p^{m-\{l\}}\to L_p)}\|\mathcal{M}f\|^{\delta/m}_{L_p}\Big\|(\Lambda^{-m}|f|)\Big(\frac{\cdot}{4}\Big)S_l\gamma_\rho\Big\|^{1-\delta/m}_{L_p}$$
$$+c\,\|\nabla_{[l]}\gamma_\rho\|_{M(H_p^{m-\{l\}}\to L_p)}\|\Lambda^{\delta-m}|f|\|_{H_p^{m-\delta}}\,.$$

Applying (3.2.16), we finally obtain

$$N_3 \le c\,\|\gamma_\rho\|^{\delta[l]/ml}_{M(H_p^m\to L_p)}\|\gamma_\rho\|^{\delta\{l\}/ml}_{M(H_p^m\to L_p)}\big(\|S_l\gamma_\rho\|_{M(H_p^m\to L_p)}$$
$$+\|\gamma_\rho\|^{[l]/l}_{M(H_p^m\to H_p^l)}\|\gamma_\rho\|^{\{l\}/l}_{M(H_p^{m-l}\to L_p)}\big)^{1-\delta/m}\,\|f\|_{L_p}\,. \tag{3.2.24}$$

Adding the estimates (3.2.22)–(3.2.24), we conclude that

$$\Big\{\int\Big(\int_0^\infty\Big[\int_{\mathcal{B}_1}|\Delta_{y\theta}u(x)||\Delta_{y\theta}(\nabla_{[l]}\gamma_\rho)(x)|\,d\theta\Big]^2 y^{-1-2\{l\}}dy\Big)^{p/2}dx\Big\}^{1/p}$$
$$\le \big(\varepsilon\|\gamma_\rho\|_{M(H_p^m\to H_p^l)}+\varepsilon\,\|S_l\gamma_\rho\|_{M(H_p^m\to L_p)}$$
$$+\,c(\varepsilon)\,\|\gamma_\rho\|_{M(H_p^{m-l}\to L_p)}\big)\|u\|_{H_p^m}. \tag{3.2.25}$$

This together with (3.2.20) and (3.2.21) makes it possible to deduce from (3.2.19) that

$$\|S_{\{l\}}(u\nabla_{[l]}\gamma_\rho)\|_{L_p} \le \big(\varepsilon\,\|\gamma_\rho\|_{M(H_p^m\to H_p^l)}+c\,\|S_l\gamma_\rho\|_{M(H_p^m\to L_p)}$$
$$+c(\varepsilon)\|\gamma_\rho\|_{M(H_p^{m-l}\to L_p)}\big)\|u\|_{H_p^m}\,.$$

Substitution of this estimate into (3.2.18) leads to (3.2.14) for γ_ρ. Reference to Lemma 3.2.1 completes the proof. □

3.2.7 Lower Bound for the Norm of a Multiplier

We prove the assertion converse to Lemma 3.2.7.

Lemma 3.2.8. *If* $\gamma \in M(H_p^m \to H_p^l)$, *then* $\gamma \in H^l_{p,\mathrm{loc}}$, $S_l\gamma \in M(H_p^m \to L_p)$, *and* $\gamma \in M(H_p^{m-l} \to L_p)$. *The following inequality holds*

$$\|S_l\gamma\|_{M(H_p^m\to L_p)} + \|\gamma\|_{M(H_p^{m-l}\to L_p)} \le c\,\|\gamma\|_{M(H_p^m\to H_p^l)}\,. \tag{3.2.26}$$

Proof. It is clear that $\gamma \in H^l_{p,\mathrm{loc}}$. The upper estimate of $\|\gamma\|_{M(H_p^{m-l}\to L_p)}$ is contained in Lemma 3.2.3.

The assertion was proved for integer l in Lemma 2.3.4. Suppose that $\{l\}>0$. For any $u \in C_0^\infty$,

$$\|S_{\{l\}}(u\nabla_{[l]}\gamma_\rho)\|_{L_p} \le \|S_l(\gamma_\rho u)\|_{L_p} + \sum_{\substack{0\le\beta<\alpha\\ |\alpha|=[l]}} \|S_{\{l\}}(D^\beta\gamma_\rho D^{\alpha-\beta}u)\|_{L_p}$$

and therefore

$$\|uS_l\gamma_\rho\|_{L_p} \le \|\gamma_\rho u\|_{H_p^l} + \sum_{\substack{0\le\beta<\alpha\\ |\alpha|\le[l]}} \|S_{\{l\}}(D^\beta\gamma_\rho D^{\alpha-\beta}u)\|_{L_p} + \||\nabla_{[l]}\gamma_\rho|S_{\{l\}}u\|_{L_p}$$

$$+\Big\{\int\Big(\int_0^\infty\Big[\int_{B_1}|\Delta_{y\theta}u(x)||\Delta_{y\theta}(\nabla_{[l]}\gamma_\rho)(x)|\,d\theta\Big]^2 y^{-1-2\{l\}}dy\Big)^{p/2}dx\Big\}^{1/p}.$$

This estimate and (3.2.17), (3.2.21), (3.2.25) give

$$\begin{aligned}\|uS_l\gamma_\rho\|_{L_p} \le{}& \big(\varepsilon\,\|S_l\gamma_\rho\|_{M(H_p^m\to L_p)}\\ &+c_1\|\gamma_\rho\|_{M(H_p^m\to H_p^l)} + c\,\|\gamma_\rho\|_{M(H_p^{m-l}\to L_p)}\big)\|u\|_{H_p^m}\\ \le{}& \big(\varepsilon\|S_l\gamma_\rho\|_{M(H_p^m\to L_p)} + c_2\,\|\gamma_\rho\|_{M(H_p^m\to H_p^l)}\big)\|u\|_{H_p^m}\end{aligned}$$

which implies that

$$\|S_l\gamma_\rho\|_{M(H_p^m\to L_p)} \le c\,\|\gamma_\rho\|_{M(H_p^m\to H_p^l)}\,.$$

It remains to use Lemma 3.2.1. □

Corollary 3.2.2. *Let* $0<l<m$ *and let* $p\in(1,\infty)$. *Then*

$$\|\gamma\|_{M(H_p^m\to H_p^l)}$$

$$\sim \sum_{j=0}^{[l]}\big(\|S_{l-j}\gamma\|_{M(H_p^{m-j}\to L_p)} + \|\nabla_{[l]-j}\gamma\|_{M(H_p^{m-j-\{l\}}\to L_p)}\big). \tag{3.2.27}$$

Proof. The upper estimate in (3.2.27) follows from Lemma 3.2.7. In view of (3.2.26) and Corollary 3.2.1, the lower estimate in (3.2.27) results from

$$\|S_{l-j}\gamma\|_{M(H_p^{m-j}\to L_p)} \le c\,\|\gamma\|_{M(H_p^{m-j}\to H_p^{l-j})} \le c\,\|\gamma\|_{M(H_p^m\to H_p^l)}$$

and

$$\|\nabla_{[l]-j}\gamma\|_{M(H_p^{m-j-\{l\}}\to L_p)} \le c\,\|\gamma\|_{M(H_p^{m-j-\{l\}}\to H_p^{[l]-j})} \le c\,\|\gamma\|_{M(H_p^m\to H_p^l)}.$$

3.2.8 Description of the Space $M(H_p^m \to H_p^l)$

Combining Lemmas 3.2.7 and 3.2.8, we obtain:

Theorem 3.2.1. *Let $m \ge l \ge 0$, $p \in (1,\infty)$. A function γ belongs to the space $M(H_p^m \to H_p^l)$ if and only if $\gamma \in H^l_{p,\mathrm{loc}}$, $S_l\gamma \in M(H_p^m \to L_p)$, and $\gamma \in M(H_p^{m-l} \to L_p)$. The relation*

$$\|\gamma\|_{M(H_p^m\to H_p^l)} \sim \|S_l\gamma\|_{M(H_p^m\to L_p)} + \|\gamma\|_{M(H_p^{m-l}\to L_p)}$$

holds.

This, together with Lemma 3.2.2, gives the following assertion.

Theorem 3.2.2. *A function γ belongs to the space $M(H_p^m \to H_p^l)$, $m \ge l \ge 0$, $p \in (1,\infty)$ if and only if $\gamma \in H^l_{p,\mathrm{loc}}$ and, for any compact set $e \subset \mathbb{R}^n$,*

$$\|S_l\gamma; e\|^p_{L_p} \le c\,C_{p,m}(e)\,,$$
$$\|\gamma; e\|^p_{L_p} \le c\,C_{p,m-l}(e)\,.$$

The relations

$$\|\gamma\|_{M(H_p^m\to H_p^l)} \sim \sup_e \Big(\frac{\|S_l\gamma; e\|_{L_p}}{[C_{p,m}(e)]^{1/p}} + \frac{\|\gamma; e\|_{L_p}}{[C_{p,m-l}(e)]^{1/p}} \Big)\,, \tag{3.2.28}$$

$$\|\gamma\|_{M(H_p^m\to H_p^l)} \sim \sup_{\{e:d(e)\le 1\}} \Big(\frac{\|S_l\gamma; e\|_{L_p}}{[C_{p,m}(e)]^{1/p}} + \frac{\|\gamma; e\|_{L_p}}{[C_{p,m-l}(e)]^{1/p}} \Big) \tag{3.2.29}$$

hold.

In particular,

$$\|\gamma\|_{MH_p^l} \sim \sup_e \frac{\|S_l\gamma; e\|_{L_p}}{[C_{p,l}(e)]^{1/p}} + \|\gamma\|_{L_\infty}\,, \tag{3.2.30}$$

$$\|\gamma\|_{MH_p^l} \sim \sup_{\{e:d(e)\le 1\}} \frac{\|S_l\gamma; e\|_{L_p}}{[C_{p,l}(e)]^{1/p}} + \|\gamma\|_{L_\infty}\,. \tag{3.2.31}$$

Remark 3.2.1. According to Stein's theorem [St1], the norm in H_p^l for $p \geq 2$, $0 < l < 1$, is equivalent to $\|Tu\|_{L_p} + \|u\|_{L_p}$, where

$$(Tu)(x) = \left(\int |u(x+y) - u(x)|^2 \frac{dy}{|y|^{n+2l}} \right)^{1/2} . \tag{3.2.32}$$

This and the inequality

$$|T(\gamma u) - u\,T\gamma| \leq \|\gamma\|_{L_\infty} Tu$$

imply that

$$\|\gamma\|_{MH_p^l} \sim \sup_e \frac{\|T\gamma; e\|_{L_p}}{[C_{p,l}(e)]^{1/p}} + \|\gamma\|_{L_\infty} ,$$

where $p \geq 2$, $0 < l < 1$.

3.2.9 Equivalent Norm in $M(H_p^m \to H_p^l)$ Involving the Norm in $L_{mp/(m-l)}$

In this subsection we obtain one more norm for the space $M(H_p^m \to H_p^l)$.

Lemma 3.2.9. *If $\gamma \in M(H_p^m \to H_p^l)$, $m > l$, $p \in (1, \infty)$ and k is an integer, $1 \leq k \leq m/(m-l)$, then $\gamma^k \in M(H_p^m \to H_p^{m-k(m-l)})$. Moreover,*

$$\|\gamma^k\|_{M(H_p^m \to H_p^{m-k(m-l)})} \leq c\, \|\gamma\|^{k\,\theta(k)}_{M(H_p^{m-l} \to L_p)} \|\gamma\|^{k(1-\theta(k))}_{M(H^m \to H_p^l)} , \tag{3.2.33}$$

where $\theta(k) = (k-1)(m-l)/2l$.

Proof. We have

$$\|\gamma^k\|_{M(H_p^m \to H_p^{m-k(m-l)})} \leq c\, \|\gamma\|_{M(H_p^{m-(k-1)(m-l)} \to H_p^{m-k(m-l)})} \times \|\gamma^{k-1}\|_{M(H_p^m \to H_p^{m-(k-1)(m-l)})} . \tag{3.2.34}$$

Suppose that (3.2.33) is proved for $k-1$. Then from (3.2.34) and the interpolation inequality

$$\|\gamma\|_{M(H_p^{m-(k-1)(m-l)} \to H_p^{m-k(m-l)})} \leq c\, \|\gamma\|^{(k-1)(m-l)/l}_{M(H_p^{m-l} \to L_p)} \|\gamma\|^{(m-k(m-l))/l}_{M(H_p^m \to H_p^l)}$$

we obtain

$$\|\gamma^k\|_{M(H_p^m \to H_p^{m-k(m-l)})} \leq c\, \|\gamma\|^{(k-1)(m-l)/l + (k-1)\theta(k-1)}_{M(H_p^m \to L_p)} \times \|\gamma\|^{m-k(m-l)/l + (k-1)(1-\theta(k-1))}_{M(H_p^m \to H_p^l)}$$

which is equivalent to (3.2.33). □

Corollary 3.2.3. *The inequality*

$$\sup_e \frac{\|\gamma; e\|_{L_{pm/(m-l)}}}{[C_{p,m}(e)]^{(m-l)/mp}} \le c\, \|\gamma\|^{1-\mu}_{M(H_p^{m-l}\to L_p)} \|\gamma\|^{\mu}_{M(H_p^m \to H_p^l)}$$

holds, where $\mu = k(1-\theta(k))$, $k = [m/(m-l)]$, $\theta(k) = (k-1)(m-l)/2l$.

Proof. By Lemma 3.2.2,

$$\| |\gamma|^{m/(m-l)} u\|_{L_p} \le c \left(\sup_e \frac{\int_e |\gamma|^{p\alpha}\, dx}{C_{p,\alpha(m-l)}(e)} \right)^{1/p} \|\gamma^k u\|_{H_p^{\alpha(m-l)}} ,$$

where α is a fractional part of $m/(m-l)$. Applying Lemmas 3.2.2 and 2.3.6, we get from the last estimate that

$$\| |\gamma|^{m/(m-l)} u\|_{L_p} \le c\, \|\gamma\|^{\alpha}_{M(H_p^{m-l}\to L_p)} \|\gamma^k u\|_{H_p^{\alpha(m-l)}} .$$

It remains to use Lemma 3.2.9. □

Theorem 3.2.3. *The equivalence relation*

$$\|\gamma\|_{M(H_p^m\to H_p^l)} \sim \sup_e \left(\frac{\|\gamma; e\|_{L_{mp/(m-l)}}}{[C_{p,m}(e)]^{(m-l)/mp}} + \frac{\|S_l\gamma; e\|_{L_p}}{[C_{p,m}(e)]^{1/p}} \right) \qquad (3.2.35)$$

holds with $p \in (1,\infty)$, $m > l \ge 0$.

Proof. The upper estimate for the norm in $M(H_p^m \to H_p^l)$ follows from Lemmas 3.2.7 and 2.3.6 which implies that

$$\sup_e \frac{\|\gamma; e\|_{L_p}}{[C_{p,m-l}(e)]^{1/p}} \le c \sup_e \frac{\|\gamma; e\|_{L_{mp/(m-l)}}}{[C_{p,m}(e)]^{(m-l)/mp}} .$$

The lower bound is deduced from Corollary 3.2.3 and Lemma 3.2.8. □

From this theorem and (3.1.14) we obtain

Corollary 3.2.4. *The relation*

$$\|\gamma\|_{M(H_p^m\to H_p^l)} \sim \sup_{\{e: d(e)\le 1\}} \left(\frac{\|\gamma; e\|_{L_{mp/(m-l)}}}{[C_{p,m}(e)]^{(m-l)/mp}} + \frac{\|S_l\gamma; e\|_{L_p}}{[C_{p,m}(e)]^{1/p}} \right)$$

holds with $p \in (1,\infty)$, $m > l \ge 0$.

3.2.10 Characterization of $M(H_p^m \to H_p^l)$, $m > l$, Involving the Norm in $L_{1,\text{unif}}$

In the case $m > l$, the second term on the right-hand sides of (3.2.28) and (3.2.29) can be replaced by the norm of γ in $L_{1,\text{unif}}$ as shown by the following assertion.

Theorem 3.2.4. *Let $0 < l < m \le n/p$, $p \in (1, \infty)$. Then*

$$\|\gamma\|_{M(H_p^m \to H_p^l)} \sim \sup_e \frac{\|S_l\gamma; e\|_{L_p}}{[C_{p,m}(e)]^{1/p}} + \|\gamma\|_{L_{1,\text{unif}}}. \tag{3.2.36}$$

This relation also holds if e is any compact subset of $\mathbb{R}^n$ with diameter less than 1.

The lower estimate for the norm in $M(H_p^m \to H_p^l)$ is a direct corollary of (3.2.28), (3.2.29), and the obvious inequality

$$\sup_e \frac{\|\gamma; e\|_{L_p}}{[C_{p,m-l}(e)]^{1/p}} \ge c \sup_x \|\gamma; \mathcal{B}_1(x)\|_{L_1}.$$

The upper estimate follows from Theorem 3.2.3 and the next assertion obtained by Verbitsky (see Sect. 2.6 in [MSh16]). This assertion is similar in nature to Lemma 2.3.9.

Lemma 3.2.10. *Let $0 < l < m \le n/p$, $p \in (1, \infty)$, and let $\gamma \in H^l_{p,\text{loc}}$. Then*

$$\sup_e \frac{\|\gamma; e\|_{L_{mp/(m-l)}}}{(C_{p,m}(e))^{(m-l)/mp}} \le c\Big(\sup_e \frac{\|S_l\gamma; e\|_{L_p}}{(C_{p,m}(e))^{1/p}} + \|\gamma\|_{L_{1,\text{unif}}}\Big), \tag{3.2.37}$$

The proof is based on several auxiliary assertions. In the first three of them we use the Poisson operator which is defined for functions $\gamma \in L_{1,\text{unif}}$ by

$$(T\gamma)(x, y) = \frac{1}{|\partial\mathcal{B}^{(n+1)}|} \int_{\mathbb{R}^n} \frac{y\,\gamma(\xi) d\xi}{(y^2 + |x - \xi|^2)^{(n+1)/2}}, \quad (x, y) \in \mathbb{R}^{n+1}_+, \tag{3.2.38}$$

where $|\partial\mathcal{B}^{(n+1)}|$ is the area of the $(n + 1)$-dimensional unit ball.

Lemmas 3.2.11–3.2.14 below are proved by Verbitsky, see Sect. 2.6 [MSh16].

Lemma 3.2.11. *For any $k = 0, 1, \ldots$ there holds the inequality*

$$|\gamma(x)| \le c\Big(\|\gamma\|_{L_{1,\text{unif}}} + \int_0^1 \Big|\frac{\partial^{k+1}(T\gamma)(x, y)}{\partial y^{k+1}}\Big| y^k dy\Big). \tag{3.2.39}$$

Proof. The following equality is readily checked by integration by parts

$$\gamma(x) = (T\gamma)(x, 1) - \frac{\partial(T\gamma)(x, 1)}{\partial y} + \frac{1}{2}\frac{\partial^2(T\gamma)(x, 1)}{\partial y^2} - \ldots + \frac{(-1)^k}{k!}\frac{\partial^k(T\gamma)(x, 1)}{\partial y^k}$$

$$+\frac{(-1)^{k+1}}{k!}\int_0^1 \frac{\partial^{k+1}(T\gamma)(x,y)}{\partial y^{k+1}} y^k\,dy. \tag{3.2.40}$$

Now we show that for any $k = 0, 1, \ldots$ the inequality

$$\Big\|\frac{\partial^k(T\gamma)(\cdot,1)}{\partial y^k}\Big\|_{L_\infty} \le c\,\|\gamma\|_{L_{1,\mathrm{unif}}} \tag{3.2.41}$$

holds.

Let $k > 0$. The Poisson kernel

$$P(x,y) = y(|x|^2 + y^2)^{-(n+1)/2}$$

obeys the estimates

$$\Big|\frac{\partial^k P(x,y)}{\partial y^k}\Big| \le c\,(|x| + y)^{-n-k} \tag{3.2.42}$$

(see [St2], Ch. 5, Sect. 4). Hence

$$\Big|\frac{\partial^k(T\gamma)(x,1)}{\partial y^k}\Big| = \Big|\int \frac{\partial^k P(t,1)}{\partial y^k}\gamma(x-t)dt\Big|$$

$$\le c\int \frac{|\gamma(x-t)|dt}{(|t|+1)^{n+k}} \le c\Big(\int_{|t|\le 1}|\gamma(x-t)|dt + \int_{|t|\ge 1}\frac{|\gamma(x-t)|dt}{|t|^{n+k}}\Big)$$

$$\le c\Big(\|\gamma\|_{L_{1,\mathrm{unif}}} + \int_1^\infty r^{-n-k-1}dr\int_{|t|\le r}|\gamma(x-t)|dt\Big).$$

Note that for $r \ge 1$

$$\int_{|t|\le r}|\gamma(x-t)|dt \le c\,r^n\|\gamma\|_{L_{1,\mathrm{unif}}}.$$

Therefore,

$$\Big|\frac{\partial^k(T\gamma)(x,1)}{\partial y^k}\Big| \le c\,\|\gamma\|_{L_{1,\mathrm{unif}}}\Big(1 + \int_1^\infty r^{-k-1}dr\Big).$$

Combining (3.2.40) and (3.2.41), we complete the proof for $k > 0$.

The case $k = 0$ is treated in a similar way with the help of the estimate $P(x,1) \le c\,(|x|+1)^{-n-1}$. □

Lemma 3.2.12. *Let $\gamma \in W^{[l]}_{1,\mathrm{loc}}$ and let $k = [l] + 1$. Then*

$$\Big(\int_0^\infty \Big|\frac{\partial^k(T\gamma)(x,y)}{\partial y^k}\Big|^2 y^{1-2\{l\}}dy\Big)^{1/2} \le c\,(S_l\gamma)(x).$$

Proof. First, consider the case $k = 1$, $l \in (0,1)$. The identity

$$\int \frac{\partial P(t,y)}{\partial y} dt = 0$$

(see [St2], Ch. 5, Sect. 4) implies that

$$\frac{\partial (T\gamma)(x,y)}{\partial y} = \int \frac{\partial P(t,y)}{\partial y} \gamma(x-t)dt = \int \frac{\partial P(t,y)}{\partial y} \big(\gamma(x-t) - \gamma(x)\big)dt.$$

Using (3.2.42), we get

$$\Big|\frac{\partial (T\gamma)(x,y)}{\partial y}\Big| \le c \int \frac{|\gamma(x-t) - \gamma(x)|}{(|t|+y)^{n+1}} dt.$$

Consequently,

$$\int_0^\infty \Big|\frac{\partial (T\gamma)(x,y)}{\partial y}\Big|^2 y^{1-2l} dy \le c \int_0^\infty y^{1-2l} \Big(\int \frac{|\gamma(x-t) - \gamma(x)|}{(|t|+y)^{n+1}} dt\Big)^2 dy$$

$$\le c \int_0^\infty y^{-(1+2l+2n)} dy \Big(\int_{|t|\le y} |\gamma(x-t) - \gamma(x)| dt\Big)^2$$

$$+ c \int_0^\infty y^{1-2l} \Big(\int_{|t|\ge y} \frac{|\gamma(x-t) - \gamma(x)|}{(|t|+y)^{n+1}} dt\Big)^2 dy = A_1 + A_2.$$

We have

$$A_1 = c \int_0^\infty y^{-(1+2l+2n)} \Big(\int_{|\tau|\le 1} |\gamma(x-\tau y) - \gamma(x)| y^n d\tau\Big)^2 dy$$

$$= c \int_0^\infty y^{-(1+2l)} \Big(\int_{|\tau|\le 1} |\gamma(x-\tau y) - \gamma(x)|\, d\tau\Big)^2 dy \le c\big((S_l\gamma)(x)\big)^2.$$

To find a majorant for A_2, we rewrite it as follows

$$A_2 = c \int_0^\infty y^{1-2l} \Big(\int_{r\ge y} r^{-n-2} dr \int_{y\le |t|\le r} |\gamma(x-t) - \gamma(x)| dt\Big)^2 dy.$$

Applying the Hardy inequality, we get

$$A_2 \le c \int_0^\infty r^{-2n-2l-1} \Big(\int_{|t|\le r} |\gamma(x-t) - \gamma(x)| dt\Big)^2 dr$$

$$= c \int_0^\infty r^{-2l-1} \Big(\int_{|\tau|\le 1} |\gamma(x-\tau r) - \gamma(x)| d\tau\Big)^2 dr \le c\,\big((S_l\gamma)(x)\big)^2.$$

Thus, for $l \in (0,1)$

$$\int_0^\infty \Big|\frac{\partial (T\gamma)(x,y)}{\partial y}\Big|^2 y^{1-2l} dy \le c\,\big((S_l\gamma)(x)\big)^2.$$

Next, let $k = 2m$, $m = 1, 2, \ldots$. Since $T\gamma$ is a harmonic function in $\mathbb{R}^{n+1}_+$,

$$\frac{\partial^2 (T\gamma)(x,y)}{\partial y^2} = -\sum_{j=1}^{n} \frac{\partial^2 (T\gamma)(x,y)}{\partial x_j^2}.$$

Therefore

$$\frac{\partial^k (T\gamma)(x,y)}{\partial y^k} = (-1)^m \sum_{i_1,\ldots,i_m=1}^{n} \frac{\partial^{2m} (T\gamma)(x,y)}{\partial x_{i_1}^2 \ldots \partial x_{i_m}^2}. \tag{3.2.43}$$

Using the identity

$$\int \frac{\partial P(t,y)}{\partial t_i} dt = 0, \qquad i = 1, \ldots, n,$$

and the estimate

$$\Big| \frac{\partial P(x,y)}{\partial x_i} \Big| \le c\,(|x| + y)^{-n-1}$$

(see [St2], Ch. 5, Sect. 4), we get

$$\Big|\frac{\partial^k (T\gamma)(x,y)}{\partial y^k}\Big| = \Big| \sum_{i_1,\ldots,i_m=1}^{n} \int \frac{\partial P(t,y)}{\partial t_{i_1}} \Big(\frac{\partial^{[l]}\gamma(x-t)}{\partial x_{i_1} \partial x_{i_2}^2 \ldots \partial x_{i_m}^2} - \frac{\partial^{[l]}\gamma(x)}{\partial x_{i_1} \partial x_{i_2}^2 \ldots \partial x_{i_m}^2} \Big) dt$$

$$\le c \int \frac{|\nabla_{[l]}\gamma(x-t) - \nabla_{[l]}\gamma(x)|}{(|t| + y)^{n+1}} dt.$$

We complete the proof in the same way as for $l \in (0,1)$.

For $k = 2m+1$, $m = 1, 2, \ldots$, we use the identity

$$\frac{\partial^k (T\gamma)(x,y)}{\partial y^k} = (-1)^m \sum_{i_1,\ldots,i_m=1}^{n} \frac{\partial^{2m+1} (T\gamma)(x,y)}{\partial y \partial x_{i_1}^2 \ldots \partial x_{i_m}^2}$$

$$= (-1)^m \sum_{i_1,\ldots,i_m=1}^{n} \int \frac{\partial P(t,y)}{\partial y} \Big(\frac{\partial^{[l]}\gamma(x-t)}{\partial x_{i_1}^2 \ldots \partial x_{i_m}^2} - \frac{\partial^{[l]}\gamma(x)}{\partial x_{i_1}^2 \ldots \partial x_{i_m}^2} \Big) dt$$

which implies that

$$\Big|\frac{\partial^k (T\gamma)(x,y)}{\partial y^k}\Big| \le c \int \frac{|\nabla_{[l]}\gamma(x-t) - \nabla_{[l]}\gamma(x)|}{(|t| + y)^{n+1}} dt.$$

Again we complete the proof in the same way as for $l \in (0,1)$. □

In the next two assertions we use the notation

$$K = \sup_{x \in \mathbb{R}^n, r \in (0,1]} r^{m-n/p} \| S_l \gamma; \mathcal{B}_r(x) \|_{L_p}. \tag{3.2.44}$$

Lemma 3.2.13. *Let $\gamma \in W_{1,\mathrm{loc}}^{[l]}$, $y \in (0,1]$, and let $k = [l]+1$. Then*

$$\Big|\frac{\partial^k (T\gamma)(x,y)}{\partial y^k}\Big| \le c\,K y^{\{l\}-m-1}.$$

Proof. Let $r/2 < y \le r$, $r \in (0,1]$. By Lemma 3.2.12,

$$\int_{\mathcal{B}_r(x)} \Big(\int_0^\infty \Big|\frac{\partial^k (T\gamma)(t,y)}{\partial y^k}\Big|^2 y^{1-2\{l\}} dy\Big)^{p/2} dt \le c\,K^p r^{n-mp}. \tag{3.2.45}$$

Applying the mean value theorem for harmonic functions and then the Cauchy inequality, we obtain

$$\begin{aligned}\Big|\frac{\partial^k (T\gamma)(x,y)}{\partial y^k}\Big| &\le c\,r^{-n-1} \int_{\mathcal{B}_r(x)} \int_{r/2}^r \Big|\frac{\partial^k (T\gamma)(t,\eta)}{\partial \eta^k}\Big| d\eta\, dt \\ &\le c\,r^{-n-1/2} \int_{\mathcal{B}_r(x)} \Big(\int_{r/2}^r \Big|\frac{\partial^k (T\gamma)(t,\eta)}{\partial \eta^k}\Big|^2 d\eta\Big)^{1/2} dt \\ &\le c\,r^{-n-1+\{l\}} \int_{\mathcal{B}_r(x)} \Big(\int_{r/2}^r \Big|\frac{\partial^k (T\gamma)(t,\eta)}{\partial \eta^k}\Big|^2 \eta^{1-2\{l\}} d\eta\Big)^{1/2} dt.\end{aligned}$$

Using (3.2.45) and the Hölder inequality, we find that

$$\Big|\frac{\partial^k (T\gamma)(x,y)}{\partial y^k}\Big| \le c\,r^{-n-1+\{l\}}\, r^{n(p-1)/p}\, K\, r^{(n-mp)/p} \le c\,K\, y^{\{l\}-m-1}.$$

The proof is complete. □

Lemma 3.2.14. *Let $\gamma \in W_{1,\mathrm{loc}}^{[l]}$, $0 < l < m \le n/p$. Then, for almost all $x \in \mathbb{R}^n$,*

$$|\gamma(x)| \le c\,\big(K^{l/m}((S_l\gamma)(x))^{(m-l)/m} + \|\gamma\|_{L_{1,\mathrm{unif}}}\big), \tag{3.2.46}$$

where K is introduced in (3.2.44).

Proof. We use Lemma 3.2.11 with $k = [l]+1$. Let

$$\varphi(y) = \begin{cases} |\partial^{k+1}(T\gamma)(x,y)/\partial y^{k+1}| & \text{for } 0 < y \le 1, \\ 0 & \text{for } y \ge 1. \end{cases}$$

Then, for any $R > 0$,

$$\begin{aligned}\int_0^1 \Big|\frac{\partial^{k+1}(T\gamma)(x,y)}{\partial y^{k+1}}\Big| y^k dy &= \int_0^\infty \varphi(y)\, y^k\, dy \\ &= \int_0^R \varphi(y)\, y^k\, dy + \int_R^\infty \varphi(y)\, y^k\, dy.\end{aligned}$$

Applying the Cauchy inequality to the first term and using Lemma 3.2.13 to estimate the second one, we find that

$$\int_0^\infty \varphi(y) y^k dy$$

$$\le c\Big(\Big(\int_0^R \varphi(y)^2 y^{1-2\{l\}} dy\Big)^{1/2}\Big(\int_0^R y^{2k+2\{l\}-1} dy\Big)^{1/2} + K\int_R^\infty y^{k+\{l\}-1-m} dy\Big)$$

$$\le c\Big(\Big(\int_0^\infty \varphi(y)^2 y^{1-2\{l\}}\Big)^{1/2} R^l + KR^{l-m}\Big).$$

Putting here

$$R = K^{1/m}\Big(\int_0^\infty \varphi(y)^2 y^{1-2\{l\}} dy\Big)^{-1/2m},$$

we arrive at

$$\int_0^\infty \varphi(y) y^k dy \le c\, K^{l/m}\Big(\int_0^\infty \varphi(y)^2 y^{1-2\{l\}} dy\Big)^{(m-l)/2m}.$$

Consequently,

$$|\gamma(x)| \le c\Big(K^{l/m}\Big(\int_0^1 \Big|\frac{\partial^{k+1}(T\gamma)(x,y)}{\partial y^{k+1}}\Big|^2 y^{1-2\{l\}} dy\Big)^{(m-l)/2m} + \|\gamma\|_{L_{1,\mathrm{unif}}}\Big).$$

Reference to Lemma 3.2.12 completes the proof. □

Proof of Lemma 3.2.10. Let K be defined by (3.2.44). By Lemma 3.2.14,

$$\int_e |\gamma(x)|^{mp/(m-l)} dx \le c\Big(K^{lp/(m-l)}\int_e |(S_l\gamma)(x)|^p dx + \|\gamma\|_{L_{1,\mathrm{unif}}}^{mp/(m-l)} \mathrm{mes}_n e\Big),$$

which together with the obvious estimate $\mathrm{mes}_n e \le C_{p,m}(e)$ implies that

$$\frac{\|\gamma; e\|_{L_{mp/(m-l)}}}{(C_{p,m}(e))^{(m-l)/mp}} \le c\big(K^{l/m} Q^{(m-l)/m} + \|\gamma\|_{L_{1,\mathrm{unif}}}\big),$$

where

$$Q = \sup_e \frac{\|S_l\gamma; e\|_{L_p}}{(C_{p,m}(e))^{1/p}}.$$

Since $K \le Q$, we complete the proof. □

Corollary 3.2.5. *Let $0 < l < m$ and let $p \in (1,\infty)$. Then*

$$\|\gamma\|_{M(H_p^m \to H_p^l)} \sim \|S_l\gamma\|_{M(H_p^m \to L_p)} + \|\gamma\|_{L_{1,\mathrm{unif}}}. \tag{3.2.47}$$

For $m = l$ the norm $\|\gamma\|_{L_{1,\mathrm{unif}}}$ should be replaced by $\|\gamma\|_{L_\infty}$.

Proof. The result follows by Theorem 3.2.4 and Lemma 3.2.2.

3.2.11 The Space $M(H_p^m \to H_p^l)$ for $mp > n$

Theorem 3.2.5. *If $mp > n$, $p \in (1, \infty)$, then*

$$\|\gamma\|_{M(H_p^m \to H_p^l)} \sim \|\gamma\|_{H^l_{p,\mathrm{unif}}}.$$

Proof. The required lower bound for the norm in $M(H_p^m \to H_p^l)$ follows from the inequality

$$\|\gamma\eta_z\|_{H_p^l} \le c\,\|\gamma\|_{M(H_p^m \to H_p^l)}\|\eta_z\|_{H_p^m}\,.$$

Let us obtain the upper bound. Since H_p^m is imbedded into L_∞, we have $C_{p,m}(e) \sim 1$ for any compact set e with diameter $d(e)$ not exceeding 1. Therefore, it follows from Theorem 3.2.4 that

$$\|\gamma\|_{M(H_p^m \to H_p^l)} \sim \sup_{z \in \mathbb{R}^n} \|S_l\gamma; \mathcal{B}_{1/2}(z)\|_{L_p} + \|\gamma\|_{L_{1,\mathrm{unif}}}.$$

We have

$$\|S_l\gamma; \mathcal{B}_{1/2}(z)\|_{L_p} \le \|S_l(\gamma\eta_z); \mathcal{B}_{1/2}(z)\|_{L_p} + \|S_l[\gamma(1-\eta_z)]; \mathcal{B}_{1/2}(z)\|_{L_p}\,.$$

The first norm on the right-hand side does not exceed $\|\gamma\eta_z\|_{H_p^l}$. By Theorem 3.1.1 the second one is not greater than

$$c \sup_{x \in \mathcal{B}_{1/2}(z)} \Big(\int_{1/2}^{\infty} \Big[y^{-n}\int_{|s|<y} |\nabla_{[l]}[\gamma(x+s)(1-\eta_z(x+s))]|\,ds\Big]^2 y^{-1-2\{l\}}\,dy\Big)^{1/2}$$

$$\le c_1 \sup_{x \in \mathbb{R}^n} \|\gamma; \mathcal{B}_1(x)\|_{W_p^{[l]}} \le c_2 \sup_{x \in \mathbb{R}^n} \|\gamma\eta_x\|_{H_p^l}\,.$$

Thus the upper estimate for $\|\gamma\|_{M(H_p^m \to H_p^l)}$ follows. □

Remark 3.2.2. The coincidence of MH_p^m and $H^m_{p,\mathrm{unif}}$ for $mp > n$ is a result by Strichartz [Str].

3.3 One-Sided Estimates for the Norm in $M(H_p^m \to H_p^l)$

We present here some lower and, separately, upper bounds for the norm in $M(H_p^m \to H_p^l)$, $mp \le n$, which do not contain the capacity and which follow from the characterization of multipliers in $M(H_p^m \to H_p^l)$ obtained in the previous section.

3.3.1 Lower Estimate for the Norm in $M(H_p^m \to H_p^l)$ Involving Morrey Type Norms

The next assertion follows directly from Proposition 3.1.4 and Theorem 3.2.2.

Proposition 3.3.1. *Let* $0 < l < m$. *If* $mp < n$, *then*

$$\|\gamma\|_{M(H_p^m \to H_p^l)}$$
$$\geq c \sup_{\substack{x \in \mathbb{R}^n \\ r \in (0,1)}} \left(r^{m-n/p} \|S_l \gamma; \mathcal{B}_r(x)\|_{L_p} + r^{m-l-n/p} \|\gamma; \mathcal{B}_r(x)\|_{L_p} \right) \tag{3.3.1}$$

and, if $mp = n$, *then*

$$\|\gamma\|_{M(H_p^m \to H_p^l)}$$
$$\geq c \sup_{\substack{x \in \mathbb{R}^n \\ r \in (0,1)}} \left((\log 2r^{-1})^{1-1/p} \|S_l \gamma; \mathcal{B}_r(x)\|_{L_p} + r^{-l} \|\gamma; \mathcal{B}_r(x)\|_{L_p} \right). \tag{3.3.2}$$

For $m = l$ *the second term on the right-hand sides of* (3.3.1) *and* (3.3.2) *should be replaced by* $\|\gamma\|_{L_\infty}$.

Remark 3.3.1. Let $p \in (1, \infty)$ and let $\lambda \in (0, n)$. By the Morrey space $\mathcal{L}_{p,\lambda}$ one means the set of functions in $\mathbb{R}^n$ with the finite norm

$$\|f\|_{\mathcal{L}_{p,\lambda}} = \sup_{x \in \mathbb{R}^n, r > 0} r^{-\lambda/p} \|f; \mathcal{B}_r(x)\|_{L_p}. \tag{3.3.3}$$

Using this norm, we can rewrite estimate (3.3.1), where $mp < n$, as

$$\|\gamma\|_{M(H_p^m \to H_p^l)} \geq c\left(\|S_l \gamma\|_{\mathcal{L}_{p,n-mp}} + \|\gamma\|_{L_{1,\mathrm{unif}}} \right). \tag{3.3.4}$$

3.3.2 Upper Estimate for the Norm in $M(H_p^m \to H_p^l)$ Involving Marcinkiewicz Type Norms

The next assertion is a corollary of Theorem 3.2.4 and estimates (3.1.5), (3.1.6).

Proposition 3.3.2. *If* $mp < n$ *and* $m > l$, *then*

$$\|\gamma\|_{M(H_p^m \to H_p^l)}$$
$$\leq c\Big(\sup_{\{e: d(e) \leq 1\}} (\mathrm{mes}_n e)^{m/n - 1/p} \|S_l \gamma; e\|_{L_p} + \|\gamma\|_{L_{1,\mathrm{unif}}} \Big) \tag{3.3.5}$$

and, if $mp = n$, $m > l$, *then*

$$\|\gamma\|_{M(H_p^m \to H_p^l)}$$
$$\leq c\Big(\sup_{\{e: d(e) \leq 1\}} (\log(2^n / \mathrm{mes}_n e))^{(p-1)/p} \|S_l \gamma; e\|_{L_p} + \|\gamma\|_{L_{1,\mathrm{unif}}} \Big), \tag{3.3.6}$$

where $d(e)$ *is the diameter of* e.

In case $m = l$ *the norm in* $L_{1,\mathrm{unif}}$ *on the right-hand sides of* (3.3.5) *and* (3.3.6) *should be replaced by* $\|\gamma\|_{L_\infty}$.

In connection with (3.3.5) we make the following remark.

Remark 3.3.2. By the Marcinkiewicz space $\mathfrak{M}_\alpha$ one means the linear set of functions in a domain $\Omega \subset \mathbb{R}^n$ for which

$$\sup_{0<t<\infty} t\,[m(t)]^\alpha < \infty\,, \tag{3.3.7}$$

where $\alpha \in (0,1)$ and

$$m(t) = \text{mes}_n\{x \in \Omega : |f(x)| \geq t\}.$$

We denote the left-hand side of (3.3.7) by $\|f;\Omega\|_{\mathfrak{M}_\alpha^*}$ and we set

$$\|f;\Omega\|_{\mathfrak{M}_\alpha} = \sup_{e\subset\Omega} \frac{\int_e |f|\,dx}{(\text{mes}_n e)^{1-\alpha}}\,.$$

It is known that

$$(1-\alpha)\|f;\Omega\|_{\mathfrak{M}_\alpha} \leq \|f;\Omega\|_{\mathfrak{M}_\alpha^*} \leq \|f;\Omega\|_{\mathfrak{M}_\alpha} \tag{3.3.8}$$

(cf. [KZPS], 8.3, Ch. 2). In fact, for all e,

$$\int_e |f|\,dx \leq \|f;\Omega\|_{\mathfrak{M}_\alpha}(\text{mes}_n e)^{1-\alpha}$$

and, in particular,

$$\int_{|f|\geq t} |f|\,dx \leq \|f;\Omega\|_{\mathfrak{M}_\alpha}[m(t)]^{1-\alpha}\,.$$

On the other hand,

$$\int_{|f|\geq t} |f|\,dx \geq t\,m(t)\,.$$

Consequently,

$$t\,[m(t)]^\alpha \leq \|f;\Omega\|_{\mathfrak{M}_\alpha}$$

and the right inequality (3.3.8) is proved.

Now let $(0,\text{mes}_n e) \ni \mu \to F(\mu)$ be a non-increasing rearrangement of $|f|$, i.e. a non-increasing function which satisfies

$$m(t) = \text{mes}_1(\{\mu : F(\mu) \geq t\})\,.$$

It is clear that

$$\int_e |f|\,dx \leq \int_0^{\text{mes}_n e} F(\mu)\,d\mu\,.$$

Moreover,

$$F(\mu) \leq \mu^{-\alpha}\|f;\Omega\|_{\mathfrak{M}_\alpha^*}\,.$$

Hence

$$\int_e |f|\,dx \leq \|f;\Omega\|_{\mathfrak{M}_\alpha^*}\int_0^{\text{mes}_n e} \mu^{-\alpha}\,d\mu = \|f;\Omega\|_{\mathfrak{M}_\alpha^*}\frac{(\text{mes}_n e)^{1-\alpha}}{1-\alpha}$$

and the left estimate (3.3.8) is proved.

Now it is clear that (3.3.5) can be written in terms of Marcinkiewicz spaces. For $l < m < n/p$ this inequality is equivalent to

$$\|\gamma\|_{M(H_p^m \to H_p^l)} \le c\big(\sup_{x\in\mathbb{R}^n} \|S_l\gamma; \mathcal{B}_1(x)\|_{\mathfrak{M}_{m/n}} + \|\gamma\|_{L_{1,\mathrm{unif}}}\big). \tag{3.3.9}$$

For $l = m < n/p$ a similar result takes the form

$$\|\gamma\|_{MH_p^l} \le c\big(\sup_{x\in\mathbb{R}^n} \|S_l\gamma; \mathcal{B}_1(x)\|_{\mathfrak{M}_{l/n}} + \|\gamma\|_{L_\infty}\big). \tag{3.3.10}$$

We complete this subsection by stating the two-sided estimate of the norm in $M(H_p^m \to H_p^l)$ unifying inequalities (3.3.4) and (3.3.9):

$$\begin{aligned} c_1\big(\|S_l\gamma\|_{\mathcal{L}_{p,n-mp}} + \|\gamma\|_{L_{1,\mathrm{unif}}}\big) &\le \|\gamma\|_{M(H_p^m \to H_p^l)} \\ &\le c_2\big(\sup_{x\in\mathbb{R}^n} \|S_l\gamma; \mathcal{B}_1(x)\|_{\mathfrak{M}_{m/n}} + \|\gamma\|_{L_{1,\mathrm{unif}}}\big). \end{aligned} \tag{3.3.11}$$

Similarly,

$$\begin{aligned} c_1\big(\|S_l\gamma\|_{\mathcal{L}_{p,n-lp}} + \|\gamma\|_{L_\infty}\big) &\le \|\gamma\|_{MH_p^l} \\ &\le c_2\big(\sup_{x\in\mathbb{R}^n} \|S_l\gamma; \mathcal{B}_1(x)\|_{\mathfrak{M}_{l/n}} + \|\gamma\|_{L_\infty}\big). \end{aligned} \tag{3.3.12}$$

These estimates mean that the space of multipliers $M(H_p^m \to H_p^l)$ is situated between Morrey-Sobolev type and Marcinkiewicz-Sobolev spaces with norms of the same space dimension $m - l$.

Remark 3.3.3. An estimate similar to (3.3.9), in which $p = 2$, $0 < l < 1$ and the function $T\gamma$, defined by (3.2.32), plays the role of $S_l\gamma$, was given by Hirschman [Hi1]. In the case $l \in (0, 1)$, $lp < n$, inequality (3.3.9) was proved by Strichartz [Str] with the use of interpolation methods.

3.3.3 Upper Estimates for the Norm in $M(H_p^m \to H_p^l)$ Involving Norms in $H_{n/m}^l$

Theorem 3.3.1. *(i) If $lp < n$ and $\gamma \in H_{n/l,\mathrm{unif}}^l \cap L_\infty$, then $\gamma \in MH_p^l$ and*

$$\|\gamma\|_{MH_p^l} \le c\left(\|\gamma\|_{H_{n/l,\mathrm{unif}}^l} + \|\gamma\|_{L_\infty}\right). \tag{3.3.13}$$

(ii) If $mp < n$, $l < m$ and $\gamma \in H_{n/m,\mathrm{unif}}^l$, then $\gamma \in M(H_p^m \to H_p^l)$ and

$$\|\gamma\|_{M(H_p^m \to H_p^l)} \le c\,\|\gamma\|_{H_{n/m,\mathrm{unif}}^l}.$$

Proof. Let $\eta \in C_0^\infty(\mathcal{B}_1)$, $\eta = 1$ on $\mathcal{B}_{1/2}$ and $\eta_z(x) = \eta(x - z)$. By Theorem 3.1.2,

$$\|\gamma\|_{M(H_p^m \to H_p^l)} \le c \sup_{z\in\mathbb{R}^n} \|\eta_z\gamma\|_{M(H_p^m \to H_p^l)}.$$

This and (4.3.96) imply that the norm of γ in $M(H_p^m \to H_p^l)$ has the majorant

$$c \sup_{z \in \mathbb{R}^n} \left(\|S_l(\eta_z \gamma)\|_{L_{n/m}} + \|\eta_z \gamma\|_{L_{n/(m-l)}} \right) .$$

Now, using Theorem 3.1.1, we derive the estimate

$$\|\gamma\|_{M(H_p^m \to H_p^l)} \le c\,(\|\gamma\|_{H^l_{n/m,\mathrm{unif}}} + \|\gamma\|_{L_{n/(m-l),\mathrm{unif}}})$$

which coincides with (3.3.13) for $m = l$. Since by the Sobolev theorem $H^l_{n/m} \subset L_{n/(m-l)}$ for $m > l$, we have

$$\|\gamma\|_{L_{n/(m-l),\mathrm{unif}}} \le c\,\|\gamma\|_{H^l_{n/m,\mathrm{unif}}} .$$

The proof is complete. □

Remark 3.3.4. Estimate (3.3.13) was first obtained by Polking [Pol1] with a different proof.

3.4 Upper Estimates for the Norm in $M(H_p^m \to H_p^l)$ by Norms in Besov Spaces

Let $\{\mu\} > 0$. We define the space $B^\mu_{q,\infty}$ of functions in $\mathbb{R}^n$ with the norm

$$\|v\|_{B^\mu_{q,\infty}} = \sup_{h \in \mathbb{R}^n} |h|^{-\{\mu\}} \|\Delta_h \nabla_{[\mu]} v\|_{L_q} + \|v\|_{W_q^{[\mu]}} . \tag{3.4.1}$$

In the present section we find sufficient conditions for the inclusion $\gamma \in M(H_p^m \to H_p^l)$ formulated in terms of the space $B^\mu_{q,\infty}$ (see Theorem 3.4.1).

3.4.1 Auxiliary Assertions

Lemma 3.4.1. *Let $q \ge 1$, $\{\mu\} > 0$, $\mu q < n$. Further, let $v \in W_q^{[\mu]}(\mathcal{B}_1)$. Then*

$$\sup_{e \subset \mathcal{B}_1} \frac{\|v; e\|_{L_q}}{(\mathrm{mes}_n e)^{\mu/n}} \le c \left(\sup_{e \subset \mathcal{B}_1} \frac{\|\nabla_{[\mu]} v; e\|_{L_q}}{(\mathrm{mes}_n e)^{\{\mu\}/n}} + \|v; \mathcal{B}_1\|_{L_q} \right) . \tag{3.4.2}$$

Proof. By (2.2.8), for any integer $l < m$ we have

$$(\mathrm{mes}_n e)^{(m-l)/n-1/q} \|\gamma; e\|_{L_q}$$

$$\le c\Big(\|\gamma; \mathcal{B}_1\|_{L_q} + (\mathrm{mes}_n e)^{(m-l)/n-1/q} \Big(\int_e \Big(\int_{|z|<2} \frac{|\nabla_l \gamma(x+z)|}{|z|^{n-l}} dz \Big)^q dx \Big)^{1/q} \Big).$$

We estimate the integral over the ball $|z| < 2$ by the sum

$$\int_{\mathcal{B}^{(0)}} + \sum_{j=0}^{N} \int_{\mathcal{B}^{(j+1)} \backslash \mathcal{B}^{(j)}},$$

where

$$\mathcal{B}^{(j)} = \{z : |z| \le 2^j (\text{mes}_n e)^{1/n}\}$$

and $2^N (\text{mes}_n e)^{1/n} \le 1$. By Minkowski's inequality

$$\Big(\int_e \Big(\int_{\mathcal{B}^{(0)}} \dots dz\Big)^q dx\Big)^{1/q} \le \int_{\mathcal{B}^{(0)}} \frac{dz}{|z|^{n-l}} \Big(\int_e |\nabla_l \gamma(x+z)|^q dx\Big)^{1/q}.$$

Let $mq \le n$. We introduce the notation

$$s(\gamma) = \sup_{\{e:d(e)\le 1\}} \frac{\|\nabla_l \gamma; e\|_{L_q}}{(\text{mes}_n e)^{1/q - m/n}}.$$

Clearly, the right-hand side of the last inequality does not exceed

$$s(\gamma)(\text{mes}_n e)^{1/q-m/n} \int_{\mathcal{B}^{(0)}} \frac{dz}{|z|^{n-l}} \le c\, s(\gamma)(\text{mes}_n e)^{1/q-(m-l)/n}.$$

Next we estimate the integral over the spherical layer $\mathcal{B}^{(j+1)} \backslash \mathcal{B}^{(j)}$:

$$\int_{\mathcal{B}^{(j+1)} \backslash \mathcal{B}^{(j)}} \dots dz \le c(2^j (\text{mes}_n e)^{1/n})^{l-n/q} \Big(\int_{\mathcal{B}^{(j+1)} \backslash \mathcal{B}^{(j)}} |\nabla_l \gamma(x+z)|^q dz\Big)^{1/q}$$

$$\le c\, s(\gamma)(2^j (\text{mes}_n e)^{1/n}))^{l-m}.$$

Consequently

$$\Big(\int_e \Big(\sum_{j=0}^{N} \int_{\mathcal{B}^{(j+1)} \backslash \mathcal{B}^{(j)}} \dots dz\Big)^q dx\Big)^{1/q}$$

$$\le c\, s(\gamma) \sum_{j=0}^{N} (2^j (\text{mes}_n e)^{1/n})^{l-m} (\text{mes}_n e)^{1/q} \le c\, s(\gamma)(\text{mes}_n e)^{1/q-(m-l)/n}.$$

Thus

$$(\text{mes}_n e)^{(m-l)/n-1/q} \|\gamma; e\|_{L_q}$$

$$\le c\Big(\|\gamma; \mathcal{B}_1\|_{L_q} + \sup_{e \subset \overline{\mathcal{B}}} (\text{mes}_n e)^{m/n-1/q} \|\nabla_l \gamma; e\|_{L_q}\Big).$$

To complete the proof it remains to put here $m = n/q - \{\mu\}$ and $l = [\mu]$. □

Lemma 3.4.2. *Let $q \geq 1$, $\mu > 0$, $\{\mu\} > 0$, $v \in W_q^{[\mu]}(\mathcal{B}_2)$. Then*

$$\sup_{e \subset \mathcal{B}_1} (\text{mes}_n e)^{-\{\mu\}/n} \|\nabla_{[\mu]} v; e\|_{L_q}$$
$$\leq c \left[\sup_{h \in \mathcal{B}_1} |h|^{-\{\mu\}} \|\Delta_h \nabla_{[\mu]} v; \mathcal{B}_1\|_{L_q} + \sup_{x \in \mathcal{B}_1, 0<r<1} r^{-\mu} \|v; \mathcal{B}_r(x)\|_{L_q} \right],$$

where $(\Delta_h w)(x) = w(x+h) - w(x)$.

Proof. Let

$$Q \in C_0^\infty(\mathcal{B}_1) \qquad \text{and} \quad \int Q\, dx = 1.$$

Further, let $\rho \in (0,1)$, and let e be an arbitrary compact set in $\mathcal{B}_1$. We have

$$|D^\alpha v(x)| \leq \left| \rho^{-n} \int Q(h/\rho)(D^\alpha v(x+h) - D^\alpha v(x))\, dh \right|$$
$$+ \rho^{-n-[\mu]} \left| \int (D^\alpha Q)(h/\rho) v(x+h)\, dh \right|$$

for any multi-index α of order $[\mu]$. Consequently,

$$\int_e |D^\alpha v|^q dx \leq c \Big[\rho^{-qn} \int_e \Big(|Q(h/\rho)|^{1/q'} |Q(h/\rho)|^{1/q} |D^\alpha v(x+h) - D^\alpha v(x)| dh \Big)^q dx$$
$$+ (\text{mes}_n e) \rho^{-q[\mu]-n} \sup_{x \in \mathcal{B}_1} \int_{\mathcal{B}_\rho(x)} |v(h)|^q\, dh \Big].$$

Applying Hölder's inequality to the first term on the right-hand side, we obtain

$$\int_e |D^\alpha v|^q\, dx \leq c\, \rho^{q\{\mu\}} \Big[\sup_{h \in \mathcal{B}_\rho} |h|^{-q\{\mu\}} \int_e |D^\alpha v(x+h) - D^\alpha v(x)|^q dx$$
$$+ (\text{mes}_n e) \rho^{-n} \sup_{x \in \mathcal{B}_1, r \in (0,1)} r^{-\mu q} \|v; \mathcal{B}_r(x)\|_{L_q}^q \Big].$$

It remains to put $\rho = 2^{-1} (\text{mes}_n e)^{1/n}$ and to multiply both sides of the last inequality by $\rho^{-q\{\mu\}}$. □

The following lemma contains a well known integral representation of a function formulated in terms of its higher-order differences (see [BIN], Ch. 2, Sect.7.8).

Lemma 3.4.3. *Let v be a function in L_q, $1 \leq q < \infty$. Then, for almost all $z \in \mathbb{R}^n$,*

$$v(z) = \sum_{i=1}^n \int_0^\infty \frac{d\sigma}{\sigma^{n+2}} \int_{\mathbb{R}^1} M\Big(\frac{t}{\sigma}\Big) dt \int_{\mathbb{R}^n} \Omega_i\Big(\frac{y-z}{\sigma}\Big) \Delta_{te_i}^s v(z+y+te_i) dy \quad (3.4.3)$$

where $M \in C_0^\infty(0,1)$, $\Omega_i \in C_0^\infty((0,1)^n)$, s is an arbitrary integer and $\Delta_{te_i}^s$ is the difference of order s in the direction of the unit vector e_i.

Proof. Let $K \in C_0^\infty(0,(s+1)^{-1})$ and let

$$\int_{\mathbb{R}^1} K(t)\,dt = 1.$$

We put

$$L(t) = (s+1)\sum_{j=0}^{s} \frac{(-1)^j}{(1+j)^2}\binom{s}{j} K\left(\frac{t}{1+j}\right). \tag{3.4.4}$$

Since

$$\int_{\mathbb{R}^1} \sum_{j=0}^{s} \frac{(-1)^s}{(1+j)^2}\binom{s}{j} K\left(\frac{t}{1+j}\right)\,dt = \int_0^1 (1-t)^s\,dt = (s+1)^{-1},$$

it follows that

$$\int_{\mathbb{R}^1} L(t)\,dt = 1.$$

We introduce the function Φ:

$$\Phi(\tau) = \int_{\mathbb{R}^1} L(\tau-\eta)L(\eta)\,d\eta$$

and set

$$v_\sigma(z) = \int_{\mathbb{R}^n} v(z+y)\prod_{j=1}^{n} \sigma^{-1}\Phi\Big(\frac{y_j}{\sigma}\Big)\,dy\,, \quad z\in\mathbb{R}^n, \sigma>0\,.$$

The inclusion $v\in L_q$, $1\le q<\infty$, implies that $v_\sigma(z)\to 0$ as $\sigma\to\infty$. Hence

$$v(z) = -\int_0^\infty \frac{\partial}{\partial\sigma} v_\sigma(z)\,d\sigma$$

for almost all $z\in\mathbb{R}^n$. We have

$$\begin{aligned}\frac{\partial}{\partial\sigma}v_\sigma(z) &= \int_{\mathbb{R}^n} v(z+y)\frac{\partial}{\partial\sigma}\Big(\prod_{j=1}^{n}\sigma^{-1}\Phi\Big(\frac{y_j}{\sigma}\Big)\Big)\,dy\\ &= \sum_{i=1}^{n}\int_{\mathbb{R}^n} v(z+y)\Big(\prod_{j\ne i}\sigma^{-1}\Phi\Big(\frac{y_j}{\sigma}\Big)\Big)\frac{\partial}{\partial\sigma}\Big(\sigma^{-1}\Phi\Big(\frac{y_i}{\sigma}\Big)\Big)\,dy.\end{aligned} \tag{3.4.5}$$

From the definition of Φ we obtain

$$\begin{aligned}\frac{\partial}{\partial\sigma}\Big(\sigma^{-1}\Phi\Big(\frac{y_i}{\sigma}\Big)\Big) &= \frac{\partial}{\partial\sigma}\int_{\mathbb{R}^1}\Big(\frac{1}{\sigma}L\Big(\frac{y_i-\tau}{\sigma}\Big)\Big)\Big(\frac{1}{\sigma}L\Big(\frac{\tau}{\sigma}\Big)\Big)\,d\tau\\ &= 2\int_{\mathbb{R}^1}\frac{1}{\sigma}L\Big(\frac{y_i-\tau}{\sigma}\Big)\frac{\partial}{\partial\sigma}\Big(\frac{1}{\sigma}L\Big(\frac{\tau}{\sigma}\Big)\Big)\,d\tau\\ &= -2\sigma^{-2}\int_{\mathbb{R}^1} L\Big(\frac{y_i}{\sigma}-t\Big)(tL(t))'\,dt\\ &= 2\sigma^{-3}\int_{\mathbb{R}^1}\frac{\tau}{\sigma}L\Big(\frac{\tau}{\sigma}\Big)\,L'\Big(\frac{y_i-\tau}{\sigma}\Big)\,d\tau\,.\end{aligned}$$

Substituting the above expression for $\partial(\sigma^{-1}\Phi(y_i/\sigma))/\partial\sigma$ into (3.4.5), we obtain

$$\frac{\partial}{\partial\sigma}v_\sigma(z) = -2\sum_{i=1}^{n}\sigma^{-n-2}\int_{\mathbb{R}^1}\frac{t}{\sigma}L\Big(\frac{t}{\sigma}\Big)\,dt \times\int_{\mathbb{R}^n} v(z+y)\Big(\prod_{j\neq i}\Phi\Big(\frac{y_i}{\sigma}\Big)\Big)L'\Big(\frac{y_i-\tau}{\sigma}\Big)\,dy$$
$$= \sigma^{-n-2}\sum_{i=1}^{n}\int_{\mathbb{R}^1}\frac{t}{\sigma}L\Big(\frac{t}{\sigma}\Big)\,dt\int_{\mathbb{R}^n} v(z+y+te_i)\,\Omega_i\Big(\frac{y}{\sigma}\Big)\,dy,$$

where

$$\Omega_i(y) = 2\prod_{j\neq i}\Phi(y_j)L'(y_i), \quad \text{supp}\ \Omega_i\subset(0,1)^n.$$

Thus

$$v(z) = \sum_{i=1}^{n}\int_0^\infty \sigma^{-n-2}d\sigma\int_{\mathbb{R}^n}\Omega_i\Big(\frac{y}{\sigma}\Big)dy\int_{\mathbb{R}^1} v(z+y+te_i)\frac{t}{\sigma}L\Big(\frac{t}{\sigma}\Big)dt. \quad (3.4.6)$$

Using definition (3.4.4) of the function L, we write the last integral over $\mathbb{R}^1$ as

$$(s+1)\int_{\mathbb{R}^1} v(z+y+te_i)\sum_{j=0}^{s}\frac{(-1)^j}{1+j}\binom{s}{j}\frac{t}{\sigma(1+j)}K\Big(\frac{t}{\sigma(1+j)}\Big)\frac{dt}{1+j}$$
$$= (-1)^s(s+1)\int_{\mathbb{R}^1}\sum_{j=0}^{s}(-1)^{s-j}\binom{s}{j}v(z+y+(1+j)\tau e_i)\frac{\tau}{\sigma}K\Big(\frac{\tau}{\sigma}\Big)\,d\tau$$
$$= \int_{\mathbb{R}^1}\Delta^s_{te_i}v(z+y+te_i)M\Big(\frac{t}{\sigma}\Big)\,dt,$$

where $M(t) = (-1)^s(s+1)tK(t)$. Hence, using (3.4.6), we obtain (3.4.3). □

3.4.2 Properties of the Space $B^\mu_{q,\infty}$

Proposition 3.4.1. (cf. [BIN], Ch. 4, Sect. 18.15). *The norm (3.4.1) is equivalent to the norm*

$$\|v\|^{(1)}_{B^\mu_{q,\infty}} = \sum_{i=1}^{n}\sup_{t\in\mathbb{R}^1}|t|^{-\mu}\|\Delta^s_{te_i}v\|_{L_q} + \|v\|_{L_q},$$

where $s>\mu$, $1\le q<\infty$ and $\Delta^s_{te_i}$ is the difference of order s in the direction of the unit vector e_i.

Proof. Let $v \in B^{\mu}_{q,\infty}$. Putting $k = [\mu]$ and $m = s - [\mu]$ in the formula

$$\Delta^{k+m}_{te_i} v(x) = t^k \underbrace{\int_0^1 \cdots \int_0^1}_{k} \Delta^m_{te_i} \frac{\partial^k v}{\partial x_i^k}(x + t(\xi_1 + \cdots + \xi_k)e_i)\, d\xi_1 \ldots, d\xi_k$$

and applying Minkowski's inequality, we get

$$|t|^{-\mu} \|\Delta^s_{te_i} v\|_{L_q} \leq |t|^{-\{\mu\}} \Big\| \Delta^{s-[\mu]}_{te_i} \frac{\partial^{[\mu]}}{\partial x_i^{[\mu]}} v \Big\|_{L_q}$$

$$\leq 2^{s-[\mu]-1} |t|^{-\{\mu\}} \Big\| \Delta_{te_i} \frac{\partial^{[\mu]}}{\partial x_i^{[\mu]}} v \Big\|_{L_q}.$$

This implies that $\|v\|^{(1)}_{B^{\mu}_{q,\infty}} \leq c\, \|v\|_{B^{\mu}_{q,\infty}}$.

Now we derive the converse inequality. We show that the finiteness of the norm $\|v\|^{(1)}_{B^{\mu}_{q,\infty}}$ implies the inclusion $v \in W^{[\mu]}_q$. Let v_σ be the mean value of v defined in the proof of Lemma 3.4.3. We have

$$v_\sigma(x) - v_\varepsilon(x) = \int_\varepsilon^\sigma \frac{\partial}{\partial \tau} v_\tau(x)\, d\tau\,, \quad 0 < \varepsilon < \sigma,\ x \in \mathbb{R}^n\,.$$

Using the expression for $\partial v_\tau(x)/\partial\tau$ borrowed from the just mentioned proof, we arrive at the identity

$$v_\varepsilon(x) = v_\sigma(x)$$
$$+ \int_\varepsilon^\sigma \frac{d\tau}{\tau^{n+2}} \int_{\mathbb{R}^1} M\Big(\frac{t}{\tau}\Big) dt \sum_{i=1}^n \int_{\mathbb{R}^n} \Omega_i\Big(\frac{y}{\tau}\Big) \Delta^s_{te_i} v(x+y+te_i) dy. \qquad (3.4.7)$$

Differentiating (3.4.7) with $k \leq [\mu]$, we find that

$$\nabla_k v_\varepsilon = \nabla_k v_\sigma(x)(t-1)^k \int_\varepsilon^\sigma \frac{d\tau}{\tau^{n+2+k}} \int_{\mathbb{R}^1} M\Big(\frac{t}{\tau}\Big)\, dt$$
$$\times \sum_{i=1}^n \int_{\mathbb{R}^n} (\nabla_k \Omega_i)\Big(\frac{y}{\tau}\Big) \Delta^s_{te_i} v(x+y+te_i)\, dy \qquad (3.4.8)$$

which together with Minkowski's inequality yields

$$\|\nabla_k v_\varepsilon - \nabla_k v_\sigma\|_{L_q} \leq c \sum_{i=1}^n \int_\varepsilon^\sigma \tau^{-2-k}\, d\tau \int_{\mathbb{R}^1} M\Big(\frac{t}{\tau}\Big) \|\Delta^s_{te_i} v\|_{L_q}\, dt$$
$$\leq \sum_{i=1}^n \Big(\sup_{t \in \mathbb{R}^1} |t|^{-\mu} \|\Delta^s_{te_i} v\|_{L_q}\Big) \int_0^\sigma \tau^{-2-k}\, d\tau \int_{\mathbb{R}^1} \Big| M\Big(\frac{t}{\tau}\Big)\Big| |t|^\mu\, dt$$
$$\leq c\, \sigma^{\mu-k} \|v\|^{(1)}_{B^{\mu}_{q,\infty}}\,.$$

The latter and the L_q-convergence $v_\sigma \to v$ as $\sigma \to +0$ imply the existence of any distributional derivative $D^\alpha v \in L_q$, $|\alpha| \leq [\mu]$. Also,

$$\|\nabla_k v_\varepsilon\|_{L_q} \le \|\nabla_k v_\sigma\|_{L_q} + c\,\sigma^{\mu-k}\|v\|^{(1)}_{B^\mu_{q,\infty}}\,.$$

Passing here to the limit as $\varepsilon \to +0$, we get

$$\|\nabla_k v\|_{L_q} \le \|\nabla_k v_\sigma\|_{L_q} + c\,\sigma^{\mu-k}\|v\|^{(1)}_{B^\mu_{q,\infty}} \le c\,\sigma^{-k}\|v\|_{L_q} + c\,\sigma^{\mu-k}\|v\|^{(1)}_{B^\mu_{q,\infty}}\,.$$

Putting $\sigma = 1$, we arrive at the estimate

$$\|v\|_{W_q^{[\mu]}} \le c\,\|v\|^{(1)}_{B^\mu_{q,\infty}}.$$

Since $\nabla_k v \in L_q$, it follows that $\nabla_k v_\sigma(x) \to 0$ as $\sigma \to \infty$. We also have $\nabla_k v_\varepsilon(x) \to \nabla_k v(x)$ as $\varepsilon \to 0$ for almost all $x \in \mathbb{R}^n$. Therefore, by (3.4.7) we find that

$$\nabla_k v(x) = (-1)^k \int_0^\infty \frac{d\tau}{\tau^{n+2+k}} \int_{\mathbb{R}^1} M\Big(\frac{t}{\tau}\Big)dt \sum_{i=1}^n \int_{\mathbb{R}^n} (\nabla_k \Omega_i)\Big(\frac{y}{\tau}\Big) \Delta^s_{te_i}(x+y+te_i)dy$$

for almost all $x \in \mathbb{R}^n$. Hence, for any $\lambda > 0$ and $k = [\mu]$, we have

$$\Delta_{\lambda e_j}\nabla_k v(x) \tag{3.4.9}$$

$$= (-1)^k \sum_{i=1}^n \int_{\mathbb{R}^n} \Delta_{\lambda e_j}\Delta_{te_i} v(x+y+te_i)dy \int_{\mathbb{R}^1} dt \int_0^\lambda M\Big(\frac{t}{\tau}\Big)(\nabla_k \Omega_i)\Big(\frac{y}{\tau}\Big)\frac{d\tau}{\tau^{n+2+k}}$$

$$+(-1)^{k+1}\lambda \sum_{i=1}^n \int_0^1 d\xi \int_{\mathbb{R}^n} \Delta^s_{te_i} v(x+y+te_i+\xi e_j) \int_{\mathbb{R}^1} dt \int_\lambda^\infty M\Big(\frac{t}{\tau}\Big)\psi\Big(\frac{y}{\tau}\Big)\frac{d\tau}{\tau^{n+3+k}},$$

where

$$\psi(z) = \nabla_k \frac{\partial}{\partial z_j}\Omega_i(z).$$

To derive equality (3.4.9), we made the change of variables $z = x + y$ in the second summand and then applied the difference $\Delta_{\lambda e_j}$ to the kernel $(\nabla_k \Omega_i)\left(\frac{z-x}{\tau}\right)$, using the formula

$$(\nabla_k \Omega_i)\Big(\frac{z-x-\lambda e_j}{\tau}\Big) - (\nabla_k \Omega_i)\Big(\frac{z-x}{\tau}\Big) = -\lambda \int_0^1 \psi\Big(\frac{z-x-\lambda\xi e_j}{\tau}\Big)\,d\xi.$$

After that we made the reverse change of variables $y = z - x - \lambda\xi e_j$. Since $M \in C_0^\infty(0,1)$ and $\Omega_i \in C_0^\infty((0,1)^n)$, $i = 1,\dots,n$, it follows for $\alpha = \max\{|y_1|,\dots,|y_n|,t\}$ and $0 < t < \lambda$, that

$$\Big|\int_0^\lambda \tau^{-n-2-k} M\Big(\frac{t}{\tau}\Big)\nabla_k \Omega_i\Big(\frac{y}{\tau}\Big)\,d\tau\Big|$$

$$\le c\int_\alpha^\infty \tau^{-n-2-k}\,d\tau \le c\,(|y|+t)^{-n-1-k}. \tag{3.4.10}$$

If $t > \lambda$ then the integral on the left-hand side of (3.4.10) is equal to zero. In a similar way we get

$$\left| \int_{\lambda}^{\infty} \tau^{-n-3-k} M\Big(\frac{t}{\tau}\Big)\psi\Big(\frac{y}{\tau}\Big)\, d\tau \right| \le c\,(|y| + t + \lambda)^{-n-2-k} \tag{3.4.11}$$

provided that $t > \delta\lambda$ with some $\delta > 0$. In the case $t \le \delta\lambda$, the integral on the left-hand side of (3.4.10) is equal to zero. Using (3.4.8)–(3.4.11), we find that

$$\begin{aligned} |\Delta_{\lambda e_j} \nabla_k v(x)| \le c \sum_{i=1}^{n} \int_0^{\lambda} dt \int_{\mathbb{R}^n} \frac{|\Delta^s_{te_i} \Delta_{\lambda e_j} v(x + y + te_i)}{(|y| + t)^{n+1+k}}\, dy \\ + c \sum_{i=1}^{n} \lambda \int_{\delta\lambda}^{\infty} dt \int_{\mathbb{R}^n} \frac{\int_0^1 |\Delta^s_{te_i} v(x + y + te_i + \lambda\xi e_j)| d\xi}{(|y| + t)^{n+2+k}}\, dy\, . \end{aligned}$$

Applying Minkowski's inequality we obtain

$$\begin{aligned} \lambda^{-\{\mu\}} \|\Delta_{\lambda e_j} \nabla_k v\|_{L_q} \le c \sum_{i=1}^{n} \lambda^{-\{\mu\}} \int_0^{\lambda} \|\Delta_{te_i} v\|_{L_q} t^{-1-k}\, dt \\ + c \sum_{i=1}^{n} \lambda^{1-\{\mu\}} \int_{\delta\lambda}^{\infty} \|\Delta_{te_i} v\|_{L_q} t^{-2-k}\, dt \le c \sum_{i=1}^{n} \sup_{t>0} t^{-\mu} \|\Delta^s_{te_i} v\|_{L_q} . \end{aligned}$$

Thus

$$\sup_{\lambda>0} \lambda^{-\{\mu\}} \|\Delta_{\lambda e_j} \nabla_k v\|_{L_q} \le c\, \|v\|^{(1)}_{B^{\mu}_{q,\infty}}\,, \quad j = 1, \ldots, n\, . \tag{3.4.12}$$

Next we note that

$$\|\Delta_\eta v\|_{L_q} \le \sum_{j=1}^{n} \|\Delta_{|\eta_j| e_j} v\|_{L_q}\,, \quad \eta \in \mathbb{R}^n\, .$$

Therefore

$$|\eta|^{-\{\mu\}} \|\Delta_\eta \nabla_k v\|_{L_q} \le \sum_{j=1}^{n} |\eta_j|^{-\{\mu\}} \|\Delta_{|\eta_j| e_j} \nabla_k v\|_{L_q}\, .$$

By (3.4.12), the last sum does not exceed $c\|v\|^{(1)}$. The result follows. □

The proposition just proved and the definition of the norms $\|\cdot\|$ and $\|\cdot\|^{(1)}$ imply that the norm in $B^{\mu}_{q,\infty,\mathrm{unif}}$ is equivalent to either of the following two norms:

$$\sup_{x\in\mathbb{R}^n, h\in\mathcal{B}_1} |h|^{-\{\mu\}} \|\Delta_h \nabla_{[\mu]} v; \mathcal{B}_1(x)\|_{L_q} + \sup_{x\in\mathbb{R}^n} \|v; \mathcal{B}_1(x)\|_{L_q}\,,$$

$$\sum_{i=1}^{n} \sup_{x\in\mathbb{R}^n, |t|<1} |t|^{-\mu} \|\Delta^s_{te_i} v; \mathcal{B}_1(x)\|_{L_q} + \sup_{x\in\mathbb{R}^n} \|v; \mathcal{B}_1(x)\|_{L_q}\, .$$

Next we present an assertion which is due to V. P. Il'in (personal communication).

Lemma 3.4.4. *Let v be a function in L_q with compact support and let $q \geq 1$, $\mu > 0$, $\mu q < n$. Then*

$$\sup_{x\in\mathbb{R}^n, r>0} r^{-\mu}\|v; \mathcal{B}_r(x)\|_{L_q} \leq c\sum_{i=1}^{n} \sup_{t\in\mathbb{R}^1} |t|^{-\mu}\|\Delta_{te_i}^s v\|_{L_q}\,. \tag{3.4.13}$$

Proof. It follows from (3.4.3) that

$$|v(z)| \leq \sum_{i=1}^{n}\int_0^\infty \frac{d\sigma}{\sigma^{n+2}}\int_{\mathbb{R}^1}\Big|M\Big(\frac{t}{\sigma}\Big)\Big|\,dt\int|\Delta_{te_i}^s v(y+te_i)|\Big|\Omega\Big(\frac{y-z}{\sigma}\Big)\Big|\,dy,$$

where $M \in C_0^\infty(0,1)$ and $\Omega \in C_0^\infty((0,1)^n)$.

By U_i we denote the i-th term on the right-hand side. Let us represent U_i as the sum $V_i + W_i$ of two integrals over σ so that the integration in V_i is taken over $\sigma \in [0,r]$. By Minkowski's inequality,

$$\|V_i; \mathcal{B}_r(x)\|_{L_q} \leq \int_0^r \frac{d\sigma}{\sigma^{n+2}}\int_{\mathbb{R}^1}\Big|M\Big(\frac{t}{\sigma}\Big)\Big|\,dt$$
$$\times\Big\|\int|\Delta_{te_i}^s v(y+te_i)|\Big|\Omega\Big(\frac{y-z}{\sigma}\Big)\Big|\,dy; \mathcal{B}_r(x)\Big\|_{L_q}.$$

Applying Minkowski's inequality once more, we obtain

$$\Big\|\int|\Delta_{te_i}^s v(y+te_i)|\Big|\Omega\Big(\frac{y-z}{\sigma}\Big)\Big|\,dy; \mathcal{B}_r(x)\Big\|_{L_q} \leq c\,\sigma^n\|\Delta_{te_i}^s v\|_{L_q}.$$

Therefore,

$$\|V_i; \mathcal{B}_r(x)\|_{L_q} \leq c\int_0^r \frac{d\sigma}{\sigma^{n+2}}\sigma^n\int_0^\sigma \tau^\mu\,d\tau\sup_{t\in\mathbb{R}^1}|t|^{-\mu}\|\Delta_{te_i}^s v\|_{L_q}$$
$$\leq c\,r^\mu\sup_{t\in\mathbb{R}^1}|t|^{-\mu}\|\Delta_{te_i}^s v\|_{L_q}\,. \tag{3.4.14}$$

By Hölder's inequality,

$$\sup_{z\in\mathcal{B}_r(x)}|W_i(z)| \leq c_1\int_r^\infty \frac{d\sigma}{\sigma^{n+2}}\int_0^\sigma \sigma^{n/q'}\Big[\sup_{t\in\mathbb{R}^1}|t|^{-\mu}\|\Delta_{te_i}^s v\|_{L_q}\Big]\tau^\mu\,d\tau$$
$$= c\,r^{\mu-n/q}\sup_{t\in\mathbb{R}^1}|t|^{-\mu}\|\Delta_{te_i}^s v\|_{L_q}\,.$$

Consequently,

$$\|W_i; \mathcal{B}_r(x)\|_{L_q} \leq c\,r^{n/q}\sup_{z\in\mathcal{B}_r(x)}|W_i(z)| \leq c\,r^\mu\sup_{t\in\mathbb{R}^1}|t|^{-\mu}\|\Delta_{te_i}^s v\|_{L_q}\,.$$

This, together with (3.4.14), implies (3.4.13). □

The next assertion follows immediately from Lemmas 3.4.1 - 3.4.4.

Corollary 3.4.1. *Let* $\{\mu\} > 0$, $q \geq 1$, $\mu q \leq n$ *and let* $v \in W^{[\mu]}_{q,\mathrm{loc}}$. *Then*

$$\sup_{\{e:d(e)\leq 1\}} (\mathrm{mes}_n e)^{-\mu/n}\big((\mathrm{mes}_n e)^{[\mu]/n}\|\nabla_{[\mu]}v; e\|_{L_q} + \|v; e\|_{L_q}\big)$$

$$\leq \begin{cases} c\,\|v\|_{B^{\mu}_{q,\infty,\mathrm{unif}}} & \text{for } \mu q < n\,,\\ c\,(\|v\|_{B^{\mu}_{q,\infty,\mathrm{unif}}} + \|v\|_{L_\infty}) & \text{for } \mu q = n\,, \end{cases}$$

where $d(e)$ *is the diameter of a compact set* e.

3.4.3 Estimates for the Norm in $M(H^m_p \to H^l_p)$ by the Norm in $B^{\mu}_{q,\infty}$

The following assertion is the main result of the present section.

Theorem 3.4.1. *Let* $q \geq p$, $\mu = n/q - m + l$, $\mu > l$, $\{\mu\} > 0$.

(*i*) *If* $\gamma \in B^{\mu}_{q,\infty,\mathrm{unif}} \cap L_\infty$, *then* $\gamma \in MH^l_p$ *and*

$$\|\gamma\|_{MH^l_p} \leq c\Big(\sup_{x\in\mathbb{R}^n, h\in\mathcal{B}_1} |h|^{-\{\mu\}}\|\Delta_h\nabla_{[\mu]}\gamma; \mathcal{B}_1(x)\|_{L_q} + \|\gamma\|_{L_\infty}\Big). \qquad (3.4.15)$$

(*ii*) *If* $\gamma \in B^{\mu}_{q,\infty,\mathrm{unif}}$, *then* $\gamma \in M(H^m_p \to H^l_p)$ *and*

$$\|\gamma\|_{M(H^m_p\to H^l_p)}$$
$$\leq c\big(\sup_{x\in\mathbb{R}^n, h\in\mathcal{B}_1} |h|^{-\{\mu\}}\|\Delta_h\nabla_{[\mu]}\gamma; \mathcal{B}_1(x)\|_{L_q} + \sup_{x\in\mathbb{R}^n}\|\gamma; \mathcal{B}_1(x)\|_{L_q}\big). \qquad (3.4.16)$$

Proof. Let $m \geq l$. It is sufficient to assume that the difference $\varepsilon = n/q - m$ is small, since the general case follows by interpolation between the pairs $\{H^{m-l}_p, L_p\}$ and $\{H^{n/q-\varepsilon}_p, H^{\mu-\varepsilon}_p\}$ (cf. (3.2.1)). Thus we assume that $1 - \{l\} > n/q - m > 0$.

We introduce the function Φ_x on $(0,\infty)$ with the values

$$\Phi_x(y) = \int_{\mathcal{B}_y} |\nabla_{[l]}\gamma(x+z) - \nabla_{[l]}\gamma(x)|\,dz\,.$$

Clearly,

$$\|S_l\gamma; e\|^p_{L_p} = \int_e\Big(\int_0^\infty [\Phi_x(y)]^2 y^{-1-2(\{l\}+n)}\,dy\Big)^{p/2}dx,$$

where e is a compact set with $d(e) \leq 1$. Since Φ_x is an increasing function, the internal integral on the right-hand side is dominated by

$$(\{l\}+n)\int_0^\infty \frac{\Phi_x(y)\,dy}{y^{1+\{l\}+n}}\int_y^\infty \Phi_x(t)\frac{dt}{t^{1+\{l\}+n}} = \frac{\{l\}+n}{2}\Big(\int_0^\infty \Phi_x(y)\frac{dy}{y^{1+\{l\}+n}}\Big)^2.$$

We write the last integral as the sum of two integrals $i_1(x)+i_2(x)$, of which the first is over the semi-axis $y>|e|^{1/n}$, where $|e|=\text{mes}_n e$. We have

$$\int_e i_1(x)^p\,dx \le \int_e \Big\{\int_{|e|^{1/n}}^\infty \Big(y^{-n}\int_{\mathcal{B}_y} |\nabla_{[l]}\gamma(x+z)|\,dz + |\nabla_{[l]}\gamma(x)|\Big)\frac{dy}{y^{1+\{l\}}}\Big\}^p\,dx.$$

This and Minkowski's inequality imply that

$$\int_e i_1(x)^p dx \le \Big\{\int_{|e|^{1/n}}^\infty \Big(\int_e\Big[y^{-n}\int_{\mathcal{B}_y}|\nabla_{[l]}\gamma(x+z)|dz+|\nabla_{[l]}\gamma(x)|\Big]^p dx\Big)^{1/p}\frac{dy}{y^{1+\{l\}}}\Big\}^p.$$

Therefore

$$\int_e i_1(x)^p dx \le c\Big\{\int_{|e|^{1/n}}^\infty \Big(\int_e\Big[y^{-n}\int_{\mathcal{B}_y}|\nabla_{[l]}\gamma(x+z)|^p dz+|\nabla_{[l]}\gamma(x)|^p\Big]dx\Big)^{1/p}\frac{dy}{y^{1+\{l\}}}\Big\}^p$$

$$\le c\,|e|^{1-p/q}\Big\{\int_{|e|^{1/n}}^\infty \Big[y^{-n}\int_{\mathcal{B}_y}\Big(\int_e|\nabla_{[l]}\gamma(x+z)|^q\,dx\Big)^{1/q}dz$$

$$+\Big(\int_e|\nabla_{[l]}\gamma(x)|^q\,dx\Big)^{1/q}\Big]\frac{dy}{y^{1+\{l\}}}\Big\}^p.$$

By $N(\gamma)$ we denote the right-hand sides of (3.4.15) and (3.4.16). According to Corollary 3.4.1, where $\mu=n/q-m+l$,

$$\Big(\int_e|\nabla_{[l]}\gamma(x+z)|^q\,dx\Big)^{1/q} \le c\,|e|^{1/q-m/n}N(\gamma)\,.$$

Consequently,

$$\int_e i_1(x)^p\,dx \le c\,|e|^{1-mp/n}N(\gamma)^p\,. \tag{3.4.17}$$

Applying Minkowski's inequality, we obtain

$$\int_e i_2(x)^p\,dx = \int_e\Big(\int_0^{|e|^{1/n}}\int_{\mathcal{B}_y}|\nabla_{[l]}\gamma(x+z)-\nabla_{[l]}\gamma(x)|\,dz\frac{dy}{y^{1+\{l\}+n}}\Big)^p\,dx$$

$$\le\Big\{\int_0^{|e|^{1/n}}\int_{\mathcal{B}_y}\Big(\int_e|\nabla_{[l]}\gamma(x+z)-\nabla_{[l]}\gamma(x)|^p\,dx\Big)^{1/p}dz\frac{dy}{y^{1+\{l\}+n}}\Big\}^p.$$

By Hölder's inequality the last expression does not exceed

$$|e|^{1-p/q}\Big\{\int_0^{|e|^{1/n}}\int_{\mathcal{B}_y}\Big(\int_e|\nabla_{[l]}\gamma(x+z)-\nabla_{[l]}\gamma(x)|^q\,dx\Big)^{1/q}dz\frac{dy}{y^{1+\{l\}+n}}\Big\}^p.$$

Since

$$|z|^{q(m-\{l\})-n}\int_e|\nabla_{[l]}\gamma(x+z)-\nabla_{[l]}\gamma(x)|^q\,dx \le c\,N(\gamma)^q\,,$$

it follows that

$$\int_e i_2(x)^p\,dx \le c\,|e|^{1-p/q} N(\gamma)^p \Big(\int_0^{|e|^{1/n}} \int_{\mathcal{B}_y} |z|^{-m+\{l\}+n/q}\,dz \frac{dy}{y^{1+\{l\}+n}}\Big)^p$$
$$= c\,|e|^{1-pm/n} N(\gamma)^p\,. \tag{3.4.18}$$

Adding together (3.4.17) and (3.4.18), we arrive at

$$\|S_l\gamma; e\|^p_{L_p} \le c\,|e|^{1-pm/n} N(\gamma)^p.$$

By Hölder's inequality and Corollary 3.4.1 with $\mu = n/q - m + l$ we obtain

$$\|\gamma; e\|_{L_p} \le |e|^{1/p-1/q} \|\gamma; e\|_{L_q} \le c\,|e|^{1/p-1/q} |e|^{1/q-(m-l)/n} N(\gamma)\,. \tag{3.4.19}$$

Reference to Proposition 3.3.2 completes the proof. □

3.4.4 Estimate for the Norm of a Multiplier in $MH^l_p(\mathbb{R}^1)$ by the q-Variation

Hirschman [Hi2] obtained the following sufficient conditions for a function γ to belong to the class MW^l_2 on a unit circumference C: γ is bounded and has a finite q-variation $\mathrm{Var}_q(\gamma)$ for some q, $2 < q < 1/l$.

Here q-variation is defined by

$$\mathrm{Var}_q(\gamma) = \sup\Big(\sum_{j=0}^{m-1} |\gamma(t_{j+1}) - \gamma(t_j)|^q\Big)^{1/q}, \tag{3.4.20}$$

with the supremum taken over all partitions of the circumference C by points t_j.

Using Theorem 3.4.1, one may easily derive a sufficient condition for a function to belong to $MH^l_p(\mathbb{R}^1)$ which, for $p = 2$, coincides with Hirschman's condition up to the change of $\mathbb{R}^1$ for C.

We define the local q-variation of a function γ defined on $\mathbb{R}^1$ by (3.4.20), with the supremum taken over all choices of a finite number of points $t_0 < t_1 < \cdots < t_m$ on any interval σ of unit length.

Since

$$\int_\sigma |\gamma(t+h) - \gamma(t)|^q\,dt \le c\,|h|[\mathrm{Var}_q(\gamma)]^q,$$

we arrive at the following assertion.

Corollary 3.4.2. *Let $n = 1$, $q \ge p$, $lq < 1$. If $\gamma \in L_\infty$ and $\mathrm{Var}_q(\gamma) < \infty$, then $\gamma \in MH^l_p$ and*

$$\|\gamma\|_{MH^l_p} \le c\,(\|\gamma\|_{L_\infty} + \mathrm{Var}_q(\gamma))\,.$$

3.5 Miscellaneous Properties of Multipliers in $M(H_p^m \to H_p^l)$

The following assertion is a generalization of Proposition 2.7.1.

Proposition 3.5.1. *If $k \in [0, l]$, then $M(H_p^m \to H_p^l) \subset M(H_p^{m-l+k} \to H_p^k)$ and*

$$\|\gamma\|_{M(H_p^{m-l+k} \to H_p^k)} \le c \, \|\gamma\|_{M(H_p^m \to H_p^l)} \, .$$

Proof. The imbedding $M(H_p^m \to H_p^l) \subset M(H_p^{m-l} \to L_p)$ and the corresponding inequality for the norms were proved in Lemma 3.2.3. It remains to use the interpolation inequality

$$\|\gamma\|_{M(H_p^{m-l+k} \to H_p^k)} \le c \, \|\gamma\|^{k/l}_{M(H_p^m \to H_p^l)} \|\gamma\|^{(l-k)/l}_{M(H_p^{m-l} \to L_p)}$$

which is a particular case of estimate (3.2.1). □

Proposition 3.5.2. (*i*) *If $mp > n$, $1 < q < \infty$, $0 \le k < l$ and $k \le l + n(1/q - 1/p)$, then $M(H_p^m \to H_p^l) \subset M(H_q^{m-l+k} \to H_q^k)$ and*

$$\|\gamma\|_{M(H_q^{m-l+k} \to H_q^k)} \le c \, \|\gamma\|_{M(H_p^m \to H_p^l)} \, . \tag{3.5.1}$$

(*ii*) *If $mp = n$, $0 \le k \le l$, $q > 1$ and $k < l - m + n/q$, then $M(H_p^m \to H_p^l) \subset M(H_q^{m-l+k} \to H_q^k)$ and inequality* (3.5.1) *holds.*

Proof. (*i*) According to Theorem 3.2.5, $M(H_p^m \to H_p^l) = H^l_{p,\mathrm{unif}}$. Since $mp > n$, it follows that

$$M(H_p^m \to H_p^l) \subset H^l_{n/m,\mathrm{unif}} \quad \text{for } m > l$$

and

$$MH_p^m \subset H^m_{n/m,\mathrm{unif}} \cap L_\infty \quad \text{for } m = l.$$

This, together with Propostion 3.3.1, implies that

$$M(H_p^m \to H_p^l) \subset M(H_q^m \to H_q^l)$$

for any $q \in (1, n/m)$. Applying Proposition 3.5.1 and interpolating with respect to q between $1 + \varepsilon$ and p (cf. (3.2.1)), we obtain the inclusion $M(H_p^m \to H_p^l) \subset M(H_q^{m-l+k} \to H_q^k)$ for all $k \in [0, l]$, $q \in (1, p]$.

For $q \in (p, \infty)$ we put $s = l + n(1/q - 1/p)$. It is clear that $s < l$ and $q(s + m - l) > n$. By Theorem 3.2.5 and the Sobolev imbedding theorem,

$$M(H_p^m \to H_p^l) = H^l_{p,\mathrm{unif}} \subset H^s_{q,\mathrm{unif}}.$$

Since $q(s + m - l) > n$, we can apply Theorem 3.2.5 once more to obtain $M(H_p^m \to H_p^l) \subset M(H_q^{m-l+s} \to H_q^s)$. Further, by Proposition 3.5.1, $M(H_p^m \to H_p^l) \subset M(H_q^{m-l+k} \to H_q^k)$ for all $k \le s$, $q \in [p, \infty)$.

(*ii*) By Theorem 3.2.5, $M(H_p^m \to H_p^l) \subset H_{n/m,\text{unif}}^l$ for $m > l$ and $MH_p^m \subset H_{n/m,\text{unif}}^m \cap L_\infty$. According to the Sobolev imbedding theorem,

$$H_{n/m,\text{unif}}^l \subset H_{n/r,\text{unif}}^k, \qquad \text{where} \quad r = m - l + k, \;\; k < l.$$

This and Proposition 3.3.1 imply that

$$M(H_p^m \to H_p^l) \subset M(H_q^{m-l+k} \to H_q^k)$$

for any $q \in (1, n/r)$. □

Next we present an imbedding theorem for the space MH_p^m.

Proposition 3.5.3. *The space* MH_p^m *is imbedded into* MH_q^k, $k \leq m, 1 < q < \infty$, *provided that*

(*i*) $mp > n$, $k \leq m + n(1/q - 1/p)$,

(*ii*) $mp \leq n$, $k < mp/q$.

Proof. For $mp \geq n$ the assertion is proved in Propostion 3.5.2. Let $mp < n$. We start with the case $q \geq p$. If $\gamma \in MH_p^m$ then $\gamma \in L_\infty$ and hence $\gamma \in ML_r$ for any $r \in (1, \infty)$. The interpolation between H_p^m and L_r with $r \in [q, \infty)$ (cf. inequality (3.2.1)) implies the imbedding $MH_p^m \subset MH_q^k$ with $0 < k < mp/q$, $q \geq p$. Now suppose that $q < p$, $k \leq m$ and $\gamma \in MH_p^m$. Then by Theorem 3.2.2

$$\sup_e \frac{\|S_m\gamma; e\|_{L_p}^p}{C_{p,m}(e)} < \infty\,.$$

This and Lemma 2.3.6 yield

$$\sup_e \frac{\|S_m\gamma; e\|_{L_q}^q}{C_{p,mq/p}(e)} < \infty\,.$$

Applying the inequality

$$C_{p,mq/p}(e) \leq c\, C_{q,m}(e) \qquad \text{with} \quad q < p$$

(see Adams, Meyers [AM]), we find that

$$\sup_e \frac{\|S_m\gamma; e\|_{L_q}^q}{C_{q,m}(e)} < \infty\,.$$

Since $\gamma \in L_\infty$, we get $\gamma \in MH_q^m$ by Theorem 3.2.2. Again, interpolating between H_q^m and L_q, we obtain $\gamma \in MH_q^k$ with $k \leq m$, $q < p$. The result follows.

Remark 3.5.1. For $mp \geq n$ this proposition was obtained by Strichartz [Str]. In the case $mp < n$ the result proved in his paper is incomplete. Above we have presented a stronger statement communicated to us by Verbitsky who also showed that these embeddings cannot be improved. For $q \leq p$, $k \leq m$ the latter can be easily verified by an example of lacunary series. The proof of the exactness is much more difficult in the case $q > p$, $k \leq mp/q$; in particular, for $k = mp/q$ it can be performed by using Blaschke products.

Next we give a simple sufficient condition for a function to belong to the space $M(H_p^m(\mathbb{R}^n) \to H_p^l(\mathbb{R}^n))$ which is formulated in terms of $M(H_p^m(\mathbb{R}^1) \to H_p^l(\mathbb{R}^1))$.

Proposition 3.5.4. *If for all $j = 1, \ldots, n$*

$$\operatorname*{ess\ sup}_{(x_1,\ldots,x_{j-1},x_{j+1},\ldots,x_n)} \|\gamma(x_1, \ldots, x_{j-1}, \cdot, x_{j+1}, \ldots, x_n; \mathbb{R}^1\|_{M(H_p^m \to H_p^l)} < \infty,$$

then $\gamma \in M(H_p^m(\mathbb{R}^n) \to H_p^l(\mathbb{R}^n))$.

Proof. It can be easily verified that the functions

$$\frac{(1+|\xi|^2)^{l/2}}{\sum_{j=1}^{n}(1+\xi_j^2)^{l/2}}, \qquad \left(\frac{1+\xi_k^2}{1+|\xi|^2}\right)^{l/2}, \qquad k = 1, 2, \ldots, n\ ,$$

satisfy the condition of Lemma 1.2.1. Therefore

$$\|u\|_{H_p^l} \sim \sum_{j=1}^{n} \left\| \left(1 - \frac{\partial^2}{\partial x_j^2}\right)^{l/2} u \right\|_{L_p} . \tag{3.5.2}$$

The conclusion is obvious. □

From (3.5.2) we immediately obtain

Proposition 3.5.5. *If a function γ depends only on variables $x_1, \ldots, x_s$, $s < n$, then $\gamma \in M(H_p^m(\mathbb{R}^n) \to H_p^l(\mathbb{R}^n))$ if and only if*

$$\gamma \in M(H_p^m(\mathbb{R}^s) \to H_p^l(\mathbb{R}^s)) .$$

Moreover,

$$\|\gamma; \mathbb{R}^n\|_{M(H_p^m \to H_p^l)} \sim \|\gamma; \mathbb{R}^s\|_{M(H_p^m \to H_p^l)} .$$

Proof. The upper bound for the norm in $M(H_p^m(\mathbb{R}^n) \to H_p^l(\mathbb{R}^n))$ follows directly from Proposition 3.5.4. Let $\eta \in C_0^\infty(\mathbb{R}^{n-s})$, $\eta \neq 0$. By (3.5.2) for $v \in H_p^m(\mathbb{R}^s)$ we have

$$\|\gamma v; \mathbb{R}^s\|_{H_p^l} \leq c\,\|\gamma\eta v; \mathbb{R}^n\|_{H_p^l} \leq c\,\|\gamma; \mathbb{R}^n\|_{M(H_p^m \to H_p^l)}\|\eta v; \mathbb{R}^n\|_{H_p^m}$$
$$\leq c\,\|\gamma; \mathbb{R}^n\|_{M(H_p^m \to H_p^l)}\|v; \mathbb{R}^s\|_{H_p^m} \,,$$

which yields the lower bound for the norm in $M(H_p^m(\mathbb{R}^n) \to H_p^l(\mathbb{R}^n))$. □

Corollary 3.5.1. *The characteristic function χ of the half-space $\mathbb{R}^n_+$ belongs to $MH_p^l(\mathbb{R}^n)$ if and only if $lp < 1$.*

Proof. By Proposition 3.5.5 we may limit consideration to the case $n = 1$. According to Theorem 3.4.1, for $lp < 1$

$$\|\chi; \mathbb{R}^1\|_{MH_p^l} \leq c\Big(\sup_{x\in\mathbb{R}^1, |h|<1} |h|^{-1/p}\|\Delta_h\chi; \mathcal{B}_1(x)\|_{L_p} + 1\Big) .$$

Since the right-hand side is bounded, $\chi \in MH_p^l(\mathbb{R}^1)$.

Now let $lp = 1$. By η we denote a function in $C_0^\infty(\mathbb{R}^1)$ which is equal to one in a neighborhood of the point O. Since $(F\chi)(\xi) = i\xi^{-1} + \pi\delta(\xi)$, it follows that

$$[F(\eta\chi)](\xi) = i\xi^{-1} + O(|\xi|^2)$$

for large $|\xi|$. Therefore

$$\Lambda^l(\eta\chi) = F^{-1}(1+\xi^2)^{l/2}F(\eta\chi) = c\,|x|^{-l}\text{sgn } x + O(1)$$

for small $|x|$ and hence $\chi \notin H^l_{p,\text{loc}}(\mathbb{R}^1)$. Consequently, $\chi \notin MH_p^l(\mathbb{R}^1)$. □

Remark 3.5.2. Corollary 3.5.1 is well known. For $p = 2$ it was proved by Hirschman [Hi2]. The case $p \in (1, \infty)$ was considered by Shamir [Sha] and Strichartz [Str]. An analogous assertion for the space W_p^l is given by Lions and Magenes [LiM1]. The conditions for the function χ to belong to classes of multipliers in various functional spaces have been studied by Triebel [Tr1], [Tr2], and [Tr4], Frazier and Jawerth [FrJ], Franke [Fr], Gulisashvili [Gu1], [Gu2], Runst and Sickel [RS].

Remark 3.5.3. Corollary 3.5.1 for $n = 1$ and Proposition 3.5.4 immediately imply that the characteristic function of any convex open subset of $\mathbb{R}^n$ is a multiplier in $H_p^l(\mathbb{R}^n)$ for $lp < 1$. It is even sufficient to assume that the multiplicity of the intersection of the set with almost any straight line parallel to one coordinate axis is bounded.

We state some simple properties of the class MH_p^l.

Proposition 3.5.6. *The estimate*

$$\|\gamma\|_{L_\infty} \leq \|\gamma\|_{MH_p^l}$$

holds.

Proof. This estimate is deduced in precisely the same way as (2.7.1).

The following assertion complements Proposition 2.7.5, where the case $\{l\} = 0$ is considered.

Proposition 3.5.7. *Let $\{l\} > 0$, $\gamma \in MH_p^l$ and let σ be a segment of the real axis such that $\gamma(x) \in \sigma$ for almost all $x \in \mathbb{R}^n$. Further, let $f \in C^{[l],1}(\sigma)$. Then $f(\gamma) \in MH_p^l$ and the estimate*

$$\|f(\gamma)\|_{MH_p^l} \leq c \sum_{j=0}^{[l]+1} \|f^{(j)}; \sigma\|_{L_\infty} \|\gamma\|_{MH_p^l}^j$$

holds.

Proof. Let $l \in (0,1)$. By Theorem 3.1.1,

$$\|uf(\gamma)\|_{H_p^l} \leq c\left(\|S_l(uf(\gamma))\|_{L_p} + \|uf(\gamma)\|_{L_p}\right)$$

for all $u \in C_0^\infty$. Since

$$\begin{aligned} S_l(uf(\gamma))) &\leq |u| S_l f(\gamma) + \|f(\gamma)\|_{L_\infty} S_l u \\ &\leq |u| \|f'; \sigma\|_{L_\infty} S_l \gamma + \|f(\gamma)\|_{L_\infty} S_l u\,, \end{aligned}$$

we have

$$\|uf(\gamma)\|_{H_p^l} \leq c(\|f'\|_{L_\infty} \|S_l \gamma\|_{M(H_p^l \to L_p)} + \|f(\gamma)\|_{L_\infty}) \|u\|_{H_p^l}.$$

This, together with Lemma 3.2.8, implies the required estimate

$$\|f(\gamma)\|_{MH_p^l} \leq c(\|f'\|_{L_\infty} \|\gamma\|_{MH_p^l} + \|f(\gamma)\|_{L_\infty})\,.$$

It remains to proceed by induction on $[l]$ (cf. the proof of Proposition 2.7.5). □

This and Proposition 3.5.6 imply

Corollary 3.5.2. *If $\gamma \in MH_p^l$, $\{l\} > 0$, and $\|\gamma^{-1}\|_{L_\infty} < \infty$, then $\gamma^{-1} \in MH_p^l$ and*

$$\|\gamma^{-1}\|_{MH_p^l} \leq c \|\gamma^{-1}\|_{L_\infty}^{[l]+2} \|\gamma\|_{MH_p^l}^{[l]+1}\,.$$

3.6 Spectrum of Multipliers in H_p^l and $H_{p'}^{-l}$

3.6.1 Preliminary Information

We recall certain definitions and facts of the operator spectral theory (see for example, [DS], Ch. VII, Sect. 3.4 and 5.4).

Let X be a complex Banach space and let A be a bounded linear operator on X.

Definition 3.6.1. *The set of complex values λ for which the operator $(\lambda I - A)^{-1}$ exists, is defined on the whole of X and is bounded, is called the resolvent set $\rho(A)$ of the operator A. The complement of $\rho(A)$ is called the spectrum $\sigma(A)$ of A.*

It is known that the resolvent set $\rho(A)$ is open and that the function $(\lambda I - A)^{-1}$ is analytic on $\rho(A)$.

Definition 3.6.2. *The value $r(A) = \sup|\sigma(A)|$ is called the spectral radius of the operator A.*

The Gelfand formula holds:

$$r(A) = \lim_{m\to\infty} \sqrt[m]{\|A^m\|}. \tag{3.6.1}$$

Definition 3.6.3. *The operator A is called quasinilpotent if*

$$\lim_{m\to\infty} \sqrt[m]{\|A^m\|} = 0.$$

The next three definitions give a classification of points of the spectrum.

Definition 3.6.4. *The set of points $\lambda \in \sigma(A)$ such that the mapping $\lambda I - A$ is not one-to-one is called the pointwise spectrum and is denoted by $\sigma_\rho(A)$. In other words, $\lambda \in \sigma_p(A)$ if and only if there exists a nontrivial solution $u \in X$ of the equation $(\lambda I - A)u = 0$. Elements of σ_p are called eigenvalues.*

Definition 3.6.5. *The set of numbers $\lambda \in \sigma(A)$ for which the mapping $\lambda I - A$ is one-to-one and the range of $\lambda I - A$ is not dense in X is called the residual spectrum and is denoted by $\sigma_r(A)$.*

Definition 3.6.6. *The set of numbers $\lambda \in \sigma(A)$ for which the mapping $\lambda I - A$ is one-to-one and the range of $\lambda I - A$ is dense in X but does not coincide with X is called the continuous spectrum of A and is denoted by $\sigma_c(A)$.*

It is clear that the sets $\sigma_p(A)$, $\sigma_r(A)$ and $\sigma_c(A)$ are disjoint. By the Banach theorem on isomorphism, the condition $(\lambda I - A)X \neq X$ in Definition 3.6.5 is unnecessary and therefore

$$\sigma(A) = \sigma_p(A) \cup \sigma_r(A) \cup \sigma_c(A)\,. \tag{3.6.2}$$

Let A^* be the operator adjoint of A. Definitions 3.6.4–3.6.6 imply that

$$\sigma_r(A) \subset \overline{\sigma_p(A^*)} \subset \sigma_r(A) \cup \sigma_p(A)\,, \tag{3.6.3}$$

where the bar denotes complex conjugation.

3.6.2 Facts from Nonlinear Potential Theory

Definition 3.6.7. *If E is any subset of $\mathbb{R}^n$, then the numbers*

$$\underline{C_{p,m}}(E) = \sup\{C_{p,m}(e) : e \subset E, e \text{ is a compact set}\}$$

and

$$\overline{C_{p,m}}(E) = \inf\{\underline{C_{p,m}}(G) : G \supset E, G \text{ is an open set}\}$$

are called the inner and outer capacities of E.

We formulate some known properties of these capacities (see [AH], Ch. 2).
1. If the set $e \subset \mathbb{R}^n$ is compact, then $\overline{C_{p,m}}(e) = C_{p,m}(e)$.
2. If $E_1 \subset E_2 \subset \mathbb{R}^n$, then

$$\underline{C_{p,m}}(E_1) \leq \underline{C_{p,m}}(E_2)$$

and

$$\overline{C_{p,m}}(E_1) \leq \overline{C_{p,m}}(E_2).$$

3. Let $\{E_k\}_{k=1}^{\infty}$ be a sequence of sets in $\mathbb{R}^n$, $E = \bigcup_k E_k$. Then

$$\overline{C_{p,m}}(E) \leq \sum_{k=1}^{\infty} \overline{C_{p,m}}(E_k).$$

4. Any analytic (in particular, any Borel) subset E of the space $\mathbb{R}^n$ is measurable with respect to the capacity $C_{p,m}$ (i.e. $\overline{C_{p,m}}(E) = \underline{C_{p,m}}(E)$).

If the inner and the outer capacities of the set E are equal, then their value is called the capacity of E and is denoted $C_{p,m}(E)$.

By $V_{p,m}\mu$ we mean the nonlinear Bessel potential of a measure μ, i.e.

$$V_{p,m}\mu = J_m(J_m\mu)^{p'-1}.$$

The potential $V_{p,m}\mu$ satisfies the following 'rough' maximum principle (cf. [MH1], [MH2], and [AH], Sect. 2.6).

Proposition 3.6.1. *There exists a constant $\mathfrak{M}$ depending only on n, p, m, and such that*

$$(V_{p,m}\mu)(x) \leq \mathfrak{M} \sup\{(V_{p,m}\mu)(x) : x \in \text{supp } \mu\}\ . \tag{3.6.4}$$

The following assertion contains basic properties of the so-called capacitary measure (see [MH1], [MH2], and [AH], Sect. 2.5).

Proposition 3.6.2. *Let E be a subset of $\mathbb{R}^n$. If $\overline{C_{p,m}}(E) < \infty$, then there exists a unique measure μ_E with the properties:*
1. $\|J_m\mu_E\|_{L'_p}^{p'} = \overline{C_{p,m}}(E)$,
2. $(V_{p,m}\mu_E)(x) \geq 1$ *(p,m)-quasi everywhere in E.*

(Here '(p,m)-quasi everywhere' means 'everywhere except for a set of zero outer capacity $\overline{C_{p,m}}$'.)

3. $\operatorname{supp}\mu_E \subset \bar{E}$,
4. $\mu_E(\bar{E}) = \overline{C_{p,m}}(E)$,
5. $(V_{p,m}\mu_E)(x) \le 1$ *for all* $x \in \operatorname{supp}\mu_E$.

The measure μ_E is called the capacitary measure of the set E and $V_{p,m}\mu_E$ is called the capacitary potential of the set E.

In addition we notice that the capacity $C_{p,m}(e)$ may be defined by

$$C_{p,m}(e) = \sup\{\mu(e) : \operatorname{supp}\mu \subset e \text{ and } (V_{p,m}\mu)(x) \le 1 \text{ on } \operatorname{supp}\mu\}$$

(see [AH], Sect. 2.5).

3.6.3 Main Theorem

In the present and subsequent subsections the role of the space X is played by H_p^l and $H_{p'}^{-l}$; the multipliers are considered as an operator A.

Corollaries 2.7.1, 3.5.2 and the imbedding $MH_p^l \subset L_\infty$ immediately imply:

Corollary 3.6.1. *A number λ belongs to the spectrum of a multiplier $\gamma \in MH_p^l$ if and only if $(\gamma-\lambda)^{-1} \notin L_\infty$ or, which is equivalent, for any $\varepsilon > 0$ the set $\{x : |\gamma(x)-\lambda| < \varepsilon\}$ has positive n-dimensional measure.*

Since the adjoint operator of $\gamma \in MH_p^l$ is the multiplier $\bar{\gamma}$ in $H_{p'}^{-l}$, Corollary 3.6.1 implies:

Corollary 3.6.2. *A number λ belongs to the spectrum of $\gamma \in MH_{p'}^{-l}$ if and only if $(\gamma - A)^{-1} \notin L_\infty$.*

From Corollaries 3.6.1 and 3.6.2 we obtain that the spectral radius $r(\gamma)$ of a multiplier γ in H_p^l or in $H_{p'}^{-l}$ is equal to $\|\gamma\|_{L_\infty}$.

This and (3.6.1) imply that

$$\lim_{m\to\infty} \sqrt[m]{\|\gamma^m\|_{MH_p^l}} = \|\gamma\|_{L_\infty}.$$

Thus the only quasinilpotent multiplier is zero. In other words, the algebra MH_p^l is semisimple.

This is a generalization of results obtained in [DH] for $p=2$, $2l<1$.

The main theorem of the present section contains a description of the decomposition (3.6.2) for multipliers in H_p^l and $H_{p'}^{-l}$. Before we pass to its formulation, we present certain auxiliary definitions and results.

Definition 3.6.8. *A function u is called (p,l)-refined if for any $\varepsilon > 0$ one can find an open set ω such that $C_{p,l}(\omega) < \varepsilon$ and u is continuous on $\mathbb{R}^n\backslash\omega$.*

For proofs of the next assertions see [MH2].

Proposition 3.6.3. *For any $u \in H^l_{p,loc}$ there exists a (p,l)-refined Borel function which coincides with u almost everywhere.*

Proposition 3.6.4. *If two (p,l)-refined functions u_1 and u_2 are equal almost everywhere then they are equal (p,l)-quasi everywhere.*

Henceforth in this section all the functions are assumed to be (p,l)-refined and Borel.

The following assertion is proved for integer l in [Maz14] and for fractional l in [APo] for compact sets. The passage to arbitrary sets does not need new arguments if one uses Proposition 3.6.2.

Proposition 3.6.5. *Let $E \subset \mathbb{R}^n$. The capacity $\overline{C_{p,l}}(E)$ is equivalent to the set function*

$$\inf\{\|v\|^p_{H^l_p} : v \in \mathfrak{M}(E)\},$$

where $\mathfrak{M}(E)$ is the collection of (p,l)-refined functions equal to one (p,l)-quasi everywhere on E and satisfying the inequalities $0 \le v \le 1$.

Definition 3.6.9. *A set $E \subset \mathbb{R}^n$ is called the set of uniqueness for the space H^l_p if the conditions $u \in H^l_p$, $u(x) = 0$ for (p,l)-quasi all $x \in \mathbb{R}^n \backslash E$ imply that $u = 0$.*

A description of sets of uniqueness for H^l_p is given in [Hed3] and [Pol2]. The first result of such a kind for $H^{1/2}_2$ on a circumference is due to Ahlfors and Beurling [AB].

Proposition 3.6.6. (see [Hed3]). *Let E be a Borel subset of $\mathbb{R}^n$. The following conditions are equivalent:*

(i) E is the set of uniqueness for H^l_p;

(ii) $\underline{C_{p,l}}(G\backslash E) = C_{p,l}(G)$ for any open set G;

(iii) for almost all x

$$\lim_{\rho \to 0} \rho^{-n} C_{p,l}(\mathcal{B}_\rho(x)\backslash E) > 0.$$

If $lp > n$ then E is the set of uniqueness if and only if it has no interior points.

Now we state a theorem which gives a characteristic of the sets $\sigma_p(\gamma)$, $\sigma_r(\gamma)$ and $\sigma_c(\gamma)$ for a multiplier γ in H^l_p or $H^{-l}_{p'}$.

Theorem 3.6.1. (*i*) *Let $\gamma \in MH^l_p$ and $\lambda \in \sigma(\gamma)$.*

1. *$\lambda \in \sigma_p(\gamma)$ if and only if the set $Z_\lambda = \{x : \gamma(x) = \lambda\}$ does not satisfy any of conditions (i)–(iii) of Propostion 3.6.6.*
2. *$\lambda \in \sigma_r(\gamma)$ if and only if the set Z_λ satisfies any one of the conditions of Proposition 3.6.6 and $C_{p,l}(Z_\lambda) > 0$.*
3. *$\lambda \in \sigma_c(\gamma)$ if and only if $C_{p,l}(Z_\lambda) = 0$.*

(*ii*) *Let $\gamma \in MH^{-l}_{p'}$ and $\lambda \in \sigma(\gamma)$.*

1. $\lambda \in \sigma_p(\gamma)$ *if and only if* $C_{p,l}(Z_\lambda) > 0$.
2. $\lambda \in \sigma_c(\gamma)$ *if and only if* $C_{p,l}(Z_\lambda) = 0$ *(hence the set* $\sigma_r(\gamma)$ *is empty).*

An obvious corollary of Propositions 3.6.3 and 3.6.4 is the following assertion.

Lemma 3.6.1. *Let* γ *be a* (p,l)*-refined function in* MH_p^l*. The equation* $(\gamma - \lambda)u = 0$ *has a nontrivial solution in* H_p^l *if and only if there exists a* (p,l)*-refined non-zero function in* H_p^l *vanishing* (p,l)*-quasi everywhere outside* Z_λ*.*

This lemma shows that part $(i)1$ of the theorem immediately follows from Proposition 3.6.6. The proof of the other assertions of the theorem is contained in the next subsection.

3.6.4 Proof of Theorem 3.6.1

Below we shall use the following assertion.

Lemma 3.6.2. *Let* γ *be a* (p,l)*-refined function in* MH_p^l *and let* $Z_0 = \{x : \gamma(x) = 0\}$*. If* $C_{p,l}(Z_0) = 0$*, then the set* γH_p^l *is dense in* H_p^l*.*

Proof. Let $f \in C_0^\infty$ and let

$$N_\tau = \{x \in \text{supp}\, f : |\gamma(x)| \le \tau\}.$$

By ε we denote a small positive number and by ω we mean an open set with $C_{p,l}(\omega) < \varepsilon$ and such that γ is continuous on $\mathbb{R}^n \backslash \omega$. Let G stand for a neighborhood of the set $N_0 \backslash \omega$ with $C_{p,l}(G) < \varepsilon$.

We note that $N_\tau \backslash \omega \subset G$ for small enough $\tau > 0$. In fact, if for any $\tau > 0$ there exists a point $x_\tau \in N_\tau \backslash \omega$ which is not contained in G then, by continuity of γ outside ω, the limit point x_0 of the family $\{x_\tau\}$ is in $N_0 \backslash \omega$, contrary to the definition of G.

Consequently, $C_{p,l}(N_\tau \backslash \omega) < \varepsilon$ for small values of τ and

$$C_{p,l}(N_\tau) \le C_{p,l}(N_\tau \backslash \omega) + C_{p,l}(\omega) < 2\varepsilon\ .$$

Thus, $C_{p,l}(N_\tau) \to 0$ as $\tau \to 0$.

By $\{w_\tau\}_{\tau > 0}$ we denote a family of functions in $\mathfrak{M}(N_\tau)$ such that

$$\lim_{\tau \to 0} \|w_\tau\|_{H_p^l} = 0$$

(see Proposition 3.6.5). Further, we put

$$u_{\tau,\delta} = (1 - w_\tau)\bar{\gamma} f / (\gamma\bar{\gamma} + \delta),$$

where $\delta > 0$. Since

$$(1 - w_\tau) f \in H_p^l, \ \ \bar{\gamma} \in MH_p^l, \ \ \gamma\bar{\gamma} \in MH_p^l, \text{ and } \gamma\bar{\gamma} + \delta \ge \delta,$$

it follows that $u_{\tau,\delta} \in H_p^l$. We have

$$f - \gamma u_{\tau,\delta} = w_\tau f + \delta(1-w_\tau)f/(\gamma\bar{\gamma}+\delta)\,.$$

Let φ be a smooth increasing function on $[0,\infty)$, $\varphi(0) = \tau^2/4$, $\varphi(\tau) = t$ for $t > \tau^2/2$. Since $1 - w_\tau = 0$ (p,l)-quasi everywhere on N_τ,

$$f - \gamma u_{\tau,\delta} = w_\tau f + \delta(1-w_\tau)f/[\varphi(\gamma\bar{\gamma})+\delta]\,.$$

Using the inequality

$$\varphi(\gamma\bar{\gamma}) + \delta > \tau^2/4,$$

we obtain from Proposition 2.7.5 and Corollary 3.5.2 that the norm

$$\|[\varphi(\gamma\bar{\gamma})+\delta]^{-1}\|_{MH_p^l}$$

is uniformly bounded with respect to δ. Therefore

$$\|f - \gamma u_{\tau,\delta}\|_{H_p^l} \le \|w_\tau f\|_{H_p^l} + \delta k(\tau),$$

where $k(\tau)$ does not depend on δ. We put $\delta(\tau) = \tau/k(\tau)$. Then

$$\|f - \gamma u_{\tau,\delta(\tau)}\|_{H_p^l} \le c\,\|w_\tau\|_{H_p^l} + \tau$$

and hence $\gamma u_{\tau,\delta(\tau)} \to f$ as $\tau \to 0$ in H_p^l. □

In the next three propositions γ is a (p,l)-refined function from MH_p^l.

Proposition 3.6.7. *A number λ is contained in the pointwise spectrum of a multiplier γ in $H_{p'}^{-l}$ if and only if $C_{p,l}(Z_\lambda) > 0$.*

Proof. Sufficiency. Let R be so large that $C_{p,l}(Z_\lambda \cap \mathcal{B}_R) > 0$ and let μ be the capacitary measure of $Z_\lambda \cap \mathcal{B}_R$. Note that, whatever the (p,l)-refined function $u \in H_p^l$, we have $u(x)(\gamma(x) - \lambda) = 0$ for (p,l)-quasi all $x \in Z_\lambda \cap \mathcal{B}_R$. By Proposition 3.6.2, the last equality holds μ-almost everywhere. Therefore,

$$\int u(\gamma - \lambda)\,d\mu = 0.$$

In other words, $(\gamma - \lambda)\mu = 0$. Since

$$\|\mu\|_{H_{p'}^{-l}}^{p'} = \|J_l\mu\|_{L_{p'}}^{p'} = C_{p,l}(Z_\lambda \cap \mathcal{B}_R) < \infty\,,$$

we conclude that $\lambda \in \sigma_p(\gamma)$.

Necessity. Let $\lambda \in \sigma_p(\gamma)$. Then there exists a distribution $T \in H_{p'}^{-l}$, $T \neq 0$ such that $(\gamma - \lambda)T = 0$. Therefore $(T, (\gamma-\lambda)u) = 0$ for all $u \in H_p^l$ and the set $(\gamma - \lambda)H_p^l$ is not dense in H_p^l. The result follows by application of Lemma 3.6.2. □

Proposition 3.6.8. *A number λ is contained in the residual spectrum of a multiplier in H_p^l if and only if $\lambda \notin \sigma_p(\gamma)$ and $C_{p,l}(Z_\lambda) > 0$.*

Proof. Sufficiency. Since $C_{p,l}(Z_\lambda) > 0$ it follows by Proposition 3.6.7 that $\bar{\lambda}$ is an eigenvalue of the multiplier $\bar{\gamma}$ in $H_{p'}^{-l}$. This fact and (3.6.3) imply that

$$\lambda \in \sigma_r(\gamma) \cup \sigma_p(\gamma) = \sigma_r(\gamma).$$

Necessity. Let $\lambda \in \sigma_r(\gamma)$. By (3.6.3), $\bar{\lambda}$ is an eigenvalue of the multiplier $\bar{\gamma}$ in $H_{p'}^{-l}$. Hence, according to Proposition 3.6.7, $C_{p,l}(Z_\lambda) > 0$. □

Proposition 3.6.9. *The multiplier γ in $H_{p'}^{-l}$ has no residual spectrum.*

Proof. Let $\lambda \in \sigma_r(\gamma)$. By (3.6.3), $\bar{\lambda}$ is an eigenvalue of $\bar{\gamma}$ in H_p^l. This and the first assertion in Theorem 3.6.1 part (*i*) imply that $C_{p,l}(G \backslash Z_\lambda) < C_{p,l}(G)$ for some open set $G \subset \mathbb{R}^n$. Since

$$C_{p,l}(Z_\lambda) > C_{p,l}(G) - C_{p,l}(G \backslash Z_\lambda),$$

we have $C_{p,l}(Z_\lambda) > 0$. According to Proposition 3.6.7, this means that $\lambda \in \sigma_p(\gamma)$. Thus we arrive at a contradiction. □

Thus the statements of Theorem 3.6.1 concerning the pointwise and residual spectrum are proved. The characterization of the continuous spectrum obviously follows from these criteria and the relation (3.6.2).

3.7 The Space $M(h_p^m \to h_p^l)$

In this section we assume that $mp < n$, $p \in (1, \infty)$.

Using Sobolev's theorem on the imbedding $h_p^m \subset L_q$ with $q = mp/(n - mp)$, $n > mp$, one can easily prove that any function $\psi \in C_0^\infty$ belongs to the space Mh_p^m of multipliers in h_p^m. This immediately implies that the norm in Mh_p^m of the function ψ_r with values $\psi_r(x) = \psi(x/r)$, $r > 0$, does not depend on r.

This enables one to proceed as in the proofs of Lemmas 3.2.7 and 3.2.8, with γ_ρ replaced by $\psi_r \gamma_\rho$, where $\psi \in C_0^\infty$, $\psi = 1$ on $\mathcal{B}_1$, passing to the limit as $\rho \to 0$, $r \to \infty$ at the final step.

Thus we are led to the relation for the norm in $M(h_p^m \to h_p^l)$ with $0 \leq l \leq m < n/p$:

$$\|\gamma\|_{M(h_p^m \to h_p^l)} \sim \sup_e \left(\frac{\|S_l \gamma; e\|_{L_p}}{[c_{p,m}(e)]^{1/p}} + \frac{\|\gamma; e\|_{L_p}}{[c_{p,m-l}(e)]^{1/p}} \right). \tag{3.7.1}$$

The next two assertions are proved in the same way as Lemmas 2.8.1 and 2.8.2 with W replaced by H and w replaced by h.

Lemma 3.7.1. *(i) The inequality*

$$\|\gamma\|_{M(H_p^m \to L_p)} \leq \|\gamma\|_{M(h_p^m \to L_p)} \tag{3.7.2}$$

holds.

(ii) Let $\rho > 0$ and let $\gamma \in M(h_p^m \to L_p)$. Then

$$\lim_{\rho \to 0} \|\rho^{-m}\, \gamma(\cdot/\rho)\|_{M(H_p^m \to L_p)} = \|\gamma\|_{M(h_p^m \to L_p)}. \tag{3.7.3}$$

(iii) The function $\gamma \in M(h_p^m \to L_p)$ satisfies

$$\|\gamma\|_{M(h_p^m \to L_p)} \geq c \sup_{\substack{x \in \mathbb{R}^n \\ r>0}} r^{m-n/p} \|\gamma; \mathcal{B}_r(x)\|_{L_p}. \tag{3.7.4}$$

Lemma 3.7.2. *(i) Let $m \geq l$ and let $\gamma \in M(h_p^m \to h_p^l)$. Then*

$$\|\gamma\|_{M(H_p^m \to H_p^l)} \leq c\, \|\gamma\|_{M(h_p^m \to h_p^l)}. \tag{3.7.5}$$

(ii) The inequality

$$\liminf_{\rho \to 0} \|\rho^{l-m}\, \gamma(\cdot/\rho)\|_{M(H_p^m \to H_p^l)} \geq \|\gamma\|_{M(h_p^m \to h_p^l)} \tag{3.7.6}$$

holds.

Now we give a description of the space $M(h_p^m \to h_p^l)$.

Theorem 3.7.1. *Let $mp < n$, $m \geq l$, $p \in (1, \infty)$. Then $\gamma \in M(h_p^m \to h_p^l)$ if and only if $\gamma \in h_{p,\mathrm{loc}}^l$,*

$$S_l \gamma \in M(h_p^m \to L_p), \tag{3.7.7}$$

and

$$\begin{aligned} &\gamma \in L_\infty(\mathbb{R}^n) && \text{for} \quad m = l, \\ &\lim_{r \to \infty} r^{-n} \|\gamma; \mathcal{B}_r\|_{L_1} = 0 && \text{for} \quad m > l. \end{aligned} \tag{3.7.8}$$

The norm in the space $M(h_p^m \to h_p^l)$, $m > l$, is subject to the equivalence relation

$$\|\gamma\|_{M(h_p^m \to h_p^l)} \sim \|S_l \gamma\|_{M(h_p^m \to L_p)}. \tag{3.7.9}$$

For $m = l$ the norm $\|\gamma\|_{L_\infty}$ should be added to the right-hand side of this relation.

The equivalence relation

$$\|\gamma\|_{M(h_p^m \to h_p^l)}$$

$$\sim \sum_{j=0}^{[l]} \left(\|S_{l-j}\gamma\|_{M(h_p^{m-j} \to L_p)} + \|\nabla_{[l]-j}\gamma\|_{M(h_p^{m-\{l\}-j} \to L_p)} \right) \tag{3.7.10}$$

holds.

Proof. We replace γ by $\rho^{l-m}\gamma(\cdot/\rho)$ in (3.2.27). Then (3.7.10) follows from Lemmas 3.7.1 and 3.7.2 as $\rho \to 0$.

We take γ as $\rho^{l-m}\gamma(\cdot/\rho)$ in (3.2.47) to obtain

$$\|\rho^{l-m}\,\gamma(\cdot/\rho)\|_{M(H_p^m\to H_p^l)}$$
$$\leq c\big(\|S_l(\rho^{l-m}\,\gamma(\cdot/\rho))\|_{M(H_p^m\to L_p)} + \sup_{\substack{x\in\mathbb{R}^n\\ R>0}} R^{m-l-n}\|\rho^{l-m}\gamma(\cdot/\rho);\mathcal{B}_R(x)\|_{L_1}\big).$$

Since the second term on the right-hand side is equal to

$$\sup_{\substack{x\in\mathbb{R}^n\\ r>0}} r^{m-l-n}\|\gamma;\mathcal{B}_r(x)\|_{L_1},$$

and the first term tends to $\|S_l\gamma\|_{M(h_p^m\to L_p)}$ as $\rho\to 0$ by (3.7.3), reference to (3.7.6) gives

$$\|\gamma\|_{M(h_p^m\to h_p^l)} \leq c\big(\|S_l\gamma\|_{M(h_p^m\to L_p)} + \sup_{\substack{x\in\mathbb{R}^n\\ r>0}} r^{m-l-n}\|\gamma;\mathcal{B}_r(x)\|_{L_1}\big). \quad (3.7.11)$$

It remains to remove the second term on the right-hand side in the case $m > l$.

We use the inequality

$$|\gamma(x)| \leq c\,\big(\sup_{x\in\mathbb{R}^n, r\in(0,1)} r^{m-\frac{n}{p}}\|S_l\gamma;\mathcal{B}_r(x)\|_{L_p}\big)^{\frac{l}{m}}\big((S_l\gamma)(x)\big)^{1-\frac{l}{m}}, \quad (3.7.12)$$

which follows from (3.2.46) with the term $\|\gamma\|_{L_{1,\mathrm{unif}}}$ omitted on the right-hand side due to condition (3.7.8). Integrating (3.7.12) over an arbitrary ball $\mathcal{B}_r(x)$, we arrive at

$$\|\gamma;\mathcal{B}_r(x)\|_{L_1} \leq c\,\big(\sup_{\substack{x\in\mathbb{R}^n\\ r\in(0,1)}} r^{m-n/p}\|S_l\gamma;\mathcal{B}_r(x)\|_{L_p}\big)^{l/m}\|S_l\gamma^{1-l/m};\mathcal{B}_r(x)\|_{L_1}.$$

By Hölder's inequality

$$r^{m-l-n}\|\gamma;\mathcal{B}_r(x)\|_{L_1}$$
$$\leq c\,\big(\sup_{\substack{x\in\mathbb{R}^n\\ r\in(0,1)}} r^{m-\frac{n}{p}}\|S_l\gamma;\mathcal{B}_r(x)\|_{L_p}\big)^{\frac{l}{m}}\big(r^{m-\frac{n}{p}}\|S_l\gamma;\mathcal{B}_r(x)\|_{L_p}\big)^{1-\frac{l}{m}}. \quad (3.7.13)$$

The result follows by (3.7.11). □

Remark 3.7.1. By Theorem 3.1.4, condition (3.7.7) can be formulated in four ways as follows:

(*i*) for all compact sets $e\subset\mathbb{R}^n$

$$\|S_l\gamma;e\|^p_{L_p} \leq C\,c_{p,m}(e),$$

(*ii*) for all cubes Q in $\mathbb{R}^n$

$$\int \left(I_m(\chi_Q S_l\gamma)^p(x)\right)^{p'} dx \le C \int_Q (S_l\gamma)^p dx,$$

(*iii*) for almost all $x \in \mathbb{R}^n$

$$I_m\left(I_m(S_l\gamma)^p\right)^{p'}(x) \le C\, I_m(S_l\gamma)^p(x),$$

(*iv*) for every dyadic cube P_0 in $\mathbb{R}^n$

$$\sum_{P\subseteq P_0} \left[(\mathrm{mes}_n P)^{(m-n)/n} \int_P (S_l\gamma(x))^p dx\right]^{p'} \mathrm{mes}_n P \le C \int_{P_0} (S_l\gamma(x))^p dx,$$

where the sum is taken over all dyadic cubes P contained in P_0.

3.8 Positive Homogeneous Multipliers

In this section we give a description of elements in the spaces $M(h_p^m \to h_p^l)$ and $M(H_p^m \to H_p^l)$ which have the form $|x|^{l-m} f(x/|x|)$.

3.8.1 The Space $M(H_p^m(\partial\mathcal{B}_1) \to H_p^l(\partial\mathcal{B}_1))$

Let $\{U_i\}$ be a finite covering of $\partial\mathcal{B}_1$ by open sets with small diameters, and let $\{\varphi_i\}$ be a family of diffeomorphisms $U_i \to \mathbb{R}^{n-1}$ which form a smooth structure on $\partial\mathcal{B}_1$. Further, let $\{\nu_i\}$ be a smooth partition of unity on $\partial\mathcal{B}_1$ subordinate to the covering $\{U_i\}$.

We say that a function v on $\partial\mathcal{B}_1$ is in the space $H_p^l(\partial\mathcal{B}_1)$ if

$$(\nu_i \circ v) \circ \varphi_i^{-1} \in H_p^l(\mathbb{R}^{n-1}) \qquad \text{for all } i.$$

We equip $H_p^l(\partial\mathcal{B}_1)$ with the norm

$$\|v;\partial\mathcal{B}_1\|_{H_p^l} = \Big(\sum_i \|(\nu_i v)\circ\varphi_i^{-1};\mathbb{R}^{n-1}\|_{H_p^l}^p\Big)^{1/p}.$$

It is well known that the passage from one collection $\{U_i, \varphi_i, \nu_i\}$ to another leads to an equivalent norm.

Proposition 3.8.1. *A function f is contained in the space $M(H_p^m(\partial\mathcal{B}_1) \to H_p^l(\partial\mathcal{B}_1))$ if and only if*

$$(\nu_i \circ f)\circ\varphi_i^{-1} \in M(H_p^m(\mathbb{R}^{n-1}) \to H_p^l(\mathbb{R}^{n-1}))$$

for all i. Moreover,

$$\|f;\partial\mathcal{B}_1\|_{M(H_p^m\to H_p^l)} \sim \max_i \|(\nu_i f)\circ\varphi_i^{-1};\mathbb{R}^{n-1}\|_{M(H_p^m\to H_p^l)}\,. \tag{3.8.1}$$

Proof. For any $v \in H_p^m(\partial\mathcal{B}_1)$ we have

$$\|fv; \partial\mathcal{B}_1\|_{H_p^l} \le \Big(\sum_{i,j} \|(\nu_i f \nu_j \zeta_i v) \circ \varphi_i^{-1}; \mathbb{R}^{n-1}\|_{H_p^l}^p\Big)^{1/p}$$

$$\le \sup_i \|(\nu_i f) \circ \varphi_i^{-1}; \mathbb{R}^{n-1}\|_{M(H_p^m \to H_p^l)} \Big(\sum_{i,j} \|(\nu_j \zeta_i v) \circ \varphi_i^{-1}; \mathbb{R}^{n-1}\|_{H_p^m}^p\Big)^{1/p},$$

where $\zeta_i \in C_0^\infty(U_i)$ and $\zeta_i \nu_i = \nu_i$. Since the mapping

$$\varphi_j \varphi_i^{-1} : \varphi_j(U_j \cap U_i) \to \varphi_i(U_j \cap U_i)$$

is infinitely differentiable, we have

$$\|(\nu_j \zeta_i v) \circ \varphi_i^{-1}; \mathbb{R}^{n-1}\|_{H_p^m} = \|(\nu_j \zeta_i v) \circ \varphi_j^{-1} \varphi_j \varphi_i^{-1}; \mathbb{R}^{n-1}\|_{H_p^m}$$

$$\le c\, \|(\nu_j v) \circ \varphi_j^{-1}; \mathbb{R}^{n-1}\|_{H_p^m},$$

and thus the required upper estimate for the norm in $M(H_p^m(\partial\mathcal{B}_1) \to H_p^l(\partial\mathcal{B}_1))$ follows.

On the other hand, for any $w \in H_p^m(\mathbb{R}^{n-1})$

$$\|[(\nu_i f) \circ \varphi_i^{-1}]w; \mathbb{R}^{n-1}\|_{H_p^l} = \|[\nu_i f(w \circ \varphi_i)] \circ \varphi_i^{-1}; \mathbb{R}^{n-1}\|_{H_p^l}$$

$$\le \|\nu_i f(w \circ \varphi_i); \partial\mathcal{B}_1\|_{H_p^l} \le \|f; \partial\mathcal{B}_1\|_{M(H_p^m \to H_p^l)} \|\nu_i(w \circ \varphi_i); \partial\mathcal{B}_1\|_{H_p^m}.$$

It is clear that the last norm does not exceed

$$\Big(\sum_j \|[\nu_j \nu_i(w \circ \varphi_i)] \circ \varphi_j^{-1}; \mathbb{R}^{n-1}\|_{H_p^m}^p\Big)^{1/p}$$

$$\le c\Big(\sum_j \|[\nu_j \nu_i(w \circ \varphi)] \circ \varphi_i^{-1}; \mathbb{R}^{n-1}\|_{H_p^m}^p\Big)^{1/p} \le c\, \|w; \mathbb{R}^{n-1}\|_{H_p^m}$$

and the lower estimate for the norm in $M(H_p^m(\partial\mathcal{B}_1) \to H_p^l(\partial\mathcal{B}_1))$ follows. □

We note that $H_p^l(\partial\mathcal{B}_1)$ can be supplied with an equivalent norm using the operator $(1-\delta)^{l/2}$ where δ is the Beltrami operator on the sphere. Namely,

$$\|v; \partial\mathcal{B}_1\|_{H_p^l} \sim \|(1-\delta)^{l/2} v; \partial\mathcal{B}_1\|_{L_p}. \tag{3.8.2}$$

It is essentially a consequence of the following property established in [Se]: $(1-\delta)^{l/2}$ is the pseudo-differential operator with symbol $|\xi|^l$ (see also [Sh]).

3.8.2 Other Normalizations of the Spaces h_p^m and H_p^m

The normalizations in the title of this subsection are given below in Lemma 3.8.1 and Corollary 3.8.1.

Let $0 \le l \le m$, $p > 1$, $mp < n$. Then the well known Hardy type inequality

$$\left\| \frac{u}{|x|^{m-l}} \right\|_{h_p^l} \le c \, \|u\|_{h_p^m} \tag{3.8.3}$$

is valid. It means that $|x|^{l-m} \in M(h_p^m \to h_p^l)$ and it follows, for instance, from (3.7.1) together with the inequalities

$$\int_e \frac{dx}{|x|^{mp}} \le c \, (\text{mes}_n e)^{1-mp/n} ,$$

$$\big(S_l |y|^{l-m}\big)(x) \le c \, |x|^{-m} .$$

Here, the second inequality results from the easily verified estimate

$$\int_{\mathcal{B}_1} \big| \, |x+\theta y|^{l-m} - |x|^{l-m} \big| \, d\theta \le c \min\{y, \, |x|\} |x|^{l-m-1} .$$

Using (3.8.3), we may equip the space h_p^m, $mp < n$, with an equivalent norm. Let

$$G_k = \{x : 2^{k-1} < |x| < 2^{k+1}\}, \qquad k = 0, \pm 1, \ldots ,$$

and let $\{\psi_k\}$ be a partition of unity on $\mathbb{R}^n \backslash \{0\}$ subject to the covering $\{G_k\}$. We suppose that $|D^\alpha \psi_k| \le c_\alpha 2^{-k|\alpha|}$.

Lemma 3.8.1. *Let $p > 1$, $mp < n$. The relation*

$$\|u\|_{h_p^m} \sim \Big(\sum_{k=-\infty}^{\infty} \|\psi_k u\|_{h_p^m}^p \Big)^{1/p} \tag{3.8.4}$$

holds.

Proof. First we show that the proof of (3.8.4) easily reduces to the case $0 \le m \le 1$. Suppose that $m \ge 1$ and that the assertion is proved for $m-1$. One can readily check that, for integer s, the norms of the functions $|x|^\alpha D^\alpha \psi_k$ in Mh_p^s are uniformly bounded with respect to k. The same is true for fractional s by interpolation. Therefore,

$$\begin{aligned} &\big| \, \|\psi_k \nabla u\|_{h_p^{m-1}} - \|\nabla(\psi_k u)\|_{h_p^{m-1}} \big| \\ &\le \|\sigma_k u \nabla \psi_k\|_{h_p^{m-1}} \le c \, \| |x|^{-1} \sigma_k u\|_{h_p^{m-1}} , \end{aligned} \tag{3.8.5}$$

where

$$\sigma_k \in C_0^\infty(G_k), \quad \sigma_k \psi_k = \psi_k, \quad |D^\alpha \sigma_k| \le c_\alpha 2^{-k|\alpha|}.$$

Consequently,

$$\|\nabla u\|^p_{h_p^{m-1}} \le c \sum_{k=-\infty}^{\infty} \|\psi_k \nabla u\|^p_{h_p^{m-1}}$$
$$\le c \sum_{k=-\infty}^{\infty} \big(\|\nabla(\psi_k u)\|^p_{h_p^{m-1}} + \||x|^{-1}\sigma_k u\|^p_{h_p^{m-1}}\big).$$

This and (3.8.3) imply

$$\|u\|^p_{h_p^m} \le c \sum_{k=-\infty}^{\infty} \big(\|\psi_m u\|^p_{h_p^m} + \|\sigma_k u\|^p_{h_p^m}\big).$$

Since $\{\psi_k\}$ is a partition of unity and $\|\sigma_k\|_{Mh_p^m} \le \text{const}$, the upper estimate for the norm $\|u\|_{h_p^m}$ follows.

Next we derive the lower bound. By (3.8.5) we have

$$\sum_{k=-\infty}^{+\infty} \|\psi_k u\|^p_{h_p^m} \le c \sum_{k=-\infty}^{\infty} \big(\|\psi_k \nabla u\|^p_{h_p^{m-1}} + \||x|^{-1}\sigma_k u\|^p_{h_p^{m-1}}\big).$$

Replacing σ_k by ψ_k in the last norm and using the induction hypothesis, we obtain that the right-hand side does not exceed

$$c\,\big(\|u\|^p_{h_p^m} + \||x|^{-1}u\|^p_{h_p^{m-1}}\big).$$

It remains to make use of (3.8.3).

In the case $m = 0$ the relation (3.8.4) is trivial. Let $0 < m < 1$. It is clear that

$$\|S_m u\|^p_{L_p} \sim \sum_{k=-\infty}^{\infty} \|\psi_k S_m u\|^p_{L_p}\,. \tag{3.8.6}$$

The definition of S_m and Minkowski's inequality imply that

$$\big|\psi_k(x)(S_m u)(x) - (S_m(\psi_k u))(x)\big|$$
$$\le \Big(\int_0^\infty \Big[\int_{\mathcal{B}_y} |u(x+h)||\psi_k(x+h) - \psi_k(x)|\,dh\Big]^2 y^{-1-2n-2m}\,dy\Big)^{1/2}$$
$$\le c \int |u(z)|\,|\psi_k(z) - \psi_k(x)|\,dz \Big(\int_{|x-z|}^{\infty} y^{-1-2n-2m} dy\Big)^{1/2}$$
$$\le c \int |u(z)| \frac{|\psi_k(z) - \psi_k(x)|}{|z-x|^{n+m}}\,dz\,.$$

By $A(x)$ we denote the right-hand side and put $g_k = G_{k-1} \cup G_k \cup G_{k+1}$. Since $\operatorname{supp} \psi_k \subset G_k$, it follows that

$$A(x) \le \begin{cases} c\,2^{-k(n+m)} \displaystyle\int_{G_k} |u(z)|\,dz & \text{if } |x| < 2^{k-2}\,, \\ c\,|x|^{-(n+m)} \displaystyle\int_{G_k} |u(z)|\,dz & \text{if } |x| > 2^{k+2}\,, \\ c\Big(2^{-k} \displaystyle\int_{\mathcal{B}_{2|x|}} \frac{|u(z)|\,dz}{|z-x|^{n+m-1}} + \int_{\mathbb{R}^n\setminus\mathcal{B}_{2|x|}} \frac{|u(z)|\,dz}{|z|^{n+m}}\Big) & \text{if } x \in g_k\,. \end{cases}$$

Therefore,

$$\|\psi_k S_m u - S_m(\psi_k u)\|^p_{L_p} \le c\,2^{k(n-pm-pn)}\Big(\int_{G_k} |u(z)|\,dz\Big)^p$$

$$+c\,2^{-kp}\int_{g_k}\Big(\int_{\mathcal{B}_{2|x|}} \frac{|u(z)|\,dz}{|z-x|^{n+m-1}}\Big)^p dx + c\int_{g_k}\Big(\int_{\mathbb{R}^n\setminus\mathcal{B}_{2|x|}} \frac{|u(z)|dz}{|z|^{n+m}}\Big)^p dx.$$

The first term on the right-hand side does not exceed $c\,2^{-kpm}\|u;G_k\|^p_{L_p}$. The second one is majorized by

$$c\,2^{-kp}\Big(\max_x \int_{\mathcal{B}_{2^{k+3}}} \frac{d\zeta}{|\zeta-x|^{n+m-1}}\Big)^p \int_{\mathcal{B}_{2^{k+3}}} |u(z)|^p\,dz \le c\,2^{-kpm}\int_{\mathcal{B}_{2^{k+3}}} |u(z)|^p\,dz.$$

The third term can be written as

$$c\int_{2^{k-2}}^{2^{k+2}} r^{n-1}\Big(\int_{2r}^{\infty} \frac{v(\rho)}{\rho^{m+1}}\,d\rho\Big)^p dr,$$

where $v(\rho)$ is the mean value of $|u|$ on $\partial\mathcal{B}_\rho$. Summing over k and using the one-dimensional Hardy inequality, we arrive at

$$\sum_{k=-\infty}^{+\infty} \|\psi_k S_m u - S_m(\psi_k u)\|^p_{L_p}$$

$$\le c\,\|\,|x|^{-m}u\|^p_{L_p} \sim \sum_{k=-\infty}^{+\infty} \|\,|x|^{-m}\psi_k u\|^p_{L_p}\,. \tag{3.8.7}$$

Now, (3.8.7) and (3.8.3) imply that

$$\|S_m u\|^p_{L_p} \le c\sum_{k=-\infty}^{+\infty} \|S_m(\psi_k u)\|^p_{L_p}\,.$$

Thus the upper bound for the norm in h^m_p follows.

Also, by (3.8.7) we have

$$\sum_{k=-\infty}^{+\infty} \|S_m(\psi_k u)\|^p_{L_p} \le \|S_m u\|^p_{L_p} + \|\,|x|^{-m}u\|^p_{L_p}$$

Applying (3.8.3) once again, we complete the proof. □

Relation (3.8.4) and the equivalence

$$\|u\|_{H_p^m} \sim \|u\|_{h_p^m} + \|u\|_{L_p}$$

imply the result similar to the last lemma for the space H_p^m.

Corollary 3.8.1. *Let $p > 1$ and $mp < n$. Then*

$$\|u\|_{H_p^m} \sim \Big(\sum_{k=-\infty}^{+\infty} \|\psi_k u\|_{H_p^m}^p \Big)^{1/p}.$$

A direct corollary of Lemma 3.8.1 is the following assertion on the normalization of the space $M(h_p^m \to h_p^l)$, $pm < n$, which will be used in Subsect. 3.8.3.

Corollary 3.8.2. *If $p > 1$ and $pm < n$, then*

$$\|\gamma\|_{M(h_p^m \to h_p^l)} \sim \sup_{-\infty<k<\infty} \|\psi_k \gamma\|_{M(h_p^m \to h_p^l)} \,. \tag{3.8.8}$$

3.8.3 Positive Homogeneous Elements of the Spaces $M(h_p^m \to h_p^l)$ and $M(H_p^m \to H_p^l)$

We state and prove the main result of this section.

Theorem 3.8.1. *Let $p > 1$ and $pm < n$. The function*

$$x \to \gamma(x) = |x|^{l-m} f(x/|x|)$$

is contained in any of the spaces $M(h_p^m \to h_p^l)$ and $M(H_p^m \to H_p^l)$ if and only if

$$f \in M(H_p^m(\partial\mathcal{B}_1) \to H_p^l(\partial\mathcal{B}_1)).$$

Moreover,

$$\|\gamma\|_{M(h_p^m \to h_p^l)} \sim \|\gamma\|_{M(H_p^m \to H_p^l)} \sim \|f; \partial\mathcal{B}_1\|_{M(H_p^m \to H_p^l)} \,.$$

Proof. We start with the space $M(h_p^m \to h_p^l)$. Let $\zeta \in C_0^\infty(-1/2, 2)$, $\zeta(t) = 1$ for $t \in (1, 3/2)$ and $\zeta_k(x) = \zeta(2^{-k}|x|)$. Since the norms of the functions ζ_k and ψ_k in Mh_p^l are uniformly bounded with respect to k, it follows from (3.8.8) that

$$\|\gamma\|_{M(h_p^m \to h_p^l)} \sim \sup_{-\infty<k<\infty} \|\zeta_k \gamma\|_{M(h_p^m \to h_p^l)} \,. \tag{3.8.9}$$

Using the homogeneity of the norm in h_p^k with respect to the similarity transform and the homogeneity of γ we conclude that the norm on the right-hand side of (3.8.9) does not depend on k. Therefore,

$$\|\gamma\|_{M(h_p^m \to h_p^l)} \sim \|\zeta\gamma\|_{M(h_p^m \to h_p^l)} \,. \tag{3.8.10}$$

Let $\{\nu_i\}$ be a partition of unity on $\partial\mathcal{B}_1$ subordinate to the covering of $\partial\mathcal{B}_1$ by a family $\{U_i\}$ of coordinate neighborhoods on $\partial\mathcal{B}_1$. Further, let $\{\varphi_i\}$ be the family of diffeomorphisms used in the definition of the space $H_p^l(\partial\mathcal{B}_1)$. Since $(S_l\nu_i)(x) = O(|x|^{-l})$, we have $\nu_i \in Mh_p^l$ (see (3.7.1)). From this inclusion and from (3.8.10) it follows that

$$\|\gamma\|_{M(h_p^m \to h_p^l)} \sim \max_i \|\nu_i\zeta\gamma\|_{M(h_p^m \to h_p^l)} \,. \tag{3.8.11}$$

We introduce the local coordinates $y = (y', y_n)$, where $y_n = |x|$ and $y' = \varphi_i(x/|x|)$, on the set $\{x : 1/2 < |x| < 2, x/|x| \in U_i\}$. Since the mapping $\Phi_i : x \to y$ is a diffeomorphism, we have

$$\|\nu_i\zeta\gamma\|_{M(h_p^m \to h_p^l)} \sim \|(\nu_i\zeta\gamma) \circ \Phi_i^{-1}\|_{M(h_p^m \to h_p^l)} \,. \tag{3.8.12}$$

Next, we note that

$$\big((\nu_i\zeta\gamma) \circ \Phi_i^{-1}\big)(y) = \zeta(y_n)\, y_n^{l-m}\, \big((\nu_i f) \circ \varphi_i^{-1}\big)(y')$$

and that the function $y \to \zeta(y_n)\, y_n^{l-m}$ is smooth and has a compact support. Therefore,

$$\begin{aligned}\|\nu_i\zeta\gamma\|_{M(h_p^m \to h_p^l)} &\le c\, \|(\nu_i f) \circ \Phi_i^{-1}\|_{M(h_p^m \to h_p^l)} \\ &\le c\, \|(\nu_i f) \circ \varphi_i^{-1}; \mathbb{R}^{n-1}\|_{M(h_p^m \to h_p^l)} \,. \end{aligned} \tag{3.8.13}$$

By the relation

$$\|u\|_{h_p^l} \sim \sum_{i=1}^{n} \Big\| \Big|\frac{\partial}{\partial y_i}\Big|^l u \Big\|_{L_p}$$

(see the proof of (3.5.2)), every function $y \to v(y')$ with $v \in C_0^\infty(\mathbb{R}^{n-1})$ satisfies the inequality

$$\|(\nu_i f) \circ \varphi_i^{-1} v; \mathbb{R}^{n-1}\|_{h_p^l} \le c\, \|(\nu_i\zeta\gamma) \circ \Phi_i^{-1} v\zeta; \mathbb{R}^n\|_{h_p^l} \,.$$

The right-hand side, obviously, does not exceed

$$\begin{aligned} &c\, \|(\nu_i\zeta\gamma) \circ \Phi_i^{-1}\|_{M(h_p^m \to h_p^l)}\, \|v\zeta; \mathbb{R}^n\|_{h_p^m} \\ &\le c\, \|(\nu_i\zeta\gamma) \circ \Phi_i^{-1}\|_{M(h_p^m \to h_p^l)}\, \|v; \mathbb{R}^{n-1}\|_{h_p^m} \,. \end{aligned}$$

Thus

$$\|(\nu_i f) \circ \varphi_i^{-1}; \mathbb{R}^{n-1}\|_{M(h_p^m \to h_p^l)} \le c\, \|(\nu_i\zeta\gamma) \circ \Phi_i^{-1}\|_{M(h_p^m \to h_p^l)}$$

which together with (3.8.12) and (3.8.13) leads to

$$\|\nu_i\zeta\gamma\|_{M(h_p^m \to h_p^l)} \sim \|(\nu_i f) \circ \Phi_i^{-1}; \mathbb{R}^{n-1}\|_{M(h_p^m \to h_p^l)} \,. \tag{3.8.14}$$

Comparing (3.8.14), (3.8.1) and (3.8.11), we complete the proof for the space $M(h_p^m \to h_p^l)$.

Now we turn to the space $M(H_p^m \to H_p^l)$. For all $u \in C_0^\infty$ we have

$$\|\gamma u\|_{h_p^l} \leq c\,\|\gamma\|_{M(H_p^m \to H_p^l)}(\|u\|_{h_p^m} + \|u\|_{L_p})\,.$$

Putting here $u(x) = w(ax)$, where a is an arbitrary positive number, and using the positive homogeneity of γ, we find that, for any $w \in C_0^\infty$,

$$\|\gamma w\|_{h_p^l} \leq c\,\|\gamma\|_{M(H_p^m \to H_p^l)}(\|w\|_{h_p^m} + a^{-m}\|w\|_{L_p})\,.$$

Passing to the limit as $a \to \infty$, we arrive at

$$\|\gamma\|_{M(h_p^m \to h_p^l)} \leq c\,\|\gamma\|_{M(H_p^m \to H_p^l)}\,. \tag{3.8.15}$$

Now we derive the converse estimate. Let $\{\eta_j\}$ be the sequence of functions defined in Theorem 3.1.2. Since $M(h_p^m \to h_p^l) \subset M(h_p^{m-l} \to L_p)$, it follows that

$$\begin{aligned}
\|\gamma u\eta_j\|_{H_p^l} &\leq c\,(\|\gamma u\eta_j\|_{h_p^l} + \|\gamma u\eta_j\|_{L_p}) \\
&\leq c\,(\|\gamma\|_{M(h_p^m \to h_p^l)}\,\|u\eta_j\|_{h_p^m} + \|\gamma\|_{M(h_p^m \to h_p^l)}\|u\eta_j\|_{h_p^{m-l}}) \\
&\leq c\,\|\gamma\|_{M(h_p^m \to h_p^l)}\,\|u\eta_j\|_{H_p^m}\,.
\end{aligned}$$

Therefore, using Theorem 3.1.2, we obtain

$$\|\gamma u\|_{H_p^l} \leq c\,\|\gamma\|_{M(h_p^m \to h_p^l)}\|u\|_{H_p^m}\,.$$

Thus

$$\|\gamma\|_{M(H_p^m \to H_p^l)} \leq c\,\|\gamma\|_{M(h_p^m \to h_p^l)}$$

which together with (3.8.15) shows that the norms of γ in $M(H_p^m \to H_p^l)$ and in $M(h_p^m \to h_p^l)$ are equivalent. The theorem is proved. □

4

The Space $M(B_p^m \to B_p^l)$ with $p > 1$

4.1 Introduction

In the present chapter we give necessary and sufficient conditions for a function to be a multiplier acting from one Besov space $B_p^m(\mathbb{R}^n)$ into another $B_p^l(\mathbb{R}^n)$, where $0 < l \le m$ and $p \in (1, \infty)$ (see Sect. 4.3).

Let $l = k + \alpha$, where $\alpha \in (0, 1]$ and k is a nonnegative integer. Further, as before, let

$$\Delta_h^{(2)} u(x) = u(x + 2h) - 2u(x + h) + u(x)$$

and

$$(\mathfrak{D}_{p,l} u)(x) = \Big(\int_{\mathbb{R}^n} |\Delta_h^{(2)} \nabla_k u(x)|^p |h|^{-n-p\alpha} dh \Big)^{1/p}, \tag{4.1.1}$$

where ∇_k stands for the gradient of order k, i.e.

$$\nabla_k u = \{\partial_{x_1}^{\alpha_1} \dots \partial_{x_n}^{\alpha_n}\}, \qquad \alpha_1 + \dots + \alpha_n = k.$$

The Besov space $B_p^l(\mathbb{R}^n)$ is introduced as the completion of $C_0^\infty(\mathbb{R}^n)$ in the norm

$$\|u; \mathbb{R}^n\|_{B_p^l} = \|\mathfrak{D}_{p,l} u; \mathbb{R}^n\|_{L_p} + \|u; \mathbb{R}^n\|_{L_p}. \tag{4.1.2}$$

Let $\{l\}$ and $[l]$ denote the fractional and integer parts of a positive number l and let

$$(D_{p,l} u)(x) = \Big(\int_{\mathbb{R}^n} |\Delta_h \nabla_{[l]} u(x)|^p |h|^{-n-p\{l\}} dh \Big)^{1/p},$$

where $\Delta_h v(x) = v(x + h) - v(x)$. The fractional Sobolev space W_p^l is defined as the closure of C_0^∞ in the norm $\|D_{p,l} u\|_{L_p} + \|u\|_{L_p}$. As before, we omit $\mathbb{R}^n$ in the notation of norms, spaces, and in the range of integration.

For $\{l\} > 0$ the spaces B_p^l and W_p^l have the same elements and their norms are equivalent since

$$(2 - 2^{\{l\}}) D_{p,l} u \le \mathfrak{D}_{p,l} u \le (2 + 2^{\{l\}}) D_{p,l} u, \tag{4.1.3}$$

V.G. Maz'ya, T.O. Shaposhnikova, *Theory of Sobolev Multipliers*,
Grundlehren der mathematischen Wissenschaften 337,

which follows directly from the identity

$$2[u(x+h) - u(x)] = -[u(x+2h) - 2u(x+h) + u(x)] + [u(x+2h) - u(x)].$$

Similarly to the case of Sobolev spaces in Sect. 2.1, one can show that the space $M(B_p^m \to B_p^l)$ is trivial provided that $m < l$.

We formulate the main result of this chapter.

Theorem 4.1.1. *Let* $0 < l \le m$, $p \in (1, \infty)$, *and let* $\gamma \in B^l_{p,\mathrm{loc}}$. *A function* γ *belongs to* $M(B_p^m \to B_p^l)$ *if and only if* $\gamma \in B^l_{p,\mathrm{loc}}$, $\mathfrak{D}_{p,l}\gamma \in M(B_p^m \to L_p)$, *and either* $\gamma \in L_{1,\mathrm{unif}}$ *for* $m > l$ *or* $\gamma \in L_\infty$ *for* $m = l$. *The equivalence relation*

$$\|\gamma\|_{M(B_p^m \to B_p^l)} \sim \sup_e \frac{\|\mathfrak{D}_{p,l}\gamma; e\|_{L_p}}{[C_{p,m}(e)]^{1/p}} + \begin{cases} \|\gamma\|_{L_{1,\mathrm{unif}}}, & m > l, \\ \|\gamma\|_{L_\infty}, & m = l \end{cases} \tag{4.1.4}$$

holds, where e *is an arbitrary compact set in* $\mathbb{R}^n$.

The relation (4.1.4) *remains valid if the condition* $d(e) \le 1$ *is added, where* $d(e)$ *is the diameter of* e.

For $mp > n$ the statement of the above theorem can be simplified. Namely, the relation (4.1.4) is equivalent to

$$\|\gamma\|_{M(B_p^m \to B_p^l)} \sim \|\gamma\|_{B^l_{p,\mathrm{unif}}} \quad \text{for } m \ge l, \tag{4.1.5}$$

and for $lp > n$

$$\|\gamma\|_{MB_p^l} \sim \|\mathfrak{D}_{p,l}\gamma\|_{L_{p,\mathrm{unif}}} + \|\gamma\|_{L_\infty}. \tag{4.1.6}$$

Various upper estimates for the norm in $M(B_p^m \to B_p^l)$ are derived in Sects. 4.4 and 4.5.

The following assertion concerning the composition $\varphi(\gamma)$, where $\gamma \in M(W_p^m \to W_p^l)$, $0 < l < 1$, $p > 1$, is proved in the last Sect. 4.6. If a function φ satisfies the Hölder condition

$$|\varphi(t+\tau) - \varphi(t)| \le A\,|\tau|^\rho, \quad |\tau| < 1,$$

with $\rho \in (0,1)$, then $\varphi(\gamma) \in M(W_p^{m-l+r} \to W_p^r)$ for any $r \in (0, l\rho)$.

4.2 Properties of Besov Spaces

4.2.1 Survey of Known Results

We start with three classical properties of the space B_p^l.

Proposition 4.2.1. (see [St2], Sect. 5.1) *The equivalence relation*

$$\|u\|_{B_p^k} \sim \|\Lambda^\alpha u\|_{B_p^{k-\alpha}}, \tag{4.2.1}$$

holds, where $p \in (1,\infty)$, $\alpha \in (0,k)$, *and* $\Lambda = (1-\Delta)^{1/2}$.

Proposition 4.2.2. (see [St2], Sect. 5.3) *The inequalities*

$$\|u\|_{H^l_p} \le c\,\|u\|_{B^l_p} \quad \text{for} \quad 1 < p \le 2$$

and

$$\|u\|_{B^l_p} \le c\,\|u\|_{H^l_p} \quad \text{for} \quad 2 \le p < \infty$$

hold.

Proposition 4.2.3. (see [Bes]) *Let* $\mathbb{R}^n = \mathbb{R}^m \times \mathbb{R}^k = \{x = (y, z) : y \in \mathbb{R}^m, z \in \mathbb{R}^k\}$. *The relation*

$$\|u\|_{B^l_p} \sim \|u\|_{L_p} + \Big(\int_{\mathbb{R}^k}\int_{\mathbb{R}^m} \|\Delta^{(2)}_\eta \nabla_k u(\cdot, z); \mathbb{R}^m\|^p_{L_p} \frac{d\eta\, dz}{|\eta|^{m+p\alpha}}\Big)^{1/p}$$

$$+\Big(\int_{\mathbb{R}^m}\int_{\mathbb{R}^k} \|\Delta^{(2)}_\zeta \nabla_k u(y, \cdot); \mathbb{R}^k\|^p_{L_p} \frac{d\zeta\, dy}{|\zeta|^{k+p\alpha}}\Big)^{1/p} \tag{4.2.2}$$

holds, where $l = k + \alpha > 0$ *and* $p \in [1, \infty)$. *Similarly, for noninteger* l,

$$\|u\|_{W^l_p} \sim \|u\|_{L_p} + \Big(\int_{\mathbb{R}^k}\int_{\mathbb{R}^m} \|\Delta_\eta \nabla_{[l]} u(\cdot, z); \mathbb{R}^m\|^p_{L_p} \frac{d\eta\, dz}{|\eta|^{m+p\{l\}}}\Big)^{1/p}$$

$$+\Big(\int_{\mathbb{R}^m}\int_{\mathbb{R}^k} \|\Delta_\zeta \nabla_{[l]} u(y, \cdot); \mathbb{R}^k\|^p_{L_p} \frac{d\zeta\, dy}{|\zeta|^{k+p\{l\}}}\Big)^{1/p}. \tag{4.2.3}$$

Analogous norms appear under decomposition of $\mathbb{R}^n$ *into more than two factors.*

Now we recall well known trace properties of Besov spaces (see [Usp] or [Bur], Sect. 5.4).

Proposition 4.2.4. *Let* $U \in C_0^\infty(\mathbb{R}^{n+s})$ *be any extension of* $u \in C_0^\infty(\mathbb{R}^n)$ *onto the space* $\mathbb{R}^{n+s} = \{z = (x, y) : x \in \mathbb{R}^n,\ y \in \mathbb{R}^s\}$.

(*i*) *If* $p \in (1, \infty)$ *and* $l > 0$, *then*

$$\|u; \mathbb{R}^n\|_{B^l_p} \sim \inf_{\{U\}} \|U; \mathbb{R}^{n+s}\|_{H^{l+s/p}_p}. \tag{4.2.4}$$

(*ii*) *If* $p \in [1, \infty)$ *and* $l > 0$, *then*

$$\|u; \mathbb{R}^n\|_{B^l_p} \sim \inf_{\{U\}} \|U; \mathbb{R}^{n+s}\|_{B^{l+s/p}_p}. \tag{4.2.5}$$

(*iii*) *If* $p \in [1, \infty)$ *and* $l > 0$, *then*

$$\|u; \mathbb{R}^n\|_{B^l_p} \sim \inf_{\{U\}} \Big(\int_{\mathbb{R}^{n+s}} |y|^{p(1-\{l\})-s}\big(|\nabla_{[l]+1} U|^p + |U|^p\big)\, dz\Big)^{1/p}. \tag{4.2.6}$$

Another classical property of B^l_p is the following Sobolev type imbedding result.

Proposition 4.2.5. (see [Bes], [Tr3], Sect. 2.8). *The inequality*

$$\|u\|_{L_q} \le c\,\|u\|_{B_p^l}$$

holds with $p > 1$ *and*

$$\begin{aligned} q = pn/(n-pl) \quad &\text{if} \quad n > pl, \\ q \in [p, \infty) \quad &\text{if} \quad n = pl, \\ q \in [p, \infty] \quad &\text{if} \quad n < pl. \end{aligned}$$

The next assertion follows from (4.2.4) and Theorem 3.1.2.

Proposition 4.2.6. *Let* $\{\eta_j\}_{j\ge 0}$ *be the same sequence as in Theorem* 3.1.2. *Then*

$$\|u\|_{B_p^l} \sim \Big(\sum_{j\ge 0} \|u\eta_j\|^p_{B_p^l}\Big)^{1/p}.$$

One can prove either by the Banach isomorphism theorem or directly that the equivalence relation

$$\|u\|_{B_p^l} \sim \|\mathfrak{D}_{p,l} u\|_{L_p} \tag{4.2.7}$$

holds for functions $u \in B_p^l$ with supports in the unit ball. This gives

Proposition 4.2.7. *Let* $u \in B_p^m$ *and* $\operatorname{supp} u \subset \mathcal{B}_\delta$. *Then for any* $k \in (0, l]$

$$\delta^k \|\mathfrak{D}_{p,l} u\|_{L_p} + \|u\|_{L_p} \le c\,\delta^l\, \|\mathfrak{D}_{p,l} u\|_{L_p}. \tag{4.2.8}$$

This and Proposition 4.2.6 imply

Corollary 4.2.1. *Let* $\{\mathcal{B}_\delta^{(j)}\}$ *be a covering of* $\mathbb{R}^n$ *by open balls of radius* $\delta \in (0,1)$ *with finite multiplicity depending only on* n. *Further, let* $u^{(j)} \in W_p^l$ *and* $\operatorname{supp} u^{(j)} \subset \mathcal{B}_\delta^{(j)}$. *Then*

$$\|\sum_j u^{(j)}\|^p_{W_p^l} \le c \sum_j \|D_{p,l} u^{(j)}\|^p_{L_p}.$$

4.2.2 Properties of the Operators $\mathfrak{D}_{p,l}$ and $D_{p,l}$

We start this section with a composition property of the operator $D_{p,l}$.

Lemma 4.2.1. *For any* α, $\beta > 0$ *with* $\alpha + \beta < 1$ *the inequality*

$$\|D_{p,\alpha} D_{p,\beta} u\|_{L_p} \le c\,\|D_{p,\alpha+\beta} u\|_{L_p}$$

holds.

Proof. Let $t \in \mathbb{R}^n$ and let $u_t(x) = u(x+t)$. We have

$$|(D_{p,\beta}u)(x) - (D_{p,\beta}u_t)(x)| \le \Big(\int |\Delta_h\big(u(x) - u_t(x)\big)|^p \frac{dh}{|h|^{n+p\beta}}\Big)^{1/p}.$$

Therefore,

$$\|D_{p,\alpha}D_{p,\beta}u\|^p_{L_p} \le \iiint |\Delta_h\big(u(x) - u_t(x)\big)|^p \frac{dhdtdx}{|h|^{n+p\beta}|t|^{n+p\alpha}}.$$

We write the integral over $\mathbb{R}^{3n}$ on the right-hand side as the sum of two integrals, one of which is taken over $|h| \le |t|$ and does not exceed

$$\int dx \int \frac{|\Delta_h u(x)|^p}{|h|^{n+p\beta}} dh \int_{|t|\ge|h|} \frac{dt}{|t|^{n+p\alpha}}$$

$$+ \int \frac{dh}{|h|^{n+p\beta}} \int_{|t|\ge|h|} \frac{dt}{|t|^{n+p\alpha}} \int |\Delta_h u_t(x)|^p dx.$$

Clearly, the second term coincides with the first one which in its turn is equal to

$$c\int dx \int |\Delta_h u(x)|^p \frac{dh}{|h|^{n+(\alpha+\beta)p}} = c\,\|D_{p,\alpha+\beta}u\|^p_{L_p}.$$

The integral over the set $|h| > |t|$ is estimated in the same way. The result follows. □

We proceed with elementary upper pointwise estimates for $\mathfrak{D}_{p,l}$ and $D_{p,l}$.

Lemma 4.2.2. *For any positive $l > 0$ and $m > 0$ the inequalities*

$$(\mathfrak{D}_{p,l}u)(x) \le \big(J_m \mathfrak{D}_{p,l}\Lambda^m u\big)(x), \tag{4.2.9}$$

$$(D_{p,l}u)(x) \le \big(J_m D_{p,l}\Lambda^m u\big)(x) \tag{4.2.10}$$

hold with J_m and Λ^m defined in Sects. 1.2.1 *and* 3.1, *respectively.*

Proof. Let $f = \Lambda^m u$ and let $l = k + \alpha$, $\alpha \in (0,1]$. Clearly,

$$(\mathfrak{D}_{p,l}u)(x) = (\mathfrak{D}_{p,l}J_m f)(x)$$

$$= \Big(\int \Big|\int \big(G_m(x-\xi+2h) - 2G_m(x-\xi+h) + G_m(x-\xi)\big)\nabla_k f(\xi)d\xi\Big|^p \frac{dh}{|h|^{n+p\alpha}}\Big)^{1/p}$$

$$= \Big(\int \Big|\int G_m(x-\xi)\big(\nabla_k f(\xi+2h) - 2\nabla_k f(\xi+h) + \nabla_k f(\xi)\big)d\xi\Big|^p \frac{dh}{|h|^{n+p\alpha}}\Big)^{1/p}.$$

Using Minkowski's inequality, we obtain $\mathfrak{D}_{p,l}u \le J_m\mathfrak{D}_{p,l}f$. An analogous estimate for $D_{p,l}u$ with $\{l\} > 0$ does not need a separate proof in view of (4.1.3). □

4.2.3 Pointwise Estimate for Bessel Potentials

Henceforth we need the following modification of Hedberg's inequality (1.2.17).

Lemma 4.2.3. *Let $\mathcal{M}$ be the Hardy–Littlewood maximal operator in $\mathbb{R}^n$ and let $J_r^{(n+s)}$ denote the Bessel potential in $\mathbb{R}^{n+s}$, $s \geq 1$. Then, for any nonnegative function $f \in L_p(\mathbb{R}^{n+s})$, $p > 1$, and for almost all $x \in \mathbb{R}^n$*

$$(J^{(n+s)}_{r\theta+s/p}f)(x,0) \leq c\big((J^{(n+s)}_{r+s/p}f)(x,0)\big)^\theta \big(\mathcal{M}F(x)\big)^{1-\theta}, \tag{4.2.11}$$

where $F(x) = \|f(x,\cdot);\mathbb{R}^s\|_{L_p}$ and $0 < \theta < 1$.

Proof. Let $\delta \in (0,1]$ and let

$$E_\delta(x) = \{\zeta = (\xi,\eta) : \xi \in \mathbb{R}^n,\ \ \eta \in \mathbb{R}^s,\ \ (x-\xi)^2 + \eta^2 > \delta^2\}.$$

We express the potential on the left-hand side of (4.2.11) as the sum of two integrals one of which is over $E_\delta(x)$. Let $r\theta < n+s/p'$. Then in view of (1.2.4) we have

$$\int_{\mathbb{R}^{n+s}\setminus E_\delta(x)} G_{r\theta+s/p}(x-\xi,\eta)\, f(\xi,\eta)\, d\xi\, d\eta$$

$$\leq c\int_{\mathcal{B}_\delta(x)} F(\xi)\Big(\int_{|\eta|<\delta}\frac{d\eta}{(|x-\xi|+|\eta|)^{p'q}}\Big)^{1/p'} d\xi, \tag{4.2.12}$$

where $p' = p/(p-1)$ and $q = n - r\theta + s/p'$.

One checks directly that

$$\Big(\int_{|\eta|<\delta}\frac{d\eta}{(|x-\xi|+|\eta|)^{p'q}}\Big)^{1/p'} \leq \begin{cases} c\,|x-\xi|^{r\theta-n} & r\theta < n,\\ c\log\dfrac{2\delta}{|x-\xi|}, & r\theta = n,\\ c\,\delta^{r\theta-n}, & r\theta > n.\end{cases} \tag{4.2.13}$$

In the case $r\theta < n$ it follows from (4.2.13) and (1.2.19) that the right-hand side of (4.2.12) does not exceed

$$c\int_{\mathcal{B}_\delta(x)}\frac{F(\xi)\,d\xi}{|x-\xi|^{n-r\theta}} \leq c\,\delta^{r\theta}(\mathcal{M}F)(x).$$

The same estimate follows from (4.2.12), (4.2.13) for $r\theta \geq n$.

Now let $r\theta = n + s/p'$. By (4.2.13) and (1.2.4) with $t = \delta$, the left-hand side of (4.2.12) is not greater than

$$c\int_{\mathcal{B}_\delta(x)} F(\xi)\Big(\int_{|\eta|<\delta} |\log c\,(|x-\xi|+|\eta|)^{p'}\,d\eta\Big)^{1/p'} d\xi$$

$$\leq c\,(1+|\log\delta|)\,\delta^{s/p'}\int_{\mathcal{B}_\delta(x)} F(\xi)\,d\xi \leq c\,(1+|\log\delta|)\,\delta^{r\theta}(\mathcal{M}F)(x).$$

If $r\theta > n + s/p'$, the right-hand side of (4.2.12) is dominated by

$$c\,\delta^{s/p'} \int_{\mathcal{B}_\delta(x)} F(\xi)\, d\xi \le c\,\delta^{n+s/p'} (\mathcal{M}F)(x).$$

Thus, for $\delta \le 1$

$$\int_{\mathbb{R}^{n+s}\setminus E_\delta(x)} G_{r\theta+s/p}(x-\xi,\eta)\, f(\xi,\eta)\, d\xi\, d\eta$$

$$\le \begin{cases} c\,\delta^{r\theta} (\mathcal{M}F)(x), & r\theta < n + s/p', \\ c\,(1 + |\log\delta|)\, \delta^{r\theta} (\mathcal{M}F)(x), & r\theta = n + s/p', \\ c\,\delta^{n+s/p'} (\mathcal{M}F)(x), & r\theta > n + s/p'. \end{cases} \tag{4.2.14}$$

Next we estimate the integral over $E_\delta(x)$. In the case $r\theta < n + s/p'$ we have

$$\int_{E_\delta(x)} G_{r\theta+s/p}(x-\xi,\eta)\, f(\xi,\eta)\, d\xi\, d\eta \le c \int_{E_\delta(x)\setminus E_1(x)} \frac{f(\xi,\eta)\, d\xi\, d\eta}{(|x-\xi| + |\eta|)^{n-r\theta+s/p'}}$$

$$+ c \int_{E_1(x)} e^{-\sqrt{(x-\xi)^2+\eta^2}} \frac{f(\xi,\eta)\, d\xi\, d\eta}{(|x-\xi|+|\eta|)^{(n+1-r\theta+s/p')/2}}$$

$$\le c\,\delta^{-r(1-\theta)} \int_{E_\delta(x)\setminus E_1(x)} \frac{f(\xi,\eta)\, d\xi\, d\eta}{(|x-\xi|+|\eta|)^{n-r+s/p'}}$$

$$+ c\,\delta^{-r(1-\theta)/2} \int_{E_1(x)} e^{-\sqrt{(x-\xi)^2+\eta^2}} \frac{f(\xi,\eta)\, d\xi\, d\eta}{(|x-\xi|+|\eta|)^{(n+1-r+s/p')/2}}$$

$$\le c\,\delta^{-r(1-\theta)} (J^{(n+s)}_{r+s/p} f)(x,0).$$

The case $r\theta \ge n + s/p'$ is treated the same way. As a result we have

$$\int_{E_\delta(x)} G_{r\theta+s/p}(x-\xi,\eta)\, f(\xi,\eta)\, d\xi\, d\eta$$

$$\le \begin{cases} c\,\delta^{-r(1-\theta)} (J^{(n+s)}_{r+s/p} f)(x,0), & r\theta < n + s/p', \\ c\,(1 + |\log\delta|)\, (J^{(n+s)}_{r+s/p} f)(x,0), & r\theta = n + s/p', \\ c\,(J^{(n+s)}_{r+s/p} f)(x,0), & r\theta > n + s/p'. \end{cases} \tag{4.2.15}$$

Further, we recall that $G_r(z) = O(e^{-c|z|})$ for $|z| > 1$ to obtain

$$\int_{E_1(x)} G_{r\theta+s/p}(x-\xi,\eta)\, f(\xi,\eta)\, d\xi\, d\eta \le c \int_{E_1(x)} F(\xi) e^{-c|x-\xi|} d\xi$$

$$= c \sum_{j=0}^{\infty} \int_{2^j < |x-\xi| < 2^{j+1}} e^{-c|x-\xi|} F(\xi)\, d\xi \le c \sum_{j=0}^{\infty} e^{-c2^j}\, 2^{nj} (\mathcal{M}F)(x) = c\,(\mathcal{M}F)(x).$$

Combining this with (4.2.14), where $\delta = 1$, we arrive at

$$\big(J^{(n+s)}_{r\theta+s/p} f\big)(x,0) \le c\,(\mathcal{M}F)(x). \tag{4.2.16}$$

In the case $r\theta > n + s/p'$ we have

$$\int_{\mathbb{R}^{n+s}\setminus E_1(x)} G_{r\theta+s/p}(x-\xi,\eta)\, f(\xi,\eta)\, d\xi\, d\eta$$

$$\le \int_{\mathbb{R}^{n+s}\setminus E_1(x)} f(\xi,\eta)\, d\xi\, d\eta \le c\,(J^{(n+s)}_{r\theta+s/p} f)(x,0)$$

which together with (4.2.15) yields

$$\big(J^{(n+s)}_{r\theta+s/p} f\big)(x,0) \le c\,\big(J^{(n+s)}_{r+s/p} f\big)(x,0). \tag{4.2.17}$$

For $r\theta < n + s/p'$ estimates (4.2.14) –(4.2.16) imply that

$$\big(J^{(n+s)}_{r\theta+s/p} f\big)(x,0) \le c\,\big(\delta^{r\theta}(\mathcal{M}F)(x) + \delta^{-r(1-\theta)}(J^{(n+s)}_{r+s/p} f)(x,0)\big)$$

for all $\delta \in (0,\infty)$. Minimizing the right-hand side with respect to δ, we arrive at (4.2.11).

If $r\theta = n + s/p'$, then by (4.2.14) and (4.2.15)

$$(J^{(n+s)}_{r\theta+s/p} f)(x,0) \le c\,(1 + |\log\delta|)\big(\delta^{r\theta}(\mathcal{M}F)(x) + (J^{(n+s)}_{r+s/p} f)(x,0)\big) \tag{4.2.18}$$

for all $\delta \in (0,1)$. In the case

$$(\mathcal{M}F)(x) \le (J^{(n+s)}_{r+s/p} f)(x,0),$$

by (4.2.16) we have

$$\big(J^{(n+s)}_{r\theta+s/p} f\big)(x,0) \le c\,\big(J^{(n+s)}_{r+s/p} f\big)(x,0) \tag{4.2.19}$$

and (4.2.11) follows from (4.2.16) and (4.2.19). Let

$$(\mathcal{M}F)(x) > (J^{(n+s)}_{r+s/p} f)(x,0).$$

We put

$$\delta^{r\theta} = (J^{(n+s)}_{r+s/p} f)(x,0)/(\mathcal{M}F)(x)$$

in (4.2.18). Then

$$\big(J^{(n+s)}_{r\theta+s/p} f\big)(x,0) \le c\,(1 + |\log\delta|)(J^{(n+s)}_{r+s/p}) f(x,0)$$

$$= c\,(1 + |\log\delta|)\delta^{r\theta(1-\theta)}\big((J^{(n+s)}_{r+s/p} f)(x,0)\big)^{\theta}\big((\mathcal{M}F)(x)\big)^{1-\theta}.$$

Since $\delta \in (0,1]$, inequality (4.2.11) is proved.

In the case $r\theta > n + s/p'$ estimate (4.2.11) results from (4.2.16) and (4.2.17). $\square$

4.3 Proof of Theorem 4.1.1

4.3.1 Estimate for the Product of First Differences

While deriving estimates involving the second difference of the product of two functions, we need to estimate the product of their first differences, which is based on the next lemma.

Lemma 4.3.1. *For $p \in (1, \infty)$, $\delta \in (0, 1)$ and any integer $k \geq 1$, we have*

$$\Big(\int\int |\varDelta_h \gamma(x) \varDelta_h u(x)|^p \frac{dhdx}{|h|^{n+p}}\Big)^{1/p} \leq c \sup_e \frac{\|\mathfrak{D}_{p,\delta}\gamma; e\|_{L_p}}{[C_{p,k-1+\delta}(e)]^{1/p}} \|u\|_{B_p^k}. \quad (4.3.1)$$

Proof. Let $U \in C_0^\infty(\mathbb{R}^{n+s})$ be an extension of a function $u \in B_p^k(\mathbb{R}^n)$ to $\mathbb{R}^{n+s}$ with k subject to $k < n + s/p'$. By f we denote the function $\varLambda^{k+s/p}U$. Then

$$u(x) = \int_{\mathbb{R}^{n+s}} G_{k+s/p}(x - \xi, \eta) f(\xi, \eta) d\xi d\eta.$$

We shall use the properties of the function G_r listed in Sect. 1.2.1. Let us introduce the sets

$$\mathcal{N}_1 = \{(\xi, \eta) : 4|h| < |x - \xi| + |\eta| < 1\},$$

$$\mathcal{N}_2 = \{(\xi, \eta) : |x - \xi| + |\eta| < \min(1, 4|h|)\},$$

$$\mathcal{N}_3 = \{(\xi, \eta) : |x - \xi| + |\eta| > \max(1, 4|h|)\},$$

$$\mathcal{N}_4 = \{(\xi, \eta) : 4|h| > |\xi - x| + |\eta| > 1\}.$$

It is clear that

$$|\varDelta_h u(x)| \leq \int_{\mathbb{R}^{n+s}} \big|G_{k+s/p}(x - \xi + h, \eta) - G_{k+s/p}(x - \xi, \eta)\big| \, |f(\xi, \eta)| d\xi d\eta.$$

We represent the last integral as the sum of four integrals over $\mathcal{N}_1, \ldots, \mathcal{N}_4$ and estimate each of them. First,

$$\int_{\mathcal{N}_1} \leq c\,|h| \int_{\mathcal{N}_1} t_\theta^{-(n-k+1+s/p')} |f(\xi, \eta)| d\xi d\eta,$$

where $t_\theta^2 = (x + \theta h - \xi)^2 + \eta^2$, $\theta \in (0, 1)$. Since $t_\theta^2 \geq c\big((x - \xi)^2 + \eta^2\big)$ on $\mathcal{N}_1$, it follows that

$$\int_{\mathcal{N}_1} \leq c\,|h|^{1-\delta} \int_{\mathcal{N}_1} t_0^{-(n-k+1+s/p'-\delta)} |f(\xi, \eta)| d\xi d\eta$$

$$\leq c\,|h|^{1-\delta} \big(\varLambda^{-(k-1+s/p+\delta)} |f|\big)(x, 0). \quad (4.3.2)$$

The integral over $\mathcal{N}_2$ is dominated by

$$\int_{\mathcal{N}_2} (t_1^{-(n-k+s/p')} + t_0^{-(n-k+s/p')})|f(\xi,\eta)|d\xi d\eta$$

$$\leq c\,|h|^{1-\delta} \int_{\mathcal{N}_2} (t_1^{-(n-k+s/p'+1-\delta)} + t_0^{-(n-k+s/p'+1-\delta)}|f(\xi,\eta)|d\xi d\eta.$$

Consequently,

$$\int_{\mathcal{N}_2} \leq c|h|^{1-\delta}\big((\Lambda^{-(k-1+s/p+\delta)}|f|)(x+h,0)+(\Lambda^{-(k-1+s/p+\delta)}|f|)(x,0)\big). \quad (4.3.3)$$

Using (2.8.5), we obtain

$$\int_{\mathcal{N}_3} \leq c\,|h| \int_{\mathcal{N}_3} e^{-t_0/2}|f(\xi,\eta)|d\xi d\eta.$$

Hence,

$$\int_{\mathcal{N}_3} \leq c\,|h|^{1-\delta} \int_{\mathcal{N}_3} t_0^{(k-n-s/p'-1+\delta)/2} e^{-t_0/4}|f(\xi,\eta)|d\xi d\eta$$

$$\leq c\,|h|^{1-\delta}(\Lambda^{-(k-1+s/p+\delta)}|F|)(\frac{x}{4},0), \quad (4.3.4)$$

where $F(\xi,\eta) = f(4\xi,4\eta)$. In a similar way we find that

$$\int_{\mathcal{N}_4} \leq c \int_{\mathcal{N}_4} (e^{-t_1/2} + e^{-t_0/2})|f(\xi,\eta)|d\xi d\eta$$

$$\leq c\,|h|^{1-\delta}\int_{\mathcal{N}_4} (t_1^{(k-n-s/p'-1+\delta)/2}e^{-t_1/4}+t_0^{(k-n-s/p'-1+\delta)/2}e^{-t_0/4})|f(\xi,\eta)|d\xi d\eta,$$

and consequently

$$\int_{\mathcal{N}_4} \leq c\,|h|^{1-\delta}\Big((\Lambda^{-(k-1+s/p+\delta)}|F|)(\frac{x+h}{4},0) + (\Lambda^{-(k-1+s/p+\delta)}|F|)(\frac{x}{4},0)\Big).$$

Adding the last inequality to (4.3.2)–(4.3.4) and noting that $G_r(az) \geq G_r(z)$ for any constant $a < 1$, we conclude that

$$|\Delta_h u(x)| \leq c\,|h|^{1-\delta}\Big((\Lambda^{-(k-1+s/p+\delta)}|F|)(\frac{x+h}{4},0)+(\Lambda^{-(k-1+s/p+\delta)}|F|)(\frac{x}{4},0)\Big).$$

Hence,

$$\int\int |\Delta_h\gamma(x)\Delta_h u(x)|^p \frac{dhdx}{|h|^{n+p}}$$

$$\leq c\int \Big((\Lambda^{-(k-1+s/p+\delta)}|F|)(\frac{x}{4},0)\Big)^p \int |\Delta_h\gamma(x)|^p \frac{dhdx}{|h|^{n+p\delta}}$$

$$= c_1 \int ((\Lambda^{-(k-1+s/p+\delta)}|F|)(x,0))^p \int |\Delta_h\gamma(4x)|^p \frac{dhdx}{|h|^{n+p\delta}}.$$

By (4.3.12),

$$\int\int |\Delta_h\gamma(x)\Delta_h u(x)|^p \frac{dhdx}{|h|^{n+p}}$$

$$\leq c\sup_e \frac{\int_e\int |\Delta_h\gamma(4x)|^p \frac{dhdx}{|h|^{n+p\delta}}}{C_{p,k-1+\delta}(e)} \|(\Lambda^{-(k-1+s/p+\delta)}|F|)(\cdot,0)\|^p_{W_p^{k-1+\delta}}. \quad (4.3.5)$$

Clearly,

$$\int_e\int |\Delta_h\gamma(4x)|^p \frac{dhdx}{|h|^{n+p\delta}} = c\int_{4e}\int |\Delta_h\gamma(x)|^p \frac{dhdx}{|h|^{n+p\delta}} \quad (4.3.6)$$

and

$$C_{p,k-1+\delta}(e) \geq c\, C_{p,k-1+\delta}(4e), \quad (4.3.7)$$

where $4e = \{x : x/4 \in e\}$. Note also that

$$\|(\Lambda^{-(k-1+s/p+\delta)}|F|)(\cdot,0)\|_{W_p^{k-1+\delta}} \leq c\,\|F;\mathbb{R}^{n+s}\|_{L_p}$$

$$= c\,4^{-(n+s)}\|f;\mathbb{R}^{n+s}\|_{L_p} = c\,4^{-(n+s)}\|\Lambda^{k+s/p}U;\mathbb{R}^{n+s}\|_{L_p}.$$

This, together with (4.3.5)–(4.3.7), implies that

$$\int\int |\Delta_h\gamma(x)\Delta_h u(x)|^p \frac{dhdx}{|h|^{n+p}}$$

$$\leq c\,\sup_e \frac{\|\mathfrak{D}_{p,\delta}\gamma; e\|^p_{L_p}}{C_{p,k-1+\delta}(e)} \|\Lambda^{k+s/p}U;\mathbb{R}^{n+s}\|^p_{L_p}.$$

Minimizing the last norm over all extensions U, we complete the proof. □

4.3.2 Trace Inequality for B_p^k, $p > 1$

Lemma 4.3.2. *Let μ be a measure in $\mathbb{R}^n$, $p \in (1,\infty)$, $k \in (0,\infty)$. The best constant C in the inequality*

$$\int |u|^p\, d\mu \leq C\,\|u\|^p_{B_p^k}, \qquad u \in C_0^\infty, \quad (4.3.8)$$

is equivalent to

$$\sup \frac{\mu(e)}{C_{p,k}(e)}, \quad (4.3.9)$$

where e is an arbitrary compact set of positive capacity $C_{p,k}(e)$.

Proof. Let $U \in C_0^\infty(\mathbb{R}^{n+1})$ be an arbitrary extension of $u \in C_0^\infty(\mathbb{R}^n)$ to $\mathbb{R}^{n+1}$. By Theorem 3.1.4,

$$\int_{\mathbb{R}^n} |u|^p \, d\mu \le c \sup_{e \subset \mathbb{R}^n} \frac{\mu(e)}{C^{(n+1)}_{p,k+1/p}(e)} \|U; \mathbb{R}^{n+1}\|^p_{H_p^{k+1/p}}.$$

In the present proof we use the notation $C^{(n+1)}_{p,k+1/p}(e)$ temporarily in order to stress that the functions in the definition (1.2.6) of the capacity are given on $\mathbb{R}^{n+1}$ instead of $\mathbb{R}^n$. Minimizing the right-hand side over all extensions of u and applying the relation

$$C_{p,k}(e) \sim C^{(n+1)}_{p,k+1/p}(e), \tag{4.3.10}$$

which follows from the definition of capacity and the equivalence

$$\|u; \mathbb{R}^n\|_{B_p^k} \sim \inf_{\{U\}} \|U; \mathbb{R}^{n+1}\|_{H_p^{k+1/p}}, \tag{4.3.11}$$

we obtain the required upper bound for C. The lower bound is obvious. □

Remark 4.3.1. The above result can be found in [Maz9] and [Maz11]. For its generalizations to Besov spaces with three indices see [Wu] and [AX].

By Lemma 4.3.2,

$$\|\gamma\|_{M(B_p^k \to L_p)} \sim \sup_e \frac{\|\gamma; e\|_{L_p}}{[C_{p,k}(e)]^{1/p}} \tag{4.3.12}$$

for $p \in (1, \infty)$. Hence, in view of Lemma 3.2.2,

$$\|\gamma\|_{M(B_p^k \to L_p)} \sim \|\gamma\|_{M(H_p^k \to L_p)} \sim \sup_{e, \operatorname{diam}(e) \le 1} \frac{\|\gamma; e\|_{L_p}}{[C_{p,k}(e)]^{1/p}}. \tag{4.3.13}$$

By Proposition 3.1.4 containing the estimates for the capacity of a ball, one obtains the following relations from (4.3.12):

if $pk > n$, then

$$\|\gamma\|_{M(B_p^k \to L_p)} \sim \|\gamma\|_{L_{p,\mathrm{unif}}}; \tag{4.3.14}$$

if $pk < n$, then

$$\|\gamma\|_{M(B_p^k \to L_p)} \ge c \sup_{x \in \mathbb{R}^n, r \in (0,1)} r^{k-n/p} \|\gamma; \mathcal{B}_r(x)\|_{L_p}; \tag{4.3.15}$$

if $pk = n$, then

$$\|\gamma\|_{M(B_p^k \to L_p)} \ge c \sup_{x \in \mathbb{R}^n, r \in (0,1)} \Big(\log \frac{2}{r}\Big)^{(p-1)/p} \|\gamma; \mathcal{B}_r(x)\|_{L_p}. \tag{4.3.16}$$

By Propositions 3.1.2, 3.1.3 the following upper estimates for the norm in $M(B_p^k \to L_p)$ hold:

if $pk < n$, then

$$\|\gamma\|_{M(B_p^k \to L_p)} \leq c \sup_{e, \mathrm{diam}(e) \leq 1} (\mathrm{mes}_n e)^{k/n-1/p} \|\gamma; e\|_{L_p}; \tag{4.3.17}$$

if $pk = n$, then

$$\|\gamma\|_{M(B_p^k \to L_p)} \leq c \sup_{e, \mathrm{diam}(e) \leq 1} \Big(\log \frac{2^n}{\mathrm{mes}_n e}\Big)^{(p-1)/p} \|\gamma; e\|_{L_p}. \tag{4.3.18}$$

4.3.3 Auxiliary Assertions Concerning $M(B_p^m \to B_p^l)$

We start with inequalities for mollifiers of multipliers.

Lemma 4.3.3. *Let γ_ρ denote a mollifier of a function γ which is defined as*

$$\gamma_\rho(x) = \rho^{-n} \int K\big(\rho^{-1}(x-\xi)\big)\gamma(\xi)d\xi,$$

where $K \in C_0^\infty(\mathcal{B}_1)$, $K \geq 0$, and $\|K\|_{L_1} = 1$. The inequalities

$$\|\gamma_\rho\|_{M(B_p^m \to B_p^l)} \leq \|\gamma\|_{M(B_p^m \to B_p^l)} \leq \liminf_{\rho \to 0} \|\gamma_\rho\|_{M(B_p^m \to B_p^l)}, \tag{4.3.19}$$

$$\|\gamma_\rho\|_{M(B_p^m \to L_p)} \leq \|\gamma\|_{M(B_p^m \to L_p)} \leq \liminf_{\rho \to 0} \|\gamma_\rho\|_{M(B_p^m \to L_p)}, \tag{4.3.20}$$

and

$$\sup_e \frac{\|\mathfrak{D}_{p,l}\gamma_\rho; e\|_{L_p}}{[C_{p,m}(e)]^{1/p}} \leq \sup_e \frac{\|\mathfrak{D}_{p,l}\gamma; e\|_{L_p}}{[C_{p,m}(e)]^{1/p}} \tag{4.3.21}$$

hold.

Proof. Let $u \in C_0^\infty$. Clearly,

$$\|\gamma_\rho u\|_{B_p^l} = \Big(\int\int \Big|\rho^{-n} \int K\Big(\frac{\xi}{\rho}\Big) \nabla_{l-1,x} \Delta_h^{(2)} (\gamma(x-\xi)u(x))d\xi\Big|^p \frac{dh}{|h|^{n+1}} dx\Big)^{1/p}$$
$$+ \Big(\int\Big|\int \rho^{-n} K\Big(\frac{\xi}{\rho}\Big) \gamma(x-\xi)u(x)d\xi\Big|^p dx\Big)^{1/p}. \tag{4.3.22}$$

By Minkowski's inequality,

$$\|\mathfrak{D}_{p,l}(\gamma_\rho u)\|_{L_p} \leq \rho^{-n} \int K\Big(\frac{\xi}{\rho}\Big) \|\mathfrak{D}_{p,l}\big(\gamma(\cdot - \xi)u(\cdot)\big)\|_{L_p} d\xi.$$

This and (4.3.22) imply that

$$\|\gamma_\rho u\|_{B_p^l} \leq \rho^{-n} \int K\Big(\frac{\xi}{\rho}\Big) \|\gamma(\cdot - \xi)u(\cdot)\|_{B_p^l} d\xi.$$

Since

$$\|\gamma(\cdot - \xi)u(\cdot)\|_{B_p^l} \leq \|\gamma\|_{M(B_p^m \to B_p^l)} \|u\|_{B_p^m},$$

the left inequality (4.3.19) follows. One can prove the right inequality (4.3.19) duplicating the argument in Lemma 2.3.1. The proof of (4.3.20) is obvious.

To derive the estimate (4.3.21) we use Minkowski's inequality once more to obtain

$$\frac{\|\mathfrak{D}_{p,l}\gamma_\rho; e\|_{L_p}}{[C_{p,m}(e)]^{1/p}} \leq \frac{\int K(z)\Big(\int_e (\mathfrak{D}_{p,l}\gamma(x - \rho z))^p dx\Big)^{1/p} dz}{[C_{p,m}(e)]^{1/p}}$$

$$\leq \frac{\int_{\mathcal{B}_1} K(z)\Big(\int_E (\mathfrak{D}_{p,l}\gamma(\xi))^p d\xi\Big)^{1/p} dz}{[C_{p,m}(E)]^{1/p}} \leq \|K\|_{L_1} \sup_e \frac{\|\mathfrak{D}_{p,l}\gamma; e\|_{L_p}}{[C_{p,m}(e)]^{1/p}}$$

where $E = \{x - \rho z : x \in e, z \in \mathcal{B}_1\}$. The proof is complete. □

We use the interpolation properties

$$B_p^{m-k} = \Big(B_p^m,\, H_p^{m-l}\Big)_{k/l,p} \tag{4.3.23}$$

and

$$B_p^{m-k} = \Big(B_p^m,\, B_p^{m-l}\Big)_{k/l,p}, \tag{4.3.24}$$

where $l < k < m$ (see, [T], Th. 2.4.2). In particular, (4.3.24) implies that

$$\|\gamma\|_{MB_p^r} \leq c\, \|\gamma\|_{MB_p^\sigma}^\theta\, \|\gamma\|_{MB_p^\tau}^{1-\theta}, \tag{4.3.25}$$

where $p \in (1, \infty)$, $\sigma > \tau > 0$, $0 < \theta < 1$, and $r = \theta\sigma + (1 - \theta)\tau$. It follows from (4.3.12) and (4.3.24) that $\gamma \in M(B_p^m \to B_p^l) \cap M(B_p^{m-l} \to L_p)$ implies that $\gamma \in M(B_p^{m-k} \to B_p^{l-k})$ for $0 < k < l$. Moreover,

$$\|\gamma\|_{M(B_p^{m-k} \to B_p^{l-k})} \leq c\, \|\gamma\|_{M(B_p^m \to B_p^l)}^{1-k/l} \|\gamma\|_{M(B_p^{m-l} \to L_p)}^{k/l} \tag{4.3.26}$$

for $0 < k < l < m$ and

$$\|\gamma\|_{MB_p^{l-k}} \leq c\, \|\gamma\|_{MB_p^l}^{1-k/l}\, \|\gamma\|_{L_\infty}^{k/l} \tag{4.3.27}$$

for $0 < k < l$.

4.3.4 Lower Estimates for the Norm in $M(B_p^m \to B_p^l)$

The following is the main result of this subsection.

Lemma 4.3.4. *Let $0 < l \le m$ and $p \in (1, \infty)$. Then*

$$\|\gamma\|_{L_\infty} \le \|\gamma\|_{MB_p^l} \qquad \textit{for } \; m = l \tag{4.3.28}$$

and

$$\|\gamma\|_{M(B_p^{m-l} \to L_p)} \le c\,\|\gamma\|_{M(B_p^m \to B_p^l)} \qquad \textit{for } \; m > l. \tag{4.3.29}$$

Proof. Let $u \in B_p^l$ and let N be a positive integer. Clearly,

$$\|\gamma^N u\|_{L_p}^{1/N} \le \|\gamma^N u\|_{B_p^l}^{1/N} \le \|\gamma\|_{MB_p^l}\, \|u\|_{B_p^l}^{1/N}.$$

Passing to the limit as $N \to \infty$ we arrive at (4.3.28).

Now suppose that $0 < l < m$. Let γ_ρ be the mollification of $\gamma \in M(B_p^m \to B_p^l)$. By Lemma 4.3.3, it suffices to prove (4.3.29) for γ_ρ. To simplify the notation we write γ in place of γ_ρ.

We consider two cases: $m \ge 2l$ and $2l > m > l$. Assume first that $m \ge 2l$. In view of Lemma 4.2.4, there exists $U \in H_p^{m-l+1/p}(\mathbb{R}^{n+1})$ which is an extension of the function $u \in B_p^{m-l}(\mathbb{R}^n)$ to $\mathbb{R}^{n+1}$ such that

$$\|U; \mathbb{R}^{n+1}\|_{H_p^{m-l+1/p}} \le c\,\|u; \mathbb{R}^n\|_{B_p^{m-l}}. \tag{4.3.30}$$

By the same lemma, the converse estimate

$$\|u; \mathbb{R}^n\|_{B_p^{m-l}} \le c\,\|U; \mathbb{R}^{n+1}\|_{H_p^{m-l+1/p}} \tag{4.3.31}$$

holds for all extensions U. Let us take U as the Bessel potential $J_{m-l+1/p}^{(n+1)} f$ with density $f \in L_p(\mathbb{R}^{n+1})$. By Lemma 4.2.3,

$$|u(x)| \le c\left((J_{m+1/p}^{(n+1)}|f|)(x,0)\right)^{(m-l)/l} \left(\mathcal{M}F(x)\right)^{l/m},$$

where $F(x) = \|f(x,\cdot); \mathbb{R}^1\|_{L_p}$. Therefore,

$$\|\gamma u\|_{L_p} \le c\,\|f; \mathbb{R}^{n+1}\|_{L_p}^{l/m}\, \|\, |\gamma|^{l/(m-l)} (J_{m+1/p}^{(n+1)}|f|)(\cdot,0)\|_{L_p}^{(m-l)/m}.$$

According to Lemma 4.3.2, the right-hand side does not exceed

$$c\,\|f; \mathbb{R}^{n+1}\|_{L_p}^{\frac{l}{m}}\, \|\gamma\, (J_{m+1/p}^{(n+1)}|f|)(\cdot,0)\|_{B_p^l}^{1-\frac{l}{m}} \sup_e \left(\frac{\displaystyle\int_e |\gamma|^{\frac{pl}{m-l}}\, dx}{C_{p,l}(e)} \right)^{\frac{m-l}{mp}}. \tag{4.3.32}$$

Setting $\varphi = |\gamma|^{1/(m-l)}$, $\nu = l$, $\mu = m - l$ in Lemma 2.3.6, which is valid for all ν and μ such that $0 < \mu < \nu$, we find that in the case $m \ge 2l$ the supremum in (4.3.32) is dominated by

$$c\left(\sup_e \frac{\displaystyle\int_e |\gamma|^p dx}{C_{p,m-l}(e)}\right)^{l/mp} \le c\,\|\gamma\|_{M(B_p^{m-l} \to L_p)}^{l/m}.$$

Therefore, by (4.3.32) we obtain

$$\|\gamma u\|_{L_p} \leq c\|f;\mathbb{R}^{n+1}\|_{L_p}^{\frac{l}{m}} \|\gamma\|_{M(B_p^m \to B_p^l)}^{1-\frac{l}{m}} \|J_{m+1/p}^{(n+1)}|f|(\cdot,0)\|_{B_p^m}^{1-\frac{l}{m}} \|\gamma\|_{M(B_p^{m-l}\to L_p)}^{\frac{l}{m}}.$$

Using first (4.3.31) and then the equality $\|u\|_{H_p^k} = \|\Lambda^k u\|_{L_p}$ and (4.3.30), we obtain

$$\|J_{m+1/p}^{(n+1)}|f|(\cdot,0)\|_{B_p^m} \leq c\,\|J_{m+1/p}^{(n+1)}|f|;\mathbb{R}^{n+1}\|_{H_p^{m+1/p}} = c\,\|f;\mathbb{R}^{n+1}\|_{L_p}$$

$$= c\,\|U;\mathbb{R}^{n+1}\|_{H_p^{m-l+1/p}} \leq c\,\|u;\mathbb{R}^n\|_{B_p^{m-l}}.$$

Thus,

$$\|\gamma u\|_{L_p} \leq c\,\|\gamma\|_{M(B_p^{m-l}\to L_p)}^{l/m} \|\gamma\|_{M(B_p^m\to B_p^l)}^{(m-l)/m} \|u\|_{B_p^{m-l}},$$

which implies (4.3.29) for $m \geq 2l$.

Suppose that $2l > m > l$. Let μ be an arbitrary positive number less than $m-l$. By (4.3.26) with $k = l-\mu$,

$$\|\gamma\|_{M(B_p^{m-l+\mu}\to B_p^\mu)} \leq c\,\|\gamma\|_{M(B_p^{m-l}\to L_p)}^{(l-\mu)/l} \|\gamma\|_{M(B_p^m\to B_p^l)}^{\mu/l}.$$

Since $m-l+\mu > 2\mu$, it follows from the first part of the proof that (4.3.29) holds with m and l replaced by $m-l+\mu$ and μ, respectively, i.e.

$$\|\gamma\|_{M(B_p^{m-l}\to L_p)} \leq c\,\|\gamma\|_{M(B_p^{m-l+\mu}\to B_p^\mu)}.$$

Consequently,

$$\|\gamma\|_{M(B_p^{m-l}\to L_p)} \leq c\,\|\gamma\|_{M(B_p^{m-l}\to L_p)}^{(l-\mu)/l} \|\gamma\|_{M(B_p^m\to B_p^l)}^{\mu/l}$$

and (4.3.29) is proved for $2l > m > l$ as well. □

By Lemma 4.3.4 and (4.3.12), the following assertion holds.

Corollary 4.3.1. *Let $\gamma \in M(B_p^m \to B_p^l)$, $0 < l < m$. Then*

$$\sup_e \frac{\|\gamma;e\|_{L_p}}{[C_{p,m-l}(e)]^{1/p}} \leq c\,\|\gamma\|_{M(B_p^m\to B_p^l)}.$$

Lemma 4.3.4 in combination with (4.3.26) and (4.3.27) implies:

Corollary 4.3.2. *Let $\gamma \in M(B_p^m \to B_p^l)$, $0 < l \leq m$. Then*

$$\gamma \in M(B_p^{m-k} \to B_p^{l-k}), \quad 0 < k < l,$$

and

$$\|\gamma\|_{M(B_p^{m-k}\to B_p^{l-k})} \leq c\,\|\gamma\|_{M(B_p^m\to B_p^l)}.$$

The following assertion contains an estimate for derivatives of a multiplier.

Lemma 4.3.5. *Let* $\gamma \in M(B_p^m \to B_p^l)$, $0 < l \le m$. *Then*

$$D^\alpha \gamma \in M(B_p^m \to B_p^{l-|\alpha|})$$

for any multi-index α *of order* $|\alpha| \le l$ *and*

$$\|D^\alpha \gamma\|_{M(B_p^m \to B_p^{l-|\alpha|})} \le c\, \|\gamma\|_{M(B_p^m \to B_p^l)}.$$

Proof. It suffices to consider the case $|\alpha| = 1$, $l \ge 1$. Clearly,

$$\begin{aligned}\|u\nabla\gamma\|_{B_p^{l-1}} &\le \|u\gamma\|_{B_p^l} + \|\gamma\nabla u\|_{B_p^{l-1}} \\ &\le \big(\|\gamma\|_{M(B_p^m \to B_p^l)} + \|\gamma\|_{M(B_p^{m-1} \to B_p^{l-1})}\big)\|u\|_{B_p^m}.\end{aligned}$$

Hence, using Corollary 4.3.2, we find that

$$\|u\nabla\gamma\|_{B_p^{l-1}} \le c\, \|\gamma\|_{M(B_p^m \to B_p^l)}\|u\|_{B_p^m}$$

which completes the proof. □

Corollary 4.3.3. *Let* $\gamma \in M(B_p^m \to B_p^l)$, $0 < l \le m$. *Then, for any* $\varepsilon > 0$ *and every multi-index* α *of order* $|\alpha| \le l$, $D^\alpha\gamma \in M(B_p^{m-l-|\alpha|} \to L_p)$, *and the inequality*

$$\|D^\alpha \gamma\|_{M(B_p^{m-l+|\alpha|} \to L_p)} \le \varepsilon\, \|\gamma\|_{M(B_p^m \to B_p^l)} + c(\varepsilon)\, \|\gamma\|_{M(B_p^m \to L_p)}$$

holds.

Proof. The result follows by Lemma 4.3.5 and inequality (4.3.26).

Lemma 4.3.4 and Corollary 4.3.3 imply the following assertion.

Corollary 4.3.4. *Let* $\gamma \in M(B_p^m \to B_p^l)$, $0 < l \le m$. *Then, for any multi-index* α *of order* $|\alpha| \le l$, $D^\alpha\gamma \in M(B_p^{m-l-|\alpha|} \to L_p)$, *and the inequality*

$$\|D^\alpha \gamma\|_{M(B_p^{m-l+|\alpha|} \to L_p)} \le c\, \|\gamma\|_{M(B_p^m \to B_p^l)}$$

holds.

4.3.5 Proof of Necessity in Theorem 4.1.1

In this section we derive the inequalities

$$\sup_e \frac{\|\mathfrak{D}_{p,l}\gamma; e\|_{L_p}}{[C_{p,m}(e)]^{1/p}} + \sup_{x\in\mathbb{R}^n} \|\gamma; \mathcal{B}_1(x)\|_{L_p} \le c\,\|\gamma\|_{M(B_p^m \to B_p^l)}, \quad m > l, \tag{4.3.33}$$

and

$$\sup_e \frac{\|\mathfrak{D}_{p,l}\gamma; e\|_{L_p}}{[C_{p,l}(e)]^{1/p}} + \|\gamma\|_{L_\infty} \le c\,\|\gamma\|_{MB_p^l}. \tag{4.3.34}$$

The core of the proof is the following assertion.

Lemma 4.3.6. *Let $\gamma \in M(B_p^m \to B_p^l)$, where $0 < l \le m$ and $p \in (1, \infty)$. Then*

$$\sup_e \frac{\|\mathfrak{D}_{p,l}\gamma; e\|_{L_p}}{[C_{p,m}(e)]^{1/p}} \le c\, \|\gamma\|_{M(B_p^m \to B_p^l)}. \tag{4.3.35}$$

Proof. We use induction on l and start by showing that (4.3.35) holds for $l \in (0, 1]$.

(*i*) Let $l \in (0, 1)$. We have

$$\begin{aligned} \|uD_{p,l}\gamma\|_{L_p} &\le c\big(\|\gamma u\|_{B_p^l} + \|\gamma D_{p,l}u\|_{L_p}\big) \\ &\le c\big(\|\gamma\|_{M(B_p^m \to B_p^l)} \|u\|_{B_p^l} + \|\gamma D_{p,l}u\|_{L_p}\big). \end{aligned} \tag{4.3.36}$$

Consider first the case $m = l$. Clearly,

$$\|\gamma D_{p,l}u\|_{L_p} \le \|\gamma\|_{L_\infty} \|u\|_{B_p^l}$$

which together with (4.3.36) and (4.3.28) gives

$$\|uD_{p,l}\gamma\|_{L_p} \le c\, \|\gamma\|_{MB_p^l} \|u\|_{B_p^l}.$$

Therefore,

$$\|D_{p,l}\gamma\|_{M(B_p^l \to L_p)} \le c\, \|\gamma\|_{MB_p^l}$$

and, in view of (4.3.12), we obtain (4.3.35).

Suppose now that $l < m$. By (4.2.9),

$$\|\gamma D_{p,l}u\|_{L_p} \le \|\gamma\|_{M(B_p^{m-l} \to L_p)} \|J_{m-l} D_{p,l} \Lambda^{m-l} u\|_{B_p^{m-l}}. \tag{4.3.37}$$

By Lemma 4.2.1, the last norm does not exceed

$$c\, \|D_{p,l}\Lambda^{m-l}u\|_{L_p} \le c\, \|\Lambda^{m-l}u\|_{B_p^l} \le c\, \|u\|_{B_p^m}$$

which in combination with (4.3.37) implies that

$$\|\gamma D_{p,l}u\|_{L_p} \le c\, \|\gamma\|_{M(B_p^{m-l} \to L_p)} \|u\|_{B_p^m}. \tag{4.3.38}$$

Using (4.3.36), (4.3.38) and Lemma 4.3.4, we arrive at

$$\|uD_{p,l}\gamma\|_{L_p} \le c\, \|\gamma\|_{M(B_p^m \to B_p^l)} \|u\|_{B_p^m}.$$

Thus,

$$\|D_{p,l}\gamma\|_{M(B_p^m \to L_p)} \le c\, \|\gamma\|_{M(B_p^m \to B_p^l)}$$

which together with (4.3.12) gives (4.3.35).

(*ii*) Let $l = 1$. In view of the identity

$$\Delta_h^{(2)}(\gamma u) = \gamma \Delta_h^{(2)} u + u \Delta_h^{(2)} \gamma + \Delta_{2h}\gamma \Delta_{2h} u - 2\Delta_h \gamma \Delta_h u \tag{4.3.39}$$

one has

$$\| u\mathfrak{D}_{p,1}\gamma\|_{L_p} \le \|\gamma u\|_{B^1_p} + \|\gamma\mathfrak{D}_{p,1}u\|_{L_p}$$
$$+ 4\Big(\int\int |\Delta_h\gamma(x)\Delta_h u(x)|^p|h|^{-n-p}dhdx\Big)^{1/p} \tag{4.3.40}$$

for any $u \in C_0^\infty$.

We proceed separately for $m = 1$ and $m > 1$. Let first $m = 1$. Using (4.3.1) with $k = 1$ and $\delta \in (0,1)$ together with (4.3.40) and (4.3.28), we find that

$$\|u\mathfrak{D}_{p,1}\gamma\|_{L_p} \le c\left(\|\gamma\|_{MB^1_p} + \sup_e \frac{\|\mathfrak{D}_{p,\delta}\gamma; e\|_{L_p}}{[C_{p,\delta}(e)]^{1/p}}\right)\|u\|_{B^1_p}. \tag{4.3.41}$$

In view of part (i) of this proof, the last supremum is majorized by $c\,\|\gamma\|_{MB^\delta_p}$. Hence (4.3.41) leads to the inequality

$$\sup_e \frac{\|\mathfrak{D}_{p,1}\gamma; e\|_{L_p}}{[C_{p,1}(e)]^{1/p}} \le c\,(\|\gamma\|_{MB^1_p} + \|\gamma\|_{MB^\delta_p}). \tag{4.3.42}$$

Since by Corollary 4.3.2

$$\|\gamma\|_{MB^\delta_p} \le c\,\|\gamma\|_{MB^1_p},$$

we arrive at (4.3.35) for $m = l = 1$.

Next we estimate the right-hand side of (4.3.40) for $m > 1$. By (4.2.9), its second term is majorized by

$$\|\gamma J_{m-1}\mathfrak{D}_{p,1}\Lambda^{m-1}u\|_{L_p} \le c\,\|\gamma\|_{M(B^{m-1}_p\to L_p)}\|J_{m-1}\mathfrak{D}_{p,1}\Lambda^{m-1}u\|_{B^{m-1}_p}$$
$$\le c\,\|\gamma\|_{M(B^{m-1}_p\to L_p)}\|\mathfrak{D}_{p,1}\Lambda^{m-1}u\|_{L_p}$$
$$\le c\,\|\gamma\|_{M(B^{m-1}_p\to L_p)}\|\Lambda^{m-1}u\|_{B^1_p} \le c\,\|\gamma\|_{M(B^m_p\to B^1_p)}\|u\|_{B^m_p}. \tag{4.3.43}$$

The last inequality in this chain follows from (4.2.1) and (4.3.29). We estimate the third term on the right-hand side of (4.3.40) using (4.3.1) with $k = m > 1$ and (4.3.35) with $l = \delta < 1$. Then this term does not exceed

$$c\sup_e \frac{\|\mathfrak{D}_{p,\delta}\gamma; e\|_{L_p}}{[C_{p,m-1+\delta}(e)]^{1/p}}\|u\|_{B^m_p} \le c\,\|\gamma\|_{M(B^{m-1+\delta}_p\to B^\delta_p)}\|u\|_{B^m_p}. \tag{4.3.44}$$

Furthermore, by Corollary 4.3.2

$$\|\gamma\|_{M(B^{m-1+\delta}_p\to B^\delta_p)} \le c\,\|\gamma\|_{M(B^m_p\to B^1_p)}.$$

Hence the third term on the right-hand side of (4.3.40) is dominated by

$$c\,\|\gamma\|_{M(B^m_p\to B^1_p)}\|u\|_{B^m_p}.$$

This along with (4.3.40) and (4.3.43) implies that

$$\|u\mathfrak{D}_{p,1}\gamma\|_{L_p} \le c\,\|\gamma\|_{M(B_p^m\to B_p^1)}\|u\|_{B_p^m}$$

and thus (4.3.35) holds for $l=1$.

(*iii*) Suppose that l is positive and integer, and that the lemma is proved for $\gamma \in M(B_p^m \to B_p^k)$, where k is any positive integer not exceeding $l-1$. Applying (4.3.39), we find that

$$\|u\mathfrak{D}_{p,l}\gamma\|_{L_p} \le \|\gamma u\|_{B_p^l} + c\sum_{j=0}^{l-1}\|\,|\nabla_j\gamma|\mathfrak{D}_{p,l-j}u\|_{L_p} + c\sum_{j=1}^{l-1}\|\,|\nabla_j u|\mathfrak{D}_{p,l-j}\gamma\|_{L_p}$$

$$+c\sum_{j=0}^{l-1}\Big(\int\!\!\int|\Delta_h\nabla_j\gamma(x)|^p|\Delta_h\nabla_{l-1-j}u|^p|h|^{-n-p}dhdx\Big)^{1/p}. \qquad (4.3.45)$$

By (4.2.9) with $\alpha=l-j$ and $\beta=m-l+j$, we have

$$(\mathfrak{D}_{p,l-j}u)(x) \le (J_{m-l+j}\mathfrak{D}_{p,l-j}\Lambda^{m-l+j}u)(x).$$

Therefore, for $j=1,\dots,l-1$ and $m\ge l$,

$$\|\,|\nabla_j\gamma|\mathfrak{D}_{p,l-j}u\|_{L_p} \le c\,\|\nabla_j\gamma\|_{M(B_p^{m-l+j}\to L_p)}\|J_{m-l+j}\mathfrak{D}_{p,l-j}\Lambda^{m-l+j}u\|_{B_p^{m-l+j}}$$

$$\le c\,\|\nabla_j\gamma\|_{M(B_p^{m-l+j}\to L_p)}\|\mathfrak{D}_{p,l-j}\Lambda^{m-l+j}u\|_{L_p}. \qquad (4.3.46)$$

According to (4.2.1),

$$\|\mathfrak{D}_{p,l-j}\Lambda^{m-l+j}u\|_{L_p} \le \|\Lambda^{m-l+j}u\|_{B_p^{l-j}} \le c\,\|u\|_{B_p^m}. \qquad (4.3.47)$$

By Corollary 4.3.4,

$$\|\nabla_j\gamma\|_{M(B_p^{m-l+j}\to L_p)} \le c\,\|\gamma\|_{M(B_p^m\to B_p^l)}, \quad j=1,\dots,l-1,\ \ m\ge l. \quad (4.3.48)$$

For $j=0$ by Lemma 4.3.4 we obtain

$$\|\gamma\mathfrak{D}_{p,l}u\|_{L_p} \le \|\gamma\|_{M(B_p^m\to B_p^l)}\|u\|_{B_p^m}. \qquad (4.3.49)$$

Unifying (4.3.46)–(4.3.49), we find that for all $j=0,\dots,l-1$ and $1\le l\le m$,

$$\|\,|\nabla_j\gamma|\mathfrak{D}_{p,l-j}u\|_{L_p} \le c\,\|\gamma\|_{M(B_p^m\to B_p^l)}\|u\|_{B_p^m}. \qquad (4.3.50)$$

For $j=1,\dots,l-1$ we have

$$\|\,|\nabla_j u|\mathfrak{D}_{p,l-j}\gamma\|_{L_p} \le c\sup_e \frac{\|\mathfrak{D}_{p,l-j}\gamma;e\|_{L_p}}{[C_{p,m-j}(e)]^{1/p}}\|u\|_{B_p^m}. \qquad (4.3.51)$$

From the induction assumption and Corollary 4.3.2 it follows that for $m \geq l$ one has

$$\sup_e \frac{\|\mathfrak{D}_{p,l-j}\gamma; e\|_{L_p}}{[C_{p,m-j}(e)]^{1/p}} \leq c\,\|\gamma\|_{M(B_p^{m-j}\to B_p^{l-j})} \leq c\,\|\gamma\|_{M(B_p^m\to B_p^l)} \tag{4.3.52}$$

which together with (4.3.51) implies that

$$\|\,|\nabla_j u|\mathfrak{D}_{p,l-j}\gamma\|_{L_p} \leq c\,\|\gamma\|_{M(B_p^m\to B_p^l)}\|u\|_{B_p^m}, \quad j=1,\dots,l-1. \tag{4.3.53}$$

Next we estimate the last sum in (4.3.45). Let $\delta \in (0,1)$ be such that $m+\delta$ is a noninteger. By (4.3.1) with γ replaced by $\nabla_j\gamma$, u replaced by $\nabla_{l-1-j}u$, and $k = m-l+j+1$, each term of the last sum in (4.3.45) does not exceed

$$c \sup_e \frac{\|\mathfrak{D}_{p,j+\delta}\gamma; e\|_{L_p}}{[C_{p,m-l+j+\delta}(e)]^{1/p}}\|\nabla_{l-1-j}u\|_{B_p^{m-l+j+1}}. \tag{4.3.54}$$

By the induction assumption and Corollary 4.3.2 this implies that

$$\Big(\int\!\!\int |\Delta_h\nabla_j\gamma(x)|^p|\Delta_h\nabla_{l-1-j}u|^p|h|^{-n-p}dhdx\Big)^{1/p}$$

$$\leq c\,\|\gamma\|_{M(B_p^{m-l+j+\delta}\to B_p^{j+\delta})}\|u\|_{B_p^m} \leq c\,\|\gamma\|_{M(B_p^m\to B_p^l)}\|u\|_{B_p^m}. \tag{4.3.55}$$

Combining this with (4.3.53) and (4.3.51), we obtain from (4.3.45)

$$\|u\mathfrak{D}_{p,l}\gamma\|_{L_p} \leq c\,\|\gamma\|_{M(B_p^m\to B_p^l)}\|u\|_{B_p^m} \tag{4.3.56}$$

and thus (4.3.35) follows for all integer l.

(*iv*) Now let l be a noninteger. Suppose that

$$\sup_e \frac{\|\mathfrak{D}_{p,l}\gamma; e\|_{L_p}}{[C_{p,m}(e)]^{1/p}} \leq c\,\|\gamma\|_{M(B_p^m\to B_p^l)}$$

for all noninteger $l \in (0,N)$, where N is an integer. Let $N < l < N+1$. In view of the equivalence $\mathfrak{D}_{p,l}\gamma \sim D_{p,l}\gamma$ we have

$$\|uD_{p,l}\gamma\|_{L_p} \leq \|\gamma u\|_{B_p^l} + c\sum_{j=0}^{N}\|\,|\nabla_j\gamma|D_{p,l-j}u\|_{L_p}$$

$$+\,c\sum_{j=1}^{N}\|\,|\nabla_j u|D_{p,l-j}\gamma\|_{L_p}. \tag{4.3.57}$$

Let $t \in (0, m-l+j)$ if $m > l$ or $m = l$, $j > 0$ and let $t = 0$ if $m = l$ and $j = 0$. By (4.2.10) with $\alpha = l-j$ and $\beta = t$ one has

$$(D_{p,l-j}u)(x) \leq (J_t D_{p,l-j}\Lambda^t u)(x).$$

Using (4.2.1), we find

$$\| \, |\nabla_j \gamma| D_{p,l-j} u\|_{L_p} \le \|\nabla_j \gamma\|_{M(W_p^{m-l+j} \to L_p)} \|J_t D_{p,l-j} \Lambda^t u\|_{W_p^{m-l+j}}$$
$$\le c \, \|\nabla_j \gamma\|_{M(B_p^{m-l+j} \to L_p)} \|D_{p,l-j} \Lambda^t u\|_{W_p^{m-l+j-t}}. \tag{4.3.58}$$

By definition of the operator $D_{p,l}$ and the space W_p^l,

$$\|D_{p,l-j} v\|_{W_p^{m-l+j-t}} = \|D_{p,m-l+j-t} D_{p,\{l\}} \nabla_{[l-j]} v\|_{L_p} + \|D_{p,l-j} v\|_{L_p}.$$

We use Lemma 4.2.1 with $\alpha = m - l + j - t$, $\beta = \{l\}$ assuming t to be so close to $m - l + j$ that $0 < m - t - [l] + j < 1$. Then

$$\| \, D_{p,m-l+j-t} D_{p,\{l\}} \nabla_{[l]-j} v\|_{L_p}$$
$$\le c \, \|D_{p,m-t-[l]-j} \nabla_{[l]-j} v\|_{L_p} \le c \, \|v\|_{W_p^{m-t}}. \tag{4.3.59}$$

We may also choose t in such a way that $m-t$ is a noninteger so that $W_p^{m-t} = B_p^{m-t}$. Then (4.3.57) together with (4.3.58) and (4.3.59), where $v = \Lambda^t u$, and Corollary 4.3.4 imply that

$$\| \, |\nabla_j \gamma| D_{p,l-j} u\|_{L_p} \le c \, \|\nabla_j \gamma\|_{M(B_p^{m-l+j} \to L_p)} \|\Lambda^t u\|_{B_p^{m-t}}$$
$$\le c \, \|\gamma\|_{M(B_p^m \to B_p^l)} \|u\|_{B_p^m}. \tag{4.3.60}$$

By the induction hypothesis, we have

$$\| \, |\nabla_j u| D_{p,l-j} \gamma\|_{L_p} \le c \sup_e \frac{\|D_{p,l-j} \gamma ; e\|_{L_p}}{[C_{p,m-j}(e)]^{1/p}} \|\nabla_j u\|_{B_p^{m-j}}$$
$$\le c \, \|\gamma\|_{M(B_p^{m-j} \to B_p^{l-j})} \|u\|_{B_p^m} \tag{4.3.61}$$

for $j = 1, \ldots, N$ which, together with Corollary 4.3.2, implies that

$$\| \, |\nabla_j u| D_{p,l-j} \gamma\|_{L_p} \le c \, \|\gamma\|_{M(B_p^m \to B_p^l)} \|u\|_{B_p^m}.$$

This together with (4.3.60) and (4.3.57) leads to

$$\|u D_{p,l} \gamma\|_{L_p} \le c \, \|\gamma\|_{M(B_p^m \to B_p^l)} \|u\|_{B_p^m}.$$

The proof is complete. □

The following simple corollary contains the required lower estimate of the norm in $M(B_p^m \to B_p^l)$ in Theorem 4.1.1. It also finishes the proof of necessity in Theorem 4.1.1.

Corollary 4.3.5. *Let $\gamma \in M(B_p^m \to B_p^l)$, where $0 < l \le m$ and $p \in (1, \infty)$. Then* (4.3.33) *and* (4.3.34) *hold.*

Proof. Since $\gamma \in M(B_p^m \to B_p^l)$, it follows that

$$\|\gamma\eta\|_{L_p} \leq \|\gamma\|_{M(B_p^m \to B_p^l)} \|\eta\|_{B_p^m}$$

for any $\eta \in C_0^\infty(\mathcal{B}_2(x))$, $\eta = 1$ on $\mathcal{B}_1(x)$, where x is an arbitrary point of $\mathbb{R}^n$. Therefore,

$$\sup_{x \in \mathbb{R}^n} \|\gamma; \mathcal{B}_1(x)\|_{L_p} \leq c \, \|\gamma\|_{M(B_p^m \to B_p^l)}.$$

The result follows by combining this inequality with Lemma 4.3.6. □

The next corollary contains one more lower estimate for the norm in the space $M(B_p^m \to B_p^l)$.

Corollary 4.3.6. *Let $\gamma \in M(B_p^m \to B_p^l)$, where $0 < l \leq m$, $p \in (1, \infty)$. Then for any $k = 0, \ldots, [l]$, if l is a noninteger, and for any $k = 0, \ldots, l-1$, if l is an integer, the inclusion $\mathfrak{D}_{p,l-k}\gamma \in M(B_p^{m-k} \to L_p)$ holds and*

$$\|\mathfrak{D}_{p,l-k}\gamma\|_{M(B_p^{m-k} \to L_p)} \leq c \, \|\gamma\|_{M(B_p^m \to B_p^l)}.$$

Proof. By Corollaries 4.3.5 and 4.3.2,

$$\sup_e \frac{\|\mathfrak{D}_{p,l-k}\gamma; e\|_{L_p}}{[C_{p,m-k}(e)]^{1/p}} \leq c \, \|\gamma\|_{M(B_p^{m-k} \to B_p^{l-k})} \leq c \, \|\gamma\|_{M(B_p^m \to B_p^l)}. \tag{4.3.62}$$

It remains to make use of (4.3.12). □

4.3.6 Proof of Sufficiency in Theorem 4.1.1

The aim of this section is to prove the upper estimate for $\|\gamma\|_{M(B_p^m \to B_p^l)}$ in (4.1.4).

Lemma 4.3.7. *Let $\gamma \in B_{p,loc}^l$, $p \in (1, \infty)$. Then for $m > l$*

$$\|\gamma\|_{M(B_p^m \to B_p^l)} \leq c \sup_{e, \mathrm{diam}(e) \leq 1} \Big(\frac{\|\mathfrak{D}_{p,l}\gamma; e\|_{L_p}}{[C_{p,m}(e)]^{1/p}} + \frac{\|\gamma; e\|_{L_p}}{[C_{p,m-l}(e)]^{1/p}} \Big). \tag{4.3.63}$$

For $m = l$ the second term should be replaced by $\|\gamma\|_{L_\infty}$.

Proof. It follows from the finiteness of the right-hand side of (4.3.63) that $\gamma \in L_{1,\mathrm{unif}}$. Let γ_ρ denote a mollifier of γ with radius ρ. Since $\gamma \in L_{1,\mathrm{unif}}$ we see that all derivatives of γ_ρ are bounded. Hence $\gamma_\rho \in M(B_p^m \to B_p^l)$.

For integer l we find by (4.3.45) that

$$\|\gamma_\rho u\|_{B_p^l} \leq c \Big(\sum_{j=0}^{l-1} \| \, |\nabla_j \gamma_\rho| \mathfrak{D}_{p,l-j} u\|_{L_p} + \sum_{j=0}^{l-1} \| \, |\nabla_j u| \mathfrak{D}_{p,l-j} \gamma_\rho\|_{L_p}$$

$$+ \sum_{j=0}^{l-1} \Big(\int\!\!\int |\Delta_h \nabla_j \gamma_\rho(x)|^p |\Delta_h \nabla_{l-1-j} u|^p |h|^{-n-p} dh dx \Big)^{1/p} \Big). \tag{4.3.64}$$

By Corollary 4.3.4,

$$\|\nabla_j \gamma_\rho\|_{M(B_p^{m-l+j} \to L_p)} \leq c\, \|\gamma_\rho\|_{M(B_p^{m-l+j+\alpha} \to B_p^\alpha)} \tag{4.3.65}$$

for any $\alpha \in (0, 1)$. In view of (4.3.26), for $m > l$ the right-hand side in (4.3.65) does not exceed

$$c\, \|\gamma_\rho\|^{(l-\alpha)/l}_{M(B_p^{m-l} \to L_p)} \|\gamma_\rho\|^{\alpha/l}_{M(B_p^m \to B_p^l)}.$$

Combining this fact with (4.3.46) and (4.3.47), we obtain

$$\|\, |\nabla_j \gamma_\rho| \mathfrak{D}_{p,l-j} u\|_{L_p}$$

$$\leq \big(\varepsilon \|\gamma_\rho\|_{M(B_p^m \to B_p^l)} + c(\varepsilon) \|\gamma_\rho\|_{M(B_p^{m-l} \to L_p)}\big) \|u\|_{B_p^m}, \tag{4.3.66}$$

where $j = 0, \dots, l-1$, and ε is an arbitrary positive number.

In case $m = l$ inequalities (4.3.65) and (4.3.27) imply that

$$\|\nabla_j \gamma_\rho\|_{M(B_p^j \to L_p)} \leq c\, \|\gamma_\rho\|^{(l-j)/l}_{L_\infty} \|\gamma_\rho\|^{j/l}_{MB_p^l}.$$

Unifying this estimate with (4.3.46) and (4.3.47) for $m = l$, we obtain

$$\|\, |\nabla_j \gamma_\rho| \mathfrak{D}_{p,l-j} u\|_{L_p} \leq \big(\varepsilon \|\gamma_\rho\|_{MB_p^l} + c(\varepsilon) \|\gamma_\rho\|_{L_\infty}\big) \|u\|_{B_p^l}. \tag{4.3.67}$$

It follows from (4.3.51), (4.3.52), and (4.3.26), (4.3.27) that for $j > 0$

$$\|\, |\nabla_j u| \mathfrak{D}_{p,l-j} \gamma_\rho\|_{L_p}$$

$$\leq \big(\varepsilon \|\gamma_\rho\|_{M(B_p^m \to B_p^l)} + c(\varepsilon) \|\gamma_\rho\|_{M(B_p^{m-l} \to L_p)}\big) \|u\|_{B_p^m}, \tag{4.3.68}$$

if $m > l$, and

$$\|\, |\nabla_j u| \mathfrak{D}_{p,l-j} \gamma_\rho\|_{L_p} \leq \big(\varepsilon \|\gamma_\rho\|_{MB_p^l} + c(\varepsilon) \|\gamma_\rho\|_{L_\infty}\big) \|u\|_{B_p^l}, \tag{4.3.69}$$

if $m = l$.

The third sum on the right-hand side of (4.3.64) is estimated by using (4.3.55) and (4.3.26), (4.3.27). It has the same majorant as the right-hand side of (4.3.68) for $m > l$ or (4.3.69) for $m = l$. Thus, for $m > l$ we find that

$$\|\gamma_\rho u\|_{B_p^l} \leq \Big(\varepsilon \|\gamma_\rho\|_{M(B_p^m \to B_p^l)} + c(\varepsilon) \|\gamma_\rho\|_{M(B_p^{m-l} \to L_p)}$$

$$+ c \sup_{e, \operatorname{diam}(e) \leq 1} \frac{\|\mathfrak{D}_{p,l} \gamma_\rho; e\|_{L_p}}{[C_{p,m}(e)]^{1/p}}\Big) \|u\|_{B_p^m}. \tag{4.3.70}$$

Similarly, for $m = l$,

$$\|\gamma_\rho u\|_{B_p^l} \le \Big(\varepsilon \|\gamma_\rho\|_{MB_p^l} + c(\varepsilon)\|\gamma_\rho\|_{L_\infty}$$

$$+ c \sup_{e, \mathrm{diam}(e)\le 1} \frac{\|\mathfrak{D}_{p,l}\gamma_\rho; e\|_{L_p}}{[C_{p,l}(e)]^{1/p}}\Big)\|u\|_{B_p^l}. \tag{4.3.71}$$

For noninteger l the following estimate, simpler than (4.3.64), holds:

$$\|\gamma_\rho u\|_{B_p^l} \le c\Big(\sum_{j=0}^{[l]} \|\,|\nabla_j \gamma_\rho| D_{p,l-j} u\|_{L_p} + \sum_{j=0}^{[l]} \|\,|\nabla_j u| D_{p,l-j}\gamma_\rho\|_{L_p}\Big)$$

Combining (4.3.60) with Corollary 4.3.4 and (4.3.26), (4.3.27), we arrive at (4.3.66) and (4.3.67) in the same way as for integer l. We also note that (4.3.61) and (4.3.26) for $m > l$ and (4.3.27) for $m = l$ imply (4.3.68) and (4.3.69) for noninteger l. Reference to (4.3.12) and Lemma 4.3.3 completes the proof. □

The required upper estimate of $\|\gamma\|_{MB_p^l}$ in (4.1.4) is obtained in Lemma 4.3.7. In order to show that the second term in the right-hand side of (4.3.63) can be replaced by $\|\gamma\|_{L_{1,\mathrm{unif}}}$ for $m > l$, we need several auxiliary assertions.

Let $(T\gamma)(x, y)$ denote the Poisson integral of a function $\gamma \in L_{1,\mathrm{unif}}$ defined by (3.2.38).

Lemma 4.3.8. *Let l be a noninteger and let $\gamma \in W_{1,\mathrm{loc}}^{[l]}$. Then*

$$\Big(\int_0^\infty \Big|\frac{\partial^{[l]+1}(T\gamma)(x,y)}{\partial y^{[l]+1}}\Big|^p y^{p-1-p\{l\}} dy\Big)^{1/p} \le c\,(D_{p,l}\gamma)(x).$$

Proof. Our argument is similar to that used in the proof of Lemma 3.2.12. We start with the inequality

$$\Big|\frac{\partial^{[l]+1}(T\gamma)(x,y)}{\partial y^{[l]+1}}\Big| \le c\int \frac{|\nabla_{[l]}\gamma(x-\xi) - \nabla_{[l]}\gamma(x)|}{(|\xi| + y)^{n+1}} d\xi \tag{4.3.72}$$

derived in the proof of Lemma 3.2.12. Hence

$$\int_0^\infty \Big|\frac{\partial^{[l]+1}\gamma(x,y)}{\partial y^{[l]+1}}\Big|^p y^{p-1-p\{l\}} dy$$

$$\le c\int_0^\infty y^{p(1-\{l\})}\Big(\int \frac{|\nabla_{[l]}\gamma(x-\xi) - \nabla_{[l]}\gamma(x)|}{y^{n+1}}\Big(1 + \frac{|\xi|}{y}\Big)^{-n-1} d\xi\Big)^p \frac{dy}{y}$$

$$= c\int_0^\infty \Big(\int \frac{|\nabla_{[l]}\gamma(x-\xi) - \nabla_{[l]}\gamma(x)|}{|\xi|^{n+\{l\}}}\Big(\frac{|\xi|}{y}\Big)^{n+\{l\}}\Big(1 + \frac{|\xi|}{y}\Big)^{-n-1} d\xi\Big)^p \frac{dy}{y}. \tag{4.3.73}$$

Introducing spherical coordinates, we write the last expression as

$$c\int_0^\infty \Big(\int_0^\infty f\big(\frac{t}{y}\big)g(t,x)\frac{dt}{t}\Big)^p \frac{dy}{y},$$

where

$$f(s) = s^{n+\{l\}}(1+s)^{-n-1}$$

and

$$g(t,x) = t^{-\{l\}} \int_{\partial\mathcal{B}_1} |\nabla_{[l]}\gamma(x+t\theta) - \nabla_{[l]}\gamma(x)|d\theta.$$

Clearly,

$$\int_0^\infty \Big(\int_0^\infty f\big(\frac{t}{y}\big)g(t,x)\frac{dt}{t}\Big)^p \frac{dy}{y} = \int_0^\infty \Big(\int_0^\infty f(s)g(sy,x)\frac{ds}{s}\Big)^p \frac{dy}{y}. \qquad (4.3.74)$$

By Minkowski's inequality the expression on the right-hand side does not exceed

$$\Big(\int_0^\infty \Big(\int_0^\infty (f(s))^p (g(sy,x))^p \frac{dy}{y}\Big)^{1/p} \frac{ds}{s}\Big)^p$$

$$= \Big(\int_0^\infty f(s)\Big(\int_0^\infty (g(\tau,x))^p \frac{d\tau}{\tau}\Big)^{1/p}\frac{ds}{s}\Big)^p = \Big(\int_0^\infty f(s)\frac{ds}{s}\Big)^p \int_0^\infty (g(\tau,x))^p \frac{d\tau}{\tau}.$$

We deduce from the definition of f that

$$\int_0^\infty f(s)\frac{ds}{s} \le \int_0^1 s^{n+\{l\}}\frac{ds}{s} + \int_1^\infty s^{\{l\}-1}\frac{ds}{s} < \infty. \qquad (4.3.75)$$

Therefore, (4.3.73)–(4.3.75) imply the estimate

$$\int_0^\infty \Big|\frac{\partial^{[l]+1}(T\gamma)(x,y)}{\partial y^{[l]+1}}\Big|^p y^{p-1-p\{l\}}dy \le c\int_0^\infty (g(\tau,x))^p \frac{d\tau}{\tau}.$$

It remains to note that

$$\int_0^\infty (g(\tau,x))^p \frac{d\tau}{\tau} = \int_0^\infty \tau^{-p\{l\}}\Big(\int_{\partial\mathcal{B}_1} |\nabla_{[l]}\gamma(\tau\theta+x) - \nabla_{[l]}\gamma(x)|d\theta\Big)^p \frac{d\tau}{\tau}$$

$$\le \int_0^\infty \int_{\partial\mathcal{B}_1} |\nabla_{[l]}\gamma(\tau\theta+x) - \nabla_{[l]}\gamma(x)|^p d\theta \frac{d\tau}{\tau^{1+p\{l\}}}$$

$$\le c\int \frac{|\nabla_{[l]}\gamma(x+h) - \nabla_{[l]}\gamma(x)|^p}{|h|^{n+p\{l\}}}dh.$$

The proof is complete. □

The following two lemmas are similar to Lemmas 3.2.13 and 3.2.14.

Lemma 4.3.9. *Let $\gamma \in W^{[l]}_{1,loc}$, $y \in (0,1]$. Then*

$$\Big|\frac{\partial^{[l]+1}(T\gamma)(x,y)}{\partial y^{[l]+1}}\Big| \le c\, y^{\{l\}-m-1} \sup_{x\in\mathbb{R}^n, r\in(0,1)} r^{m-n/p}\|D_{p,l}\gamma;\mathcal{B}_r(x)\|_{L_p}.$$

Proof. We introduce the notation

$$K = \sup_{x\in\mathbb{R}^n, r\in(0,1)} r^{m-n/p}\|D_{p,l}\gamma;\mathcal{B}_r(x)\|_{L_p}. \tag{4.3.76}$$

Let $r \in (0,1]$. By Lemma 4.3.8,

$$\int_{\mathcal{B}_r(x)}\int_0^\infty \Big|\frac{\partial^{[l]+1}(T\gamma)(t,y)}{\partial y^{[l]+1}}\Big|^p y^{p-1-p\{l\}} dy\, dt \le c\, K^p r^{n-mp}. \tag{4.3.77}$$

Applying the mean value theorem for harmonic functions, we find for $r/2 < y < 2r/3$ that

$$\Big|\frac{\partial^{[l]+1}(T\gamma)(x,y)}{\partial y^{[l]+1}}\Big| \le c\, r^{-n-1}\int_{\mathcal{B}_r(x)}\int_{r/4}^{r}\Big|\frac{\partial^{[l]+1}(T\gamma)(t,\eta)}{\partial \eta^{[l]+1}}\Big|\, d\eta dt.$$

By Hölder's inequality the right-hand side is dominated by

$$c\, r^{\{l\}-1-n/p}\Big(\int_{\mathcal{B}_r(x)}\int_{r/4}^{r}\Big|\frac{\partial^{[l]+1}(T\gamma)(t,\eta)}{\partial \eta^{[l]+1}}\Big|^p \eta^{p-1-p\{l\}} d\eta dt\Big)^{1/p}$$

which by (4.3.77) does not exceed $c\, r^{\{l\}-m-1}K$. The result follows. □

Lemma 4.3.10. *Let $\gamma \in W^{[l]}_{1,loc}$. Then for almost all $x \in \mathbb{R}^n$ the inequality*

$$|\gamma(x)| \le c\Big(\Big(\sup_{\substack{x\in\mathbb{R}^n,\\ r\in(0,1)}} r^{m-n/p}\|D_{p,l}\gamma;\mathcal{B}_r(x)\|_{L_p}\Big)^{l/m}(D_{p,l}\gamma(x))^{(m-l)/m} + \|\gamma\|_{L_{1,\text{unif}}}\Big)$$

holds.

Proof. We put

$$v(y) = \begin{cases} \Big|\dfrac{\partial^{[l]+1}(T\gamma)(x,y)}{\partial y^{[l]+1}}\Big| & \text{for } 0 < y \le 1,\\[2ex] 0 & \text{for } y > 1.\end{cases}$$

Then, for any $R > 0$

$$\int_0^1\Big|\frac{\partial^{[l]+1}(T\gamma)(x,y)}{\partial y^{[l]+1}}\Big| y^{[l]}dy = \int_0^\infty v(y)y^{[l]}dy = \int_0^R v(y)y^{[l]}dy + \int_R^\infty v(y)y^{[l]}dy.$$

Applying Hölder's inequality, we find that

$$\int_0^R v(y) y^{[l]} dy \le c\, R^l \Big(\int_0^R (v(y))^p y^{p-p\{l\}-1} dy \Big)^{1/p}.$$

By Lemma 4.3.9,

$$\Big| \frac{\partial^{[l]+1} (T\gamma)(x,y)}{\partial y^{[l]+1}} \Big| \le c\, K y^{\{l\}-m-1},$$

where K is defined by (4.3.76). Hence

$$\int_0^\infty v(y) y^{[l]} dy \le c\Big(R^l \Big(\int_0^\infty (v(y))^p y^{p-p\{l\}-1} dy \Big)^{1/p} + R^{l-m} K \Big).$$

Putting here

$$R = K^{1/m} \Big(\int_0^\infty v(y)^p y^{p-p\{l\}-1} dy \Big)^{-1/pm},$$

we arrive at

$$\int_0^\infty v(y) y^{[l]} dy \le c\, K^{l/m} \Big(\int_0^\infty v(y)^p y^{p-p\{l\}-1} dy \Big)^{(m-l)/pm}.$$

Combining this inequality with (3.2.39) for $k = [l]$ we arrive at

$$|\gamma(x)| \le c\, \Big(K^{l/m} \Big(\int_0^\infty v(y)^p y^{p-p\{l\}-1} dy \Big)^{(m-l)/pm} + \|\gamma\|_{L_{1,\mathrm{unif}}} \Big).$$

Reference to Lemma 4.3.8 completes the proof. $\square$

Now, we are in a position to prove the main result of this section.

Lemma 4.3.11. *Let* $0 < l < m$, $p \in (1, \infty)$. *Then*

$$\|\gamma\|_{M(B_p^m \to B_p^l)} \le c\Big(\sup_{e, \mathrm{diam}(e) \le 1} \frac{\|\mathfrak{D}_{p,l}\gamma; e\|_{L_p}}{[C_{p,m}(e)]^{1/p}} + \|\gamma\|_{L_{1,\mathrm{unif}}} \Big). \tag{4.3.78}$$

Proof. By (2.3.18) with $\varphi = |\gamma_\rho|^{1/(m-l)}$, $\lambda = m - l$, and $\mu = m - \varepsilon$, where ε is a positive number less than l such that both $l - \varepsilon$ and $m - \varepsilon$ are nonintegers, we find that

$$\sup_e \frac{\int_e |\gamma_\rho|^p(x) dx}{C_{p,m-l}(e)} \le c \sup_e \left(\frac{\int_e |\gamma_\rho|^{\frac{m-\varepsilon}{m-l} p}(x) dx}{C_{p,m-\varepsilon}(e)} \right)^{\frac{m-l}{m-\varepsilon}}. \tag{4.3.79}$$

Using Lemma 4.3.10 with l replaced by $l - \varepsilon$ and m replaced by $m - \varepsilon$, we obtain

$$\int_e |\gamma_\rho|^{\frac{(m-\varepsilon)p}{m-l}} dx \le c\Big(\Big(\sup_{x\in\mathbb{R}^n, r\in(0,1)} r^{m-\varepsilon-\frac{n}{p}} \|D_{p,l-\varepsilon}\gamma_\rho; \mathcal{B}_r(x)\|_{L_p}\Big)^{\frac{(l-\varepsilon)p}{m-l}}$$

$$\times \int_e |(D_{p,l-\varepsilon}\gamma_\rho)(x)|^p dx + \|\gamma_\rho\|_{L_{1,\mathrm{unif}}}^{\frac{(m-\varepsilon)p}{m-l}} \operatorname{mes}_n e\Big).$$

Hence

$$\left(\frac{\int_e |\gamma_\rho|^{\frac{(m-\varepsilon)p}{m-l}}(x)dx}{C_{p,m-\varepsilon}(e)}\right)^{\frac{m-l}{(m-\varepsilon)p}} \le c\Big\{\Big(\sup_{\substack{x\in\mathbb{R}^n,\\ r\in(0,1)}} r^{m-\varepsilon-\frac{n}{p}} \|D_{p,l-\varepsilon}\gamma_\rho; \mathcal{B}_r(x)\|_{L_p}\Big)^{\frac{l-\varepsilon}{m-\varepsilon}}$$

$$\times \left(\sup_e \frac{\|D_{p,l-\varepsilon}\gamma_\rho; e\|_{L_p}}{[C_{p,m-\varepsilon}(e)]^{1/p}}\right)^{\frac{m-l}{m-\varepsilon}} + \|\gamma_\rho\|_{L_{1,\mathrm{unif}}}\Big\}. \tag{4.3.80}$$

By Corollary 4.3.2,

$$\sup_e \frac{\|D_{p,l-\varepsilon}\gamma_\rho; e\|_{L_p}}{[C_{p,m-\varepsilon}(e)]^{1/p}} \le c\,\|\gamma_\rho\|_{M(W_p^{m-\varepsilon}\to W_p^{l-\varepsilon})}$$

$$= c\,\|\gamma_\rho\|_{M(B_p^{m-\varepsilon}\to B_p^{l-\varepsilon})} \le c\,\|\gamma_\rho\|_{M(B_p^m\to B_p^l)}.$$

Thus, the left-hand side of (4.3.80) has the majorant

$$c\Big(\Big(\sup_{\substack{x\in\mathbb{R}^n,\\ r\in(0,1)}} r^{m-\varepsilon-\frac{n}{p}} \|D_{p,l-\varepsilon}\gamma_\rho; \mathcal{B}_r(x)\|_{L_p}\Big)^{\frac{l-\varepsilon}{m-\varepsilon}} \|\gamma_\rho\|_{M(B_p^m\to B_p^l)}^{\frac{m-l}{m-\varepsilon}} + \|\gamma_\rho\|_{L_{1,\mathrm{unif}}}\Big)$$

which together with (4.3.79) implies the inequality

$$\sup_e \left(\frac{\int_e |\gamma_\rho|^p(x)dx}{C_{p,m-l}(e)}\right)^{1/p} \le c(\delta) \sup_{\substack{x\in\mathbb{R}^n,\\ r\in(0,1)}} r^{m-\varepsilon-\frac{n}{p}} \|D_{p,l-\varepsilon}\gamma_\rho; \mathcal{B}_r(x)\|_{L_p}$$

$$+\delta\|\gamma_\rho\|_{M(B_p^m\to B_p^l)} + c\,\|\gamma_\rho\|_{L_{1,\mathrm{unif}}}, \tag{4.3.81}$$

where δ is an arbitrary positive number.

Next we show that

$$\sup_{x\in\mathbb{R}^n, r\in(0,1)} r^{m-\varepsilon-\frac{n}{p}} \|D_{p,l-\varepsilon}\gamma_\rho; \mathcal{B}_r(x)\|_{L_p}$$

$$\le c(\sigma) \sup_{x\in\mathbb{R}^n, r\in(0,1)} r^{m-\frac{n}{p}} \|\mathfrak{D}_{p,l}\gamma_\rho; \mathcal{B}_r(x)\|_{L_p} + \sigma\,\|\gamma_\rho\|_{M(B_p^m\to B_p^l)} \tag{4.3.82}$$

where σ is an arbitrary positive number. We note that by (4.1.3) $D_{p,l-\varepsilon}\gamma_\rho$ can be replaced by $\mathfrak{D}_{p,l-\varepsilon}\gamma_\rho$. Let ω denote a positive number to be chosen later. Further, let $k=l-1$ and $\lambda=1$ for integer l, and let $k=[l]$ and $\lambda=\{l\}$ for noninteger l. We then have

$$\int_{\mathcal{B}_r(x)} dy \int_{\mathcal{B}_{\omega r}} \frac{|\nabla_k \gamma_\rho(y+2h) - 2\nabla_k \gamma_\rho(y+h) + \nabla_k \gamma_\rho(y)|^p}{|h|^{n+p(\lambda-\varepsilon)}} dh$$

$$\le (\omega r)^{p\varepsilon} \int_{\mathcal{B}_r(x)} dy \int_{\mathcal{B}_{\omega r}} \frac{|\nabla_k \gamma_\rho(y+2h) - 2\nabla_k \gamma_\rho(y+h) + \nabla_k \gamma_\rho(y)|^p}{|h|^{n+p\lambda}} dh$$

$$\le (\omega r)^{p\varepsilon} \|\mathfrak{D}_{p,l}\gamma_\rho; \mathcal{B}_r(x)\|_{L_p}^p. \tag{4.3.83}$$

Also,

$$\int_{\mathcal{B}_r(x)} dy \int_{\mathbb{R}^n \backslash \mathcal{B}_{\omega r}} \frac{|\nabla_k \gamma_\rho(y+2h) - 2\nabla_k \gamma_\rho(y+h) + \nabla_k \gamma_\rho(y)|^p}{|h|^{n+p(\lambda-\varepsilon)}} dh$$

$$\le c\Big(\int_{\mathcal{B}_r(x)} dy \int_{\mathbb{R}^n \backslash \mathcal{B}_{\omega r}} \frac{|\nabla_k \gamma_\rho(y+2h)|^p}{|h|^{n+p(\lambda-\varepsilon)}} dh + \int_{\mathcal{B}_r(x)} dy \int_{\mathbb{R}^n \backslash \mathcal{B}_{\omega r}} \frac{|\nabla_k \gamma_\rho(y+h)|^p}{|h|^{n+p(\lambda-\varepsilon)}} dh$$

$$+ (\omega r)^{p(\varepsilon-\lambda)} \|\nabla_k \gamma_\rho; \mathcal{B}_r(x)\|_{L_p}^p\Big). \tag{4.3.84}$$

Further, we have

$$\int_{\mathcal{B}_r(x)} dy \int_{\mathbb{R}^n \backslash \mathcal{B}_{\omega r}} \frac{|\nabla_k \gamma_\rho(y+2h)|^p}{|h|^{n+p(\lambda-\varepsilon)}} dh$$

$$\le \int_{\mathbb{R}^n \backslash \mathcal{B}_{\omega r}} \frac{dh}{|h|^{n+p(\lambda-\varepsilon)}} \int_{\mathcal{B}_r(x+2h)} |\nabla_k \gamma_\rho(z)|^p dz$$

$$\le c\,\omega^{p(\varepsilon-\lambda)} r^{n-pm+p\varepsilon} \sup_{x\in\mathbb{R}^n, r\in(0,1)} r^{p(m-\lambda)-n} \|\nabla_k \gamma_\rho; \mathcal{B}_r(x)\|_{L_p}^p.$$

By (4.3.14)–(4.3.16) the last supremum is dominated by

$$c\,\|\nabla_k \gamma_\rho\|_{M(W_p^{m-\lambda} \to L_p)}^p$$

which by Corollary 4.3.4 does not exceed $c\,\|\gamma_\rho\|_{M(B_p^m \to B_p^l)}^p$.

Clearly, the second term on the right-hand side of (4.3.84) is estimated in the same way. Similarly, the third term does not exceed

$$c\,\omega^{p(\varepsilon-\lambda)} r^{n-pm+p\varepsilon} \|\gamma_\rho\|_{M(B_p^m \to B_p^l)}^p.$$

Hence

$$\int_{\mathcal{B}_r(x)} dy \int_{\mathbb{R}^n \backslash \mathcal{B}_{\omega r}} \frac{|\nabla_k \gamma_\rho(y+2h) - 2\nabla_k \gamma_\rho(y+h) + \nabla_k \gamma_\rho(y)|^p}{|h|^{n+p(\lambda-\varepsilon)}} dh$$

$$\le \; c\,\omega^{p(\varepsilon-\lambda)} r^{n-pm+p\varepsilon} \|\gamma_\rho\|_{M(B_p^m \to B_p^l)}^p. \tag{4.3.85}$$

From (4.3.83) and (4.3.85) we obtain

$$r^{m-\varepsilon-n/p}\|D_{p,l-\varepsilon}\gamma_\rho\|_{L_p}\le c\big(\omega^\varepsilon r^{m-n/p}\|\mathfrak{D}_{p,l}\gamma_\rho;\mathcal{B}_r(x)\|_{L_p}+\omega^{\varepsilon-\lambda}\|\gamma_\rho\|_{M(B_p^m\to B_p^l)}\big).$$

Setting $\sigma=c\omega^{\varepsilon-\lambda}$, we arrive at (4.3.82).

By (4.3.14)–(4.3.16) and (4.3.82),

$$\sup_{x\in\mathbb{R}^n,r\in(0,1)} r^{m-\varepsilon-\frac{n}{p}}\|D_{p,l-\varepsilon}\gamma_\rho;\mathcal{B}_r(x)\|_{L_p}$$

$$\le c(\sigma)\sup_e\frac{\|\mathfrak{D}_{p,l}\gamma_\rho;e\|_{L_p}}{[C_{p,m}(e)]^{1/p}}+\sigma\|\gamma_\rho\|_{M(B_p^m\to B_p^l)},$$

which together with (4.3.81) and Lemma 4.3.7 gives

$$\|\gamma_\rho\|_{M(B_p^m\to B_p^l)}\le c\Big(\sup_e\frac{\|\mathfrak{D}_{p,l}\gamma_\rho;e\|_{L_p}}{[C_{p,m}(e)]^{1/p}}+\|\gamma_\rho\|_{L_{1,\mathrm{unif}}}\Big). \tag{4.3.86}$$

It remains to estimate the right-hand side of (4.3.86) by Lemma 4.3.3, and to use the equivalence relation

$$C_{p,m}(e)\sim\sum_{j\ge1}C_{p,m}(e\cap\mathcal{B}^{(j)}),$$

where $\{\mathcal{B}^{(j)}\}_{j\ge0}$ is a covering of $\mathbb{R}^n$ by balls of diameter one with multiplicity depending only on n (see Proposition 3.1.5). The result follows. □

Combining the statements of Lemmas 4.3.6 and 4.3.11, we complete the proof of Theorem 4.1.1.

The next assertion contains a modified version of Theorem 4.1.1.

Corollary 4.3.7. *Let $0<l<m$ and let $p\in(1,\infty)$. Then for noninteger l*

$$\|\gamma\|_{M(B_p^m\to B_p^l)}$$

$$\sim\sum_{j=0}^{[l]}\big(\|D_{p,l-j}\gamma\|_{M(B_p^{m-j}\to L_p)}+\|\nabla_{[l]-j}\gamma\|_{M(B_p^{m-j-\{l\}}\to L_p)}\big), \tag{4.3.87}$$

and for integer l

$$\|\gamma\|_{M(B_p^m\to B_p^l)}$$

$$\sim\sum_{j=0}^{l-1}\|\mathfrak{D}_{p,l-j}\gamma\|_{M(B_p^{m-j}\to L_p)}+\sum_{j=1}^{l}\|\nabla_{l-j}\gamma\|_{M(B_p^{m-j}\to L_p)}. \tag{4.3.88}$$

Also,

$$\|\gamma\|_{M(B_p^m\to B_p^l)}\sim\|D_{p,l}\gamma\|_{M(B_p^m\to L_p)}+\|\gamma\|_{L_1,\mathrm{unif}}. \tag{4.3.89}$$

For $m=l$ the norm $\|\gamma\|_{L_1,\mathrm{unif}}$ should be replaced by $\|\gamma\|_{L_\infty}$.

Proof. The upper estimate in (4.3.87) follows from Theorem 4.1.1 and the equivalence relation (4.3.12). By Corollaries 4.3.6 and 4.3.4, the lower bound in (4.3.87) results from

$$\|D_{p,l-j}\gamma\|_{M(B_p^{m-j}\to L_p)} \le c\,\|\gamma\|_{M(B_p^{m-j}\to B_p^{l-j})} \le c\,\|\gamma\|_{M(B_p^m\to H_p^l)}$$

and

$$\|\nabla_{[l]-j}\gamma\|_{M(B_p^{m-j-\{l\}}\to L_p)} \le c\,\|\gamma\|_{M(B_p^{m-j-\{l\}}\to B_p^{[l]-j})} \le c\,\|\gamma\|_{M(B_p^m\to B_p^l)}.$$

□

Remark 4.3.2. It follows from Remark 3.1.3 that the supremum on the right-hand side of (4.1.4) is equivalent to each of the suprema

$$\sup_{\{Q\}} \frac{\|J_m \chi_Q (\mathfrak{D}_{p,l}\gamma)^p; Q\|_{L_{p/(p-1)}}}{\|\mathfrak{D}_{p,l}\gamma; Q\|_{L_p}^{p-1}}, \tag{4.3.90}$$

where $\{Q\}$ is the collection of all cubes, χ_Q is the characteristic function of Q, and

$$\sup_{x\in\mathbb{R}^n} \frac{J_m(J_m(\mathfrak{D}_{p,l}\gamma)^p)^{p/(p-1)}(x)}{J_m(\mathfrak{D}_{p,l}\gamma)^p(x)}. \tag{4.3.91}$$

Adding to (4.3.90) and (4.3.91) the norms $\|\gamma\|_{L_{1,\mathrm{unif}}}$ for $m > l$ and $\|\gamma\|_{L_\infty}$ for $m = l$, we arrive at two non-capacitary necessary and sufficient conditions for $\gamma \in M(B_p^m \to B_p^l)$.

4.3.7 The Case $mp > n$

For $mp > n$ Theorem 4.1.1 has a simpler formulation.

Corollary 4.3.8. *Let $0 < l \le m$, $mp > n$, and $p \in (1,\infty)$. Then*

$$\|\gamma\|_{M(B_p^m\to B_p^l)} \sim \sup_{x\in\mathbb{R}^n} \left(\|\mathfrak{D}_{p,l}\gamma; \mathcal{B}_1(x)\|_{L_p} + \|\gamma; \mathcal{B}_1(x)\|_{L_p}\right). \tag{4.3.92}$$

For $m = l$ the second term on the right-hand side can be replaced by $\|\gamma\|_{L_\infty}$.

Proof. The lower estimate of $\|\gamma\|_{M(B_p^m\to B_p^l)}$ follows from the relation

$$C_{p,m}(e) \sim 1 \tag{4.3.93}$$

wich holds for $mp > n$ and e with $\mathrm{diam}(e) \le 1$, and from Corollary 4.3.5. The upper estimate results from

$$\|\gamma\|_{M(B_p^m\to B_p^l)} \le \|\gamma\|_{MB_p^l} \le c\,\Big(\sup_{e,\mathrm{diam}(e)\le 1} \|\mathfrak{D}_{p,l}\gamma; e\|_{L_p} + \|\gamma\|_{L_\infty}\Big)$$

$$\le c\,\sup_{x\in\mathbb{R}^n}\left(\|\mathfrak{D}_{p,l}\gamma; \mathcal{B}_1(x)\| + \|\gamma; \mathcal{B}_1(x)\|_{L_p}\right).$$

The proof is complete. □

Remark 4.3.3. One can easily verify that the right-hand side of (4.3.92) is equivalent to the norm of γ in $B^l_{p,\mathrm{unif}}$. Hence $M(B^m_p \to B^l_p)$ is isomorphic to $B^l_{p,\mathrm{unif}}$ for $0 < l \le m$, $mp > n$, $p \in (1, \infty)$.

4.3.8 Lower and Upper Estimates for the Norm in $M(B^m_p \to B^l_p)$

Here we present some lower and, separately, upper bounds for the norm in $M(B^m_p \to B^l_p)$, $mp \le n$, which do not involve the capacity and which follow from the characterization of multipliers in $M(B^m_p \to B^l_p)$. The next assertion stems directly from Proposition 3.1.4 and Theorem 4.1.1.

Proposition 4.3.1. *Let $0 < l < m$. If $mp < n$, then*

$$\|\gamma\|_{M(B^m_p \to B^l_p)} \ge c\Big(\sup_{\substack{x\in\mathbb{R}^n \\ r\in(0,1)}} r^{m-n/p} \|\mathfrak{D}_{p,l}\gamma; \mathcal{B}_r(x)\|_{L_p} + \|\gamma\|_{L_{1,\mathrm{unif}}}\Big) \tag{4.3.94}$$

and, if $mp = n$, then

$$\|\gamma\|_{M(B^m_p \to B^l_p)}$$
$$\ge c\Big(\sup_{\substack{x\in\mathbb{R}^n \\ r\in(0,1)}} (\log 2r^{-1})^{1-1/p} \|\mathfrak{D}_{p,l}\gamma; \mathcal{B}_r(x)\|_{L_p} + \|\gamma\|_{L_{1,\mathrm{unif}}}\Big). \tag{4.3.95}$$

For $m = l$ the second term on the right-hand side of (4.3.94) *and* (4.3.95) *should be replaced by $\|\gamma\|_{L_\infty}$.*

Finally we formulate a corollary of Theorem 4.1.1 and Propositions 3.1.2 and 3.1.3.

Proposition 4.3.2. *Let $0 < l < m$. If $mp < n$, then*

$$\|\gamma\|_{M(B^m_p \to B^l_p)}$$
$$\le c\Big(\sup_{\{e: d(e)\le 1\}} (\mathrm{mes}_n e)^{m/n-1/p} \|\mathfrak{D}_{p,l}\gamma; e\|_{L_p} + \|\gamma\|_{L_{1,\mathrm{unif}}}\Big) \tag{4.3.96}$$

and, if $mp = n$, then

$$\|\gamma\|_{M(B^m_p \to B^l_p)}$$
$$\le c\Big(\sup_{\{e: d(e)\le 1\}} (\log(2^n/\mathrm{mes}_n e))^{1/p'} \|\mathfrak{D}_{p,l}\gamma; e\|_{L_p} + \|\gamma\|_{L_{1,\mathrm{unif}}}\Big). \tag{4.3.97}$$

For $m = l$ the second term on the right-hand side of (4.3.96) *and* (4.3.97) *should be replaced by $\|\gamma\|_{L_\infty}$.*

4.4 Sufficient Conditions for Inclusion into $M(W_p^m \to W_p^l)$ with Noninteger m and l

It may be of use to compare the contents of this section with sufficient conditions for inclusion into the class $M(H_p^m \to H_p^l)$ obtained in 3.4. Here similar conditions are found for $M(W_p^m \to W_p^l)$, $\{m\} > 0$, $\{l\} > 0$. They are formulated in terms of the spaces $B^\mu_{q,\infty}$ (cf. (4.4.2)), $B^l_{q,p}$ (cf. (4.4.3)) and $H^l_{n/m}$ (cf. (4.4.4).

By $B^s_{q,\theta}$ we denote the space of functions in $\mathbb{R}^n$ having the finite norm

$$\|u\|_{B^s_{q,\theta}} = \Big(\int \|\Delta_h \nabla_{[s]} u\|^\theta_{L_q} |h|^{-n-\theta\{s\}}\, dh\Big)^{1/\theta} + \|u\|_{W_q^{[s]}}, \tag{4.4.1}$$

where $\{s\} > 0$, q, $\theta \geq 1$.

4.4.1 Conditions Involving the Space $B^\mu_{q,\infty}$

We prove an assertion analogous to Theorem 3.4.1 and formulated in terms of the space $B^s_{q,\theta,\mathrm{unif}}$.

It is clear that

$$\|u\|_{B^s_{q,\theta,\mathrm{unif}}} \sim \sup_{x\in\mathbb{R}^n} \Big[\Big(\int_{\mathcal{B}_1} \|\Delta_h \nabla_{[s]} u; \mathcal{B}_1(x)\|^\theta_{L_q} |h|^{-n-\theta\{s\}}\, dh\Big)^{1/\theta} + \|u; \mathcal{B}_1(x)\|_{W_q^{[s]}}\Big].$$

Theorem 4.4.1. *Let $q \geq p > 1$, $\{m\} > 0$, $\{l\} > 0$, $\mu = n/q - m + l$, $\mu > l$, and $\{\mu\} > 0$.*

(i) If $\gamma \in B^\mu_{q,\infty,\mathrm{unif}} \cap L_\infty$ then $\gamma \in MW_p^l$ and

$$\|\gamma\|_{MW_p^l} \leq c\,(\|\gamma\|_{B^\mu_{q,\infty,\mathrm{unif}}} + \|\gamma\|_{L_\infty}). \tag{4.4.2}$$

(ii) If $\gamma \in B^\mu_{q,\infty,\mathrm{unif}}$, then $\gamma \in M(W_p^m \to W_p^l)$ and

$$\|\gamma\|_{M(W_p^m \to W_p^l)} \leq c\,\|\gamma\|_{B^\mu_{q,\infty,\mathrm{unif}}}\,. \tag{4.4.3}$$

Proof. Let $m \geq l$. It suffices to assume that the difference $\varepsilon = \mu - l$ is small, since the general case follows by interpolation between the pairs $\{W_p^{m-l}, L_p\}$ and $\{W_p^{n/q-\varepsilon}, W_p^{\mu-\varepsilon}\}$ (see (4.3.26)). Thus we assume that $1 + [l] > \mu > l$.

Let e be a compact set in $\mathbb{R}^n$, $d(e) \leq 1$, and let $|e| = \mathrm{mes}_n e$. We have

$$\|D_{p,l}\gamma; e\|^p_{L_p} = \int \frac{dh}{|h|^{n+p\{l\}}} \int_e |\nabla_{[l]}\gamma(x+h) - \nabla_{[l]}\gamma(x)|^p\, dx\,. \tag{4.4.4}$$

We express the integral over $\mathbb{R}^n$ as the sum of two integrals $i_1 + i_2$, the first being taken over the exterior of the ball $\{h : |h| < |e|^{1/n}\}$. Obviously,

$$i_1 \le c \int_{|h|>|e|^{1/n}} \frac{dh}{|h|^{n+p\{l\}}} \Big(\int_e |\nabla_{[l]}\gamma(x+h)|^p \, dx + \int_e |\nabla_{[l]}\gamma(x)|^p \, dx \Big)$$
$$\le c\,|e|^{1-p/q} \int_{|h|>|e|^{1/n}} \frac{dh}{|h|^{n+p\{l\}}} \Big(\int_e |\nabla_{[l]}\gamma(x+h)|^q \, dx + \int_e |\nabla_{[l]}\gamma(x)|^q \, dx \Big)^{p/q}.$$

Hence, using Corollary 3.4.1, we obtain

$$i_1 \le c\,|e|^{1-p/q+\{\mu\}p/n} \int_{|h|>|e|^{1/n}} \frac{dh}{|h|^{n+p\{l\}}} N(\gamma)^p \,, \tag{4.4.5}$$

where $N(\gamma)$ is the right-hand side of either (4.4.2) or (4.4.3). Since $[\mu] = [l]$, we have $\{\mu\} - \{l\} = \mu - l$ and consequently

$$i_1 \le c\,|e|^{1-mp/n} N(\gamma)^p \,.$$

By Hölder's inequality,

$$i_2 \le |e|^{1-p/q} \int_{|h|<|e|^{1/n}} \frac{dh}{|h|^{n+p\{l\}}} \Big(\int_e |\nabla_{[l]}\gamma(x+h) - \nabla_{[l]}\gamma(x)|^q \, dx \Big)^{p/q}$$
$$\le |e|^{1-p/q} \int_{|h|\le|e|^{1/n}} \frac{dh}{|h|^{n+p(\{l\}-\{\mu\})}} N(\gamma)^p = c\,|e|^{1-pm/n} N(\gamma)^p \,.$$

Combining this result with (4.4.4) and (4.4.5), we have

$$\|D_{p,l}\gamma; e\|_{L_p} \le c\,|e|^{1/p-m/n} N(\gamma).$$

In addition, according to (3.4.19),

$$\|\gamma; e\|_{L_p} \le c\,|e|^{1/p-(m-l)/n} N(\gamma) \tag{4.4.6}$$

for $m > l$. It remains to refer to Proposition 4.3.2. □

Using the same arguments as in the proof of Corollary 3.4.2, we obtain:

Corollary 4.4.1. *Let $n = 1$, $q \ge p$ and $lq < 1$. If $\gamma \in L_\infty$ and $\mathrm{Var}_q(\gamma) < \infty$, then $\gamma \in MW_p^l$ and*

$$\|\gamma\|_{MW_p^l} \le c\,(\|\gamma\|_{L_\infty} + \mathrm{Var}_q(\gamma)) \,.$$

We give one more sufficient condition for a function to belong to the space $M(B_p^m \to B_p^l)$ for noninteger m and l in the case $mp = n$. We introduce the semi-norm

$$\langle\gamma\rangle = \sup_{y\in\mathbb{R}^n} \sup_{h\in\mathcal{B}_{1/2}} |h|^{-\{l\}} \log(1/|h|) \|\varDelta_h \nabla_{[l]}\gamma; \mathcal{B}_1(y)\|_{L_p} \,.$$

Theorem 4.4.2. *Let* $\{m\} > 0$, $\{l\} > 0$, $p > 1$.

(*i*) *If* $lp = n$, $\gamma \in L_\infty$ *and* $\langle\gamma\rangle < \infty$, *then* $\gamma \in MW_p^l$ *and*

$$\|\gamma\|_{MW_p^l} \leq c\,(\langle\gamma\rangle + \|\gamma\|_{L_\infty})\,. \tag{4.4.7}$$

(*ii*) *If* $mp = n$, $\gamma \in L_{p,\mathrm{unif}}$ *and* $\langle\gamma\rangle < \infty$, *then* $\gamma \in M(W_p^m \to W_p^l)$ *for* $l < m$ *and*

$$\|\gamma\|_{M(W_p^m \to W_p^l)} \leq c\,(\langle\gamma\rangle + \|\gamma\|_{L_{p,\mathrm{unif}}})\,. \tag{4.4.8}$$

Proof. By $Q(\gamma)$ we denote either of the right-hand sides of (4.4.7) and (4.4.8). We express the integral over $\mathbb{R}^n$ in (4.4.4) as the sum $i_1 + i_2$ of two integrals, the first being taken over the exterior of the ball

$$K = \{h : |h| < c_0|e|^{1/n}(\log 2^n/|e|)^{(p-1)/p\{l\}}\}\,,$$

where $|e| = \mathrm{mes}_n e$ and c_0 is a small positive constant depending on n and p. By Corollary 3.4.1 with $q = p$ and $\mu = n/p = l$,

$$\begin{aligned} i_1 &\leq c\int_{\mathbb{R}^n\setminus K}\frac{dh}{|h|^{n+p\{l\}}}\Big(\int_e |\nabla_{[l]}\gamma(x+h)|^p\,dx + \int_e |\nabla_{[l]}\gamma(x)|^p\,dx\Big) \\ &\leq c\,|e|^{\{l\}p/n}\int_{\mathbb{R}^n\setminus K}\frac{dh}{|h|^{n+p\{l\}}}Q(\gamma)^p \leq c(\log 2^n/|e|)^{1-p}Q(\gamma)^p\,. \end{aligned} \tag{4.4.9}$$

Moreover,

$$i_2 \leq \langle\gamma\rangle^p\int_K \frac{dh}{|h|^n(\log 1/|h|)^p}\,.$$

The integral on the right-hand side does not exceed $c\,(\log 2^n/|e|)^{1-p}$. This estimate, together with (4.4.4) and (4.4.9), yields

$$(\log 2^n/|e|)^{(p-1)/p}\|D_{p,l}\gamma; e\|_{L_p} \leq c\,Q(\gamma).$$

In addition, by (4.4.6)

$$|e|^{-l/n}\|\gamma; e\|_{L_p} \leq c\,Q(\gamma).$$

Reference to Proposition 4.3.1 completes the proof. □

4.4.2 Conditions Involving the Fourier Transform

We start with the following known characterization of the space $B_{2,\infty}^s$ (see, for example, [Tr3]).

Lemma 4.4.1. *The relation*

$$\|u\|_{B_{2,\infty}^s} \sim \sup_{R>1} R^s\|Fu; \mathcal{B}_{2R}\setminus\mathcal{B}_R\|_{L_2} + \|u\|_{L_2} \tag{4.4.10}$$

holds.

Proof. The lower bound for the norm in $B^s_{2,\infty}$ is obtained as follows:

$$\begin{aligned}\sup_h |h|^{-2\{s\}} \|\Delta_h \nabla_{[s]} u\|^2_{L_2} &= c \sup_{\rho>0} \rho^{-2\{s\}} \int_{\partial\mathcal{B}_1} \|\Delta_{\rho\theta} \nabla_{[s]} u\|^2_{L_2}\, d\sigma_\theta \\ &= c \sup_{\rho>0} \rho^{-2\{s\}} \iint_{\partial\mathcal{B}_1} |\xi|^{2[s]} |Fu(\xi)|^2 \sin^2 \frac{\rho(\theta,\xi)}{2}\, d\sigma_\theta d\xi \\ &\geq c \sup_{\rho>0} \rho^{2(1-\{s\})} \int_{\rho^{-1}>|\xi|>(2\rho)^{-1}} |\xi|^{2[s]+1} |Fu(\xi)|^2\, d\xi \\ &\geq c \sup_{\rho>0} R^{2s} \int_{\mathcal{B}_{2R}\setminus\mathcal{B}_R} |Fu(\xi)|^2\, d\xi\, .\end{aligned}$$

On the other hand,

$$\begin{aligned}&\sup_h |h|^{-2\{s\}} \|\Delta_h \nabla_{[s]} u\|^2_{L_2} \leq c \sup_h |h|^{-2\{s\}} \int_{|\xi||h|>1} |\xi|^{2[s]} |Fu(\xi)|^2 d\xi \\ &+ c \sup_h |h|^{2(1-\{l\})} \int_{|\xi||h|<1} |\xi|^{2[s]+2} |Fu(\xi)|^2\, d\xi\, .\end{aligned} \tag{4.4.11}$$

The first term on the right-hand side does not exceed

$$\begin{aligned}&c \sup_h |h|^{-2s} \sum_{j=1}^{\infty} 4^{j[s]} \int_{2^j>|\xi||h|>2^{j-1}} |Fu(\xi)|^2\, d\xi \\ &\leq c \sum_{j=1}^{\infty} 4^{-j\{s\}} \sup_R R^{2s} \int_{\mathcal{B}_{2R}\setminus\mathcal{B}_R} |Fu(\xi)|^2\, d\xi\, .\end{aligned}$$

The second term on the right-hand side of (4.4.11) is majorized by

$$\begin{aligned}&c \sup_h |h|^{-2s} \sum_{j=0}^{\infty} 4^{-j([s]+1)} \int_{2^{-j-1}<|\xi||h|<2^{-j}} |Fu(\xi)|^2\, d\xi \\ &\leq c \sum_{j=0}^{\infty} 4^{-j(1-\{s\})} \sup_R R^s \int_{\mathcal{B}_{2R}\setminus\mathcal{B}_R} |Fu(\xi)|^2\, d\xi\, .\end{aligned}$$

The proof is complete. □

In the same way one may prove:

Lemma 4.4.2. *The norms*

$$\sup_{y\in\mathbb{R}^n} \sup_{h\in\mathcal{B}_{1/2}} |h|^{-\{l\}} \log(1/|h|) \|\Delta_h \nabla_{[l]} u; \mathcal{B}_1(y)\|_{L_2} + \|u\|_{L_2}$$

and

$$\sup_{R>2} R^l \log R\, \|Fu; \mathcal{B}_{2R}\setminus\mathcal{B}_R\|_{L_2} + \|u\|_{L_2}$$

are equivalent.

Theorems 4.4.1, 4.4.2 and Lemmas 4.4.1, 4.4.2 imply:

Theorem 4.4.3. *(i) If $1 < p \le 2$, $n/2 > m$, m, l are noninteger, $m > l$ and*

$$(F\gamma)(\xi) = O((1+|\xi|)^{m-l-n}),$$

then $\gamma \in M(W_p^m \to W_p^l)$.

(ii) If $1 < p \le 2$, $n/2 > l$, l is a noninteger, $\gamma \in L_\infty$ and

$$(F\gamma)(\xi) = O((1+|\xi|)^{-n}),$$

then $\gamma \in MW_p^l$.

(iii) If n is odd, $2l = n$, $\gamma \in L_\infty$ and

$$(F\gamma)(\xi) = O(|\xi|^{-n}(\log|\xi|)^{-1})$$

for $|\xi| \ge 2$, then $\gamma \in MW_2^l$.

4.4.3 Conditions Involving the Space $B^l_{q,p}$

The condition $\gamma \in B^\mu_{q,\infty,\mathrm{unif}}$ in Theorem 4.4.1 requires the 'number of derivatives' μ to exceed l. In this subsection we obtain sufficient conditions for a function to belong to the class $M(W_p^m \to W_p^l)$ in terms of the space $B^l_{q,p,\mathrm{unif}}$. We recall that diminishing the exponent θ leads to narrowing of $B^\mu_{q,\theta}$, and diminishing of μ leads to expansion of this space. So new sufficient conditions are not comparable with the conditions of Theorem 4.4.1.

Theorem 4.4.4. *Let $\{m\} > 0$, $\{l\} > 0$, $p > 1$.*

(i) Let $q \in [n/l, \infty]$ for $pl < n$ and $q \in (p, \infty]$ for $lp = n$. If $\gamma \in B^l_{q,p,\mathrm{unif}} \cap L_\infty$, then $\gamma \in MW_p^l$ and

$$\|\gamma\|_{MW_p^l} \le c\Big(\sup_{x\in\mathbb{R}^n}\Big(\int_{\mathcal{B}_1}\|\Delta_h\nabla_{[l]}\gamma;\mathcal{B}_1(x)\|^p_{L_q}|h|^{-n-p\{l\}}\,dh\Big)^{1/p} + \|\gamma\|_{L_\infty}\Big). \quad (4.4.12)$$

(ii) Let $m > l$, $q \in [n/m, \infty]$ for $mp < n$ and $q \in (p, \infty)$ for $mp = n$. If $\gamma \in B^l_{q,p,\mathrm{unif}}$, then $\gamma \in M(W_p^m \to W_p^l)$ and

$$\|\gamma\|_{M(W_p^m\to W_p^l)} \le c\sup_{x\in\mathbb{R}^n}\Big(\Big(\int_{\mathcal{B}_1}\|\Delta_h\nabla_{[l]}\gamma;\mathcal{B}_1(x)\|^p_{L_q}|h|^{-n-p\{l\}}\,dh\Big)^{1/p} + \|\gamma;\mathcal{B}_1(x)\|_{L_p}\Big). \quad (4.4.13)$$

Proof. Proposition 4.2.6 implies that

$$\|\gamma\|_{M(W_p^m\to W_p^l)} \le c\sup_{x\in\mathbb{R}^n}\|\eta_x\gamma\|_{M(B_p^m\to B_p^l)},$$

where $\eta \in C_0^\infty(\mathcal{B}_1)$, $\eta = 1$ on $\mathcal{B}_{1/2}$ and $\eta_x(y) = \eta(x-y)$. Therefore it suffices to obtain (4.4.12), (4.4.13) under the assumption that the diameter of supp γ does not exceed 1.

Let $e \subset \mathbb{R}^n$, $d(e) \le 1$. We have

$$\begin{aligned}\|D_{p,l}\gamma; e\|_{L_p}^p &\le \int \|\Delta_h \nabla_{[l]}\gamma; e\|_{L_p}^p \frac{dh}{|h|^{n+p\{l\}}}\\ &\le (\mathrm{mes}_n e)^{1-p/q} \int \|\Delta_h \nabla_{[l]}\gamma; e\|_{L_q}^p \frac{dh}{|h|^{n+p\{l\}}}\\ &\le (\mathrm{mes}_n e)^{1-p/q} \sup_{x\in\mathbb{R}^n} \int \|\Delta_h \nabla_{[l]}\gamma; \mathcal{B}_1(x)\|_{L_q}^p \frac{dh}{|h|^{n+p\{l\}}}\,.\end{aligned}$$

Reference to Proposition 4.3.2 completes the proof. □

Putting $q = \infty$ in (4.4.12), we obtain a simple condition for a function γ to belong to the class MW_p^l (and hence to $M(W_p^m \to W_p^l)$) formulated in terms of the modulus of continuity ω of the vector-function $\nabla_{[l]}\gamma$:

$$\int_0 \left[\frac{\omega(t)}{t^{\{l\}+1/p}}\right]^p dt < \infty\,. \tag{4.4.14}$$

The last theorem contains the condition $lp \le n$. Nevertheless, (4.4.14) ensures that $\gamma \in MW_p^l$ for $lp > n$, since in that case $MW_p^l = W_{p,\mathrm{unif}}^l \supset B_{\infty,p}^l$.

We show that even a rough condition (4.4.14) is the best possible in some sense.

Example 4.4.1. Let ω be a continuous increasing function on $[0,1]$ satisfying the inequalities

$$\delta \int_0^1 \frac{\omega(t)}{t^2}\,dt + \int_0^\delta \frac{\omega(t)}{t}\,dt \le c\,\omega(\delta)\,, \qquad 1 > \omega(\delta) \ge c\,\delta. \tag{4.4.15}$$

Further, let

$$\int_0^1 [\omega(t)t^{-\{l\}-1/p}]^p\,dt = \infty.$$

We construct a function γ on $\mathbb{R}^n$ such that

1. the modulus of continuity of the vector-function $\nabla_{[l]}\gamma$ does not exceed $c\,\omega$, where $c = $ const;
2. $\gamma \notin W_{p,\mathrm{unif}}^l$ and hence $\gamma \notin M(W_p^m \to W_p^l)$.

We put

$$\gamma(x) = \prod_{i=1}^n \eta(x_i) \sum_{k=1}^\infty e^{-[l]k}\omega(e^{-k})\sin(e^k x_1)\,, \tag{4.4.16}$$

where $\eta \in C_0^\infty(-2\pi, 2\pi)$, $\eta = 1$ on $(-\pi,\pi)$, $0 < \eta \le 1$.

For small enough $|h|$ we have

$$|\nabla_{[l]}\gamma(x+h) - \nabla_{[l]}\gamma(x)|$$

$$\leq c\Big(|h|\sum_{k=1}^{\infty}\omega(e^{-k}) + |h|\sum_{k\leq\log|h|^{-1}}\omega(e^{-k})e^k + \sum_{k>\log|h|^{-1}}\omega(e^k)\Big)$$

which, together with (4.4.15), gives

$$|\nabla_{[l]}\gamma(x+h) - \nabla_{[l]}\gamma(x)| \leq c\,\omega(|h|)\,. \tag{4.4.17}$$

Further,

$$\|\gamma\|^p_{W_p^l} \geq c\int_{\mathbb{R}^n}\int_{\mathbb{R}^1}\left|\Delta_{te_1}\frac{\partial^{[l]}\gamma}{\partial x_1^{[l]}}\right|^p\frac{dt}{t^{1+p\{l\}}}\,. \tag{4.4.18}$$

We set

$$f(x_1) = \sum_{k=1}^{\infty} e^{-[l]k}\omega(e^{-k})e^{ie^k x_1}. \tag{4.4.19}$$

By virtue of (4.4.18) we have

$$\|f\|^p_{W_p^l} \geq c\,\|\mathrm{Im} f^{[l]};(-\pi,\pi)\|^p_{W_p^{\{l\}}} \geq c\,\|f^{[l]};(-\pi,\pi)\|^p_{W_p^{\{l\}}}.$$

It is clear that

$$\Delta_t f^{[l]}(x_1) = \sum_{k=1}^{\infty} e^{-[l]k}\omega(e^{-k})(e^{ie^k t}-1)e^{ie^k x_1}.$$

According to a known property of lacunary trigonometric series (see [Zy], v. 1, Th. 8.20),

$$\|\Delta_t f^{[l]};(-\pi,\pi)\|^2_{L_p} \sim \sum_{k=1}^{\infty} e^{-2[l]k}(\omega(e^{-k})\sin(e^k t/2))^2\,.$$

Therefore

$$\|\gamma\|^p_{W_p^l} \geq c\int_0^{\pi}[\omega(e^{-k(t)})\sin(e^{k(t)}t/2)]^p\frac{dt}{t^{1+p\{l\}}}\,,$$

where $k(t) = [\log 2t^{-1}]$. Finally,

$$\|\gamma\|^p_{W_p^l} \geq c\int_0^1\left[\frac{\omega(t)}{t^{\{l\}+1/p}}\right]^p dt = \infty\,.$$

4.5 Conditions Involving the Space $H^l_{n/m}$

Here we obtain an upper bound for the norm in $M(W^m_p \to W^l_p)$ with noninteger m and l, $p \geq 2$, and $mp < n$, by the norm in the Bessel potential space $H^l_{n/m,\mathrm{unif}}$.

Theorem 4.5.1. *The estimates hold:*

$$\|\gamma\|_{M(W^m_p \to W^l_p)} \leq c\, \|\gamma\|_{H^l_{n/m,\mathrm{unif}}}\,, \tag{4.5.1}$$

where $m > l$, $\{m\} > 0$, $\{l\} > 0$, $mp < n$ and $p \geq 2$, and

$$\|\gamma\|_{MW^l_p} \leq c\,(\|\gamma\|_{H^l_{n/l,\mathrm{unif}}} + \|\gamma\|_{L_\infty})\,, \tag{4.5.2}$$

where $\{l\} > 0$, $lp < n$ and $p \geq 2$.

Proof. It suffices to derive (4.5.1) and (4.5.2) under the assumption that the diameter of supp γ does not exceed 1 (cf. the beginning of the proof of Theorem 4.4.4).

We use the inequality

$$\|D_{p,l}\gamma\|_{L_{n/m}} \leq c\, \|\gamma\|_{H^l_{n/m}},$$

where $p \geq 2$, $n > p(m - \{l\})$, see Polking [Pol1]. Moreover, by Theorem 3.1.3,

$$\|\gamma\|_{L_{n/(m-l)}} \leq c\, \|\gamma\|_{H^l_{n/m}},$$

where $m > l$. It remains to apply the estimate

$$\|\gamma\|_{M(B^m_p \to B^l_p)} \leq c \sup_{x \in \mathbb{R}^n} (\|D_{p,l}\gamma; \mathcal{B}_1(x)\|_{L_{n/m}} + \|\gamma; \mathcal{B}_1(x)\|_{L_{n/(m-l)}})$$

which follows from (4.3.96). □

Possibly, it is of interest to compare the last theorem with Theorem 3.3.1, according to which the right-hand sides of (4.5.1) and (4.5.2) majorize the norms of MH^l_p and $M(H^m_p \to H^l_p)$ for any $p \in (1, \infty)$.

We show that the condition $p \geq 2$ in the theorem of this subsection cannot be omitted.

Example 4.5.1. Let us consider the function γ defined by (4.4.16). Since, for $q \in (1, \infty)$,

$$\|\gamma\|_{H^l_q} \sim \sum_{j=1}^{n} \Big\| \Big(1 - \frac{\partial^2}{\partial x_j^2}\Big)^{l/2} \gamma \Big\|_{L_q}$$

(see the proof of Proposition 3.5.4), it follows that

$$\|\gamma\|_{H_q^l} \sim \|\eta \operatorname{Im} f; \mathbb{R}^1\|_{H_q^l},$$

where η and f are functions defined in Example 4.4.1.

It is known that $\eta \operatorname{Im} f \in H_q^l(\mathbb{R}^1)$ if and only if the function $\operatorname{Im} f$ belongs to the space H_q^l on the unit circumference C and

$$\|\eta \operatorname{Im} f; \mathbb{R}^1\|_{H_q^l} \sim \|\operatorname{Im} f; C\|_{H_q^l} = \Big(\int_{-\pi}^{\pi} \Big|\Big(1 - \frac{d^2}{d\theta^2}\Big)^{l/2} \operatorname{Im} f(\theta)\Big|^q d\theta\Big)^{1/q}.$$

(We omit a standard but rather tedious proof of this fact.) Therefore

$$\|\gamma\|_{H_q^l} \sim \Big(\sum_{k=1}^{\infty} e^{-2[l]k}(1+e^{2k})^{l/2}[\omega(e^{-k})]^2\Big)^{1/2}$$

and consequently $\gamma \in H_q^l$ if and only if

$$\int_0^1 \Big(\frac{\omega(t)}{t^{\{l\}}}\Big)^2 \frac{dt}{t} < \infty. \tag{4.5.3}$$

From Theorem 3.3.1 and from the imbedding $M(H_p^m \to H_p^l) \subset H_{p,\mathrm{unif}}^l$ we obtain that (4.5.3) is equivalent to $\gamma \in M(H_p^m \to H_p^l)$.

It was shown in Example 4.4.1 that $\gamma \in M(W_p^m \to W_p^l)$ if and only if

$$\int_0^1 \Big(\frac{\omega(t)}{t^{\{l\}}}\Big)^p \frac{dt}{t} < \infty. \tag{4.5.4}$$

If the theorem of this subsection were true for $p < 2$, we would have implication (4.5.3) $\Rightarrow$ (4.5.4), which is obviously wrong.

What is more, we see that for $p < 2$ one cannot give sufficient conditions for inclusion into $M(W_p^m \to W_p^l)$, $\{l\} > 0$ stated in terms of H_q^l.

4.6 Composition Operator on $M(W_p^m \to W_p^l)$

According to a theorem by Hirschman [Hi1], the composition $\varphi(\gamma)$ of $\varphi \in C^{0,\rho}$, $\rho \in (0,1]$, and of a multiplier γ in the space W_2^l, $l \in (0,1)$, represents a multiplier in W_2^r, where $r \in (0, l\rho)$ if $\rho < 1$ and $r = l$ if $\rho = 1$. The case $\rho = 1$ was considered earlier by Beurling [Beu].

The purpose of this section is to present a generalization of Hirschman's result, obtained in [MSh8].

Theorem 4.6.1. *Let $\gamma \in M(W_p^m \to W_p^l)$, $m \ge l$, $0 < l < 1$, $p > 1$. Further, let φ be a function defined on $\mathbb{R}^1$ if $\operatorname{Im} \gamma = 0$ or on $\mathbb{C}^1$ if γ is complex-valued. Suppose that $\varphi(0) = 0$ and for all t and τ, $|\tau| < 1$, the inequality*

$$|\varphi(t+\tau) - \varphi(t)| \le A\,|\tau|^\rho \quad \text{with } \rho \in (0,1]$$

is valid. Then

$$\varphi(\gamma) \in M(W_p^{m-l+r} \to W_p^r),$$

where $r \in (0, l\rho)$ if $\rho < 1$ and $r = l$ if $\rho = 1$. The following estimate holds:

$$\|\varphi(\gamma)\|_{M(W_p^{m-l+r} \to W_p^r)} \leq c\, A(\|\gamma\|^{\rho}_{M(W_p^m \to W_p^l)} + \|\gamma\|_{M(W_p^m \to W_p^l)}) \,.$$

Proof. First we note that for all t and τ

$$|\varphi(t+\tau) - \varphi(t)| \leq A\,(|\tau|^{\rho} + |\tau|) \,. \tag{4.6.1}$$

By (4.6.1), the inclusions $\varphi(\gamma) \in L_{1,\mathrm{unif}}$ and $\varphi(\gamma) \in L_\infty$ follow from $\gamma \in L_{1,\mathrm{unif}}$ and $\gamma \in L_\infty$, respectively.

Consider the case $\rho = 1$. We have

$$\begin{aligned}\|\varphi(\gamma)u\|_{W_p^l} &= \|D_{p,l}[\varphi(\gamma)u]\|_{L_p} + \|\varphi(\gamma)u\|_{L_p}\\ &\leq \|uD_{p,l}\varphi(\gamma)\|_{L_p} + \|\varphi(\gamma)D_{p,l}u\|_{L_p} + \|\varphi(\gamma)u\|_{L_p}.\end{aligned}$$

Using (4.6.1), we see that the sum on the right-hand side does not exceed

$$2A(\|uD_{p,l}\gamma\|_{L_p} + \|\gamma D_{p,l}u\|_{L_p} + \|\gamma u\|_{L_p}).$$

It is clear that

$$\|\gamma D_{p,l}u\|_{L_p} \leq \|D_{p,l}(\gamma u)\|_{L_p} + \|uD_{p,l}\gamma\|_{L_p}.$$

Hence

$$\|\varphi(\gamma)u\|_{W_p^l} \leq 2A(2\|uD_{p,l}\gamma\|_{L_p} + \|\gamma u\|_{W_p^l}) \,.$$

Applying (4.3.12), we get

$$\|\varphi(\gamma)u\|_{W_p^l} \leq 2A\left(c \sup_e \frac{\|D_{p,l}\gamma; e\|_{L_p}}{[C_{p,m}(e)]^{1/p}} + \|\gamma\|_{M(W_p^m \to W_p^l)}\right)\|u\|_{W_p^m}$$

which together with Theorem 4.1.1, gives

$$\|\varphi(\gamma)\|_{M(W_p^m \to W_p^l)} \leq c\,A\,\|\gamma\|_{M(W_p^m \to W_p^l)}.$$

Now let $0 < \rho < 1$. Let us write the integral

$$\iint |\varphi(\gamma(x))u(x) - \varphi(\gamma(y))u(y)|^p |x-y|^{-n-pr}\, dydx$$

as the sum of two integrals of which one is taken over the set

$$\mathfrak{M} = \{(x,y) : |\gamma(y)| \leq |\gamma(x)|\}.$$

It is sufficient to estimate the integral over $\mathfrak{M}$ which obviously does not exceed

$$c\Big(\iint_{\mathfrak{M}} |u(x)|^p|\varphi(\gamma(x)) - \varphi(\gamma(y))|^p|x-y|^{-n-pr}\,dydx$$

$$+\iint_{\mathfrak{M}} |\varphi(\gamma(y))|^p|u(x)-u(y)|^p|x-y|^{-n-pr}\,dydx\,. \tag{4.6.2}$$

We introduce two sets

$$\mathfrak{M}_1(x) = \{y : |\gamma(y)| \le |\gamma(x)|,\ \ |\gamma(x)-\gamma(y)| \le 1\}$$

and

$$\mathfrak{M}_2(x) = \{y : |\gamma(y)| \le |\gamma(x)|,\ \ |\gamma(x)-\gamma(y)| > 1\}.$$

It follows from (4.6.1) that

$$\int_{\mathfrak{M}_1(x)} |\varphi(\gamma(x)) - \varphi(\gamma(y))|^p|x-y|^{-n-pr}\,dy$$
$$\le c\,A^p \int_{\mathfrak{M}_1(x)} |\gamma(x)-\gamma(y)|^{p\rho}|x-y|^{-n-pr}\,dy$$
$$\le c\,A^p\Big(2^{p\rho}|\gamma(x)|^{p\rho}\int_{|x-y|\ge\delta_x}|x-y|^{-n-pr}\,dy$$
$$+\int_{|x-y|<\delta_x}|\gamma(x)-\gamma(y)|^{p\rho}|x-y|^{-n-pr}\,dy\Big)\,, \tag{4.6.3}$$

where δ_x is a nonnegative function to be chosen later. Estimating the last integral by the Hölder inequality, we find that it is dominated by

$$\Big(\int_{|x-y|<\delta_x}|\gamma(x)-\gamma(y)|^p|x-y|^{-n-pl}\,dy\Big)^\rho$$
$$\times\Big(\int_{|x-y|<\delta_x}|x-y|^{-n-p(r-l\rho)/(1-\rho)}\,dy\Big)^{1-\rho}$$
$$\le c(D_{p,l}\gamma(x))^{p\rho}\delta_x^{p(l\rho-r)}\,. \tag{4.6.4}$$

Using (4.6.3) and (4.6.4), we have

$$\int_{\mathfrak{M}_1(x)} |\varphi(\gamma(x)) - \varphi(\gamma(y))|^p|x-y|^{-n-pr}\,dy$$
$$\le c\,A^p\Big(|\gamma(x)|^{p\rho}\delta_x^{-pr} + D_{p,l}\gamma(x))^{p\rho}\delta_x^{p(\rho l-r)}\Big)\,.$$

Minimizing the right-hand side over δ_x, we obtain

$$\int_{\mathfrak{M}_1(x)} |\varphi(\gamma(x)) - \varphi(\gamma(y))|^p|x-y|^{-n-pr}\,dy$$
$$\le c\,A^p|\gamma(x)|^{p(\rho-r/l)}(D_{p,l}\gamma(x))^{pr/l}\,.$$

By (4.6.1),

$$\int_{\mathfrak{M}_2(x)} |\varphi(\gamma(x)) - \varphi(\gamma(y))|^p |x-y|^{-n-pr}\, dy$$
$$\leq c\, A^p \int_{\mathfrak{M}_2(x)} |\gamma(x) - \gamma(y)|^p |x-y|^{-n-pr}\, dy \leq c\, A^p (D_{p,r}\gamma(x))^p\, .$$

Consequently, the first integral in (4.6.2) is not greater than

$$\int |u(x)|^p |\gamma(x)|^{p(\rho - r/l)} (D_{p,l}\gamma(x))^{pr/l}\, dx + \int |u(x)|^p (D_{p,r}\gamma(x))^p\, dx\, . \quad (4.6.5)$$

Let $m \neq l$. We set $v = \Lambda^{-s}|\Lambda^s u|$, where $\Lambda = (-\Delta + 1)^{1/2}$, $s = m - l + r - \varepsilon$ and ε is a small positive number. The properties of the kernel G_r, given in Sect. 3.2.5, and the Hölder inequality imply that

$$v \leq c\, (\Lambda^r v)^{(l-r)/l} (\Lambda^{r-l} v)^{r/l}\, . \quad (4.6.6)$$

Hence the first term in (4.6.5) does not exceed

$$c \int (\Lambda^{r-l} v D_{p,l}\gamma)^{pr/l} (\Lambda^r v)^{p(1-r/l)} |\gamma|^{p(\rho - r/l)}\, dx$$
$$\leq c\Big(\int (\Lambda^{r-l} v)^p (D_{p,l}\gamma(x))^p\, dx\Big)^{r/l} \Big(\int (\Lambda^r v)^p |\gamma|^{p(l\rho - r)/(l-r)}\, dx\Big)^{1-r/l}$$
$$\leq c_1 \Big(\sup_e \frac{\|D_{p,l}\gamma; e\|^p_{L_p}}{C_{p,m}(e)}\Big)^{r/l} \|\Lambda^{r-l} v\|^{pr/l}_{W_p^m}$$
$$\times \Big(\sup_e \frac{\||\gamma|^{(lp-r)/(l-r)}; e\|^p_{L_p}}{C_{p,m-l}(e)}\Big)^{1-r/l} \|\Lambda^r v\|^{p(1-r/l)}_{W_p^{m-l}}.$$

Here we used the Hölder inequality and (4.3.12). Furthermore,

$$\|\Lambda^{r-l} v\|_{W_p^m} = \|\Lambda^{\varepsilon - m}\, |\Lambda^{m-l+r-\varepsilon} u|\|_{W_p^m} \leq c \||\Lambda^{m-l+r-\varepsilon} u|\|_{W_p^\varepsilon}$$
$$\leq c\, \|\Lambda^{m-l+r-\varepsilon} u\|_{W_p^\varepsilon} \leq c_1 \|u\|_{W_p^{m-l+r}}$$

(see (4.3.57)) and similarly

$$\|\Lambda^r v\|_{W_p^{m-l}} = \|\Lambda^{\varepsilon - m + l}\, |\Lambda^{m-l+r-\varepsilon} u|\|_{W_p^{m-l}} \leq c\, \||\Lambda^{m-l+r-\varepsilon} u|\|_{W_p^\varepsilon}$$
$$\leq c_1 \|u\|_{W_p^{m-l+r}}.$$

Now the first term in (4.6.5) is majorized by

$$c\Big(\sup_e \frac{\|D_{p,l}\gamma; e\|^p_{L_p}}{C_{p,m}(e)}\Big)^{r/l} \Big(\sup_e \frac{\||\gamma|^{(l\rho - r)/(l-r)}; e\|^p_{L_p}}{C_{p,m-l}(e)}\Big)^{1-r/l} \|u\|^p_{W_p^{m-l+r}}\, . \quad (4.6.7)$$

The same bound can be obtained for $m = l$, if we put $v = \Lambda^{-r}|\Lambda^r u|$ and apply the inequality

$$v \le c(\mathcal{M}\Lambda^r v)^{(l-r)/l}(\Lambda^{r-l}v)^{r/l}$$

instead of (4.6.6) (see Lemma 1.2.5).

Taking into account that

$$\begin{aligned}\frac{\||\gamma|^{(l\rho-r)/(l-r)}; e\|_{L_p}^p}{C_{p,m-l}(e)} &\le \Big(\frac{\|\gamma; e\|_{L_p}^p}{C_{p,m-l}(e)}\Big)^{(l\rho-r)/(l-r)}\Big(\frac{\mathrm{mes}_n e}{C_{p,m-l}(e)}\Big)^{l(1-\rho)/(l-r)}\\ &\le \Big(\frac{\|\gamma; e\|_{L_p}^p}{C_{p,m-l}(e)}\Big)^{(l\rho-r)/(l-r)},\end{aligned}$$

we obtain from Corollary 4.3.1 and Lemma 4.3.6 that (4.6.7), and consequently the first term in (4.6.5), do not exceed

$$c\,\|\gamma\|^{p\rho}_{M(W_p^m \to W_p^l)}\|u\|^p_{W_p^{m-l+r}}\,.$$

In view of Corollary 4.3.1 and the imbedding of $M(W_p^m \to W_p^l)$ into $M(W_p^{m-l+r} \to W_p^r)$, the second term in (4.6.5) is estimated in the following way:

$$\begin{aligned}\int |u(x)|^p (D_{p,l}\gamma(x))^p\,dx &\le c\sup_e \frac{\|D_{p,l}\gamma; e\|^p_{L_p}}{C_{p,m-l+r}(e)}\|u\|^p_{W_p^{m-l+r}}\\ &\le c_1\|\gamma\|^p_{M(W_p^m \to W_p^l)}\|u\|^p_{W_p^{m-l+r}}.\end{aligned}$$

The proof is complete. □

5

The Space $M(B_1^m \to B_1^l)$

A complete description of the space $M(B_1^m(\mathbb{R}^n) \to B_1^l(\mathbb{R}^n))$, $0 \le l \le m$, is given in Sect. 5.3 for integer l and in Sect. 5.4 for noninteger l. Sections 5.1 and 5.2 are auxiliary. In Sect. 5.5 we survey some results on multipliers in Besov, BMO, and related function spaces.

5.1 Trace Inequality for Functions in $B_1^l(\mathbb{R}^n)$

Let $l = k + \alpha$, where $\alpha \in (0,1]$ and k is a nonnegative integer. In concert with (4.1.1) and (4.1.2) we use the notations

$$(\mathfrak{D}_{1,l}u)(x) = \int_{\mathbb{R}^n} |\Delta_h^{(2)} \nabla_k u(x)|\, |h|^{-n-\alpha} dh, \tag{5.1.1}$$

and

$$\|u; \mathbb{R}^n\|_{B_1^l} = \|\mathfrak{D}_{1,l}u; \mathbb{R}^n\|_{L_1} + \|u; \mathbb{R}^n\|_{L_1}. \tag{5.1.2}$$

Let us adopt the notation $z = (x, y)$ and $\zeta = (\xi, \eta)$ for points of $\mathbb{R}^{n+s}$, where $x, \xi \in \mathbb{R}^n$ and $y, \eta \in \mathbb{R}^s$. Further, let $\mathcal{B}_r^{(d)}(q)$ be a d-dimensional ball with centre $q \in \mathbb{R}^d$. If $d = n$, we write $\mathcal{B}_r(q)$ instead of $\mathcal{B}_r^{(n)}(q)$.

In this section we prove the following assertion.

Theorem 5.1.1. [Maz11] *The best constant* K_1 *in*

$$\int_{\mathbb{R}^n} |u| d\mu \le K \, \|\mathfrak{D}_{1,l}u; \mathbb{R}^n\|_{L_1} \tag{5.1.3}$$

is equivalent to

$$Q = \sup_{x \in \mathbb{R}^n, \rho > 0} \rho^{l-n} \mu(\mathcal{B}_\rho(x)).$$

5.1.1 Auxiliary Facts

Lemma 5.1.1. *Let g be an open subset of $\mathbb{R}^{n+s}$ with compact closure and smooth boundary ∂g such that*

$$\int_{\mathcal{B}_r^{(n+s)}(z)\cap g} |\eta|^\alpha d\zeta \Big(\int_{\mathcal{B}_r^{(n+s)}(z)} |\eta|^\alpha d\zeta \Big)^{-1} = 1/2, \tag{5.1.4}$$

where $\alpha > -s$. Then

$$\int_{\mathcal{B}_r^{(n+s)}(z)\cap \partial g} |\eta|^\alpha d\sigma(\zeta) \geq c\, r^{n+s-1}(r+|y|)^\alpha, \tag{5.1.5}$$

where σ is the $(n+s-1)$-dimensional area.

The proof is based on the following lemma.

Lemma 5.1.2. *Let $\alpha > -s$ for $s > 1$ and $0 \geq \alpha > -1$ for $s = 1$. Then for any $v \in C^\infty(\overline{\mathcal{B}_r^{(n+s)}})$ there exists a constant V such that*

$$\int_{\mathcal{B}_r^{(n+s)}} |v(\zeta) - V||\eta|^\alpha d\zeta \leq c\, r \int_{\mathcal{B}_r^{(n+s)}} |\eta|^\alpha |\nabla v(\zeta)| d\zeta. \tag{5.1.6}$$

Proof. It suffices to derive (5.1.6) for $r = 1$. We adopt the notation $\mathcal{B}_1^{(s)} \times \mathcal{B}_1 = Q$. By $R(\zeta)$ we denote the distance from the point $\zeta \in \partial Q$ to the origin i.e. $R(\zeta) = (1+|\zeta|^2)^{1/2}$ for $|\eta| = 1, |\xi| < 1$, and $R(\zeta) = (1+|\eta|^2)^{1/2}$ for $|\xi| = 1, |\eta| < 1$. Since $\mathcal{B}_1^{(n+s)}$ is the bi-Lipschitz image of Q under the mapping $\zeta \to \zeta/R(\zeta)$, we can deduce (5.1.6) from the inequality

$$\int_Q |v(\zeta) - V||\eta|^\alpha d\zeta \leq c \int_Q |\nabla v(\zeta)||\eta|^\alpha d\zeta. \tag{5.1.7}$$

Let us show that (5.1.7) holds. Since

$$(s+\alpha)|\eta|^\alpha = \operatorname{div}(|\eta|^\alpha \eta),$$

we find by integrating by parts on the left-hand side of (5.1.7) that it does not exceed

$$(s+\alpha)^{-1}\Big(\int_Q |\nabla v||\eta|^{\alpha+1} d\zeta + \int_{\mathcal{B}_1^{(n)}} d\xi \int_{\partial \mathcal{B}_1^{(s)}} |v(\zeta) - V| ds(\eta) \Big). \tag{5.1.8}$$

We put $T = \mathcal{B}_1 \times (\mathcal{B}_1^{(s)} \backslash \mathcal{B}_{1/2}^{(s)})$. Let $s > 1$. The second term in (5.1.8) is not greater than

$$c \int_T |\nabla v| d\zeta + c \int_T |v - V| d\zeta.$$

Hence, taking the mean value of v in T as V, we get (5.1.7) from (5.1.8). If $s = 1$ then the set T has two components: $T_+ = \mathcal{B}_1 \times (1/2, 1)$ and $T_- = \mathcal{B}_1 \times (-1, -1/2)$. The same arguments as for the case $s > 1$ lead to

$$\int_{\mathcal{B}_1} |v(\xi, \pm 1) - V_\pm|\, d\xi \le c \int_{T_\pm} |\nabla v(\zeta)| d\zeta \le c \int_Q |\nabla v(\zeta)||\eta|^\alpha d\zeta,$$

where $V_\pm$ is the mean value of v in $T_\pm$. It remains to be noted that

$$|V_+ - V_-| \le c \int_{\mathcal{B}_1} d\xi \int_{-1}^1 \left|\frac{\partial v}{\partial \eta}\right| d\eta \le c \int_Q |\nabla v(\zeta)||\eta|^\alpha d\zeta$$

for $\alpha \le 0$. Thus, for $s = 1$, inequality (5.1.7) follows with V_+ or V_- in place of V.

Proof of Lemma 5.1.1. For the sake of brevity, let $\mathcal{B} = \mathcal{B}_r^{(n+s)}(z)$. We let v in (5.1.6) be the mollification χ_ρ of the characteristic function of g. Then the left-hand side is bounded from below by the sum

$$|1 - V| \int_{e_1} |\eta|^\alpha d\zeta + |V| \int_{e_0} |\eta|^\alpha d\zeta,$$

where $e_i = \{z \in \mathcal{B} : \chi_\rho(z) = i\}, i = 0, 1$.

Let ϵ be an arbitrarily small positive number. By (5.1.4),

$$(\frac{1}{2} - \epsilon)(|1 - V| + |V|) \int_{\mathcal{B}} |\eta|^\alpha d\zeta \le c\, r \int_{\mathcal{B}} |\eta|^\alpha |\nabla \chi_\rho(\zeta)| d\zeta$$

for sufficiently small ρ. Consequently

$$\frac{1}{2} \int_{\mathcal{B}} |y|^\alpha d\zeta \le c\, r \limsup_{\rho \to +0} \int_{\mathcal{B}} |\eta|^\alpha |\nabla_\zeta \chi_\rho(\zeta)| d\zeta = c\, r \int_{\mathcal{B} \cap \partial g} |\eta|^\alpha d\sigma(\zeta).$$

(The last equality can be derived from Corollary 1.1.1.) It remains to note that

$$\int_{\mathcal{B}} |y|^\alpha d\zeta \ge c\, r^{n+s} (r + |y|)^\alpha.$$

□

Lemma 5.1.3. *Let ν be a measure in $\mathbb{R}^{n+s}$ and let $\alpha > -s$. The best constant K_1 in*

$$\int_{\mathbb{R}^{n+s}} |U| d\nu \le K_1 \int_{\mathbb{R}^{n+s}} |y|^\alpha |\nabla_z U| dz, \qquad U \in C_0^\infty(\mathbb{R}^{n+s}), \tag{5.1.9}$$

is equivalent to

$$Q_1 = \sup_{z, \rho > 0} (\rho + |y|)^{-\alpha} \rho^{1-n+s} \nu(\mathcal{B}_\rho^{(n+s)}(z)). \tag{5.1.10}$$

Proof. 1. First let $m > 1$ or $0 \ge \alpha > -1, m = 1$. According to Proposition 1.1.1,

$$K_1 = \sup_g \frac{\nu(g)}{\displaystyle\int_{\partial g} |y|^\alpha d\sigma},$$

where g is an arbitrary open subset of $\mathbb{R}^{n+s}$ with compact closure and smooth boundary. We show that for any g there exists a covering of g by a sequence of balls $\mathcal{B}_{\rho_i}^{(n+s)}(z_i)$, $i = 1, 2, \cdots$, such that

$$\sum_i \rho_i^{n+s-1}(\rho_i + |y_i|)^\alpha \le c \int_{\partial g} |y|^\alpha d\sigma. \tag{5.1.11}$$

Every point $z \in g$ is the centre of a ball $\mathcal{B}_r^{(n+s)}(z)$ for which (5.1.4) holds. In fact, the ratio on the left-hand side of (5.1.4) is a continuous function in r, equal to one for small values of r and tending to zero as $r \to \infty$. By Lemma 1.1.3 there exists a sequence of disjoint balls $\mathcal{B}_{r_i}^{(n+s)}(z_i)$ such that

$$g \subset \bigcup_{i=1}^{\infty} \mathcal{B}_{3r_i}^{(n+s)}(z_i).$$

Lemma 5.1.1 implies that

$$\int_{\mathcal{B}_{r_i}^{(n+s)}(z_i)\cap\partial g} |y|^\alpha d\sigma \ge c\, r_i^{n+s-1}(r_i + |y_i|)^\alpha.$$

Consequently $\{\mathcal{B}_{3r_i}^{(n+s)}(z_i)\}_{i\ge 1}$ is the required covering. Obviously,

$$\nu(g) \le \sum_i \nu(\mathcal{B}_{3r_i}^{(n+s)}(z_i)) \le Q_1 \sum_i r_i^{n+s-1}(r_i + |y_i|^\alpha)$$

$$\le c\, Q_1 \int_{\partial g} |y|^\alpha d\sigma.$$

Thus $K_1 \le cQ_1$.

2. Let $s = 1$ and $\alpha > 0$. We construct a covering of the set $\{\zeta : \eta \ne 0\}$ by balls $\mathcal{B}^{(j)}$ such that the radius ρ_j of $\mathcal{B}^{(j)}$ is equal to the distance from $\mathcal{B}^{(j)}$ to the hyperplane $\{\zeta : \eta = 0\}$. By $\{\varphi_j\}$ we denote a partition of unity subordinate to the covering $\{\mathcal{B}^{(j)}\}$ with $|\nabla\varphi_j| \le c\,\rho_j^{-1}$ (see [St2], Ch. VI, §1). Using the present assertion for $\alpha = 0$, we get

$$\int_{\mathbb{R}^{n+1}} |\varphi_j u| d\nu \le c \sup_{\rho, z} \rho^{-n} \nu_j(\mathcal{B}_\rho^{(n+1)}(z)) \|\nabla(\varphi_j u); \mathbb{R}^{n+1}\|_{L_1},$$

where ν_j is the restriction of the measure ν to $\mathcal{B}^{(j)}$. It is clear that

$$\sup_{\rho, z} \rho^{-n} \nu_j(\mathcal{B}_\rho^{(n+1)}(z)) \le c \sup_{\rho \le \rho_j, z \in \mathcal{B}^{(j)}} \rho^{-n} \nu(\mathcal{B}_\rho^{(n+1)}(z)).$$

Therefore,

$$\int_{\mathbb{R}^{n+1}} |\varphi_j u| d\nu$$
$$\leq c \sup_{\rho \leq \rho_j, z \in \mathcal{B}^{(j)}} (\rho + \rho_j)^{-\alpha} \rho^{-n} \nu(\mathcal{B}_\rho^{(n+1)}(z)) \int_{\mathbb{R}^{n+1}} |\nabla(\varphi_j u)||\eta|^\alpha d\zeta.$$

Summing over j, we find that

$$\int_{\mathbb{R}^{n+1}} |u| d\nu \leq c K_1 \Big(\int_{\mathbb{R}^{n+1}} |\nabla u||\eta|^\alpha d\zeta + \int_{\mathbb{R}^{n+1}} |u||\eta|^\alpha d\zeta \Big).$$

Since

$$\int_{\mathbb{R}^{n+1}} |u||\eta|^{\alpha-1} d\zeta \leq \alpha^{-1} \int_{\mathbb{R}^{n+1}} |\nabla u||\eta|^\alpha d\zeta$$

for $\alpha > 0$, we also have $K_1 \leq c Q_1$ for $s = 1$, $\alpha > 0$.

3. To obtain the converse estimate, we put $U(\xi) = \varphi(\rho^{-1}(\zeta - z))$ into (5.1.9), where $\varphi \in C_0^\infty(\mathcal{B}_2^{(n+s)})$, $\varphi = 1$ on $\mathcal{B}_1^{(n+s)}$. We notice that

$$\int_{\mathcal{B}_{2\rho}^{(n+s)}(z)} |\eta|^\alpha |\nabla_\zeta U| d\zeta \leq c\, \rho^{-1} \int_{\mathcal{B}_{2\rho}^{(n+s)}(z)} |\eta|^\alpha d\zeta \leq c\, \rho^{n+s-1} (\rho + |y|)^\alpha.$$

The proof is complete. □

Corollary 5.1.1. *Let ν be a measure in $\mathbb{R}^n$ and let $\alpha > -s$. The best constant in (5.1.9) is equivalent to*

$$\sup_{x \in \mathbb{R}^n, \rho > 0} \rho^{1-n-s-\alpha} \nu(\mathcal{B}_\rho^{(n+s)}(z)).$$

To prove this assertion, it suffices to note that the value Q_1 defined in (5.1.10) is equivalent to the last supremum if supp $\nu \subset \mathbb{R}^n$.

5.1.2 Main Result

Now we are in a position to prove Theorem 5.1.1.

The estimates $K_1 \geq c\, Q$ and $K_2 \geq c\, Q$ can be be obtained quite simply. It suffices to put $u(\xi) = \varphi(\rho^{-1}(x - \xi))$ into (5.1.3), where $\varphi \in C_0^\infty(\mathcal{B}_2)$ and $\varphi = 1$ on $\mathcal{B}_1$, and to note that

$$\int_{\mathbb{R}^n} |u| d\mu \geq \mu(\mathcal{B}_\rho(x)), \quad \|\mathfrak{D}_{1,l} u; \mathbb{R}^n\|_{L_1} = c\, \rho^{n-l}.$$

Now we obtain the estimates $K \leq cQ$. Let $l \in (0, 1)$. According to Corollary 5.1.1,

$$\int_{\mathbb{R}^n} |u| d\mu \leq c\, Q \int_{\mathbb{R}^{n+1}} |y|^{-l} |\nabla U| dz, \tag{5.1.12}$$

where $U \in C_0^\infty(\mathbb{R}^{n+1})$ is an arbitrary extension of $u \in C_0^\infty(\mathbb{R}^n)$ to $\mathbb{R}^{n+1}$. If $l = 1$, then by Theorem 1.1.1

$$\int_{\mathbb{R}^n} |u| d\mu \leq c Q \int_{\mathbb{R}^{n+1}} |\nabla_2 U| dz. \tag{5.1.13}$$

It is known (see [Usp]) that

$$\|\mathfrak{D}_{1,l} u; \mathbb{R}^n\|_{L_1} \sim \inf_U \int_{\mathbb{R}^{n+1}} |y|^{-l} |\nabla U| dz, \quad l \in (0,1),$$

$$\|\mathfrak{D}_{1,1} u; \mathbb{R}^n\|_{L_1} \sim \inf_U \int_{\mathbb{R}^{n+1}} |\nabla_2 U| dz.$$

Hence, minimizing the right-hand sides of (5.1.12) and (5.1.13) over all extensions U, we arrive at

$$\int_{\mathbb{R}^n} |u| d\mu \leq c Q \|\mathfrak{D}_{1,l} u; \mathbb{R}^n\|_{L_1}, \qquad l \in (0,1].$$

Suppose that the estimate $K \leq cQ$ is proved under the condition $l \in (m-2, m-1]$, where m is an integer, $m \geq 2$. Duplicating the argument used in part *(ii)* of the proof of Theorem 1.1.1, we obtain the required estimate for $l \in (m-1, m]$. □

Remark 5.1.1. It follows from Theorem 5.1.1 that (5.1.3) with $l > n$ is valid only in the trivial case $\mu = 0$, and for $l = n$ if and only if the measure μ is finite.

Theorem 5.1.2. *The best constant K_0 in*

$$\int_{\mathbb{R}^n} |u| d\mu \leq K_0 \|u; \mathbb{R}^n\|_{B_1^l}$$

is equivalent to

$$Q_0 = \sup_{x \in \mathbb{R}^n, \rho \in (0,1)} \rho^{l-n} \mu(\mathcal{B}_\rho(x)).$$

Proof. The estimate $K_0 \geq c\, Q_0$ can be obtained in the same way as the estimate $K \geq c\, Q$ in Theorem 5.1.1. To prove the converse inequality, we use the sequence $\{\eta_j\}_{j \geq 0}$ defined in Theorem 3.1.2. We apply Theorem 5.1.1 to the integral

$$\int_{\mathbb{R}^n} |\eta_j u| d\mu_j,$$

where μ_j is the restriction of μ to the support of η_j. Then

$$\int_{\mathbb{R}^n} |u| d\mu \leq c \sum_j \int_{\mathbb{R}^n} |\eta_j u| d\mu_j \leq c Q_0 \sum_j \|\mathfrak{D}_{1,l}(\eta_j u)\|_{L_1}.$$

Since

$$\|u;\mathbb{R}^n\|_{B_1^l} \sim \sum_j \|\eta_j u;\mathbb{R}^n\|_{B_1^l},$$

the last sum does not exceed $c\,\|u;\mathbb{R}^n\|_{B_1^l}$. □

Remark 5.1.2. It is clear that

$$Q_0 = \sup_{x\in\mathbb{R}^n} \mu(\mathcal{B}_1(x))$$

for $l \geq n$.

5.2 Properties of Functions in the Space $B_1^k(\mathbb{R}^n)$

5.2.1 Trace and Imbedding Properties

We start with the statement of a well-known trace and extension result.

Lemma 5.2.1. [Usp] *Suppose that $m \geq 1$.*

(i) Let U be an arbitrary function in the space $W_1^{l+1}(\mathbb{R}_+^{n+1})$. Then the limit

$$u(x) = \lim_{\rho\to 0} U(x,\rho)$$

exists for almost all $x \in \mathbb{R}^n$, the function u belongs to the space $B_1^l(\mathbb{R}^n)$, and

$$\|u;\mathbb{R}^n\|_{B_1^l} \leq c\,\|U;\mathbb{R}_+^{n+1}\|_{W_1^{l+1}}.$$

(ii) Let Tu denote the action of the Poisson operator on the function $u \in L_{1,\mathrm{unif}}(\mathbb{R}^n)$ defined by (3.2.38). *Then*

$$\|Tu;\mathbb{R}_+^{n+1}\|_{W_1^{l+1}} \leq c\,\|u;\mathbb{R}^n\|_{B_1^l}.$$

The next lemma contains an interpolation inequality for functions in $B_1^l(\mathbb{R}^n)$.

Lemma 5.2.2. *Let $u \in B_1^l(\mathbb{R}^n)$, where l is an integer, $l \geq 1$. Then for $j = 0,\dots,l-1$*

$$\|u;\mathbb{R}^n\|_{B_1^{l-j}} \leq c\,\|u;\mathbb{R}^n\|_{B_1^l}^{(l-j)/l}\,\|u;\mathbb{R}^n\|_{L_1}^{j/l}. \tag{5.2.1}$$

Proof. We introduce the function

$$(\mathcal{D}_s^{(q)}u)(x) = \int_{\mathbb{R}^n} \frac{|\Delta_h^{(q)}u(x)|}{|h|^{n+s}}\,dh \tag{5.2.2}$$

with any integer $q > s$, where $\Delta_h^{(q)} u(x)$ is the difference of order q defined by

$$\Delta_h^{(q)} u(x) = \sum_{i=0}^{q} \binom{q}{i} (-1)^i \, u(x + (q-i)h).$$

Given $s > 0$, the equivalence relation

$$\|u; \mathbb{R}^n\|_{B_1^s} \sim \|\mathcal{D}_s^{(q)} u; \mathbb{R}^n\|_{L_1} + \|u; \mathbb{R}^n\|_{L_1} \tag{5.2.3}$$

holds for all values of q greater than s (see [Tr4], Sect. 3.5.3). Let $q > l$. Adding together the two inequalities

$$\int_{\mathbb{R}^n} \int_{\mathcal{B}} \frac{|\Delta_h^{(q)} u(x)|}{|h|^{n+l-j}} \, dh \, dx \le \int_{\mathbb{R}^n} \int_{\mathcal{B}} \frac{|\Delta_h^{(q)} u(x)|}{|h|^{n+l}} \, dh \, dx$$

and

$$\int_{\mathbb{R}^n} \int_{\mathbb{R}^n \setminus \mathcal{B}} \frac{|\Delta_h^{(q)} u(x)|}{|h|^{n+l-j}} \, dh \, dx \le c \, \|u; \mathbb{R}^n\|_{L_1(\mathbb{R}^n)},$$

we find that

$$\begin{aligned} \|\mathcal{D}_{l-j}^{(q)} u; \mathbb{R}^n\|_{L_1} &= \int_{\mathbb{R}^n} \int_{\mathbb{R}^n} \frac{|\Delta_h^{(q)} u(x)|}{|h|^{n+l-j}} \, dh \, dx \\ &\le c \left(\|\mathcal{D}_l^{(q)} u; \mathbb{R}^n\|_{L_1} + \|u; \mathbb{R}^n\|_{L_1} \right). \end{aligned} \tag{5.2.4}$$

By (5.2.3),

$$\|\mathfrak{D}_{1,l-j} u; \mathbb{R}^n\|_{L_1} \le c \left(\|\mathcal{D}_{l-j}^{(q)} u; \mathbb{R}^n\|_{L_1} + \|u; \mathbb{R}^n\|_{L_1} \right).$$

This, together with (5.2.4) and (5.2.3), leads to

$$\|\mathfrak{D}_{1,l-j} u; \mathbb{R}^n\|_{L_1} \le c \left(\|\mathfrak{D}_{1,l} u; \mathbb{R}^n\|_{L_1} + \|u; \mathbb{R}^n\|_{L_1} \right). \tag{5.2.5}$$

The interpolation inequality (5.2.1) follows from (5.2.5) by dilation. □

Let $\mathcal{B}$ denote the unit ball centered at the origin and let l be a positive integer. We introduce the space $B_1^l(\mathcal{B})$ of functions on $\mathcal{B}$ with finite norm

$$\|u; \mathcal{B}\|_{B_1^l} = \sum_{j=0}^{l-1} \|\nabla_j u; \mathcal{B}\|_{L_1} + \sum_{j=0}^{l-1} \int_{\mathcal{B}} \int_{\mathcal{B}} |(\Delta_y^{(2)} \nabla_j u)(x)| \frac{dx \, dy}{|x-y|^{n+1}}.$$

A local variant of inequality (5.2.1) is contained in the next statement.

Corollary 5.2.1. *Let $u \in B_1^l(\mathcal{B})$. Then, for any $j = 0, \ldots, l-1$*

$$\|u; \mathcal{B}\|_{B_1^{l-j}} \le c \, \|u; \mathcal{B}\|_{B_1^l}^{(l-j)/l} \, \|u; \mathcal{B}\|_{L_1}^{j/l}. \tag{5.2.6}$$

Proof. It is well known (see [Tr4], Sect. 4.5) that u can be extended onto $\mathbb{R}^n$ so that

$$\|u;\mathbb{R}^n\|_{B_1^l} \leq c\,\|u;\mathcal{B}\|_{B_1^l} \tag{5.2.7}$$

and

$$\|u;\mathbb{R}^n\|_{L_1} \leq c\,\|u;\mathcal{B}\|_{L_1}. \tag{5.2.8}$$

These inequalities, combined with Lemma 5.2.2, give (5.2.6). □

We need the following Hardy-type inequality.

Lemma 5.2.3. *Let $u \in B_1^l(\mathbb{R}^n)$, where l is an integer, $1 \leq l < n$. Then*

$$\int_{\mathbb{R}^n} |x|^{-l}\,|u(x)|\,dx \leq c\,\|u;\mathbb{R}^n\|_{B_1^l}. \tag{5.2.9}$$

Proof. Let $U \in W_1^{l+1}(\mathbb{R}_+^{n+1})$ be an arbitrary extension of u. We have

$$\int_{\mathbb{R}^n} |x|^{-l}\,|u(x)|\,dx = \frac{2^l-1}{l}\int_0^\infty \frac{dr}{r^{l+1}}\int_{\mathcal{B}_{2r}\setminus\mathcal{B}_r} |u(x)|\,dx. \tag{5.2.10}$$

To estimate the right-hand side of (5.2.10), we use the standard trace inequality

$$\int_{\mathcal{B}_{2r}\setminus\mathcal{B}_r} |u(x)|\,dx \leq c\int_{\mathcal{G}_{2r}\setminus\mathcal{G}_r} \big(r^{-1}|U(z)| + |\nabla U(z)|\big)dz,$$

where $\mathcal{G}_r = \mathcal{B}_r^{(n+1)} \cap \mathbb{R}_+^{n+1}$. Together with (5.2.10), this inequality implies that

$$\int_{\mathbb{R}^n} |x|^{-l}\,|u(x)|\,dx \leq c\int_{\mathbb{R}_+^{n+1}} \Big(\frac{|U(z)|}{|z|} + |\nabla U(z)|\Big)\frac{dz}{|z|^l}. \tag{5.2.11}$$

Iterating the Hardy-type inequality

$$\int_{\mathbb{R}_+^{n+1}} |\nabla_j U(z)|\frac{dz}{|z|^{l+1-j}} \leq c\int_{\mathbb{R}_+^{n+1}} |\nabla_{j+1} U(z)|\frac{dz}{|z|^{l-j}}$$

with $j = 0,\dots,l-1$, we find that the right-hand side in (5.2.11) is dominated by

$$c\int_{\mathbb{R}_+^{n+1}} |\nabla_{l+1} U(z)|dz.$$

Taking into account that u is the trace of U on $\mathbb{R}^n$ and using part (i) of Lemma 5.2.1, we complete the proof. □

The next lemma contains two more inequalities for intermediate derivatives of functions given on the ball $\mathcal{B}_r$. The integral over $(\mathcal{B}_r)^2$ stands for the double integral over $\mathcal{B}_r$.

Lemma 5.2.4. *Let l be a positive integer and let $j = 0, \dots, l-1$. Then for any $r \in (0,1]$*

$$r^{j-l}\|\nabla_j u; \mathcal{B}_r\|_{L_1} \le c\Big(\int_{(\mathcal{B}_r)^2} |(\Delta^{(2)}\nabla_{l-1}u)(x,y)| \frac{dx\,dy}{|x-y|^{n+1}} + r^{-l}\|u; \mathcal{B}_r\|_{L_1}\Big) \tag{5.2.12}$$

and

$$r^{j+1-l}\int_{(\mathcal{B}_r)^2} |(\Delta^{(2)}\nabla_j u)(x,y)| \frac{dx\,dy}{|x-y|^{n+1}} \le c\Big(\int_{(\mathcal{B}_r)^2} |(\Delta^{(2)}\nabla_{l-1}u)(x,y)| \frac{dx\,dy}{|x-y|^{n+1}} + r^{-l}\|u; \mathcal{B}_r\|_{L_1}\Big), \tag{5.2.13}$$

where

$$(\Delta^{(2)}v)(x,y) = v(x) - 2v\Big(\frac{x+y}{2}\Big) + v(y).$$

Proof. By dilation, the proof reduces to the case $r = 1$. It is well known that for $j = 1, \dots, l-2$

$$\|\nabla_j u; \mathcal{B}\|_{L_1} \le c\,(\|\nabla_{l-1}u; \mathcal{B}\|_{L_1} + \|u; \mathcal{B}\|_{L_1}). \tag{5.2.14}$$

Hence, it suffices to prove (5.2.12) for $j = l-1$. We introduce the function

$$\varphi \in C_0^\infty(\mathcal{B}) \quad \text{subject to} \quad \int_{\mathbb{R}^n} \varphi(y)dy = 1.$$

We have

$$\nabla_{l-1}u(x) = \int_{\mathcal{B}} \varphi(y)\nabla_{l-1}u(x)dy$$

$$= \int_{\mathcal{B}} \varphi(y)\Delta^{(2)}\nabla_{l-1}u(x,y)dy + \int_{\mathcal{B}} \varphi(y)\big(2\nabla_{l-1}u\big(\frac{x+y}{2}\big) - \nabla_{l-1}u(y)\big)dy.$$

Integrating by parts in the last integral, we obtain

$$\nabla_{l-1}u(x) = \int_{\mathcal{B}} \varphi(y)\Delta^{(2)}\nabla_{l-1}u(x,y)dy + (-1)^{l-1}\int_{\mathcal{B}} \big(2^{2-l}u\big(\frac{x+y}{2}\big) - u(y)\big)(\nabla_{l-1}\varphi)(y)dy. \tag{5.2.15}$$

Therefore,

$$\int_{\mathcal{B}} |\nabla_{l-1}u(x)|dx \le \int_{\mathcal{B}}\Big|\int_{\mathcal{B}} \varphi(y)\big(\Delta^{(2)}\nabla_{l-1}u\big)(x,y)dy\Big|dx + c\,\|u; \mathcal{B}\|_{L_1}.$$

Since the right-hand side does not exceed

$$c\Big(\int_{\mathcal{B}}\int_{\mathcal{B}}\big|\big(\Delta^{(2)}\nabla_{l-1}u\big)(x,y)\big| \frac{dydx}{|x-y|^{n+1}} + \|u; \mathcal{B}\|_{L_1}\Big),$$

we arrive at (5.2.12) with $j = l-1$. The proof of (5.2.12) is complete. Finally, (5.2.13) results from the definition of the space $B_1^l(\mathcal{B})$ and inequalities (5.2.6) and (5.2.12). □

5.2.2 Auxiliary Estimates for the Poisson Operator

We deal with the Poisson operator T defined by (3.2.38).

Lemma 5.2.5. *Let* $\gamma \in W^{l-1}_{1,\mathrm{loc}}(\mathbb{R}^n)$. *Then*

$$\int_0^\infty \Big| \frac{\partial^{l+1}(T\gamma)}{\partial y^{l+1}} \Big| dy \le c\,(\mathfrak{D}_{1,l}\gamma)(x), \tag{5.2.16}$$

where $(\mathfrak{D}_l\gamma)(x)$ *is defined by* (5.1.1).

Proof. For every n-dimensional multi-index α with $|\alpha| = 2$,

$$D_x^\alpha (T\gamma)(x,y) = y^{-n-2} \int_{\mathbb{R}^n} (D^\alpha \zeta)\Big(\frac{\xi - x}{y}\Big)\gamma(\xi)d\xi$$

$$= y^{-n-2} \int_{\mathbb{R}^n} (D^\alpha \zeta)\Big(\frac{h}{y}\Big)\gamma(x+h)dh = y^{-n-2} \int_{\mathbb{R}^n} \zeta_{0,\alpha}\Big(\frac{h}{y}\Big)\Delta_h^{(2)}\gamma(x)dh, \tag{5.2.17}$$

where

$$\zeta_{0,\alpha} = \frac{1}{2}(D^\alpha \zeta)(\xi).$$

The last equality in (5.2.17) holds because $D^\alpha\zeta$ is even and satisfies

$$\int_{\mathbb{R}^n} D^\alpha \zeta(t)dt = 0.$$

Since

$$(T\gamma)(x,y) = \frac{1}{2} y^{-n} \int_{\mathbb{R}^n} \zeta\Big(\frac{h}{y}\Big)\Delta_h^{(2)}\gamma(x)dh + \gamma(x), \tag{5.2.18}$$

it follows for every n-dimensional multi-index β with $|\beta| = 1$ that

$$\frac{\partial}{\partial y} D_x^\beta (T\gamma)(x,y) = \frac{1}{2} \int_{\mathbb{R}^n} \frac{\partial}{\partial y}\Big(y^{-n-1}(D_y^\beta\zeta)\Big(\frac{h}{y}\Big)\Big)\Delta_h^{(2)}\gamma(x)dh$$

$$= y^{-n-2} \int_{\mathbb{R}^n} \zeta_{0,\beta}\Big(\frac{h}{y}\Big)\Delta_h^{(2)}\gamma(x)dh, \tag{5.2.19}$$

where

$$\zeta_{0,\beta} = \frac{-1}{2}\big((n+1+\langle \xi, \nabla\rangle)D^\beta\zeta\big)(\xi).$$

Suppose that $l \ge 2$. Let $\tau = \alpha + \delta$, where $|\tau| = l+1$, $|\alpha| = 2$, $|\delta| = l-1$. By (5.2.17),

$$D_x^\tau (T\gamma)(x,y) = y^{-n-2} \int_{\mathbb{R}^n} \zeta_{0,\alpha}\big(\frac{h}{y}\big)\Delta_h^{(2)}(D^\delta\gamma)(x)dh. \tag{5.2.20}$$

Next, let $\tau = \beta + \delta$, where $|\tau| = l$, $|\beta| = 1$, $|\delta| = l - 1$. By (5.2.19),

$$\frac{\partial}{\partial y} D_x^\tau (T\gamma)(x,y) = y^{-n-2} \int_{\mathbb{R}^n} \zeta_{0,\beta}\Big(\frac{h}{y}\Big) \Delta_h^{(2)} (D^\delta \gamma)(x) dh. \tag{5.2.21}$$

Suppose that $l+1$ is even, then the harmonicity of $T\gamma$ implies that

$$\frac{\partial^{l+1}}{\partial y^{l+1}} (T\gamma)(x,y) = (-\Delta_x)^{(l+1)/2} (T\gamma)(x,y).$$

Hence, by (5.2.20),

$$\Big|\frac{\partial^{l+1}}{\partial y^{l+1}} (T\gamma)(x,y)\Big| \le c\, y^{-n-2} \int_{\mathbb{R}^n} \zeta_1\Big(\frac{h}{y}\Big)\, |(\Delta_h^{(2)} \nabla_{l-1}\gamma)(x)| dh, \tag{5.2.22}$$

where

$$0 < \zeta_1(\xi) \le c\,(1+|\xi|)^{-n-3}. \tag{5.2.23}$$

If $l+1$ is odd, then we have by harmonicity of $T\gamma$

$$\frac{\partial^{l+1}}{\partial y^{l+1}} (T\gamma)(x,y) = \frac{\partial}{\partial y} (-\Delta_x)^{l/2} (T\gamma)(x,y).$$

This, together with (5.2.21), gives

$$\Big|\frac{\partial^{l+1}}{\partial y^{l+1}} (T\gamma)(x,y)\Big| \le c\, y^{-n-2} \int_{\mathbb{R}^n} \zeta_2\Big(\frac{h}{y}\Big)\, |(\Delta_h^{(2)} \nabla_{l-1}\gamma)(x)| dh, \tag{5.2.24}$$

where

$$0 < \zeta_2(\xi) \le c\,(1+|\xi|)^{-n-2}. \tag{5.2.25}$$

Hence,

$$\begin{aligned}\int_0^\infty \Big|\frac{\partial^{l+1}(T\gamma)}{\partial y^{l+1}}\Big| dy &\le c \int_0^\infty dy \int_{\mathbb{R}^n} \frac{|(\Delta_h^{(2)} \nabla_{l-1}\gamma)(x)|}{(y+|h|)^{n+2}} dh \\ &= \frac{c}{n+1} \int_{\mathbb{R}^n} \frac{|(\Delta_h^{(2)} \nabla_{l-1}\gamma)(x)|}{|h|^{n+1}} dh\end{aligned}$$

which completes the proof. □

Lemma 5.2.6. *Suppose that* $\gamma \in W_{1,\mathrm{loc}}^{l-1}(\mathbb{R}^n)$ *and let*

$$N = \sup_{\substack{x\in\mathbb{R}^n \\ r\in(0,1)}} r^{m-n} \|\mathfrak{D}_{1,l}\gamma; \mathcal{B}_r(x)\|_{L_1}. \tag{5.2.26}$$

Then, for any $y \in (0,1]$

$$\Big|\frac{\partial^{l+1}(T\gamma)(x,y)}{\partial y^{l+1}}\Big| \le c N\, y^{-m-1}.$$

Proof. By Lemma 5.2.5,

$$\int_{\mathcal{B}_r(x)} dz \int_0^\infty \Big|\frac{\partial^{l+1}(T\gamma)(z,y)}{\partial y^{l+1}}\Big|\, dy \le c\, N\, r^{n-m} \tag{5.2.27}$$

for $r \in (0,1)$. Let $r/2 < y \le r$. Applying the mean value theorem for harmonic functions, we find that

$$\Big|\frac{\partial^{l+1}(T\gamma)(x,y)}{\partial y^{l+1}}\Big| \le \frac{c}{r^{n+1}} \int_{\mathcal{B}_r(x)} dz \int_{r/2}^r \Big|\frac{\partial^{l+1}(T\gamma)(z,\eta)}{\partial \eta^{l+1}}\Big| d\eta.$$

By (5.2.27), the right-hand side is dominated by $c\, N\, r^{-1-m}$. The proof is complete. □

The next assertion is based mainly on two previous lemmas.

Corollary 5.2.2. *Let $0 < l < m \le n$ and let $\gamma \in W_{1,\mathrm{loc}}^{l-1}(\mathbb{R}^n)$. Then, for all $x \in \mathbb{R}^n$*

$$|\gamma(x)| \le c\left(N^{l/m}\big((\mathfrak{D}_{1,l}\gamma)(x)\big)^{(m-l)/m} + \|\gamma\|_{L_{1,\mathrm{unif}}}\right).$$

Proof. Introducing the notation

$$\varphi(y) = \begin{cases} |\partial^{l+1}(T\gamma)(x,y)/\partial y^{l+1}| & \text{for } 0 < y \le 1, \\ 0 & \text{for } y > 1, \end{cases}$$

for any $R > 0$, we have

$$\int_0^1 \Big|\frac{\partial^{l+1}(T\gamma)(x,y)}{\partial y^{l+1}}\Big|\, y^l dy = \int_0^\infty \varphi(y)\, y^l\, dy \le R^l \int_0^R \varphi(y) dy + \int_R^\infty \varphi(y) y^l dy.$$

By Lemma 5.2.5, the first term on the right-hand side is majorized by $cR^l(D_{1,l}\gamma)(x)$ and, by Lemma 5.2.6,

$$\int_R^\infty \varphi(y) y^l dy \le c\, N \int_R^\infty y^{l-m-1} dy = c\, N\, R^{l-m}.$$

Choosing R as

$$R = N^{1/m}\big((\mathfrak{D}_{1,l}\gamma)(x)\big)^{-1/m},$$

we arrive at the inequality

$$\int_0^\infty \varphi(y) y^l dy \le c\, N^{l/m}\big((\mathfrak{D}_{1,l}\gamma)(x)\big)^{(m-l)/m}$$

which together with (3.2.39) completes the proof. □

Corollary 5.2.3. *Suppose that* $\gamma \in W_{1,\mathrm{loc}}^{l-1}(\mathbb{R}^n)$. *For any integer* $l \geq 1$ *and any* $z \in \mathbb{R}^n$

$$r^{m-n-l}\|\gamma; \mathcal{B}_r(z)\|_{L_1} \leq c\Big(\sup_{\rho\in(0,1)} \rho^{m-n}\|\mathfrak{D}_{1,l}\gamma; \mathcal{B}_r(z)\|_{L_1} + \|\gamma\|_{L_{1,\mathrm{unif}}}\Big). \quad (5.2.28)$$

Proof. By Corollary 5.2.2,

$$r^{m-n-l}\|\gamma; \mathcal{B}_r(z)\|_{L_1}$$

$$\leq c\left(N^{l/m} r^{m-n-l}\int_{\mathcal{B}_r(z)} ((\mathfrak{D}_{1,l}\gamma)(x))^{(m-l)/m}dx + r^{m-l}\|\gamma; \mathbb{R}^n\|_{L_{1,\mathrm{unif}}}\right)$$

which does not exceed

$$c\left(N^{l/m}\left(r^{m-n}\int_{\mathcal{B}_r(z)} (\mathfrak{D}_{1,l}\gamma)(x)dx\right)^{(m-l)/m} + \|\gamma; \mathbb{R}^n\|_{L_{1,\mathrm{unif}}}\right)$$

by Hölder's inequality. Using (5.2.26), we complete the proof. □

In the next two lemmas, we return to the Poisson operator T.

Lemma 5.2.7. *Let* $r < y < 1$. *Then, for any integer* $k \geq 1$

$$|\nabla_k(T\gamma)(x,y)| \leq c\, r^{l-m-k} \sup_{\substack{z\in\mathbb{R}^n\\ \rho\in(0,1)}} \rho^{m-n-l}\|\gamma; \mathcal{B}_\rho(z)\|_{L_1}. \quad (5.2.29)$$

Proof. By (3.2.38),

$$|\nabla_k(T\gamma)(x,y)| \leq c\int_{\mathbb{R}^n} \frac{|\gamma(\xi)|}{(|x-\xi|+y)^{n+k}}d\xi. \quad (5.2.30)$$

We have

$$\int_{\mathcal{B}_r(x)} \frac{|\gamma(\xi)|}{(|x-\xi|+y)^{n+k}}d\xi \leq c\, y^{-n-k}\int_{\mathcal{B}_r(x)} |\gamma(\xi)|d\xi$$

$$\leq c\, r^{l-m-k} \sup_{\substack{z\in\mathbb{R}^n\\ \rho\in(0,1)}} \rho^{m-n-l}\|\gamma; \mathcal{B}_\rho(z)\|_{L_1}. \quad (5.2.31)$$

Also,

$$\int_{\mathbb{R}^n\setminus\mathcal{B}_r(x)} \frac{|\gamma(\xi)|}{(|x-\xi|+y)^{n+k}}d\xi \leq \int_{\mathbb{R}^n\setminus\mathcal{B}_r(x)} \frac{|\gamma(\xi)|}{|x-\xi|^{n+k}}d\xi$$

$$\leq c\, r^{-n}\int_{\mathbb{R}^n\setminus\mathcal{B}_{2r}(x)} dz\int_{\mathcal{B}_r(z)} \frac{|\gamma(\xi)|}{|x-\xi|^{n+k}}d\xi.$$

Since $|\xi - x| > |z-x|/2$, it follows that the right-hand side of the last inequality does not exceed

$$c\,r^{l-m} \int_{\mathbb{R}^n \setminus \mathcal{B}_{2r}(x)} \frac{dt}{|t-x|^{n+k}} \sup_{\substack{x \in \mathbb{R}^n \\ \rho \in (0,1)}} \rho^{m-n-l} \|\gamma; \mathcal{B}_\rho(x)\|_{L_1}.$$

Therefore,

$$\int_{\mathbb{R}^n \setminus \mathcal{B}_r(x)} \frac{|\gamma(\xi)|}{(|x-\xi|+y)^{n+k}} d\xi \le c\,r^{l-m-k} \sup_{\substack{z \in \mathbb{R}^n \\ \rho \in (0,1)}} \rho^{m-n-l} \|\gamma; \mathcal{B}_\rho(z)\|_{L_1}. \quad (5.2.32)$$

Now, (5.2.29) results by combining inequalities (5.2.31), (5.2.32), and (5.2.30). □

Lemma 5.2.8. *Let $y > 1$ and let $k \ge 0$. Then*

$$|\nabla_k (T\gamma)(x,y)| \le c\,y^{-k} \|\gamma\|_{L_{1,\mathrm{unif}}}. \quad (5.2.33)$$

Proof. First, observe that

$$\int_{\mathcal{B}_y(x)} \frac{|\gamma(\xi)|\, d\xi}{(|x-\xi|+y)^{n+k}} \le c\,y^{-n-k} \int_{\mathcal{B}_y(x)} |\gamma(\xi)| d\xi \le c y^{-k} \|\gamma\|_{L_{1,\mathrm{unif}}}. \quad (5.2.34)$$

Clearly,

$$\int_{\mathbb{R}^n \setminus \mathcal{B}_y(x)} \frac{|\gamma(\xi)|}{(|x-\xi|+y)^{n+k}} d\xi \le c \int_{\mathbb{R}^n \setminus \mathcal{B}_2(x)} dz \int_{\mathcal{B}_1(z)} \frac{|\gamma(\xi)|}{|x-\xi|^{n+k}} d\xi. \quad (5.2.35)$$

Since $|\xi - x| > |z-x|/2$, the right-hand side of (5.2.35) is dominated by

$$c\,\|\gamma\|_{L_{1,\mathrm{unif}}} \int_{\mathbb{R}^n \setminus \mathcal{B}_2(x)} \frac{dz}{|z-x|^{n+k}}.$$

Therefore,

$$\int_{\mathbb{R}^n \setminus \mathcal{B}_y(x)} \frac{|\gamma(\xi)|}{(|x-\xi|+y)^{n+k}} d\xi \le c\,y^{-k} \|\gamma\|_{L_{1,\mathrm{unif}}}. \quad (5.2.36)$$

Combining the inequalities (5.2.34) and (5.2.36), and then using (5.2.30), we complete the proof. □

5.3 Descriptions of $M(B_1^m \to B_1^l)$ with Integer l

In this section we give two different characterizations of the space $M(B_1^m \to B_1^l)$ with integer l.

5.3.1 A Norm in $M(B_1^m \to B_1^l)$

As before, in the following theorem the integral over $(\mathcal{B}_r^{(n)}(z))^2$ stands for the double integral over $\mathcal{B}_r^{(n)}(z)$.

Theorem 5.3.1. *Let l be an integer and let $m \geq l \geq 1$. The equivalence relation holds:*

$$\|\gamma\|_{M(B_1^m \to B_1^l)} \sim$$

$$\sup_{\substack{z\in\mathbb{R}^n \\ r\in(0,1)}} r^{m-n}\Big(\int_{(\mathcal{B}_r(z))^2} |(\varDelta^{(2)}\nabla_{l-1}\gamma)(x,y)|\frac{dx\,dy}{|x-y|^{n+1}} + r^{-l}\|\gamma;\mathcal{B}_r(z)\|_{L_1}\Big). \quad (5.3.1)$$

Proof. We use the norm

$$\|v;\mathcal{B}_r\|_{B_1^l} = \sum_{j=0}^{l-1} r^{j-l}\|\nabla_j v;\mathcal{B}_r\|_{L_1}$$

$$+\sum_{j=0}^{l-1} r^{j+1-l}\int_{(\mathcal{B}_r)^2} |\varDelta_y^{(2)}\nabla_j v(x)|\frac{dx\,dy}{|x-y|^{n+1}} \quad (5.3.2)$$

defined for a positive integer l and $r\in(0,1)$. Lemma 5.2.4 implies that

$$\|v;\mathcal{B}_r\|_{B_1^l} \sim \int_{(\mathcal{B}_r)^2} |(\varDelta_y^{(2)}\nabla_{l-1}v)(x)|\frac{dx\,dy}{|x-y|^{n+1}} + r^{-l}\|v;\mathcal{B}_r\|_{L_1}. \quad (5.3.3)$$

By dilation in (5.2.6) we obtain

$$\|v;\mathcal{B}_r\|_{B_1^{l-j}} \leq c\,\|v;\mathcal{B}_r\|_{B_1^l}^{1-j/l}\|v;\mathcal{B}_r\|_{L_1}^{j/l} \quad (5.3.4)$$

for any $j=0,\dots,l-1$.

By (5.3.3), the required relation (5.3.1) can be written as

$$\|\gamma\|_{M(B_1^m \to B_1^l)} \sim \sup_{\substack{z\in\mathbb{R}^n \\ r\in(0,1)}} r^{m-n}\|\gamma;\mathcal{B}_r(z)\|_{B_1^l}. \quad (5.3.5)$$

From Lemma 5.2.3 and (5.3.3), we see that

$$\|v;\mathcal{B}_r\|_{B_1^l} \leq c\,\|v;\mathbb{R}^n\|_{B_1^l} \quad (5.3.6)$$

for $l<n$. Let $u(y)=\eta\left(\frac{y-x}{r}\right)$, where $r\in(0,1)$ for $m<n$, and $r=1$ for $m\geq n$, and $\eta\in C_0^\infty(\mathcal{B}_2)$, $\eta=1$ on $\mathcal{B}_1$. Setting this u into the inequality

$$\|\gamma\,u\|_{B_1^l} \leq \|\gamma\|_{M(B_1^m\to B_1^l)}\|u\|_{B_1^m} \quad (5.3.7)$$

and using (5.3.6) with $v=\gamma u$, we have

$$\|\gamma;\mathcal{B}_r(x)\|_{B_1^l} \leq c\,r^{n-m}\|\gamma;\mathbb{R}^n\|_{M(B_1^m\to B_1^l)} \quad (5.3.8)$$

for any $x \in \mathbb{R}^n$. The required lower estimate for the norm $\|\gamma; \mathbb{R}^n\|_{M(B_1^m \to B_1^l)}$ follows from (5.3.3).

Now we obtain the upper estimate for the norm $\|\gamma; \mathbb{R}^n\|_{M(B_1^m \to B_1^l)}$. As before, let $T\gamma$ stand for the Poisson integral of γ. For any $U \in W_1^{m+1}(\mathbb{R}_+^{n+1})$, we have by Lemma 5.2.1 that

$$\|\gamma\, u; \mathbb{R}^n\|_{B_1^l} \le c\,\|(T\gamma)U; \mathbb{R}_+^{n+1}\|_{W_1^{l+1}}, \tag{5.3.9}$$

where $u(x) = U(x, 0)$. Let $X = (x, y) \in \mathbb{R}_+^{n+1}$ and let

$$\mathcal{G}_r(X) = \mathcal{B}_r^{(n+1)}(X) \cap \mathbb{R}_+^{n+1}.$$

By Theorem 2.4.2, for any integer $l \in [0, m)$,

$$\|\varGamma; \mathbb{R}_+^{n+1}\|_{M(W_1^{m+1} \to W_1^{l+1})} \sim$$

$$\sup_{\substack{X \in \mathbb{R}_+^{n+1} \\ r \in (0,1)}} r^{m-n} \|\nabla_{l+1}\varGamma; \mathcal{G}_r(X)\|_{L_1} + \sup_{X \in \mathbb{R}_+^{n+1}} \|\varGamma; \mathcal{G}_1(X)\|_{L_1}. \tag{5.3.10}$$

The first supremum in (5.3.10) can be replaced by

$$\sup_{X \in \mathbb{R}_+^{n+1}} \|\nabla_{l+1}\varGamma; \mathcal{G}_1(X)\|_{L_1}$$

in the case $m \ge n$. Furthermore,

$$\|\varGamma; \mathbb{R}_+^{n+1}\|_{MW_1^{l+1}} \sim \sup_{\substack{X \in \mathbb{R}_+^{n+1} \\ r \in (0,1)}} r^{l-n} \|\nabla_{l+1}\varGamma; \mathcal{G}_r(X)\|_{L_1} + \|\varGamma; \mathbb{R}_+^{n+1}\|_{L_\infty}. \tag{5.3.11}$$

This relation and (5.3.9) give

$$\|\gamma\, u; \mathbb{R}^n\|_{B_1^l} \le c\, K_{m,l} \|U; \mathbb{R}_+^{n+1}\|_{W_1^{m+1}}, \tag{5.3.12}$$

where

$$K_{m,l} = \sup_{\substack{X \in \mathbb{R}_+^{n+1} \\ r \in (0,1)}} r^{m-n} \|\nabla_{l+1}(T\gamma); \mathcal{G}_r(X)\|_{L_1} + \sup_{X \in \mathbb{R}_+^{n+1}} \|T\gamma; \mathcal{G}_1(X)\|_{L_1}. \tag{5.3.13}$$

We introduce one more notation

$$k_{m,l} := \sup_{\substack{z \in \mathbb{R}^n \\ r \in (0,1)}} r^{m-n} \int_{(\mathcal{B}_r(z))^2} |\varDelta^{(2)} \nabla_{l-1}\gamma(x, y)| \frac{dx\, dy}{|x - y|^{n+1}} \tag{5.3.14}$$

and intend to show that

$$K_{m,l} \le c\,\big(k_{m,l} + \sup_{z \in \mathbb{R}^n} \|\gamma; \mathcal{B}_1^{(n)}(z)\|_{L_1}\big). \tag{5.3.15}$$

Then the upper estimate for $\|\gamma; \mathbb{R}^n\|_{M(B_1^m \to B_1^l)}$ follows from (5.3.12) by Lemma 5.2.1 and the arbitrariness of U.

Let us justify (5.3.15). When estimating $\|\nabla_{l+1}(T\gamma); \mathcal{G}_r(X_0)\|_{L_1}$, where $X_0 \in \mathbb{R}^{n+1}_+$, it suffices to take $X_0 = (0, y_0)$. Suppose first that $y_0 > 2$. Then, by Lemma 5.2.8,

$$r^{m-n}\|\nabla_{l+1}(T\gamma); \mathcal{G}_r(X_0)\|_{L_1} \le c\,\|\gamma; \mathbb{R}^n\|_{L_{1,\mathrm{unif}}}.$$

For $2 > y_0 \ge 2r$, from Lemma 5.2.7 we have

$$r^{m-n}\|\nabla_{l+1}(T\gamma); \mathcal{G}_r(X_0)\|_{L_1} \le c \sup_{\substack{x \in \mathbb{R}^n \\ \rho \in (0,1)}} \rho^{m-n-l}\|\gamma; \mathcal{B}_\rho^{(n)}(x)\|_{L_1}.$$

Given any $r \in (0,1)$, it remains to estimate the norm $\|\nabla_{l+1}(T\gamma); \mathcal{G}_r(X_0)\|_{L_1}$ for $y_0 < 2r$.

For any even $k \ge 2$ and $|\sigma| = l + 1 - k$, the harmonicity of $T\gamma$ in $\mathbb{R}^{n+1}_+$ implies that

$$\frac{\partial^k}{\partial y^k} D_x^\sigma (T\gamma)(x,y) = D_x^\sigma (-\Delta_x)^{k/2}(T\gamma)(x,y).$$

This together with (5.2.20) gives

$$\left|\frac{\partial^k}{\partial y^k} D_x^\sigma (T\gamma)(x,y)\right| \le c\, y^{-n-2} \int_{\mathbb{R}^n} \zeta_1\left(\frac{h}{y}\right) |(\Delta_h^{(2)} \nabla_{l-1}\gamma)(x)|dh, \tag{5.3.16}$$

where ζ_1 obeys (5.2.23). Similarly, for any odd $k \ge 3$

$$\frac{\partial^k}{\partial y^k} D_x^\sigma (T\gamma)(x,y) = \frac{\partial}{\partial y} D_x^\sigma (-\Delta_x)^{(k-1)/2}(T\gamma)(x,y).$$

Using (5.2.21), we have

$$\left|\frac{\partial^k}{\partial y^k} D_x^\sigma (T\gamma)(x,y)\right| \le c\, y^{-n-2} \int_{\mathbb{R}^n} \zeta_2\left(\frac{h}{y}\right) |(\Delta_h^{(2)} \nabla_{l-1}\gamma)(x)|dh \tag{5.3.17}$$

with ζ_2 satisfying (5.2.25).

Introducing the notation

$$J_1 := \int_{\mathcal{G}_r(X_0)} y^{-n-2} dx dy \int_{\mathcal{B}_y} \zeta_2\left(\frac{h}{y}\right) |\Delta_h^{(2)}(\nabla_{l-1}\gamma)(x)|dh, \tag{5.3.18}$$

we deduce from (5.2.25) that for $y_0 < 2r$

$$\begin{aligned} J_1 &\le c \int_0^{3r} y^{-n-2} dy \int_{\mathcal{B}_{2r}} dx \int_{\mathcal{B}_y} |\Delta_h^{(2)} \nabla_{l-1}\gamma)(x)|dh \\ &= c \int_{\mathcal{B}_{3r}} |h|^{-n-1} dh \int_{\mathcal{B}_{2r}} |(\Delta_h^{(2)} \nabla_{l-1}\gamma)(x)|dx. \end{aligned}$$

Therefore,

$$J_1 \le c\, r^{n-m}\, k_{m,l} \tag{5.3.19}$$

with $k_{m,l}$ given by (5.3.14).

Let

$$J_2 := \int_{\mathcal{G}_r(X_0)} \int_{\mathbb{R}^n\backslash\mathcal{B}_y} \zeta_2\big(\frac{h}{y}\big)\, |(\Delta_h^{(2)}\nabla_{l-1}\gamma)(x)|dh\, y^{-n-2} dxdy. \tag{5.3.20}$$

By (5.3.17) and (5.2.25), we have for the inner integral over $\mathbb{R}^n\backslash\mathcal{B}_y^{(n)}$

$$\int_{\mathbb{R}^n\backslash\mathcal{B}_y} \zeta_2\big(\frac{h}{y}\big)\, |(\Delta_h^{(2)}\nabla_{l-1}\gamma)(x)|dh \le c\, y^{n+2} \int_{\mathbb{R}^n\backslash\mathcal{B}_y} \frac{|(\Delta_h^{(2)}\nabla_{l-1}\gamma)(x)|dh}{|h|^{n+2}}.$$

We write the integral on the right-hand side as the sum of two integrals, one taken over $\mathcal{B}_{2r}\backslash\mathcal{B}_y$ and another over $\mathbb{R}^n\backslash\mathcal{B}_{2r}$. We see that

$$\int_0^{3r} dy \int_{\mathcal{B}_r} |(\Delta_h^{(2)}\nabla_{l-1}\gamma)(x)|dx \int_{\mathcal{B}_{2r}\backslash\mathcal{B}_y} \frac{dh}{|h|^{n+2}}$$

$$\le \int_{\mathcal{B}_r} dx \int_{\mathcal{B}_{2r}} \frac{|(\Delta_h^{(2)}\nabla_{l-1}\gamma)(x)|dh}{|h|^{n+1}} \le c\, r^{n-m}\, k_{m,l}. \tag{5.3.21}$$

Also,

$$\int_0^{3r} dy \int_{\mathcal{B}_r^{(n)}} dx \int_{\mathbb{R}^n\backslash\mathcal{B}_{2r}} \frac{|(\Delta_h^{(2)}\nabla_{l-1}\gamma)(x)|dh}{|h|^{n+2}} \le c\,(I + I_+ + I_-), \tag{5.3.22}$$

where

$$I = \int_0^{3r} dy \int_{\mathcal{B}_r^{(n)}} |\nabla_{l-1}\gamma(x)|\, dx \int_{\mathbb{R}^n\backslash\mathcal{B}_{2r}} \frac{dh}{|h|^{n+2}}$$

and

$$I_\pm = \int_0^{3r} dy \int_{\mathcal{B}_r} \int_{\mathbb{R}^n\backslash\mathcal{B}_{2r}} \frac{|\nabla_{l-1}\gamma(x \pm h)|dh}{|h|^{n+2}} dx.$$

Clearly,

$$I \le c\, r^{-1} \int_{\mathcal{B}_r} |\nabla_{l-1}\gamma(x)|dx. \tag{5.3.23}$$

Hence,

$$I \le c\, r^{n-m} \Big(\sup_{\substack{z\in\mathbb{R}^n \\ \rho\in(0,1)}} \rho^{m-n-1} \int_{\mathcal{B}_\rho(z)} |\nabla_{l-1}\gamma(x)|dx \Big). \tag{5.3.24}$$

Obviously,

$$I_\pm \le c\, r^{1-n} \int_{\mathcal{B}_r^{(n)}} \int_{\mathbb{R}^n \setminus \mathcal{B}_{2r}} \int_{\mathcal{B}_r(\xi)} \frac{|\nabla_{l-1}\gamma(x \pm h)|dh}{|h|^{n+2}}\, d\xi\, dx.$$

In view of the estimate $|\xi| \le r + |h| < \frac{1}{2}|\xi| + |h|$, this implies that

$$I_\pm \le c\, r^{1-n} \int_{\mathbb{R}^n \setminus \mathcal{B}_{2r}} \int_{\mathcal{B}_r(\xi)} \int_{\mathcal{B}_r} |\nabla_{l-1}\gamma(x \pm h)|\, dx\, dh \frac{d\xi}{|\xi|^{n+2}}, \tag{5.3.25}$$

and therefore

$$I_\pm \le c\, r^{n-m} \Big(\sup_{\substack{z \in \mathbb{R}^n \\ \rho \in (0,1)}} \rho^{m-n-1} \int_{\mathcal{B}_\rho(z)} |\nabla_{l-1}\gamma(x)|dx \Big). \tag{5.3.26}$$

Combining (5.3.21), (5.3.23), and (5.3.26), we conclude that J_2 defined by (5.3.20) is subject to the inequality

$$J_2 \le c\, r^{n-m} \Big(k_{l,m} + \sup_{\substack{z \in \mathbb{R}^n \\ \rho \in (0,1)}} \rho^{m-n-1} \int_{\mathcal{B}_\rho(z)} |\nabla_{l-1}\gamma(x)|dx \Big). \tag{5.3.27}$$

Together with (5.3.19), this leads to

$$r^{m-n} \|\nabla_{l+1}(T\gamma); \mathcal{G}_r(X_0)\|_{L_1}$$

$$\le c \Big(k_{l,m} + \sup_{\substack{z \in \mathbb{R}^n \\ \rho \in (0,1)}} \rho^{m-n-1} \int_{\mathcal{B}_\rho(z)} |\nabla_{l-1}\gamma(x)|dx \Big). \tag{5.3.28}$$

It remains to show that

$$\sup_{X_0 \in \mathbb{R}^{n+1}_+} \|T\gamma; \mathcal{G}_1(X_0)\|_{L_1} \le c \sup_{x \in \mathbb{R}^n} \|\gamma; \mathcal{B}_1(x)\|_{L_1}. \tag{5.3.29}$$

If $y_0 \ge 2$, this inequality stems directly from (5.2.33). Let $y_0 < 2$. Clearly,

$$\|T\gamma; \mathcal{G}_1(X_0)\|_{L_1} \le \int_0^3 \int_{\mathcal{B}} \int_{\mathcal{B}_y(x)} \zeta\Big(\frac{\xi - x}{y}\Big) |\gamma(\xi)| d\xi\, dx \frac{dy}{y^n}$$

$$+ \int_0^3 dy \int_{\mathcal{B}_1^{(n)}} \int_{\mathbb{R}^n \setminus \mathcal{B}_y(x)} \zeta\Big(\frac{\xi - x}{y}\Big) |\gamma(\xi)| d\xi\, dx \frac{dy}{y^n}. \tag{5.3.30}$$

The first term on the right-hand side does not exceed

$$\int_{\mathcal{B}} \zeta(t)\, dt \int_0^3 \int_{\mathcal{B}} |\gamma(x + ty)|\, dx dy \le c \sup_{z \in \mathbb{R}^n} \|\gamma; \mathcal{B}_1(z)\|_{L_1}. \tag{5.3.31}$$

Since ζ is the Poisson kernel, the second term in (5.3.30) is dominated by

$$c\int_0^3 \int_{\mathcal{B}} \int_{\mathbb{R}^n \setminus \mathcal{B}_y} \frac{|\gamma(x+h)|\, dh}{(y+|h|)^{n+1}} dx\, ydy$$

$$\leq c\int_0^3 \int_{\mathcal{B}} \int_{\mathbb{R}^n \setminus \mathcal{B}_{2y}} d\xi \int_{\mathcal{B}_y(\xi)} \frac{|\gamma(x+h)|\, dh}{|h|^{n+1}} dx \frac{dy}{y^{n-1}}. \qquad (5.3.32)$$

In view of the inequality $|h| > |\xi|/2$ which is valid since $|\xi| \leq r+|h| < \frac{1}{2}|\xi|+|h|$, the right-hand side in (5.3.32) is majorized by

$$c\int_0^3 \int_{\mathbb{R}^n \setminus \mathcal{B}_{2y}} \int_{\mathcal{B}_y(\xi)} \int_{\mathcal{B}} |\gamma(x+h)| dx dh \frac{d\xi}{|\xi|^{n+1}} \frac{dy}{y^{n-1}} \leq c \sup_{z\in\mathbb{R}^n} \|\gamma; (\mathcal{B}_1(z))\|_{L_1}.$$

Combining the last estimate with (5.3.31) and (5.3.32), we arrive at (5.3.29).

Now, adding the inequalities (5.3.19), (5.3.27), and (5.3.29), we conclude that the value $K_{l,m}$ defined by (5.3.13) satisfies

$$K_{l,m} \leq c\Big(k_{l,m} + \sup_{\substack{z\in\mathbb{R}^n \\ r\in(0,1)}} r^{m-n-1} \int_{\mathcal{B}_r(z)} |\nabla_{l-1}\gamma(x)| dx + \sup_{z\in\mathbb{R}^n} \|\gamma; \mathcal{B}_1(z)\|_{L_1}\Big). \quad (5.3.33)$$

Estimating the second term on the right-hand side by Lemma 5.2.4, we arrive at (5.3.15). The result follows for $l < m$. For $m = l$, instead of (5.3.29), we use the maximum principle

$$\|T\gamma; \mathbb{R}^{n+1}_+\|_{L_\infty} \leq \|\gamma; \mathbb{R}^n\|_{L_\infty}.$$

The proof of Theorem 5.3.1 is complete. □

Remark 5.3.1. It is obvious that for $m = l$ the relation (5.3.1) can be written as

$$\|\gamma\|_{MB_1^l} \sim$$

$$\sup_{\substack{z\in\mathbb{R}^n \\ r\in(0,1)}} r^{m-n} \int_{(\mathcal{B}_r(z))^2} |(\varDelta^{(2)}\nabla_{l-1}\gamma)(x,y)| \frac{dx\, dy}{|x-y|^{n+1}} + \|\gamma\|_{L_\infty}. \qquad (5.3.34)$$

Corollary 5.3.1. *Suppose that $\gamma \in M(B_1^m(\mathbb{R}^n) \to B_1^l(\mathbb{R}^n))$. Then $\nabla_j\gamma \in M(B_1^{m-j}(\mathbb{R}^n) \to B_1^{l-j}(\mathbb{R}^n))$.*

Proof. This follows directly from Theorem 5.3.1 and Lemma 5.2.4.

5.3.2 Description of $M(B_1^m \to B_1^l)$ Involving $\mathfrak{D}_{1,l}$

Now we give another description of the space $M(B_1^m(\mathbb{R}^n) \to B_1^l(\mathbb{R}^n))$.

Theorem 5.3.2. *Let l be an integer and let $m \geq l \geq 1$. Then the equivalence relation*

$$\|\gamma\|_{M(B_1^m \to B_1^l)} \sim \sup_{\substack{z \in \mathbb{R}^n \\ r \in (0,1)}} r^{m-n} \|\mathfrak{D}_{1,l}\gamma; \mathcal{B}_r(z)\|_{L_1} + \|\gamma\|_{L_{1,\mathrm{unif}}} \tag{5.3.35}$$

holds. If $m = l$, then

$$\|\gamma\|_{MB_1^l} \sim \sup_{\substack{z \in \mathbb{R}^n \\ r \in (0,1)}} r^{m-n} \|\mathfrak{D}_{1,l}\gamma; \mathcal{B}_r(z)\|_{L_1} + \|\gamma\|_{L_\infty}. \tag{5.3.36}$$

For $m \geq n$ and $m > l$,

$$\|\gamma\|_{M(B_1^m \to B_1^l)} \sim \sup_{z \in \mathbb{R}^n} \big(\|\mathfrak{D}_{1,l}\gamma; \mathcal{B}_1(z)\|_{L_1} + \|\gamma; (\mathcal{B}_1(z))\|_{L_1}\big) \tag{5.3.37}$$

which, in its turn, is equivalent to $\|\gamma; \mathbb{R}^n\|_{B^l_{1,\mathrm{unif}}}$.

Proof. The desired lower estimate for the norm $\|\gamma\|_{M(B_1^m \to B_1^l)}$ follows from (5.3.8) and the estimate

$$\|\mathfrak{D}_{1,l}\gamma; \mathcal{B}_r(z)\|_{L_1} \leq c \sup_{\xi \in \mathbb{R}^n} \|\gamma; \mathcal{B}_r(\xi)\|_{B_1^l}, \tag{5.3.38}$$

which holds for all $z \in \mathbb{R}^n$ and $r \in (0, 1]$. In order to justify (5.3.38), it suffices to check that

$$\int_{\mathcal{B}_r(z)} \int_{\mathbb{R}^n \setminus \mathcal{B}_r} |(\Delta_h^{(2)} \nabla_{l-1}\gamma)(x)| \frac{dh}{|h|^{n+1}} dx \leq c r^{-1} \sup_{\xi \in \mathbb{R}^n} \|\nabla_{l-1}\gamma; \mathcal{B}_r(\xi)\|_{L_1}. \tag{5.3.39}$$

Clearly,

$$\int_{\mathcal{B}_r(z)} \int_{\mathbb{R}^n \setminus \mathcal{B}_r} |\nabla_{l-1}\gamma(x)| \frac{dh}{|h|^{n+1}} dx \leq c r^{-1} \sup_{\xi \in \mathbb{R}^n} \|\nabla_{l-1}\gamma; \mathcal{B}_r(\xi)\|_{L_1}. \tag{5.3.40}$$

Also,

$$\int_{\mathcal{B}_r(z)} \int_{\mathbb{R}^n \setminus \mathcal{B}_r} |\nabla_{l-1}\gamma(x \pm h)| \frac{dh}{|h|^{n+1}} dx$$
$$\leq \frac{c}{r^n} \int_{\mathcal{B}_r(z)} \int_{\mathbb{R}^n \setminus \mathcal{B}_{2r}} \int_{\mathcal{B}_r(\xi)} |\nabla_{l-1}\gamma(x \pm h)| \frac{dh}{|h|^{n+1}} d\xi \, dx.$$

Since $|\xi| < r + |h|$, it follows that $|h| > |\xi|/2$ and, therefore, the right-hand side of the last inequality is dominated by

$$c \int_{\mathbb{R}^n \setminus \mathcal{B}_{2r}} \int_{\mathcal{B}_r(\xi)} |\nabla_{l-1}\gamma(x \pm h)| dh \frac{d\xi}{|\xi|^{n+1}} \leq c r^{-1} \sup_{z \in \mathbb{R}^n} \|\nabla_{l-1}\gamma; \mathcal{B}_r(z)\|_{L_1}$$

which together with (5.3.40) implies (5.3.39).

To get the required upper estimate for $\|\gamma\|_{M(B_1^m \to B_1^l)}$, we combine (5.3.33) with Lemma 5.2.4 and Corollary 5.2.3 to conclude that

$$K_{m,l} \leq c\Big(\sup_{\substack{z \in \mathbb{R}^n \\ r \in (0,1)}} r^{m-n} \|\mathfrak{D}_{1,l}\gamma; \mathcal{B}_r(z)\|_{L_1} + \|\gamma\|_{L_{1,\text{unif}}}\Big).$$

Using this inequality in (5.3.12), the result follows.

For $m \geq n$, the right-hand side in (5.3.5) is obviously equivalent to $B^l_{1,\text{unif}}(\mathbb{R}^n)$. The proof is complete. □

5.3.3 $M(B_1^m(\mathbb{R}^n) \to B_1^l(\mathbb{R}^n))$ as the Space of Traces

We use the notation $\mathbb{R}^{n+1}_+ = \{(x, y) : x \in \mathbb{R}^n, y > 0\}$ and $\mathbb{R}^n = \partial\mathbb{R}^{n+1}_+$. By $W_1^k(\mathbb{R}^{n+1}_+)$ with integer k we mean the space of functions defined on $\mathbb{R}^{n+1}_+$ with finite norm

$$\|U; \mathbb{R}^{n+1}_+\|_{W_1^k} = \|\nabla_k U; \mathbb{R}^{n+1}_+\|_{L_1} + \|U; \mathbb{R}^{n+1}_+\|_{L_1}.$$

The next theorem shows that $M(B_1^m(\mathbb{R}^n) \to B_1^l(\mathbb{R}^n))$, with integer m and l, is the space of traces on $\mathbb{R}^n$ of functions in $M(W_1^{m+1}(\mathbb{R}^{n+1}_+) \to W_1^{l+1}(\mathbb{R}^{n+1}_+))$.

Theorem 5.3.3. *Let m and l be integers, $m \geq l \geq 1$.*

(i) Suppose that

$$\gamma \in M(B_1^m(\mathbb{R}^n) \to B_1^l(\mathbb{R}^n)).$$

Then the Dirichlet problem

$$\Delta\Gamma = 0 \text{ on } \mathbb{R}^{n+1}_+, \quad \Gamma|_{\mathbb{R}^n} = \gamma$$

has a unique solution in $M(W_1^{m+1}(\mathbb{R}^{n+1}_+) \to W_1^{l+1}(\mathbb{R}^{n+1}_+))$ and the estimate

$$\|\Gamma; \mathbb{R}^{n+1}_+\|_{M(W_1^{m+1} \to W_1^{l+1})} \leq c\,\|\gamma; \mathbb{R}^n\|_{M(B_1^m \to B_1^l)} \tag{5.3.41}$$

holds.

(ii) Suppose that

$$\Gamma \in M(W_1^{m+1}(\mathbb{R}^{n+1}_+) \to W_1^{l+1}(\mathbb{R}^{n+1}_+)).$$

If γ is the trace of Γ on $\mathbb{R}^n$, then

$$\gamma \in M(B_1^m(\mathbb{R}^n) \to B_1^l(\mathbb{R}^n))$$

and the estimate

$$\|\gamma; \mathbb{R}^n\|_{M(B_1^m \to B_1^l)} \leq c\,\|\Gamma; \mathbb{R}^{n+1}_+\|_{M(W_1^{m+1} \to W_1^{l+1})} \tag{5.3.42}$$

holds.

Proof. (*i*) Suppose that $\gamma \in M(B_1^m(\mathbb{R}^n) \to B_1^l(\mathbb{R}^n))$. Then by Theorem 5.3.1, the right-hand side in (5.3.1) is finite. Taking into account (5.3.15), we conclude that $K_{m,l}$ defined in (5.3.13) is finite. Then reference to the equivalence relation (5.3.10) completes the proof of part (*i*).

(*ii*) Let $U \in W_1^{m+1}(\mathbb{R}_+^{n+1})$ and $U(x,0) = u(x)$. Clearly, by part (i) of Lemma 5.2.1,

$$\|\gamma\, u; \mathbb{R}^n\|_{B_1^l} \le c\, \|\Gamma U; \mathbb{R}_+^{n+1}\|_{W_1^{l+1}}$$

$$\le c\, \|\Gamma; \mathbb{R}_+^{n+1}\|_{M(W_1^{m+1} \to W_1^{l+1})} \|U; \mathbb{R}_+^{n+1}\|_{W_1^{m+1}}.$$

Minimizing the right-hand side over all extensions U of u and using part (ii) of Lemma 5.2.1, we complete the proof.

5.3.4 Interpolation Inequality for Multipliers

From Theorem 5.1.2 one readily obtains

Corollary 5.3.2. *Let* $0 < s < n$*, then*

$$\|\gamma; \mathbb{R}^n\|_{M(B_1^s \to L_1)} \sim \sup_{\substack{x \in \mathbb{R}^n \\ r \in (0,1)}} r^{s-n} \|\gamma; \mathcal{B}_r(x)\|_{L_1}. \tag{5.3.43}$$

Let $s \ge n$*, then*

$$\|\gamma; \mathbb{R}^n\|_{M(B_1^s \to L_1)} \sim \sup_{x \in \mathbb{R}^n} \|\gamma; \mathcal{B}_1(x)\|_{L_1}. \tag{5.3.44}$$

Theorem 5.3.4. *Let* m *and* l *be integers,* $m \ge l > 0$*, and let* $j = 0, \dots, l-1$*. Then*

$$\|\gamma\|_{M(B_1^{m-j} \to B_1^{l-j})}$$
$$\le c\, \|\gamma\|^{1-j/l}_{M(B_1^m \to B_1^l)} \|\gamma\|^{j/l}_{M(B_1^{m-l} \to L_1)}. \tag{5.3.45}$$

Proof. By (5.3.4),

$$\|u; \mathcal{B}_r(x)\|_{B_1^{l-j}} \le c\, \|u; \mathcal{B}_r(x)\|^{1-j/l}_{B_1^l} \|\gamma; \mathcal{B}_r(x)\|^{j/l}_{L_1}.$$

Hence,

$$\sup_{\substack{x \in \mathbb{R}^n \\ r \in (0,1)}} r^{m-j-n} \|u; \mathcal{B}_r(x)\|_{B_1^{l-j}}$$

$$\le c \Big(\sup_{\substack{x \in \mathbb{R}^n \\ r \in (0,1)}} r^{m-n} \|u; \mathcal{B}_r(x)\|_{B_1^{l-j}} \Big)^{1-j/l} \Big(\sup_{\substack{x \in \mathbb{R}^n \\ r \in (0,1)}} r^{m-l-n} \|\gamma; \mathcal{B}_r(x)\|_{L_1} \Big)^{j/l}.$$

It remains to apply Theorem 5.3.1 and Corollary 5.3.2.

5.4 Description of the Space $M(B_1^m \to B_1^l)$ with Noninteger l

Before we pass to a complete characterization of the space of multipliers $M(W_1^m(\mathbb{R}^n) \to W_1^l(\mathbb{R}^n))$ for noninteger l, we prove an assertion similar to Lemma 5.2.4.

Lemma 5.4.1. *Let l be a positive noninteger, and let $j = 0, \dots, [l]$. Then for any $r \in (0, 1]$,*

$$r^{j-l}\|\nabla_j u; \mathcal{B}_r\|_{L_1}$$
$$\leq c\Big(\int_{(\mathcal{B}_r)^2} |\nabla_{[l]}u(x) - \nabla_{[l]}u(y)|\frac{dx\,dy}{|x-y|^{n+\{l\}}} + r^{-l}\|u; \mathcal{B}_r\|_{L_1}\Big) \tag{5.4.1}$$

and

$$r^{j-[l]}\int_{(\mathcal{B}_r)^2} |\nabla_j u(x) - \nabla_j u(y)|\frac{dx\,dy}{|x-y|^{n+\{l\}}}$$
$$\leq c\Big(\int_{(\mathcal{B}_r)^2} |\nabla_{[l]}u(x) - \nabla_{[l]}u(y)|\frac{dx\,dy}{|x-y|^{n+\{l\}}} + r^{-l}\|u; \mathcal{B}_r\|_{L_1}\Big). \tag{5.4.2}$$

Proof. We apply the same argument as in the proof of Lemma 5.2.4 with (5.2.15) replaced by the identity

$$\nabla_{[l]}u(x) = \int_{\mathcal{B}} \varphi(y)\big(\nabla_{[l]}u(x) - \nabla_{[l]}u(y)\big)dy$$
$$+(-1)^{[l]}\int_{\mathcal{B}} u(y)(\nabla_{[l]}\varphi)(y)dy. \tag{5.4.3}$$

□

Theorem 5.4.1. *Let l be a noninteger and let $m \geq l \geq 0$. The equivalence relation*

$$\|\gamma\|_{M(W_1^m \to W_1^l)} \sim$$
$$\sup_{\substack{z\in\mathbb{R}^n\\ r\in(0,1)}} r^{m-n}\Big(\int_{(\mathcal{B}_r(z))^2} |\nabla_{[l]}\gamma(x) - \nabla_{[l]}\gamma(y)|\frac{dx\,dy}{|x-y|^{n+\{l\}}} + r^{-l}\|\gamma; \mathcal{B}_r(z)\|_{L_1}\Big)$$

holds. For $m \geq n$

$$\|\gamma\|_{M(W_1^m \to W_1^l)} \sim \|\gamma\|_{W^l_{1,\mathrm{unif}}}.$$

Proof. We use the norm

$$\|v; \mathcal{B}_r\|_{W_1^l} = \sum_{j=0}^{[l]} r^{j-l}\|\nabla_j v; \mathcal{B}_r\|_{L_1}$$
$$+\sum_{j=0}^{[l]} r^{j-[l]}\int_{(\mathcal{B}_r)^2} |\nabla_j v(x) - \nabla_j v(y)|\frac{dx\,dy}{|x-y|^{n+\{l\}}} \tag{5.4.4}$$

defined for positive and noninteger l, and $r \in (0,1]$. Lemma 5.4.1 implies that

$$\|v; \mathcal{B}_r\|_{W_1^l} \sim \int_{(\mathcal{B}_r)^2} |\nabla_{[l]} v(x) - \nabla_{[l]} v(y)| \frac{dx\, dy}{|x-y|^{n+\{l\}}} + r^{-l} \|v; \mathcal{B}_r\|_{L_1}. \quad (5.4.5)$$

Making dilation in (5.2.6) with noninteger l, we obtain

$$\|v; \mathcal{B}_r\|_{W_1^{l-j}} \le c \, \|v; \mathcal{B}_r\|_{W_1^l}^{1-j/l} \|v; \mathcal{B}_r\|_{L_1}^{j/l} \quad (5.4.6)$$

for any $j = 0, \dots, l-1$, and hence

$$\|v; \mathcal{B}_r\|_{W_1^{l-j}} \le c \, \|v; \mathcal{B}_r\|_{W_1^l}. \quad (5.4.7)$$

By (5.4.5), the equivalence required in the theorem can be written as

$$\|\gamma; \mathbb{R}^n\|_{M(W_1^m \to W_1^l)} \sim \sup_{z \in \mathbb{R}^n} \|\gamma; \mathcal{B}_1(z)\|_{W_1^l}. \quad (5.4.8)$$

For $m \ge n$ the right-hand side of (5.4.8) becomes

$$\sup_{\substack{z \in \mathbb{R}^n \\ r \in (0,1)}} r^{m-n} \|\gamma; \mathcal{B}_r(z)\|_{W_1^l} \sim \|\gamma; \mathbb{R}^n\|_{W_{1,\mathrm{unif}}^l}.$$

From Lemma 5.2.3 and (5.4.5), we obtain

$$\|v; \mathcal{B}_r\|_{W_1^l} \le c \, \|v; \mathbb{R}^n\|_{W_1^l} \quad (5.4.9)$$

for $l < n$. Let $u(y) = \eta\left(\frac{y-x}{r}\right)$, where $r \in (0,1)$ for $m < n$ and $r = 1$ for $m \ge n$, and $\eta \in C_0^\infty(\mathcal{B}_2)$, $\eta = 1$ on $\mathcal{B}_1$. Setting this u into the inequality

$$\|\gamma\, u\|_{W_1^l} \le \|\gamma\|_{M(W_1^m \to W_1^l)} \|u\|_{W_1^m} \quad (5.4.10)$$

and using (5.4.9) with $v = \gamma u$, we have

$$\|\gamma; \mathcal{B}_r(x)\|_{W_1^l} \le c\, r^{n-m} \|\gamma; \|_{M(W_1^m \to W_1^l)} \quad (5.4.11)$$

for any $x \in \mathbb{R}^n$. The required lower estimate for the norm $\|\gamma\|_{M(W_1^m \to W_1^l)}$ follows from (5.4.5).

Now we obtain the upper estimate for the norm $\|\gamma\|_{M(W_1^m \to W_1^l)}$. We have

$$\|D_{1,l}(\gamma u)\|_{L_1}$$

$$\le c \sum_{j=0}^{[l]} \left(\| \, |\nabla_j u| \, D_{1,l-j}\gamma\|_{L_1} + \| \, |\nabla_j \gamma| \, D_{1,l-j} u\|_{L_1} \right). \quad (5.4.12)$$

By Theorem 5.1.2

$$\| \, |\nabla_j u| \, D_{1,l-j}\gamma\|_{L_1}$$

$$\le c \sup_{\substack{x \in \mathbb{R}^n \\ r \in (0,1)}} r^{m-j-n} \|D_{1,l-j}\gamma; \mathcal{B}_r(x)\|_{L_1} \|\nabla_j u\|_{W_1^{m-j}}. \quad (5.4.13)$$

Note that the estimate

$$\|D_{1,l}\gamma; \mathcal{B}_r(z)\|_{L_1} \le c \sup_{\xi \in \mathbb{R}^n} \|\gamma; \mathcal{B}_r(\xi)\|_{W_1^l}, \tag{5.4.14}$$

holds for all $z \in \mathbb{R}^n$ and $r \in (0,1]$. To justify this estimate, it suffices to check that

$$\int_{\mathcal{B}_r^{(n)}(z)} \int_{\mathbb{R}^n \setminus \mathcal{B}_r} |\nabla_{[l]}\gamma(x+h) - \nabla_{[l]}\gamma(x)| \frac{dh}{|h|^{n+\{l\}}} dx$$
$$\le c\, r^{-\{l\}} \sup_{\xi \in \mathbb{R}^n} \|\nabla_{[l]}\gamma; \mathcal{B}_r(\xi)\|_{L_1} \tag{5.4.15}$$

which is proved in the same way as (5.3.39) with obvious changes. By (5.4.15),

$$r^{-j}\|D_{1,l-j}\gamma; \mathcal{B}_r(x)\|_{L_1} \le c\, r^{-j-\{l\}} \sup_{\xi \in \mathbb{R}^n} \|\nabla_{[l]-j}\gamma; \mathcal{B}_r(\xi)\|_{L_1}$$
$$\le c \sup_{\xi \in \mathbb{R}^n} \|\gamma; \mathcal{B}_r(\xi)\|_{W_1^l}.$$

This together with (5.4.13) implies that

$$\|\, |\nabla_j u|\, D_{1,l-j}\gamma\|_{L_1} \le c \sup_{\substack{\xi \in \mathbb{R}^n \\ r \in (0,1)}} r^{m-n} \|\gamma; \mathcal{B}_r(\xi)\|_{W_1^l} \|u\|_{W_1^m} \tag{5.4.16}$$

for $m < n$, and

$$\|\, |\nabla_j u|\, D_{1,l-j}\gamma\|_{L_1} \le c \sup_{z \in \mathbb{R}^n} \|\gamma; \mathcal{B}_1(z)\|_{W_1^l} \|u\|_{W_1^m} \tag{5.4.17}$$

for $m \ge n$.

Since the proof of (4.3.60) is also valid for $p = 1$, we have

$$\|\, |\nabla_j \gamma|\, D_{1,l-j} u\|_{L_1} \le c\, \|\nabla_j \gamma\|_{M(W_1^{m-l+j} \to L_1)} \|u\|_{W_1^m}$$

which together with Theorem 5.1.2 gives

$$\|\, |\nabla_j \gamma|\, D_{1,l-j} u\|_{L_1} \le c \sup_{\substack{x \in \mathbb{R}^n \\ r \in (0,1)}} r^{m-l+j-n} \|\nabla_j \gamma; \mathcal{B}_r(x)\|_{L_1} \|u\|_{W_1^m}.$$

This estimate along with Lemma 5.4.1 implies that

$$\|\, |\nabla_j \gamma|\, D_{1,l-j} u\|_{L_1} \le c \sup_{\substack{z \in \mathbb{R}^n \\ r \in (0,1)}} r^{m-n} \|\gamma; \mathcal{B}_r(z)\|_{W_1^l} \|u\|_{W_1^m} \tag{5.4.18}$$

for $m < n$, and

$$\|\, |\nabla_j \gamma|\, D_{1,l-j} u\|_{L_1} \le c \sup_{z \in \mathbb{R}^n} \|\gamma; \mathcal{B}_1(z)\|_{W_1^l} \|u\|_{W_1^m} \tag{5.4.19}$$

for $m \ge n$. Substituting (5.4.16)–(5.4.19) into (5.4.12), we complete the proof. □

5.5 Further Results on Multipliers in Besov and Other Function Spaces

5.5.1 Peetre's Imbedding Theorem

As early as in 1976 Peetre showed that, in order for a function γ be a multiplier in $B^l_{p,\theta}$, it suffices that γ belongs to $B^l_{\infty,\theta}$ (see [Pe2]). He used a Littlewood-Paley decomposition of a function u as the sum of elementary functions u_n such that the Fourier transform of u_n is supported by the dyadic annulus of width $\sim 2^n$. In this section we reproduce Peetre's result.

The space $B^l_{p,\theta}$ with $p \geq 1$ defined in Sect. 4.4 can be supplied with other norms (see [Pe2], [Tr3]). For example, for $l \in (0,1)$ one may put

$$\|u\|_{B^l_{p,\theta}} = \begin{cases} \left(\displaystyle\int_0^\infty \left(t^{-1}\omega_p(t,u)\right)^\theta \frac{dt}{t}\right)^{1/\theta} + \|u\|_{L_p} & \text{for } \theta < \infty, \\ \displaystyle\sup_{t>0} t^{-1}\omega_p(t,u) + \|u\|_{L_p} & \text{for } \theta = \infty, \end{cases}$$

where

$$\omega_p(t,u) = \sup_{|h|\leq t} \left(\int_{\mathbb{R}^n} |u(x+h) - u(x)|^p \, dx\right)^{1/p}.$$

Another frequently used norm in $B^l_{p,\theta}$ introduced by Peetre has the following definition. Let $\varphi \in \mathcal{S}$, where $\mathcal{S}$ is the space of rapidly decreasing functions. Further let

(1) $\operatorname{supp}\varphi = \{\xi : 2^{-1} \leq |\xi| \leq 2\}$,
(2) $\varphi(\xi) > 0$ for $2^{-1} < |\xi| < 2$,
(3) $\displaystyle\sum_{k=-\infty}^{\infty} \varphi(2^{-k}\xi) = 1, \quad \xi \neq 0.$

Let Ψ and φ_k be given by

$$F\Psi(\xi) = 1 - \sum_{k\geq 1} \varphi(2^{-k}\xi); \quad F\varphi_k(\xi) = \varphi(2^{-k}\xi), \quad -\infty < k < +\infty, \tag{5.5.1}$$

where F is the Fourier transform. Then the norm (the quasi-norm for $0<\theta<1$) in $B^l_{p,\theta}$ is equivalent to

$$\|u\|^{(1)}_{B^l_{p,\theta}} = \begin{cases} \left(\displaystyle\sum_{k\geq 1} \left(2^{kl}\|u * \varphi_k\|_{L_p}\right)^\theta\right)^{1/\theta} + \|u * \Psi\|_{L_p} & \text{for } \theta < \infty, \\ \displaystyle\sup_{k\geq 1} 2^{kl}\,\|u * \varphi_k\|_{L_p} + \|u * \Psi\|_{L_p} & \text{for } \theta = \infty \end{cases}$$

(see, for example, [Pe2], [Tr3]). Using the right-hand sides of these equalities, one can define the spaces $B^l_{p,\theta}$ for all $l \in \mathbb{R}^1$ and for $p \in (0,1)$.

Theorem 5.5.1. [Pe2] *If $l > 0$, $p \in [1,\infty]$, and $\theta \in (0,\infty]$, then $B^l_{\infty,\theta} \subset MB^l_{p,\theta}$.*

Proof. Let $u \in B^l_{p,\theta}$ and let $\gamma \in B^l_{\infty,\theta}$. We put $U = u * \Psi$, $u_k = u * \varphi_k$, and similarly $\Gamma = \gamma * \Psi$, $\gamma_k = \gamma * \varphi_k$. Then

$$u = U + \sum_{k\geq 1} u_k, \qquad \gamma = \Gamma + \sum_{k\geq 1} \gamma_k.$$

For the product $g = \gamma\, u$ we have the decomposition

$$g = \Gamma U + \Gamma \sum_{j\geq 1} u_j + U \sum_{k\geq 1} \gamma_k + \sum_{k\geq 1}\sum_{j\geq 1} \gamma_k\, u_j. \tag{5.5.2}$$

Clearly,

$$\|g * \Psi\|_{L_p} \leq c\,\|g\|_{L_p}.$$

In order to estimate the norm

$$\Big(\sum_{m\geq 1} \big(2^{ml}\, \|g * \varphi_m\|_{L_p}\big)^\theta\Big)^{1/\theta},$$

we consider only the functions $(\gamma_k\, u_j) * \varphi_m$ because other terms in (5.5.2) can be estimated in a similar way. Let us introduce the spherical layer

$$G_m = \{\xi : 2^{m-1} < |\xi| < 2^{m+1}\}.$$

Note that

$$F^{-1}\big((\gamma_k\, u_j) * \varphi_m\big) = \big(F^{-1}\gamma_k * F^{-1}u_j\big)F^{-1}\varphi_m$$

and

$$\operatorname{supp} F^{-1}\gamma_k \subset G_k, \quad \operatorname{supp} F^{-1}u_j \subset G_j, \quad \operatorname{supp} F^{-1}\varphi_m \subset G_m.$$

Therefore, if $(\gamma_k\, u_j) * \varphi_m$ is not identical to zero, we have $2^{m-1} \leq 2^{j+1} + 2^{k+1}$. Hence it is sufficient to estimate the norms

$$\mathcal{M}_m := \Big\| \sum_{j\geq m-3}\sum_{k\geq 1} (\gamma_k\, u_j) * \varphi_m \Big\|_{L_p}$$

and

$$\mathcal{L}_m := \Big\| \sum_{k\geq m-3}\sum_{j\geq 1} (\gamma_k\, u_j) * \varphi_m \Big\|_{L_p}.$$

We have

$$\mathcal{M}_m = \Big\| \sum_{j\geq m-3} \Big(u_j \sum_{k\geq 1} \gamma_k\Big) * \varphi_m \Big\|_{L_p}$$

$$\leq c \Big\|\sum_{k\geq 1} \gamma_k\Big\|_{L_\infty} \sum_{j\geq m-3} \|u_j\|_{L_p} \leq c\,\|\gamma\|_{L_\infty} \sum_{j\geq m-3} \|u_j\|_{L_p}.$$

This implies the estimate

$$\mathcal{M}^\theta_m \leq c\,\|\gamma\|^\theta_{L_\infty} \sup_{j\geq m-3} \big(2^{s\theta j}\, \|u_j\|^\theta_{L_p}\big) 2^{-s\theta m} \tag{5.5.3}$$

for any $s \in (0, l)$. Hence

$$\sum_{m\geq 4} 2^{l\theta m} \mathcal{M}_m^\theta \leq c\,\|\gamma\|_{L_\infty}^\theta \sum_{m\geq 4} 2^{(l-s)\theta m} \sum_{j\geq m-3} 2^{s\theta j}\|u_j\|_{L_p}^\theta$$

$$\leq c\,\|\gamma\|_{L_\infty}^\theta \sum_{m\geq 1} 2^{l\theta m}\|u_m\|_{L_p}^\theta.$$

The norm $\mathcal{L}_m$ satisfies

$$\mathcal{L}_m = \Big\| \sum_{k\geq m-3} \big(\gamma_k \sum_{j\geq 1} u_j\big) * \varphi_m \Big\|_{L_p}$$

$$\leq c\,\Big\|\sum_{j\geq 1} u_j\Big\|_{L_p} \sum_{k\geq m-3} \|\gamma_k\|_{L_\infty} \leq c\,\|u\|_{L_p} \sum_{k\geq m-3} \|\gamma_k\|_{L_\infty}.$$

Hence, by the same argument as in the case of the norm $\mathcal{M}_m$, we conclude that

$$\sum_{m\geq 4} 2^{l\theta m} \mathcal{M}_m^\theta \leq c\,\|u\|_{L_p}^\theta \sum_{m\geq 1} 2^{l\theta m}\|\gamma_m\|_{L_\infty}^\theta.$$

Finally, we obtain

$$\|\gamma u\|_{B^l_{p,\theta}} \leq c\left(\|\gamma\|_{B^l_{\infty,\theta}}\|u\|_{L_p} + \|\gamma\|_{L_\infty}\|u\|_{B^l_{p,\theta}}\right).$$

The proof is complete. □

5.5.2 Related Results on Multipliers in Besov and Triebel-Lizorkin Spaces

We begin an overview of further results with Triebel's theorem, similar to Theorem 5.5.1, on multipliers in $B^l_{p,\theta}$ and in the Triebel-Lizorkin space $F^l_{p,\theta}$ with $l \in \mathbb{R}^1$ and positive p and θ.

The norm (quasi-norm) in $F^l_{p,\theta}$ is defined by

$$\|u\|_{F^l_{p,\theta}} = \Big\| \Big(\sum_{k\geq 1} \big(2^{kl}\,|u * \varphi_k|\big)^\theta\Big)^{1/\theta} \Big\|_{L_p} + \|u * \Psi\|_{L_p}.$$

(For properties of these spaces see [Tr4] and [RS].)

Theorem 5.5.2. [Tr4] *(i) If $l \in \mathbb{R}^1$, $p \in (0,\infty]$, $\theta \in (0,\infty]$, and $\rho > \max\{l, -l+n/p\}$, then $B^\rho_{\infty,\infty} \subset MB^l_{p,\theta}$.*

(ii) If $l \in \mathbb{R}^1$, $p \in (0,\infty]$, $\theta \in (0,\infty]$, and $\rho > \max\{l, -l+n/\min(p,\theta)\}$, then $B^\rho_{\infty,\infty} \subset MF^l_{p,\theta}$.

Note that $F^l_{p,2}$ coincides with the space of Bessel potentials H^l_p for $p > 1$ and in this case part (ii) of Theorem 5.5.2 is contained in part (i) of Theorem 3.4.1.

Peetre's approach, based on the decomposition (5.5.2) and sometimes called the paraproduct algorithm, was used in the study of multipliers in Besov and Triebel-Lizorkin spaces in [Sic1]–[Sic3], [Tr4], Sect. 2.8, [Jo], [RS], [Mar1]–[Mar3], [Yam], [Yu], [SS], [KoS], [Ger]. In [SS] Sickel and Smirnov showed that

$$MB^l_{p,\theta} = B^l_{p,\theta,\mathrm{unif}}$$

for any p, θ such that $1 \le p \le \theta$ and $l > n/p$, whereas Bourdaud [Bo] demonstrated that

$$MB^l_{p,\theta} \neq B^l_{p,\theta,\mathrm{unif}}$$

for $1 \le \theta < p \le \infty$, $l > n/p$.

Multipliers preserving the Besov spaces $B^0_{\infty,1}$ and $B^0_{\infty,\infty}$ were characterized by Koch and Sickel [KoS]. In particular, they found the equivalence relation

$$\|\gamma\|_{MB^0_{\infty,\infty}} \sim \|\gamma\|_{L_\infty} + \|\gamma\|_{F^0_{\infty,1}} + \sup_{k\ge 0}(k+1)\,\|\gamma * \varphi_k\|_{L_\infty}$$

with φ_k defined in (5.5.1). We also mention that $MB^0_{p,\theta} \neq L_\infty$ unless $p=\theta=2$, according to Frazier and Jawerth [FrJ].

The imbeddings of the form $B^s_{p,\theta}$, $F^s_{p,\theta} \subset M(X)$, where $X = B^{s_0}_{p_0,\theta_0}$ or $X = F^{s_0}_{p_0,\theta_0}$ were studied in [RS], Ch. 4. A description of $M(w^m_2 \to w^{-k}_2)$, where m and k belong to $(-n/2, n/2)$ and $m \neq k$, can be found in [Ger].

Netrusov [Net] gave a characterization of multipliers in Triebel-Lizorkin spaces $F^l_{p,\theta}$, $p \le \theta \le \infty$, and Besov spaces $B^l_{p,\infty}$, where $0 < p \le 1$. Different characterizations of $MF^l_{p,\theta}$ were obtained by Sickel [Sic3] in the case

$$0 < p < \infty, \quad 0 < \theta \le \infty, \quad l > n\max\{0,\, p^{-1} - 1,\, \theta^{-1} - 1\}.$$

We note that, since $B^1_{1,\infty} = BV$ (see [Pe2], p. 164, and [Gu1]), it follows from Theorem 2.9.3 that

$$\|\gamma\|_{MB^1_{1,\infty}} \sim \sup_{\substack{x\in\mathbb{R}^n,\\ r\in(0,1)}} r^{1-n} \mathrm{var}\, \nabla\gamma(\mathcal{B}_r(x)). \tag{5.5.4}$$

The following description of the space $M(B^l_{p,1} \to B^l_{p,\infty})$, where $p \in (1,\infty)$ and $0 < l \le 1/p$, was given by Gulisashvili [Gu1], [Gu2].

Theorem 5.5.3. *A function γ belongs to the space $M(B^l_{p,1} \to B^l_{p,\infty})$, $p \in (1,\infty)$, $0 < l \le 1/p$, if and only if $\gamma \in L_\infty$ and*

$$\int_{\mathcal{B}_r(x)} |\Delta_h \gamma(t)|^p dt \le c\,|h|^{pl}\, r^{n-pl}$$

for all balls $\mathcal{B}_r(x)$, $r \in (0,1)$ and any $h \in \mathbb{R}^n$.

The relation

$$\|\gamma\|_{M(B^l_{p,1}\to B^l_{p,\infty})} \sim \sup_{\substack{x\in\mathbb{R}^n, r\in(0,1)\\ h\in\mathbb{R}^n\setminus\{0\}}} \Big(\frac{1}{|h|^{pl}\, r^{n-pl}} \int_{\mathcal{B}_r(x)} |\Delta_h \gamma(t)|^p dt\Big)^{1/p} + \|u\|_{L_\infty}$$

holds.

As a corollary of this result one obtains the following condition for the inclusion of χ_E into $M(B^{1/p}_{p,1} \to B^{1/p}_{p,\infty})$, where χ_E is the characteristic function of a Lebesgue measurable set E in $\mathbb{R}^n$. (See Sect. 2.9 for the notations s and ∂^*.)

Corollary 5.5.1. [Gu1] *The inclusion $\chi_E \in M(B^{1/p}_{p,1} \to B^{1/p}_{p,\infty})$ holds if and only if E is a set with local finite perimeter and*

$$s(\mathcal{B}_r(x)\cap \partial^* E) \le c\, r^{n-1}$$

for all balls $\mathcal{B}_r(x)$ with $r \in (0,1)$.

Comparing this assertion with Corollary 2.9.1 and (5.5.4), we see that χ_E belongs to $M(B^{1/p}_{p,1} \to B^{1/p}_{p,\infty})$ and $M(B^1_{1,1} \to B^1_{1,\infty})$ simultaneously.

5.5.3 Multipliers in *BMO*

The space BMO of functions with bounded mean oscillation (see [JN], [Cam], [F1], [Ste2], [Ja1], [Ja2], and elsewhere) plays an important role in modern analysis. It is situated between $B^0_{\infty,1}$ and $B^0_{\infty,\infty}$. This space is defined as follows. Let $Q_r(x)$ be the cube in $\mathbb{R}^n$ with side length r centered at x whose sides are parallel to coordinate axes. By $f(Q)$ we denote the mean value of f on a cube Q, that is,

$$f(Q) = \frac{1}{\text{mes}_n Q} \int_Q f(x)dx.$$

Further, we introduce the mean oscillation of f on Q by

$$\mathcal{O}(f,Q) = \frac{1}{\text{mes}_n Q} \int_Q |f(x) - f(Q)|dx.$$

By BMO we denote the space of functions integrable on $\mathbb{R}^n$ and such that

$$\sup_{x\in\mathbb{R}^n, r\in(0,1)} \mathcal{O}(f, Q_r(x)) < \infty.$$

Endowed with the norm

$$\|f\|_{BMO} = \|f\|_{L_1} + \sup_{x\in\mathbb{R}^n, r\in(0,1/2)} \mathcal{O}(f, Q_r(x)),$$

BMO becomes a Banach space.

We can include BMO into the family of spaces BMO_φ of locally integrable functions with the finite norm

$$\|f\|_{L_1} + \sup_{x\in\mathbb{R}^n, r\in(0,1/2)} \frac{\mathcal{O}(f, Q_r(x))}{\varphi(r)},$$

where φ ia a positive nondecreasing function on $(0, 1/2)$.

The following theorem, containing a description of the space $M(BMO)$ of multipliers in BMO, is due to Stegenga [Ste1] and Janson [Ja1].

Theorem 5.5.4. *The space $M(BMO)$ coincides with $BMO_{|\log r|^{-1}} \cap L_\infty$.*

Proof. We begin by deriving the following useful estimate for functions in BMO:

$$|f(Q_r) - f(Q_{1/2})| \le c\,\|f\|_{BMO}\,|\log r|, \tag{5.5.5}$$

where $Q_\rho = Q_\rho(x)$. It is clear that the left-hand side of this inequality does not exceed

$$\int_r^{1/2} \Big|\frac{d}{d\rho} f(Q_\rho)\Big| d\rho \le \int_r^{1/2} \frac{d\rho}{\rho^n} \int_{\partial Q_\rho} |f(y) - f(Q_\rho)| ds_y$$

$$\le c_1 \int_r^{1/2} \frac{d\rho}{\rho^{n+1}} \int_{\rho/2}^{\rho} dr \int_{\partial Q_\rho} |f(y) - f(Q_\rho)| ds_y.$$

Therefore,

$$|f(Q_r) - f(Q_{1/2})| \le c \int_r^{1/2} \frac{d\rho}{\rho^{n+1}} \int_{Q_\rho} |f(y) - f(Q_\rho)| dy,$$

which implies (5.5.5). By (5.5.5),

$$|f(Q_r)| \le 2^n \|f\|_{L_1} + c\,\|f\|_{BMO} |\log r|. \tag{5.5.6}$$

Let $\gamma \in BMO_{|\log r|^{-1}} \cap L_\infty$. Then

$$r^{-n} \int_{Q_r} |(\gamma f)(y) - \gamma(Q_r) f(Q_r)| dy$$

$$\le r^{-n} \int_{Q_r} |\gamma(x)|\, |f(x) - f(Q_r)| dy + r^{-n} \int_{Q_r} |f(Q_r)|\, |\gamma(y) - \gamma(Q_r)| dy$$

$$\le \|\gamma\|_{L_\infty} \mathcal{O}(f, Q_r) + |f(Q_r)|\, \mathcal{O}(\gamma, Q_r).$$

From (5.5.6) it follows that the last sum is dominated by

$$c\,|\log r| \big(\|\gamma\|_{L_\infty} + \|\gamma\|_{BMO_{|\log r|^{-1}}}\big) \|f\|_{BMO}.$$

It remains to note that

$$\mathcal{O}(f, Q_r) \leq \frac{2}{\text{mes}_n Q_r} \int_{Q_r} |f(y) - a| dy$$

for any number a. Hence $\gamma \in M(BMO)$.

Now let us show that $M(BMO) \subset BMO_{|\log r|^{-1}} \cap L_\infty$. Obviously,

$$\|\gamma\|_{L_\infty} = \lim_{N\to\infty} \|\gamma^N f\|_{L_1}^{1/N} \leq \lim_{N\to\infty} \inf \|\gamma^N f\|_{BMO}^{1/N} \leq \|\gamma\|_{M(BMO)}. \tag{5.5.7}$$

We see that

$$\frac{1}{\text{mes}_n Q_r} \int_{Q_r} |\gamma(y) - \gamma(Q_r)|\, |f(y)| dy \leq \mathcal{O}(\gamma f, Q_r) + 2\|\gamma\|_{L_\infty} \mathcal{O}(f, Q_r).$$

This together with (5.5.7) and the inequality

$$\mathcal{O}(\gamma f, Q_r) \leq \|\gamma\|_{M(BMO)} \|f\|_{BMO}$$

implies that

$$\frac{1}{\text{mes}_n Q_r} \int_{Q_r} |\gamma(y) - \gamma(Q_r)|\, |f(y)| dy \leq 3\, \|\gamma\|_{M(BMO)} \|f\|_{BMO}.$$

Setting here

$$f(y) = \eta(y) \log[x - y],$$

where $\eta \in C_0^\infty(Q_r)$ and $\eta = 1$ on $Q_{1/2}$, we obtain

$$\mathcal{O}(\gamma, Q_r) \leq c\,|\log r|^{-1}.$$

The proof is complete. □

The following general result has a slightly more complicated proof (see [Ja1]).

Theorem 5.5.5. *Let the function $\varphi(r)\, r^{-1}$ be almost decreasing in the sense that*

$$\varphi(\rho)\, \rho^{-1} \leq c\, \varphi(r)\, r^{-1} \qquad \text{for } \rho \geq r.$$

Then

$$M(BMO_\varphi) = BMO_\psi \cap L_\infty,$$

where

$$\psi(r) = \varphi(r) \Big(\int_r^1 \varphi(t) \frac{dt}{t} \Big)^{-1}.$$

Concerning the space $M(BMO_\varphi)$ for $\mathbb{R}^n$ and for general domains, see the series of papers by Bloom [Blo], Nakai [Na1], [Na2], Nakai and Yabuta [NY1], [NY2], Yabuta [Ya].

Using the duality of the Hardy space H^1 and BMO (see [St2]), Janson [Ja1] proved the coincidence of spaces of multipliers in H^1 and BMO. In other words, $MH^1 = BMO_{|\log r|^{-1}} \cap L_\infty$.

6

Maximal Algebras in Spaces of Multipliers

6.1 Introduction

Let A be a subset of a Banach function space. Then A is called a multiplication algebra if for all u and v in A their product uv belongs to A and there exists a constant c such that

$$\|uv\| \le c\,\|u\|\,\|v\|.$$

Let l be an integer. For $lp \le n$ and $p \in (1, \infty)$, or for $l < n$ and $p = 1$, the space W_p^l contains unbounded functions which are certainly not multipliers in W_p^l (see, for example, (2.7.1)). Hence the space W_p^l is not a multiplication algebra for the values of p, l, and n given above. It is not difficult to describe the maximal algebra contained in W_p^l. If $u \in A$, then for any $N = 1, 2, \dots$

$$\|u^N\|_{L_p}^{1/N} \le \|u^N\|_{W_p^l}^{1/N} \le c\,\|u\|_{W_p^l}.$$

Consequently, $A \subset W_p^l \cap L_\infty$. On the other hand, it is well known that the intersection $W_p^l \cap L_\infty$ is a multiplication algebra. In fact, for all u and v in $W_p^l \cap L_\infty$,

$$\|\nabla_l(uv)\|_{L_p} \le c \sum_{k=0}^{l} \|\,|\nabla_k u|\,|\nabla_{l-k} v|\,\|_{L_p} \le c \sum_{k=0}^{l} \|\nabla_k u\|_{L_{pl/k}}\,\|\nabla_{l-k} v\|_{L_{pl/(l-k)}}$$
$$\le c \sum_{k=0}^{l} \|u\|_{L_\infty}^{(l-k)/l}\,\|u\|_{W_p^l}^{k/l}\,\|v\|_{L_\infty}^{k/l}\,\|v\|_{W_p^l}^{(l-k)/l}.$$

Here we have used the Gagliardo - Nirenberg inequality

$$\|\nabla_j u\|_{L_{pl/j}} \le c\,\|u\|_{L_\infty}^{(l-j)/l}\,\|u\|_{W_p^l}^{j/l}, \quad j = 1, \dots, l-1.$$

(see [Gag2] and [Nir]). Thus the space $W_p^l \cap L_\infty$ is the maximal algebra contained in W_p^l.

V.G. Maz'ya, T.O. Shaposhnikova, *Theory of Sobolev Multipliers*,
Grundlehren der mathematischen Wissenschaften 337,

Since by Sobolev's theorem $W^l_p \subset L_\infty$ for $lp > n$, $p \in (1, \infty)$ or for $l \geq n$, $p = 1$, it follows that W^l_p is a multiplication algebra for the indicated values of p and l. Obviously, this known assertion follows from (2.2.5) and (2.3.29) as well.

Let us turn to the question of Banach algebras in spaces of multipliers $M(W^m_p \to W^l_p)$. Obviously, this question is trivial for MW^l_p which is a Banach algebra itself, but this is not the case for spaces of multipliers mapping one Sobolev space into a different one.

In Sect. 6.3 we show that the maximal Banach algebra $A^{m,l}_p$, imbedded in the space of multipliers $M(W^m_p \to W^l_p)$ which map the Sobolev space W^m_p to W^l_p with noninteger m and l, $m > l$ and $p \in [1, \infty)$, is isomorphic to $M(W^m_p \to W^l_p) \cap L_\infty$.

It is proved in Sect. 6.4 that the maximal Banach algebra $\mathcal{A}^{m,l}_p$, imbedded in the space of multipliers acting between Bessel potential spaces $M(H^m_p \to H^l_p)$, is isomorphic to $M(H^m_p \to H^l_p) \cap L_\infty$. Precise descriptions of the imbeddings $A^{m,l}_p \subset A^{\mu,\lambda}_p$ and $\mathcal{A}^{m,l}_p \subset \mathcal{A}^{\mu,\lambda}_p$ are given in Sect. 6.5.

6.2 Pointwise Interpolation Inequalities for Derivatives

6.2.1 Inequalities Involving Derivatives of Integer Order

The aim of this subsection is the following inequality (see [Kal] and [MSh17]).

Lemma 6.2.1. *Let* $0 < k < m$. *Then*

$$|\nabla_k u(x)| \leq c\, (\mathcal{M}u(x))^{\frac{m-k}{m}} (\mathcal{M}\nabla_m u(x))^{\frac{k}{m}}. \tag{6.2.1}$$

Proof. Let η be a function in the ball $\mathcal{B}_1$ with Lipschitz derivatives of order $m-2$ and which vanishes on $\partial\mathcal{B}_1$ together with all these derivatives. Also let

$$\int_{\mathcal{B}_1} \eta(y)\, dy = 1. \tag{6.2.2}$$

We need the Sobolev integral representation:

$$v(0) = \sum_{|\beta|<m-k} t^{-n} \int_{\mathcal{B}_t} \frac{(-y)^\beta}{\beta!} \partial^\beta v(y) \eta(y/t)\, dy$$

$$+(-1)^{m-k}(m-k) \sum_{|\alpha|=m-k} \int_{\mathcal{B}_t} \frac{y^\alpha}{\alpha!} \partial^\alpha v(y) \int_{|y|/t}^\infty \eta\Big(\rho \frac{y}{|y|}\Big) \rho^{n-1} d\rho \frac{dy}{|y|^n} \tag{6.2.3}$$

(see [Maz14], Sec. 1.5.1).

Setting here $v = \partial^\gamma u$ with an arbitrary multi-index γ of order k and integrating by parts in the first integral, we arrive at the identity

$$\partial^\gamma u(0) = (-1)^k t^{-n} \int_{\mathcal{B}_t} u(y) \sum_{|\beta|<m-k} \frac{1}{\beta!} \partial^{\beta+\gamma} \left(y^\beta \eta(y/t)\right) dy$$

$$+ \sum_{|\alpha|=m-k} (-1)^{m-k}(m-k) \int_{\mathcal{B}_t} \frac{y^\alpha}{\alpha!} \partial^{\alpha+\gamma} u(y) \int_{|y|/t}^{\infty} \eta\Big(\rho\frac{y}{|y|}\Big) \rho^{n-1} d\rho \frac{dy}{|y|^n}. \quad (6.2.4)$$

Hence

$$|\nabla_k u(0)| \le c_1 t^{-n-k} \int_{\mathcal{B}_t} |u(y)|\, dy + c_2 \int_{\mathcal{B}_t} |\nabla_m u(y)| \frac{dy}{|y|^{n-m+k}}. \quad (6.2.5)$$

If $m-k \ge n$, the second integral does not exceed

$$t^{m-k-n} \int_{\mathcal{B}_t} |\nabla_m u(y)|\, dy\,.$$

In the case $m-k<n$ the second integral in (6.2.5) equals

$$t^{m-k-n} \int_{\mathcal{B}_t} |\nabla_m u(y)|\, dy + (n-m+k) \int_0^t \frac{d\tau}{\tau^{n-m+k+1}} \int_{\mathcal{B}_\tau} |\nabla_m u(y)|\, dy\,.$$

Therefore,

$$\int_{\mathcal{B}_t} |\nabla_m u(y)| \frac{dy}{|y|^{n-m+k}} \le \frac{n}{m-k} t^{m-k} \sup_{\tau \le t} \tau^{-n} \int_{\mathcal{B}_\tau} |\nabla_m u(y)|\, dy. \quad (6.2.6)$$

Thus, for any $t>0$,

$$|\nabla_k u(0)| \le c_3\, t^{-k} \mathcal{M} u(0) + c_4\, t^{m-k} \mathcal{M} \nabla_m u(0) \quad (6.2.7)$$

which implies (6.2.1). □

6.2.2 Inequalities Involving Derivatives of Fractional Order

Lemma 6.2.2. *Let k and l be integers, and let m be a noninteger, $0 \le l \le k < m$. If $u \in W^{[m]}_{p,\mathrm{loc}}$, then*

$$|\nabla_k u(x)| \le c \left((\mathcal{M}\nabla_l u)(x)\right)^{\frac{m-k}{m-l}} \left((D_{p,m} u)(x)\right)^{\frac{k-l}{m-l}} \quad (6.2.8)$$

for almost all $x \in \mathbb{R}^n$.

Proof. It suffices to prove (6.2.8) for $l=0$ and $x=0$. Let η be a function in the ball $\mathcal{B}_1$ with Lipschitz derivatives of order $m-2$ and which vanishes on $\partial\mathcal{B}_1$ together with all these derivatives. We assume also that η is subject to (6.2.2). Let t be an arbitrary positive number to be chosen later.

Replacing m by $[m]$ in (6.2.3) and setting there $v = \partial^\gamma u$ with an arbitrary multi-index γ of order k and then integrating by parts in the first integral, we arrive at the identity

$$\partial^\gamma u(0) = (-1)^k t^{-n} \int_{\mathcal{B}_t} u(y) \sum_{|\beta|<[m]-k} \frac{1}{\beta!} \partial^{\beta+\gamma}(y^\beta \eta(y/t))\, dy \tag{6.2.9}$$

$$+ \sum_{|\alpha|=[m]-k} (-1)^{[m]-k}([m]-k) \int_{\mathcal{B}_t} \frac{y^\alpha}{\alpha!} \partial^{\alpha+\gamma} u(y) \int_{|y|/t}^\infty \eta(\rho \frac{y}{|y|}) \rho^{n-1} d\rho \frac{dy}{|y|^n} .$$

Hence, for $k < [m]$ we have

$$|\nabla_k u(0)| \le c\Big(t^{-k} \mathcal{M}u(0) + t^{[m]-k} |\nabla_{[m]} u(0)|$$

$$+ \int_{\mathcal{B}_t} \frac{|\nabla_{[m]} u(y) - \nabla_{[m]} u(0)|}{|y|^{n-[m]+k}} dy\Big) . \tag{6.2.10}$$

Hölder's inequality implies that

$$\int_{\mathcal{B}_t} \frac{|\nabla_{[m]} u(y) - \nabla_{[m]} u(0)|}{|y|^{n-[m]+k}} dy \le c\, t^{m-k} (D_{p,m} u)(0) . \tag{6.2.11}$$

Let γ be an arbitrary multi-index of order $[m]$. The identity

$$\partial^\gamma u(0) = t^{-n} \int_{\mathcal{B}_t} \eta(\frac{y}{t}) \partial^\gamma u(y) dy + t^{-n} \int_{\mathcal{B}_t} \eta(\frac{y}{t}) \left[\partial^\gamma u(0) - \partial^\gamma u(y)\right] dy$$

gives

$$|\nabla_{[m]} u(0)| \le t^{-n-[m]} \Big| \int_{\mathcal{B}_t} u(y) (\nabla_{[m]} \eta)(\frac{y}{t})\, dy \Big|$$

$$+ t^{\{m\}} \Big(\int_{\mathcal{B}_t} |\eta(y)|^q |y|^{(\frac{n}{p}+\{m\})q} dy \Big)^{1/q} (D_{p,m} u)(0) , \tag{6.2.12}$$

where $p^{-1} + q^{-1} = 1$. Combining (6.2.10)–(6.2.12), we arrive at

$$|\nabla_k u(0)| \le c \left(t^{-k} (\mathcal{M}u)(0) + t^{m-k} (D_{p,m} u)(0) \right) . \tag{6.2.13}$$

Minimization of the right-hand side in t completes the proof. □

Remark 6.2.1. Note that one can replace $D_{p,m}u$ on the right-hand sides of (6.2.11) and (6.2.12) by $D^{(r)}_{p,m} u$, where

$$(D^{(r)}_{p,m} u)(x) = \Big(\int_{\mathcal{B}_r} |\nabla_{[m]} u(x+h) - \nabla_{[m]} u(x)|^p \frac{dh}{|h|^{n+p\{s\}}} \Big)^{1/p} . \tag{6.2.14}$$

Hence, by (6.2.10),

$$|\nabla_k u(0)| \le c \left(t^{-k} (\mathcal{M}u)(0) + t^{m-k} (D^{(t)}_{p,m} u)(0) \right) . \tag{6.2.15}$$

Lemma 6.2.2 implies the following assertion.

Corollary 6.2.1. *Let k be an integer and let m be a noninteger, $0 < k < m$. Then*

$$|\nabla_k u(x)| \le c\, \|u\|_{L_\infty}^{\frac{m-k}{m}} \big((D_{p,m}u)(x)\big)^{\frac{k}{m}} \tag{6.2.16}$$

for almost all $x \in \mathbb{R}^n$.

In the following lemma we derive a multiplicative inequality involving the function (6.2.14) and use the notation of integral with a bar for the mean value.

Lemma 6.2.3. *Let s and l be positive nonintegers, $s < l$, and let $1 \le p < \infty$.*

If either $\{s\} \ge \{l\}$ or $0 < s < 1$, then there exists a positive constant $c = c(s, l, n, p)$ such that

$$(D^{(r)}_{p,s}u)(x) \le c\Big(\fint_{\mathcal{B}_r(x)} |u(y)|dy\Big)^{\frac{l-s}{l}} \Big(r^{-l}\fint_{\mathcal{B}_r(x)} |u(y)|dy + D^{(r)}_{p,l}u(x)\Big)^{\frac{s}{l}} \tag{6.2.17}$$

for almost all $x \in \mathbb{R}^n$.

Proof. Let $s \ge 1$. It is enough to prove (6.2.17) for $x = 0$. Since $s < l$ and $\{s\} \ge \{l\}$, it follows that $[s] \le [l] - 1$. We have

$$\begin{aligned}&\Big(\int_{\mathcal{B}_1} |\nabla_{[s]}u(h) - \nabla_{[s]}u(0)|^p \frac{dh}{|h|^{n+p\{s\}}}\Big)^{1/p}\\ &\le \Big(\int_{\mathcal{B}_1} \Big|\nabla_{[s]}u(h) - \sum_{\{\alpha:|\alpha|\le[l]-1-[s]\}} \frac{h^\alpha}{\alpha!} D^\alpha \nabla_{[s]}u(0)\Big|^p \frac{dh}{|h|^{n+p(l-[s])}}\Big)^{1/p}\\ &+ c\sum_{i=[s]+1}^{[l]-1} |\nabla_i u(0)|.\end{aligned} \tag{6.2.18}$$

The last sum should be omitted if $[s] = [l] - 1$. For $[s] \le [l] - 1$ the identity

$$\begin{aligned}&\nabla_{[s]}u(h) - \sum_{\{\alpha:|\alpha|\le[l]-1-[s]\}} \frac{h^\alpha}{\alpha!} D^\alpha \nabla_{[s]}u(0)\\ &= ([l]-[s]) \int_0^1 \sum_{\{\alpha:|\alpha|=[l]-[s]\}} \frac{h^\alpha}{\alpha!} \big(D^\alpha\nabla_{[s]}u(th)\\ &\qquad -D^\alpha\nabla_{[s]}u(0)\big)(1-t)^{[l]-[s]-1}dt\end{aligned} \tag{6.2.19}$$

implies that the integral on the right-hand side of (6.2.18) does not exceed

$$c\Big(\int_{\mathcal{B}_1}\Big(\int_0^1 |\nabla_{[l]}u(th) - \nabla_{[l]}u(0)|(1-t)^{[l]-[s]-1}dt\Big)^p \frac{dh}{|h|^{n+p\{l\}}}\Big)^{1/p}.$$

By Minkowski's inequality this last expression is dominated by

$$\int_0^1 \Big(\int_{\mathcal{B}_1} |\nabla_{[l]} u(th) - \nabla_{[l]} u(0)|^p \frac{dh}{|h|^{n+p\{l\}}}\Big)^{1/p} (1-t)^{[l]-[s]-1} dt$$

$$= c\, D^{(1)}_{p,l} u(0). \tag{6.2.20}$$

It follows from (6.2.5) that for every $i \leq [l]-1$

$$|\nabla_i u(0)| \leq c_1 \int_{\mathcal{B}_1} |u(h)| dh + c_2 \int_{\mathcal{B}_1} |\nabla_{[l]} u(h)| \frac{dh}{|h|^{n-[l]+i}}$$

$$\leq c_1 \int_{\mathcal{B}_1} |u(h| dh + c_2 \int_{\mathcal{B}_1} \frac{|\nabla_{[l]} u(h) - \nabla_{[l]} u(0)|}{|h|^{n-[l]+i}} dh + c_3\, |\nabla_{[l]} u(0)|. \tag{6.2.21}$$

Let η be a function in the ball $\mathcal{B}_1$ with Lipschitz derivatives of order $l-2$ and which vanishes on $\partial \mathcal{B}_1$ together with all these derivatives. We also assume that η satisfies (6.2.2). For any multi-index β of order $[l]$ the identity

$$D^\beta u(0) = \int_{\mathcal{B}_1} \eta(h) D^\beta u(h) dh + \int_{\mathcal{B}_1} \eta(h) \big(D^\beta u(0) - D^\beta u(h)\big) dh$$

and Hölder's inequality imply that

$$|\nabla_{[l]} u(0)| \leq c_1 \int_{\mathcal{B}_1} |u(h)| dh$$

$$+ c_2 \Big(\int_{\mathcal{B}_1} |\eta(h)|^q |h|^{(\frac{n}{p}+\{l\})q} dh\Big)^{1/q} D^{(1)}_{p,l} u(0), \tag{6.2.22}$$

where $p^{-1}+q^{-1}=1$. Using Hölder's inequality again, we obtain

$$\int_{\mathcal{B}_1} \frac{|\nabla_{[l]} u(h) - \nabla_{[l]} u(0)|}{|h|^{n-[l]+i}} dh \leq c\, D^{(1)}_{p,l} u(0)$$

which together with (6.2.22) and (6.2.21) results in

$$|\nabla_i u(0)| \leq c_1 \int_{\mathcal{B}_1} |u(h)| dh + c_2\, D^{(1)}_{p,l} u(0), \qquad i \leq [l]-1. \tag{6.2.23}$$

Combining this inequality with (6.2.18) and (6.2.20), we arrive at

$$D^{(1)}_{p,s} u(0) \leq c_1 \int_{\mathcal{B}_1} |u(h)| dh + c_2\, D^{(1)}_{p,l} u(0) \tag{6.2.24}$$

which, after the dilation $h \to h/r$, becomes

$$D^{(r)}_{p,s} u(0) \leq c\Big(r^{-s} \fint_{\mathcal{B}_r} |u(h)| dh + r^{l-s} D^{(r)}_{p,l} u(0)\Big). \tag{6.2.25}$$

Now, one has the alternatives: either

$$\fint_{\mathcal{B}_r} |u(h)|dh \le r^l D^{(r)}_{p,l} u(0)$$

and

$$D^{(r)}_{p,s} u(0) \le c \left(\fint_{\mathcal{B}_r} |u(h)|dh \right)^{1-\frac{s}{l}} \left(D^{(r)}_{p,l} u(0) \right)^{\frac{s}{l}},$$

or

$$\fint_{\mathcal{B}_r} |u(h)|dh > r^l D^{(r)}_{p,l} u(0)$$

and

$$D^{(r)}_{p,s} u(0) \le c\, r^{-s} \fint_{\mathcal{B}_r} |u(h)|dh.$$

Therefore, (6.2.17) holds.

For $0 < s < 1$ the inequality (6.2.17) is, obviously, valid. □

Corollary 6.2.2. *Let s and l be positive nonintegers such that $s < l$ and let $p \ge 1$.*

(i) If either $\{s\} \ge \{l\}$ or $0 < s < 1$, then there exists a positive constant $c = c(s, l, n, p)$ such that

$$(D_{p,s} u)(x) \le c \left(\mathcal{M} u(x) \right)^{\frac{l-s}{l}} \left(D_{p,l} u(x) \right)^{\frac{s}{l}} \tag{6.2.26}$$

for any $u \in W^l_{p,loc}$ and almost all $x \in \mathbb{R}^n$.

(ii) If $s > 1$ and $\{s\} < \{l\}$, then even the rougher inequality

$$(D_{p,s} u)(x) \le c \, \|u\|^{\frac{l-s}{l}}_{L_\infty} \left(D_{p,l} u(x) \right)^{\frac{s}{l}}$$

does not hold.

Proof. (*i*) From (6.2.17) we deduce that

$$(D^{(r)}_{p,s} u)(x) \le c \left(\mathcal{M} u(x) \right)^{\frac{l-s}{l}} \left(r^{-l} \mathcal{M} u(x) + D^{(r)}_{p,l} u(x) \right)^{\frac{s}{l}}.$$

The result follows by passing to the limit as $r \to \infty$.

(*ii*) Let $s > 1$ and $\{s\} < \{l\}$. Suppose that the inequality

$$(D_{p,s} u)(x) \le c \, \|u\|^{\frac{\{l\}-\{s\}}{[s]+\{l\}}}_{L_\infty} \left(D_{p,[s]+\{l\}} u(x) \right)^{\frac{s}{[s]+\{l\}}} \tag{6.2.27}$$

holds. According to part (i),

$$(D_{p,[s]+\{l\}} u)(x) \le c \, \|u\|^{\frac{[l]-[s]}{l}}_{L_\infty} \left(D_{p,l} u(x) \right)^{\frac{[s]+\{l\}}{l}}$$

which together with (6.2.27) results in (6.2.17). Hence it is enough to disprove (6.2.27).

We set $x = 0$ and assume that supp $u \subset \mathcal{B}_2 \backslash \overline{\mathcal{B}_1}$. Then (6.2.27) implies that

$$\Big(\int_{\mathcal{B}_2 \backslash \mathcal{B}_1} |\nabla_{[s]} u(h)|^p dh\Big)^{1/p} \leq c\, \|u\|_{L_\infty}$$

which, obviously, does not hold for all u. This completes the proof. □

Remark 6.2.2. For $p = \infty$ the inequality (6.2.17) becomes

$$\sup_y \frac{|\nabla_{[s]} u(y) - \nabla_{[s]} u(x)|}{|y-x|^{\{s\}}} \leq c\, (\mathcal{M}u(x))^{\frac{l-s}{l}} \Big(\sup_y \frac{|\nabla_{[l]} u(y) - \nabla_{[l]} u(x)|}{|y-x|^{\{l\}}}\Big)^{\frac{s}{l}},$$

where $s < l$ and either $\{s\} \leq \{l\}$ or $0 < s < 1$.

6.3 Maximal Banach Algebra in $M(W_p^m \to W_p^l)$

6.3.1 The Case $p > 1$

Theorem 6.3.1. *Let $m \geq l \geq 0$, and let $p \in (1, \infty)$. The maximal Banach algebra $A_p^{m,l}$ imbedded into*

$$M(W_p^m \to W_p^l)$$

is isomorphic to the space

$$M(W_p^m \to W_p^l) \cap L_\infty. \tag{6.3.1}$$

The estimate

$$\|\gamma_1 \gamma_2\|_{M(W_p^m \to W_p^l)}$$
$$\leq c\, \big(\|\gamma_1\|_{L_\infty} \|\gamma_2\|_{M(W_p^m \to W_p^l)} + \|\gamma_2\|_{L_\infty} \|\gamma_1\|_{M(W_p^m \to W_p^l)}\big)$$

holds.

Proof. The statement is trivial for $m = l$, since MW_p^l is an algebra and is imbedded into L_∞ (see (2.3.15) and (4.3.28)).

Let $A_p^{m,l}$ be a Banach subalgebra of $M(W_p^m \to W_p^l)$ and let c be a constant such that

$$\|\gamma\|_{M(W_p^m \to W_p^l)} \leq c\, \|\gamma\|_{A_p^{m,l}}$$

for all $\gamma \in A_p^{m,l}$. For every $N = 1, 2 \ldots$ and all $\gamma \in A_p^{m,l}$, $u \in W_p^m$ we have

$$\|\gamma^N u\|_{L_p}^{1/N} \leq \|\gamma^N u\|_{W_p^l}^{1/N} \leq \|\gamma^N\|_{M(W_p^m \to W_p^l)}^{1/N} \|u\|_{W_p^m}^{1/N}$$

$$\leq (c\, \|\gamma^N\|_{A_p^{m,l}})^{1/N} \|u\|_{W_p^m}^{1/N} \leq c^{1/N} \|\gamma\|_{A_p^{m,l}} \|u\|_{W_p^m}^{1/N}.$$

Passing to the limit as $N \to \infty$, we obtain that $\gamma \in L_\infty$ and

$$\|\gamma\|_{L_\infty} \le \|\gamma\|_{A_p^{m,l}}. \tag{6.3.2}$$

Consider first the case of integer l. Suppose that γ_1 and γ_2 belong to (6.3.1). Then, for all $u \in W_p^m$

$$\|\nabla_l (\gamma_1\gamma_2 u)\|_{L_p} \le c\Big(\|\gamma_1\|_{L_\infty} \|\nabla_l (\gamma_2 u)\|_{L_p} + \|\gamma_2\|_{L_\infty} \sum_{h=1}^{l} \| \, |\nabla_h\gamma_1| \, |\nabla_{l-h} u| \, \|_{L_p}$$

$$+ \sum_{h=1}^{l-1}\sum_{k=1}^{l-h} \| \, |\nabla_h\gamma_1| \, |\nabla_k\gamma_2| \, |\nabla_{l-h-k} u| \, \|_{L_p}\Big). \tag{6.3.3}$$

The first term on the right-hand side is majorized by

$$c\,\|\gamma_1\|_{L_\infty} \|\gamma_2\|_{M(W_p^m\to W_p^l)} \|u\|_{W_p^m} \,.$$

To estimate the second term, we recall that, if $\gamma \in M(W_p^m \to W_p^l)$, then for any $j = 0, \dots, l$,

$$\nabla_j\gamma \in M(W_p^m \to W_p^{l-j}) \subset M(W_p^{m-l+j} \to L_p)$$

and the estimate

$$\|\nabla_j\gamma\|_{M(W_p^{m-l+j}\to L_p)} \le c\,\|\gamma\|_{M(W_p^m\to W_p^l)} \tag{6.3.4}$$

holds (see (2.3.13)). Therefore, the second term on the right-hand side of (6.3.3) is not greater than

$$c\,\|\gamma_2\|_{L_\infty} \|\gamma_1\|_{M(W_p^m\to W_p^l)} \|u\|_{W_p^m} \,.$$

To estimate the remaining terms on the right-hand side of (6.3.3), we use the inequality

$$|\nabla_h\gamma(x)| \le c\,\|\gamma\|_{L_\infty}^{\frac{k}{h+k}} (\mathcal{M}\nabla_{h+k}\gamma(x))^{\frac{h}{k+h}}$$

stemming from (6.2.1). Hence

$$\| \, |\nabla_h\gamma_1| \, |\nabla_k\gamma_2| \, |\nabla_{l-h-k} u| \, \|_{L_p}$$
$$\le c\,\|\gamma_1\|_{L_\infty}^{\frac{k}{h+k}} \|\gamma_2\|_{L_\infty}^{\frac{h}{h+k}} \| \, (\mathcal{M}\nabla_{h+k}\gamma_1)^{\frac{h}{h+k}} (\mathcal{M}\nabla_{h+k}\gamma_2)^{\frac{k}{h+k}} |\nabla_{l-h-k} u| \, \|_{L_p}$$
$$\le c\|\gamma_1\|_{L_\infty}^{\frac{k}{h+k}} \|\gamma_2\|_{L_\infty}^{\frac{h}{h+k}} \|(\mathcal{M}\nabla_{h+k}\gamma_1)|\nabla_{l-h-k} u| \, \|_{L_p}^{\frac{h}{h+k}} \|(\mathcal{M}\nabla_{h+k}\gamma_2)|\nabla_{l-h-k} u| \, \|_{L_p}.$$

By (2.3.22) we have

$$\|\mathcal{M}\gamma\|_{M(W_p^s\to L_p)} \le c\,\|\gamma\|_{M(W_p^s\to L_p)} \,. \tag{6.3.5}$$

This inequality and (6.3.4) give

$$\| \, |\nabla_h \gamma_1| \, |\nabla_k \gamma_2| \, |\nabla_{l-h-k} u| \, \|_{L_p}$$
$$\le c \, \|\gamma_1\|_{L_\infty}^{\frac{k}{h+k}} \, \|\gamma_2\|_{L_\infty}^{\frac{h}{h+k}} \, \| \, |\nabla_{h+k}\gamma_1| \, |\nabla_{l-h-k} u| \, \|_{L_p}^{\frac{h}{h+k}} \, \| \, |\nabla_{h+k}\gamma_2| \, |\nabla_{l-h-k} u| \, \|_{L_p}^{\frac{k}{h+k}}$$
$$\le c \, \|\gamma_1\|_{L_\infty}^{\frac{k}{h+k}} \, \|\gamma_2\|_{L_\infty}^{\frac{h}{h+k}} \, \|\gamma_1\|_{M(W_p^m \to W_p^l)}^{\frac{h}{h+k}} \, \|\gamma_2\|_{M(W_p^m \to W_p^l)}^{\frac{k}{h+k}} \, \|u\|_{W_p^m}$$

which completes the proof for integer l.

Let l be a positive noninteger. Suppose that γ_1 and γ_2 belong to (6.3.1). Then, for all $u \in W_p^m$

$$\begin{aligned} \|D_{p,l}(\gamma_1\gamma_2 u)\|_{L_p} &= \|D_{p,\{l\}} \nabla_{[l]}(\gamma_1\gamma_2 u)\|_{L_p} \\ &\le c \sum_{|\alpha|+|\beta|+|\sigma|=[l]} \|D_{p,\{l\}}(D^\alpha\gamma_1 D^\beta\gamma_2 D^\sigma u)\|_{L_p} \\ &\le c \sum_{|\alpha|+|\beta|+|\sigma|=[l]} (A_{\alpha,\beta,\sigma} + B_{\alpha,\beta,\sigma} + C_{\alpha,\beta,\sigma}), \end{aligned} \tag{6.3.6}$$

where

$$A_{\alpha,\beta,\sigma} = \|(D^\alpha\gamma_1)\,(D^\beta\gamma_2)\,D_{p,\{l\}} D^\sigma u\|_{L_p},$$
$$B_{\alpha,\beta,\sigma} = \|(D^\alpha\gamma_1)\,(D_{p,\{l\}} D^\beta\gamma_2)\,D^\sigma u\|_{L_p},$$

and $C_{\alpha,\beta,\sigma}$ is given by

$$\Big(\int\!\!\int |D^\alpha\gamma_1(x)|^p |D^\beta\gamma_2(x+h)|^p |D^\sigma u(x+h) - D^\sigma u(x)|^p \frac{dh dx}{|h|^{n+p\{l\}}}\Big)^{1/p}. \tag{6.3.7}$$

To estimate $A_{\alpha,\beta,\sigma}$, we use the inequality

$$|\nabla_i \varphi(x)| \le c\, \|\varphi\|_{L_\infty}^{\frac{k-i}{k}} (\mathcal{M}\nabla_k\varphi(x))^{\frac{i}{k}} \quad \text{a.e. in } \mathbb{R}^n \tag{6.3.8}$$

with $0 \le i \le k$, which follows directly from (6.2.1). We have

$$A_{\alpha,\beta,\sigma} \le c \|\gamma_1\|_{L_\infty}^{1-\frac{|\alpha|}{[l]-|\sigma|}} \|\gamma_2\|_{L_\infty}^{1-\frac{|\beta|}{[l]-|\sigma|}}$$
$$\times \Big(\int\!\!\int (\mathcal{M}\nabla_{[l]-|\sigma|}\gamma_1(x))^{\frac{p|\alpha|}{[l]-|\sigma|}} (\mathcal{M}\nabla_{[l]-|\sigma|}\gamma_2(x))^{\frac{p|\beta|}{[l]-|\sigma|}}$$
$$\times |D^\sigma u(x+h) - D^\sigma u(x)|^p \frac{dh dx}{|h|^{n+p\{l\}}}\Big)^{1/p}.$$

By Hölder's inequality, the double integral on the right-hand side is dominated by

$$c \Big(\int\!\!\int (\mathcal{M}\nabla_{[l]-|\sigma|}\gamma_1(x))^p |D^\sigma u(x+h) - D^\sigma u(x)|^p \frac{dh dx}{|h|^{n+p\{l\}}}\Big)^{\frac{|\alpha|}{([l]-|\sigma|)p}}$$
$$\times \Big(\int\!\!\int (\mathcal{M}\nabla_{[l]-|\sigma|}\gamma_2(x))^p |D^\sigma u(x+h) - D^\sigma u(x)|^p \frac{dh dx}{|h|^{n+p\{l\}}}\Big)^{\frac{|\beta|}{([l]-|\sigma|)p}}$$

which equals

$$c \,\|(\mathcal{M}\nabla_{[l]-|\sigma|}\gamma_1) D_{p,|\sigma|+\{l\}} u\|_{L_p}^{\frac{|\alpha|}{[l]-|\sigma|}}$$

$$\times \|(\mathcal{M}\nabla_{[l]-|\sigma|}\gamma_2) D_{p,|\sigma|+\{l\}} u\|_{L_p}^{\frac{|\beta|}{[l]-|\sigma|}}. \tag{6.3.9}$$

This expression does not exceed

$$c \,\|\mathcal{M}\nabla_{[l]-|\sigma|}\gamma_1\|_{M(W_p^{m-|\sigma|-\{l\}} \to L_p)}^{\frac{|\alpha|}{[l]-|\sigma|}} \|u\|_{W_p^m}^{\frac{|\alpha|}{[l]-|\sigma|}}$$

$$\times \|\mathcal{M}\nabla_{[l]-|\sigma|}\gamma_2\|_{M(W_p^{m-|\sigma|-\{l\}} \to L_p)}^{\frac{|\beta|}{[l]-|\sigma|}} \|u\|_{W_p^m}^{\frac{|\beta|}{[l]-|\sigma|}}. \tag{6.3.10}$$

We estimate (6.3.10) using (6.3.5) combined with the inequality

$$\|\nabla_j \gamma\|_{M(W_p^{m-l+j} \to L_p)} \le c \,\|\gamma\|_{M(W_p^m \to W_p^l)} \tag{6.3.11}$$

(see (2.2.11)). Since $|\alpha| + |\beta| + |\sigma| = [l]$, we obtain from (6.3.9)– (6.3.11)

$$A_{\alpha,\beta,\sigma} \le c \,\|\gamma_1\|_{L_\infty}^{\frac{|\beta|}{|\alpha|+|\beta|}} \|\gamma_2\|_{L_\infty}^{\frac{|\alpha|}{|\alpha|+|\beta|}}$$

$$\times \|\gamma_1\|_{M(W_p^m \to W_p^l)}^{\frac{|\alpha|}{|\alpha|+|\beta|}} \|\gamma_2\|_{M(W_p^m \to W_p^l)}^{\frac{|\beta|}{|\alpha|+|\beta|}} \|u\|_{W_p^m}.$$

Now, by Hölder's inequality

$$A_{\alpha,\beta,\sigma} \le c\big(\|\gamma_1\|_{L_\infty} \|\gamma_2\|_{M(W_p^m \to W_p^l)}$$

$$+ \|\gamma_2\|_{L_\infty} \|\gamma_1\|_{M(W_p^m \to W_p^l)}\big) \|u\|_{W_p^m}. \tag{6.3.12}$$

To estimate $B_{\alpha,\beta,\sigma}$ defined by (6.3.7), we use (6.3.8) with $\varphi = D^\alpha \gamma_1$ and apply Lemma 6.2.3 with $\varphi = \gamma_2$ and $s = \{l\} + |\beta|$. Then

$$B_{\alpha,\beta,\sigma} \le c \,\|\gamma_1\|_{L_\infty}^{1-\frac{|\alpha|}{l-|\sigma|}} \|\gamma_2\|_{L_\infty}^{1-\frac{|\beta|+\{l\}}{l-|\sigma|}}$$

$$\times \|(D_{p,l-|\sigma|}\gamma_1)^{\frac{|\alpha|}{l-|\sigma|}} (D_{p,l-|\sigma|}\gamma_2)^{\frac{|\beta|+\{l\}}{l-|\sigma|}} D^\sigma u\|_{L_p}.$$

By Hölder's inequality, the last norm is dominated by

$$\|D_{p,l-|\sigma|}\gamma_1 D^\sigma u\|_{L_p}^{\frac{|\alpha|}{l-|\sigma|}} \|D_{p,l-|\sigma|}\gamma_2 D^\sigma u\|_{L_p}^{\frac{|\beta|+\{l\}}{l-|\sigma|}}.$$

Since

$$\|D_{p,l-j}\gamma\|_{M(W_p^{m-j} \to L_p)} \le c \,\|\gamma\|_{M(W_p^m \to W_p^l)}$$

(see (4.3.87)), it follows for $i = 1, 2$ that

$$\|D_{p,l-|\sigma|}\gamma_i D^\sigma u\|_{L_p} \le \|D_{p,l-|\sigma|}\gamma_i\|_{M(W_p^{m-|\sigma|} \to L_p)} \|u\|_{W_p^m}$$

$$\le c \,\|\gamma_i\|_{M(W_p^m \to W_p^l)} \|u\|_{W_p^m}.$$

Hence,

$$B_{\alpha,\beta,\sigma} \leq c\, \|\gamma_1\|_{L_\infty}^{\frac{|\beta|+\{l\}}{|\alpha|+|\beta|+\{l\}}} \|\gamma_2\|_{L_\infty}^{\frac{|\alpha|+\{l\}}{|\alpha|+|\beta|+\{l\}}}$$
$$\times \|\gamma_1\|_{M(W_p^m \to W_p^l)}^{\frac{|\alpha|}{|\alpha|+|\beta|+\{l\}}} \|\gamma_2\|_{M(W_p^m \to W_p^l)}^{\frac{|\beta|+\{l\}}{|\alpha|+|\beta|+\{l\}}} \|u\|_{W_p^m}.$$

Therefore, by Hölder's inequality, $B_{\alpha,\beta,\sigma}$ has the same majorant (6.3.12) as $A_{\alpha,\beta,\sigma}$.

In order to estimate $C_{\alpha,\beta,\sigma}$, we use (6.3.8). Then

$$C_{\alpha,\beta,\sigma} \leq c\, \|\gamma_1\|_{L_\infty}^{1-\frac{|\alpha|}{[l]-|\sigma|}} \|\gamma_2\|_{L_\infty}^{1-\frac{|\beta|}{[l]-|\sigma|}}$$
$$\times \Big(\int\!\!\int (\mathcal{M}\nabla_{[l]-|\sigma|}\gamma_1(x))^{\frac{p|\alpha|}{[l]-|\sigma|}} (\mathcal{M}\nabla_{[l]-|\sigma|}\gamma_2(x+h))^{\frac{p|\beta|}{[l]-|\sigma|}}$$
$$\times \frac{|D^\sigma u(x+h) - D^\sigma u(x)|^p}{|h|^{n+p\{l\}}} dh dx\Big)^{1/p}.$$

By Hölder's inequality, the double integral on the right-hand side is not greater than

$$c\Big(\int\!\!\int (\mathcal{M}\nabla_{[l]-|\sigma|}\gamma_1(x))^p |D^\sigma u(x+h) - D^\sigma u(x)|^p \frac{dh dx}{|h|^{n+p\{l\}}}\Big)^{\frac{|\alpha|}{([l]-|\sigma|)p}}$$
$$\times\Big(\int\!\!\int (\mathcal{M}\nabla_{[l]-|\sigma|}\gamma_2(x+h))^p |D^\sigma u(x+h) - D^\sigma u(x)|^p \frac{dh dx}{|h|^{n+p\{l\}}}\Big)^{\frac{|\beta|}{([l]-|\sigma|)p}}$$

which coincides with (6.3.9). The above estimate of (6.3.9) implies (6.3.12) with $A_{\alpha,\beta,\sigma}$ replaced by $C_{\alpha,\beta,\sigma}$. This completes the proof. □

Theorems 6.3.1 and 4.1.1 imply

Corollary 6.3.1. *The maximal Banach algebra in* $M(W_p^m \to W_p^l)$, $m \geq l$, $p \in (1,\infty)$, *consists of functions* $\gamma \in W_{p,loc}^l$ *with finite norm*

$$\sup_{\substack{e \subset \mathbb{R}^n \\ \mathrm{diam}(e) \leq 1}} \frac{\|D_{p,l}\gamma; e\|_{L_p}}{(C_{p,m}(e))^{1/p}} + \|\gamma\|_{L_\infty}. \tag{6.3.13}$$

In the case $mp > n$ *the norm* (6.3.13) *can be simplified as*

$$\|D_{p,l}\gamma\|_{L_{p,\mathrm{unif}}} + \|\gamma\|_{L_\infty}.$$

6.3.2 Maximal Banach Algebra in $M(W_1^m \to W_1^l)$

Theorem 6.3.2. *Let* $m \geq l \geq 0$. *The maximal Banach algebra* $A_1^{m,l}$ *imbedded into* $M(W_1^m \to W_1^l)$ *is isomorphic to the space*

$$M(W_1^m \to W_1^l) \cap L_\infty. \tag{6.3.14}$$

The estimate

$$\|\gamma_1\gamma_2\|_{M(W_1^m \to W_1^l)} \le c\,\big(\|\gamma_1\|_{L_\infty}\|\gamma_2\|_{M(W_1^m\to W_1^l)} + \|\gamma_2\|_{L_\infty}\|\gamma_1\|_{M(W_1^m\to W_1^l)}\big)$$

holds.

Proof. The imbedding $A_1^{m,l} \subset L_\infty$ is proved in the same way as in Theorem 6.3.1, where the case $p > 1$ is considered.

(*i*) Let l be integer. Suppose that γ_1 and γ_2 belong to (6.3.14). Then

$$\|\gamma_1\gamma_2 u\|_{W_1^l} \le c \sum_{j=0}^{l} \|\,|\nabla_j(\gamma_1\gamma_2)|\,|\nabla_{l-j}u|\|_{L_1}.$$

We show that

$$\nabla_j(\gamma_1\gamma_2) \in M(W_1^{m-l+j} \to L_1), \quad j = 0, \dots, l, \tag{6.3.15}$$

which implies the result.

Since the intersection $W_1^j(\mathcal{B}_1) \cap L_\infty(\mathcal{B}_1)$ is an algebra, we have

$$\|\gamma_1\gamma_2; \mathcal{B}_1\|_{W_1^j} \le c\,\big(\|\gamma_1; \mathcal{B}_1\|_{L_\infty}\|\gamma_2; \mathcal{B}_1\|_{W_1^j} + \|\gamma_2; \mathcal{B}_1\|_{L_\infty}\|\gamma_1; \mathcal{B}_1\|_{W_1^j}\big).$$

Hence, for any $r > 0$

$$r^j \int_{\mathcal{B}_r} |\nabla_j(\gamma_1\gamma_2)(x)|dx \le c\,\big(\|\gamma_1; \mathcal{B}_r\|_{L_\infty}(r^j\|\nabla_j\gamma_2; \mathcal{B}_r\|_{L_1} + r^n\|\gamma_2; \mathcal{B}_r\|_{L_1})$$
$$+ \|\gamma_2; \mathcal{B}_r\|_{L_\infty}(r^j\|\nabla_j\gamma_1; \mathcal{B}_r\|_{L_1} + r^n\|\gamma_1; \mathcal{B}_r\|_{L_1})\big).$$

This inequality, taken for $r \in (0,1)$, along with (2.2.3) gives

$$r^{m-l+j-n}\|\nabla_j(\gamma_1\gamma_2); \mathcal{B}_r\|_{L_1} \le c\,\big(\|\gamma_1; \mathcal{B}_r\|_{L_\infty}\|\gamma_2\|_{M(W_1^m\to W_1^l)}$$
$$+\|\gamma_2; \mathcal{B}_r\|_{L_\infty}\|\gamma_1\|_{M(W_1^m\to W_1^l)}\big)$$

which implies (6.3.15).

(*ii*) Let l be a noninteger. We have

$$\int_{\mathcal{B}_1}\int_{\mathcal{B}_1} |\nabla_{[l]}(\gamma_1(x)\gamma_2(x)) - \nabla_{[l]}(\gamma_1(y)\gamma_2(y))|\frac{dxdy}{|x-y|^{n+\{l\}}}$$
$$\le \sum_{k=0}^{[l]}\Big(\int_{\mathcal{B}_1}\int_{\mathcal{B}_1} |\nabla_k\gamma_1(x)|\,|\nabla_{[l]-k}\gamma_2(x) - \nabla_{[l]-k}\gamma_2(y)|\frac{dxdy}{|x-y|^{n+\{l\}}}$$
$$+\int_{\mathcal{B}_1}\int_{\mathcal{B}_1} |\nabla_{[l]-k}\gamma_2(y)|\,|\nabla_k\gamma_1(x) - \nabla_k\gamma_1(y)|\frac{dxdy}{|x-y|^{n+\{l\}}}\Big). \tag{6.3.16}$$

By Hölder's inequality, the right-hand side does not exceed

$$c\sum_{i=0}^{1}\sum_{k=0}^{[l]}\|\nabla_k\gamma_{1+i};\mathcal{B}_1\|_{L_{\frac{l}{k}}}$$

$$\times\Big(\int_{\mathcal{B}_1}\Big(\int_{\mathcal{B}_1}|\nabla_{[l]-k}\gamma_{2-i}(x)-\nabla_{[l]-k}\gamma_{2-i}(y)|\frac{dy}{|x-y|^{n+\{l\}}}\Big)^{\frac{l}{l-k}}dx\Big)^{\frac{l-k}{l}}. \quad (6.3.17)$$

Duplicating the argument leading from (6.2.25) to (6.2.17), we conclude that (6.2.15) implies the estimate

$$|\nabla_k u(0)|\le c\,(\|u;\mathcal{B}_t\|_{L_\infty})^{1-\frac{k}{l}}\,(t^l\|u;\mathcal{B}_t\|_{L_\infty}+D^{(t)}_{p,l}u(0))^{\frac{k}{l}}. \quad (6.3.18)$$

For $t=1$ and u replaced by γ it becomes

$$|\nabla_k\gamma(x)|$$

$$\le c\,\|\gamma;\mathcal{B}_1\|^{1-\frac{k}{l}}_{L_\infty}\Big(\int_{\mathcal{B}_1}|\nabla_{[l]}\gamma(x)-\nabla_{[l]}\gamma(y)|\frac{dy}{|x-y|^{n+\{l\}}}+\|\gamma;\mathcal{B}_1\|_{L_\infty}\Big)^{\frac{k}{l}}. \quad (6.3.19)$$

Furthermore, by Lemma 6.2.3

$$\int_{\mathcal{B}_1}|\nabla_{[l]-k}\gamma(x)-\nabla_{[l]-k}\gamma(y)|\frac{dy}{|x-y|^{n+\{l\}}}$$

$$\le c\|\gamma;\mathcal{B}_1\|^{\frac{k}{l}}_{L_\infty}\Big(\int_{\mathcal{B}_1}|\nabla_{[l]}\gamma(x)-\nabla_{[l]}\gamma(y)|\frac{dy}{|x-y|^{n+\{l\}}}+\|\gamma;\mathcal{B}_1\|_{L_1}\Big)^{\frac{l-k}{l}}. \quad (6.3.20)$$

Using (6.3.17)–(6.3.20) in (6.3.16), we find that

$$\int_{\mathcal{B}_1}\int_{\mathcal{B}_1}|\nabla_{[l]}(\gamma_1(x)\gamma_2(x))-\nabla_{[l]}(\gamma_1(y)\gamma_2(y))|\frac{dxdy}{|x-y|^{n+\{l\}}}$$

$$\le c\sum_{i=0}^{1}\sum_{k=0}^{[l]}\|\gamma_{1+i};\mathbb{R}^n\|^{1-\frac{k}{l}}_{L_\infty}\Big(\int_{\mathcal{B}_1}\int_{\mathcal{B}_1}|\nabla_{[l]}\gamma_{1+i}(x)-\nabla_{[l]}\gamma_{1+i}(y)|\frac{dy}{|x-y|^{n+\{l\}}}$$

$$+\|\gamma_{1+i};\mathcal{B}_1\|_{L_1}\Big)^{\frac{k}{l}}\|\gamma_{2-i};\mathbb{R}^n\|^{\frac{k}{l}}_{L_\infty}$$

$$\times\Big(\int_{\mathcal{B}_1}\int_{\mathcal{B}_1}|\nabla_{[l]}\gamma_{2-i}(x)-\nabla_{[l]}\gamma_{2-i}(y)|\frac{dy}{|x-y|^{n+\{l\}}}+\|\gamma_{2-i};\mathcal{B}_1\|_{L_1}\Big)^{1-\frac{k}{l}}.$$

By the dilation $x\to x/r$ we obtain

$$r^{m-n}\int_{\mathcal{B}_r}\int_{\mathcal{B}_r}|\nabla_{[l]}(\gamma_1(x)\gamma_2(x))-\nabla_{[l]}(\gamma_1(y)\gamma_2(y))|\frac{dxdy}{|x-y|^{n+\{l\}}}$$

$$\leq c \sum_{i=0}^{1} \sum_{k=0}^{[l]} \|\gamma_{1+i}\|_{L_\infty}^{1-\frac{k}{l}} \Big(r^{m-n} \int_{\mathcal{B}_r} \int_{\mathcal{B}_r} |\nabla_{[l]}\gamma_{1+i}(x) - \nabla_{[l]}\gamma_{1+i}(y)| \frac{dy}{|x-y|^{n+\{l\}}}$$

$$+ r^{m-l} \|\gamma_{1+i}; \mathcal{B}_r\|_{L_1} \Big)^{\frac{k}{l}} \|\gamma_{2-i}\|_{L_\infty}^{\frac{k}{l}}$$

$$\times \Big(r^{m-n} \int_{\mathcal{B}_r} \int_{\mathcal{B}_r} |\nabla_{[l]}\gamma_{2-i}(x) - \nabla_{[l]}\gamma_{2-i}(y)| \frac{dy}{|x-y|^{n+\{l\}}} + r^{m-l} \|\gamma_{2-i}; \mathcal{B}_r\|_{L_1} \Big)^{1-\frac{k}{l}}.$$

Using this estimate for $r \in (0,1)$ together with Theorem 5.4.1, we arrive at

$$r^{m-n} \|D_{1,l}^{(r)}(\gamma_1\gamma_2); \mathcal{B}_r\|_{L_1} \leq$$

$$c \sum_{i=0}^{1} \sum_{k=0}^{[l]} \|\gamma_{1+i}\|_{L_\infty}^{1-\frac{k}{l}} \|\gamma_{1+i}\|_{M(W_1^m \to W_1^l)}^{\frac{k}{l}} \|\gamma_{2-i}\|_{L_\infty}^{\frac{k}{l}} \|\gamma_{2-i}\|_{M(W_1^m \to W_1^l)}^{1-\frac{k}{l}}$$

$$\leq \|\gamma_1\|_{L_\infty} \|\gamma_2\|_{M(W_1^m \to W_1^l)} + \|\gamma_2\|_{L_\infty} \|\gamma_1\|_{M(W_1^m \to W_1^l)}$$

which, along with Theorem 5.4.1 and the obvious inequality

$$r^{m-l-n} \|\gamma_1\gamma_2; \mathcal{B}_r(x)\|_{L_1} \leq c\, \|\gamma_1\|_{L_\infty} \|\gamma_2\|_{M(W_1^m \to W_1^m)},$$

completes the proof. □

The following assertion resulting from Theorem 6.3.2 involves the norm $\|\!|\cdot|\!\|$ defined by (5.4.4) and possessing the property (5.4.5).

Corollary 6.3.2. *Let $m \geq l \geq 0$. The maximal Banach algebra in $M(W_1^m \to W_1^l)$, consists of functions $\gamma \in W_{1,\mathrm{loc}}^l$ with finite norm*

$$\sup_{\substack{x \in \mathbb{R}^n \\ r \in (0,1)}} r^{m-n} \|\!|\gamma; \mathcal{B}_r(x)|\!\|_{W_1^l} + \|\gamma\|_{L_\infty}. \tag{6.3.21}$$

In the case $m \geq n$ the norm (6.3.21) *can be simplified as*

$$\sup_{x \in \mathbb{R}^n} \|\!|\gamma; \mathcal{B}_1(x)|\!\|_{W_1^l} + \|\gamma\|_{L_\infty}.$$

6.4 Maximal Algebra in Spaces of Bessel Potentials

6.4.1 Pointwise Inequalities Involving the Strichartz Function

We start with an inequality similar to (6.2.16).

Lemma 6.4.1. *Let k and r be an integer and noninteger, respectively, with $0 < k < r$ and let $\varphi \in W^{[r]}_{p,\mathrm{loc}}$. There exists a positive constant $c = c(k, r, n)$ such that*

$$|\nabla_k \varphi(x)| \le c\,(\mathcal{M}\varphi(x))^{\frac{r-k}{r}}\,(S_r\varphi(x))^{\frac{k}{r}} \tag{6.4.1}$$

for almost all $x \in \mathbb{R}^n$.

Proof. We use the inequality

$$|\nabla_k \varphi(0)| \le c\Big(\int_{\mathcal{B}_1} |\varphi(y)|dy + \int_{\mathcal{B}_1} |\nabla_{[r]}\varphi(y)| \frac{dy}{|y|^{n-[r]+k}}\Big) \tag{6.4.2}$$

(see (6.2.5)). Clearly, the right-hand side is majorized by

$$c\Big(\mathcal{M}\varphi(0) + \int_{\mathcal{B}_1} |\nabla_{[r]}\varphi(y) - \nabla_{[r]}\varphi(0)| \frac{dy}{|y|^{n-[r]+k}} + |\nabla_{[r]}\varphi(0)|\Big). \tag{6.4.3}$$

Using the notation

$$\psi(y) = |\nabla_{[r]}\varphi(y) - \nabla_{[r]}\varphi(0)|,$$

we find that the second term in (6.4.3) is equal to

$$(n-[r]-k)\int_0^1 t^{-n+[r]-k-1}\int_{\mathcal{B}_t}\psi(z)dzdt + \int_{\mathcal{B}_1}\psi(z)dz.$$

This sum is dominated by

$$c\Big(S_r\varphi(0) + \int_{\mathcal{B}_2}(2-|z|)\psi(z)dz\Big).$$

We have

$$\int_{\mathcal{B}_2}(2-|z|)\psi(z)dz = \int_0^2\int_{\mathcal{B}_t}\psi(z)dzdt$$

$$\le c\Big(\int_0^2 t^{-2n-1-2\{r\}}\Big(\int_{\mathcal{B}_t}\psi(z)dz\Big)^2 dt\Big)^{1/2} \le c\,S_r\varphi(0). \tag{6.4.4}$$

Therefore,

$$|\nabla_k\varphi(0)| \le c\,\big(\mathcal{M}\varphi(0) + S_r\varphi(0) + |\nabla_{[r]}\varphi(0)|\big). \tag{6.4.5}$$

In order to estimate the third term in (6.4.5), we use the identity

$$\nabla_{[r]}\varphi(0) = \int_{\mathcal{B}_1}\eta(y)\nabla_{[r]}\varphi(y)dy + \int_{\mathcal{B}_1}\eta(y)\big(\nabla_{[r]}\varphi(0) - \nabla_{[r]}\varphi(y)\big)dy, \tag{6.4.6}$$

where $\eta \in C_0^\infty(\mathcal{B}_1)$ is such that

$$\int_{\mathcal{B}_1}\eta(y)dy = 1.$$

Clearly, the first term on the right-hand side of (6.4.6) is majorized by $c\,\mathcal{M}\varphi(0)$. The second term is estimated by

$$c \int_{\mathcal{B}_2} (2 - |z|)\, \psi(z) dz$$

and does not exceed $c\, S_r\varphi(0)$ by (6.4.4). Therefore,

$$|\nabla_{[r]}\varphi(0)| \le c\,\big(\mathcal{M}\varphi(0) + S_r\varphi(0)\big). \tag{6.4.7}$$

Combining (6.4.7) with (6.4.5), we arrive at

$$|\nabla_k\varphi(0)| \le c\,\big(\mathcal{M}\varphi(0) + S_r\varphi(0)\big).$$

Now we obtain by dilation $x \to x/\rho$ that

$$|\nabla_k\varphi(0)| \le c\big(\rho^{-k}\mathcal{M}\varphi(0) + \rho^{r-k} S_r\varphi(0)\big).$$

The result follows by minimization of the right-hand side in ρ. □

Lemma 6.4.2. *Let k and l be an integer and noninteger, respectively, with $0 < k < l$, and let $\varphi \in W^{[l]}_{p,\mathrm{loc}}$. There exists a positive constant $c = c(k,l,n)$ such that*

$$S_{l-k}\varphi(x) \le c\,\|\varphi\|^{\frac{k}{l}}_{L_\infty} (S_l\varphi(x))^{\frac{l-k}{l}} \tag{6.4.8}$$

for almost all $x \in \mathbb{R}^n$.

Proof. Clearly,

$$\int_0^2 \Big(\int_{\mathcal{B}_1} |(\nabla_{[l]-k}\varphi)(\theta y) - (\nabla_{[l]-k}\varphi)(0)| d\theta\Big)^2 \frac{dy}{y^{1+2\{l\}}}$$

$$\le \int_0^2 \Big(\int_{\mathcal{B}_1} \Big|(\nabla_{[l]-k}\varphi)(\theta y) - \sum_{\{\alpha:|\alpha|\le k-1\}} \frac{y^\alpha\theta^\alpha}{\alpha!} (D^\alpha\nabla_{[l]-k}\varphi)(0)\Big| d\theta\Big)^2 \frac{dy}{y^{1+2\{l\}}}$$
$$+ \sum_{i=0}^{k-1} |\nabla_{[l]-i}\varphi(0)|^2. \tag{6.4.9}$$

The difference in the integral over $\mathcal{B}_1$ on the right-hand side is equal to

$$k \int_0^1 \sum_{\{\alpha:|\alpha|=k\}} \frac{y^\alpha\theta^\alpha}{\alpha!} \Big((D^\alpha\nabla_{[l]-k}\varphi)(\tau\theta y) - (D^\alpha\nabla_{[l]-k}\varphi)(0)\Big)(1-\tau)^{k-1} d\tau.$$

Using this and Minkowski's inequality, we see that the first term on the right-hand side of (6.4.9) is dominated by

$$c\Big(\int_0^1 \Big(\int_0^\infty \Big(\int_{\mathcal{B}_1} |(\nabla_{[l]}\varphi)(\tau\theta y) - (\nabla_{[l]}\varphi)(0)| d\theta\Big)^2 \frac{dy}{y^{1+2\{l\}}}\Big)^{1/2} d\tau\Big)^2$$
$$= c\,(S_l\varphi(0))^2. \tag{6.4.10}$$

By Lemma 6.4.1,

$$|\nabla_{[l]-i}\varphi(0)| \le c\,\|\varphi\|_{L_\infty}^{\frac{\{l\}+i}{l}} \left(S_l\varphi(0)\right)^{\frac{[l]-i}{l}}, \quad i = 0, \dots, [l], \tag{6.4.11}$$

which together with (6.4.9) and (6.4.10) gives

$$\int_0^2 \Big(\int_{\mathcal{B}_1} |(\nabla_{[l]-k}\varphi)(\theta y) - -(\nabla_{[l]-k}\varphi)(0)|d\theta\Big)^2 \frac{dy}{y^{1+\{l\}}} \le c\left(\|\varphi\|_{L_\infty} + S_l\varphi(0)\right)^2. \tag{6.4.12}$$

Obviously,

$$\int_2^\infty \Big(\int_{\mathcal{B}_1} |(\nabla_{[l]-k}\varphi)(\theta y) - (\nabla_{[l]-k}\varphi)(0)|d\theta\Big)^2 \frac{dy}{y^{1+\{l\}}} \le c\Big(|(\nabla_{[l]-k}\varphi)(0)|^2 + \int_2^\infty \Big(\int_{B_y} |\nabla_{[l]-k}\varphi(z)|dz\Big)^2 \frac{dy}{y^{1+2\{l\}+2n}}\Big). \tag{6.4.13}$$

The second term on the right-hand side does not exceed

$$\int_2^\infty \Big(\int_{\mathcal{B}_{y+1}} d\xi \int_{\mathcal{B}_1(\xi)} |\nabla_{[l]-k}\varphi(z)|dz\Big)^2 \frac{dy}{y^{1+2\{l\}+2n}}. \tag{6.4.14}$$

Since by (6.2.5)

$$\int_{\mathcal{B}_1(\xi)} |\nabla_{[l]-k}\varphi(z)|dz \le c\int_{\mathcal{B}_1(\xi)} \left(|\nabla_{[l]}\varphi(z)| + |\varphi(z)|\right)dz,$$

the expression (6.4.14) is dominated by

$$\begin{aligned} &c\Big(\int_2^\infty \Big(\int_{\mathcal{B}_{y+1}} d\xi \int_{\mathcal{B}_1(\xi)} |\nabla_{[l]}\varphi(z)|dz\Big)^2 \frac{dy}{y^{1+2\{l\}+2n}} + \|\varphi\|^2_{L_\infty}\Big) \\ &\le c\Big(\int_2^\infty \Big(\int_{\mathcal{B}_{y+2}} |\nabla_{[l]}\varphi(z)|dz\Big)^2 \frac{dy}{y^{1+2\{l\}+2n}} + \|\varphi\|^2_{L_\infty}\Big) \\ &\le c\left(S_l\varphi(0) + |\nabla_{[l]}\varphi(0)| + \|\varphi\|_{L_\infty}\right)^2. \end{aligned}$$

Hence, we find by (6.4.13) and (6.4.11) that

$$\int_2^\infty \Big(\int_{\mathcal{B}_1} |(\nabla_{[l]-k}\varphi)(\theta y) - (\nabla_{[l]-k}\varphi)(0)|d\theta\Big)^2 \frac{dy}{y^{1+2\{l\}}} \le c\left(\|\varphi\|_{L_\infty} + S_l\varphi(0)\right)^2.$$

Using (6.4.12), we obtain

$$S_{l-k}\varphi(x) \le c\left(\|\varphi\|_{L_\infty} + S_l\varphi(x)\right).$$

The result follows by dilation as in Lemma 6.4.1. □

6.4.2 Banach Algebra $\mathcal{A}_p^{m,l}$

Theorem 6.4.1. *The maximal Banach algebra $\mathcal{A}_p^{m,l}$ imbedded into $M(H_p^m \to H_p^l)$, $m \geq l$, is isomorphic to the space*

$$M(H_p^m \to H_p^l) \cap L_\infty. \tag{6.4.15}$$

The estimate

$$\|\gamma_1\gamma_2\|_{M(H_p^m\to H_p^l)} \leq c\big(\|\gamma_1\|_{L_\infty}\|\gamma_2\|_{M(H_p^m\to H_p^l)} + \|\gamma_2\|_{L_\infty}\|\gamma_1\|_{M(H_p^m\to H_p^l)}\big)$$

holds.

Proof. Let $\mathcal{A}_p^{m,l}$ be a subset of $M(H_p^m \to H_p^l)$ and let $\gamma \in \mathcal{A}_p^{m,l}$. The inequality

$$\|\gamma\|_{L_\infty} \leq \|\gamma\|_{\mathcal{A}_p^{m,l}} \tag{6.4.16}$$

is proved in the same way as (6.3.2).

Suppose that γ_1 and γ_2 belong to the space (6.4.15). For any $u \in H_p^m$,

$$\|S_l(\gamma_1\gamma_2 u)\|_{L_p} = \|S_{\{l\}}\nabla_{[l]}(\gamma_1\gamma_2 u)\|_{L_p} \leq c \sum_{|\alpha|+|\beta|+|\sigma|=[l]} \|S_{\{l\}}(D^\alpha\gamma_1 D^\beta\gamma_2 D^\sigma u)\|_{L_p}$$

$$\leq c \sum_{|\alpha|+|\beta|+|\sigma|=[l]} \big(\mathcal{A}_{\alpha,\beta,\sigma} + \mathcal{B}_{\alpha,\beta,\sigma} + \mathcal{C}_{\alpha,\beta,\sigma}\big),$$

where

$$\mathcal{A}_{\alpha,\beta,\sigma} = \|(D^\alpha\gamma_1)\,(D^\beta\gamma_2)\,S_{\{l\}}D^\sigma u\|_{L_p}, \tag{6.4.17}$$

$$\mathcal{B}_{\alpha,\beta,\sigma} = \|(D^\alpha\gamma_1)\,(S_{\{l\}}D^\beta\gamma_2)\,D^\sigma u\|_{L_p}, \tag{6.4.18}$$

$$\mathcal{C}_{\alpha,\beta,\sigma} = \Big(\int |D^\alpha\gamma_1(x)|^p \Big(\int_0^\infty \Big(\int_{\mathcal{B}_1} |D^\beta\gamma_2(x+\theta y)|\,|D^\sigma u(x+\theta y) - D^\sigma u(x)|d\theta\Big)^2 \frac{dy}{y^{1+2\{l\}}}\Big)^{p/2} dx\Big)^{1/p}. \tag{6.4.19}$$

Applying Lemma 6.4.1, we obtain

$$\mathcal{A}_{\alpha,\beta,\sigma} \leq c\|\gamma_1\|_{L_\infty}^{1-\frac{|\alpha|}{[l]-|\sigma|}}\|\gamma_2\|_{L_\infty}^{1-\frac{|\beta|}{[l]-|\sigma|}}$$

$$\times\|\big(\mathcal{M}\nabla_{[l]-|\sigma|}\gamma_1\big)^{\frac{|\alpha|}{[l]-|\sigma|}}\big(\mathcal{M}\nabla_{[l]-|\sigma|}\gamma_2\big)^{\frac{|\beta|}{[l]-|\sigma|}} S_{\{l\}+|\sigma|}u\|_{L_p}. \tag{6.4.20}$$

By Hölder's inequality the last norm is dominated by

$$c\,\|\big(\mathcal{M}\nabla_{[l]-|\sigma|}\gamma_1\big)S_{\{l\}+|\sigma|}u\|_{L_p}^{\frac{|\alpha|}{[l]-|\sigma|}}\,\|\big(\mathcal{M}\nabla_{[l]-|\sigma|}\gamma_2\big)\,S_{\{l\}+|\sigma|}u\|_{L_p}^{\frac{|\beta|}{[l]-|\sigma|}}. \tag{6.4.21}$$

Since by Minkowski's inequality

$$S_{\{l\}}v \le \Lambda^{|\sigma|+\{l\}-m} S_{\{l\}} \Lambda^{m-|\sigma|-\{l\}} v,$$

it follows that for $\gamma \in M(H_p^{m-|\sigma|-\{l\}} \to L_p)$

$$\|\gamma S_{\{l\}}v\|_{L_p} \le \|\gamma\|_{M(H_p^{m-|\sigma|-\{l\}} \to L_p)} \|\Lambda^{|\sigma|+\{l\}-m} S_{\{l\}} \Lambda^{m-|\sigma|-\{l\}} v\|_{H_p^{m-|\sigma|-\{l\}}}$$

$$\le c\,\|\gamma\|_{M(H_p^{m-|\sigma|-\{l\}} \to L_p)} \|v\|_{H_p^{m-|\sigma|}}.$$

Putting here $\gamma = \mathcal{M}\nabla_{[l]-|\sigma|}\gamma_i$, $i = 1, 2$, and $v = \nabla_{|\sigma|}u$, we find that (6.4.21) does not exceed

$$c\,\|\mathcal{M}\nabla_{[l]-|\sigma|}\gamma_1\|^{\frac{|\alpha|}{[l]-|\sigma|}}_{M(H_p^{m-|\sigma|-\{l\}} \to L_p)} \|u\|^{\frac{|\alpha|}{[l]-|\sigma|}}_{H_p^m}$$

$$\times \|\mathcal{M}\nabla_{[l]-|\sigma|}\gamma_2\|^{\frac{|\beta|}{[l]-|\sigma|}}_{M(H_p^{m-|\sigma|-\{l\}} \to L_p)} \|u\|^{\frac{|\beta|}{[l]-|\sigma|}}_{H_p^m}. \tag{6.4.22}$$

We use (6.3.5) and the estimate

$$\|\nabla_j \gamma\|_{M(H_p^{m-l+j} \to L_p)} \le c\,\|\gamma\|_{M(H_p^m \to H_p^l)} \tag{6.4.23}$$

(see Corollary 3.2.1). Then, using the equality $|\alpha| + |\beta| + |\sigma| = [l]$, we obtain that (6.4.22) does not exceed

$$c\,\|\gamma_1\|^{\frac{|\alpha|}{|\alpha|+|\beta|}}_{M(H_p^m \to H_p^l)} \|\gamma_2\|^{\frac{|\beta|}{|\alpha|+|\beta|}}_{M(H_p^m \to H_p^l)} \|u\|_{H_p^m}. \tag{6.4.24}$$

Now, by (6.4.20) and Hölder's inequality,

$$\mathcal{A}_{\alpha,\beta,\sigma} \le c\big(\|\gamma_1\|_{L_\infty} \|\gamma_2\|_{M(H_p^m \to H_p^l)} + \|\gamma_2\|_{L_\infty} \|\gamma_1\|_{M(H_p^m \to H_p^l)}\big) \|u\|_{H_p^m}. \tag{6.4.25}$$

To estimate $B_{\alpha,\beta,\sigma}$, defined by (6.4.18), we apply Lemma 6.4.1 to the function $D^\alpha\gamma_1$ and Lemma 6.4.2 to the function $S_{\{l\}}D^\beta\gamma_2$. Then, by Hölder's inequality,

$$\mathcal{B}_{\alpha,\beta,\sigma} \le c\,\|\gamma_1\|^{1-\frac{|\alpha|}{l-|\sigma|}}_{L_\infty} \|\gamma_2\|^{1-\frac{|\beta|+\{l\}}{l-|\sigma|}}_{L_\infty} \|\big(S_{l-|\sigma|}\gamma_1\big)^{\frac{|\alpha|}{l-|\sigma|}} \big(S_{l-|\sigma|}\gamma_2\big)^{\frac{|\beta|+\{l\}}{l-|\sigma|}} D^\sigma u\|_{L_p}$$

$$\le c\,\|\gamma_1\|^{1-\frac{|\alpha|}{l-|\sigma|}}_{L_\infty} \|\gamma_2\|^{1-\frac{|\beta|+\{l\}}{l-|\sigma|}}_{L_\infty} \|\big(S_{l-|\sigma|}\gamma_1\big) D^\sigma u\|^{\frac{|\alpha|}{l-|\sigma|}}_{L_p} \|\big(S_{l-|\sigma|}\gamma_2\big) D^\sigma u\|^{\frac{|\beta|+\{l\}}{l-|\sigma|}}_{L_p}.$$

By Lemma 3.2.8, we have for $i = 1, 2$

$$\|\big(S_{l-|\sigma|}\gamma_i\big) D^\sigma u\|_{L_p} \le \|S_{l-|\sigma|}\gamma_i\|_{M(H_p^{m-|\sigma|} \to L_p)} \|u\|_{H_p^m}$$

$$\le c\,\|\gamma_i\|_{M(H_p^m \to H_p^l)} \|u\|_{H_p^m}.$$

Hence $\mathcal{B}_{\alpha,\beta,\sigma}$ has the same majorant (6.4.25) as $\mathcal{A}_{\alpha,\beta,\sigma}$.

In order to estimate $\mathcal{C}_{\alpha,\beta,\sigma}$ defined by (6.4.19), we use Lemma 6.4.1 and get

$$\mathcal{C}_{\alpha,\beta,\sigma} \leq c\,\|\gamma_1\|_{L_\infty}^{1-\frac{|\alpha|}{[l]-|\sigma|}} \|\gamma_2\|_{L_\infty}^{1-\frac{|\beta|}{[l]-|\sigma|}} K_{\alpha,\beta,\sigma},$$

where

$$K_{\alpha,\beta,\sigma} = \Big(\int (\mathcal{M}\nabla_{[l]-|\sigma|}\gamma_1(x))^{\frac{p|\alpha|}{[l]-|\sigma|}}$$

$$\times\int_0^\infty\Big(\int_{\mathcal{B}_1}(\mathcal{M}\nabla_{[l]-|\sigma|}\gamma_2(x+\theta y))^{\frac{p|\beta|}{[l]-|\sigma|}}|\nabla_{|\sigma|}u(x+\theta y)-\nabla_{|\sigma|}u(x)|d\theta\Big)^2\frac{dy}{y^{1+2\{l\}}}\Big)^{\frac{p}{2}}dx\Big)^{\frac{1}{p}}.$$

By Hölder's inequality, $K_{\alpha,\beta,\sigma}$ is dominated by (6.4.22) which, as was shown above, has the majorant (6.4.24). Therefore, (6.4.25) holds with $A_{\alpha,\beta,\sigma}$ replaced by $C_{\alpha,\beta,\sigma}$. This completes the proof. □

Corollary 6.4.1. *The maximal Banach algebra $\mathcal{A}_p^{m,l}$ in $M(H_p^m \to H_p^l)$, $m \geq l$, $1 < p < \infty$, is the set of functions $\gamma \in H_{p,\mathrm{loc}}^l$ such that*

$$\sup_{\substack{e\subset R^n\\ \mathrm{diam}(e)\leq 1}} \frac{\|S_l\gamma; e\|_{L_p}}{(C_{p,m}(e))^{1/p}} + \|\gamma\|_{L_\infty} < \infty. \tag{6.4.26}$$

In the case $mp > n$ this condition can be simplified as

$$\|S_l\gamma\|_{L_{p,\mathrm{unif}}} + \|\gamma\|_{L_\infty} < \infty.$$

6.5 Imbeddings of Maximal Algebras

In this section we deal with the imbeddings $A_p^{m,l} \subset A_p^{\mu,\lambda}$ and $\mathcal{A}_p^{m,l} \subset \mathcal{A}_p^{\mu,\lambda}$. We fix an arbitrary μ and find the maximum value of λ for which the imbeddings hold. Since the best value of λ is equal to l when $\mu \geq m$, we can restrict ourselves to $\mu < m$.

The next theorem contains a complete characterization of the imbedding $A_p^{m,l} \subset A_p^{\mu,\lambda}$. The corresponding assertion relating the algebras $\mathcal{A}_p^{m,l}$ is stated and proved in exactly the same way with $D_{p,l}$ replaced by S_l in the proof.

Theorem 6.5.1. *Let $m, l, m\theta, l\theta$ be nonintegers for any $\theta \in (0,1)$, $m \geq l$, and let $p \in (1,\infty)$. The following imbeddings hold:*

(i) if $pm \leq n$, then $A_p^{m,l} \subset A_p^{m\theta,l\theta}$,

(ii) if $pm\theta > n$, then $A_p^{m,l} \subset A_p^{m\theta,\min\{m\theta,l\}}$,

(iii) if $pm\theta = n$, then

$$A_p^{m,l} \subset A_p^{m\theta,m\theta} \quad \text{for } m\theta < l$$

and

$$A_p^{m,l} \subset A_p^{m\theta,l-\epsilon} \quad \text{for } m\theta \geq l$$

with an arbitrary small $\epsilon > 0$,
(iv) if $pm\theta < n < pm$, *then*

$$A_p^{m,l} \subset A_p^{m\theta,m\theta} \qquad \text{for } pl > n$$

and

$$A_p^{m,l} \subset A_p^{m\theta,m\theta lp/n} \quad \text{for } pl \leq n.$$

All these imbeddings are best possible.

Proof. (*i*) Since the multiplication by $\gamma \in A_p^{m,l}$ continuously maps L_p to L_p and W_p^m to W_p^l, the imbedding $A_p^{m,l} \subset A_p^{m\theta,l\theta}$ results by complex interpolation (see [Tr4], Sec. 2.4.7).

In the cases (*ii*)–(*iv*) we have $pm > n$. Thus, by Corollary 6.2.2,

$$A_p^{m,l} = (W_{p,\mathrm{unif}}^l \cap L_\infty).$$

(*ii*) Since $pm\theta > n$, Corollary 6.2.2 implies that $A_p^{m\theta,\lambda} = (W_{p,\mathrm{unif}}^\lambda \cap L_\infty)$ for $\lambda \leq m\theta$. The result follows from the imbedding $W_{p,\mathrm{unif}}^l \subset W_{p,\mathrm{unif}}^{\min\{m\theta,l\}}$.

(*iii*) If $pm\theta = n$ and $m\theta < l$, then $pl > n$ and by Remark 4.3.3 we have $W_{p,\mathrm{unif}}^l = MW_p^l$. This last space is imbedded into $MW_p^{m\theta} = A_p^{m\theta,m\theta}$ by the complex interpolation between W_p^l and L_p. Hence $A_p^{m,l} \subset A_p^{m\theta,m\theta}$.

Let $pm\theta = n$ and $m\theta \geq l$. By Corollary 6.2.2, a norm in $A_p^{m\theta,l-\epsilon}$ can be given by

$$\sup_{\substack{e \subset \mathbb{R}^n \\ \mathrm{diam}(e) \leq 1}} \frac{\|D_{p,l-\epsilon}\gamma; e\|_{L_p}}{(C_{p,m\theta}(e))^{1/p}} + \|\gamma\|_{L_\infty}.$$

Since (3.1.6) shows for $pm\theta = n$ that

$$C_{p,m\theta}(e) \geq c\,(\mathrm{mes}_n e)^{\epsilon/n}$$

with an arbitrary $\epsilon > 0$, it follows that

$$\|\gamma\|_{A_p^{m\theta,l-\epsilon}} \leq c\left(\|\gamma\|_{W_{q,\mathrm{unif}}^{l-\epsilon}} + \|\gamma\|_{L_\infty}\right)$$

with $q = pn/(n - p\epsilon)$. It remains to use the Sobolev imbedding $W_{p,\mathrm{unif}}^l \subset W_{q,\mathrm{unif}}^{l-\epsilon}$.

(*iv*) Let $pm\theta < n < pm$ and $pl > n$. Then $A_p^{m,l} = W_{p,\mathrm{unif}}^l = MW_p^l$. Since $l > m\theta$, we have

$$MW_p^l \subset MW_p^{m\theta} = A_p^{m\theta,m\theta}.$$

The result follows.

Now let $pm\theta < n < pm$ and $lp \leq n$. By (3.1.6), we obtain from Corollary 6.2.2 that

$$\|\gamma;\mathbb{R}^n\|_{A_p^{m\theta,m\theta lp/n}} \le c\Big(\sup_{\substack{e\subset\mathbb{R}^n\\ \mathrm{diam}(e)\le 1}} \frac{\|D_{p,m\theta lp/n}\gamma;e\|_{L_p}}{(\mathrm{mes}_n e)^{(n-pm\theta)/np}} + \|\gamma\|_{L_\infty}\Big)$$
$$\le c\,\big(\|\gamma\|_{W^{m\theta lp/n}_{n/m\theta,\mathrm{unif}}} + \|\gamma\|_{L_\infty}\big). \tag{6.5.1}$$

Suppose first that $\{m\theta lp/n\}>0$. Let $H^t_p(\mathbb{R}^{n+1}_+)$ denote the Bessel potential space of functions defined on $\mathbb{R}^{n+1}_+ = \{(x,y): x\in\mathbb{R}^n, y>0\}$. The space $W^t_p(\mathbb{R}^n)$ is the space of traces on $\mathbb{R}^n$ of functions in $H_p^{t+1/p}(\mathbb{R}^{n+1}_+)$, where $\{t\}>0$ and $p\in(1,\infty)$. This and the Gagliardo-Nirenberg type inequality

$$\|\Gamma;\mathbb{R}^{n+1}_+\|_{H^{m\theta(lp+1)/n}_{n/m\theta,\mathrm{unif}}} \le c\,\|\Gamma;\mathbb{R}^{n+1}_+\|^{pm\theta/n}_{H^{l+1/p}_{p,\mathrm{unif}}}\|\Gamma;\mathbb{R}^{n+1}_+\|^{1-pm\theta/n}_{L_\infty}$$

(see [AF], Lemma 3.4) imply that the right-hand side of (6.5.1) is dominated by

$$c\,\big(\|\gamma\|^{pm\theta/n}_{W^l_{p,\mathrm{unif}}}\|\gamma\|^{1-pm\theta/n}_{L_\infty} + \|\gamma\|_{L_\infty}\big) \le c\,\|\gamma\|_{A_p^{m,l}}.$$

Hence the imbedding $A_p^{m,l}\subset A_p^{m\theta,m\theta lp/n}$ is valid.

To obtain the same imbedding for integer $m\theta lp/n$ we use Hölder's inequality and the estimate

$$|\nabla_k u(x)| \le c\,\|u\|^{1-k/l}_{L_\infty}(D_{p,l}u(x))^{k/l}, \quad k<l, \tag{6.5.2}$$

valid by Corollary 6.2.1. Then the right-hand side of (6.5.1) is dominated by

$$\|\gamma\|^{pm\theta/n}_{W^l_{p,unif}}\|\gamma;\|^{1-pm\theta/n}_{L_\infty}.$$

We now show that the imbedding $A_p^{m,l}\subset A_p^{m\theta,\lambda}$ with λ given in (i)–(iv) cannot be improved. Let

$$\gamma_\mu(x) = \exp(i|x|^{-\mu}) \tag{6.5.3}$$

with $\mu>0$. From the equivalence relations

$$|\nabla_{[l]}\gamma_\mu(x)| \sim |x|^{-[l](\mu+1)}$$

and

$$|\nabla_{[l]}\gamma_\mu(x+h) - \nabla_{[l]}\gamma_\mu(x)| \sim \frac{\min\{|h|,|x|^{1+\mu}\}}{|x|^{([l]+1)(1+\mu)}},$$

where $|x|$ is sufficiently small, it follows that

$$D_{p,l}\gamma_\mu(x) \sim |x|^{-l(\mu+1)} \tag{6.5.4}$$

for $|x|<1$. Furthermore, $D_{p,l}\gamma_\mu(x)$ is bounded for $|x|\ge 1$. Now, by Corollary 6.3.1,

$$\gamma_\mu \in A_p^{m,l} \Longleftrightarrow \begin{cases} (\mu+1)l \le m & \text{if } pm < n, \\ (\mu+1)l < m & \text{if } pm = n, \\ p(\mu+1)l < n & \text{if } pm > n. \end{cases} \tag{6.5.5}$$

We conclude by (6.5.5) that in the case $pm \le n$

$$\gamma \in A_p^{m,l} \Longleftrightarrow \gamma \in A_p^{m\theta, l\theta} \quad \text{for } \theta \in (0,1)$$

which shows that the imbedding (i) is sharp.

Since the imbedding (*ii*) is equivalent to

$$W^l_{p,\mathrm{unif}} \cap L_\infty \subset W^{\min\{m\theta, l\}}_{p,\mathrm{unif}} \cap L_\infty$$

and, obviously, $\lambda \le m\theta$ in $A_p^{m,l} \subset A_p^{m\theta,\lambda}$, it follows that (*ii*) cannot be improved.

We turn to the imbeddings (*iii*). The optimality of the first one (corresponding to $m\theta < l$) is obvious. Let $m\theta \ge l$. We show that $A_p^{m,l}$ is not imbedded into $A_p^{m\theta,l}$.

Let $\mu \ge 0$ and $pl(\mu+1) = n$. We introduce the function

$$\Gamma_{\mu,\delta}(x) = \eta(x) \exp\big(i|x|^{-\mu}(\log|x|^{-1})^{-\delta}\big),$$

where $\delta > -1$ and η is a function in $C_0^\infty(\mathbb{R}^n)$ with support in a small neighbourhood of the origin, equal to 1 near the origin. For $\mu > 0$ direct calculations imply that

$$|\nabla_{[l]} \Gamma_{\mu,\delta}(x)| \sim |x|^{-[l](\mu+1)} (\log|x|^{-1})^{-[l]\delta}$$

and

$$|\nabla_{[l]} \Gamma_{\mu,\delta}(x+h) - \nabla_{[l]} \Gamma_{\mu,\delta}(x)| \sim \frac{\min\{|h|, |x|^{1+\mu}(\log|x|^{-1})^\delta\}}{|x|^{([l]+1)(1+\mu)} (\log|x|^{-1})^{([l]+1)\delta}},$$

where $|x|$ is sufficiently small. Therefore, for small $|x|$

$$D_{p,l} \Gamma_{\mu,\delta}(x) \sim |x|^{-l(\mu+1)} (\log|x|^{-1})^{-l\delta}. \tag{6.5.6}$$

Analogously,

$$|\nabla_{[l]} \Gamma_{0,\delta}(x)| \sim |x|^{-[l]} (\log|x|^{-1})^{-[l](\delta+1)}$$

and

$$|\nabla_{[l]} \Gamma_{0,\delta}(x+h) - \nabla_{[l]} \Gamma_{0,\delta}(x)| \sim \frac{\min\{|h|, |x|(\log|x|^{-1})^{\delta+1}\}}{|x|^{[l]+1} (\log|x|^{-1})^{[l]+1)(\delta+1)}}$$

for small $|x|$. Hence,

$$D_{p,l} \Gamma_{0,\delta}(x) \sim |x|^{-l} (\log|x|^{-1})^{-l(\delta+1)}. \tag{6.5.7}$$

Now it is straightforward that

$$\Gamma_{\mu,\delta} \in A_p^{m,l} = (W_{p,\mathrm{unif}}^l \cap L_\infty)$$

if and only if $pl\delta > 1$ for $\mu > 0$, and $pl(\delta+1) > 1$ for $\mu = 0$. On the other hand, by Corollary 6.2.2 and Proposition 3.1.4,

$$\|\Gamma_{\mu,\delta}\|_{A_p^{m\theta,l}} \geq c\,(\log \rho^{-1})^{(p-1)/p} \|D_{p,l}\Gamma_{\mu,\delta}; \mathcal{B}_\rho\|_{L_p}$$

for small $\rho > 0$. Applying (6.5.6) and (6.5.7), we obtain

$$\|\Gamma_{\mu,\delta}\|_{A_p^{m\theta,l}} \geq c\,(\log \rho^{-1})^{(p-1)/p+(1-pl\delta)/p} \quad \text{for } \mu > 0$$

and

$$\|\Gamma_{0,\delta}\|_{A_p^{m\theta,l}} \geq c\,(\log \rho^{-1})^{(p-1)/p+(1-pl(\delta+1))/p}.$$

This, obviously, implies that $\Gamma_{\mu,\delta} \in A_p^{m,l}$ and $\Gamma_{\mu,\delta} \notin A_p^{m\theta,l}$ if $1 > l\delta > 1/p$ for $\mu > 0$, and $\Gamma_{0,\delta} \in A_p^{m,l}$ and $\Gamma_{0,\delta} \notin A_p^{m\theta,l}$ if $1 > l(\delta+1) > 1/p$. The result follows.

We pass to (iv). It suffices to consider only the case $pl < n$. Assume that

$$A_p^{m,l} \subset A_p^{m\theta,\lambda} \quad \text{with} \quad \lambda = m\theta pl/n(1-\epsilon)$$

for some $\epsilon > 0$. We choose μ to satisfy $pl(\mu+1) > n(1-\epsilon)$. Then the function γ_μ, introduced by (6.5.3), belongs to $W_{p,\mathrm{unif}}^l = A_p^{m,l}$. On the other hand, by Corollary 6.2.2 and (6.5.4),

$$\|\gamma_\mu\|_{A_p^{m\theta,\lambda}} \geq c\,\rho^{m\theta - n/p} \|D_{p,\lambda}\gamma_\mu; \mathcal{B}_\rho\|_{L_p} \geq c\,\rho^{m\theta-\lambda(\mu+1)}$$

for $\rho < 1$. Since

$$m\theta - \lambda(\mu+1) = m\theta(1 - pl(\mu+1)/n(1-\epsilon)) < 0,$$

we have $\gamma_\mu \notin A_p^{m\theta,\lambda}$. The proof is complete. □

The following assertion is an analogue of Theorem 6.5.1 for $p = 1$.

Theorem 6.5.2. *Let $m, l, m\theta, l\theta$ be nonintegers for any $\theta \in (0,1)$, $m \geq l$. The following imbeddings are valid:*

(i) if $m < n$, then $A_1^{m,l} \subset A_1^{m\theta,l\theta}$,

(ii) if $m\theta > n$, then $A_1^{m,l} \subset A_1^{m\theta,\min\{m\theta,l\}}$,

(iii) if $m\theta < n < m$, then

$$A_1^{m,l} \subset A_1^{m\theta,m\theta} \quad \text{for } l > n$$

and

$$A_1^{m,l} \subset A_1^{m\theta,m\theta l/n} \quad \text{for } l < n.$$

All these imbeddings are best possible.

Proof. (*i*) By (6.2.17) and Hölder's inequality, for all $\theta \in (0,1)$

$$\sup_{\substack{x\in\mathbb{R}^n\\ r\in(0,1)}} r^{m\theta-n}\|D^{(r)}_{1,l\theta}\gamma;\mathcal{B}_r(x)\|_{L_1}$$

$$\leq c\,\|\gamma\|^{1-\theta}_{L_\infty}\sup_{\substack{x\in\ R^n\\ r\in(0,1)}}(r^{m-n}\|D^{(r)}_{1,l}\gamma;\mathcal{B}_r(x)\|_{L_1})^{\theta}.$$

Hence $A^{m,l}_1 \subset A^{m\theta,l\theta}_1$ and

$$\|\gamma\|_{A^{m\theta,l\theta}_1} \leq c\,\|\gamma\|^{1-\theta}_{L_\infty}\|\gamma\|^{\theta}_{A^{m,l}_1}.$$

In the cases (*ii*), (*iii*) we have $m\theta \geq n$. Therefore, by Corollary 6.2.2, $A^{m,l}_1 = W^l_{1,\mathrm{unif}} \cap L_\infty$.

(*ii*) Since $m\theta > n$, Corollary 6.2.2 implies that

$$A^{m\theta,\lambda}_1 = W^{\lambda}_{1,\mathrm{unif}} \cap L_\infty \quad \text{for} \quad \lambda \leq m\theta.$$

The result follows from the imbedding

$$W^l_{1,\mathrm{unif}} \subset W^{\min\{m\theta,l\}}_{1,\mathrm{unif}}.$$

(*iii*) Let $m\theta < n < m$ and $l > n$. Then

$$A^{m,l}_1 = W^l_{1,\mathrm{unif}} = MW^l_1.$$

Since $l > m\theta$, we have

$$MW^l_1 \subset MW^{m\theta}_1 = A^{m\theta,m\theta}_1.$$

The result follows.

Now let $m\theta < n < m$ and $l < n$. By Corollary 6.3.2,

$$\|\gamma\|_{A^{m\theta,m\theta l/n}_1} \leq c\Big(\sup_{\substack{x\in\mathbb{R}^n\\ r\in(0,1)}} r^{m\theta-n}\|D^{(r)}_{1,m\theta l/n}\gamma;\mathcal{B}_r(x)\|_{L_1} + \|\gamma\|_{L_\infty}\Big). \tag{6.5.8}$$

If $m\theta l/n$ is integer, then the imbedding $A^{m,l}_1 \subset A^{m\theta,m\theta l/n}_1$ follows by the same argument as for $p > 1$ (see part (*iv*) of Theorem 6.5.1). Suppose that $m\theta l/n$ is a noninteger and $\{m\theta l/n\} \geq \{l\}$. Then, by Lemma 6.2.3 and Hölder's inequality, (6.5.8) is dominated by

$$c\,\|\gamma\|^{m\theta/n}_{W^l_{1,\mathrm{unif}}}\|\gamma\|^{1-m\theta/n}_{L_\infty},$$

and we arrive at the imbedding $A^{m,l}_1 \subset A^{m\theta,m\theta l/n}_1$.

Let $\{m\theta l/n\} < \{l\}$. Then

$$\|\gamma\|_{W^{m\theta l/n}_{n/m\theta,\mathrm{unif}}} \leq c\,\|\gamma\|^{m\theta/n}_{W^l_{1,\mathrm{unif}}}\|\gamma\|^{1-m\theta/n}_{L_\infty}.$$

One can show that the imbedding $A_1^{m,l} \subset A_1^{m\theta,\lambda}$ with λ given in (i)–(iii) cannot be improved by using the same argument as in the proof of the sharpness of (i), (ii), and (iv) in Theorem 6.5.1 with $p = 1$. □

Remark 6.5.1. The conditions on parameters of integrability and smoothness for concrete function spaces to be multiplication algebras, were studied by many authors, see Strichartz [Str], Peetre [Pe2], Hertz [Her], Bennet and Gilbert [BG], Johnson [Jo], Triebel [Tr1], [Tr2], [Tr4], Zolesio [Zo], Bliev [Bl1], [Bl2], Kalyabin [K1]–[K3], Marschall [Mar3], Benchekroun and Benkirane [BB], Ali Mehmeti and Nicaise [AN], Runst [Ru].

Considerable attention was paid to the description of the range of a product of two or more functions in Sobolev type spaces (see Maz'ya [Maz6], Amann [Am], Hanouzet [Ha], Valent [Va], Franke [Fr] Johnsen [Jo], Miyachi [Mi], Runst and Sickel [RS], Sickel [Sic1], Sickel and Triebel [ST], Sickel and Youssfi [SY], Dacorogna and Moser [DM1], Ye [Ye], Runst and Youssfi [RY], Youssfi [Yo], Drihem and Moussai [DM2] et al.

7

Essential Norm and Compactness of Multipliers

In this chapter we study elements of the space $M(W_p^m \to W_p^l)$, where $p \in [1, \infty)$, and m and l are arbitrary, integer and noninteger, with $m \geq l \geq 0$. As usual, we omit $\mathbb{R}^n$ in notations of spaces, norms, and integrals.

By $\operatorname{ess} \|\gamma\|_{M(W_p^m \to W_p^l)}$ we denote the essential norm of the operator of multiplication by $\gamma \in M(W_p^m \to W_p^l)$, that is,

$$\inf_{\{T\}} \|\gamma - T\|_{W_p^m \to W_p^l},$$

where $\{T\}$ is the set of compact operators $W_p^m \to W_p^l$.

As before,

$$(D_{p,l}\gamma)(x) = |\nabla_l \gamma(x)|$$

for integer l and

$$(D_{p,l}\gamma)(x) = \Big(\int \frac{|\Delta_h \nabla_{[l]} u(x)|^p}{|h|^{n+p\{l\}}} dh\Big)^{1/p}$$

for noninteger l.

The main results are sharp two-sided estimates for $\operatorname{ess} \|\gamma\|_{M(W_p^m \to W_p^l)}$. We formulate the main result concerning $m > l$.

Theorem 7.0.3. *Let $\gamma \in M(W_p^m \to W_p^l)$, where $m > l \geq 0$.*

(i) If $p \in (1, \infty)$ and $mp \leq n$, then

$$\operatorname{ess} \|\gamma\|_{M(W_p^m \to W_p^l)}$$

$$\sim \lim_{\delta \to 0} \Big(\sup_{\{e: d(e) \leq \delta\}} \frac{\|D_{p,l}\gamma; e\|_{L_p}}{(C_{p,m}(e))^{\frac{1}{p}}} + \sup_{\substack{x \in \mathbb{R}^n \\ \rho \leq \delta}} \rho^{m-l-\frac{n}{p}} \|\gamma; \mathcal{B}_\rho(x)\|_{L_p} \Big)$$

$$+ \lim_{r \to \infty} \Big(\sup_{\substack{e \subset \mathbb{R}^n \setminus \mathcal{B}_r \\ d(e) \leq 1}} \frac{\|D_{p,l}\gamma; e\|_{L_p}}{(C_{p,m}(e))^{\frac{1}{p}}} + \sup_{x \in \mathbb{R}^n \setminus \mathcal{B}_r} \|\gamma; \mathcal{B}_1(x)\|_{L_p} \Big), \tag{7.0.1}$$

V.G. Maz'ya, T.O. Shaposhnikova, *Theory of Sobolev Multipliers*,
Grundlehren der mathematischen Wissenschaften 337,

where $d(e)$ is the diameter of a compact set $e \subset \mathbb{R}^n$.

(ii) If $m < n$, then

$$\operatorname{ess} \|\gamma\|_{M(W_1^m \to W_1^l)}$$

$$\sim \lim_{\delta \to 0} \sup \delta^{m-n} \sup_{x \in \mathbb{R}^n} \big(\|D_{1,l}\gamma; \mathcal{B}_\delta(x)\|_{L_1} + \delta^{-l} \|\gamma; \mathcal{B}_\delta(x)\|_{L_1} \big)$$

$$+ \lim_{|x| \to \infty} \sup \big(\sup_{r \in (0,1)} r^{m-n} \|D_{1,l}\gamma; \mathcal{B}_r(x)\|_{L_1} + \|\gamma; \mathcal{B}_1(x)\|_{L_1} \big). \tag{7.0.2}$$

(iii) If $mp > n$, $p \in (1, \infty)$, then

$$\operatorname{ess} \|\gamma\|_{M(W_p^m \to W_p^l)} \sim \lim_{|x| \to \infty} \sup \|\gamma; \mathcal{B}_1(x)\|_{W_p^l}. \tag{7.0.3}$$

Theorem 7.0.4. *Let $\gamma \in MW_p^l$.*

(i) If $p \in (1, \infty)$ and $lp \le n$, then

$$\operatorname{ess} \|\gamma\|_{MW_p^l} \sim \lim_{\delta \to 0} \sup_{\{e: d(e) \le \delta\}} \frac{\|D_{p,l}\gamma; e\|_{L_p}}{(C_{p,l}(e))^{\frac{1}{p}}}$$

$$+ \lim_{r \to \infty} \sup_{\substack{e \subset \mathbb{R}^n \setminus \mathcal{B}_r \\ d(e) \le 1}} \frac{\|D_{p,l}\gamma; e\|_{L_p}}{(C_{p,l}(e))^{\frac{1}{p}}} + \|\gamma\|_{L_\infty}. \tag{7.0.4}$$

(ii) If $l < n$, then

$$\operatorname{ess} \|\gamma\|_{MW_1^l} \sim \lim_{\delta \to 0} \sup \delta^{l-n} \sup_{x \in \mathbb{R}^n} \|D_{1,l}\gamma; \mathcal{B}_\delta(x)\|_{L_1}$$

$$+ \lim_{|x| \to \infty} \sup \sup_{r \in (0,1)} r^{l-n} \|D_{1,l}\gamma; \mathcal{B}_r(x)\|_{L_1} + \|\gamma\|_{L_\infty}.$$

(iii) If $m \ge n$, then

$$\operatorname{ess} \|\gamma\|_{MW_1^l} \sim \lim_{|x| \to \infty} \sup \|\gamma; \mathcal{B}_1(x)\|_{W_p^l} + \|\gamma\|_{L_\infty}. \tag{7.0.5}$$

Clearly, if multipliers have compact supports, the above equivalence relations for the essential norm are simplified, since all terms containing either $r \to \infty$ or $|x| \to \infty$ vanish.

As simple corollaries we obtain characterizations of the space $\mathring{M}(W_p^m \to W_p^l)$, $m > l$, of compact multipliers. We note also that Sect. 7.2.7 contains one-sided estimates for the essential norm of a multiplier which do not involve capacities.

7.1 Auxiliary Assertions

In this chapter we use the following cutoff functions.

Definition 7.1.1. *Let $x \in \mathbb{R}^n$, $\delta \in (0,1)$, and let η be a function in $C_0^\infty[0,2)$, equal to* 1 *on* $[0,1]$. *Furthermore, we assume that $0 \le \eta \le 1$. We set*

$$\mathbb{R}^n \ni y \to \eta_{\delta,x}(y) = \eta\Big(\frac{|y-x|}{\delta}\Big).$$

Definition 7.1.2. *Let η be the same as in Definition* 7.1.1. *We put*

$$\mathbb{R}^n \ni y \to \mu_{\delta,x}(y) = \eta\Big(\frac{2\log\delta}{\log|y-x|}\Big), \qquad \delta \in (0,1/2).$$

We also adopt the notation $\eta_\delta = \eta_{\delta,0}$ and $\mu_\delta = \mu_{\delta,0}$.

Definition 7.1.3. *Let $\zeta_r(y) = \zeta(y/r)$, where $r > 1$, $\zeta \in C^\infty(\mathbb{R}^n)$, $\zeta = 0$ for $y \in \mathcal{B}_1$ and $\zeta(y) = 1$ for $y \in \mathbb{R}^n \backslash \mathcal{B}_2$. Furthermore, let $0 \le \zeta \le 1$.*

As usual, by $W_p^l(\mathcal{B}_r)$ we denote the space of functions with the finite norm

$$\|u;\mathcal{B}_r\|_{W_p^l} = \begin{cases} \displaystyle\sum_{j=0}^{l} \|\nabla_j u;\mathcal{B}\|_{L_p} & \text{for} \quad \{l\}=0, \\ \displaystyle\sum_{j=0}^{[l]} \Big(\int_{\mathcal{B}_r}\int_{\mathcal{B}_r} |\nabla_j u(x)-\nabla_j u(y)|^p \frac{dxdy}{|x-y|^{n+p\{l\}}}\Big)^{1/p} + \|u;\mathcal{B}_r\|_{W_p^{[l]}} \\ \text{for} \quad \{l\}>0. \end{cases}$$

We introduce one more norm in $W_p^l(\mathcal{B}_r)$ depending on $r \in (0,1)$. Namely, we set

$$|\!|\!| u;\mathcal{B}_r |\!|\!|_{W_p^l} = \begin{cases} \displaystyle\sum_{j=0}^{l} r^{j-l}\|\nabla_j u;\mathcal{B}_r\|_{L_p} & \text{for} \quad \{l\}=0, \\ \displaystyle\sum_{j=0}^{[l]} r^{j-[l]}\Big(\int_{\mathcal{B}_r}\int_{\mathcal{B}_r} |\nabla_j u(x)-\nabla_j u(y)|^p \frac{dxdy}{|x-y|^{n+p\{l\}}}\Big)^{1/p} \\ \displaystyle +\sum_{j=0}^{[l]} r^{j-l}\|\nabla_j u;\mathcal{B}_r\|_{L_p} & \text{for} \quad \{l\}>0. \end{cases}$$

It is clear that the last norm is invariant under dilation.

We present some properties of the norm $|\!|\!| u;\mathcal{B}_r |\!|\!|_{W_p^l}$ which will be used henceforth.

Lemma 7.1.1. *If l is a positive noninteger, then*

$$\|D_{p,l}u;\mathcal{B}_r\|_{L_p} \le c \sup_{x\in\mathbb{R}^n} |\!|\!| u;\mathcal{B}_r |\!|\!|_{W_p^l}.$$

Proof. It suffices to estimate

$$\Big(\int_{\mathcal{B}_{r/2}(z)}\int_{\mathbb{R}^n\setminus\mathcal{B}_{r/2}(z)} |\nabla_{[l]}u(x)-\nabla_{[l]}u(y)|^p\frac{dxdy}{|x-y|^{n+p\{l\}}}\Big)^{1/p},$$

where z is an arbitrary point of the ball $\mathcal{B}_r$. This value does not exceed

$$\Big(\int_{\mathcal{B}_{r/2}(z)} |\nabla_{[l]}u(y)|^p\,dy\int_{\mathbb{R}^n\setminus\mathcal{B}_{r/2}(z)}\frac{dx}{|x-y|^{n+p\{l\}}}\Big)^{1/p}$$

$$+\Big(\int_{\mathbb{R}^n\setminus\mathcal{B}_{r/2}(z)} |\nabla_{[l]}u(x)|^p\,dx\int_{\mathcal{B}_{r/2}(z)}\frac{dy}{|x-y|^{n+p\{l\}}}\Big)^{1/p}$$

$$\le c\,r^{-\{l\}}\|\nabla_{[l]}u;\mathcal{B}_r(z)\|_{L_p}+c\Big(\int_{\mathbb{R}^n\setminus\mathcal{B}_{r/2}(z)}\frac{|\nabla_{[l]}u(x)|^p\,dx}{|x-z|^{n+p\{l\}}}\Big)^{1/p}.$$

The second term on the right-hand side does not exceed

$$c\Big(\int_{\mathbb{R}^n\setminus\mathcal{B}_{r/2}(z)}\frac{1}{|x-\xi|^{n+p\{l\}}}\int_{\mathcal{B}_r(\xi)}|\nabla_{[l]}u(x)|^p\,dxd\xi\Big)^{1/p},$$

which is dominated by

$$c\,r^{-\{l\}}\sup_{x\in\mathbb{R}^n}\|u;\mathcal{B}_r\|_{W_p^l}.$$

The proof is complete. □

Next we formulate three well-known properties of the norm $\|u;\mathcal{B}_r\|_{W_p^m}$, leaving their proof to the reader as an exercise.

Lemma 7.1.2. *If $\varphi\in C_0^\infty(\mathcal{B}_r)$ and $|\nabla_k\varphi|\le c\,r^{-k}$, $k=0,1,\dots,m$, then for all $u\in W_p^m(\mathcal{B}_r)$*

$$\|\varphi\, u;\mathbb{R}^n\|_{W_p^m}\le c\,\|u;\mathcal{B}_r\|_{W_p^m}\quad\text{for}\quad r\le 1$$

and

$$\|\varphi\, u;\mathbb{R}^n\|_{W_p^m}\le c\,\|u;\mathcal{B}_r\|_{W_p^k}\quad\text{for}\quad r>1.$$

Lemma 7.1.3. *Let $u\in W_p^m(\mathcal{B}_\delta)$. There exists a polynomial P of degree $[m]$ of the form*

$$P(u;x)=\sum_\beta\Big(\frac{x}{\delta}\Big)^\beta\delta^{-n}\int_{\mathcal{B}_\delta}\varphi_\beta\Big(\frac{y}{\delta}\Big)u(y)\,dy,$$

where $\varphi\in C_0^\infty(\mathcal{B}_1)$, and such that

$$\|u-P(u;\cdot);\mathcal{B}_\delta\|_{W_p^m}\le c\int_{\mathcal{B}_\delta}\int_{\mathcal{B}_\delta}|\nabla_{[m]}u(x)-\nabla_{[m]}u(y)|^p\frac{dxdy}{|x-y|^{n+p\{m\}}}.$$

Lemma 7.1.4. *The inequality*

$$\|u; \mathcal{B}_r\|_{W_p^s} \le c\, \|u; \mathcal{B}_r\|_{W_p^k}^{s/k} \|u; \mathcal{B}_r\|_{L_p}^{1-s/k}, \qquad 0 < s < k, \tag{7.1.1}$$

holds.

From (7.1.1) we immediately obtain

Corollary 7.1.1. *If k is a noninteger, then*

$$\|u; \mathcal{B}_r\|_{W_p^k} \sim \Big(\int_{\mathcal{B}_r}\int_{\mathcal{B}_r} |\nabla_{[k]}u(x) - \nabla_{[k]}u(y)|^p \frac{dxdy}{|x-y|^{n+p\{k\}}}\Big)^{1/p} + r^{-k}\|u; \mathcal{B}_r\|_{L_p}.$$

If k is an integer, then

$$\|u; \mathcal{B}_r\|_{W_p^k} \sim \|\nabla_k u; \mathcal{B}_r\|_{L_p} + r^{-k}\|u; \mathcal{B}_r\|_{L_p}.$$

Using the Hardy-type inequality

$$\|u\,|x|^{-k}\|_{L_p} \le c\,\|u\|_{W_p^k}, \qquad kp < n,$$

(which, in particular, easily follows from (4.3.17)), we deduce

Lemma 7.1.5. *If $kp < n$, then*

$$\|u; \mathcal{B}_r\|_{W_p^k} \le c\,\|u\|_{W_p^k}$$

with a constant c independent of r.

Next we prove some technical lemmas.

Lemma 7.1.6. *Let $\varphi \in C_0^\infty(\mathcal{B}_\delta)$ with $\delta < 1$, and let $|\nabla_k \varphi| \le c\,\delta^{-k}$, $k = 0, 1, \dots, [l]+1$. Then*

$$\|\varphi\|_{MW_p^l} \le c,$$

where $lp < n$ and $p \in [1, \infty)$, or $l = n$ and $p = 1$.

Proof. Let $u \in C_0^\infty(\mathbb{R}^n)$. According to Lemma 7.1.2 and Corollary 7.1.1,

$$\|\varphi\, u\|_{W_p^l} \le c\,\|u; \mathcal{B}_{2\delta}\|_{W_p^l}$$

$$\le c\Big(\Big(\int_{\mathcal{B}_{2\delta}}\int_{\mathcal{B}_{2\delta}} |\nabla_{[l]}u(x) - \nabla_{[l]}u(y)|^p \frac{dx\,dy}{|x-y|^{n+p\{l\}}}\Big)^{\frac{1}{p}} + \|u; \mathcal{B}_{2\delta}\|_{L_{\frac{pn}{n-lp}}}\Big). \tag{7.1.2}$$

Now the result follows from Proposition 4.2.5. □

Lemma 7.1.7. *If $u \in C_0^\infty$ and $lp < n$, then*

$$\sup_{x \in \mathbb{R}^n} \|\eta_{\delta,x}\, u\|_{W_p^l} \to 0 \quad \text{as } \delta \to 0, \tag{7.1.3}$$

where $\eta_{\delta,x}$ is the function in Definition 7.1.1.

Proof. This assertion follows from (7.1.2).

Lemma 7.1.8. *Let* $\{l\} > 0$. *Then*

$$|\nabla_j \mu_\delta(z)| \leq c\, |\log|z|\,|^{-1}\, |z|^{-j} \tag{7.1.4}$$

and

$$\int_{\mathcal{B}_\delta} \frac{|\nabla_j \mu_\delta(z) - \nabla_j \mu_\delta(y)|^p}{|z-y|^{n+p\{l\}}} dy \leq c_j\, |\log|z|\,|^{-p}\, |z|^{-p(\{l\}+j)}, \tag{7.1.5}$$

where $z \in \mathcal{B}_\delta$, $j = 0, 1, \ldots$

Proof. Estimate (7.1.4) is obvious. We prove (7.1.5). Since

$$|D^\alpha \mu_\delta(z)| \leq |z|^{-|\alpha|} \sum_{k=1}^{|\alpha|} \sigma_k \Big(\frac{2\log\delta}{\log|z|}\Big)\, (2\log\delta)^{-k},$$

with $\sigma_k \in C_0^\infty(-1,1)$, it follows that

$$|\nabla_j \mu_\delta(z) - \nabla_j \mu_\delta(y)| \leq \begin{cases} c\,|\log\delta|^{-1}|z-y|\,|z|^{-j-1} \\ \text{if } |z|/2 \leq |y| \leq |z|; \\ \\ c\,|\log\delta|^{-1}\big(\max\{|z|,|y|\}\big)^{-j} \quad \text{if } j>0 \\ \text{and either } |y| < |z|/2 \text{ or } |z| < |y|/2; \\ \\ c\,|\log\delta|^{-1}\Big|\log\frac{|z|}{|y|}\Big| \qquad \text{if } j = 0 \\ \text{and either } |y| < |z|/2 \text{ or } |z| < |y|/2. \end{cases} \tag{7.1.6}$$

These estimates imply that

$$\int_{\mathcal{B}_\delta} \frac{|\nabla_j \mu_\delta(z) - \nabla_j \mu_\delta(y)|^p}{|z-y|^{n+p\{l\}}} dy \leq c\, |\log\delta|^{-p}\, |z|^{-p(\{l\}+j)}$$

which is equivalent to (7.1.4) for $z \in \mathcal{B}_\delta \backslash \mathcal{B}_{\delta^3}$ by the second and the third estimates in (7.1.6). Let $z \in \mathcal{B}_{\delta^3}$. Then

$$\int_{\mathcal{B}_\delta} \frac{|\nabla_j \mu_\delta(z) - \nabla_j \mu_\delta(y)|^p}{|z-y|^{n+p\{l\}}} dy \leq c \int_{\mathcal{B}_\delta \backslash \mathcal{B}_{\delta^2}} \frac{|\delta_{0j} - \nabla_j \mu_\delta(y)|^p}{|y|^{n+\{l\}}} dy$$

$$\leq c_j\, |\log\delta|^{-p} \int_{\mathcal{B}_\delta \backslash \mathcal{B}_{\delta^2}} \frac{|y|^{-jp}}{|y|^{n+\{l\}}} dy = c_j |\log\delta|^{-p}\, \delta^{-3p(j+\{l\})},$$

where δ_{0j} is the Kroneker delta. Putting here $|z| = \delta^3$ and noting that $t^{3(\{l\}+j)}|\log t|$ increases near $t = 0$, we arrive at (7.1.5) for $z \in \mathcal{B}_{\delta^3}$. □

Lemma 7.1.9. *Let* $\nabla\psi \in C_0^\infty(\mathcal{B}_1)$ *and* $\psi_r(y) = \psi(y/r)$, $r > 1$. *Then*

$$\|\psi_r\|_{MW_p^l} \leq 1 + c\,r^{-\sigma},$$

where $\sigma = l$ *if* $0 < l < 1$ *and* $\sigma = 1$ *if* $l \geq 1$.

Proof. The assertion is obvious for integer l. Let $\{l\} > 0$. Then

$$\|\psi_r\, u\|_{W_p^l} \leq \|\psi_r\, D_{p,l} u\|_{L_p} + \|\psi_r\, u\|_{L_p}$$

$$+ \sum_{\substack{|\alpha|+|\beta|=[l],\\ |\alpha|>0}} \|D^\alpha\psi_r\, D^\beta u\|_{W_p^{\{l\}}} + \|\nabla_{[l]} u\, D_{p,\{l\}}\psi_r\|_{L_p}. \tag{7.1.7}$$

Note that

$$D^\alpha\psi_r = r^{-|\alpha|}(D^\alpha\psi)_r \quad \text{and} \quad D_{p,\{l\}}\psi_r = r^{-\{l\}}(D_{p,\{l\}}\psi)_r.$$

Since the function $(D^\alpha\psi)_r$ is bounded together with all its derivatives and since $(D_{p,\{l\}}\psi)_r$ is uniformly bounded with respect to r, it follows that

$$\|D^\alpha\psi_r\, D^\beta u\|_{W_p^{\{l\}}} \leq c\,r^{-|\alpha|}\|u\|_{W_p^{|\beta|+\{l\}}}$$

and

$$\|\nabla_{[l]} u\, D_{p,\{l\}}\psi_r\|_{L_p} \leq c\,r^{-\{l\}}\|u\|_{W_p^l}. \tag{7.1.8}$$

Combining (7.1.7) and (7.1.8), we complete the proof. □

Lemma 7.1.10. *Let* ψ *and* ψ_r *be the functions defined in Lemma* 7.1.9. *Further, let* $\psi = 0$ *in the ball* $\mathcal{B}_{1/2}$. *Then, for any* $u \in W_p^l$,

$$\lim_{r\to\infty} \|\psi_r\, u\|_{W_p^l} = 0.$$

Proof. The result follows from the inequality

$$\|\psi_r\, u\|_{W_p^l} \leq \|\psi_r\, D_{p,l} u\|_{L_p} + \|\psi_r\, u\|_{L_p} + c\,r^{-\sigma}\|u\|_{W_p^l}$$

established in the proof of Lemma 7.1.9. □

Lemma 7.1.11. *Let* $mp < n$, $k \in [0, m]$. *Further, let* e *be a compact subset of the ball* $\mathcal{B}_r$, $r \in (0,1)$. *Then*

$$C_{p,k}(e) \leq c\,r^{(m-k)p}C_{p,m}(e). \tag{7.1.9}$$

Proof. Let $u \in C_0^\infty$ with $u \geq 1$ on e. We have

$$[C_{p,k}(e)]^{1/p} \leq \|\eta_r u\|_{W_p^k} = \|D_{p,k}(\eta_r u)\|_{L_p} + \|\eta_r u\|_{L_p}.$$

By inequality (4.2.8),

$$\|\eta_r u\|_{L_p} \le c\, r^m \, \|D_{p,m}(\eta_r u)\|_{L_p}$$

and therefore

$$[C_{p,k}(e)]^{1/p} \le c\, r^{m-k} \, \|\eta_r u\|_{W_p^m}.$$

This and Lemmas 7.1.6 and 7.1.9 imply that

$$[C_{p,k}(e)]^{1/p} \le c_1 \, r^{m-k} \, \|u\|_{W_p^m}.$$

Minimizing the right-hand side, we arrive at (7.1.9). □

7.2 Two-Sided Estimates for the Essential Norm. The Case $m > l$

7.2.1 Estimates Involving Cutoff Functions

Lemma 7.2.1. *The estimate*

$$\limsup_{r\to\infty} \|\zeta_r \gamma\|_{M(W_p^m \to W_p^l)} \le c \operatorname{ess} \|\gamma\|_{M(W_p^m \to W_p^l)}$$

holds, where $m \ge l$, $p \ge 1$.

Proof. Let $\varepsilon > 0$ and let $T = T(\gamma, \varepsilon)$ be a compact operator such that

$$\|\gamma - T\| \le \operatorname{ess} \|\gamma\|_{M(W_p^m \to W_p^l)} + \varepsilon\,.$$

Then, for all $u \in W_p^m$,

$$\|\gamma u - Tu\|_{W_p^l} \le (\operatorname{ess} \|\gamma\|_{M(W_p^m \to W_p^l)} + \varepsilon)\|u\|_{W_p^m}\,. \tag{7.2.1}$$

Let S be the unit ball in W_p^m centered at the origin and let $\{v_k\}$ be a finite ε-net in TS. Without loss of generality we assume that $v_k \in C_0^\infty$. It is clear that $\zeta_r v_k = 0$ for sufficiently large r. Therefore,

$$\|\zeta_r \gamma u\|_{W_p^l} = \|\zeta_r(\gamma u - v_k)\|_{W_p^l} \le \|\zeta_r(\gamma u - Tu)\|_{W_p^l} + \|\zeta_r(Tu - v_k)\|_{W_p^l}\,.$$

From this inequality and from Lemma 7.1.9 we obtain

$$\|\zeta_r \gamma u\|_{W_p^l} \le c\,(\operatorname{ess} \|\gamma\|_{M(W_p^m \to W_p^l)} + \varepsilon)\|u\|_{W_p^m}\,.$$

The result follows. □

Theorem 7.2.1. *Let $m > 1$ and let either $lp < n$ and $p > 1$, or $l \le n$ and $p = 1$. Then the equivalence relation*

$$\text{ess}\,\|\gamma\|_{M(W_p^m \to W_p^l)} \sim \limsup_{\delta\to 0}\ \sup_{x\in\mathbb{R}^n} \|\eta_{\delta,x}\gamma\|_{M(W_p^m\to W_p^l)} + \limsup_{r\to\infty} \|\zeta_r\gamma\|_{M(W_p^m\to W_p^l)} \tag{7.2.2}$$

holds.

Proof. (*i*) *The lower bound for the essential norm.* We use the notation introduced in the proof of Lemma 7.2.1. For any $u \in S$,

$$\|\eta_{\delta,x}\gamma u\|_{W_p^l} \le \|\eta_{\delta,x}(\gamma u - v_k)\|_{W_p^l} + \|\eta_{\delta,x} v_k\|_{W_p^l}$$

$$\le \|\eta_{\delta,x}(\gamma u - Tu)\|_{W_p^l} + \|\eta_{\delta,x}(Tu - v_k)\|_{W_p^l} + \|\eta_{\delta,x} v_k\|_{W_p^l}\,.$$

From this inequality and Lemma 7.1.6 we get

$$\|\eta_{\delta,x}\gamma u\|_{W_p^l} \le c\,(\,\text{ess}\,\|\gamma\|_{M(W_p^m\to W_p^l)} + 2\varepsilon) + \varepsilon$$

which, together with reference to Lemma 7.2.1, completes the proof of the lower bound for $\text{ess}\,\|\gamma\|_{M(W_p^m\to W_p^l)}$.

(*ii*) *The upper bound for the essential norm.* We choose δ and r so that the following estimates hold:

$$\sup_{x\in\mathbb{R}^n} \|\eta_{2\delta,x}\gamma\|_{M(W_p^m\to W_p^l)} \le \limsup_{\delta\to 0}\ \sup_{x\in\mathbb{R}^n} \|\eta_{\delta,x}\gamma\|_{M(W_p^m\to W_p^l)} + \varepsilon,$$

$$\|\zeta_r\gamma\|_{M(W_p^m\to W_p^l)} \le \limsup_{r\to\infty} \|\zeta_r\gamma\|_{M(W_p^m\to W_p^l)} + \varepsilon\,. \tag{7.2.3}$$

By $\{K_\delta^{(j)}\}$ we denote a finite covering of the ball $\mathcal{B}_{2r}$ by open balls with radius δ and centers x_j. We can choose the balls $K_\delta^{(j)}$ so that the multiplicity of the covering of $\mathcal{B}_{2r}$ by the balls $K_{2\delta}^{(j)}$ depends only on n. Let $\{\varphi^{(j)}\}$ be a smooth partition of unity subordinate to $\{K_\delta^{(j)}\}$ and such that

$$|\nabla_k \varphi^{(j)}| \le c\,\delta^{-k}, \quad k = 0, 1, \ldots.$$

Given any $u \in W_p^m(\mathcal{B}_\delta)$, we use the polynomials introduced in Lemma 7.1.3. Let $P^{(j)} = P^{(j)}(u;\cdot)$ be such polynomials constructed for the balls $K_{2\delta}^{(j)}$.

Further, let $\Gamma = (1 - \zeta_r)\gamma$ and let T_* be the finite-dimensional operator defined by

$$(T_* u)(x) = \Gamma(x) \sum_j \varphi^{(j)}(x) P^{(j)}(u; x)\,. \tag{7.2.4}$$

We have

$$\|(\gamma - T_*)u\|_{W_p^l} \leq \|(\Gamma - T_*)u\|_{W_p^l} + \|\zeta_r \gamma u\|_{W_p^l} . \tag{7.2.5}$$

Since

$$(\Gamma - T_*)u = \sum_j \Gamma\, \eta_{2\delta,x_j}\, \varphi^{(j)}(u - P^{(j)}),$$

it follows from Corollary 4.2.1 that

$$\|(\Gamma - T_*)u\|^p_{W_p^l} \leq \sum_j \|\Gamma\, \eta_{2\delta,x_j}\, \varphi^{(j)}(u - P^{(j)})\|^p_{W_p^l}$$

$$\leq c \sup_j \|\Gamma\, \eta_{2\delta,x_j}\|^p_{M(W_p^m \to W_p^l)} \sum_j \|\varphi^{(j)}(u - P^{(j)})\|^p_{W_p^m} . \tag{7.2.6}$$

Using Lemmas 7.1.2 and 7.1.3, we obtain that the last sum does not exceed

$$c \sum_j |\!|\!| u - P^{(j)}; K_{2\delta}^{(j)} |\!|\!|^p_{W_p^m} \leq c_1 \|D_{p,m} u\|^p_{L_p} . \tag{7.2.7}$$

We further note that, by Lemma 7.1.9,

$$\|\Gamma\, \eta_{2\delta,x_j}\|_{M(W_p^m \to W_p^l)} \leq c\, \|\gamma\, \eta_{2\delta,x_j}\|_{M(W_p^m \to W_p^l)} .$$

This and inequalities (7.2.3), (7.2.5)–(7.2.7) imply that

$$\|\gamma - T_*\|_{W_p^m \to W_p^l} \tag{7.2.8}$$

$$\leq c\Big(\limsup_{\delta \to 0} \sup_{x \in \mathbb{R}^n} \|\eta_{\delta,x}\gamma\|_{M(W_p^m \to W_p^l)} + \limsup_{r \to \infty} \|\zeta_r \gamma\|_{M(W_p^m \to W_p^l)} + \varepsilon\Big) .$$

□

Remark 7.2.1. Relation (7.2.2) fails for $lp > n$. In fact, let $\gamma = 1$. It is clear that $1 \in M(W_p^m \to W_p^l)$. On the other hand, Remark 4.3.3 implies that

$$\lim_{\delta \to 0} \|\eta_{\delta,x}\|_{M(W_p^m \to W_p^l)} = \infty .$$

7.2.2 Estimate Involving Capacity (The Case $mp < n$, $p > 1$)

The following theorem presents one more relation for the essential norm.

Theorem 7.2.2. *If $mp < n$ and $p \in (1, \infty)$, then*

$$\text{ess}\, \|\gamma\|_{M(W_p^m \to W_p^l)} \sim \lim_{\delta \to 0} \sup_{\{e:\, d(e) \leq \delta\}} \Big(\frac{\|\gamma; e\|_{L_p}}{[C_{p,m-l}(e)]^{1/p}} + \frac{\|D_{p,l}\gamma; e\|_{L_p}}{[C_{p,m}(e)]^{1/p}} \Big)$$

$$+ \lim_{r \to \infty} \sup_{\{e \subset \mathbb{R}^n \setminus \mathcal{B}_r:\, d(e) \leq 1\}} \Big(\frac{\|\gamma; e\|_{L_p}}{[C_{p,m-l}(e)]^{1/p}} + \frac{\|D_{p,l}\gamma; e\|_{L_p}}{[C_{p,m}(e)]^{1/p}} \Big). \tag{7.2.9}$$

Proof. We limit consideration to the case of noninteger l, since for integer l the proof is analogous and slightly simpler.

(*i*) *The lower bound for the essential norm.* We introduce the notation

$$f_k(\gamma; e) = \frac{\|D_{p,k}\gamma; e\|_{L_p}}{[C_{p,m-l+k}(e)]^{1/p}}$$

for noninteger k, $0 < k \le l$, and

$$f_k(\gamma; e) = \frac{\|\nabla_k \gamma; e\|_{L_p}}{[C_{p,m-l+k}(e)]^{1/p}}$$

for integer k, $0 \le k \le l$. Clearly, for any compact set e with $d(e) \le \delta$,

$$f_0(\gamma; e) \le \sup_{x\in\mathbb{R}^n} \sup_{e\subset\mathcal{B}_\delta(x)} f_0(\gamma; e) \le \sup_{x\in\mathbb{R}^n} \sup_{e\subset\mathcal{B}_\delta(x)} f_0(\eta_{\delta,x}\gamma; e)\,.$$

This, together with Corollary 4.3.1, implies that

$$f_0(\gamma; e) \le c \sup_{x\in\mathbb{R}^n} \|\eta_{\delta,x}\gamma\|_{M(W_p^m\to W_p^l)}\,.$$

Applying Theorem 7.2.1, we obtain

$$f_0(\gamma; e) \le c\,(\operatorname{ess} \|\gamma\|_{M(W_p^m\to W_p^l)} + \varepsilon). \tag{7.2.10}$$

Now we turn to the bound for $f_l(\gamma; e)$. For $e \subset \mathcal{B}_\delta(x)$ we have

$$\|D_{p,l}\gamma; e\|_{L_p}^p = \int_e dy \int_{\mathcal{B}_{2\delta}(x)} \frac{|\nabla_{[l]}(\gamma(y)\eta_{2\delta,x}(y)) - \nabla_{[l]}(\gamma(z)\eta_{2\delta,x}(z))|^p}{|y-z|^{n+p\{l\}}}\, dz$$
$$+ \int_e dy \int_{\mathbb{R}^n\setminus\mathcal{B}_{2\delta}(x)} |\nabla_{[l]}\gamma(y) - \nabla_{[l]}\gamma(z)|^p \frac{dz}{|y-z|^{n+p\{l\}}}\,.$$

Therefore,

$$\|D_{p,l}\gamma; e\|_{L_p} \le c\Big[\|D_{p,l}(\eta_{2\delta,x}\gamma); e\|_{L_p}$$
$$+\Big(\int_e |\nabla_{[l]}\gamma(y)|^p dh \int_{\mathbb{R}^n\setminus\mathcal{B}_{2\delta}(x)} \frac{dz}{|y-z|^{n+p\{l\}}}\Big)^{1/p}$$
$$+\Big(\int_{\mathbb{R}^n\setminus\mathcal{B}_{2\delta}(x)} |\nabla_{[l]}\gamma(y)|^p dy \int_e \frac{dz}{|y-z|^{n+p\{l\}}}\Big)^{1/p}\Big]. \tag{7.2.11}$$

The second term on the right-hand side does not exceed

$$c_1\delta^{-\{l\}}\|\nabla_{[l]}\gamma; e\|_{L_p} \le c_2\,\delta^{-\{l\}}\|\nabla_{[l]}(\gamma\eta_{2\delta,x}); e\|_{L_p}\,. \tag{7.2.12}$$

Since $e \subset \mathcal{B}_\delta(x)$, the third term is not greater than

$$c\,(\mathrm{mes}_n e)^{1/p}\Big(\int_{\mathbb{R}^n\backslash\mathcal{B}_{2\delta}(x)} \frac{|\nabla_{[l]}\gamma(z)|^p dz}{|z-x|^{n+p\{l\}}}\Big)^{1/p}$$

$$\le c_1(\mathrm{mes}_n e)^{1/p}\Big(\int_{\mathbb{R}^n\backslash\mathcal{B}_{2\delta}(x)} |z-x|^{-n-p\{l\}}\delta^{-n}\int_{\mathcal{B}_\delta(z)} |\nabla_{[l]}\gamma(\xi)|^p d\xi\, dz\Big)^{1/p}$$

$$\le c_2\,(\mathrm{mes}_n e)^{1/p} \sup_{z\in\mathbb{R}^n} \|\nabla_{[l]}(\eta_{2\delta,z}\gamma);\mathcal{B}_\delta(z)\|_{L_p}\delta^{-\{l\}-n/p}\,. \tag{7.2.13}$$

Therefore,

$$f_l(\gamma;e) \le c\Big[f_l(\eta_{2\delta,x}\gamma;e) + \delta^{-\{l\}}\Big(\frac{C_{p,m-\{l\}}(e)}{C_{p,m}(e)}\Big)^{1/p} f_{[l]}(\eta_{2\delta,x}\gamma;e)$$
$$+\Big(\frac{\delta^{-mp}\mathrm{mes}_n e}{C_{p,m}(e)}\Big)^{1/p} \sup_{x\in\mathbb{R}^n}\ \sup_{e\subset\mathcal{B}_\delta(x)} f_{[l]}(\eta_{2\delta,x}\gamma;e)\Big]\,. \tag{7.2.14}$$

Now from (7.2.14) and Lemma 7.1.11 we obtain

$$f_l(\gamma;e) \le c \sup_{x\in\mathbb{R}^n}\ \sup_{e\subset\mathcal{B}_\delta(x)} [f_l(\eta_{2\delta,x}\gamma;e) + f_{[l]}(\eta_{2\delta,x}\gamma;e)]$$

which, together with Proposition 4.3.1 and Theorem 7.2.1, implies that

$$f_l(\gamma;e) \le \sup_{x\in\mathbb{R}^n} \|\eta_{2\delta,x}\gamma\|_{M(W_p^m\to W_p^l)} \le c_1\big(\mathrm{ess}\,\|\gamma\|_{M(W_p^m\to W_p^l)} + \varepsilon\big). \tag{7.2.15}$$

Combining (7.2.10) and (7.2.16), we arrive at the inequality

$$\lim_{\delta\to 0}\ \sup_{\{e:\ d(e)\le\delta\}}\Big(\frac{\|\gamma;e\|_{L_p}}{[C_{p,m-l}(e)]^{1/p}} + \frac{\|D_{p,l}\gamma;e\|_{L_p}}{[C_{p,m}(e)]^{1/p}}\Big)$$
$$\le c\,\mathrm{ess}\,\|\gamma\|_{M(W_p^m\to W_p^l)}\,. \tag{7.2.16}$$

Let $e \subset \mathbb{R}^n\backslash\mathcal{B}_{3r}$. It is clear that $f_0(\gamma;e) \le f_0(\zeta_r\gamma;e)$ and, by Theorems 4.1.1 and 7.2.1, the estimate (7.2.10) holds if r is sufficiently large.

Let us estimate $f_l(\gamma;e)$. We have

$$\|D_{p,l}\gamma;e\|_{L_p}^p = \int_e dy \int_{\mathbb{R}^n\backslash\mathcal{B}_{2r}} \frac{|\nabla_{[l]}(\gamma(y)\zeta_r(y)) - \nabla_{[l]}(\gamma(z)\zeta_r(z))|^p}{|y-z|^{n+p\{l\}}}\,dz$$
$$+\int_e dy\int_{\mathcal{B}_{2r}} \frac{|\nabla_{[l]}\gamma(y) - \nabla_{[l]}\gamma(z)|^p}{|y-z|^{n+p\{l\}}}\,dz\,.$$

Consequently,

$$\|D_{p,l}\gamma;e\|_{L_p} \le c\Big[\|D_{p,l}(\zeta_r\gamma);e\|_{L_p} + \Big(\int_e |\nabla_{[l]}\gamma(y)|^p dy\int_{\mathcal{B}_{2r}}\frac{dz}{|y-z|^{n+p\{l\}}}\Big)^{1/p}$$
$$+\Big(\int_{\mathcal{B}_{2r}} |\nabla_{[l]}\gamma(z)|^p dz\int_e\frac{dy}{|y-z|^{n+p\{l\}}}\Big)^{1/p}\Big]\,. \tag{7.2.17}$$

The second term on the right-hand side of (7.2.17) does not exceed

$$cr^{-\{l\}}\Big(\int_e |\nabla_{[l]}\gamma(y)|^p dy\Big)^{1/p} \le c_1 r^{-\{l\}} f_{[l]}(\gamma;e)\big[C_{p,m-\{l\}}(e)\big]^{1/p}$$

which, by Lemma 7.1.11 and Theorem 4.1.1, is not greater than

$$c_2\, r^{-\{l\}}\|\gamma\|_{M(W_p^m\to W_p^l)}[C_{p,m}(e)]^{1/p}.$$

Let us get a similar estimate for the third term on the right-hand side of (7.2.17). We have

$$\Big(\int_{\mathcal{B}_{2r}} |\nabla_{[l]}\gamma(z)|^p dz \int_e \frac{dy}{|y-z|^{n+p\{l\}}}\Big)^{1/p}$$

$$\le c\, r^{-n/p-\{l\}}(\mathrm{mes}_n e)^{1/p}\|\nabla_{[l]}\gamma;\mathcal{B}_{2r}\|_{L_p} \le c\, f_{[l]}(\gamma;\mathcal{B}_{2r})(\mathrm{mes}_n e)^{1/p} r^{-m}\,.$$

Therefore, the third term is majorized by

$$cr^{-m}\|\gamma\|_{M(W_p^m\to W_p^l)}[C_{p,m}(e)]^{1/p}\,.$$

Finally,

$$f_l(\gamma;e) \le c\big(f_l(\zeta_r\gamma;e) + (r^{-\{l\}} + r^{-m})\|\gamma\|_{M(W_p^m\to W_p^l)}\big)\,. \qquad (7.2.18)$$

Taking the supremum with respect to e on both sides of this inequality and making $r\to\infty$, we arrive at

$$\lim_{r\to\infty}\ \sup_{e\subset\mathbb{R}^n\setminus\mathcal{B}_r} f_l(\gamma;e) \le c\,\lim_{r\to\infty}\ \sup_e f_l(\zeta_r\gamma;e)\,.$$

Combining this estimate with (7.2.10) and Theorem 7.2.1, we conclude that

$$\lim_{r\to\infty}\ \sup_{e\subset\mathbb{R}^n\setminus\mathcal{B}_r}\Big(\frac{\|\gamma;e\|_{L_p}}{[C_{p,m-l}(e)]^{1/p}} + \frac{\|D_{p,l}\gamma;e\|_{L_p}}{[C_{p,m}(e)]^{1/p}}\Big) \le c\,\mathrm{ess}\,\|\gamma\|_{M(W_p^m\to W_p^l)}. \quad (7.2.19)$$

Adding (7.2.16) and (7.2.19), we obtain the required estimate for the essential norm.

(ii) The upper bound for the essential norm. Let e be an arbitrary compact set in $\mathbb{R}^n$. We have

$$\|D_{p,l}(\eta_{\delta,x}\gamma);e\|_{L_p} \le c\Big[\sum_{j=0}^{[l]} \||\nabla_j(\eta_{2\delta,x}\gamma)|D_{p,l-j}\eta_{\delta,x};e\,\|_{L_p}$$

$$+\sum_{j=0}^{[l]}\Big(\int_{\mathbb{R}^n}|\nabla_j\eta_{\delta,x}(y)|^p dy\int_e \frac{|\nabla_{[l]-j}(\eta_{2\delta,x}(y)\gamma(y)) - \nabla_{[l]-j}(\eta_{2\delta,x}(z)\gamma(z))|^p}{|y-z|^{n+p\{l\}}}dz\Big)^{1/p}$$

$$+\|D_{p,l}(\eta_{\delta,x}\eta_{2\delta,x}\gamma);e\setminus\mathcal{B}_{4\delta}(x)\|_{L_p}\Big]. \qquad (7.2.20)$$

The obvious estimate

$$D_{p,l-j}\eta_{\delta,x} \le c\,\delta^{j-l}$$

and Lemma 7.1.11 imply that

$$\begin{aligned}\| |\nabla_j(\eta_{2\delta,x}\gamma)| D_{p,l-j}\eta_{\delta,x}; e\|_{L_p} &\le c\,\delta^{j-l} f_j(\eta_{2\delta,x}\gamma; e)[C_{p,m-l+j}(e)]^{1/p} \\ &\le c f_j(\eta_{2\delta,x}\gamma; e)[C_{p,m}(e)]^{1/p}.\end{aligned}$$

By Theorem 4.1.1,

$$f_j(\eta_{2\delta,x}\gamma; e) \le c\,\|\eta_{2\delta,x}\gamma\|_{M(W_p^{m-l+j}\to W_p^j)}. \tag{7.2.21}$$

Hence, from Lemma 7.1.6 and Corollary 4.3.7, we obtain

$$\begin{aligned}f_j(\eta_{2\delta,x}\gamma; e) &\le c \sup_{\xi\in\mathbb{R}^n} \|\eta_{\delta,\xi}\eta_{2\delta,x}\gamma\|_{M(W_p^{m-l+j}\to W_p^j)} \\ &\le c \sup_{\xi\in\mathbb{R}^n} \|\eta_{\delta,\xi}\gamma\|_{M(W_p^{m-l+j}\to W_p^j)} \\ &\le \varepsilon \sup_{\xi\in\mathbb{R}^n} \|\eta_{\delta,\xi}\gamma\|_{M(W_p^m\to W_p^l)} + c(\varepsilon) \sup_{\xi\in\mathbb{R}^n} \|\eta_{\delta,\xi}\gamma\|_{M(W_p^{m-l}\to L_p)}.\end{aligned}$$

Thus the first sum in (7.2.20) does not exceed

$$\big(\varepsilon\,\|\eta_{\delta,x}\gamma\|_{M(W_p^m\to W_p^l)} + c(\varepsilon)\|\eta_{\delta,x}\gamma\|_{M(W_p^{m-l}\to L_p)}\big)[C_{p,m}(e)]^{1/p}. \tag{7.2.22}$$

The j-th term in the second sum on the right-hand side of (7.2.20) is majorized by

$$\begin{aligned}c\,\delta^{-j}\|D_{p,l-j}(\eta_{2\delta,x}\gamma); e\|_{L_p} &\le c\,\delta^{-j} f_{l-j}(\eta_{2\delta,x}\gamma; e)[C_{p,m-j}(e)]^{1/p} \\ &\le c\, f_{l-j}(\eta_{2\delta,x}\gamma; e)[C_{p,m}(e)]^{1/p}.\end{aligned}$$

Using the same arguments as when estimating $f_j(\gamma; e)$, we get

$$f_{l-j}(\eta_{2\delta,x}\gamma; e) \le \varepsilon \sup_{\xi\in\mathbb{R}^n} \|\eta_{\delta,\xi}\gamma\|_{M(W_p^m\to W_p^l)} + c(\varepsilon) \sup_{\xi\in\mathbb{R}^n} \|\eta_{\delta,\xi}\gamma\|_{M(W_p^{m-l}\to L_p)}$$

for $j = 1, \dots, [l]$. Therefore, the second sum on the right-hand side of (7.2.20) does not exceed

$$\big(\varepsilon \sup_{x\in\mathbb{R}^n} \|\eta_{\delta,x}\gamma\|_{M(W_p^m\to W_p^l)} + c(\varepsilon) \sup_{\xi\in\mathbb{R}^n} \|\eta_{\delta,\xi}\gamma\|_{M(W_p^{m-l}\to L_p)} + c f_l(\gamma; e)\big)[C_{p,m}(e)]^{1/p}.$$

Now we give a bound for

$$\|D_{p,l}(\eta_{\delta,x}\eta_{2\delta,x}\gamma); e\backslash\mathcal{B}_{4\delta}(x)\|_{L_p}.$$

By Hölder's inequality and the estimate for the capacity (3.1.2), we find that for $z \in \mathcal{B}_{2\delta}(x)$

$$\int_{e\backslash\mathcal{B}_{4\delta}(x)} |y-z|^{-n-p\{l\}}dy$$
$$\leq c\,(\text{mes}_n e)^{(n-mp)/n}\Big(\int_{e\backslash\mathcal{B}_{4\delta}(x)} |y-z|^{-(n+p\{l\})n/mp}dy\Big)^{mp/n}$$
$$\leq c\,C_{p,m}(e)\,\delta^{p(m-\{l\})-n}.$$

Consequently,

$$\|D_{p,l}(\eta_{\delta,x}\eta_{2\delta,x}\gamma); e\backslash\mathcal{B}_{4\delta}(x)\|^p_{L_p}$$
$$\leq c\sum_{j=0}^{[l]}\int_{\mathcal{B}_\delta(x)}|\nabla_j(\eta_{2\delta,x}(z)\gamma(z))|^p dz\,\delta^{(j-[l])p}\int_{e\backslash\mathcal{B}_{4\delta}(x)}|y-z|^{-p\{l\}-n}dy$$
$$\leq c\,C_{p,m}(e)\sum_{j=0}^{[l]}\delta^{(m-j+l)p-n}\int_{\mathcal{B}_{2\delta}(x)}|\nabla_j(\eta_{2\delta,x}(z)\gamma(z))|^p dz$$
$$\leq c\,C_{p,m}(e)\sum_{j=0}^{[l]}f_j(\eta_{2\delta,x}\gamma;\mathcal{B}_{2\delta}(x))\,.$$

Following the same lines as when estimating $f_j(\eta_{2\delta,x}\gamma; e)$, we conclude that the third term on the right-hand side of (7.2.20) does not exceed (7.2.22). Substituting the derived estimates into (7.2.20), we arrive at

$$f_l(\eta_{\delta,x}\gamma;e) \leq c\Big(\varepsilon\sup_{\xi\in\mathbb{R}^n}\|\eta_{\delta,\xi}\gamma\|_{M(W^m_p\to W^l_p)}$$
$$+\,c(\varepsilon)\sup_{\xi\in\mathbb{R}^n}\|\eta_{\delta,\xi}\gamma\|_{M(W^{m-l}_p\to L_p)}+f_l(\gamma;e)\Big),$$

which implies that

$$\|\eta_{\delta,x}\gamma\|_{M(W^m_p\to W^l_p)}$$
$$\leq c\,\big(\varepsilon\sup_{\xi\in\mathbb{R}^n}\|\eta_{\delta,\xi}\gamma\|_{M(W^m_p\to W^l_p)}+c(\varepsilon)\sup_{\xi\in\mathbb{R}^n}\|\eta_{\delta,\xi}\gamma\|_{M(W^{m-l}_p\to L_p)}$$
$$+\sup_{e\subset\mathcal{B}_{2\delta}(x)}f_0(\gamma;e)+\sup_{e\subset\mathcal{B}_{4\delta}(x)}f_l(\gamma;e)\big).$$

Taking the supremum over x on both sides, we find that

$$\sup_{x\in\mathbb{R}^n}\|\eta_{\delta,x}\gamma\|_{M(W^m_p\to W^l_p)}$$
$$\leq c\,\big(\sup_{\{e:\,d(e)\leq 4\delta\}}f_0(\gamma;e)+\sup_{\{e:\,d(e)\leq 8\delta\}}f_l(\gamma;e)\big). \tag{7.2.23}$$

Now let $e \subset \mathbb{R}^n$ with $d(e) \leq 1$, and let r be a sufficiently large positive number. We have

$$\|D_{p,l}(\zeta_{3r}\gamma); e\|_{L_p}$$
$$\leq c\Big[\sum_{j=0}^{[l]} \||\nabla_j\gamma| D_{p,l-j}\zeta_{3r}; e\backslash\mathcal{B}_r\|_{L_p}$$
$$+ \sum_{j=0}^{[l]}\Big(\int_{\mathbb{R}^n} |\nabla_j\zeta_{3r}(y)|^p dy \int_{e\backslash\mathcal{B}_r} \frac{|\nabla_{[l]-j}\gamma(y) - \nabla_{[l]-j}\gamma(z)|^p}{|y-z|^{n+p\{l\}}} dy\, dz\Big)^{1/p}$$
$$+ \|D_{p,l}(\zeta_{3r}\gamma); e\cap\mathcal{B}_r\|_{L_p}\Big]. \tag{7.2.24}$$

The first sum on the right-hand side does not exceed

$$c\sum_{j=0}^{[l]} r^{j-l} f_j(\gamma; e\backslash\mathcal{B}_r)[C_{p,m-l+j}(e\backslash\mathcal{B}_r)]^{1/p}$$
$$\leq c\, r^{-\{l\}} \|\gamma\|_{M(W_p^m\to W_p^l)}[C_{p,m}(e\backslash\mathcal{B}_r)]^{1/p}.$$

The j-th term on the right-hand side of (7.2.24) is majorized by

$$c\, r^{-j}\|D_{p,l-j}\gamma; e\backslash\mathcal{B}_r\|_{L_p} \leq c\, r^{-j} f_{l-j}(\gamma; e\backslash\mathcal{B}_r)[C_{p,m-j}(e\backslash\mathcal{B}_r)]^{1/p}.$$

Hence the sum in question is dominated by

$$c\,(f_l(\gamma; e\backslash\mathcal{B}_r) + r^{-1}\|\gamma\|_{M(W_p^m\to W_p^l)})[C_{p,m}(e)]^{1/p}.$$

Further, we estimate the last term on the right-hand side of (7.2.24). We have

$$\|D_{p,l}(\zeta_{3r}\gamma); e\cap\mathcal{B}_r\|^p_{L_p}$$
$$\leq \sum_{j=0}^{[l]} \int_{\mathbb{R}^n\backslash\mathcal{B}_{3r}} |\nabla_j\gamma(z)|^p dz\, r^{(j-[l])p} \int_{e\cap\mathcal{B}_r} |y-z|^{-n-p\{l\}} dy$$
$$\leq c\,\mathrm{mes}_n e \sum_{j=0}^{[l]} \int_{\mathbb{R}^n\backslash\mathcal{B}_{3r}} \frac{|\nabla_j\gamma(z)|^p}{|z|^{n+p\{l\}}} dz\, r^{(j-[l])p}.$$

We note that

$$\int_{\mathbb{R}^n\backslash\mathcal{B}_{3r}} |\nabla_j\gamma(z)|^p \frac{dz}{|z|^{n+p\{l\}}} \leq r^{-p\{l\}} \sup_{\xi\in\mathbb{R}^n\backslash\mathcal{B}_{2r}} \|\nabla_j\gamma; \mathcal{B}_1(\xi)\|^p_{L_p}$$
$$\leq c\, r^{-p\{l\}} \|\gamma\|^p_{M(W_p^m\to W_p^l)}.$$

Therefore,

$$\|D_{p,l}(\zeta_{3r}\gamma); e \cap \mathcal{B}_r\|_{L_p} \le c\,(\mathrm{mes}_n e)^{1/p} r^{-\{l\}} \|\gamma\|_{M(W_p^m \to W_p^l)}\,.$$

Consequently,

$$\|D_{p,l}(\zeta_{3r}\gamma); e\|_{L_p} \le c\,\big(f_l(\gamma; e\backslash \mathcal{B}_r) + r^{-\{l\}} \|\gamma\|_{M(W_p^m \to W_p^l)}\big)\,[C_{p,m}(e)]^{1/p}\,.$$

Using Theorem 4.1.1, we obtain the estimate

$$\|\zeta_{3r}\gamma\|_{M(W_p^m \to W_p^l)} \le c_1 \sup_{e \subset \mathbb{R}^n \backslash \mathcal{B}_r, d(e) \le 1} \big(f_l(\gamma; e) + f_0(\gamma; e)\big)$$

$$+ c_2 r^{-\{l\}} \|\gamma\|_{M(W_p^m \to W_p^l)}. \tag{7.2.25}$$

This inequality, together with (7.2.23) and Theorem 7.2.1, implies the required upper bound for the essential norm. □

7.2.3 Estimates Involving Capacity (The Case $mp = n$, $p > 1$)

Theorem 7.2.3. *The relation* (7.2.9) *also holds for* $mp = n$, $p > 1$.

Proof. As in the proof of Theorem 7.2.2, we need to consider only the more difficult case of noninteger l, $0 < l < m$.

(*i*) *The lower bound for the essential norm.* Let e be a compact set in $\mathbb{R}^n$ with $d(e) \le \delta \le 1/2$. The argument leading to (7.2.10) applies equally for obtaining the required estimate of $f_0(\gamma; e)$.

For $f_l(\gamma; e)$ we have the estimates (7.2.11)–(7.2.13). The right-hand side of (7.2.12) does not exceed

$$c\,\delta^{-\{l\}} (\mathrm{mes}_n e)^{\{l\}/n} \|\nabla_{[l]}(\gamma\eta_{2\delta,x}); e\|_{L_{pn/(n-p\{l\})}}$$
$$\le c\,\frac{|\log \delta|^{(p-1)/p}}{|\log \mathrm{mes}_n e|^{(p-1)/p}} \|\nabla_{[l]}(\gamma\eta_{2\delta,x}); e\|_{L_{pn/(n-p\{l\})}}\,.$$

The expression on the right-hand side here is not greater than

$$c\,[C_{p,m}(e)]^{1/p} \sup_{x \in \mathbb{R}^n} \frac{\|D_{p,l}(\gamma\eta_{2\delta,x}); \mathcal{B}_{4\delta}(x)\|_{L_p}}{[C_{p,m}(\mathcal{B}_{4\delta}(x))]^{1/p}}\,. \tag{7.2.26}$$

Similarly, the right-hand side of (7.2.13) does not exceed

$$c\,(\mathrm{mes}_n e)^{1/p} |\log \mathrm{mes}_n e|^{(p-1)/p} [C_{p,m}(e)]^{1/p} \delta^{-n/p} \sup_{z \in \mathbb{R}^n} \|D_{p,l}(\gamma\eta_{2\delta,z}); \mathcal{B}_{4\delta}(z)\|_{L_p}$$

which, in turn, is majorized by (7.2.26). Thus

$$f_l(\gamma; e) \le c \sup_{x \in \mathbb{R}^n} f_l(\gamma\eta_{2\delta,x}; \mathcal{B}_{4\delta}(x))\,.$$

Combining this estimate with Theorem 4.1.1 and Theorem 7.2.1, we get (7.2.15). Consequently we arrive at (7.2.16).

The proof of (7.2.18) holds for $mp = n$ as well. Thus the required lower bound for the essential norm is obtained.

(ii) The upper bound for the essential norm. We take the relation (7.2.2) as the basis of the proof. Since $\eta_{\delta^2,x}\mu_{\delta,x} = \eta_{\delta^2,x}$, then, by Lemma 7.1.6,

$$\|\eta_{\delta^2,x}\gamma\|_{M(W_p^m \to W_p^l)} \le c\,\|\mu_{\delta,x}\gamma\|_{M(W_p^m \to W_p^l)}\,.$$

Our aim is to prove the estimate

$$\|\mu_{\delta,x}\gamma\|_{M(W_p^m \to W_p^l)} \le c \sup_{e\subset \mathcal{B}_{6\delta}(x)} \big(f_0(\gamma;e) + f_l(\gamma;e)\big)\,. \tag{7.2.27}$$

Let e be a compact set in $\mathbb{R}^n$ with $d(e) < 1/2$. We have

$$\begin{aligned}
&\|D_{p,l}(\mu_{\delta,x}\gamma);e\|_{L_p}\\
&\quad\le c\Big[\sum_{j=0}^{[l]} \||\nabla_j\gamma| D_{p,l-j}\mu_{\delta,x}; e\cap \mathcal{B}_{2\delta}(x)\|_{L_p}\\
&\qquad+\sum_{j=1}^{[l]}\Big(\int |\nabla_j\mu_{\delta,x}(y)|^p\,dy \int_{e\cap \mathcal{B}_{2\delta}(x)} \frac{|\nabla_{[l]-j}\gamma(y)-\nabla_{[l]-j}\gamma(z)|^p}{|y-z|^{n+p\{l\}}}\,dz\Big)^{1/p}\\
&\qquad+\|D_{p,l}(\mu_{\delta,x}\gamma); e\backslash \mathcal{B}_{2\delta}(x)\|_{L_p} + \|\mu_{\delta,x}D_{p,l}\gamma;e\|_{L_p}\Big]\,.
\end{aligned} \tag{7.2.28}$$

Applying Lemma 7.1.8, we find that

$$\begin{aligned}
&\||\nabla_j\gamma| D_{p,l-j}\mu_{\delta,x}; e\cap \mathcal{B}_{2\delta}(x)\|_{L_p}\\
&\qquad\le c\,\||\nabla_j\gamma|\,|\log r|^{-1}\, r^{j-l}; e\cap \mathcal{B}_{2\delta}(x)\|_{L_p}\,,
\end{aligned} \tag{7.2.29}$$

where $r(z) = |z-x|$. By Hölder's inequality the right-hand side does not exceed

$$c\,\|\nabla_j\gamma; \mathcal{B}_{2\delta}(x)\|_{L_{pn/(n-p(l-j))}}\||\log r|^{-1}\, r^{j-l}\,; e\cap \mathcal{B}_{2\delta}(x)\|_{L_{n/(l-j)}}\,.$$

Since the function $|\log t|^{-1}t^{j-l}$ decreases near $t=0$, the maximum value of the integral

$$\int_E \frac{dz}{|z|^n|\log|z||^{n/(l-j)}}$$

over all sets E with prescribed small $\mathrm{mes}_n E$ is attained for the ball centered at $z=0$. Consequently,

$$\||\log r|^{-1} r^{j-l}; e\cap \mathcal{B}_{2\delta}(x)\|_{L_{n/(l-j)}} \le c\,|\log \mathrm{mes}_n(e\cap \mathcal{B}_{2\delta}(x))|^{(l-j-n)/n}. \tag{7.2.30}$$

We further note that

$$\|\nabla_j\gamma; \mathcal{B}_{2\delta}(x)\|_{L_{np/(n-p(l-j))}} \le c\,\big(\|D_{p,l}\gamma; \mathcal{B}_{2\delta}(x)\|_{L_p} + \delta^{-l}\|\gamma; \mathcal{B}_{2\delta}(x)\|_{L_p}\big).$$

From (7.2.29) and two last inequalities we obtain

$$\begin{aligned}
&\| |\nabla_j \gamma| D_{p,l-j}\mu_{\delta,x}; e \cap \mathcal{B}_{2\delta}(x)\|_{L_p} \\
&\quad \le c\,|\log\delta|^{(p(l-j)-n)/np}\big(|\log\delta|^{(1-p)/p} f_l(\gamma;\mathcal{B}_{2\delta}(x)) + f_0(\gamma;\mathcal{B}_{2\delta}(x))\big) \\
&\qquad \times |\log \mathrm{mes}_n(e \cap \mathcal{B}_{2\delta}(x))|^{(1-p)/p}.
\end{aligned}$$

Applying Proposition 4.3.1, we arrive at

$$\begin{aligned}
&\| |\nabla_j \gamma| D_{p,l-j}\mu_{\delta,x}; e \cap \mathcal{B}_{2\delta}(x)\|_{L_p} \\
&\quad \le c\,\big(f_l(\gamma;\mathcal{B}_{2\delta}(x)) + f_0(\gamma;\mathcal{B}_{2\delta}(x))\big)[C_{p,m}(e)]^{1/p}. \qquad (7.2.31)
\end{aligned}$$

The general term of the second sum on the right-hand side of (7.2.28) is equal to

$$\Big(\int \frac{dh}{|h|^{n+p\{l\}}} \int_{e\cap\mathcal{B}_{2\delta}(x)} |\Delta_h \nabla_{[l]-j}\gamma(z)|^p |\nabla_j \mu_{\delta,x}(z+h)|^p dz\Big)^{1/p}.$$

Since $\operatorname{supp}\mu_{\delta,x} \subset \mathcal{B}_\delta(x)$, the last expression does not exceed

$$\| |\nabla_j \mu_{\delta,x}; e\cap\mathcal{B}_{2\delta}(x)|_{L_{n/j}} \Big(\int_{\mathcal{B}_{3\delta}(x)} \|\Delta_h \nabla_{[l]-j}\gamma; \mathcal{B}_{2\delta}(x)\|^p_{L_{np/(n-jp)}} \frac{dh}{|h|^{n+p\{l\}}}\Big)^{1/p}.$$

Using Lemma 7.1.8 and applying the same argument as in the proof of (7.2.30), we conclude that

$$\|\nabla_j \mu_{\delta,x}; e\cap\mathcal{B}_{2\delta}(x)\|_{L_{n/j}} \le c\,\Big|\log \mathrm{mes}_n(e\cap\mathcal{B}_{2\delta}(x))\Big|^{(j-n)/n}. \qquad (7.2.32)$$

It is known that the space $B^{s_1}_{q_1,p}$ is imbedded continuously into $B^{s_2}_{q_2,p}$ with $s_1 - n/q_1 = s_2 - n/q_2$, $1 < q_1 < q_2 < \infty$ (see [Bes]). This, in particular, implies that

$$\begin{aligned}
&\int_{\mathcal{B}_{3\delta}(x)} \|\Delta_h \nabla_{[l]-j}\gamma; \mathcal{B}_{2\delta}(x)\|^p_{L_{np/(n-jp)}} \frac{dh}{|h|^{n+p\{l\}}} \le c\,\|\gamma;\mathcal{B}_{6\delta}(x)\|^p_{W^l_p} \\
&\qquad \sim c\,\big(\|D_{p,l}\gamma;\mathcal{B}_{6\delta}(x)\|_{L_p} + \delta^{-l}\|\gamma;\mathcal{B}_{6\delta}(x)\|_{L_p}\big)^p. \qquad (7.2.33)
\end{aligned}$$

From (7.2.32) and (7.2.33) we obtain that the general term of the second sum on the right-hand side of (7.2.28) is dominated by

$$\begin{aligned}
&c\,|\log\delta|^{(pj-n)/np}\big(|\log\delta|^{(1-p)/p} f_l(\gamma;\mathcal{B}_{6\delta}(x)) \\
&\quad + f_0(\gamma;\mathcal{B}_{6\delta}(x))\big)|\log \mathrm{mes}_n(e\cap\mathcal{B}_{2\delta}(x))|^{(1-p)/p}.
\end{aligned}$$

This, together with Proposition 3.1.3, yields

$$\begin{aligned}
&\Big(\int |\nabla_j \mu_{\delta,x}(y)|^p\,dy \int_{e\cap\mathcal{B}_{2\delta}(x)} \frac{|\nabla_{[l]-j}\gamma(y) - \nabla_{[l]-j}\gamma(z)|^p}{|y-z|^{n+p\{l\}}}\,dz\Big)^{1/p} \\
&\quad \le c\,(f_l(\gamma;\mathcal{B}_{6\delta}(x)) + f_0(\gamma;\mathcal{B}_{6\delta}(x)))[C_{p,m}(e)]^{1/p}. \qquad (7.2.34)
\end{aligned}$$

Let us estimate the norm $\|D_{p,l}(\mu_{\delta,x}\gamma); e\backslash\mathcal{B}_{2\delta}(x)\|_{L_p}$, which is obviously equal to

$$\Big(\int_{e\backslash\mathcal{B}_{2\delta}(x)} dy \int_{\mathcal{B}_\delta(x)} \frac{|\nabla_{[l]}(\mu_{\delta,x}\gamma)(z)|^p}{|y-z|^{n+p\{l\}}}\, dz\Big)^{1/p}$$

$$= \Big(\int_{\mathcal{B}_\delta(x)} |\nabla_{[l]}(\mu_{\delta,x}\gamma)(z)|^p\, dz \int_{e\backslash\mathcal{B}_{2\delta}(x)} \frac{dy}{|y-z|^{n+p\{l\}}}\Big)^{1/p}.$$

It is clear that

$$\int_{e\backslash\mathcal{B}_{2\delta}(x)} \frac{dy}{|y-z|^{n+p\{l\}}} \le \min\{\delta^{-n-p\{l\}}\mathrm{mes}_n e,\ \delta^{-p\{l\}}\}$$

$$\le \frac{|\log\delta|^{p-1}}{|\log \mathrm{mes}_n e|^{p-1}\,\delta^{p\{l\}}}.$$

Moreover, by Lemma 4.2.7,

$$\int_{\mathcal{B}_\delta(x)} |\nabla_{[l]}(\mu_{\delta,x}\gamma)(z)|^p\, dz$$

$$\le \delta^{p\{l\}} \int_{\mathcal{B}_\delta(x)}\int_{\mathcal{B}_\delta(x)} \frac{|\nabla_{[l]}(\mu_{\delta,x}\gamma)(y)-\nabla_{[l]}(\mu_{\delta,x}\gamma)(z)|^p}{|y-z|^{n+p\{l\}}}\, dz\, dy\,.$$

Consequently,

$$\|D_{p,l}(\mu_{\delta,x}\gamma); e\backslash\mathcal{B}_{2\delta}(x)\|^p_{L_p}$$

$$\le c\,\frac{|\log\delta|^{p-1}}{|\log \mathrm{mes}_n e|^{p-1}}\Big(\|D_{p,l}\gamma;\mathcal{B}_\delta(x)\|^p_{L_p} + \sum_{j=0}^{[l]} \||\nabla_j\gamma| D_{p,l-j}\mu_{\delta,x};\mathcal{B}_\delta(x)\|^p_{L_p}$$

$$+\sum_{j=1}^{[l]} \int_{\mathcal{B}_\delta(x)} |\nabla_j\mu_{\delta,x}(y)|^p\, dy \int_{\mathcal{B}_\delta(x)} \frac{|\nabla_{[l]-j}\gamma(y)-\nabla_{[l]-j}\gamma(z)|^p}{|y-z|^{n+p\{l\}}}\, dz\Big).$$

Putting $e=\mathcal{B}_\delta(x)$ in (7.2.31) and (7.2.34), we see that either of the two last sums may be estimated from above by

$$c\,|\log\delta|^{1-p}\big(f_l(\gamma;\mathcal{B}_{6\delta}(x)) + f_0(\gamma;\mathcal{B}_{6\delta}(x))\big)^p.$$

Therefore,

$$\|D_{p,l}(\mu_{\delta,x}\gamma); e\backslash\mathcal{B}_{2\delta}(x)\|^p_{L_p} \le c\,(f_l(\gamma;\mathcal{B}_{6\delta}(x)) + f_0(\gamma;\mathcal{B}_{6\delta}(x)))^p C_{p,m}(e)\,.$$

Setting this, together with (7.2.31) and (7.2.34), into (7.2.28), we arrive at (7.2.27).

The estimate

$$\limsup_{r\to\infty} \|\zeta_r\gamma\|_{M(W^m_p\to W^l_p)} \le c\lim_{r\to\infty} \sup_{e\subset\mathbb{R}^n\backslash\mathcal{B}_r,\, d(e)\le 1} (f_l(\gamma;e)+f_0(\gamma;e))$$

was obtained at the end of the proof of Theorem 7.2.2. □

7.2.4 Proof of Theorem 7.0.3

To obtain the lower estimate for $\text{ess}\|\gamma\|_{M(W_p^m \to W_p^l)}$ we use Theorem 7.2.2 and the upper estimates for the capacity of a ball (see Proposition 3.1.4). Then

$$\sup_{\{e:d(e)\le\delta\}} \frac{\|\gamma; e\|_{L_p}}{[C_{p,m-l}(e)]^{1/p}} \ge \sup_{\rho\le\delta} \frac{\|\gamma; \mathcal{B}_\rho(x)\|_{L_p}}{[C_{p,m-l}(\mathcal{B}_\rho(x))]^{1/p}}$$

$$= c \sup_{\rho\le\delta} \rho^{m-l-n/p} \|\gamma; \mathcal{B}_\rho(x)\|_{L_p} \ge c \sup_{\rho\le\delta} \rho^{m-l-n} \|\gamma; \mathcal{B}_\rho(x)\|_{L_1},$$

where x is an arbitrary point of $\mathbb{R}^n$. Also,

$$\sup_{\{e\subset\mathbb{R}^n\setminus\mathcal{B}_r : d(e)\le 1\}} \frac{\|\gamma; e\|_{L_p}}{[C_{p,m-l}(e)]^{1/p}} \ge \sup_{x\in\mathbb{R}^n\setminus\mathcal{B}_r} \|\gamma; \mathcal{B}_1(x)\|_{L_p}.$$

To prove the upper estimate in part (i) of Theorem 7.0.3, we show that

$$\sup_{\{e:d(e)\le\delta\}} \frac{\|\gamma; e\|_{L_p}}{[C_{p,m-l}(e)]^{1/p}}$$

$$\le c\Big(\sup_{\{e:d(e)\le\delta\}} \frac{\|D_{p,l}\gamma; e\|_{L_p}}{[C_{p,m}(e)]^{1/p}} + \sup_{x\in\mathbb{R}^n} \delta^{m-l-n} \|\gamma; \mathcal{B}_\delta(x)\|_{L_1}\Big) \tag{7.2.35}$$

for $pm \le n$. In view of Corollary 4.3.1 and Lemma 4.3.11 we have

$$\sup_{\{e:d(e)\le 1\}} \frac{\|\gamma; e\|_{L_p}}{[C_{p,m-l}(e)]^{1/p}}$$

$$\le c\Big(\sup_{\{e:d(e)\le 1\}} \frac{\|D_{p,l}\gamma; e\|_{L_p}}{[C_{p,m}(e)]^{1/p}} + \sup_{x\in\mathbb{R}^n} \|\gamma; \mathcal{B}_1(x)\|_{L_1}\Big). \tag{7.2.36}$$

In the last inequality we replace e by $\delta^{-1}E$ where $d(E) \le \delta$ and introduce $\Gamma(\cdot) = \gamma(\delta^{-1}\cdot)$. Then (7.2.36) becomes

$$\sup_{\{E:d(E)\le\delta\}} \frac{\delta^{-n/p}\|\Gamma; E\|_{L_p}}{[C_{p,m-l}(\delta^{-1}E)]^{1/p}}$$

$$\le c\Big(\sup_{\{E:d(E)\le\delta\}} \frac{\delta^{l-n/p}\|D_{p,l}\Gamma; E\|_{L_p}}{[C_{p,m}(\delta^{-1}E)]^{1/p}} + \sup_{y\in\mathbb{R}^n} \delta^{-n} \|\Gamma; \mathcal{B}_\delta(y)\|_{L_1}\Big). \tag{7.2.37}$$

For $mp < n$, by Corollary 3.1.1 we have

$$C_{p,m-l}(\delta^{-1}E) \le c\, \delta^{p(m-l)-n} C_{p,m-l}(E)$$

and

$$C_{p,m}(\delta^{-1}E) \ge c\, \delta^{pm-n} C_{p,m}(E).$$

For $mp = n$ the last inequality should be replaced by the estimate

$$C_{p,m}(\delta^{-1}E) \geq c\, C_{p,m}(E),$$

which holds since the capacity is a non-decreasing set function. Thus, for $mp \leq n$

$$\sup_{\{E:d(E)\leq\delta\}} \frac{\delta^{-(m-l)}\|\Gamma;E\|_{L_p}}{[C_{p,m-l}(E)]^{1/p}}$$

$$\leq c\Big(\sup_{\{E:d(E)\leq\delta\}} \frac{\delta^{-(m-l)}\|D_{p,l}\Gamma;E\|_{L_p}}{[C_{p,m}(E)]^{1/p}} + \sup_{y\in\mathbb{R}^n} \delta^{-n}\|\Gamma;\mathcal{B}_\delta(y)\|_{L_1}\Big)$$

and (7.2.35) follows. □

7.2.5 Sharpening of the Lower Bound for the Essential Norm in the Case $m > l$, $mp \leq n$, $p > 1$

In addition to Theorems 7.2.2 and 7.2.3 we prove the following result.

Theorem 7.2.4. *If $m > l$, $mp \leq n$, and $p > 1$, then*

$$\text{ess}\,\|\gamma\|_{M(W_p^m\to W_p^l)}$$

$$\geq c\lim_{\delta\to 0}\sup_{\{e:\, d(e)\leq\delta\}}\sum_{j=0}^{[l]}\Big(\frac{\|D_{p,l-j}\gamma;e\|_{L_p}}{[C_{p,m-j}(e)]^{1/p}} + \frac{\|\nabla_j\gamma;e\|_{L_p}}{[C_{p,m-l+j}(e)]^{1/p}}\Big)$$

$$+ c\lim_{r\to\infty}\sup_{\{e\subset\mathbb{R}^n\setminus\mathcal{B}_r:\, d(e)\leq 1\}}\sum_{j=0}^{[l]}\Big(\frac{\|D_{p,l-j}\gamma;e\|_{L_p}}{[C_{p,m-j}(e)]^{1/p}} + \frac{\|\nabla_j\gamma;e\|_{L_p}}{[C_{p,m-l+j}(e)]^{1/p}}\Big).$$

To prove this theorem, we need an auxiliary assertion.

Lemma 7.2.2. *For any multi-index α with $|\alpha| \leq l < m$,*

$$\text{ess}\,\|D^\alpha\gamma\|_{M(W_p^m\to W_p^{l-|\alpha|})} \leq c\sum_{j=0}^{|\alpha|}\text{ess}\,\|\gamma\|_{M(W_p^{m-j}\to W_p^{l-j})}.$$

Proof. It suffices to limit consideration to $|\alpha| = 1$. By T_j, $j = 0, 1$, we denote compact operators acting from W_p^{m-j} into W_p^{l-j} and such that

$$\|\gamma u - T_j u\|_{W_p^{l-j}} \leq (\text{ess}\,\|\gamma\|_{M(W_p^{m-j}\to W_p^{l-j})} + \varepsilon)\|u\|_{W_p^{m-j}}.$$

For any function $u \in W_p^m$ we have

$$\|\gamma\nabla u + u\nabla\gamma - \nabla T_0 u\|_{W_p^{l-1}} \leq (\text{ess}\,\|\gamma\|_{M(W_p^m\to W_p^l)} + \varepsilon)\|u\|_{W_p^m}.$$

Therefore,

$$\|u\nabla\gamma - \nabla T_0 u + T_1 \nabla u\|_{W_p^{l-1}}$$

$$\leq \big(\operatorname{ess} \|\gamma\|_{M(W_p^m \to W_p^l)} + \varepsilon\big)\|u\|_{W_p^m} + \|\gamma\nabla u - T_1 \nabla u\|_{W_p^{l-1}}$$

$$\leq \big(\operatorname{ess} \|\gamma\|_{M(W_p^m \to W_p^l)} + \operatorname{ess} \|\gamma\|_{M(W_p^{m-l} \to W_p^{l-1})} + 2\varepsilon\big)\|u\|_{W_p^m} .$$

Since $T = \nabla T_0 - T_1 \nabla$ is a compact operator acting from W_p^m into W_p^{l-1}, it follows that

$$\operatorname{ess} \|\nabla\gamma\|_{M(W_p^m \to W_p^{l-1})} \leq \operatorname{ess} \|\gamma\|_{M(W_p^m \to W_p^l)} + \operatorname{ess} \|\gamma\|_{M(W_p^{m-1} \to W_p^{l-1})} .$$

□

Proof of the Theorem 7.2.4. From the interpolation property (4.3.26) and from Theorem 7.2.1, we find that

$$\operatorname{ess} \|\gamma\|_{M(W_p^{m-j} \to W_p^{l-j})} \leq c \operatorname{ess} \|\gamma\|^{(l-j)/l}_{M(W_p^m \to W_p^l)} \operatorname{ess} \|\gamma\|^{j/l}_{M(W_p^{m-l} \to L_p)}$$

which, together with Lemmas 2.3.4 and 4.3.4, gives

$$\operatorname{ess} \|\gamma\|_{M(W_p^{m-j} \to W_p^{l-j})} \leq c \operatorname{ess} \|\gamma\|_{M(W_p^m \to W_p^l)} .$$

By this inequality and Lemma 7.2.2 we conclude that

$$\operatorname{ess} \|\nabla_j \gamma\|_{M(W_p^m \to W_p^{l-j})} \leq c \operatorname{ess} \|\gamma\|_{M(W_p^m \to W_p^l)} .$$

It remains to make use of Theorems 7.2.2 and 7.2.3. □

7.2.6 Estimates of the Essential Norm for $mp > n$, $p > 1$ and for $p = 1$

In the two cases mentioned, we do not need a capacity. The simplest formulation is for $mp > n$, $p \geq 1$ and for $m = n$, $p = 1$.

Theorem 7.2.5. *If $mp > n$ and $p \geq 1$, or $m = n$ and $p = 1$, and $\gamma \in M(W_p^m \to W_p^l)$, $m > l$, then*

$$\operatorname{ess} \|\gamma\|_{M(W_p^m \to W_p^l)} \sim \limsup_{|x| \to \infty} \|\gamma; \mathcal{B}_1(x)\|_{W_p^l} . \tag{7.2.38}$$

Proof. Applying Corollary 4.3.8 and Theorem 5.4.1 to the multiplier $\zeta_r\gamma$, we obtain from Lemma 7.2.1 that

$$\limsup_{r\to\infty}\sup_{x\in\mathbb{R}^n}\|\zeta_r\gamma;\mathcal{B}_1(x)\|_{W_p^l}\le c\,\mathrm{ess}\,\|\gamma\|_{M(W_p^m\to W_p^l)}$$

which is equivalent to

$$\limsup_{|x|\to\infty}\|\gamma;\mathcal{B}_1(x)\|_{W_p^l}\le c\,\mathrm{ess}\,\|\gamma\|_{M(W_p^m\to W_p^l)}\,.$$

Let us prove the converse estimate. Corollary 4.3.8 and Theorem 5.4.1 imply that

$$\|\gamma\|_{MW_p^l}\sim\|\gamma\|_{M(W_p^m\to W_p^l)}<\infty.$$

This and Lemma 7.1.2 yield

$$\|(1-\zeta_r)\gamma u\|_{W_p^l}\le\|\gamma\|_{MW_p^l}\|(1-\zeta_r)u\|_{W_p^l}\le c\,\|u;\mathcal{B}_{4r}\|_{W_p^l}\,.$$

Since any bounded subset of W_p^m is compact in $W_p^l(\mathcal{B}_{4r})$, the operator

$$(1-\zeta_r)\gamma\colon W_p^m\to W_p^l$$

is compact. Consequently, for any $r>0$,

$$\mathrm{ess}\,\|\gamma\|_{M(W_p^m\to W_p^l)}=\mathrm{ess}\,\|\zeta_r\gamma\|_{M(W_p^m\to W_p^l)}\le\|\zeta_r\gamma\|_{M(W_p^m\to W_p^l)}\,.$$

Estimating the last norm with the help of Corollary 4.3.8 and passing to the limit as $r\to\infty$, we complete the proof. □

Theorem 7.2.6. *If $l<m<n$, then*

$$\mathrm{ess}\,\|\gamma\|_{M(W_1^m\to W_1^l)}\sim\limsup_{\delta\to 0}\delta^{m-n}\sup_{x\in\mathbb{R}^n}\|\gamma;\mathcal{B}_\delta(x)\|_{W_1^l}$$
$$+\limsup_{|x|\to\infty}\sup_{r\in(0,1)}r^{m-n}\|\gamma;\mathcal{B}_r(x)\|_{W_1^l}\,. \quad (7.2.39)$$

Proof. According to Theorems 5.4.1 and 7.2.1,

$$\mathrm{ess}\,\|\gamma\|_{M(W_1^m\to W_1^l)}\sim\limsup_{\delta\to 0}\sup_{x,y\in\mathbb{R}^n}\sup_{r\in(0,1)}r^{m-n}\|\eta_{\delta,x}\gamma;\mathcal{B}_r(y)\|_{W_1^l}$$
$$+\limsup_{\rho\to\infty}\sup_{y\in\mathbb{R}^n}\sup_{r\in(0,1)}r^{m-n}\|\zeta_\rho\gamma;\mathcal{B}_r(y)\|_{W_1^l}\,.$$

Let the first term on the right-hand side be denoted by A_1 and the second one by A_2. Dealing first with A_2, we have

$$A_2\ge\limsup_{\rho\to\infty}\sup_{y\notin\mathcal{B}_{2\rho}}\sup_{r\in(0,1)}r^{m-n}\|\gamma;\mathcal{B}_r(y)\|_{W_1^l}$$
$$=\limsup_{|y|\to\infty}\sup_{r\in(0,1)}r^{m-n}\|\gamma;\mathcal{B}_r(y)\|_{W_1^l}\,.$$

An upper bound for A_2 can be obtained as follows:

$$A_2 \le \limsup_{\rho\to\infty} \sup_{y\notin\mathcal{B}_{\rho/2}} \sup_{r\in(0,1)} r^{m-n} \|\zeta_\rho\gamma; \mathcal{B}_r(y)\|_{W_1^l}$$

$$\le c \limsup_{\rho\to\infty} \sup_{y\notin\mathcal{B}_{\rho/2}} \sup_{r\in(0,1)} r^{m-n} \|\gamma; \mathcal{B}_{2r}(y)\|_{W_1^l}$$

$$\le c_1 \limsup_{|y|\to\infty} \sup_{r\in(0,1)} r^{m-n} \|\gamma; \mathcal{B}_r(y)\|_{W_1^l}\,.$$

Now we turn to estimates for A_1. We have

$$A_1 \ge \limsup_{\delta\to 0} \sup_{x\in\mathbb{R}^n} \delta^{m-n} \|\eta_{\delta,x}\gamma; \mathcal{B}_\delta(x)\|_{W_1^l}$$

$$\ge c \limsup_{\delta\to 0} \sup_{x\in\mathbb{R}^n} \delta^{m-n} \|\gamma; \mathcal{B}_{\delta/2}(x)\|_{W_1^l}\,.$$

On the other hand,

$$\sup_{r\in(0,1)} r^{m-n} \|\eta_{\delta,x}\gamma; \mathcal{B}_r(y)\|_{W_1^l}$$

$$\le \sup_{r\in(0,\delta/2)} r^{m-n} \|\eta_{\delta,x}\gamma; \mathcal{B}_r(y)\|_{W_1^l} + (2\delta)^{m-n} \sup_{r\in(\delta/2,1)} \|\eta_{\delta,x}\gamma; \mathcal{B}_r(y)\|_{W_1^l}\,.$$

The first term on the right-hand side does not exceed

$$c \sup_{r\in(0,\delta/2)} r^{m-n} \|\gamma; \mathcal{B}_{2r}(y)\|_{W_1^l}$$

and the second one is not greater than

$$c\,\delta^{m-n} \|\gamma; \mathcal{B}_{2\delta}(x)\|_{W_1^l}\,.$$

Consequently,

$$A_1 \le c \limsup_{\delta\to 0} \delta^{m-n} \sup_{x\in\mathbb{R}^n} \|\gamma; \mathcal{B}_\delta(x)\|_{W_1^l}\,.$$

□

Remark 7.2.2. It follows from Lemma 7.1.1 that (7.2.39) can be written as

$$\operatorname{ess} \|\gamma\|_{M(W_1^m\to W_1^l)} \sim \limsup_{\delta\to 0} \delta^{m-n} \sup_{x\in\mathbb{R}^n} \big(\delta^{-l}\|\gamma; \mathcal{B}_\delta(x)\|_{L_1} + \|D_{1,l}\gamma; \mathcal{B}_\delta(x)\|_{L_1}\big)$$

$$+ \limsup_{|x|\to\infty} \sup_{r\in(0,1)} r^{m-n} \big(r^{-l}\|\gamma; \mathcal{B}_r(x)\|_{L_1} + \|D_{1,l}\gamma; \mathcal{B}_r(x)\|_{L_1}\big)\,.$$

Remark 7.2.3. Let T_* be the operator defined by (7.2.4) for $pm \le n$, $p > 1$ and $m < n$, $p = 1$. For other values of p and m we put $T_* = (1 - \zeta_r)\gamma$. In the proofs of Theorems 7.2.1–7.2.6 we verified in passing the following estimates for the norm of $\gamma - T_*$ for fixed δ and r.

(*i*) If $mp < n$, $p > 1$, $m > l$, then

$$\|\gamma - T_*\|_{W_p^m \to W_p^l} \le c \sup_{\{e:\, d(e) \le 8\delta\}} (f_l(\gamma; e) + f_0(\gamma; e))$$

$$+ c\Big(\sup_{\{e \subset \mathbb{R}^n \setminus \mathcal{B}_{r/2}:\, d(e) \le 1\}} (f_l(\gamma; e) + f_0(\gamma; e)) + r^{-\{l\}} \|\gamma\|_{M(W_p^m \to W_p^l)}\Big).$$

(*ii*) If $mp = n$, $p > 1$, $m > l$, then the last estimate remains valid with $\{e\colon d(e) \le 8\delta\}$ replaced by $\{e\colon d(e) \le \delta^{1/2}\}$.

(*iii*) If $l < m < n$, then

$$\|\gamma - T_*\|_{W_1^m \to W_1^l} \le c\, \delta^{m-n} \sup_{x \in \mathbb{R}^n} \big(\delta^{-l} \|\gamma; \mathcal{B}_\delta(x)\|_{L_1} + \|D_{1,l}\gamma; \mathcal{B}_\delta(x)\|_{L_1}\big)$$

$$+ c \sup_{x \in \mathbb{R}^n \setminus \mathcal{B}_{r/2}} \sup_{\rho \in (0,1)} \big(\rho^{-l} \|\gamma; \mathcal{B}_\rho(x)\|_{L_1} + \|D_{1,l}\gamma; \mathcal{B}_\rho(x)\|_{L_1}\big).$$

(*iv*) If $mp > n$, $p > 1$ or $m \ge n$, $p = 1$, $m > l$, then

$$\|\gamma - T_*\|_{W_p^m \to W_p^l} \le c \sup_{x \in \mathbb{R}^n \setminus \mathcal{B}_{r/2}} \|\gamma; \mathcal{B}_1(x)\|_{W_p^l}.$$

From (*i*)-(*iv*), together with lower estimates for the essential norm proved in the above theorems, it follows that, given any ε, one can find r so large and δ so small that

$$\|\gamma - T_*\|_{W_p^m \to W_p^l} \le c\,(\operatorname{ess} \|\gamma\|_{M(W_p^m \to W_p^l)} + \varepsilon). \tag{7.2.40}$$

7.2.7 One-Sided Estimates for the Essential Norm

Using Theorem 7.0.3 and the lower estimates of the capacity $C_{p,m}$ (see Propositions 3.1.2 and 3.1.3), one can readily obtain upper estimates for the essential norm in $M(W_p^m \to W_p^l)$.

Theorem 7.2.7. *Let $\gamma \in M(W_p^m(\mathbb{R}^n) \to W_p^m(\mathbb{R}^n))$, where $m > l \ge 0$.*

(i) *If $p \in (1, \infty)$ and $mp < n$, then*

$$\operatorname{ess} \|\gamma\|_{M(W_p^m \to W_p^l)}$$

$$\le c_1 \lim_{\delta \to 0} \Big(\sup_{\{e: d(e) \le \delta\}} \frac{\|D_{p,l}\gamma; e\|_{L_p}}{(\operatorname{mes}_n e)^{\frac{1}{p} - \frac{m}{n}}} + \sup_{\substack{x \in \mathbb{R}^n \\ \rho \le \delta}} \rho^{m-l-\frac{n}{p}} \|\gamma; \mathcal{B}_\rho(x)\|_{L_p}\Big)$$

$$+ c_2 \lim_{r \to \infty} \Big(\sup_{\substack{e \subset \mathbb{R}^n \setminus \mathcal{B}_r \\ d(e) \le 1}} \frac{\|D_{p,l}\gamma; e\|_{L_p}}{(\operatorname{mes}_n e)^{\frac{1}{p} - \frac{m}{n}}} + \sup_{x \in \mathbb{R}^n \setminus \mathcal{B}_r} \|\gamma; \mathcal{B}_1(x)\|_{L_p}\Big). \tag{7.2.41}$$

(ii) *If $p \in (1, \infty)$ and $mp = n$, then*

$$\operatorname{ess} \|\gamma\|_{M(W_p^m \to W_p^l)}$$

$$\le c_1 \lim_{\delta\to 0}\Big(\sup_{\{e:d(e)\le\delta\}} (\log \frac{2^n}{\mathrm{mes}_n e})^{1-\frac{1}{p}} \|D_{p,l}\gamma; e\|_{L_p} + \sup_{\substack{x\in\mathbb{R}^n\\ \rho\le\delta}} \rho^{-l}\|\gamma; \mathcal{B}_\rho(x)\|_{L_p}\Big)$$

$$+c_2 \lim_{r\to\infty}\Big(\sup_{\substack{e\subset\mathbb{R}^n\backslash\mathcal{B}_r\\ d(e)\le 1}} (\log \frac{2^n}{\mathrm{mes}_n e})^{1-\frac{1}{p}} \|D_{p,l}\gamma; e\|_{L_p} + \sup_{x\in\mathbb{R}^n\backslash\mathcal{B}_r} \|\gamma; \mathcal{B}_1(x)\|_{L_p}\Big). \quad (7.2.42)$$

Restricting the suprema in Theorem 7.0.3 to balls of radii less than one and using the formulas for the capacity of a ball (see Proposition 3.1.4), we arrive at the following lower estimates for the essential norm in $M(W_p^m \to W_p^l)$.

Theorem 7.2.8. *Let $\gamma \in M(W_p^m(\mathbb{R}^n) \to W_p^m(\mathbb{R}^n))$, where $m > l \ge 0$.*
(i) If $p \in (1,\infty)$ and $mp < n$, then

$$\mathrm{ess}\,\|\gamma\|_{M(W_p^m\to W_p^l)}$$
$$\ge c_1 \lim_{\delta\to 0} \sup_{\substack{x\in\mathbb{R}^n\\ \rho\le\delta}} \rho^{m-\frac{n}{p}}\Big(\|D_{p,l}\gamma; \mathcal{B}_\rho(x)\|_{L_p} + \rho^{-l}\|\gamma; \mathcal{B}_\rho(x)\|_{L_p}\Big)$$
$$+c_2 \lim_{r\to\infty} \sup_{x\in\mathbb{R}^n\backslash\mathcal{B}_r} \|\gamma; \mathcal{B}_1(x)\|_{W_p^l}. \quad (7.2.43)$$

(ii) If $p \in (1,\infty)$ and $mp = n$, then

$$\mathrm{ess}\,\|\gamma\|_{M(W_p^m\to W_p^l)}$$
$$\ge c_1 \lim_{\delta\to 0} \sup_{\substack{x\in\mathbb{R}^n\\ \rho\le\delta}} \Big((\log(2/\rho))^{1-\frac{1}{p}}\|D_{p,l}\gamma; \mathcal{B}_\rho(x)\|_{L_p} + \rho^{-l}\|\gamma; \mathcal{B}_\rho(x)\|_{L_p}\Big)$$
$$+c_2 \lim_{r\to\infty} \sup_{x\in\mathbb{R}^n\backslash\mathcal{B}_r} \|\gamma; \mathcal{B}_1(x)\|_{W_p^l}. \quad (7.2.44)$$

Remark 7.2.4. Some one-sided estimates for essential norms of multipliers acting in pairs of Besov-Triebel-Lizorkin spaces $B_{p,q}^s(\mathbb{R}^n)$, $F_{p,q}^s(\mathbb{R}^n)$ were obtained by Edmunds and Shargorodsky in [ES]. These estimates involve norms in the same scales of spaces. The proofs are based on a certain abstract functional-analytic equivalent representation of the essential norm.

7.2.8 The Space of Compact Multipliers

Definition 7.2.1. *By $\mathring{M}(W_p^m \to W_p^l)$, $m > l$, we mean the set of functions γ such that the operator of multiplication by γ is a compact operator acting from W_p^m into W_p^l.*

Needless to say,

$$\gamma \in \mathring{M}(W_p^m \to W_p^l) \quad \text{if and only if} \quad \mathrm{ess}\,\|\gamma\|_{M(W_p^m\to W_p^l)} = 0.$$

Therefore, Theorems 7.0.3, 7.2.5, and 7.2.6 imply the following necessary and sufficient conditions for a function $\gamma \in M(W_p^m \to W_p^l)$ to belong to the class $\mathring{M}(W_p^m \to W_p^l)$.

Theorem 7.2.9. *(i) If $mp \le n$ and $p > 1$, then $\gamma \in \overset{\circ}{M}(W_p^m \to W_p^l)$ if and only if*

$$\lim_{\delta\to 0}\Big(\sup_{\{e:d(e)\le\delta\}}\frac{\|D_{p,l}\gamma;e\|_{L_p}}{(C_{p,m}(e))^{\frac{1}{p}}}+\sup_{\substack{x\in\mathbb{R}^n\\ \rho\le\delta}}\rho^{m-l-\frac{n}{p}}\|\gamma;\mathcal{B}_\rho(x)\|_{L_p}\Big)=0, \tag{7.2.45}$$

$$\lim_{r\to\infty}\Big(\sup_{\substack{e\subset\mathbb{R}^n\setminus\mathcal{B}_r\\ d(e)\le 1}}\frac{\|D_{p,l}\gamma;e\|_{L_p}}{(C_{p,m}(e))^{\frac{1}{p}}}+\sup_{x\in\mathbb{R}^n\setminus\mathcal{B}_r}\|\gamma;\mathcal{B}_1(x)\|_{L_p}\Big)=0. \tag{7.2.46}$$

(ii) Let either $mp > n$ and $p \ge 1$, or $m = n$ and $p = 1$. Then $\gamma \in \overset{\circ}{M}(W_p^m \to W_p^l)$ if and only if $\gamma \in W^l_{p,\mathrm{unif}}$ and

$$\lim_{|x|\to\infty}\|\gamma;\mathcal{B}_1(x)\|_{W_p^l}=0\,. \tag{7.2.47}$$

(iii) In the case $m < n$, a necessary and sufficient condition for $\gamma \in \overset{\circ}{M}(W_1^m \to W_1^l)$ is (7.2.47) together with

$$\lim_{\delta\to 0}\delta^{m-n}\sup_{x\in\mathbb{R}^n}\|\gamma;\mathcal{B}_\delta(x)\|_{W_1^l}=0\,. \tag{7.2.48}$$

Remark 7.2.5. From Theorems 7.2.2 and 7.2.3, we obtain another form of the compactness criteria for $\gamma \in \overset{\circ}{M}(W_p^m \to W_p^l)$ with $mp \le n$, $p > 1$:

$$\lim_{\delta\to 0}\sup_{\{e:\, d(e)\le\delta\}}\Big(\frac{\|\gamma;e\|_{L_p}}{[C_{p,m-l}(e)]^{1/p}}+\frac{\|D_{p,l}\gamma;e\|_{L_p}}{[C_{p,m}(e)]^{1/p}}\Big)=0\,, \tag{7.2.49}$$

$$\lim_{r\to\infty}\sup_{\{e\subset\mathbb{R}^n\setminus\mathcal{B}_r:\, d(e)\le 1\}}\Big(\frac{\|\gamma;e\|_{L_p}}{[C_{p,m-l}(e)]^{1/p}}+\frac{\|D_{p,l}\gamma;e\|_{L_p}}{[C_{p,m}(e)]^{1/p}}\Big)=0. \tag{7.2.50}$$

Theorem 7.2.1 immediately implies:

Theorem 7.2.10. *Let $lp < n$, $m > l$ and $p \ge 1$. Then $\gamma \in \overset{\circ}{M}(W_p^m \to W_p^l)$ if and only if*

$$\lim_{\delta\to 0}\sup_{x\in\mathbb{R}^n}\|\eta_{\delta,x}\gamma\|_{M(W_p^m\to W_p^l)}=0\,,$$

$$\lim_{r\to\infty}\|\zeta_r\gamma\|_{M(W_p^m\to W_p^l)}=0\,.$$

From this theorem combined with Propositions 4.3.1, 4.3.2 and the results in Sect. 4.4, we can get various necessary or sufficient conditions for $\gamma \in \overset{\circ}{M}(W_p^m \to W_p^l)$ which do not contain capacity.

The following theorem gives one more description of $\overset{\circ}{M}(W_p^m \to W_p^l)$ with $m > l$.

Theorem 7.2.11. *The space $\overset{\circ}{M}(W_p^m \to W_p^l)$ is the completion of C_0^∞ with respect to the norm in $M(W_p^m \to W_p^l)$.*

Proof. By Theorem 7.2.9, $C_0^\infty \subset \overset{\circ}{M}(W_p^m \to W_p^l)$. Therefore, any function in $M(W_p^m \to W_p^l)$, approximated by a sequence in C_0^∞ in the norm of $M(W_p^m \to W_p^l)$, generates a compact operator of multiplication: $W_p^m \to W_p^l$.

Further, we prove the converse assertion. Let $\gamma \in \overset{\circ}{M}(W_p^m \to W_p^l)$. According to parts (*ii*) and (*iii*) of Theorem 7.2.9, it suffices to consider the case $mp \geq n$, $p > 1$. By Theorem 7.2.1 we have

$$\lim_{r\to\infty} \|\gamma - (1-\zeta_r)\gamma\|_{M(W_p^m\to W_p^l)} = 0\,. \tag{7.2.51}$$

Let $\varGamma = (1-\zeta_r)\gamma$ and let $\varGamma_\rho$ be a mollification of $\varGamma$ with radius ρ. By T_* and $T_*^{(\rho)}$ we denote the operators given by (7.2.4) for $\varGamma$ and $\varGamma_\rho$ respectively. It follows from (7.2.8) that

$$\lim_{r\to 0} \|\gamma - T_*\|_{W_p^m\to W_p^l} = 0\,. \tag{7.2.52}$$

By (7.2.8) and Theorem 7.2.1,

$$\begin{aligned}\|\varGamma_\rho - T_*^{(\rho)}\|_{W_p^m\to W_p^l} &\leq c \limsup_{\delta\to 0} \sup_{x\in\mathbb{R}^n} \|\eta_{\delta,x}\varGamma_\rho\|_{M(W_p^m\to M_p^l)} + \varepsilon\\ &\leq c\,\mathrm{ess}\,\|\varGamma_\rho\|_{M(W_p^m\to W_p^l)} + \varepsilon = \varepsilon\,.\end{aligned} \tag{7.2.53}$$

The last equality holds since $\varGamma_\rho \in C_0^\infty$.

From the definitions of the operators T_* and $T_*^{(\rho)}$ we get

$$\|(T_* - T_*^{(\rho)})u\|_{W_p^l} \leq c(\delta, r)\|u\|_{L_p} \sum_j \|(\varGamma - \varGamma_\rho)\varphi^{(j)}\|_{W_p^l}$$

and hence

$$\|T_* - T_*^{(\rho)}\|_{W_p^m\to W_p^l} \leq c(\delta, r)\|\varGamma - \varGamma_\rho\|_{W_p^l}\,. \tag{7.2.54}$$

The right-hand side of this inequality tends to zero as $\rho \to 0$. Since

$$\begin{aligned}\|\varGamma - \varGamma_\rho\|_{M(W_p^m\to W_p^l)} \leq \|\varGamma - T_*\|_{W_p^m\to W_p^l} &+ \|\varGamma_\rho - T_*^{(\rho)}\|_{W_p^m\to W_p^l}\\ &+ \|T_* - T_*^{(\rho)}\|_{W_p^m\to W_p^l}\,,\end{aligned}$$

it follows by (7.2.52)–(7.2.54) that

$$\lim_{r\to\infty,\rho\to 0} \|\gamma - \varGamma_\rho\|_{M(W_p^m\to W_p^l)} = 0\,.$$

Since $\varGamma_\rho \in C_0^\infty$, the proof is complete. □

7.3 Two-Sided Estimates for the Essential Norm in the Case $m = l$

7.3.1 Estimate for the Maximum Modulus of a Multiplier in W_p^l by its Essential Norm

Theorem 7.3.1. *If $l > 0$ and $1 \le p < \infty$, then*

$$\|\gamma\|_{L_\infty} \le \text{ess}\,\|\gamma\|_{MW_p^l}. \tag{7.3.1}$$

Proof. Let T be a compact operator in W_p^l such that

$$\|(\gamma - T)u\|_{W_p^l} \le (\,\text{ess}\,\|\gamma\|_{MW_p^l} + \varepsilon)\|u\|_{W_p^l} \tag{7.3.2}$$

for all $u \in W_p^l$.

Let η be an arbitrary function in $C_0^\infty(Q_k)$, where Q_k is the cube $\{y\colon |y_j| < \pi k\}$ and k is an integer. We consider the sequence

$$u_N(y) = N^{-l}\exp(iNy_1)\eta(y)\,, \qquad N = 1, 2, \dots .$$

Obviously, for integer l we have

$$\|u_N\|_{W_p^l} = \|\eta\|_{L_p} + O(N^{-1}).$$

Let l be a noninteger, $0 < l < 1$. Then

$$\|u_N\|_{W_p^l} = N^{-l}\|e^{iNy_1}\eta\|_{W_p^l} = N^{-l}\|D_{p,l}e^{iNy_1}\eta\|_{L_p} + O(N^{-1})\,.$$

Clearly,

$$|D_{p,l}(e^{iNy_1}\eta) - |\eta|D_{p,l}e^{iNy_1}| \le D_{p,l}\eta.$$

Since

$$D_{p,l}e^{iNy_1} = a_l N^l,$$

where $a_l = \text{const} > 0$, it follows that

$$\|D_{p,l}(e^{iNy_1}\eta) - a_l N^l|\eta|\,\|_{L_p} = O(1)\,. \tag{7.3.3}$$

Let $l > 1$. Then

$$D_{p,l}(e^{iNy_1}\eta) = D_{p,\{l\}}(\nabla_{[l]}e^{iNy_1}).$$

We have

$$\left|D_{p,\{l\}}(\nabla_{[l]}(e^{iNy_1}\eta)) - N^{[l]}D_{p,\{l\}}(e^{iNy_1}\eta)\right| \le c\sum_{j=0}^{[l]} N^j D_{p,\{l\}}\left(e^{iNy_1}\frac{\partial^{[l]-j}\eta}{\partial y_1^{[l]-j}}\right).$$

This and (7.3.3) imply that

$$\|D_{p,l}(e^{iNy_1}\eta) - N^l a_{\{l\}}\|_{L_p} \le c\, N^{[l]}.$$

So in the case $\{l\} > 0$ we obtain

$$\|u_N\|_{W_p^l} = a_{\{l\}}\|\eta\|_{L_p} + O(N^{-\{l\}})\,. \tag{7.3.4}$$

We show that $\{u_N\}$ converges weakly to zero in $\mathring{W}_p^l(Q_k)$. Let f be any linear functional on $\mathring{W}_p^l(Q_k)$. If $p \le 2$, then the restriction of f to $\mathring{W}_2^l(Q_k)$ is a linear functional on $\mathring{W}_2^l(Q_k)$. Consequently,

$$f(u_N) = \int \Lambda^l u_N \Lambda^l \psi\, dx,$$

where $\psi \in \mathring{W}_2^l(Q_k)$. Since

$$\|\Lambda^l u_n - e^{iNy_1}\eta\|_{L_2} = O(N^{-1})$$

and the sequence

$$\int_{Q_k} e^{iNy_1}\eta(y)\Lambda^l\psi\, dy$$

tends to zero, being a sequence of Fourier coefficients of a function in $L_2(Q_k)$, it follows that $f(u_N) \longrightarrow 0$ as $N \to \infty$.

Let $p > 2$. Taking into account the imbedding of H_p^l into W_p^l, we get

$$|f(u_N)| \le c\,\|u_N\|_{H_p^l}.$$

Therefore,

$$f(u_N) = \int g\Lambda^l u_N\, dy\,,$$

where $g \in L_{p'}$. Since

$$\|\Lambda^l u_N - e^{iNy_1}\eta\|_{L_p} = O(N^{-1})\,,$$

we have

$$f(u_N) = \int_{Q_k} e^{iNy_1}\eta(y) g\, dy + O(N^{-1})\|g\|_{L_{p'}}\,.$$

Applying the Hausdorff-Young theorem (see [Zy], Ch. V.II) to the function $\eta g \in L_{p'}(Q_k)$, $p' < 2$, we conclude that $f(u_N) \longrightarrow 0$ as $N \to \infty$.

By φ we denote a function in $C_0^\infty(Q_1)$ which is equal to one on the cube $Q_{1-\delta}$, $\delta > 0$, and we set $\varphi_k(y) = \varphi(y/k)$. The compactness of the operator $\varphi_k T$ in $\mathring{W}_p^l(Q_k)$ implies that

$$\varphi_k T u_N \underset{N\to\infty}{\longrightarrow} 0 \quad \text{in } \mathring{W}_p^l(Q_k).$$

Now it follows from Lemma 7.1.9 and (7.3.2) that

$$\begin{aligned}\limsup_{N\to\infty} \|\varphi_k \gamma u_N\|_{W_p^l} &= \limsup_{N\to\infty} \|\varphi_k(\gamma - T)u_N\|_{W_p^l} \\ &\le (1+O(k^{-\delta}))\limsup_{N\to\infty} \|(\gamma - T)u_N\|_{W_p^l} \\ &\le (1+O(k^{-\delta}))\limsup_{N\to\infty} \|u_N\|_{W_p^l}(\,\mathrm{ess}\,\|\gamma\|_{MW_p^l} + \varepsilon)\end{aligned}$$

which together with (7.3.4) yields

$$\limsup_{N\to\infty} \|\varphi_k \gamma u_N\|_{W_p^l} \le (1+O(k^{-\delta}))a_{\{l\}}\|\eta\|_{L_p}(\,\mathrm{ess}\,\|\gamma\|_{MW_p^l} + \varepsilon)\,.$$

With the same arguments as in the proof of (7.3.4), we obtain

$$\lim_{N\to\infty} \|\varphi_k \gamma u_N\|_{W_p^l} = a_{\{l\}}\|\varphi_k \gamma \eta\|_{L_p}\,.$$

Thus

$$\limsup_{k\to\infty} \|\varphi_k \gamma \eta\|_{L_p} \le \|\eta\|_{L_p}\,\mathrm{ess}\,\|\gamma\|_{MW_p^l}\,.$$

Since $\varphi_k \eta = \eta$ for large values of k, and η is an arbitrary function in C_0^∞, the result follows. □

7.3.2 Estimates for the Essential Norm Involving Cutoff Functions (The Case $lp \le n$, $p > 1$)

Theorem 7.3.2. *For $lp < n$, $p \ge 1$ the relation*

$$\mathrm{ess}\,\|\gamma\|_{MW_p^l} \sim \limsup_{\delta\to 0}\sup_{x\in\mathbb{R}^n} \|\eta_{\delta,x}\gamma\|_{MW_p^l} + \limsup_{r\to\infty} \|\zeta_r \gamma\|_{MW_p^l} \tag{7.3.5}$$

holds.

The proof of this relation can be obtained by duplicating the proof of Theorem 7.2.1, where $m = l$.

Theorem 7.3.3. *If $0 < l \le 1$, $lp = n$, and $p > 1$, then*

$$\mathrm{ess}\,\|\gamma\|_{MW_p^l} \sim \|\gamma\|_{L_\infty} + \limsup_{\delta\to 0}\sup_{x\in\mathbb{R}^n} \|\eta_{\delta,x} D_{p,l}\gamma\|_{M(W_p^l\to L_p)}$$

$$+ \limsup_{r\to\infty} \|\zeta_r \gamma\|_{MW_p^l}\,. \tag{7.3.6}$$

Proof. (i) *The upper bound for the essential norm.* We choose δ and r so that

$$\sup_{x\in\mathbb{R}^n} \|\eta_{\delta,x} D_{p,l}\gamma\|_{M(W_p^l\to L_p)} \le \limsup_{\delta\to 0}\sup_{x\in\mathbb{R}^n} \|\eta_{\delta,x} D_{p,l}\gamma\|_{M(W_p^l\to L_p)} + \varepsilon\,,$$

$$\|\zeta_r \gamma\|_{MW_p^l} \le \limsup_{r\to\infty} \|\zeta_r \gamma\|_{MW_p^l} + \varepsilon\,.$$

Let Γ and T_* be the function and operator introduced in the second part of Theorem 7.2.1. By (7.2.5) it suffices to get the estimate for $\|(\Gamma - T_*)u\|_{W_p^l}$. We have

$$\|(\Gamma - T_*)u\|^p_{W_p^l} \le A + B + C\,, \tag{7.3.7}$$

where

$$A = \|\Gamma \sum_j \varphi^{(j)}(u - P^{(j)})\|^p_{L_p}\,,$$
$$B = \|\sum_j (D_{p,l}\Gamma)\eta_{2\delta,x_j}\varphi^{(j)}(u - P^{(j)})\|^p_{L_p}\,,$$
$$C = \|\Gamma D_{p,l} \sum_j \varphi^{(j)}(u - P^{(j)})\|^p_{L_p}\,.$$

By Lemma 7.1.2,

$$A \le \|\gamma\|^p_{L_\infty} \sum_j \|u - P^{(j)}\,; K_\delta^{(j)}\|^p_{L_p} \le c\,\|\gamma\|^p_{L_\infty} \|D_{p,l}u\|^p_{L_p}\,. \tag{7.3.8}$$

It follows from Lemmas 7.1.2 and 7.1.3 that

$$\begin{aligned} B &\le c \sum_j \|(D_{p,l}\Gamma)\eta_{2\delta,x_j}\varphi^{(j)}(u - P^{(j)})\|^p_{L_p} \\ &\le c \sup_j \|\eta_{2\delta,x_j} D_{p,l}\Gamma\|^p_{M(W_p^l \to L_p)} \sum_j \|\varphi^{(j)}(u - P^{(j)})\|^p_{L_p} \\ &\le c_1 \sup_j \|\eta_{2\delta,x_j} D_{p,l}\Gamma\|^p_{M(W_p^l \to L_p)} \|D_{p,l}u\|^p_{L_p}\,. \end{aligned} \tag{7.3.9}$$

Using Lemmas 7.1.1–7.1.3, we deduce that

$$\begin{aligned} C &\le c\,\|\gamma\|^p_{L_\infty} \sum_j \|\varphi^{(j)}(u - P^{(j)})\|^p_{W_p^l} \\ &\le c_1 \|\gamma\|^p_{L_\infty} \sum_j |\!|\!| u - P^{(j)}; K_\delta^{(j)} |\!|\!|^p_{W_p^l} \le c_2 \|\gamma\|^p_{L_\infty} \|u\|^p_{W_p^l} \end{aligned}$$

which together with (7.3.7)–(7.3.9) implies that

$$\|(\Gamma - T_*)u\|_{W_p^l} \le c\left(\|\gamma\|_{L_\infty} + \sup_j \|\eta_{2\delta,x_j} D_{p,l}\Gamma\|_{M(W_p^l \to L_p)}\right)\|u\|_{W_p^l}\,.$$

Lemma 7.1.9 enables one to replace Γ by γ on the right-hand side of the last inequality. The required upper estimate for the essential norm is obtained.

(ii) The lower bound for the essential norm. Let T be a compact operator in W_p^l such that

$$\|D_{p,l}(\gamma u) - D_{p,l}(Tu)\|_{L_p} \le (\,\text{ess}\,\|\gamma\|_{MW_p^l} + \varepsilon)\|u\|_{W_p^l}\,.$$

Hence

$$\|uD_{p,l}\gamma - D_{p,l}Tu\|_{L_p} \le (\,\text{ess}\,\|\gamma\|_{MW_p^l} + \varepsilon)\|u\|_{W_p^l} + \|\gamma D_{p,l}u\|_{L_p}\,.$$

It follows from the inequality

$$\|D_{p,l}v_1 - D_{p,l}v_2\|_{L_p} \le \|D_{p,l}(v_1 - v_2)\|_{L_p}$$

that the compactness of the set $\{Tu\colon u \in S\}$ in the space W_p^l, where S is the unit ball in W_p^l, implies the compactness of the set $\{D_{p,l}Tu\colon u \in S\}$ in L_p. Let the collection $\{w_k\}$ form a finite ε-net in the last set. Then, for $u \in S$,

$$\|u\eta_{\delta,x}D_{p,l}\gamma\|_{L_p} \le \|\eta_{\delta,x}(D_{p,l}Tu - w_k)\|_{L_p} + \|\eta_{\delta,x}w_k\|_{L_p}$$
$$+\,\text{ess}\,\|\gamma\|_{MW_p^l} + \varepsilon + \|\gamma\|_{L_\infty}\,.$$

Consequently, for any $x \in \mathbb{R}^n$ and for sufficiently small $\delta > 0$,

$$\|u\eta_{\delta,x}D_{p,l}\gamma\|_{L_p} \le c\,(\,\text{ess}\,\|\gamma\|_{MW_p^l} + \varepsilon)\,.$$

(Here we have used Theorem 7.2.1.) This, together with Lemma 7.2.1 and Theorem 7.3.1, implies the required lower bound for the norm $\text{ess}\,\|\gamma\|_{MW_p^l}$. □

Theorem 7.3.4. *If $l \ge 1$, $lp = n$, and $p > 1$, then*

$$\text{ess}\,\|\gamma\|_{MW_p^l} \sim \|\gamma\|_{L_\infty} + \limsup_{\delta\to 0}\,\sup_{x\in\mathbb{R}^n}\|\eta_{\delta,x}\nabla_k\gamma\|_{M(W_p^l\to W_p^{l-k})}$$
$$+\limsup_{r\to\infty}\|\zeta_r\gamma\|_{MW_p^l},$$

where $k = 1, \dots, [l]$.

Proof. (i) Upper bound for the essential norm. We have

$$\|(\varGamma - T_*)u\|^p_{W_p^l}$$
$$\le c\,\Big(\Big\|\nabla_k\Big(\varGamma\sum_j\varphi^{(j)}(u - P^{(j)})\Big)\Big\|^p_{W_p^{l-k}} + \|(\varGamma - T_*)u\|^p_{L_p}\Big). \quad (7.3.10)$$

The second term on the right-hand side does not exceed

$$c\,\|\gamma\|^p_{L_\infty}\,\|u\|^p_{W_p^l}$$

(see estimate (7.3.8)). The first term is not greater than

$$c\sum_{|\alpha|+|\beta|=k}\Big\|D^\alpha\varGamma\sum_j D^\beta[\varphi^{(j)}(u - P^{(j)})]\Big\|^p_{W_p^{l-k}}$$

$$\leq c \sum_{|\alpha|=0}^{k} \sup_j \|\eta_{2\delta,x_j} D^\alpha \Gamma\|^p_{M(W_p^{l-k+|\alpha|} \to W_p^{l-k})} \sum_j \|\varphi^{(j)}(u - P^{(j)})\|^p_{W_p^l}. \quad (7.3.11)$$

Since $p(l-k) < n$, we conclude, using Theorems 7.2.1 and 7.3.2, that the expression on the right-hand side of (7.3.11) is dominated by

$$c\Big(\sup_j \|\eta_{2\delta,x_j} \nabla_k \Gamma\|_{M(W_p^l \to W_p^{l-k})} + \sum_{j=0}^{k-1} \operatorname{ess} \|\nabla_j \Gamma\|_{(W_p^{l-k+j} \to W_p^{l-k})}\Big)^p \|u\|^p_{W_p^l},$$

which by Lemma 7.2.2 does not exceed

$$c\Big(\sup_j \|\eta_{2\delta,x_j} \nabla_k \Gamma\|_{M(W_p^l \to W_p^{l-k})} + \sum_{i=1}^{k} \operatorname{ess} \|\Gamma\|_{MW_p^{l-i}}\Big)^p \|u\|^p_{W_p^l}.$$

We obtain by interpolation that

$$\operatorname{ess} \|\Gamma\|_{MW_p^{l-i}} \leq \|\Gamma - T_*\|_{W_p^{l-i} \to W_p^{l-i}} \leq c\, \|\Gamma - T_*\|^{(l-i)/l}_{W_p^l \to W_p^l} \|\Gamma - T_*\|^{i/l}_{L_p \to L_p}$$

$$\leq \varepsilon \|\Gamma - T_*\|_{W_p^l \to W_p^l} + c(\varepsilon) \|\gamma\|_{L_\infty}, \quad (7.3.12)$$

where ε is an arbitrarily small positive number. Therefore, the right-hand side of (7.3.10) does not exceed

$$c\Big(\sup_j \|\eta_{2\delta,x_j} \nabla_k \Gamma\|_{M(W_p^l \to W_p^{l-k})} + \varepsilon \|\Gamma - T_*\|_{W_p^l \to W_p^l} + c(\varepsilon)\|\gamma\|_{L_\infty}\Big)^p \|u\|^p_{W_p^l}.$$

Choosing ε sufficiently small and applying Lemma 7.1.9, we obtain from the last inequality and from (7.3.10) that

$$\|(\Gamma - T_*)\|_{W_p^l \to W_p^l} \leq c\big(\sup_j \|\eta_{2\delta,x_j} \nabla_k \Gamma\|_{M(W_p^l \to W_p^{l-k})} + \|\gamma\|_{L_\infty}\big) \quad (7.3.13)$$

which, together with (7.2.5) and Lemma 7.1.9, gives the required upper bound for the essential norm.

(ii) The lower bound for the essential norm. By Lemma 7.2.1 and Theorem 7.3.1, it suffices to show that

$$\|\eta_{\delta,x} \nabla_k \gamma\|_{M(W_p^l \to W_p^{l-k})} \leq c\,(\operatorname{ess} \|\gamma\|_{MW_p^l} + \varepsilon) \quad (7.3.14)$$

for all $x \in \mathbb{R}^n$ and small enough $\delta > 0$.

Let T be a compact operator for which (7.3.2) holds. Then, for all $u \in W_p^l$,

$$\|\nabla_k[(\gamma - T)u]\|_{W_p^{l-k}} \leq (\operatorname{ess} \|\gamma\|_{MW_p^l} + \varepsilon)\|u\|_{W_p^l}.$$

In view of the inequality $p(l-k) < n$,

$$\|\eta_{\delta,x} \nabla_k[(\gamma - T)u]\|_{W_p^{l-k}} \leq c\,(\operatorname{ess} \|\gamma\|_{MW_p^l} + \varepsilon)\|u\|_{W_p^l}.$$

Let S be the unit ball in W^l_p. The set

$$\{v = D^\alpha Tu, |\alpha| = k \colon u \in S\}$$

is compact in W^{l-k}_p. Let $\{v_\nu\}$ be an ε-net in W^{l-k}_p for the last set. Without loss of generality we may assume that $v_\nu \in C^\infty_0$.

Since $p(l-k) < n$, we see by Lemma 7.1.7 that for small δ

$$\sup_{x\in\mathbb{R}^n} \|\eta_{\delta,x} v_\nu\|_{W^{l-k}_p} < \varepsilon$$

and hence

$$\sup_{x\in\mathbb{R}^n} \|\eta_{\delta,x} \nabla_k (Tu)\|_{W^{l-k}_p} < c\varepsilon\,.$$

Thus, for all $u \in S$,

$$\|u\eta_{\delta,x}\nabla_k\gamma\|_{W^{l-k}_p} \le c\big(\operatorname{ess}\|\gamma\|_{MW^l_p} + \sum_{\substack{|\alpha|+|\beta|=k,\\ |\alpha|>0}} \|\eta_{\delta,x} D^\alpha u D^\beta\gamma\|_{W^{l-k}_p} + \varepsilon\big)$$

$$\le c\big(\operatorname{ess}\|\gamma\|_{MW^l_p} + \sum_{\substack{|\alpha|+|\beta|=k\\ |\alpha|>0}} \|\eta_{\delta,x} D^\beta\gamma\|_{M(W^{l-|\alpha|}_p\to W^{l-k}_p)}\|D^\alpha u\|_{W^{l-|\alpha|}_p} + \varepsilon\big). \quad (7.3.15)$$

From Theorems 7.2.1 and 7.3.2, it follows that for small δ

$$\|\eta_{\delta,x} D^\beta\gamma\|_{M(W^{l-|\alpha|}_p\to W^{l-k}_p)} \le c\operatorname{ess}\|D^\beta\gamma\|_{M(W^{l-|\alpha|}_p\to W^{l-k}_p)}\,.$$

Making use of Theorem 7.3.2 and interpolating (see (7.3.12)), we obtain

$$\begin{aligned}\operatorname{ess}\|D^\beta\gamma\|_{M(W^{l-|\alpha|}_p\to W^{l-k}_p)} &\le c\sum_{j=0}^{|\beta|} \operatorname{ess}\|\gamma\|_{MW^{l-|\alpha|-j}_p}\\ &\le \varepsilon\|\gamma - T_*\|_{W^l_p\to W^l_p} + c(\varepsilon)\|\gamma\|_{L_\infty}\,.\end{aligned}$$

From this inequality, combined with (7.3.15) and the estimate

$$\|\gamma - T_*\|_{W^l_p\to W^l_p} \le c\big(\sup_j \|\eta_{2\delta,x_j}\nabla_k\gamma\|_{M(W^l_p\to W^{l-k}_p)} + \|\gamma\|_{L_\infty} + \|\zeta_r\gamma\|_{MW^l_p}\big)$$

established in the first part of the proof (see (7.3.13)), we get for $u \in S$

$$\|u\eta_{\delta,x}\nabla_k\gamma\|_{W^{l-k}_p} \le c(\varepsilon)\operatorname{ess}\|\gamma\|_{MW^l_p} + \varepsilon\sup_j \|\eta_{2\delta,x_j}\nabla_k\gamma\|_{M(W^l_p\to W^{l-k}_p)} + \varepsilon\,.$$

Inequality (7.3.14) follows, which completes the proof of the theorem. □

7.3.3 Estimates for the Essential Norm Involving Capacity (The Case $lp \le n$, $p > 1$)

For the theorem in this subsection we need the following interpolation inequality.

Lemma 7.3.1. *If $lp \le n$, $p > 1$, and $0 < \sigma < l$, then*

$$\operatorname{ess} \|\gamma\|_{MW_p^\sigma} \le c \operatorname{ess} \|\gamma\|_{MW_p^l}^{\sigma/l} \|\gamma\|_{L_\infty}^{1-\sigma/l} . \tag{7.3.16}$$

Proof. When proving any of Theorems 7.3.2–7.3.4, it was shown in passing that for some r and δ we obtain from the definition of T_* that

$$\|\gamma - T_*\|_{W_p^l \to W_p^l} \le c \operatorname{ess} \|\gamma\|_{MW_p^l} + \varepsilon \tag{7.3.17}$$

(cf. Remark 7.2.2). Moreover,

$$\|\gamma - T_*\|_{L_p \to L_p} \le c \|\gamma\|_{L_\infty} .$$

Hence, interpolating between W_p^l and L_p, we get

$$\operatorname{ess} \|\gamma\|_{MW_p^\sigma} \le \|\gamma - T_*\|_{W_p^\sigma \to W_p^\sigma} \le c (\operatorname{ess} \|\gamma\|_{MW_p^l} + \varepsilon)^{\sigma/l} \|\gamma\|_{L_\infty}^{1-\sigma/l} .$$

□

Theorem 7.3.5. *Let $lp \le n$ and $p > 1$. Then*

$$\operatorname{ess} \|\gamma\|_{MW_p^l} \sim \|\gamma\|_{L_\infty} + \lim_{\delta \to 0} \sup_{\{e:\, d(e) \le \delta\}} \frac{\|D_{p,l}\gamma; e\|_{L_p}}{[C_{p,l}(e)]^{1/p}}$$
$$+ \lim_{r \to \infty} \sup_{\{e \subset \mathbb{R}^n \setminus \mathcal{B}_r:\, d(e) \le 1\}} \frac{\|D_{p,l}\gamma; e\|_{L_p}}{[C_{p,l}(e)]^{1/p}} . \tag{7.3.18}$$

Proof. For $lp < n$ it suffices to duplicate the proof of Theorem 7.2.2, putting $m = l$.

Inequalities (7.2.18) and (7.2.25), which are also valid for $m = l \le n/p$, imply that

$$\limsup_{r \to \infty} \|\zeta_r \gamma\|_{MW_p^l}$$
$$\sim \lim_{r \to \infty} \Big(\|\gamma; \mathbb{R}^n \setminus \mathcal{B}_r\|_{L_\infty} + \sup_{\{e \subset \mathbb{R}^n \setminus \mathcal{B}_r:\, d(e) \le 1\}} \frac{\|D_{p,l}\gamma; e\|_{L_p}}{[C_{p,l}(e)]^{1/p}} \Big) . \tag{7.3.19}$$

Therefore, from Theorem 7.3.3 and Remark 7.2.2, we have (7.3.18) for $0 < l \le 1$ and $lp = n$.

Let $lp = n$ and $l > 1$. It is shown in the proof of Theorem 7.3.3 that

$$\|\eta_{\delta^2,x}\nabla\gamma\|_{M(W_p^l\to W_p^{l-1})} \le c\Big(\sup_{\{e:\, d(e)\le 2\delta\}} \frac{\|D_{p,l-1}(\nabla\gamma);e\|_{L_p}}{[C_{p,l}(e)]^{1/p}} + \frac{\|\nabla\gamma;e\|_{L_p}}{[C_{p,1}(e)]^{1/p}}\Big),$$

$$\|\zeta_r\gamma\|_{MW_p^l} \le c\Big(\sup_{\{e\subset\mathbb{R}^n\setminus\mathcal{B}_{r/2}:\, d(e)\le 1\}} \Big(\frac{\|D_{p,l-1}(\nabla\gamma);e\|_{L_p}}{[C_{p,l}(e)]^{1/p}} + \frac{\|\nabla\gamma;e\|_{L_p}}{[C_{p,1}(e)]^{1/p}}\Big) + \|\gamma\|_{L_\infty}\Big).$$

Hence, using the estimate for $\operatorname{ess}\|\gamma\|_{MW_p^l}$ and Theorem 7.3.4, we get

$$\operatorname{ess}\|\gamma\|_{MW_p^l} \le c\Big(\|\gamma\|_{L_\infty} + \lim_{\delta\to 0}\sup_{\{e:\, d(e)\le\delta\}} \frac{\|D_{p,l}\gamma;e\|_{L_p}}{[C_{p,l}(e)]^{1/p}}$$

$$+ \lim_{r\to\infty}\sup_{\{e\subset\mathbb{R}^n\setminus\mathcal{B}_r:\, d(e)\le 1\}} \frac{\|D_{p,l}\gamma;e\|_{L_p}}{[C_{p,l}(e)]^{1/p}} + \operatorname{ess}\|\gamma\|_{MW_p^1}\Big).$$

It remains to note that, by Lemma 7.3.1,

$$\operatorname{ess}\|\gamma\|_{MW_p^1} \le c\operatorname{ess}\|\gamma\|_{MW_p^l}^{1/l}\|\gamma\|_{L_\infty}^{(l-1)/l}.$$

Let us derive the lower bound for the essential norm. By Lemma 7.3.1 and Theorem 7.3.1,

$$\operatorname{ess}\|\gamma\|_{MW_p^l} \ge c\operatorname{ess}\|\gamma\|_{MW_p^{l-1}}$$

which, together with Lemma 7.2.2, gives the estimate

$$\operatorname{ess}\|\gamma\|_{MW_p^l} \ge c\operatorname{ess}\|\nabla\gamma\|_{M(W_p^l\to W_p^{l-1})}.$$

Taking into account Theorem 7.2.3, we obtain that the right-hand side of this inequality is not less than

$$c\lim_{\delta\to 0}\sup_{\{e:\, d(e)\le\delta\}} \frac{\|D_{p,l}\gamma;e\|_{L_p}}{[C_{p,l}(e)]^{1/p}}.$$

It remains to use (7.3.19) and Theorem 7.2.2. The result follows. □

7.3.4 Two-Sided Estimates for the Essential Norm in the Cases $lp > n$, $p > 1$, and $p = 1$

Theorem 7.3.6. *If $lp > n$ and $p > 1$, then*

$$\operatorname{ess}\|\gamma\|_{MW_p^l} \sim \|\gamma\|_{L_\infty} + \limsup_{|x|\to\infty}\|\gamma,\mathcal{B}_1(x)\|_{W_p^l}. \qquad (7.3.20)$$

Proof. From Corollary 4.3.8 and Lemma 7.1.9 we obtain

$$\limsup_{|x|\to\infty} \|\gamma; \mathcal{B}_1(x)\|_{W_p^l} \sim \limsup_{r\to\infty} \|\zeta_r \gamma\|_{MW_p^l} \,. \tag{7.3.21}$$

Hence the required lower bound for the essential norm follows by Lemma 7.2.1 and Theorem 7.3.1.

Now we establish the upper bound. Let Γ and T_* be the function and operator specified in the second part of Theorem 4.2.1. With the function Γ_ρ, which stands for a mollification of Γ with radius ρ, we associate the operator $T_*^{(\rho)}$ by the same rule. Using Corollary 4.3.8 for sufficiently small ρ, we find that

$$\|\Gamma - \Gamma_\rho\|_{MW_p^l} \le c\,\|\Gamma - \Gamma_\rho\|_{W_p^l} < \varepsilon \,. \tag{7.3.22}$$

Next we note that the proof of inequality (7.3.13) holds in the case $lp > n$. Replacing the numbers l and k by an integer s, $s > np$, in (7.3.13) and using Corollary 4.3.8, we arrive at

$$\|\Gamma_\rho - T_*^{(\rho)}\|_{W_p^s \to W_p^s} \le \Big(\sup_j \|\eta_{2\delta,x_j} \nabla_s \Gamma_\rho\|_{L_p} + \|\gamma\|_{L_\infty}\Big) \,.$$

This implies for small δ that

$$\|\Gamma_\rho - T_*^{(\rho)}\|_{W_p^s \to W_p^s} \le c\,\|\gamma\|_{L_\infty} \,. \tag{7.3.23}$$

The same inequality obviously holds for $s = 0$. Interpolating between L_p and W_p^s, we obtain (7.3.23) for $s = l$ which, together with (7.3.22), gives

$$\|\Gamma - T_*^{(\rho)}\|_{W_p^l \to W_p^l} \le c\,\|\gamma\|_{L_\infty} + \varepsilon \,.$$

The result follows from the last inequality and (7.2.5).

Further, we note that one may replace the operator $T_*^{(\rho)}$ by T_* in the last inequality. In fact, it follows from Lemmas 7.1.2, 7.1.3 and Corollary 4.3.1 that

$$\|u - \sum_j \varphi^{(j)} P^{(j)}\|_{W_p^l}^p \le c \sum_j \|\varphi^{(j)}(u - P^{(j)})\|_{W_p^l}^p \le c_1 \|u\|_{W_p^l}^p \,.$$

Therefore,

$$\|(T_* - T_*^{(\rho)})u\|_{W_p^l} \le \|\Gamma - \Gamma_\rho\|_{MW_p^l} \,\|\sum_j \varphi^{(j)} P^{(j)}\|_{W_p^l} \le c\,\|\Gamma - \Gamma_\rho\|_{MW_p^l} \,\|u\|_{W_p^l}$$

and it remains to use inequality (7.3.22). □

Theorem 7.3.7. *If $l < n$, then*

$$\operatorname{ess}\|\gamma\|_{MW_1^l} \sim \limsup_{\delta\to 0} \delta^{l-n} \sup_{x\in\mathbb{R}^n} \|\gamma; \mathcal{B}_\delta(x)\|_{W_1^l}$$

$$+ \limsup_{|x|\to\infty} \sup_{r\in(0,1)} r^{l-n} \|\gamma; \mathcal{B}_r(x)\|_{W_1^l}. \qquad (7.3.24)$$

The proof runs in the same way as that of Theorem 7.2.6, where one should put $m = l$ and use Theorem 7.3.2 instead of Theorem 7.2.1.

Remark 7.3.1. By Lemma 7.1.1, the equivalence relation (7.3.24) can be rewritten as

$$\operatorname{ess}\|\gamma\|_{MW_1^l} \sim \|\gamma\|_{L_\infty} + \limsup_{\delta\to 0} \delta^{l-n} \sup_{x\in\mathbb{R}^n} \|D_{1,l}\gamma; \mathcal{B}_\delta(x)\|_{L_1}$$

$$+ \limsup_{|x|\to\infty} \sup_{r\in(0,1)} r^{l-n} \|D_{1,l}\gamma; \mathcal{B}_r(x)\|_{L_1}.$$

Theorem 7.3.8. *If $l \geq n$, then*

$$\operatorname{ess}\|\gamma\|_{MW_1^l} \sim \|\gamma\|_{L_\infty} + \limsup_{|x|\to\infty} \|\gamma; \mathcal{B}_1(x)\|_{W_1^l}. \qquad (7.3.25)$$

Proof. The lower bound for the essential norm follows directly from Lemma 7.2.2 and Theorems 5.4.1 and 7.3.1.

Next we obtain the upper bound. Let $k = [l] + 1 - n$. We have

$$\|(\Gamma - T_*)u\|_{W_1^l} \leq c\Big(\big\|\nabla_k\big(\sum_j \varphi^{(j)}(u - P^{(j)})\big)\big\|_{W_1^{l-k}} + \|(\Gamma - T_*)u\|_{L_1}\Big).$$

The second term on the right-hand side does not exceed $c\,\|\gamma\|_{L_\infty}\|u\|_{W_1^l}$ (see (7.3.8)). The first one is not greater than

$$c \sum_{|\alpha|+|\beta|=k} \big\|D^\alpha\Gamma \sum_j D^\beta[\varphi^{(j)}(u - P^{(j)})]\big\|_{W_1^{l-k}}$$

$$\leq c \sum_{|\alpha|=0}^{k} \sup_j \|\eta_{2\delta,x_j} D^\alpha\Gamma\|_{M(W_1^{l-k+|\alpha|}\to W_1^{l-k})} \sum_j \|\varphi^{(j)}(u - P^{(j)})\|_{W_1^l}.$$

With the help of Lemmas 7.1.1–7.1.3 we obtain that the last norm is majorized by $c\,\|u\|_{W_1^l}$.

Now we show that

$$\limsup_{\delta\to 0} \sup_{x\in\mathbb{R}^n} \|\eta_{\delta,x}\Gamma\|_{MW_1^{l-k}} \leq c\,\|\Gamma\|_{L_\infty}. \qquad (7.3.26)$$

Since $l - k < n$, it follows by Theorems 7.3.2 and 7.3.3 that the left-hand side of (7.3.26) does not exceed

$$c\,\mathrm{ess}\,\|\Gamma\|_{MW_1^{l-k}} \sim \limsup_{\delta\to 0} \delta^{l-k-n} \sup_{x\in\mathbb{R}^n} \|\Gamma; \mathcal{B}_\delta(x)\|_{W_1^{l-k}}$$

$$\leq \limsup_{\delta\to 0} \delta^{l-n} \sup_{x\in\mathbb{R}^n} \|\Gamma; \mathcal{B}_\delta(x)\|_{W_1^l} = \|\Gamma\|_{L_\infty}\,.$$

Thus, (7.3.26) follows.

To complete the proof, it suffices to establish the equality

$$\lim_{\delta\to 0} \sup_{x\in\mathbb{R}^n} \|\eta_{\delta,x}\nabla_j\Gamma\|_{M(W_1^{l-k+j}\to W_1^{l-k})} = 0\,, \tag{7.3.27}$$

where $j = 1,\dots,k$. By Theorem 7.2.1 the left-hand side is equivalent to the essential norm of $\nabla_j\Gamma$ in $M(W_1^{l-k+j} \to W_1^{l-k})$. Since

$$\nabla_j\Gamma \in W_1^{l-j}, \qquad \mathrm{supp}\nabla_j\Gamma \subset \mathcal{B}_{2r},$$

and $l-k+j > n$, we have by Theorem 7.2.9, part (ii), that

$$\nabla_j\Gamma \in \mathring{M}(W_1^{l-k+j} \to W_1^{l-k}).$$

The equality (7.3.27) is proved, and so is the theorem. □

Remark 7.3.2. In addition to Lemma 7.3.1 we note that by Theorems 7.3.7, 7.3.8 and estimate (7.1.7), the following interpolation inequality is valid:

$$\mathrm{ess}\,\|\gamma\|_{MW_1^\sigma} \leq c\,(\mathrm{ess}\,\|\gamma\|_{MW_1^l})^{\sigma/l}\|\gamma\|_{L_\infty}^{1-\sigma/l}\,, \qquad 0<\sigma<l\,.$$

7.3.5 Essential Norm in $\mathring{M}W_p^l$

According to Theorem 7.2.11, the space of compact multipliers $\mathring{M}(W_p^m \to W_p^l)$, $m > l$, coincides with the completion of C_0^∞ with respect to the norm of the space $M(W_p^m \to W_p^l)$. Similarly, $\mathring{M}W_p^l$ denotes the completion of C_0^∞ with respect to the norm of the space MW_p^l. The following theorem shows that the essential norm in $\mathring{M}W_p^l$ is equivalent to the norm in L_∞.

Theorem 7.3.9. *If $\gamma \in \mathring{M}W_p^l$, $l \geq 0$, and $p \geq 1$, then*

$$\|\gamma\|_{L_\infty} \leq \mathrm{ess}\,\|\gamma\|_{MW_p^l} \leq c\,\|\gamma\|_{L_\infty}\,. \tag{7.3.28}$$

Proof. The left-hand estimate was obtained in Theorem 7.3.1. Let us establish the upper bound for the essential norm. Without loss of generality, we may assume that $\gamma \in C_0^\infty$.

Let $p > 1$ and $lp \geq n$. Since

$$C_{p,l}(e) \geq c\,(\mathrm{mes}_n e)^\nu,$$

where $\nu \in (0,1)$ and $d(e) \le \delta$ (see Proposition 3.1.2), we have

$$\frac{\|D_{p,l}\gamma; e\|_{L_p}}{[C_{p,l}(e)]^{1/p}} \le c\,(\mathrm{mes}_n e)^{(1-\nu)/p}$$

and

$$\lim_{\delta\to 0} \sup_{\{e:\, d(e)\le\delta\}} \frac{\|D_{p,l}\gamma; e\|_{L_p}}{[C_{p,l}(e)]^{1/p}} = 0\,.$$

For any compact set $e \subset \mathbb{R}^n\backslash\mathcal{B}_r$ with $d(e) \le 1$,

$$\frac{\|D_{p,l}\gamma; e\|_{L_p}}{[C_{p,l}(e)]^{1/p}} \le c r^{-\{l\}-n/p}(\mathrm{mes}_n e)^{(1-\nu)/p}\,.$$

Therefore,

$$\lim_{r\to\infty} \sup_{\{e\subset\mathbb{R}^n\backslash\mathcal{B}_r:\, d(e)\le 1\}} \frac{\|D_{p,l}\gamma; e\|_{L_p}}{[C_{p,l}(e)]^{1/p}} = 0\,.$$

Now the right-hand inequality in (7.3.28) follows from Theorem 7.3.5.

For $lp > n$, $p > 1$ and for $l \ge n$, $p = 1$, the result is a consequence of the equality

$$\lim_{|x|\to\infty} \|\gamma; \mathcal{B}_1(x)\|_{W_p^l} = 0$$

and Theorems 7.3.6 and 7.3.8.

Finally, for $l < n$ and $p = 1$, the desired estimate for the essential norm follows immediately from Remark 7.3.1, since for $\gamma \in C_0^\infty$

$$\lim_{\delta\to 0} \delta^{l-n} \sup_{x\in\mathbb{R}^n} \|D_{1,l}\gamma; \mathcal{B}_\delta(x)\|_{L_1} = 0$$

and

$$\lim_{|x|\to\infty} \sup_{r\in(0,1)} r^{l-n}\|D_{1,l}\gamma; \mathcal{B}_r(x)\|_{L_1} = 0\,.$$

□

We can describe the space $\mathring{M}W_p^l$ without approximation by functions in C_0^∞. The following assertion, supplementing Theorem 7.2.9, holds.

Theorem 7.3.10. *A function γ belongs to $\mathring{M}W_p^l$ if and only if γ is a continuous function vanishing at infinity and satisfying one of the conditions:*

(i) If $lp \le n$ and $p > 1$, then

$$\sup_{\{e:\, d(e)\le\delta\}} \frac{\|D_{p,l}\gamma; e\|_{L_p}}{[C_{p,l}(e)]^{1/p}} = o(1) \quad \text{as} \quad \delta \to 0 \tag{7.3.29}$$

and

$$\sup_{\{e\subset\mathbb{R}^n\backslash\mathcal{B}_r:\, d(e)\le 1\}} \frac{\|D_{p,l}\gamma; e\|_{L_p}}{[C_{p,l}(e)]^{1/p}} = o(1) \quad \text{as} \quad r \to \infty\,. \tag{7.3.30}$$

(*ii*) *If* $l \le n$, *then*

$$\sup_{x\in\mathbb{R}^n} \delta^{l-n} \|D_{1,l}\gamma; \mathcal{B}_\delta(x)\|_{L_1} = o(1) \quad as \quad \delta \to 0 \tag{7.3.31}$$

and

$$\sup_{r\in(0,1)} r^{l-n} \|D_{1,l}\gamma; \mathcal{B}_r(x)\|_{L_1} = o(1) \quad as \quad |x| \to \infty\,. \tag{7.3.32}$$

(*iii*) *If* $lp > n$ *and* $p > 1$, *or* $l \ge n$ *and* $p = 1$, *then*

$$\|D_{p,l}\gamma; \mathcal{B}_1(x)\|_{L_p} = o(1) \quad as \quad |x| \to \infty\,. \tag{7.3.33}$$

Proof. Necessity. Let $\gamma \in \mathring{M}W_p^l$ and let $\{\gamma_j\}$ be a sequence of functions in C_0^∞ approximating γ in MW_p^l. It follows from the expressions for equivalent norms in MW_p^l derived in Chap. 4 that the left-hand sides of (7.3.29)–(7.3.33), with γ replaced by $\gamma - \gamma_j$, are arbitrarily small for sufficiently large j. On the other hand, it has been shown in the proof of Theorem 7.3.9 that (7.3.29)–(7.3.33) hold for $\gamma_j \in C_0^\infty$. Consequently, they hold for γ as well.

Sufficiency. Let a function $\gamma \in C \cap MW_p^l$ satisfy one of the conditions (7.3.29)–(7.3.33) and let $\gamma(x) \to 0$ as $|x| \to \infty$. For $lp > n$, $p > 1$ and for $p = 1$ the possibility of approximation of γ by mollifications of functions $\zeta_r\gamma$, $r \to \infty$, immediately follows from the expressions for the norm in MW_p^l derived in Chap. 4. Consider the case $lp \le n$, $p > 1$. Then

$$\|\zeta_r\gamma\|_{MW_p^l} \to 0 \quad \text{as } r \to \infty$$

because of (7.3.19). Therefore, it suffices to approximate the multiplier γ with support in $\mathcal{B}_{r/2}$ for a fixed r by functions from C_0^∞.

Let γ_ρ be a mollification of γ with nonnegative kernel K and radius ρ. We introduce the operators

$$T_* = \gamma \sum_j \varphi^{(j)} P^{(j)}\,, \qquad T_*^{(\rho)} = \gamma_\rho \sum_j \varphi^{(j)} P^{(j)}.$$

Here we retain the same notation as in the definition of the operator T_* in the proof of Theorem 7.2.1. Obviously,

$$\|\gamma - \gamma_\rho\|_{MW_p^l} \le \|(\gamma - \gamma_\rho) - (T_* - T_*^{(\rho)})\|_{W_p^l \to W_p^l} + \|T_* - T_*^{(\rho)}\|_{W_p^l \to W_p^l}\,.$$

For $lp < n$ as well as for $lp = n$, $0 < l < 1$, we have

$$\begin{aligned} &\|(\gamma - \gamma_\rho) - (T_* - T_*^{(\rho)})\|_{W_p^l \to W_p^l} \\ &\quad \le c\Big(\sup_{\{e:\, d(e)\le\delta\}} \frac{\|D_{p,l}(\gamma - \gamma_\rho); e\|_{L_p}}{[C_{p,l}(e)]^{1/p}} + \|\gamma - \gamma_\rho\|_{L_\infty}\Big) \end{aligned} \tag{7.3.34}$$

(see the proof of Theorem 7.2.2, where the restriction $l < m$ is insignificant, and the proof of Theorem 7.3.3). The right-hand side of (7.3.34) does not exceed

$$c\Big(\sup_{\{e:\, d(e)\le\delta\}}\Big(\frac{\|D_{p,l}\gamma;e\|_{L_p}}{[C_{p,l}(e)]^{1/p}}+\frac{\|D_{p,l}\gamma_\rho;e\|_{L_p}}{[C_{p,l}(e)]^{1/p}}\Big)+\|\gamma-\gamma_\rho\|_{L_\infty}\Big).$$

Replacing here c by $2c$, we can omit the second term because of the estimate

$$\|D_{p,l}\gamma_\rho;e\|_{L_p}\le\int\rho^{-n}K(\xi/\rho)\|D_{p,l}\gamma;e_\xi\|_{L_p}d\xi,$$

where $e_\xi=\{x\colon x+\xi\in e\}$. Further, we note that

$$\|T_*-T_*^{(\rho)}\|_{W_p^l\to W_p^l}\le c(\delta,r)\|\gamma-\gamma_\rho;\mathcal{B}_r\|_{MW_p^l}.$$

Consequently,

$$\limsup_{\rho\to 0}\|\gamma-\gamma_\rho\|_{MW_p^l}\le 2c\sup_{\{e:\, d(e)\le\delta\}}\frac{\|D_{p,l}\gamma;e\|_{L_p}}{[C_{p,l}(e)]^{1/p}}$$

and it remains to make use of (7.3.29).

For $lp=n$, $l>1$, the proof follows the same lines provided that (7.3.34) is replaced by the estimate

$$\begin{aligned}\|(\gamma-\gamma_\rho)-(T_*-T_*^{(\rho)})\|_{W_p^l\to W_p^l}\le c\Big(&\sup_{\{e:\, d(e)\le\delta\}}\frac{\|D_{p,l}(\gamma-\gamma_\rho);e\|_{L_p}}{[C_{p,l}(e)]^{1/p}}\\&+\sup_{\{e:\, d(e)\le\delta\}}\frac{\|\nabla(\gamma-\gamma_\rho);e\|_{L_p}}{[C_{p,1}(e)]^{1/p}}+\|\gamma-\gamma_\rho\|_{L_\infty}\Big)\end{aligned}$$

(see the proof of Theorem 7.3.5). □

8

Traces and Extensions of Multipliers

8.1 Introduction

Let $\mathbb{R}^n_+$ denote the upper half-space $\{z = (x, y) : x \in \mathbb{R}^{n-1},\ y > 0\}$. We introduce the weighted Sobolev space $W_p^{s,\alpha}(\mathbb{R}^n_+)$ with the norm

$$\|(\min\{1, y\})^\alpha \nabla_s U; \mathbb{R}^n_+\|_{L_p} + \|(\min\{1, y\})^\alpha U; \mathbb{R}^n_+\|_{L_p}, \tag{8.1.1}$$

where s is nonnegative integer. We always assume that $-1 < \alpha p < p - 1$. Obviously, the usual Sobolev space $W_p^s(\mathbb{R}^n_+)$ is included here as $W_p^{s,0}(\mathbb{R}^n_+)$.

It is well known that the fractional Sobolev space $W_p^l(\mathbb{R}^{n-1})$ is the space of traces on $\mathbb{R}^{n-1}$ of functions in $W_p^{s,\alpha}(\mathbb{R}^n_+)$, where $s = [l]+1$, $\alpha = 1-\{l\}-1/p$, and $p \in (1, \infty)$ (see [Usp]). In Sects. 8.2–8.5 we show that a similar trace characterization holds for spaces of multipliers acting in a pair of fractional Sobolev spaces. To be precise, we prove that, for all noninteger m and l with $m \geq l > 0$, the multiplier space $M(W_p^m(\mathbb{R}^{n-1}) \to W_p^l(\mathbb{R}^{n-1}))$ is the space of traces on $\mathbb{R}^{n-1}$ of functions in $M(W_p^{t,\beta}(\mathbb{R}^n_+) \to W_p^{s,\alpha}(\mathbb{R}^n_+))$, where s and α are as above and $\beta = 1 - \{m\} - 1/p$, $t = [m] + 1$. Sect. 8.6 concerns traces of multipliers on the smooth boundary of a domain. The remaining Sects. 8.7–8.9 are devoted to three trace and extension theorems for multipliers preserving a certain Sobolev-type space.

8.2 Multipliers in Pairs of Weighted Sobolev Spaces in $\mathbb{R}^n_+$

We introduce the notion of (p, s, α)-capacity of a compact set $e \subset \mathbb{R}^n_+$:

$$\mathrm{C}_{p,s,\alpha}(e) = \inf\{\|U; \mathbb{R}^n_+\|^p_{W_p^{s,\alpha}} :\ U \in C_0^\infty(\overline{\mathbb{R}^n_+}),\ U \geq 1 \text{ on } e\}.$$

The following result is known (see [Maz15], Sects. 8.1, 8.2).

V.G. Maz'ya, T.O. Shaposhnikova, *Theory of Sobolev Multipliers*,
Grundlehren der mathematischen Wissenschaften 337,

Proposition 8.2.1. *Let k be a nonnegative integer, $-1 < \beta p < p-1$, and let $1 < p < \infty$. Then $\Gamma \in M(W_p^{k,\beta}(\mathbb{R}_+^n) \to W_p^{0,\alpha}(\mathbb{R}_+^n))$ if and only if*

$$\sup_{\substack{e \subset \mathbb{R}_+^n \\ d(e) \le 1}} \frac{\|(\min\{1,y\})^\alpha \Gamma; e\|_{L_p}}{(C_{p,k,\beta}(e))^{1/p}} < \infty,$$

where $d(e)$ is the diameter of e. The equivalence relation

$$\|\Gamma\|_{M(W_p^{k,\beta} \to W_p^{0,\alpha})} \sim \sup_{\substack{e \subset \mathbb{R}_+^n \\ d(e) \le 1}} \frac{\|(\min\{1,y\})^\alpha \Gamma; e\|_{L_p}}{(C_{p,k,\beta}(e))^{1/p}} \tag{8.2.1}$$

holds.

We shall use some general properties of multipliers in weighted Sobolev spaces. We start with the inequality

$$\|\Gamma\|_{M(W_p^{t-j,\beta} \to W_p^{s-j,\alpha})} \le c\, \|\Gamma\|^{(s-j)/s}_{M(W_p^{t,\beta} \to W_p^{s,\alpha})} \|\Gamma\|^{j/s}_{M(W_p^{t-s,\beta} \to W_p^{0,\alpha})}, \tag{8.2.2}$$

where $0 \le j \le s$, $-1 < \alpha p < p-1$, and $-1 < \beta p < p-1$, which follows from the interpolation property of weighted Sobolev spaces (see [Tr4], Sect. 3.4.2).

In this section and in Sects. 8.3, 8.4 we omit $\mathbb{R}_+^{n+1}$ in notations of spaces, norms, and integrals, when it causes no ambiguity. The notations $\mathcal{B}_r^{(d)}(x) = \{z \in \mathbb{R}^d : |z-x| < r\}$ and $\mathcal{B}_r(x) = \mathcal{B}_r^{(n)}(x)$ will be adopted.

The next assertion contains inequalities between multipliers and their mollifiers in x.

Lemma 8.2.1. *Let Γ_ρ denote a mollifier of a function Γ defined by*

$$\Gamma_\rho(x,y) = \rho^{-n+1} \int_{\mathbb{R}^{n-1}} K(\rho^{-1}(x-\xi))\Gamma(\xi,y)d\xi,$$

where $K \in C_0^\infty(\mathcal{B}_1^{(n-1)})$, $K \ge 0$, and $\|K; \mathbb{R}^{n-1}\|_{L_1} = 1$. Then

$$\|\Gamma_\rho\|_{M(W_p^{t,\beta} \to W_p^{s,\alpha})} \le \|\Gamma\|_{M(W_p^{t,\beta} \to W_p^{s,\alpha})} \le \liminf_{\rho \to 0} \|\Gamma_\rho\|_{M(W_p^{t,\beta} \to W_p^{s,\alpha})}. \tag{8.2.3}$$

Proof. Let $U \in C_0^\infty$. By Minkowski's inequality

$$\Big(\int_{\mathbb{R}_+^n} (\min\{1,y\})^{p\alpha} \Big|\nabla_{j,z} \int_{\mathbb{R}^{n-1}} \rho^{-n} K(\xi/\rho)\Gamma(x-\xi,y)U(x,y)d\xi\Big|^p dz\Big)^{1/p}$$

$$\le \int_{\mathbb{R}^{n-1}} \rho^{-n} K(\xi/\rho) \Big(\int_{\mathbb{R}_+^n} (\min\{1,y\})^{p\alpha} |\nabla_{j,z}(\Gamma(x,y)U(x+\xi,y))|^p dz\Big)^{1/p} d\xi,$$

where either $j = 0$ or $j = s$. Therefore,

$$\|\Gamma_\rho u\|_{W_p^{s,\alpha}} \leq \|\Gamma\|_{M(W_p^{t,\beta} \to W_p^{s,\alpha})}$$
$$\times \int_{\mathbb{R}^{n-1}} \rho^{-n} K(\xi/\rho) \Big\{ \Big(\int_{\mathbb{R}^n_+} (\min\{1,y\})^{p\beta} |\nabla_{t,z} U(x+\xi,y)|^p dz \Big)^{1/p}$$
$$+ \Big(\int_{\mathbb{R}^n_+} (\min\{1,y\})^{p\beta} |U(x+\xi,y)|^p dz \Big)^{1/p} \Big\} d\xi$$
$$\leq \|\Gamma\|_{M(W_p^{t,\beta} \to W_p^{s,\alpha})} \|U\|_{W_p^{t,\beta}}.$$

This gives the left inequality (8.2.3). The right inequality (8.2.3) follows from

$$\|\Gamma u\|_{W_p^{s,\alpha}} = \liminf_{\rho \to 0} \|\Gamma_\rho U\|_{W_p^{s,\alpha}} \leq \liminf_{\rho \to 0} \|\Gamma_\rho\|_{M(W_p^{t,\beta} \to W_p^{s,\alpha})} \|U\|_{W_p^{t,\beta}}.$$

The proof is complete. □

Lemma 8.2.2. *Let $\Gamma \in L_{p,loc}$, $p \in (1,\infty)$, $-1 < \beta p < p-1$, and let U be an arbitrary function in $C_0^\infty(\mathbb{R}^n_+)$. The best constant in the inequality*

$$\|(\min\{1,y\})^\alpha \Gamma \nabla_s U\|_{L_p} + \|(\min\{1,y\})^\alpha \Gamma U\|_{L_p} \leq C\, \|U\|_{W_p^{t,\beta}} \tag{8.2.4}$$

is equivalent to the norm $\|\Gamma\|_{M(W_p^{t-s,\beta} \to W_p^{0,\alpha})}$.

Proof. The estimate

$$C \leq c\, \|\Gamma\|_{M(W_p^{t-s,\beta} \to W_p^{0,\alpha})}$$

is obvious. To derive the converse estimate, we introduce a function $x \to \sigma$ which is positive on $[0,\infty)$ and equal to x for $x > 1$. For any $U \in C_0^\infty(\mathbb{R}^n_+)$ we have

$$U = (-\Delta)^s \big(\sigma(-\Delta)\big)^{-[l]-1} u + T(-\Delta) u,$$

where T is a function in $C_0^\infty([0,\infty))$. Since

$$(-\Delta)^s = (-1)^s \sum_{|\tau|=s} \frac{s!}{\tau!} D^{2\tau},$$

it follows from (8.2.4) and the theorem on the boundedness of convolution operators in weighted L_p spaces (see [And]) that

$$\int_{\mathbb{R}^n_+} (\min\{1,y\})^{p\alpha} |\Gamma(z) U(z)|^p dz$$
$$\leq\ c\, C \big(\|\nabla_s (\sigma(-\Delta))^{-s} U\|^p_{W_p^{t,\beta}} + \|TU\|^p_{W_p^{t,\beta}} \big) \leq c\, C \|U\|^p_{W_p^{t-s,\beta}}.$$

The proof is complete. □

8.3 Characterization of $M(W_p^{t,\beta} \to W_p^{s,\alpha})$

Here we derive necessary and sufficient conditions for a function to belong to the space $M(W_p^{t,\beta} \to W_p^{s,\alpha})$ for $p \in (1, \infty)$ with α and β satisfying

$$-1 < \alpha p < p - 1, \quad -1 < \beta p < p - 1, \quad t \geq s. \tag{8.3.1}$$

These inequalities will be assumed throughout. We start with an assertion on derivatives of multipliers.

Lemma 8.3.1. *Suppose that*

$$\Gamma \in M(W_p^{t,\beta} \to W_p^{s,\alpha}) \cap M(W_p^{t-s,\beta} \to W_p^{0,\alpha}), \quad p \in (1, \infty).$$

Then $D^\sigma \Gamma \in M(W_p^{t,\beta} \to W_p^{s-|\sigma|,\alpha})$ *for any multi-index* σ *of order* $|\sigma| \leq s$, *and*

$$\|D^\sigma \Gamma\|_{M(W_p^{t,\beta} \to W_p^{s-|\sigma|,\alpha})}$$
$$\leq \varepsilon \, \|\Gamma\|_{M(W_p^{t-s,\beta} \to W_p^{0,\alpha})} + c(\varepsilon) \, \|\Gamma\|_{M(W_p^{t,\beta} \to W_p^{s,\alpha})}, \tag{8.3.2}$$

where ε *is an arbitrary positive number.*

Proof. Let $U \in W_p^{s,\alpha}$ and let φ be an arbitrary function in C_0^∞. Applying the Leibniz formula

$$D^\sigma(\varphi U) = \sum_{\{\tau : \sigma \geq \tau \geq 0\}} \frac{\sigma!}{\tau!(\sigma - \tau)!} D^\tau \varphi D^{\sigma - \tau} U,$$

we obtain

$$\int \varphi U (-D)^\sigma \Gamma dz = \int \Gamma D^\sigma (\varphi U) dz = \sum_{\{\tau : \sigma \geq \tau \geq 0\}} \frac{\sigma!}{\tau!(\sigma - \tau)!} \Gamma D^\tau \varphi D^{\sigma - \tau} U dz$$

$$= \int \varphi \sum_{\{\beta : \sigma \geq \tau \geq 0\}} \frac{\sigma!}{\tau!(\sigma - \tau)!} (-D)^\tau (\Gamma D^{\sigma - \tau} U) dz.$$

Therefore,

$$U D^\sigma \Gamma = \sum_{\{\tau : \sigma \geq \tau \geq 0\}} \frac{\sigma!}{\tau!(\sigma - \tau)!} (D)^\tau (\Gamma (-D)^{\sigma - \tau} U),$$

which implies the estimate

$$\|U D^\sigma \Gamma\|_{W_p^{s-|\sigma|,\alpha}} \leq c \sum_{\{\tau : \sigma \geq \tau \geq 0\}} \|\Gamma D^{\sigma - \tau} U\|_{W_p^{s-|\sigma|+|\tau|,\alpha}}.$$

Hence, it suffices to prove (8.3.2) for $|\sigma| = 1$. We have

$$\begin{aligned}\|U\nabla\Gamma\|_{W_p^{s-1,\alpha}} &\le \|U\Gamma\|_{W_p^{s,\alpha}} + \|\Gamma\nabla U\|_{W_p^{s-1,\alpha}}\\ &\le \big(\|\Gamma\|_{M(W_p^{t,\beta}\to W_p^{s,\alpha})} + \|\Gamma\|_{M(W_p^{t-1,\beta}\to W_p^{s-1,\alpha})}\big)\|U\|_{W_p^{t,\beta}}.\end{aligned}$$

Estimating the norm $\|\Gamma\|_{M(W_p^{t-1,\beta}\to W_p^{s-1,\alpha})}$ by (8.2.2), we arrive at (8.3.2). □

Now we pass to two-sided estimates of the norms in $M(W_p^{t,\beta} \to W_p^{s,\alpha})$, $p \in (1,\infty)$, given in terms of the spaces $M(W_p^{k,\beta} \to W_p^{0,\alpha})$. We start with lower estimates.

Lemma 8.3.2. *Let $\Gamma \in M(W_p^{t,\beta} \to W_p^{s,\alpha})$. Then*

$$\|\nabla_s\Gamma\|_{M(W_p^{t,\beta}\to W_p^{0,\alpha})} + \|\Gamma\|_{M(W_p^{t-s,\beta}\to W_p^{0,\alpha})} \le c\,\|\Gamma\|_{M(W_p^{t,\beta}\to W_p^{s,\alpha})}. \tag{8.3.3}$$

Proof. Suppose first that $\Gamma \in M(W_p^{t-s,\beta} \to W_p^{0,\alpha})$. We have

$$\begin{aligned}\|\Gamma\nabla_s U\|_{W_p^{0,\alpha}} &\le \|\Gamma\|_{M(W_p^{t,\beta}\to W_p^{s,\alpha})}\|U\|_{W_p^{t,\beta}} + c\sum_{\substack{|\sigma|+|\tau|=s,\\ \tau\neq 0}} \|D^\sigma U D^\tau \Gamma\|_{W_p^{0,\alpha}}\\ &\le \Big(\|\Gamma\|_{M(W_p^{t,\beta}\to W_p^{s,\alpha})} + c\sum_{j=1}^{s}\|\nabla_j\Gamma\|_{M(W_p^{t-s+j}\to W_p^{0,\alpha})}\Big)\|U\|_{W_p^{t,\beta}}.\end{aligned} \tag{8.3.4}$$

By Lemma 8.3.1,

$$\begin{aligned}&\|\nabla_j\Gamma\|_{M(W_p^{t-s+j,\beta}\to W_p^{0,\alpha})}\\ &\quad\le \varepsilon\,\|\Gamma\|_{M(W_p^{t-s,\beta}\to W_p^{0,\alpha})} + c(\varepsilon)\,\|\Gamma\|_{M(W_p^{t-s+j,\beta}\to W_p^{j,\alpha})}.\end{aligned} \tag{8.3.5}$$

Estimating the last norm by (8.2.2), we obtain

$$\begin{aligned}&\|\nabla_j\Gamma\|_{M(W_p^{t-s+j,\beta}\to W_p^{0,\alpha})}\\ &\quad\le \varepsilon\,\|\Gamma\|_{M(W_p^{t-s,\beta}\to W_p^{0,\alpha})} + c(\varepsilon)\,\|\Gamma\|_{M(W_p^{t,\beta}\to W_p^{s,\alpha})}.\end{aligned}$$

Substitution of this inequality into (8.3.4) gives

$$\begin{aligned}\|\Gamma\nabla_s U\|_{W_p^{0,\alpha}} \le \Big(&\varepsilon\,\|\Gamma\|_{M(W_p^{t-s,\beta}\to W_p^{0,\alpha})}\\ &+ c(\varepsilon)\,\|\Gamma\|_{M(W_p^{t,\beta}\to W_p^{s,\alpha})}\Big)\|U\|_{W_p^{t,\beta}}.\end{aligned} \tag{8.3.6}$$

Also,

$$\|\Gamma U\|_{W_p^{0,\alpha}} \le \|\Gamma\|_{M(W_p^{t,\beta}\to W_p^{s,\alpha})}\|U\|_{W_p^{t,\beta}}. \tag{8.3.7}$$

Adding the last two estimates and applying Lemma 8.2.2, we arrive at

$$\|\Gamma\|_{M(W_p^{t-s,\beta}\to W_p^{0,\alpha})} \le \varepsilon\,\|\Gamma\|_{M(W_p^{t-s,\beta}\to W_p^{0,\alpha})} + c(\varepsilon)\,\|\Gamma\|_{M(W_p^{t,\beta}\to W_p^{s,\alpha})}.$$

Hence,

$$\|\Gamma\|_{M(W_p^{t-s,\beta}\to W_p^{0,\alpha})} \le c\,\|\Gamma\|_{M(W_p^{t,\beta}\to W_p^{s,\alpha})}. \tag{8.3.8}$$

Now we remove the assumption $\Gamma \in M(W_p^{t-s,\beta} \to W_p^{0,\alpha})$. Since $\Gamma \in M(W_p^{t,\beta} \to W_p^{s,\alpha})$, it follows that

$$\|\Gamma\eta\|_{W_p^{s,\alpha}} \le c\,\|\eta\|_{W_p^{t,\beta}}\,,$$

where $\eta \in C_0^\infty(\mathcal{B}_2^{(n)}(z))$, $\eta = 1$ on $\mathcal{B}_1^{(n)}(z)$, and z is an arbitrary point in $\mathbb{R}_+^n$. Hence $\Gamma \in W^{s,\alpha}_{p,\mathrm{unif}}(\mathbb{R}^n_+)$, which implies that for any $(n-1)$-dimensional multi-index τ the derivative $D_x^\tau \Gamma_\rho$ belongs to $W^{s,\alpha}_{p,\mathrm{unif}}(\mathbb{R}^n_+)$. Therefore, $\Gamma_\rho \in L_\infty(\mathbb{R}^n_+)$ which, in its turn, guarantees that $\Gamma_\rho \in M(W_p^{t-s,\beta} \to W_p^{0,\alpha})$. Thus, we may put Γ_ρ into (8.3.8) to obtain

$$\|\Gamma_\rho\|_{M(W_p^{t-s,\beta}\to W_p^{0,\alpha})} \le c\,\|\Gamma_\rho\|_{M(W_p^{t,\beta}\to W_p^{s,\alpha})}.$$

Letting $\rho \to 0$ and using Lemma 8.2.1, we arrive at (8.3.8) for all $\Gamma \in M(W_p^{t,\beta} \to W_p^{s,\alpha})$.

To estimate the first term on the right-hand side of (8.3.3), we combine (8.3.8) with (8.3.5) for $j = s$. □

The estimate opposite to (8.3.3) is contained in the following lemma.

Lemma 8.3.3. *Let $\Gamma \in M(W_p^{t-s,\beta} \to W_p^{0,\alpha})$ and let $\nabla_s\Gamma \in M(W_p^{t,\beta} \to W_p^{0,\alpha})$. Then $\Gamma \in M(W_p^{t,\beta} \to W_p^{s,\alpha})$ and the estimate*

$$\|\Gamma\|_{M(W_p^{t,\beta}\to W_p^{s,\alpha})} \le c\Big(\|\nabla_s\Gamma\|_{M(W_p^{t,\beta}\to W_p^{0,\alpha})} + \|\Gamma\|_{M(W_p^{t-s,\beta}\to W_p^{0,\alpha})}\Big) \tag{8.3.9}$$

holds.

Proof. By Lemma 8.3.2 and (8.2.2) we have

$$\begin{aligned}\|\nabla_j\Gamma\|_{M(W_p^{t-s+j,\beta}\to W_p^{0,\alpha})} &\le c\,\|\Gamma\|_{M(W_p^{t-s+j,\beta}\to W_p^{j,\alpha})}\\ &\le c\,\|\Gamma\|^{j/s}_{M(W_p^{t,\beta}\to W_p^{s,\alpha})}\|\Gamma\|^{1-j/s}_{M(W_p^{t-s,\beta}\to W_p^{0,\alpha})},\end{aligned} \tag{8.3.10}$$

where $j = 1, \dots, s$. For any $U \in C_0^\infty(\overline{\mathbb{R}^n_+})$,

$$\begin{aligned}\|(\min\{1,y\})^\alpha\nabla_s(\Gamma U)\|_{L_p} &\le c\sum_{j=0}^{s}\|(\min\{1,y\})^\alpha|\nabla_j\Gamma|\,|\nabla_{s-j}U|\,\|_{L_p}\\ &\le c\Big(\|\nabla_s\Gamma\|_{M(W_p^{t,\beta}\to W_p^{0,\alpha})} + \|\Gamma\|_{M(W_p^{t-s,\beta}\to W_p^{0,\alpha})}\\ &\quad+\sum_{j=1}^{s-1}\|\nabla_j\Gamma\|_{M(W_p^{t-s+j,\beta}\to W_p^{0,\alpha})}\Big)\|U\|_{W_p^{t,\beta}}.\end{aligned}$$

This and (8.3.10) imply that

$$\| (\min\{1,y\})^\alpha \nabla_s(\Gamma U)\|_{L_p}$$
$$\le c \left(\|\nabla_s \Gamma\|_{M(W_p^{t,\beta} \to W_p^{0,\alpha})} + \|\Gamma\|_{M(W_p^{t-s,\beta} \to W_p^{0,\alpha})} \right) \|U\|_{W_p^{t,\beta}}.$$

It remains to note that

$$\|(\min\{1,y\})^\alpha \Gamma U\|_{L_p} \le \|\Gamma\|_{M(W_p^{t-s,\beta} \to W_p^{0,\alpha})} \|U\|_{W_p^{t-s,\beta}}.$$

The proof is complete. □

Using Lemmas 8.3.2 and 8.3.3, we arrive at the following description of the space $M(W_p^{t,\beta}(\mathbb{R}_+^n) \to W_p^{s,\alpha}(\mathbb{R}_+^n))$.

Theorem 8.3.1. *A function Γ belongs to the space $M(W_p^{t,\beta} \to W_p^{s,\alpha})$ if and only if $\Gamma \in W_{p,\mathrm{loc}}^{s,\alpha}$, $\Gamma \in M(W_p^{t-s,\beta} \to W_p^{0,\alpha})$, and $\nabla_s \Gamma \in M(W_p^{t,\beta} \to W_p^{0,\alpha})$. Moreover,*

$$\|\Gamma\|_{M(W_p^{t,\beta} \to W_p^{s,\alpha})} \sim \|\nabla_s \Gamma\|_{M(W_p^{t,\beta} \to W_p^{0,\alpha})} + \|\Gamma\|_{M(W_p^{t-s,\beta} \to W_p^{0,\alpha})}.$$

The equivalence relation (8.2.1) enables us to reformulate Theorem 8.3.1 as follows.

Theorem 8.3.2. *A function Γ belongs to the space $M(W_p^{t,\beta} \to W_p^{s,\alpha})$ if and only if $\Gamma \in W_{p,\mathrm{loc}}^{s,\alpha}$, and, for any compact set $e \subset \mathbb{R}_+^n$,*

$$\|(\min\{1,y\})^\alpha \nabla_s \Gamma; e\|_{L_p}^p \le c\, C_{p,t,\beta}(e)$$

and

$$\|(\min\{1,y\})^\alpha \Gamma; e\|_{L_p}^p \le c\, C_{p,t-s,\beta}(e).$$

Moreover,

$$\|\Gamma\|_{M(W_p^{t,\beta} \to W_p^{s,\alpha})} \sim \sup_{\substack{e \subset \mathbb{R}_+^n \\ d(e) \le 1}} \left(\frac{\|(\min\{1,y\})^\alpha \nabla_s \Gamma; e\|_{L_p}}{(C_{p,t,\beta}(e))^{1/p}} + \frac{\|(\min\{1,y\})^\alpha \Gamma; e\|_{L_p}}{(C_{p,t-s,\beta}(e))^{1/p}} \right), \tag{8.3.11}$$

where $d(e)$ is the diameter of e.

We formulate the important particular case of Theorem 8.3.2 when $t = s$.

Corollary 8.3.1. *A function Γ belongs to the space $MW_p^{s,\alpha}$ if and only if $\Gamma \in W_{p,\mathrm{loc}}^{s,\alpha}$ and, for any compact set $e \subset \mathbb{R}_+^n$,*

$$\|(\min\{1,y\})^\alpha \nabla_s \Gamma; e\|_{L_p}^p \le c\, C_{p,s,\alpha}(e).$$

Moreover,

$$\|\Gamma\|_{MW_p^{s,\alpha}} \sim \sup_{\substack{e \subset \mathbb{R}_+^n \\ d(e) \le 1}} \frac{\|(\min\{1,y\})^\alpha \nabla_s \Gamma; e\|_{L_p}}{(C_{p,s,\alpha}(e))^{1/p}} + \|\Gamma\|_{L_\infty}. \tag{8.3.12}$$

8.4 Auxiliary Estimates for an Extension Operator

8.4.1 Pointwise Estimates for $T\gamma$ and $\nabla T\gamma$

For functions $\gamma \in L_{1,\mathrm{unif}}(\mathbb{R}^{n-1})$, we introduce the operator T by

$$(T\gamma)(x,y) = y^{1-n} \int_{\mathbb{R}^{n-1}} \zeta\Big(\frac{x-\xi}{y}\Big)\gamma(\xi)d\xi, \quad (x,y) \in \mathbb{R}^n_+, \tag{8.4.1}$$

where ζ is a continuously differentiable function defined on $\overline{\mathbb{R}^n_+}$. We assume that

$$(|z|+1)|\nabla\zeta(z)| + |\zeta(z)| \le C\,(|z|+1)^{-n} \tag{8.4.2}$$

and that

$$\int_{\mathbb{R}^{n-1}} \zeta(z)dz = 1. \tag{8.4.3}$$

Lemma 8.4.1. *Let $\gamma \in M(W_p^{m-l}(\mathbb{R}^{n-1}) \to L_p(\mathbb{R}^{n-1}))$, where $m \ge l$ and $1 < p < \infty$. Then*

$$|T\gamma(z)| + y|\nabla(T\gamma(z))| \;\le\; c\,(1+y^{l-m})\|\gamma; \mathbb{R}^{n-1}\|_{M(W_p^{m-l}\to L_p)}.$$

Proof. By (8.4.2),

$$|T\gamma(z)| + y|\nabla(T\gamma(z))|$$
$$\le c\,y^{1-n}\Big(\int_{\mathcal{B}_y^{(n-1)}(x)} |\gamma(\xi)|d\xi + y^n \int_{\mathbb{R}^{n-1}\backslash \mathcal{B}_y^{(n-1)}(x)} \frac{|\gamma(\xi)|d\xi}{|\xi-x|^n}\Big). \tag{8.4.4}$$

By Hölder's inequality,

$$\int_{\mathcal{B}_y^{(n-1)}(x)} |\gamma(\xi)|d\xi \le c y^{(n-1)(p-1)/p}\|\gamma; \mathcal{B}_y^{(n-1)}(x)\|_{L_p}. \tag{8.4.5}$$

Let $y \in (0,1)$. The right-hand side in (8.4.5) does not exceed

$$c\,y^{-m+l+n-1} \sup_{\substack{r\in(0,1)\\ x\in\mathbb{R}^{n-1}}} (1+r^{m-l-\frac{n-1}{p}})\,\|\gamma; \mathcal{B}_r^{(n-1)}(x)\|_{L_p}.$$

This, being combined with (8.4.5) and (4.3.15), implies that

$$\int_{\mathcal{B}_y^{(n-1)}(x)} |\gamma(\xi)|d\xi \le c\,y^{-m+l+n-1}\|\gamma; \mathbb{R}^{n-1}\|_{M(W_p^{m-l}\to L_p)} \tag{8.4.6}$$

for $y < 1$.

Suppose that $y > 1$. By (4.3.12),

$$\|\gamma; \mathbb{R}^{n-1}\|_{M(W_p^{m-l}\to L_p)} \ge c \sup_{y>1} \frac{\|\gamma; \mathcal{B}_y^{(n-1)}\|_{L_p}}{\big(C_{p,m-l}(\mathcal{B}_y^{(n-1)})\big)^{1/p}}$$
$$\ge c \sup_{y>1} y^{(1-n)/p}\|\gamma; \mathcal{B}_y^{(n-1)}\|_{L_p}, \tag{8.4.7}$$

because

$$C_{p,m-l}(\mathcal{B}_y^{(n-1)}(x)) \sim y^{n-1}$$

for $y > 1$. Combining (8.4.7) with (8.4.5), we have

$$y^{1-n} \int_{\mathcal{B}_y^{(n-1)}(x)} |\gamma(\xi)| d\xi \le c\, y^{-(n-1)/p} \|\gamma; \mathcal{B}_y^{(n-1)}\|_{L_p}$$

$$\le c\, \|\gamma; \mathbb{R}^{n-1}\|_{M(W_p^{m-l} \to L_p)}. \tag{8.4.8}$$

Thus, (8.4.6) and (8.4.8) give

$$y^{1-n} \int_{\mathcal{B}_y^{(n-1)}(x)} |\gamma(\xi)| d\xi \le c\,(1 + y^{l-m}) \|\gamma; \mathbb{R}^{n-1}\|_{M(W_p^{m-l} \to L_p)} \tag{8.4.9}$$

for any $y > 0$.

Now we estimate the second integral on the right-hand side of (8.4.4). Clearly,

$$\int_{\mathbb{R}^{n-1} \setminus \mathcal{B}_y^{(n-1)}(x)} \frac{|\gamma(\xi)| d\xi}{|\xi - x|^n} \le n \int_y^\infty \frac{d\rho}{\rho^{n+1}} \int_{\mathcal{B}_\rho^{(n-1)}(x)} |\gamma(\xi)| d\xi. \tag{8.4.10}$$

By Hölder's inequality the right-hand side of (8.4.10) has the majorant

$$c \int_y^\infty \rho^{-2-\frac{n-1}{p}} \|\gamma; \mathcal{B}_\rho^{(n-1)}(x)\|_{L_p} d\rho. \tag{8.4.11}$$

Let $y > 1$. Then by (8.4.7), the function (8.4.11) does not exceed

$$c\, y^{-1} \|\gamma; \mathbb{R}^{n-1}\|_{M(W_p^{m-l} \to L_p)}.$$

Suppose that $y < 1$. Then

$$\int_y^1 \rho^{-2-\frac{n-1}{p}} \|\gamma; \mathcal{B}_\rho^{(n-1)}(x)\|_{L_p} d\rho$$

$$\le\ c y^{-m+l-1} \sup_{\substack{r \in (0,1) \\ x \in \mathbb{R}^{n-1}}} \big(1 + r^{m-1-\frac{n-1}{p}}\big)\, \|\gamma; \mathcal{B}_r^{(n-1)}(x)\|_{L_p} \tag{8.4.12}$$

which is dominated by

$$c\, y^{-m+l-1} \|\gamma; \mathbb{R}^{n-1}\|_{M(W_p^{m-l} \to L_p)}$$

owing to (4.3.15). Furthermore, (8.4.7) implies that

$$\int_1^\infty \rho^{-2-\frac{n-1}{p}} \|\gamma; \mathcal{B}_\rho^{(n-1)}(x)\|_{L_p} d\rho \le c\, \|\gamma; \mathbb{R}^{n-1}\|_{M(W_p^{m-l} \to L_p)}.$$

Hence, for $y < 1$

$$\int_1^\infty \rho^{-2-\frac{n-1}{p}} \|\gamma; \mathcal{B}_\rho^{n-1}(x)\|_{L_p} d\rho \le c\,(1+y^{-m+l-1})\,\|\gamma; \mathbb{R}^{n-1}\|_{M(W_p^{m-l}\to L_p)}.$$

This, in combination with the case $y > 1$, implies that the integral (8.4.11) does not exceed

$$cy^{-1}(1+y^{l-m})\|\gamma; \mathbb{R}^{n-1}\|_{M(W_p^{m-l}\to L_p)}$$

for all $y > 0$. Thus, the result follows from (8.4.9), (8.4.10), and (8.4.4). □

8.4.2 Weighted L_p-Estimates for $T\gamma$ and $\nabla T\gamma$

Lemma 8.4.2. *Let the extension operator T be defined by (8.4.1). Suppose that $\gamma \in M(W_p^{m-l}(\mathbb{R}^{n-1}) \to L_p(\mathbb{R}^{n-1}))$, where $l \in (0,1)$, $[m] \ge 1$, and $1 < p < \infty$. Then, for $k = 1, \dots, [m]$,*

$$\Big(\int_0^1 y^{p(k-l)-1}\big(|T\gamma(z)| + y|\nabla(T\gamma)(z)|\big)^p dy\Big)^{1/p}$$
$$\le c\,\|\gamma; \mathbb{R}^{n-1}\|^{\frac{k-l}{m-l}}_{M(W_p^{m-l}\to L_p)} [(\mathcal{M}\gamma)(x)]^{\frac{m-k}{m-l}}, \tag{8.4.13}$$

where $\mathcal{M}$ is the Hardy–Littlewood maximal operator in $\mathbb{R}^{n-1}$.

Proof. Let δ be a number in (0,1] to be chosen later. We set

$$\int_0^1 y^{p(k-l)-1}\big(|T\gamma| + y|\nabla(T\gamma)(z)|\big)^p dy = \int_0^\delta \dots dy + \int_\delta^1 \dots dy.$$

By (8.4.4),

$$\int_0^\delta \dots dy \le c\int_0^\delta y^{p(k+1-l-n)-1}\Big(\int_{\mathcal{B}_y^{(n-1)}(x)} |\gamma(\xi)| d\xi\Big)^p dy$$
$$+\, c\int_0^\delta y^{p(k+1-l)-1}\Big(\int_{\mathbb{R}^{n-1}\backslash \mathcal{B}_y^{(n-1)}(x)} \frac{|\gamma(\xi)|}{|\xi - x|^n} d\xi\Big)^p dy.$$

From the definition of $\mathcal{M}$ it follows that

$$\int_0^\delta y^{p(k+1-l-n)-1}\Big(\int_{\mathcal{B}_y^{(n-1)}(x)} |\gamma(\xi)| d\xi\Big)^p dy \le c\,[(\mathcal{M}\gamma)(x)]^p \delta^{p(k-l)}. \tag{8.4.14}$$

Using (8.4.10), we obtain

$$\int_0^\delta y^{p(k+1-l)-1}\Big(\int_{\mathbb{R}^{n-1}\backslash \mathcal{B}_y^{(n-1)}(x)} \frac{|\gamma(\xi)|}{|\xi - x|^n} d\xi\Big)^p dy$$
$$\le c\,[(\mathcal{M}\gamma)(x)]^p \delta^{p(k-l)}. \tag{8.4.15}$$

Combining (8.4.14) and (8.4.15), we conclude that

$$\int_0^\delta \dots dy \le c\,[(\mathcal{M}\gamma)(x)]^p \delta^{p(k-l)}. \tag{8.4.16}$$

By Lemma 8.4.1,

$$\int_\delta^1 y^{p(k-l)-1}\big(|T\gamma| + y|\nabla(T\gamma)(z)|\big)^p dy$$
$$\le c\,\|\gamma;\mathbb{R}^{n-1}\|^p_{M(W_p^{m-l}\to L_p)}\delta^{p(k-m)}. \tag{8.4.17}$$

Adding (8.4.16) and (8.4.17), we find that

$$\int_0^1 y^{p(k-l)-1}\big(|T\gamma| + y|\nabla(T\gamma)(z)|\big)^p dy$$
$$\le c\big([(\mathcal{M}\gamma)(x)]^p\delta^{p(k-l)} + \|\gamma;\mathbb{R}^{n-1}\|^p_{M(W_p^{m-l}\to L_p)}\delta^{p(k-m)}\big).$$

The right-hand side in this inequality attains its minimum value for

$$\delta = \Big(\frac{\|\gamma;\mathbb{R}^{n-1}\|_{M(W_p^{m-l}\to L_p)}}{(\mathcal{M}\gamma)(x)}\Big)^{1/(m-l)}.$$

The proof is complete. □

Lemma 8.4.3. *Let the operator T be defined by* (8.4.1) *and let* $0 < l < 1$. *Then*

$$\int_0^1 y^{p(1-l)-1}|\nabla(T\gamma)(z)|^p dy \le c\,\big((D_{p,l}\gamma)(x)\big)^p.$$

Proof. Let $R(\xi,x) = \gamma(\xi) - \gamma(x)$. Using the identity

$$y^{-n+1}\int_{\mathbb{R}^{n-1}} \zeta\Big(\frac{\xi - x}{y}\Big)d\xi = const,$$

we have

$$\frac{\partial T\gamma}{\partial y}(x,y) = \frac{\partial}{\partial y}\Big(y^{-n+1}\int_{\mathbb{R}^{n-1}} \zeta\Big(\frac{\xi - x}{y}\Big)R(\xi,x)d\xi\Big). \tag{8.4.18}$$

Furthermore, it is clear that

$$\frac{\partial T\gamma}{\partial x_j}(x,y) = y^{-n+1}\int_{\mathbb{R}^{n-1}} R(\xi,x)\frac{\partial}{\partial x_j}\zeta\Big(\frac{\xi - x}{y}\Big)d\xi.$$

Therefore,

$$|\nabla(T\gamma)(x,y)| \le c\,y^{-n}\sum_{k=0}^{1}\int_{\mathbb{R}^{n-1}} \Big|\nabla_k\zeta\Big(\frac{\xi-x}{y}\Big)\Big|\Big(1+\frac{|\xi-x|}{y}\Big)^k |R(\xi,x)|d\xi.$$

This estimate and (8.4.2) imply that

$$\begin{aligned}|\nabla(T\gamma)(x,y)| &\le c\,y^{-n}\int_{\mathbb{R}^{n-1}}\Big(1+\frac{|\xi-x|}{y}\Big)^{-n}|R(\xi,x)|d\xi\\ &= c\,y^{-1/p}\int_{\mathbb{R}^{n-1}}\Big(\frac{|\xi-x|}{y}\Big)^{n-1/p}\Big(1+\frac{|\xi-x|}{y}\Big)^{-n}\frac{|R(\xi,x)|}{|\xi-x|^{n-1/p}}d\xi.\end{aligned}$$

Consequently,

$$\begin{aligned}&\int_0^1 y^{p(1-l)-1}|\nabla(T\gamma)(x,y)|^p dy\\ &\le c\int_0^1\Big(\int_{\mathbb{R}^{n-1}} f\Big(\frac{|\xi-x|}{y}\Big)\frac{|R(\xi,x)|}{|\xi-x|^{n-1/p}}d\xi\Big)^p y^{p(1-l)-1}\frac{dy}{y},\end{aligned}$$

where $f(\eta)=\eta^{n-1/p}(1+\eta)^{-n}$. We write the last integral over $(0,1)$ as

$$\begin{aligned}&\int_0^1\Big(\int_0^\infty f\Big(\frac{t}{y}\Big)g(t,x)\frac{dt}{t}\Big)^p y^{p(1-l)-1}\frac{dy}{y}\\ &\quad=\int_0^1\Big(\int_0^\infty f(s)g(sy,x)\frac{ds}{s}\Big)^p y^{p(1-l)-1}\frac{dy}{y},\end{aligned} \tag{8.4.19}$$

with

$$g(t,x)=t^{1/p-1}\int_{\partial\mathcal{B}_1^{(n-1)}}|R(t\theta+x,x)|d\theta.$$

By Minkowski's inequality, the right-hand side of (8.4.19) does not exceed

$$\begin{aligned}&\Big(\int_0^\infty\Big(\int_0^1 (f(s))^p(g(sy,x))^p y^{p(1-l)-1}\frac{dy}{y}\Big)^{1/p}\frac{ds}{s}\Big)^p\\ &=\Big(\int_0^\infty f(s)\Big(\int_0^s (g(\tau,x))^p\tau^{p(1-l)-1}\frac{d\tau}{\tau}\Big)^{1/p}\frac{ds}{s^{2-l-1/p}}\Big)^p\\ &\le\Big(\int_0^\infty f(s)\frac{ds}{s^{2-l-1/p}}\Big)^p\int_0^\infty (g(\tau,x))^p\tau^{p(1-l)-1}\frac{d\tau}{\tau}.\end{aligned} \tag{8.4.20}$$

Therefore,

$$\int_0^1 y^{p(1-l)-1}|\nabla(T\gamma)(x,y)|^p dy \le c\int_0^\infty (g(\tau,x))^p\tau^{p(1-l)-1}\frac{d\tau}{\tau}.$$

It remains to note that

$$\int_0^\infty (g(\tau,x))^p \tau^{p(1-l)-1} \frac{d\tau}{\tau} = \int_0^\infty \tau^{-pl} \Big(\int_{\partial \mathcal{B}_1^{(n-1)}} |\gamma(\tau\theta + x) - \gamma(x)| d\theta \Big)^p \frac{d\tau}{\tau}$$

$$\leq c \int_0^\infty \int_{\partial \mathcal{B}_1^{(n-1)}} |\gamma(\tau\theta + x) - \gamma(x)|^p d\theta \frac{d\tau}{\tau^{pl+1}} \leq c \int_{\mathbb{R}^{n-1}} \frac{|\gamma(x+h) - \gamma(x)|^p}{|h|^{pl+n-1}} dh$$

$$= c \big((D_{p,l}\gamma)(x) \big)^p.$$

The result follows. □

8.5 Trace Theorem for the Space $M(W_p^{t,\beta} \to W_p^{s,\alpha})$

Theorem 8.5.1. (*i*) *Let m and l be positive nonintegers with $m \geq l$, and let*

$$\Gamma \in M(W_p^{t,\beta}(\mathbb{R}_+^n) \;\to\; W_p^{s,\alpha}(\mathbb{R}_+^n))$$

where $t = [m] + 1$, $s = [l] + 1$, $\beta = 1 - \{m\} - 1/p$, and $\alpha = 1 - \{l\} - 1/p$. If γ is the trace of Γ on $\mathbb{R}^{n-1}$, then

$$\gamma \in M(W_p^m(\mathbb{R}^{n-1}) \;\to\; W_p^l(\mathbb{R}^{n-1}))$$

and the estimate

$$\|\gamma; \mathbb{R}^{n-1}\|_{M(W_p^m \to W_p^l)} \leq c\, \|\Gamma; \mathbb{R}_+^n\|_{M(W_p^{t,\beta} \to W_p^{s,\alpha})} \tag{8.5.1}$$

holds.

(*ii*) *Let*

$$\gamma \in M(W_p^m(\mathbb{R}^{n-1}) \to W_p^l(\mathbb{R}^{n-1})).$$

Then the Dirichlet problem

$$\Delta\Gamma = 0 \;\; on \;\; \mathbb{R}_+^n, \;\; \Gamma|_{\mathbb{R}^{n-1}} = \gamma \tag{8.5.2}$$

has a unique solution in $M(W_p^{t,\beta}(\mathbb{R}_+^n) \;\to\; W_p^{s,\alpha}(\mathbb{R}_+^n))$, where $t = [m] + 1$, $s = [l] + 1$, $\beta = 1 - \{m\} - 1/p$, and $\alpha = 1 - \{l\} - 1/p$. The estimate

$$\|\Gamma; \mathbb{R}_+^n\|_{M(W_p^{t,\beta} \to W_p^{s,\alpha})} \leq c\, \|\gamma; \mathbb{R}^{n-1}\|_{M(W_p^m \to W_p^l)} \tag{8.5.3}$$

holds.

Proof. We start with (*i*). Let $U \in W_p^{t,\beta}(\mathbb{R}_+^n)$ and let u be the trace of U on $\mathbb{R}^{n-1}$. By setting ΓU and γu instead of U and u, respectively, in the inequality

$$\|u; \mathbb{R}^{n-1}\|_{W_p^l} \leq c\, \|U; \mathbb{R}_+^n\|_{W_p^{s,\alpha}},$$

we obtain the estimate

$$\|\gamma u; \mathbb{R}^{n-1}\|_{W_p^l} \le c\, \|\Gamma; \mathbb{R}^n_+\|_{M(W_p^{t,\beta}\to W_p^{s,\alpha})} \|U; \mathbb{R}^n_+\|_{W_p^{t,\beta}}.$$

Minimizing the right-hand side over all extensions U of u, we obtain

$$\|\gamma u; \mathbb{R}^{n-1}\|_{W_p^l} \le c\, \|\Gamma; \mathbb{R}^n_+\|_{M(W_p^{t,\beta}\to W_p^{s,\alpha})} \|u; \mathbb{R}^{n-1}\|_{W_p^m}$$

which gives (8.5.1).

The proof of (ii) will be given separately for $l < 1$ and for $l > 1$ in Sects. 8.5.1 and 8.5.2. □

8.5.1 The Case $l < 1$

Our aim now is to prove that for $l < 1$ and $s = 1$ the operator T defined by (8.4.1) maps $M(W_p^m(\mathbb{R}^{n-1}) \to W_p^l(\mathbb{R}^{n-1}))$ into $M(W_p^{[m]+1,\beta}(\mathbb{R}^n_+) \to W_p^{1,\alpha}(\mathbb{R}^n_+))$ with $\alpha = 1 - l - 1/p$, $\beta = 1 - \{m\} - 1/p$, and that

$$\|T\gamma; \mathbb{R}^n_+\|_{M(W_p^{t,\beta}\to W_p^{1,\alpha})} \le c\, C\|\gamma; \mathbb{R}^{n-1}\|_{M(W_p^m\to W_p^l)}, \tag{8.5.4}$$

where C is the constant in (8.4.2).

We have

$$\begin{aligned}\|(\min\{1,y\})^\alpha \nabla(UT\gamma); \mathbb{R}^n_+\|^p_{L_p} \le{}& c\int_0^1 y^{p\alpha} \int_{\mathbf{R}^{n-1}} \big(|\nabla(T\gamma)|^p|U|^p + |T\gamma|^p|\nabla U|^p\big)\,dz \\ &+ c\int_1^\infty \int_{\mathbb{R}^{n-1}} \big(|\nabla(T\gamma)|^p|U|^p + |T\gamma|^p|\nabla U|^p\big)\,dz \\ ={}& c\int_{0<y<1} \ldots\, dz + c\int_{y>1} \ldots\, dz. \end{aligned}\tag{8.5.5}$$

The integration in the last two integrals is taken in y over $(0,1)$ and over $(1,\infty)$, respectively, and over $\mathbb{R}^{n-1}$ in x. By Lemma 8.4.1, for $y > 1$

$$y|\nabla(T\gamma)(z)| + |(T\gamma)(z)| \le c\, \|\gamma; \mathbb{R}^{n-1}\|_{M(W_p^{m-l}\to L_p)}.$$

Hence

$$\int_{y>1} \ldots\, dz \le c\, \|\gamma; \mathbb{R}^{n-1}\|^p_{M(W_p^{m-l}\to L_p)} \|U; \mathbb{R}^n_+\|^p_{W_p^{1,\alpha}}. \tag{8.5.6}$$

It remains to refer to the estimate

$$\|U; \mathbb{R}^n_+\|_{W_p^{1,\alpha}} \le c\, \|U; \mathbb{R}^n_+\|_{W_p^{t,\beta}}$$

which follows from the one-dimensional Hardy inequality.

Introducing the notation

$$\mathcal{R}_0 U(z) = U(z) - \sum_{k=0}^{[m]} \frac{\partial^k}{\partial y^k} U(x,0) \frac{y^k}{k!},$$

$$\mathcal{R}_1 U(z) = \nabla U(z) - \sum_{k=0}^{[m]-1} \frac{\partial^k}{\partial y^k} \nabla U(x,0) \frac{y^k}{k!} \quad \text{for } m > 1,$$

and

$$\mathcal{R}_1 U(z) = \nabla U(z) \quad \text{for } m < 1,$$

we have

$$\int_{0<y<1} \dots dz \le c \int_{0<y<1} y^{p(1-l)-1} \sum_{j=0}^{1} |\nabla_j (T\gamma)|^p |\mathcal{R}_{1-j} U(z)|^p dz$$
$$+ c \int_{0<y<1} y^{-pl-1} \big(|T\gamma(z)| + y|\nabla(T\gamma)(z)|\big)^p \sum_{k=1}^{[m]} |\nabla_k U(x,0)|^p y^{pk} dz$$
$$+ c \int_{0<y<1} y^{p(1-l)-1} |\nabla T\gamma(z)|^p |U(x,0)|^p dz \tag{8.5.7}$$

for $m > 1$. When $m < 1$ the second integral on the right-hand side of (8.5.7) should be omitted.

By Lemma 8.4.1, we have for $0 < y < 1$

$$|T\gamma(z)| + y|\nabla(T\gamma)(z)| \le c\, y^{l-m} \|\gamma; \mathbb{R}^{n-1}\|_{M(W_p^{m-l} \to L_p)}. \tag{8.5.8}$$

Since for $j = 0, 1$

$$|\mathcal{R}_{1-j} U(z)| \le \frac{y^{[m]+j-1}}{([m]+j-1)!} \int_0^y |\nabla_t U(x,t)| dt, \tag{8.5.9}$$

we obtain the inequality

$$\int_{0<y<1} y^{p(1-\{m\})-1} |\mathcal{R}_{1-j} U(z)|^p dz$$
$$\le c \int_{0<y<1} y^{-p\{m\}-1} \Big(\int_0^y |\nabla_{[m]+1} U(x,t)| dt \Big)^p dz.$$

By Hardy's inequality, the right-hand side does not exceed $c\, \|U; \mathbb{R}^n_+\|^p_{W_p^{[m]+1,\beta}}$. Combining this fact with (8.5.8), we obtain that the first integral on the right-hand side of (8.5.7) does not exceed

$$\|\gamma; \mathbb{R}^{n-1}\|^p_{M(W_p^{m-l} \to L_p)} \|U; \mathbb{R}^n_+\|^p_{W_p^{[m]+1,\beta}}. \tag{8.5.10}$$

Now we pass to the estimate of the second integral on the right-hand side of (8.5.7) for $k = 1, \dots, [m]$, $m > 1$. Applying Lemma 8.4.2, we find that

$$\int_{0<y<1} y^{p(k-l)-1}\big(|T\gamma(z)| + y|\nabla(T\gamma)(z)|\big)^p |\nabla_k U(x,0)|^p dz$$

$$\le c\,\|\gamma;\mathbb{R}^{n-1}\|^{p\frac{k-l}{m-l}}_{M(W_p^{m-l}\to L_p)} \int_{\mathbb{R}^{n-1}} (\mathcal{M}\gamma(x))^{p\frac{m-k}{m-l}} |\nabla_k U(x,0)|^p dx. \qquad (8.5.11)$$

The last integral is not greater than

$$\|(\mathcal{M}\gamma)^{\frac{m-k}{m-l}};\mathbb{R}^{n-1}\|^p_{M(W_p^{m-k}\to L_p)} \|\nabla_k U(\cdot,0);\mathbb{R}^{n-1}\|^p_{W_p^{m-k}}. \qquad (8.5.12)$$

Using Lemma 2.3.6 with $\lambda = m-k$, $\mu = m-l$ and Lemma 2.3.8, we find that (8.5.12) is dominated by

$$c\,\|\gamma;\mathbb{R}^{n-1}\|^{\frac{p(m-k)}{m-l}}_{M(W_p^{m-l}\to L_p)} \|U(\cdot,0);\mathbb{R}^{n-1}\|^p_{W_p^m}$$

which together with (8.5.11) implies that

$$\int_{0<y<1} y^{p(k-l)-1}\big(|T\gamma(z)| + y|\nabla(T\gamma)(z)|\big)^p |\nabla_k U(x,0)|^p dz$$

$$\le \ c\,\|\gamma;\mathbb{R}^{n-1}\|^p_{M(W_p^{m-l}\to L_p)} \|U;\mathbb{R}^n_+\|^p_{W_p^{[m]+1,\beta}}. \qquad (8.5.13)$$

By Lemma 8.4.3, the integral

$$\int_{0<y<1} y^{p(1-l)-1} |\nabla(T\gamma)(z)|^p |U(x,0)|^p dz \qquad (8.5.14)$$

does not exceed

$$c\int_{\mathbb{R}^{n-1}} (D_{p,l}\gamma(x))^p |U(x,0)|^p dx$$

$$\le c\,\|D_{p,l}\gamma;\mathbb{R}^{n-1}\|^p_{M(W_p^m\to L_p)} \|U(\cdot,0);\mathbb{R}^{n-1}\|^p_{W_p^m}$$

$$\le c\,\|\gamma;\mathbb{R}^{n-1}\|^p_{M(W_p^m\to W_p^l)} \|U;\mathbb{R}^n_+\|^p_{W_p^{[m]+1,\beta}}. \qquad (8.5.15)$$

Thus, we arrive at the inequality

$$\int_{0<y<1} y^{p\alpha} |\nabla(UT\gamma)(z)|^p dz \le c\,\|\gamma;\mathbb{R}^{n-1}\|^p_{M(W_p^m\to W_p^l)} \|U;\mathbb{R}^n_+\|^p_{W_p^{[m]+1,\beta}}.$$

It remains to estimate the integral

$$\int_{0<y<1} y^{p(1-l)-1} |(T\gamma)(z)|^p |U(z)|^p dz.$$

Clearly,

$$\int_{0<y<1} y^{p(1-l)-1} |(T\gamma)(z)|^p |U(z)|^p dz$$
$$\leq \int_{0<y<1} y^{p(1-l)-1} |(T\gamma)(z)|^p |\mathcal{R}_0 U(z)|^p dz$$
$$+ \sum_{k=0}^{[m]} \int_{0<y<1} y^{p(1-l+k)-1} |(T\gamma)(z)|^p |\nabla_k U(x,0)|^p dz. \quad (8.5.16)$$

By (8.5.8) and (8.5.9) with $j=1$ we have

$$\int_{0<y<1} y^{p(1-l)-1} |(T\gamma)(z)|^p |\mathcal{R}_0 U(z)|^p dz$$
$$\leq c \,\|\gamma; \mathbb{R}^{n-1}\|^p_{M(W_p^{m-l}\to L_p)} \int_{0<y<1} y^{p(1-\{m\})-1} \Big(\int_0^y |\nabla_{[m]+1} U(x,t)| dt \Big)^p dz$$

which, by Hardy's inequality, is dominated by (8.5.10). In view of (8.5.13),

$$\int_{0<y<1} y^{p(k-l)-1} |(T\gamma)(z)|^p |\nabla_k U(x,0)|^p dz$$
$$\leq c \,\|\gamma; \mathbb{R}^{n-1}\|^p_{M(W_p^{m-l}\to L_p)} \|U; \mathbb{R}^n_+\|^p_{W_p^{[m]+1,\beta}}.$$

Thus, we arrive at the estimate

$$\int_{0<y<1} \dots dz \leq c \,\|\gamma; \mathbb{R}^{n-1}\|^p_{M(W_p^m\to W_p^l)} \|U; \mathbb{R}^n_+\|^p_{W_p^{[m]+1,\beta}}.$$

Since the Poisson kernel satisfies condition (8.4.2), Theorem 8.5.1 with $l<1$ follows. □

8.5.2 The Case $l > 1$

Lemma 8.5.1. *Let m and l be nonintegers with $m \geq l > 1$, and let T be the extension operator* (8.4.1). *Suppose that $\gamma \in M(W_p^{m-l}(\mathbb{R}^{n-1}) \to L_p(\mathbb{R}^{n-1}))$. Then*

$$T\gamma \in M(W_p^{[m]-[l],\beta}(\mathbb{R}^n_+) \to W_p^{0,\alpha}(\mathbb{R}^n_+))$$

and

$$\|T\gamma; \mathbb{R}^n_+\|_{M(W_p^{[m]-[l],\beta}\to W_p^{0,\alpha})} \leq c\,\|\gamma; \mathbb{R}^{n-1}\|_{M(W_p^{m-l}\to L_p)}. \quad (8.5.17)$$

Proof. To begin with, let $[m] = [l]$. Then by (8.5.8)

$$\int_{0<y<1} y^{p(1-\{l\})-1} |U(z)(T\gamma)(z)|^p dz$$

$$\leq c\,\|\gamma;\mathbb{R}^{n-1}\|^p_{M(W_p^{m-l}\to L_p)} \int_{0<y<1} y^{p(1-\{m\})-1} |U(z)|^p dz$$

which gives the result.

Next suppose that $[m] \geq [l]+1$. We introduce the function

$$\mathcal{R}U = U(z) - \sum_{j=0}^{[m]-[l]-1} \frac{\partial^j U}{\partial y^j}(x,0)\frac{y^j}{j!}$$

which, clearly, satisfies

$$|\mathcal{R}U(z)| \leq \frac{y^{[m]-[l]-1}}{([m]-[l]-1)!}\int_0^y |\nabla_{[m]-[l]} U(x,t)|dt.$$

This estimate and (8.5.8) imply that

$$\int_{0<y<1} y^{p(1-\{l\})-1} |T\gamma(z)|^p |\mathcal{R}U(z)|^p dz$$

$$\leq c\,\|\gamma;\mathbb{R}^{n-1}\|^p_{M(W_p^{m-l}\to L_p)} \int_{0<y<1} y^{-p\{m\}-1}\Big(\int_0^y |\nabla_{[m]-[l]} U(x,t)|dt\Big)^p dz.$$

By Hardy's inequality, the right-hand side is majorized by

$$c\,\|\gamma;\mathbb{R}^{n-1}\|^p_{M(W_p^{m-l}\to L_p)} \|U;\mathbb{R}^n_+\|^p_{W_p^{[m]-[l],\beta}}.$$

Furthermore, by Lemma 8.4.2 with m replaced by $m-[l]$, l replaced by $\{l\}$, and $k = j+1$, we have for $j = 0,\dots,[m]-[l]-1$

$$\int_{0<y<1} y^{p(j+1-\{l\})-1} |T\gamma(z)|^p |\nabla_j U(x,0)|^p dz$$

$$\leq c\,\|\gamma;\mathbb{R}^{n-1}\|^{p\frac{j+1-\{l\}}{m-l}}_{M(W_p^{m-l}\to L_p)} \int_{\mathbb{R}^{n-1}} (\mathcal{M}\gamma(x))^{p\frac{m-[l]-j-1}{m-l}} |\nabla_j U(x,0)|^p dx. \quad (8.5.18)$$

The last integral is dominated by

$$\|(\mathcal{M}\gamma)^{p\frac{m-[l]-j-1}{m-l}};\mathbb{R}^{n-1}\|^p_{M(W_p^{m-[l]-j-1}\to L_p)} \|U(\cdot,0);\mathbb{R}^{n-1}\|^p_{W_p^{m-[l]-1}}$$

which by Lemma 2.3.6 does not exceed

$$\|\mathcal{M}\gamma;\mathbb{R}^{n-1}\|^{p\frac{m-[l]-j-1}{m-l}}_{M(W_p^{m-l}\to L_p)} \|U;\mathbb{R}^n_+\|^p_{W_p^{[m]-[l],\beta}}.$$

Therefore, using (8.5.18), we obtain

$$\int_{0<y<1} y^{p(j+1-\{l\})-1} |T\gamma(z)|^p |\nabla_j U(x,0)|^p dz$$
$$\leq c\,\|\gamma; \mathbb{R}^{n-1}\|^p_{M(W_p^{m-l} \to L_p)} \|U; \mathbb{R}^n_+\|^p_{W_p^{[m]-[l],\beta}}.$$

The result follows. □

8.5.3 Proof of Theorem 8.5.1 for $l > 1$

Suppose that Theorem 8.5.1 has been proved for $[l] = 1, \dots, \mathcal{L} - 1$, where $\mathcal{L} \geq 2$. Let $[l] = \mathcal{L}$ and let

$$\gamma \in M(W_p^m(\mathbb{R}^{n-1}) \to W_p^l(\mathbb{R}^{n-1})) \text{ for } m \geq \mathcal{L}. \tag{8.5.19}$$

Let $T\gamma$ denote the Poisson integral. Since by Theorem 4.1.1 we have

$$\gamma \in M(W_p^{m-l}(\mathbb{R}^{n-1}) \to L_p(\mathbb{R}^{n-1})),$$

it follows from Lemma 8.5.1 that

$$T\gamma \in M(W_p^{[m]-[l],\beta}(\mathbb{R}^n_+) \to W_p^{0,\alpha}(\mathbb{R}^n_+))$$

and (8.5.17) holds. Next we show that

$$\nabla_{\mathcal{L}+1}(T\gamma) \in M(W_p^{[m]+1,\beta}(\mathbb{R}^n_+) \to W_p^{0,\alpha}(\mathbb{R}^n_+)). \tag{8.5.20}$$

Using Lemma 4.3.5, we obtain

$$\frac{\partial\gamma}{\partial x_k} \in M(W_p^m(\mathbb{R}^{n-1}) \to W_p^{l-1}(\mathbb{R}^{n-1})), \quad k = 1, \dots, n-1.$$

Then, by the induction hypothesis applied to $\partial\gamma/\partial x_k$,

$$\frac{\partial}{\partial x_k}(T\gamma) = T\frac{\partial\gamma}{\partial x_k} \in M(W_p^{[m]+1,\beta}(\mathbb{R}^n_+) \to W_p^{\mathcal{L},\alpha}(\mathbb{R}^n_+)). \tag{8.5.21}$$

By Lemma 8.3.1,

$$\nabla_{\mathcal{L}}\frac{\partial}{\partial x_k}(T\gamma) \in M(W_p^{[m]+1,\beta}(\mathbb{R}^n_+) \to W_p^{0,\alpha}(\mathbb{R}^n_+)). \tag{8.5.22}$$

Using the harmonicity of $T\gamma$ and (8.5.22), we find that

$$\frac{\partial^{\mathcal{L}+1}(T\gamma)}{\partial y^{\mathcal{L}+1}} = -\frac{\partial^{\mathcal{L}-1}(\Delta_x(T\gamma))}{\partial y^{\mathcal{L}-1}} \in M(W_p^{[m]+1,\beta}(\mathbb{R}^n_+) \to W_p^{0,\alpha}(\mathbb{R}^n_+))$$

which together with (8.5.22) implies the inclusion (8.5.20). Combining this with (8.5.17), we find that $T\gamma \in M(W_p^{[m]+1,\beta}(\mathbb{R}^n_+) \to W_p^{[l]+1,\alpha}(\mathbb{R}^n_+))$. It remains to note that all the above inclusions are accompanied by the corresponding estimates. The result follows. □

8.6 Traces of Multipliers on the Smooth Boundary of a Domain

Let Ω be a bounded domain in $\mathbb{R}^n$ with smooth boundary $\partial\Omega$. It is well known that the fractional Sobolev space $W_p^l(\partial\Omega)$ is the space of traces of the weighted Sobolev space $W_p^{s,\alpha}(\Omega)$ endowed with the norm

$$\Big(\int_\Omega (\mathrm{dist}(x,\partial\Omega))^{p\alpha} \sum_{\{\tau:0\le|\tau|\le s\}} |D^\tau u|^p dx\Big)^{1/p},$$

where $\alpha = 1-\{l\}-1/p$, $s=[l]+1$ and $p\in(1,\infty)$ (see [Usp]). It is straightforward to deduce from this fact that the trace γ of the function

$$\Gamma \in M(W_p^{t,\beta}(\Omega) \to W_p^{s,\alpha}(\Omega)) \tag{8.6.1}$$

belongs to $M(W_p^m(\partial\Omega) \to W_p^l(\partial\Omega))$. Here m and l are nonintegers, $m \ge l > 0$, s and α are given above, $t=[m]+1$, and $\beta = 1-\{m\}-1/p$.

We prove that the converse assertion is also true showing that there exists an extension Γ of $\gamma \in M(W_p^m(\partial\Omega) \to W_p^m(\partial\Omega))$ subject to (8.6.1).

Theorem 8.6.1. *Let $\gamma \in M(W_p^m(\partial\Omega) \to W_p^l(\partial\Omega))$, where m and l are nonintegers, $m \ge l > 0$, and $p \in (1,\infty)$. There exists a linear extension operator*

$$\gamma \to \Gamma \in M(W_p^{t,\beta}(\Omega) \to W_p^{s,\alpha}(\Omega)),$$

where $t=[m]+1$, $s=[l]+1$, $\beta=1-\{m\}-1/p$, and $\alpha = 1-\{l\}-1/p$.

Proof. It suffices to construct an extension Γ only for γ with sufficiently small support. To be precise, we assume that $\gamma=0$ outside the ball $\mathcal{B}_\rho^{(n)}$ centered at $0\in\partial\Omega$, where ρ is small enough. We introduce a cutoff function $\varphi \in C_0^\infty(\mathcal{B}_{3\rho}^{(n)})$, equal to one on $\mathcal{B}_{2\rho}^{(n)}$. Let us define Cartesian coordinates $\zeta=(\xi,\eta)$ with the origin 0, where $\xi\in\mathbb{R}^{n-1}$ and $\eta\in\mathbb{R}^1$. Let

$$\Omega \cap \mathcal{B}_{3\rho}^{(n)} = \{\zeta : \xi \in \mathcal{B}_{3\rho}^{(n-1)},\ \ \eta > f(\xi)\},$$

where f is a smooth function. We make the change of variables $\tau : \zeta \to (x,y)$, where $x=\xi$, $y=\eta-f(\xi)$. The diffeomorphism τ maps $\Omega\cap\mathcal{B}_{3\rho}^{(n)}$ into the half space $\mathbb{R}_+^n = \{(x,y) : x\in\mathbb{R}^{n-1}, y>0\}$. Clearly, the function $\tilde\gamma = \gamma\circ\tau^{-1}$ belongs to $M(W_p^m(\mathbb{R}^{n-1}) \to W_p^l(\mathbb{R}^{n-1}))$. Its harmonic extension to $\mathbb{R}_+^n$, denoted by $\tilde\Gamma$, is in $M(W_p^{t,\beta}(\mathbb{R}_+^n) \to W_p^{s,\alpha}(\mathbb{R}_+^n))$ and satisfies the estimate (8.5.3) according to Theorem 8.5.1. Hence the function $\gamma = (\tilde\Gamma\circ\tau)\varphi$ is a desired extension. The proof is complete. □

8.7 $MW_p^l(\mathbb{R}^n)$ as the Space of Traces of Multipliers in the Weighted Sobolev Space $W_{p,\beta}^k(\mathbb{R}^{n+m})$

In this section we show that $MW_p^l(\mathbb{R}^n)$ is the space of traces on $\mathbb{R}^n$ of multipliers in weighted Sobolev spaces on $\mathbb{R}^{n+m}$ in the same way as $W_p^l(\mathbb{R}^n)$ is the space of traces of functions which belong to a weighted Sobolev space on $\mathbb{R}^{n+m}$. In Sect. 8.7.5 we give an application of this result to the first boundary value problem for an elliptic operator in $\mathbb{R}_+^{n+1}$. We prove the unique solvability of this problem in a space of multipliers on $\mathbb{R}_+^{n+1}$, assuming that the Dirichlet data belong to certain classes of multipliers on $\mathbb{R}^n$.

8.7.1 Preliminaries

Let $\mathbb{R}^{n+m} = \{z = (x, y)\colon x \in \mathbb{R}^n, y \in \mathbb{R}^m\}$, $m > 0$. For $U \in C_0^\infty(\mathbb{R}^{n+m})$ we introduce the norm

$$\langle U \rangle_{k,p,\beta} = \Big(\int_{\mathbb{R}^{n+m}} |y|^{p\beta} |\nabla_{k,z} U|^p \, dz \Big)^{1/p}.$$

By Hardy's inequality,

$$\langle U \rangle_{k-r,p,\beta-r} \le c \, \langle U \rangle_{k,p,\beta}$$

for $k > r$, $\beta > r - m/p$.

Let $W_{p,\beta}^k(\mathbb{R}^{n+m})$ denote the completion of $C_0^\infty(\mathbb{R}^{n+m})$ with respect to the norm

$$\langle U \rangle_{k,p,\beta} + \|U; \mathbb{R}^{n+m}\|_{L_p}.$$

The following known assertion (see [Usp]) gives a characterization of traces on $\mathbb{R}^n$ for functions in $W_{p,\beta}^k(\mathbb{R}^{n+m})$.

Lemma 8.7.1. *(i) Let U be an arbitrary function in $W_{p,\beta}^k(\mathbb{R}^{n+m})$, and let $l = k - \beta - m/p$ be a positive noninteger , $l < k$. Then for almost all $x \in \mathbb{R}^n$ the limit*

$$u(x) = \lim_{|y| \to 0} U(x, y) \quad \text{in} \quad L_p(\partial \mathcal{B}^{(n-1)})$$

exists. The function u belongs to the space $W_p^l(\mathbb{R}^n)$ and the estimate

$$\|u; \mathbb{R}^n\|_{W_p^l} \le c \, \|U; \mathbb{R}^{n+m}\|_{W_{p,\beta}^k} \tag{8.7.1}$$

holds.

(ii) Let l be a positive noninteger. There exists a linear bounded extension operator

$$E\colon W_p^l(\mathbb{R}^n) \ni u \to U \in W_{p,\beta}^k(\mathbb{R}^{n+m}),$$

where $k > l$ and $\beta = k - l - m/p$.

Lemma 8.7.2. *Let r be a positive noninteger and let ω be an m-tuple multi-index, $|\omega| < [r]$. Further, let*

$$R_\omega(h,x) = D^\omega\gamma(x+h) - \sum_{|\nu|<r-|\omega|} D^{\nu+\omega}\gamma(x)\frac{h^\nu}{\nu!}\,. \tag{8.7.2}$$

Then

$$\Big(\int_{\mathbb{R}^n} |h|^{p(|\omega|-r)-n}|R_\omega(h,x)|^p\,dh\Big)^{1/p} \le c\,(D_{p,r}\gamma)(x)\,. \tag{8.7.3}$$

Proof. It follows from the identity

$$R_\omega(h,x) = ([r]-|\omega|)\int_0^1 \sum_{|\nu|=[r]-|\omega|} \frac{h^\nu}{\nu!}$$
$$\times[(D^{\nu+\omega}\gamma)(x+th) - D^{\nu+\omega}\gamma(x)](1-t)^{[r]-|\omega|-1}\,dt$$

and the Minkowski inequality that the left-hand side of (8.7.3) does not exceed

$$c\int_0^1\Big(\int_{\mathbb{R}^n}|h|^{-p\{r\}-n}\sum_{|\alpha|=[r]}|(D^\alpha\gamma)(x+th)-(D^\alpha\gamma)(x)|^p\,dh\Big)^{1/p}dt$$

which is equal to $c\,(D_{p,r}\gamma)(x)$. □

8.7.2 A Property of Extension Operator

We introduce an extension operator which maps functions defined on $\mathbb{R}^n$ into functions on $\mathbb{R}^{n+m}$ by

$$(\mathcal{T}\gamma)(x,y) = \int\zeta(t)\gamma(x+|y|t)\,dt, \qquad x\in\mathbb{R}^n,\ \ y\in\mathbb{R}^m,$$

where

$$\zeta\in C^\infty(\mathbb{R}^n)\cap L(\mathbb{R}^n), \qquad \int\zeta(x)\,dx = 1.$$

Lemma 8.7.3. *Let $\{r\}>0$, $p\in[1,\infty)$ and let q be an integer with $q>r$. Further, let*

$$\int x^\alpha\zeta(x)\,dx = 0\,, \qquad 0<|\alpha|\le[r] \tag{8.7.4}$$

and

$$\mathcal{C}_{q,r} = \int(1+|x|)^r\sum_{j=0}^{q}\sup_{\partial\mathcal{B}^{(n)}_{|x|}}|\nabla_j\zeta(x)|(1+|x|)^j\,dx\,. \tag{8.7.5}$$

Then the estimate holds:

$$\Big(\int_{\mathbb{R}^m}|y|^{p(q-r)-m}|\nabla_{q,z}(\mathcal{T}\gamma)|^p\,dy\Big)^{1/p} \le c\,\mathcal{C}_{q,r}\,(D_{p,r}\gamma)(x)\,. \tag{8.7.6}$$

Proof. Let τ, σ, ρ, ω be m-tuple multi-indices such that $|\tau|+|\sigma|=q$, and also

$$\rho=0, \quad \omega=\tau \quad \text{if } |\tau|\le r$$

and

$$\rho=\tau-\omega>0, \quad |\omega|=[r] \quad \text{if } |\tau|>r.$$

We have

$$\begin{aligned} D_x^\tau D_y^\sigma \int \zeta(t)\gamma(x+|y|t)dt &= D_x^\rho D_y^\sigma \int \zeta(t) D_x^\omega \gamma(x+|y|t)dt \\ &= D_y^\sigma\Big(|y|^{-n-|\rho|}\int (D^\rho\zeta)\Big(\frac{\xi-x}{|y|}\Big)D^\omega\gamma(\xi)d\xi\Big) \\ &= D_y^\sigma\Big(|y|^{-n-|\rho|}\int (D^\rho\zeta)\Big(\frac{\xi-x}{|y|}\Big)R_\omega(\xi-x,x)d\xi\Big), \end{aligned}$$

where R_ω is the function defined by (8.7.2). Here we used the identity

$$\begin{aligned} & D_y^\sigma\Big(|y|^{-n-|\rho|}\int (D^\rho\zeta)\Big(\frac{\xi-x}{|y|}\Big)(\xi-x)^\nu d\xi\Big) \\ &\quad = D_y^\sigma\Big(|y|^{|\nu|-|\rho|}\int D^\rho\zeta(\xi)\xi^\nu d\xi\Big)=0. \end{aligned} \tag{8.7.7}$$

It is clear that

$$\begin{aligned} & \Big|D_y^\sigma\Big(|y|^{-n-|\rho|}\int (D^\rho\zeta)\Big(\frac{\xi-x}{|y|}\Big)R_\omega(\xi-x,x)\Big)\Big| \\ &\quad \le |y|^{r-|\rho|-|\sigma|-|\omega|}\varphi\Big(\frac{\xi-x}{|y|}\Big)\frac{|R_\omega(\xi-x,x)|}{|\xi-x|^{r-|\omega|+n}} \end{aligned}$$

with φ standing for a nonnegative function for which the estimate

$$\varphi(\xi)\le c\,|\xi|^{r-|\omega|+n}\sum_{i=0}^{|\sigma|}|\nabla_{i+|\rho|}\zeta(\xi)|\,(|\xi|^i+1) \tag{8.7.8}$$

holds. Since $|\rho|+|\omega|=|\tau|$ and $|\tau|+|\sigma|=q$, it follows that

$$\begin{aligned} & \int_{\mathbb{R}^m}|y|^{p(q-r)-m}|D_x^\tau D_y^\sigma(T\gamma)|^p\,dy \\ &\quad \le c\int_{\mathbb{R}^m}\Big(\int_{\mathbb{R}^n}\varphi\Big(\frac{\xi-x}{|y|}\Big)\frac{|R_\omega(\xi-x,x)|}{|\xi-x|^{r-|\omega|+n}}\,d\xi\Big)^p\frac{dy}{|y|^m}. \end{aligned}$$

Introducing spherical coordinates, we write the right-hand side in the form

$$c\int_0^\infty\frac{d\lambda}{\lambda}\Big(\int_0^\infty\int_{\partial B_1}\varphi\Big(\frac{t\theta}{\lambda}\Big)\frac{|R_\omega(t\theta,x)|}{t^{r-|\omega|}}\,\frac{dt}{t}\,d\theta\Big)^p.$$

This value is not greater than

$$c\int_0^\infty \frac{d\lambda}{\lambda}\Big(\int_0^\infty Q\Big(\frac{t}{\lambda}\Big)g(t)\frac{dt}{t}\Big)^p \le c\Big(\int_0^\infty Q(t)\frac{dt}{t}\Big)^p \int_0^\infty g(t)^p\frac{dt}{t},$$

where

$$Q(t) = \sup_{|\theta|=1} \varphi(t\theta), \qquad g(t) = t^{|\omega|-r}\int_{\partial\mathcal{B}_1} |R_\omega(t\theta, x)|d\theta.$$

Now we use (8.7.8) to obtain

$$\int_{\mathbb{R}^m} |y|^{p(q-r)-m}|D_x^\tau D_y^\sigma(T\gamma)|^p\, dy$$

$$\le c\Big(\int_{\mathbb{R}^n}|\xi|^{r-|\omega|}\sum_{i=0}^{|\sigma|}(|\xi|+1)^i \sup_{\partial\mathcal{B}_{|\xi|}}|\nabla_{i+|\rho|}\zeta(\xi)|d\xi\Big)^p\int_{\mathbb{R}^n}|h|^{p(|\omega|-r)-n}|R_\omega(h,x)|^p dh.$$

Making use of Lemma 8.7.2, we complete the proof. □

8.7.3 Trace and Extension Theorem for Multipliers

The following theorem shows that $MW_p^l(\mathbb{R}^n)$ coincides with the space of traces on $\mathbb{R}^n$ of functions in $MW_{p,\beta}^k(\mathbb{R}^{n+m})$.

Theorem 8.7.1. *(i) Let $\Gamma \in MW_{p,\beta}^k(\mathbb{R}^{n+m})$ with $\beta = k-l-m/p$, $\{l\} > 0$, and $k > l$. Further, let γ be the trace of Γ on $\mathbb{R}^n$ (its existence results from $\Gamma \in W_{p,\beta,loc}^k(\mathbb{R}^{n+m})$ and Lemma 8.7.1). Then $\gamma \in MW_p^l(\mathbb{R}^n)$ and the estimate*

$$\|\gamma;\mathbb{R}^n\|_{MW_p^l} \le c\,\|\Gamma;\mathbb{R}^{n+m}\|_{MW_{p,\beta}^k} \tag{8.7.9}$$

holds.

(ii) Let $\{l\} > 0$, $s = 0, 1, \ldots$, and let k be an integer with $k > l$. Further, let $\nabla_s\gamma \in MW_p^l(\mathbb{R}^n)$ and let $\mathcal{T}\gamma$ be the extension of γ to $\mathbb{R}^{n+m}$ defined in Sect. 8.7.2 with ζ subject to the conditions

$$\mathcal{C}_{k+s,l+s} < \infty, \tag{8.7.10}$$

and

$$\int x^\alpha\zeta(x)\,dx = 0, \qquad 0 < |\alpha| \le [l] + s \tag{8.7.11}$$

where $\mathcal{C}_{q,r}$ is given by (8.7.5). Then

$$\nabla_{s,z}(\mathcal{T}\gamma) \in MW_{p,\beta}^k(\mathbb{R}^{n+m}), \qquad \beta = k-l-m/p,$$

and

$$\|\nabla_{s,z}(\mathcal{T}\gamma);\mathbb{R}^{n+m}\|_{MW_{p,\beta}^k} \le c\,\mathcal{C}_{k+s,l+s}\|\nabla_{s,x}\gamma;\mathbb{R}^n\|_{MW_p^l}. \tag{8.7.12}$$

Proof. (*i*) Let $U \in W_{p,\beta}^k(\mathbb{R}^{n+m})$ and let $U(x,0) = u(x)$. We have

$$\|\gamma u;\mathbb{R}^n\|_{W_p^l} \le c\,\|\Gamma U;\mathbb{R}^{n+m}\|_{W_{p,\beta}^k} \le c\,\|\Gamma;\mathbb{R}^{n+m}\|_{MW_{p,\beta}^k}\|U;\mathbb{R}^{n+m}\|_{W_{p,\beta}^k}.$$

Using the second part of Lemma 8.7.1 and the arbitrariness of U, we arrive at (8.7.9).

(*ii*) Let χ be any m-tuple multi-index with $|\chi| = s$. It is clear that

$$\langle U D_y^\chi(\mathcal{T}\gamma)\rangle_{k,p,\beta} \le c \sum_{|\varepsilon|+|\mu|+|\nu|=k} \langle D_z^\nu U D_x^\mu D_y^{\chi+\varepsilon}(\mathcal{T}\gamma)\rangle_{0,p,\beta}. \quad (8.7.13)$$

For $|\mu| < s$ we have

$$D_x^\mu D_y^{\chi+\varepsilon}(\mathcal{T}\gamma)(z) = D_y^{\chi+\varepsilon}|y|^{-n}\int \zeta\Big(\frac{\xi - x}{|y|}\Big)$$

$$\times\Big[D_\xi^\mu\gamma(\xi) - \sum_{|\varkappa|\le s-|\mu|-1}\frac{(\xi-x)^\varkappa}{\varkappa!}D^{\varkappa+\mu}\gamma(x)\Big]\,d\xi \quad (8.7.14)$$

and, for $|\mu| \ge s$,

$$D_x^\mu D_y^{\chi+\varepsilon}(\mathcal{T}\gamma)(z) = D_x^{\mu_1} D_y^{\chi+\varepsilon}\Big(|y|^{-n}\int \zeta\Big(\frac{\xi-x}{|y|}\Big)D_\xi^{\mu_2}\gamma(\xi)\,d\xi\Big),$$

where $\mu = \mu_1 + \mu_2$, $|\mu_1| > 0$, and $|\mu_2| = s$. Thus, in both cases,

$$|D_x^\mu D_y^{\chi+\varepsilon}(\mathcal{T}\gamma)(z)| \le c\,K\|\nabla_s\gamma;\mathbb{R}^n\|_{L_\infty}|y|^{-|\mu|-|\varepsilon|}. \quad (8.7.15)$$

Here and in what follows we put

$$K = \mathcal{C}_{k+s,l+s}$$

with $\mathcal{C}_{k+s,l+s}$ given by (8.7.10). Hence, for $|\nu| > l$, we obtain

$$\langle D_z^\nu U D_x^\mu D_y^{\chi+\varepsilon}(\mathcal{T}\gamma)\rangle_{0,p,\beta} \le cK\|\nabla_s\gamma;\mathbb{R}^n\|_{L_\infty}\langle U\rangle_{|\nu|,p,|\nu|-l-m/p}$$

$$\le cK\|\nabla_s\gamma;\mathbb{R}^n\|_{L_\infty}\langle U\rangle_{k,p,\beta}. \quad (8.7.16)$$

Now let $|\nu| < l$. Using Taylor's formula for $D_z^\nu U$, we have

$$\langle D_z^\nu U D_x^\mu D_y^{\chi+\varepsilon}(\mathcal{T}\gamma)\rangle_{0,p,\beta}$$

$$\le \langle R_\nu D_z^\mu D_y^{\chi+\varepsilon}(\mathcal{T}\gamma)\rangle_{k,p,\beta} + \sum_{j=0}^{[l]-|\nu|}\Big(\int_{\mathbb{R}^{n+m}}|y|^{p(k-l+j)-m}$$

$$\times|D_z^\mu D_y^{\chi+\varepsilon}(\mathcal{T}\gamma)(z)|^p|(\nabla_{j,y}D_z^\nu U)(x,0)|^p\,dz\Big)^{1/p}, \quad (8.7.17)$$

where

$$R_\nu(z) = D_z^\nu U(z) - \sum_{|\tau|\le [l]-|\nu|} (D_y^\tau D_z^\nu U)(x,0)\frac{y^\tau}{\tau!}$$
$$= ([l]-|\nu|-1) \sum_{|\tau|=[l]-|\nu|+1} \frac{y^\tau}{\tau!}\int_0^1 (D_y^\tau D_z^\nu U)(x,ty)(1-t)^{[l]-|\nu|}\,dt\,.$$

By (8.7.15) and Minkowski's inequality,

$$\langle R_\nu D_z^\mu D_y^{\chi+\varepsilon}(\mathcal{T}\gamma)\rangle_{k,p,\beta}$$
$$\le cK\|\nabla_s\gamma;\mathbb{R}^n\|_{L_\infty}\Big(\int |y|^{p(1-\{l\})-m}\Big(\int_0^1 |(\nabla_{[l]+1,z}U)(x,ty)|\,dt\Big)^p dz\Big)^{1/p}$$
$$\le cK\|\nabla_s\gamma;\mathbb{R}^n\|_{L_\infty}\langle U\rangle_{[l]+1,p,1-\{l\}-m/p}$$
$$\le cK\|\nabla_s\gamma;\mathbb{R}^n\|_{L_\infty}\langle U\rangle_{k,p,\beta}\,.$$

Using Lemma 8.7.3 with $q=|\mu|+s+|\varepsilon|$ and $r=l+s-j-|\nu|$, we see that the sum on the right-hand side of (8.7.17) does not exceed

$$cK\sum_{j=0}^{[l]-|\nu|}\Big(\int_{\mathbb{R}^n}|(\nabla_{j,y}D_z^\nu U)(x,0)|^p[(D_{p,l-j-|\nu|}\nabla_s\gamma)(x)]^p\,dx\Big)^{1/p}. \tag{8.7.18}$$

By Lemma 4.3.2, for $p>1$ this is majorized by

$$cK\sum_{j=0}^{[l]-|\nu|}\sup_e\left(\frac{\int_e |D_{p,l-j-|\nu|}\nabla_s\gamma|^p\,dx}{C_{p,l-j-|\nu|}(e)}\right)^{1/p}\|(\nabla_{j+|\nu|,z}U)(\cdot,0);\mathbb{R}^n\|_{W_p^{l-j-|\nu|}}\,.$$

which by Theorem 4.1.1 is not greater than

$$cK\|\nabla_s\gamma;\mathbb{R}^n\|_{MW_p^l}\|U;\mathbb{R}^{n+m}\|_{W_{p,\beta}^k}\,. \tag{8.7.19}$$

The same estimate results in the case $p=1$ from Theorems 5.1.2 and 5.3.2. So, for $|\nu|<l$,

$$\langle D_z^\nu U D_x^\mu D_y^{\chi+\varepsilon}(\mathcal{T}\gamma)\rangle_{0,p,\beta}\le cK\|\nabla_s\gamma;\mathbb{R}^n\|_{MW_p^l}\|U;\mathbb{R}^{n+m}\|_{W_{p,\beta}^k}\,.$$

Combining this estimate with (8.7.16) and (8.7.13), we find that

$$\langle UD_y^\chi(\mathcal{T}\gamma)\rangle_{0,p,\beta}\le cK\|\nabla_s\gamma;\mathbb{R}^n\|_{MW_p^l}\|U;\mathbb{R}^{n+m}\|_{W_{p,\beta}^k}\,.$$

According to (8.7.15) with $|\mu|=|\varepsilon|=0$, we have

$$|D_y^\chi(\mathcal{T}\gamma)|\le cK\|\nabla_s\gamma;\mathbb{R}^n\|_{L_\infty}.$$

Therefore,

$$\|UD_y^\chi(\mathcal{T}\gamma);\mathbb{R}^{n+m}\|_{L_p}\le cK\|\nabla_s\gamma;\mathbb{R}^n\|_{L_\infty}\|U;\mathbb{R}^{n+m}\|_{L_p}\,.$$

The proof is complete. □

8.7.4 Extension of Multipliers from $\mathbb{R}^n$ to $\mathbb{R}_+^{n+1}$

An assertion analogous to Theorem 8.7.1 is also valid for the space of multipliers $MW_{p,\beta}^k(\mathbb{R}_+^{n+1})$, where $\mathbb{R}_+^{n+1} = \{z = (x, y) \colon x \in \mathbb{R}^n, y > 0\}$ and $W_{p,\beta}^k(\mathbb{R}_+^{n+1})$ is the completion of $C_0^\infty(\overline{\mathbb{R}_+^{n+1}})$ with respect to the norm

$$\Big(\int_{\mathbb{R}_+^{n+1}} y^{p\beta} |\nabla_{k,z} U|^p \, dz\Big)^{1/p} + \|U; \mathbb{R}_+^{n+1}\|_{L_p} .$$

If we put $m = 1$ and replace $\mathbb{R}^m$ by $\mathbb{R}_+^1 = \{y \colon y > 0\}$ in the statement of Lemma 8.7.3, then it holds without the condition (8.7.4). We only need to verify that

$$D_y^\sigma(y^{|\nu|-|\rho|}) \int D^\rho \zeta(\xi) \xi^\nu \, d\xi = 0 . \tag{8.7.20}$$

For $|\nu| < |\rho|$ after integration by parts we obtain that the integral in (8.7.20) is equal to zero. In the case $|\nu| \geq |\rho|$ the function $D_y^\sigma(y^{|\nu|-|\rho|})$ vanishes identically, since $\sigma > \sigma + [r] - q = [r] - |\tau| = [r] - |\omega| - |\rho| \geq |\nu| - |\rho|$.

The condition (8.7.11) was used only in (8.7.14), in the proof of Theorem 8.7.1. Since $|\chi| + |\varepsilon| = s + |\varepsilon| > s - |\mu| - 1 \geq |\chi|$, we conclude that (8.7.14) remains valid for $m = 1$ and $y \in \mathbb{R}_+^1$ without the condition (8.7.11). Therefore, we have the following assertion.

Theorem 8.7.2. *(i) Let $\{l\} > 0$, $\Gamma \in MW_{p,\beta}^k(\mathbb{R}_+^{n+1})$, $\beta = k - l - 1/p$, $k > l$ and $\gamma(x) = \Gamma(x, 0)$. Then $\gamma \in MW_p^l(\mathbb{R}^n)$ and*

$$\|\gamma; \mathbb{R}^n\|_{MW_p^l} \leq c \, \|\Gamma; \mathbb{R}_+^{n+1}\|_{MW_{p,\beta}^k} .$$

(ii) Let $\{l\} > 0$ and $\nabla_s \gamma \in MW_p^l(\mathbb{R}^n)$. Further, let $\mathcal{T}\gamma$ be the extension of γ to $\mathbb{R}_+^{n+1}$ defined in Sect. 8.7.2, where the function ζ is subject only to the condition (8.7.10). Then $\nabla_s(\mathcal{T}\gamma) \in MW_{p,\beta}^k(\mathbb{R}_+^{n+1})$, $k > l$, $\beta = k - l - 1/p$, and the estimate

$$\|\nabla_s(\mathcal{T}\gamma); \mathbb{R}_+^{n+1}\|_{MW_{p,\beta}^k} \leq c\, C_{k+s,l+s} \|\nabla_s \gamma; \mathbb{R}^n\|_{MW_p^l}$$

holds.

8.7.5 Application to the First Boundary Value Problem in a Half-Space

Let us consider the Dirichlet problem in the half-space $\mathbb{R}_+^{n+1}$

$$\begin{aligned} L(D)U &= 0 \qquad \text{for} \quad y \geq 0 , \\ \partial^j U / \partial y^j &= \varphi_j \qquad \text{for} \quad y = 0 , \qquad j = 0, \ldots, m-1 , \end{aligned}$$

where L is a homogeneous differential elliptic operator of order $2m$ with constant coefficients.

Theorem 8.7.3. *Let $\nabla_{m-1-j}\varphi_j \in MW_p^l(\mathbb{R}^n)$, where $0 < l < 1$, $1 \le p \le \infty$. Then there exists one and only one solution of the Dirichlet problem such that*

$$\nabla_{m-1}U \in MW^k_{p,k-l-1/p}(\mathbb{R}^{n+1}_+), \qquad k \ge 1.$$

This solution satisfies the estimate

$$\|\nabla_{m-1}U; \mathbb{R}^{n+1}_+\|_{MW^k_{p,k-l-1/p}} \le K \sum_{j=0}^{m-1} \|\nabla_{m-1-j}\varphi_j; \mathbb{R}^n\|_{MW^l_p},$$

where K is a constant which depends on L and n, p, m, k, l.

Proof. If $U \in MW^k_{p,k-l-1/p}(\mathbb{R}^{n+1}_+)$ is a solution of the homogeneous problem, then

$$\|\nabla_{m-l}U; \mathbb{R}^{n+1}_+\|_{L_\infty} < \infty$$

and hence $U = 0$ (see, for instance, [ADN1], Ch. I, §2).

According to the same reference, the existence of a solution follows from the assumption

$$\nabla_{m-1-j}\varphi_j \in L_\infty(\mathbb{R}^n), \qquad j = 0, 1, \dots, m-1,$$

and the solution satisfies the equality

$$D_x^\alpha \frac{\partial^i}{\partial y^i} U(x, y) = \sum_{j=0}^{m-1} \sum_{|\beta|=m-1-j} \int_{\mathbb{R}^n} K_{i,j,\beta}(x-\xi, y) D_\xi^\beta \varphi_j(\xi)\, d\xi,$$

where $0 \le i \le m-1$, α is any multi-index of order $m-1-i$ and $K_{i,j,\beta}(z)$ are positive homogeneous functions of order $-n$, smooth in $\mathbb{R}^{n+1}\backslash\{0\}$ and such that $K_{i,j,\beta}(x, 0) = 0$ for $x \ne 0$. These conditions imply the estimate

$$(|x|^2 + y^2)^{1/2} |\nabla_x K_{i,j,\beta}(x, y)| + |K_{i,j,\beta}(x, y)| \le c\, y(|x|^2 + y^2)^{-(n+1)/2}$$

which shows that the function $\zeta(x) = K_{i,j,\beta}(x, 1)$ satisfies (8.7.10) for $s = 0$, $0 < l < 1$. It remains to make use of Theorem 8.7.2. □

8.8 Traces of Functions in $MW_p^l(\mathbb{R}^{n+m})$ on $\mathbb{R}^n$

In this section we show that the space of restrictions of functions in $MW_p^l(\mathbb{R}^{n+m})$ ($lp > n$, $l - m/p$ is a noninteger) to $\mathbb{R}^n$ coincides with $MW_p^{l-m/p}(\mathbb{R}^n)$.

8.8.1 Auxiliary Assertions

We use the extension operator $\mathcal{T}$ defined in Sect. 8.7.2. Let us assume that the conditions (8.7.10) and (8.7.11) for $k = [l]$, $s = 1$ are fulfilled.

Lemma 8.8.1. *Let $\sigma \in (0, l]$ and let $p \in [1, \infty)$. Then*

$$\Big(\int_{2|\eta|<|y|} |\nabla_{[\sigma],y}(\mathcal{T}\gamma)(x, y+\eta) - \nabla_{[\sigma],y}(\mathcal{T}\gamma)(x,y)|^p |\eta|^{-m-p\{\sigma\}}\, d\eta\Big)^{1/p}$$
$$\le c\,C_{[l]+1,l+1}\, |y|^{-\sigma} \|\gamma; \mathbb{R}^n\|_{L_\infty}, \tag{8.8.1}$$

and

$$\Big(\int_{\mathbb{R}^n} |\nabla_{[\sigma],x}(\mathcal{T}\gamma)(x+h, y) - \nabla_{[\sigma],x}(\mathcal{T}\gamma)(x,y)|^p |h|^{-n-p\{\sigma\}}\, dh\Big)^{1/p}$$
$$\le c\,C_{[l]+1,l+1} |y|^{-\sigma} \|\gamma; \mathbb{R}^n\|_{L_\infty}, \tag{8.8.2}$$

where $C_{[l]+1,l+1}$ is defined by (8.7.10).

Proof. Let A_y and B_x denote the left-hand sides of (8.8.1) and (8.8.2). One verifies directly that

$$A_y \le \|\gamma; \mathbb{R}^n\|_{L_\infty} \Big\{ \int_{2|\eta|<|y|} \Big(\int_{\mathbb{R}^n} \Big| \nabla_{[\sigma],y} \Big[\zeta\Big(\frac{\xi - x}{|y+\eta|}\Big) |y+\eta|^{-n}$$
$$- \zeta\Big(\frac{\xi - x}{|y|}\Big) |y|^{-n} \Big] \Big|\, d\xi \Big)^p |\eta|^{-m-p\{\sigma\}}\, d\eta \Big\}$$
$$\le \|\gamma; \mathbb{R}^n\|_{L_\infty} \Big\{ \int_{2|\eta|<|y|} \frac{|\eta|^p\, d\eta}{|\eta|^{m+p\{\sigma\}}}$$
$$\times \Big(\int_0^1 dz \int_{\mathbb{R}^n} |\varphi_{\xi - x}^{(\sigma+1)}[|y| + z(|y+\eta| - |y|)]|\, d\xi \Big)^p \Big\}^{1/p},$$

where

$$\varphi_{\xi-x}^{(l)}(t) = -\frac{\partial^{[l]}}{\partial t^{[l]}} \Big(t^{-n} \zeta\Big(\frac{\xi - x}{t}\Big)\Big).$$

Therefore,

$$A_y \le c\,C_{[l]+1,l+1} \|\gamma; \mathbb{R}^n\|_{L_\infty}$$
$$\times \Big\{ \int_{2|\eta|<|y|} |\eta|^{p(1-\{\sigma\})-m}\, d\eta \Big(\int_0^1 (|y| + z(|y+\eta| - |y|))^{-[\sigma]-1}\, dz \Big)^p \Big\}^{1/p}$$
$$\le c\,C_{[l]+1,l+1} \|\gamma; \mathbb{R}^n\|_{L_\infty} |y|^{-\sigma}.$$

The inequality (8.8.1) is proved.

Making the change of variables

$$\xi - x = |y|\Xi, \qquad h = |y|H,$$

we obtain

$$B_x \leq |y|^{-\sigma}\Big\{\int_{\mathbb{R}^n}\Big(\int_{\mathbb{R}^n}|\nabla_{[\sigma],\Xi}[\zeta(\Xi+H)-\zeta(\Xi)]|\,d\Xi\Big)^p|H|^{-n-p\{\sigma\}}\,dH\Big\}^{1/p}.$$

Now we divide the exterior integral into two integrals, the first of which is over the ball $\mathcal{B}_1$. We have

$$\begin{aligned}\int_{\mathcal{B}_1} &\leq \int_{\mathcal{B}_1}\Big(\int_{\mathbb{R}^n}\Big|\sum_{i=1}^n H_i\int_0^1\frac{\partial}{\partial\Xi_i}\nabla_{[\sigma],\Xi}\zeta(\Xi+zH)\,dz\Big|\,d\Xi\Big)^p|H|^{-n-p\{\sigma\}}\,dH\\ &\leq \Big(\int_{\mathbb{R}^n}|\nabla_{[\sigma]+1}\zeta(\Xi)|\,d\Xi\Big)^p\int_{\mathcal{B}_1}|H|^{-n+(1-\{\sigma\})p}dH \leq c\,C^p_{[l]+1,l+1}\,.\end{aligned}$$

Finally,

$$\begin{aligned}\int_{\mathbb{R}^n\setminus\mathcal{B}_1} &\leq \int_{\mathbb{R}^n\setminus\mathcal{B}_1}\Big(\int_{\mathbb{R}^n}(|\nabla_{[\sigma]}\zeta(\Xi+H)|+|\nabla_{[\sigma]}\zeta(\Xi)|)\,d\Xi\Big)^p|H|^{-n-p\{\sigma\}}\,dH\\ &= 2^p\Big(\int_{\mathbb{R}^n}|\nabla_{[\sigma]}\zeta(\Xi)|\,d\Xi\Big)^p\int_{\mathbb{R}^n\setminus\mathcal{B}_1}|H|^{-n-p\{\sigma\}}\,dH \leq c\,C^p_{[l]+1,l+1}\,.\end{aligned}$$

The proof is complete. □

Let d_j denote the number of all derivatives of order j with respect to the variables $y_1,\dots,y_m$ and let $[W_p^\sigma(\mathbb{R}^n)]^{d_j}$ be the Cartesian product of d_j copies of the space $W_p^\sigma(\mathbb{R}^n)$. It is known (see [Usp] that there exists an extension operator E defined on vector-functions $(\boldsymbol{\varphi}_0,\boldsymbol{\varphi}_1,\dots,\boldsymbol{\varphi}_{[l-m/p]})$, where $\boldsymbol{\varphi}_j$ is the d_j-tuple vector-function. This operator maps into a space of scalar functions and has the following properties.

(*i*) E is a continuous operator:

$$\prod_{j=0}^{[l-m/p]}[W_p^{[l-j-m/p]}(\mathbb{R}^n)]^{d_j} \to W^k_{p,k-l}(\mathbb{R}^{n+m})\,. \tag{8.8.3}$$

(*ii*) The relation

$$(\nabla_j E\boldsymbol{\varphi})(x,0) = \boldsymbol{\varphi}_j(x) \qquad \text{with} \quad j=0,1,\dots,[l-m/p]$$

holds.

In what follows we use the fact that $W^{[l]+1}_{p,1-\{l\}}(\mathbb{R}^{n+m})$ is imbedded into $W^l_p(\mathbb{R}^{n+m})$ (see [Usp]). We also apply the Hardy-type inequality

$$\int_{\mathbb{R}^m} |y|^{-pl} |V(x,y)|^p \, dy \le c \, \|V(x,\cdot); \mathbb{R}^m\|^p_{W_p^l}, \tag{8.8.4}$$

where V is a function in $W_p^l(\mathbb{R}^m)$ subject to the conditions

$$(\nabla_j V)(x,0) = 0, \qquad j = 0,1,\dots,[l - m/p]. \tag{8.8.5}$$

Moreover, we make use of the following norm in $W_p^l(\mathbb{R}^{n+m})$:

$$\begin{aligned}
&\|U; \mathbb{R}^{n+m}\|_{W_p^l} \\
&\sim \Big(\int_{\mathbb{R}^n} dx \int_{\mathbb{R}^m} dy \int_{\mathbb{R}^m} |\nabla_{[l],y} U(x, y+\eta) - \nabla_{[l],y} U(x,y)|^p |\eta|^{-m-p\{l\}} \, d\eta\Big)^{1/p} \\
&+ \Big(\int_{\mathbb{R}^m} dy \int_{\mathbb{R}^n} dx \int_{\mathbb{R}^n} |\nabla_{[l],x} U(x+h, y) - \nabla_{[l],x} U(x,y)|^p |h|^{-n-p\{l\}} \, dh\Big)^{1/p} \\
&+ \|U; \mathbb{R}^{n+m}\|_{L_p}
\end{aligned} \tag{8.8.6}$$

(see Proposition 4.2.3).

8.8.2 Trace and Extension Theorem

Theorem 8.8.1. *(i) Let $lp > n$, $1 \le p < \infty$ and $l - m/p$ be noninteger. Further, let $\Gamma \in MW_p^l(\mathbb{R}^{n+m})$ and $\gamma(x) = \Gamma(x,0)$. Then $\gamma \in MW_p^{l-m/p}(\mathbb{R}^n)$ and*

$$\|\gamma; \mathbb{R}^n\|_{MW_p^{l-m/p}} \le c \, \|\Gamma; \mathbb{R}^{n+m}\|_{MW_p^l}. \tag{8.8.7}$$

(ii) Let the kernel ζ satisfy (8.7.10) *and* (8.7.11). *If $\gamma \in MW_p^{l-m/p}(\mathbb{R}^n)$ with $lp > m$, $1 \le p < \infty$ and noninteger $l - m/p$, then $\mathcal{T}\gamma \in MW_p^l(\mathbb{R}^{n+m})$ and*

$$\|\mathcal{T}\gamma \colon \mathbb{R}^{n+m}\|_{MW_p^l} \le c\,\mathcal{C}_{[l]+1,l+1} \|\gamma; \mathbb{R}^n\|_{MW_p^{l-m/p}}. \tag{8.8.8}$$

Proof. (i) Let $U \in W_p^l(\mathbb{R}^{n+m})$, $U(x,0) = u(x)$. We have

$$\|\gamma u; \mathbb{R}^n\|_{W_p^{l-m/p}} \le c \, \|\Gamma U; \mathbb{R}^{n+m}\|_{W_p^l} \le c \, \|\Gamma; \mathbb{R}^{n+m}\|_{MW_p^l} \|U; \mathbb{R}^{n+m}\|_{W_p^l}$$

which implies (8.8.7).

(ii) It is sufficient to assume that l is a noninteger, since for integer l the result is contained in Theorem 8.7.1.

Let $U \in W_p^l(\mathbb{R}^{n+m})$ and

$$\varphi(x) = (U(x,0), (\nabla_y(U)(x,0), \dots, (\nabla_{[l-m/p],y} U)(x,0)).$$

We introduce the function $V = U - E\varphi$, where E is the extension operator which was considered in Sect. 8.8.1. Then

$$\|U\mathcal{T}\gamma;\mathbb{R}^{n+m}\|_{W_p^l} \le \|(\mathcal{T}\gamma)E\varphi;\mathbb{R}^{n+m}\|_{W_p^l} + \|V\mathcal{T}\gamma;\mathbb{R}^{n+m}\|_{W_p^l}.$$

In view of the imbedding

$$W^{[l]+1}_{p,1-\{l\}}(\mathbb{R}^{n+m}) \subset W_p^l(\mathbb{R}^{n+m}),$$

the first term on the right-hand side does not exceed

$$c\,\|(\mathcal{T}\gamma)E\varphi;\mathbb{R}^{n+m}\|_{W^{[l]+1}_{p,1-\{l\}}}$$

which, by Theorem 8.7.1, is not greater than

$$c\,\mathcal{C}_{[l]+1,l+1}\|\gamma;\mathbb{R}^n\|_{MW_p^{l-m/p}}\|E\varphi;\mathbb{R}^{n+m}\|_{W^{[l]+1}_{p,1-\{l\}}}.$$

Since E performs the continuous mapping (8.8.3), it follows that

$$\|(\mathcal{T}\gamma)E\varphi;\mathbb{R}^{n+m}\|_{W_p^l}$$
$$\le c\,\mathcal{C}_{[l]+1,l+1}\|\gamma;\mathbb{R}^n\|_{MW_p^{l-m/p}}\sum_{j=0}^{[l-m/p]}\|(\nabla_{j,y}U)(\cdot,0);\mathbb{R}^n\|_{W_p^{l-m/p-j}}.$$

Consequently,

$$\|(\mathcal{T}\gamma)E\varphi;\mathbb{R}^{n+m}\|_{W_p^l} \le c\,\mathcal{C}_{[l]+1,l+1}\|\gamma;\mathbb{R}^n\|_{MW_p^{l-m/p}}\|U;\mathbb{R}^{n+m}\|_{W_p^l}. \quad (8.8.9)$$

Let us prove the inequality

$$\|V\mathcal{T}\gamma;\mathbb{R}^{n+m}\|_{W_p^l} \le c\,\mathcal{C}_{[l]+1,l+1}\|\gamma;\mathbb{R}^n\|_{L_\infty}\|V;\mathbb{R}^{n+m}\|_{W_p^l}. \quad (8.8.10)$$

It is easy to see that

$$\Big(\int_{\mathbb{R}^n} dx \int_{\mathbb{R}^m} dy \int_{2|\eta|>|y|} |\nabla_{[l],y}(V\mathcal{T}\gamma))(x,y+\eta)$$
$$-(\nabla_{[l],y}(V\mathcal{T}\gamma))(x,y)|^p|\eta|^{-m-p\{l\}}\,d\eta\Big)^{1/p}$$
$$\le c\Big(\int_{\mathbb{R}^{n+m}} |\nabla_{[l],y}(V\mathcal{T}\gamma)|^p|y|^{-p\{l\}}\,dz\Big)^{1/p} \quad (8.8.11)$$

which, by (8.7.15) with $s = 0$ and $\mu = 0$, does not exceed

$$c\sum_{j=0}^{[l]}\Big(\int_{\mathbb{R}^{n+m}} |\nabla_{[l]-j,y}\mathcal{T}\gamma|^p|\nabla_{j,y}V|^p|y|^{-p\{l\}}\,dz\Big)^{1/p}$$
$$\le c\,\mathcal{C}_{[l]+1,l+1}\|\gamma;\mathbb{R}^n\|_{L_\infty}\sum_{j=0}^{[l]}\Big(\int_{\mathbb{R}^{n+m}} |\nabla_{j,y}V|^p|y|^{(j-l)p}\,dz\Big)^{1/p}.$$

This fact and (8.8.4) show that the left-hand side of (8.8.11) is dominated by

$$c\,\mathcal{C}_{[l]+1,l+1}\|\gamma;\mathbb{R}^n\|_{L_\infty}\|V;\mathbb{R}^{n+m}\|_{W_p^l}.$$

The expression

$$\Big(\int_{\mathbb{R}^n}dx\int_{\mathbb{R}^m}dy\int_{2|\eta|<|y|}\big|\nabla_{[l],y}[(V\mathcal{T}\gamma)(x,y+\eta)-(V\mathcal{T}\gamma)(x,y)]\big|^p|\eta|^{-m-p\{l\}}d\eta\Big)^{1/p}$$

is majorized by

$$\begin{aligned}
&c\sum_{j=0}^{[l]}\Big(\int_{\mathbb{R}^n}dx\int_{\mathbb{R}^m}dy\int_{2|\eta|<|y|}|(\nabla_{[l]-j,y}\mathcal{T}\gamma)(x,y+\eta)|^p\\
&\times|\nabla_j(V(x,y+\eta)-V(x,y))|^p|\eta|^{-m-p\{l\}}d\eta\Big)^{1/p}\\
&+c\sum_{j=0}^{[l]}\Big(\int_{\mathbb{R}^n}dx\int_{\mathbb{R}^m}dy|(\nabla_{j,y}V)(x,y)|^p\\
&\times\int_{2|\eta|<|y|}\big|\nabla_{[l]-j,y}(\mathcal{T}\gamma(x,y+\eta)-\mathcal{T}\gamma(x,y))\big|^p|\eta|^{-m-p\{l\}}d\eta\Big)^{1/p}. \quad (8.8.12)
\end{aligned}$$

Since

$$|(\nabla_{[l]-j,y}\mathcal{T}\gamma)(x,y+\eta)|\le c\,\mathcal{C}_{[l]+1,l+1}\|\gamma;\mathbb{R}^n\|_{L_\infty}|y|^{j-[l]},$$

the first sum does not exceed

$$\begin{aligned}
&c\,\mathcal{C}_{[l]+1,l+1}\|\gamma;\mathbb{R}^n\|_{L_\infty}\sum_{j=0}^{[l]-1}\Big(\int_{\mathbb{R}^n}dx\int_{\mathbb{R}^m}|y|^{p(j-[l])}\int_{2|\eta|<|y|}|\eta|^{-m+p(1-\{l\})}\\
&\times\Big(\int_0^1|\nabla_{j+1,y}V(x,y+t\eta)|dt\Big)^p d\eta\Big)^{1/p}+c\,\mathcal{C}_{[l]+1,l+1}\|\gamma;\mathbb{R}^n\|_{L_\infty}\|V;\mathbb{R}^{n+m}\|_{W_p^l}.
\end{aligned}$$

Here we have used the relation (8.8.6). Further, we have

$$\begin{aligned}
&\int_{\mathbb{R}^n}|y|^{(j-[l])p}\int_{2|\eta|<|y|}|\eta|^{-m+p(1-\{l\})}\Big(\int_0^1|\nabla_{j+1,y}V(x,y+t\eta)|\,dt\Big)^p d\eta\,dy\\
&\le c\int_0^1 dt\int_{\mathbb{R}^m}|\eta|^{-m+p(1-\{l\})}\int_{2|\eta|<|y|}|y|^{(j-[l])p}|\nabla_{j+1,y}V(x,y+t\eta)|^p\,dy\,d\eta\\
&\le c\int_{\mathbb{R}^m}|\eta|^{-m+p(1-\{l\})}\int_{|\chi|>|\eta|}|\chi|^{(j-[l])p}|\nabla_{j+1,\chi}V(x,\chi)|^p\,d\chi\,d\eta\\
&=c\int_{\mathbb{R}^m}|\chi|^{p(j+1-l)}|\nabla_{j+1,\chi}V(x,\chi)|^p\,d\chi
\end{aligned}$$

which, according to (8.8.4), does not exceed $c\,\|V(x,\cdot);\mathbb{R}^m\|^p_{W^l_p}$ for almost all $x\in\mathbb{R}^n$. Thus, the first sum in (8.8.12) is not greater than

$$c\,\mathcal{C}_{[l]+1,l+1}\|V;\mathbb{R}^{n+m}\|_{W^l_p}.$$

Using (8.8.1), we find that the second sum in (8.8.12) has the majorant

$$c\,\mathcal{C}_{[l]+1,l+1}\|\gamma;\mathbb{R}^n\|_{L_\infty}\sum_{j=0}^{[l]}\Big(\int_{\mathbb{R}^{n+m}}|(\nabla_{j,y}V)(z)|^p|y|^{(j-l)p}\,dz\Big)^{1/p}$$

which, by (8.8.4), is dominated by

$$c\,\mathcal{C}_{[l]+1,l+1}\|\gamma;\mathbb{R}^n\|_{L_\infty}\|V;\mathbb{R}^{n+m}\|_{W^l_p}.$$

To obtain a bound for

$$\Big(\int_{\mathbb{R}^m}dy\int_{\mathbb{R}^n}dx\int_{\mathbb{R}^n}|\nabla_{[l],x}(V\mathcal{T}\gamma)(x+h,y)-(V\mathcal{T}\gamma)(x,y)]|^p|h|^{-n-p\{l\}}dh\Big)^{1/p},$$

it suffices to estimate the integrals

$$\int_{\mathbb{R}^m}dy\int_{\mathbb{R}^n}|\nabla_{j,x}V(x,y)|^p$$
$$\times\int_{\mathbb{R}^n}\big|\nabla_{[l]-j,x}[(\mathcal{T}\gamma)(x+h,y)-(\mathcal{T}\gamma)(x,y)]\big|^p|h|^{-n-p\{l\}}\,dh\,dx\,,$$

$$\int_{\mathbb{R}^m}dy\int_{\mathbb{R}^n}|[\nabla_{[l]-j,x}(\mathcal{T}\gamma)](x,y)|^p$$
$$\times\int_{\mathbb{R}^n}|\nabla_{j,x}(V(x+h,y)-V(x,y))\big|^p|h|^{-n-p\{l\}}\,dh\,dx\,.$$

The first integral is estimated by Lemma 8.8.1 and inequality (8.8.4). It does not exceed

$$c\,\big(\mathcal{C}_{[l]+1,l+1}\|\gamma;\mathbb{R}^n\|_{L_\infty}\|V;\mathbb{R}^{n+m}\|_{W^l_p}\big)^p\,.$$

The second integral is dominated by

$$c\,\mathcal{C}^p_{[l]+1,l+1}\|\gamma;\mathbb{R}^n\|^p_{L_\infty}\int_{\mathbb{R}^n}dx\int_{\mathbb{R}^n}|h|^{-n-p\{l\}}$$
$$\times\int_{\mathbb{R}^m}|y|^{p(j-\{l\})}|\nabla_{j,x}(V(x+h,y)-V(x,y))|^p dydh$$
$$\le c\,\mathcal{C}^p_{[l]+1,l+1}\|\gamma;\mathbb{R}^n\|^p_{L_\infty}\int_{\mathbb{R}^n}dx\int_{\mathbb{R}^n}|h|^{-n-p\{l\}}$$
$$\times\int_{\mathbb{R}^m}|\nabla_{[l],z}(V(x+h,y)-V(x,y))|^p\,dydh$$
$$\le c\,(\mathcal{C}_{[l]+1,l+1}\|\gamma;\mathbb{R}^n\|_{L_\infty}\|V;\mathbb{R}^{n+m}\|_{W^l_p})^p.$$

Here we have used (8.8.4) and (8.8.6). Thus inequality (8.8.10) is proved.

Putting $\gamma = 1$, $\mathcal{T}\gamma = 1$ in (8.8.9), we obtain the estimate

$$\|E\varphi; \mathbb{R}^{n+m}\|_{W_p^l} \le c\,\|U; \mathbb{R}^{n+m}\|_{W_p^l}$$

which, together with (8.8.10) and the equality $V = U - E\varphi$, shows that

$$\|V\mathcal{T}\gamma; \mathbb{R}^{n+m}\|_{W_p^l} \le c\,\mathcal{C}_{[l]+1,l+1}\|\gamma; \mathbb{R}^n\|_{L_\infty}\|U; \mathbb{R}^{n+m}\|_{W_p^l}.$$

The proof is complete. □

8.9 Multipliers in the Space of Bessel Potentials as Traces of Multipliers

The goal of this section is to show that multipliers in the space $H_p^l(\mathbb{R}^n)$ are traces of multipliers in a certain class of differentiable functions in $\mathbb{R}^{n+m}$ with a weighted mixed norm.

8.9.1 Bessel Potentials as Traces

The space $L_{p,\beta}^k(\mathbb{R}^{n+m})$ is defined as the completion of $C_0^\infty(\mathbb{R}^{n+m})$ in the norm

$$\Big(\int_{\mathbb{R}^n}\Big(\int_{\mathbb{R}^m}|y|^{2\beta}|\nabla_{k,z}U|^2dy\Big)^{p/2}dx\Big)^{1/p} + \Big(\int_{\mathbb{R}^n}\Big(\int_{\mathbb{R}^m}|y|^{2\beta}|U|^2dy\Big)^{p/2}dx\Big)^{1/p}.$$

Let the first term be denoted by $< U >_{p,\beta,k}$. For $k > r$ and $\beta > r - m/p$, by Hardy's inequality one has

$$< U >_{p,\beta-r,k-r} \le c < U >_{p,\beta,k}\,. \tag{8.9.1}$$

The following assertion shows that elements of $H_p^l(\mathbb{R}^n)$ are traces on $\mathbb{R}^n$ of functions in $L_{p,\beta}^k(\mathbb{R}^{n+m})$ (see [Sh1] for $0 < l < 1$, the general case is treated in a similar way). Below we use the spherical coordinates (ρ, ω) in $\mathbb{R}^m$: $\rho = |y|$ and $\omega = y/|y|$.

Lemma 8.9.1. *(i) Let $U \in L_{p,\beta}^k(\mathbb{R}^{n+m})$, where k is an integer, $2\beta > -m$, and $1 < p < \infty$. Then, for almost all $x \in \mathbb{R}^n$, the limit*

$$U(x,0) = \lim_{\rho\to+0}\int_{\partial B_1^{(m)}} U(x;\rho,\omega)d\omega$$

exists. Moreover, $U(\cdot,0) \in H_p^l(\mathbb{R}^n)$ with $l = k - \beta - m/2$, $\{l\} > 0$ and

$$\|U(\cdot,0); \mathbb{R}^n\|_{H_p^l} \le c\,\|U; \mathbb{R}^{n+m}\|_{L_{p,\beta}^k}. \tag{8.9.2}$$

(ii) Let $u \in H_p^l(\mathbb{R}^n)$, $l > 0$, $1 < p < \infty$. There exists a linear continuous extension operator:

$$H_p^l(\mathbb{R}^n) \ni u \to U \in L_{p,\beta}^k(\mathbb{R}^{n+m}),$$

where k is an integer, $k > l$, and $\beta = k - l - m/2$.

8.9.2 An Auxiliary Estimate for the Extension Operator $\mathcal{T}$

Here we use the notations $\mathcal{T}$, ζ, $\mathcal{C}_{q,r}$ introduced in Sect. 8.7.2. The following lemma will be applied in the next section.

Lemma 8.9.2. *For any positive and noninteger $\delta > 0$, and any integer $q > \delta$, the estimate*

$$\Big(\int_{\mathbb{R}^{n+m}} |y|^{2(q-\delta)-m} |\nabla_{q,z}(\mathcal{T}u)|^2 dy\Big)^{1/2} \le c\,\mathcal{C}_{q,\delta} S_\delta u(x) \tag{8.9.3}$$

holds, where $\mathcal{C}_{q,\delta}$ is defined by (8.7.5) *and S_δ is given by* (3.1.1).

Proof. Let τ, $\varkappa$, and μ be n-dimensional multi-indices, and let σ be an m-dimensional multi-index such that $|\tau| + |\sigma| = q$, $\varkappa = 0$, and $\mu = \tau$ if $|\tau| \le \delta$, and $\varkappa = \tau - \mu$, $|\mu| = [\delta]$, and $\mu < \tau$ if $|\tau| > \delta$. We introduce the notation

$$\mathcal{R}_\mu(h, x) = D^\mu u(x+h) - \sum_{|\nu| < [\delta] - |\mu|} D^{\mu+\nu} u(x) \frac{h^\nu}{\nu!}.$$

Using the identity

$$D_y^\sigma \Big(|y|^{-n-|\varkappa|} \int_{\mathbb{R}^n} (D^\varkappa \zeta)\Big(\frac{\xi - x}{|y|}\Big)(\xi - x)^\nu d\xi\Big)$$

$$= D_y^\sigma \Big(|y|^{|\nu|-|\varkappa|} \int_{\mathbb{R}^n} D^\varkappa \zeta(\xi) \xi^\nu d\xi\Big) = 0,$$

we obtain

$$D_x^\tau D_y^\sigma \int_{\mathbb{R}^n} \zeta(\eta) u(x + |y|\eta) d\eta = D_x^\varkappa D_y^\sigma \int_{\mathbb{R}^n} \zeta(\eta) D_x^\mu u(x + |y|\eta) d\eta$$

$$= D_y^\sigma \Big(|y|^{-n-|\varkappa|} \int_{\mathbb{R}^n} (D^\varkappa \zeta)\Big(\frac{\xi - x}{|y|}\Big) D^\mu u(\xi) d\xi\Big)$$

$$= D_y^\sigma \Big(|y|^{-n-|\varkappa|} \int_{\mathbb{R}^n} (D^\varkappa \zeta)\Big(\frac{\xi - x}{|y|}\Big) \mathcal{R}_\mu(\xi - x, x) d\xi\Big).$$

Clearly,

$$\Big| D_y^\sigma \Big(|y|^{-n-|\varkappa|} (D^\varkappa \zeta)\Big(\frac{\xi - x}{|y|}\Big) \mathcal{R}_\mu(\xi - x, x)\Big)\Big|$$

$$\le |y|^{\delta - |\varkappa| - |\sigma| - |\mu|} \varphi\Big(\frac{\xi - x}{|y|}\Big) \frac{|\mathcal{R}_\mu(\xi - x, x)|}{|\xi - x|^{\delta - |\mu| + n}},$$

where φ is a nonnegative function having the estimate

$$\varphi(\xi) \le c\,|\xi|^{\delta - |\mu| + n} \sum_{i=0}^{|\sigma|} |\nabla_{i+|\varkappa|} \zeta(\xi)|\,(|\xi|^i + 1). \tag{8.9.4}$$

Since $|\varkappa| + |\mu| = |\tau|$ and $|\tau| + |\sigma| = q$, we arrive at the inequality

$$\int_{\mathbb{R}^m} |y|^{2(q-\delta)-m} |D_x^\tau D_y^\sigma (\mathcal{T}u)|^2 dy$$
$$\le c \int_{\mathbb{R}^m} \Big(\int_{\mathbb{R}^n} \varphi\Big(\frac{\xi - x}{|y|}\Big) \frac{|\mathcal{R}_\mu(\xi - x, x)|}{|\xi - x|^{\delta - |\mu| + n}} d\xi \Big)^2 \frac{dy}{|y|^m}.$$

Passing to the spherical coordinates $t = |\xi - x|$ and $\theta = (\xi - x)t^{-1}$, we can write the right-hand side as

$$c \int_0^\infty \Big(\int_0^\infty \int_{\partial \mathcal{B}_1^{(n)}} \varphi\Big(\frac{t\theta}{\lambda}\Big) \frac{|\mathcal{R}_\mu(t\theta, x)|}{t^{\delta - |\mu|}} \frac{dt}{t} d\theta \Big)^2 \frac{d\lambda}{\lambda}.$$

This expression does not exceed

$$c \int_0^\infty \Big(\int_0^\infty Q\Big(\frac{t}{\lambda}\Big) g(t) \frac{dt}{t} \Big)^2 \frac{d\lambda}{\lambda}, \tag{8.9.5}$$

where

$$Q(t) = \sup_{\theta \in \partial \mathcal{B}_1^{(n)}} \varphi(t\theta)$$

and

$$g(t) = t^{|\mu| - \delta} \int_{\partial \mathcal{B}_1^{(n)}} |\mathcal{R}_\mu(t\theta, x)| d\theta.$$

By Minkowski's inequality, (8.9.5) is not greater than

$$c \Big(\int_0^\infty Q(t) \frac{dt}{t} \Big)^2 \int_0^\infty g(t)^2 \frac{dt}{t}.$$

This and (8.9.4) imply that

$$\Big(\int_{\mathbb{R}^m} |y|^{2(q-\delta)-m} |D_x^\tau D_y^\sigma (\mathcal{T}u)|^2 dy \Big)^{1/2}$$
$$\le c\, C_{q,\delta} \Big(\int_0^\infty t^{2(|\mu| - \delta) - 1} \Big(\int_{\partial \mathcal{B}_1^{(n)}} |\mathcal{R}_\mu(t\theta, x)| d\theta \Big)^2 dt \Big)^{1/2}. \tag{8.9.6}$$

For $0 < \delta < 1$ we have $\mu = 0$ and

$$\mathcal{R}_0(t\theta, x) = u(x + t\theta) - u(x).$$

Therefore, for such δ, the right-hand side of (8.9.6) is equal to $c\, C_{q,\delta} S_\delta u(x)$. Hence we need to consider only $\delta > 1$. Since

$$\mathcal{R}_\mu(t\theta, x) =$$
$$([\delta] - |\mu|) \int_0^1 \sum_{|\nu| = [\delta] - |\mu|} \frac{(t\theta)^\nu}{\nu!} \Big((D^{\nu+\mu} u)(x + ht\theta) - D^{\nu+\mu} u(x) \Big) (1 - h)^{[\delta] - |\mu| - 1} dh,$$

we obtain

$$\Big(\int_{\mathbb{R}^m} |y|^{2(q-\delta)-m} |\nabla_{q,z}(\mathcal{T}u)|^2 dy\Big)^{1/2}$$
$$\leq c\,\mathcal{C}_{q,\delta}\Big(\int_0^\infty t^{-2\{\delta\}-1}\Big(\int_{\partial\mathcal{B}_1^{(n)}}\int_0^1 \sum_{|\alpha|=[\delta]} |(D^\alpha u)(x+ht\theta)-D^\alpha u(x)|dh d\theta\Big)^2 dt\Big)^{1/2}.$$

By Minkowski's inequality, the right-hand side is dominated by

$$c\,\mathcal{C}_{q,\delta}\int_0^1\Big(\int_0^\infty t^{-2\{\delta\}-1}\Big(\int_{\partial\mathcal{B}_1^{(n)}} \sum_{|\alpha|=[\delta]} |(D^\alpha u)(x+ht\theta)-D^\alpha u(x)|d\theta\Big)^2 dt\Big)^{1/2} dh.$$

Making the change of variable $t \to h^{-1}\tau$ for any $h \in (0,1)$, we find that the last expression is equal to

$$c\,\mathcal{C}_{q,\delta}\int_0^1 h^{\{\delta\}}\Big(\int_0^\infty \tau^{-2\{\delta\}-1}\Big(\int_{\partial\mathcal{B}_1^{(n)}} \sum_{|\alpha|=[\delta]} |(D^\alpha u)(x+\tau\theta)-D^\alpha u(x)|d\theta\Big)^2 d\tau\Big)^{1/2} dh$$

and is not greater than $c\,\mathcal{C}_{q,\delta} S_\delta u(x)$. The proof is complete. □

8.9.3 MH_p^l as a Space of Traces

The main result of this section runs as follows.

Theorem 8.9.1. *(i) Let $\Gamma \in ML^k_{p,\beta}(\mathbb{R}^{n+m})$, where k is an integer, $1 < p < \infty$, $2\beta > -m$, and let $k-\beta-m/2$ be a positive noninteger. Then the function $\gamma = \Gamma(x,0)$ belongs to the space $MH_p^l(\mathbb{R}^n)$ with $l = k-\beta-m/2$, and the estimate*

$$\|\gamma;\mathbb{R}^n\|_{MH_p^l} \leq c\,\|\Gamma;\mathbb{R}^{n+m}\|_{ML^k_{p,\beta}}$$

holds.

(ii) Let $\gamma \in MH_p^l(\mathbb{R}^n)$, and let $\mathcal{T}\gamma$ be the extension of γ to $\mathbb{R}^{n+m}$ defined in Sect. 8.9.2 and subject to (8.7.4) with $r = l$. Then $\mathcal{T}\gamma \in ML^k_{p,\beta}(\mathbb{R}^{n+m})$ with an integer k, $k > l$, and $\beta = k-l-m/2$. Moreover,

$$\|\mathcal{T}\gamma;\mathbb{R}^{n+m}\|_{ML^k_{p,\beta}} \leq c\,\mathcal{C}_{k,l}\|\gamma;\mathbb{R}^n\|_{MH_p^l},$$

where $\mathcal{C}_{k,l}$ is defined by (8.7.5).

Proof. (i) The existence of a trace γ of the function $\Gamma \in ML^k_{p,\beta}(\mathbb{R}^{n+m})$ follows from the inclusion $\Gamma \in L^k_{p,\beta,\mathrm{loc}}(\mathbb{R}^{n+m})$ and Lemma 8.9.1. Let $U \in L^k_{p,\beta}(\mathbb{R}^{n+m})$ and let $u(x) = U(x,0)$. We have

$$\|\gamma u;\mathbb{R}^n\|_{H_p^l} \leq c\,\|\Gamma U;\mathbb{R}^{n+m}\|_{L^k_{p,\beta}} \leq c\,\|\Gamma;\mathbb{R}^{n+m}\|_{ML^k_{p,\beta}}\|U;\mathbb{R}^{n+m}\|_{L^k_{p,\beta}}.$$

The result follows from part (ii) of Lemma 8.9.1.

(*ii*) Let μ, ϵ, and ν be arbitrary multi-indices of dimensions n, m, and $n+m$, respectively. Clearly,

$$< U\mathcal{T}\gamma; \mathbb{R}^{n+m} >_{p,\beta,k} \leq c \sum_{|\nu|+|\mu|+|\epsilon|=k} < |D_z^\nu U| |D_x^\mu D_y^\epsilon (\mathcal{T}\gamma)|; \mathbb{R}^{n+m} >_{p,\beta,0} . \tag{8.9.7}$$

By $\sum^{(1)}$ and $\sum^{(2)}$ we denote the sums of those terms in (8.9.7) for which $|\nu| < l$ and $|\nu| > l$, respectively. Since

$$|D_x^\mu D_y^\epsilon (\mathcal{T}\gamma)(z)| \leq c\,\mathcal{C}_{k,l} \|\gamma; \mathbb{R}^n\|_{L_\infty} |y|^{-|\mu|-|\epsilon|} \tag{8.9.8}$$

(see (8.7.15)), we have

$$\sum\nolimits^{(2)} \leq c\,\mathcal{C}_{k,l} \|\gamma; \mathbb{R}^n\|_{L_\infty} < U; \mathbb{R}^{n+m} >_{p,\beta-k+|\nu|,|\nu|} .$$

By (8.9.1) the right-hand side does not exceed

$$c\,\mathcal{C}_{k,l} \|\gamma; \mathbb{R}^n\|_{L_\infty} < U; \mathbb{R}^{n+m} >_{p,\beta,k} .$$

Now let $|\nu| < l$. We put

$$\mathcal{Q}_\nu U(z) = D_z^\nu U(z) - \sum_{|\tau| \leq [l]-|\nu|} (D_y^\tau D_z^\nu U)(x,0) \frac{y^\tau}{\tau!}.$$

Then

$$\sum\nolimits^{(1)} \leq \Big(\int_{\mathbb{R}^n} \Big(\int_{\mathbb{R}^m} |y|^{2(k-l)-m} \sum_{|\nu|+|\mu|+|\epsilon|=k} |\mathcal{Q}_\nu U|^2 |D_x^\mu D_y^\epsilon (\mathcal{T}\gamma)|^2 dy \Big)^{\frac{p}{2}} dx \Big)^{\frac{1}{p}}$$

$$+ \Big(\int_{\mathbb{R}^n} \Big(\int_{\mathbb{R}^m} |y|^{2(k-l)-m} \sum_{|\nu|+|\mu|+|\epsilon|=k} \sum_{i=0}^{[l]-|\nu|} |y|^{2i} |D_x^\mu D_y^\epsilon (\mathcal{T}\gamma)|^2$$

$$\times |\nabla_{i,y} D_z^\nu U(x,0)|^2 dy \Big)^{\frac{p}{2}} dx \Big)^{\frac{1}{p}}.$$

Let us denote the first and the second terms on the right-hand side by A and B, respectively. Since

$$\mathcal{Q}_\nu U(z) = ([l] - |\nu| + 1) \sum_{|\tau|=[l]-|\nu|+1} \frac{y^\tau}{\tau!} \int_0^1 (D_y^\tau D_z^\nu) U(x,ty)(1-t)^{[l]-|\nu|} dt,$$

we have the estimate

$$|\mathcal{Q}_\nu U(z)| \leq c\,|y|^{[l]-|\nu|+1} \int_0^1 |(\nabla_{[l]+1,z} U)(x,ty)| dt.$$

Hence, using (8.9.8) and Minkowski's inequality, we find that A is majorized by

$$c\,\mathcal{C}_{k,l}\|\gamma;\mathbb{R}^n\|_{L_\infty}\int_0^1\Big(\int_{\mathbb{R}^n}\Big(\int_{\mathbb{R}^m}|y|^{2(1-\{l\})-m}|(\nabla_{[l]+1,z}U)(x,ty)|^2dy\Big)^{p/2}dx\Big)^{1/p}dt.$$

Therefore,

$$A\le c\,\mathcal{C}_{k,l}\|\gamma;\mathbb{R}^n\|_{L_\infty}<U>_{p,\beta,k}\,.$$

By Lemma 8.9.2 with $q=k-|\nu|$ and $\delta=l-|\nu|-i$, we find that B is not greater than

$$c\,\mathcal{C}_{k,l}\sum_{i=0}^{[l]-|\nu|}\Big(\int_{\mathbb{R}^n}|(\nabla_iD^\nu)U(x,0)|^p|S_{l-i-|\nu|}\gamma(x)|^pdx\Big)^{1/p}$$

$$\le c\,\mathcal{C}_{k,l}\sum_{i=0}^{[l]-|\nu|}\|S_{l-i-|\nu|}\gamma;\mathbb{R}^n\|_{M(H_p^{l-i-|\nu|}\to L_p)}\|\nabla_{i+|\nu|}U(\cdot,0);\mathbb{R}^n\|_{H_p^{l-i-|\nu|}},$$

which by (3.2.26) does not exceed

$$c\,\mathcal{C}_{k,l}\|\gamma;\mathbb{R}^n\|_{MH_p^l}\|U;\mathbb{R}^{n+m}\|_{L^k_{p,\beta}}. \tag{8.9.9}$$

Using the estimates obtained for A and B, we find that (8.9.9) is a majorant for the norm $<U\mathcal{T}\gamma>_{p,\beta,k}$. It remains to note that (8.9.8) with $\mu=\epsilon=0$ implies that

$$\Big(\int_{\mathbb{R}^n}\Big(\int_{\mathbb{R}^m}|y|^{2(k-l)-m}|U\mathcal{T}\gamma|^2dy\Big)^{p/2}dx\Big)^{1/p}$$

$$\le c\,\mathcal{C}_{k,l}\|\gamma;\mathbb{R}^n\|_{L_\infty}\Big(\int_{\mathbb{R}^n}\Big(\int_{\mathbb{R}^m}|y|^{2(k-l)-m}|U|^2dy\Big)^{p/2}dx\Big)^{1/p}.$$

The proof is complete. □

9

Sobolev Multipliers in a Domain, Multiplier Mappings and Manifolds

In this chapter we deal with multipliers in pairs of Sobolev spaces in a domain. Section 9.1 concerns the special Lipschitz domain G, i.e., $G = \{(x, y) \colon x \in \mathbb{R}^n, y > \varphi(x)\}$, where φ is a function satisfying the Lipschitz condition. We find necessary and sufficient conditions for a function to belong to the space $M(W_p^m(G) \to W_p^l(G))$, where m and l are integers with $0 \leq l \leq m$. In Sect. 9.2 we show that the Stein extension operator (see [St2], Ch.6, §3) maps continuously

$$M(W_p^m(G) \to W_p^l(G)) \text{ into } M(W_p^m(\mathbb{R}^n) \to W_p^l(\mathbb{R}^n)).$$

Analogous results for the space $M(W_p^m(\Omega) \to W_p^l(\Omega))$, where Ω is a bounded domain with boundary in the Lipschitz class $C^{0,1}$, are obtained in 9.3. A description of the space $ML_p^1(\Omega)$ is given, where $L_p^1(\Omega) = \{u \in L_{p,\mathrm{loc}}(\Omega) \colon \nabla u \in L_p(\Omega)\}$ and Ω is an arbitrary domain. We show that, in general, the restriction to Ω of a multiplier in $W_p^1(\mathbb{R}^n)$ is not a multiplier in $W_p^1(\Omega)$.

Further, in Sect. 9.4 we study the influence of a change of variables upon Sobolev spaces. Here we introduce classes of mappings ((p, l)-diffeomorphisms) which preserve the space W_p^l, as well as classes of non-smooth manifolds on which the space W_p^l is correctly defined. These definitions of mappings and manifolds involve spaces of multipliers. In conclusion, a change of variables $T_p^{m,l}$ acting in the pair of Sobolev spaces $W_p^m(V) \to W_p^l(U)$ is defined and investigated.

In Sect. 9.5 we give the following modification of the classical implicit function theorem (see, for example, [KP]) which involves multipliers in its statement. We consider a function u in a special Lipschitz domain G and assume that $\nabla u \in MW_p^{l-1}(G)$, $l \geq 2$, that u vanishes on ∂G, i.e., for $y = \varphi(x)$, and that the trace of $\partial u / \partial y$ on ∂G is separated from zero. We show that $\nabla \varphi \in MW_p^{l-1-1/p}(\mathbb{R}^{n-1})$.

Finally, in Sect. 9.6.2 we give a description of the space $M(\mathring{W}_p^m(\Omega) \to W_p^l(\Omega))$.

V.G. Maz'ya, T.O. Shaposhnikova, *Theory of Sobolev Multipliers*,
Grundlehren der mathematischen Wissenschaften 337,

9.1 Multipliers in a Special Lipschitz Domain

9.1.1 Special Lipschitz Domains

Let $z = (x, y)$, where $x \in \mathbb{R}^{n-1}$ and $y \in \mathbb{R}^1$. By a special Lipschitz domain we mean $G = \{z \in \mathbb{R}^n : x \in \mathbb{R}^{n-1}, y > \varphi(x)\}$, where φ is a function satisfying the Lipschitz condition

$$|\varphi(x_1) - \varphi(x_2)| \le L\,|x_1 - x_2|.$$

It is shown in [St2] (§3, Ch. 6) that there exists a function $z \to \delta^*(z)$ with the properties:

(i) $\delta^* \in C^\infty(\mathbb{R}^n \backslash \partial G)$ and, for any multi-index α,

$$|D^\alpha \delta^*| \le c_\alpha (\delta^*)^{1-|\alpha|},$$

where c_α are constants depending on L.

(ii) For all $z \in \mathbb{R}^n \backslash G$,

$$2\,[\varphi(x) - y] \le \delta^*(z) \le a\,[\varphi(x) - y]\,, \tag{9.1.1}$$

where $a = \text{const} > 2$.

We introduce the operator $\mathfrak{C}$ which performs an extension to the whole of $\mathbb{R}^n$ of a function f defined on G. Namely, if $z \in \mathbb{R}^n \backslash \overline{G}$, then we put

$$(\mathfrak{C} f)(z) = \int_1^2 f(x, y + \lambda \delta^*(z))\psi(\lambda)\,d\lambda\,, \tag{9.1.2}$$

where ψ is a function in $C([1, 2])$ such that

$$\int_1^2 \psi(\lambda)\,d\lambda = 1\,, \qquad \int_1^2 \lambda^k \psi(\lambda)\,d\lambda = 0\,, \qquad k = 1, 2, \ldots, l\,. \tag{9.1.3}$$

The operator $\mathfrak{C}$ maps $W_p^l(G)$ continuously into $W_p^l(\mathbb{R}^n)$ (see [St2], Ch.6, §3).

Let $\mathcal{B}_r = \{x \in \mathbb{R}^n : |x| < r\}$ and let K be a nonnegative function in $C_0^\infty(\mathcal{B}_1)$ with support in the cone $\{z\colon y > 2L|x|\}$. With a function f defined on $\mathbb{R}^n$ we associate its mollification with radius h,

$$[\mathfrak{K}(h) f](z) = \int_{\mathbb{R}^n} f(z + h\zeta) K(\zeta)\,d\zeta\,. \tag{9.1.4}$$

It is clear that if $z \in G$ then $[\mathfrak{K}(h) f](z)$ depends only on the values of f in G.

9.1.2 Auxiliary Assertions

Here, as in Sect. 9.1.1, G is a special Lipschitz domain in $\mathbb{R}^n$.

Lemma 9.1.1. *Let w be a measurable nonnegative function defined on G and let m and l be integers with $0 \le l < m$ and $1 < p < \infty$. The best constant C in the inequality*

$$\int_G (|\nabla_l u(z)|^p + |u(z)|^p)\, w(z)\, dz \le C\, \|u; G\|^p_{W_p^m}, \tag{9.1.5}$$

for all $u \in W_p^m$, is equivalent to

$$\mathcal{L} = \sup \frac{\displaystyle\int_e w(z)\, dz}{C_{p,m-l}(e)}, \tag{9.1.6}$$

where the supremum is taken over all compact subsets e of the domain G.

Proof. We extend w by zero to the exterior of G. Then (9.1.5) implies the same inequality with G replaced by $\mathbb{R}^n$. Now the desired lower estimate for the constant C follows immediately from Theorem 1.2.2 and Lemma 1.2.7.

Let us obtain the upper bound for C. By w_ε we denote a function which coincides with w on the set

$$\{z \in G\colon \operatorname{dist}(z, \partial G) > \varepsilon,\ \ w(z) < 1/\varepsilon\}, \quad \varepsilon > 0,$$

and vanishes elsewhere. By Lemma 1.2.7,

$$\int_{\mathbb{R}^n} (|\nabla_l v|^p + |v|^p) w_\varepsilon\, dz \le c \sup_E \frac{\displaystyle\int_E w_\varepsilon(z)\, dz}{C_{p,m-l}(E)} \|v; \mathbb{R}^n\|^p_{W_p^m} \tag{9.1.7}$$

for all $v \in C_0^\infty(\mathbb{R}^n)$, where the supremum is taken over all compact subsets of $\mathbb{R}^n$. It follows from the definition of w_ε and the monotonicity of the capacity that this supremum does not exceed (9.1.6). Let $u \in W_p^m(G)$. Then $\mathfrak{C}u \in W_p^m(\mathbb{R}^n)$. Approximating $\mathfrak{C}u$ by $C_0^\infty(\mathbb{R}^n)$-functions in the norm of the space $W_p^m(\mathbb{R}^n)$, we obtain from (9.1.7) that

$$\int_{\mathbb{R}^n} (|\nabla_l \mathfrak{C}u|^p + |\mathfrak{C}u|^p) w_\varepsilon\, dz \le c\mathcal{L}\, \|\mathfrak{C}u; \mathbb{R}^n\|^p_{W_p^m}.$$

Since $\mathfrak{C}u = u$ in G, $w_\varepsilon = 0$ in $\mathbb{R}^n \backslash G$ and the operator $\mathfrak{C}\colon W_p^m(G) \to W_p^m(\mathbb{R}^n)$ is continuous, it follows that

$$\int_{\mathbb{R}^n} (|\nabla_l u|^p + |u|^p) w_\varepsilon\, dz \le c\mathcal{L}\, \|u; G\|^p_{W_p^m}.$$

Passing to the limit on the left-hand side as $\varepsilon \to 0$, we complete the proof. $\square$

The next assertion results directly from Lemma 9.1.1 and (1.2.8).

Corollary 9.1.1. *The equivalence relation*

$$\|\gamma; G\|_{M(W_p^m \to L_p)} \sim \sup_{e\subset G} \frac{\|\gamma; e\|_{L_p}}{[C_{p,m}(e)]^{1/p}}$$

holds.

From the existence of the operator $\mathfrak{C}$ and the interpolation property of Sobolev spaces in $\mathbb{R}^n$ it follows that the spaces $W_p^k(G)$ have the same interpolation property. In particular,

$$\|\gamma; G\|_{M(W_p^{m-j} \to W_p^{l-j})} \le c\, \|\gamma; G\|^{(l-j)/l}_{M(W_p^m \to W_p^l)} \|\gamma; G\|^{j/l}_{M(W_p^{m-l} \to L_p)} . \tag{9.1.8}$$

We introduce some notation. Let γ be a function defined on G, whose distributional derivatives of order k are locally integrable with power p. We put

$$f_k(\gamma; e) = \frac{\|\nabla_k \gamma; e\|_{L_p}}{[C_{p,m-l+k}(e)]^{1/p}} , \tag{9.1.9}$$

where $m \ge l$, $0 \le k \le l$ and e is a compact subset of $\overline{G}$. Further, let

$$s_k(\gamma) = \sup_{e\subset G} f_k(\gamma; e) , \tag{9.1.10}$$

where the supremum is taken over all compact subsets of G of positive n-dimensional measure. If $m = l$, then $s_0(\gamma) = \|\gamma; G\|_{L_\infty}$.

We note that the value $s_k(\gamma)$ does not change if e is replaced in its definition by any compact subset of $\overline{G}$ of positive n-dimensional measure. In fact, for any $\varepsilon > 0$ there exists a compact set $E \subset \overline{G}$ such that

$$\sup_{e\subset \overline{G}} f_k(\gamma; e) \le (1+\varepsilon) f_k(\gamma; E) . \tag{9.1.11}$$

Let $E_\delta = \{z \in E \colon y \ge \varphi(x) + \delta\}$, $\delta > 0$. It is clear that

$$f_k(\gamma; e) \le \frac{\|\nabla_k \gamma; E\|_{L_p}}{[C_{p.m-l+k}(E_\delta)]^{1/p}} .$$

Since for small δ,

$$\|\nabla_k \gamma; E\|_{L_p} \le (1+\varepsilon) \|\nabla_k \gamma; E_\delta\|_{L_p} ,$$

it follows that

$$f_k(\gamma; E) \le (1+\varepsilon) f_k(\gamma; E_\delta)$$

which, together with (9.1.11), yields

$$\sup_{e\subset \overline{G}} f_k(\gamma; e) \le (1+\varepsilon)^2 \sup_{e\subset G} f_k(\gamma; e) .$$

It remains to make use of the arbitrariness of ε. □

Lemma 9.1.2. *Let $\mathfrak{K}(h)\gamma$ be the mollification of γ, given by (9.1.4). Then*

$$\|\mathfrak{K}(h)\gamma; G\|_{M(W_p^m \to W_p^l)} \le \|\gamma; G\|_{M(W_p^m \to W_p^l)} \le \liminf_{h\to 0} \|\mathfrak{K}(h)\gamma; G\|_{M(W_p^m \to W_p^l)}, \quad (9.1.12)$$

and

$$\liminf_{h\to 0} s_k(\mathfrak{K}(h)\gamma) \ge s_k(\gamma) \ge c\, s_k(\mathfrak{K}(h)\gamma). \quad (9.1.13)$$

The proof of (9.1.12) is the same as that of Lemma 2.3.1. Inequality (9.1.13) follows from (9.1.12) with $l = 0$, and Corollary 9.1.1.

9.1.3 Description of the Space of Multipliers

Theorem 9.1.1. *Let m and l be integers with $m \ge l \ge 0$ and $p \in (1, \infty)$. Then the space $M(W_p^m(G) \to W_p^l(G))$ consists of functions γ which are locally integrable with power p, along with their distributional derivatives up to order l, and such that $s_l(\gamma) + s_0(\gamma) < \infty$. The equivalence relation*

$$\|\gamma; G\|_{M(W_p^m \to W_p^l)} \sim s_l(\gamma) + s_0(\gamma) \quad (9.1.14)$$

holds.

The proof is similar to that of Theorem 2.3.2 except that we use Lemmas 9.1.1 and 9.1.2 as well as the interpolation inequality (9.1.8).

From Theorem 9.1.1 and the inequality

$$\|\nabla_j \gamma; G\|_{M(W_p^{m-l+j} \to L_p)} \le c\, \|\gamma; G\|^{j/l}_{M(W_p^m \to W_p^l)} \|\gamma; G\|^{1-j/l}_{M(W_p^{m-l} \to L_p)},$$

where $j = 1, \dots, l-1$ (cf. (2.3.8)), it follows that

$$\|\gamma; G\|_{M(W_p^m \to W_p^l)} \sim \sum_{j=0}^{l} s_j(\gamma). \quad (9.1.15)$$

Next we turn to the space $M(W_1^m(G) \to W_1^l(G))$.

Lemma 9.1.3. *Let G be a special Lipschitz domain and let w be a measurable function defined on G. Then the best constant C in*

$$\|wu; G\|_{L_1} \le C\|u; G\|_{W_1^m}, \qquad u \in W_1^m(G), \quad (9.1.16)$$

is equivalent to

$$N = \sup_{z\in\mathbb{R}^n, \rho\in(0,1)} \rho^{m-n} \|w; \mathcal{B}_\rho(z) \cap G\|_{L_1}.$$

Proof. We extend w by zero to the exterior of G. Then, from (9.1.16) we obtain the same inequality with G replaced by $\mathbb{R}^n$. Now the desired lower bound for the constant C follows from Theorem 2.2.3. The same theorem implies that

$$\|wv; \mathbb{R}^n\|_{L_1} \leq cN\|v; \mathbb{R}^n\|_{W_1^m}$$

for all $v \in W_1^m(\mathbb{R}^n)$. Minimizing the right-hand side over all extensions of $u \in W_1^m(G)$, we arrive at (9.1.16) with the constant cN. □

Remark 9.1.1. Obviously, replacing the condition $\rho \in (0,1)$ in the definition of N by $\rho \in (0, C)$, where C is an arbitrary positive constant, we obtain an equivalent value. The same is true if $z \in \mathbb{R}^n$ is replaced by $z \in G$.

Theorem 9.1.2. *Let G be a special Lipschitz domain and let m and l be integers with $0 \leq l \leq m$. The space $M(W_1^m(G) \to W_1^l(G))$ consists of functions γ which are locally integrable in $\overline{G}$ together with their distributional derivatives of order l and such that*

$$\|\nabla_l \gamma; \mathcal{B}_\rho(z) \cap G\|_{L_1} + \rho^{-l}\|\gamma; \mathcal{B}_\rho(z) \cap G\|_{L_1} \leq c\,\rho^{-m+n}$$

for all $z \in \mathbb{R}^n$, $\rho \in (0,1)$.

The relation

$$\|\gamma; G\|_{M(W_1^m \to W_1^l)} \sim \sup_{z\in\mathbb{R}^n, \rho\in(0,1)} \rho^{m-n} \sum_{j=0}^{l} \rho^{j-l}\|\nabla_j \gamma; \mathcal{B}_\rho(z) \cap G\|_{L_1}$$

holds. (Obviously, the same relation holds if we take $z \in G$ on the right-hand side.)

Proof. Let us substitute the function $\zeta \to u(\zeta) = \Phi((\zeta - z)/\rho)$, where $z \in \mathbb{R}^n$, $\Phi \in C_0^\infty(\mathcal{B}_2)$, $\Phi = 1$ on $\mathcal{B}_1$, and $\rho \in (0,1)$, into the inequality

$$\|\gamma u; G\|_{W_1^l} \leq \|\gamma; G\|_{M(W_1^m \to W_1^l)}\|u; G\|_{W_1^m}\,. \tag{9.1.17}$$

Since, for $j = 0, 1, \ldots, l-1$,

$$\rho^{j-l}\|\nabla_j(\gamma u); \mathcal{B}_{2\rho}(z) \cap G\|_{L_1} \leq c\,\|\nabla_l(\gamma u); \mathcal{B}_{2\rho}(z) \cap G\|_{L_1}\,,$$

it follows from (9.1.17) that

$$\rho^{j-l}\|\nabla_j \gamma; \mathcal{B}_\rho(z) \cap G\|_{L_1} \leq \|\gamma; G\|_{M(W_1^m \to W_1^l)}\,\rho^{-m+n}\,.$$

Thus the required lower estimate for the norm in $M(W_1^m(G) \to W_1^l(G))$ is obtained.

The upper estimate results from the obvious inequality

$$\|\gamma u; G\|_{W_1^l} \leq c \sum_{0\leq k+j\leq l} \|\,|\nabla_j u|\,|\nabla_k \gamma|; G\|_{L_1}$$

and the above lemma. □

The following theorem shows that the description of the space $M(W_p^m(G)\to W_p^l(G))$ is especially simple if either $mp > n$ and $p > 1$, or $m \geq n$ and $p = 1$. This theorem can be derived from Theorems 9.1.1, 9.1.2, but we present a direct proof.

Theorem 9.1.3. *Let G be a special Lipschitz domain and let m and l be integers with $0 \leq l \leq m$. Further, let either $mp > n$ and $p > 1$, or $m \geq n$ and $p = 1$.*

Then the space $M(W_p^m(G) \to W_p^l(G))$ consists of functions γ which are locally integrable with power p in $\overline{G}$ together with their distributional derivatives of order l and such that $\|\gamma; \mathcal{B}_1(z) \cap G\|_{W_p^l} \leq const$ for any $z \in \mathbb{R}^n$. Moreover,

$$\|\gamma; G\|_{M(W_p^m\to W_p^l)} \sim \sup_{z\in\mathbb{R}^n} \|\gamma; \mathcal{B}_1(z) \cap G\|_{W_p^l}.$$

(Obviously, the same relation holds if we take $z \in G$ on the right-hand side above.)

Proof. Putting the function $\zeta \to u(\zeta) = \varphi(\zeta - z)$, where $z \in \mathbb{R}^n$, $\varphi \in C_0^\infty(\mathcal{B}_2)$, and $\varphi = 1$ on $\mathcal{B}_1$, into

$$\|\gamma u; G\|_{W_p^l} \leq \|\gamma; G\|_{M(W_p^m\to W_p^l)} \|u; G\|_{W_p^m},$$

we obtain

$$\|\gamma; \mathcal{B}_1 \cap G\|_{W_p^l} \leq c\, \|\gamma; G\|_{M(W_p^m\to W_p^l)}.$$

By $\{\mathcal{B}^{(j)}\}_{j\geq 1}$ we denote a covering of $\mathbb{R}^n$ by unit balls with a finite multiplicity which depends only on n. We have

$$\begin{aligned}\|\gamma u; \mathcal{B}^{(j)} \cap G\|_{W_p^l} &\leq c \sum_{i=0}^{l} \| \, |\nabla_i u| \, |\nabla_{l-i}\gamma|; \mathcal{B}^{(j)} \cap G\|_{L_p} \\ &\leq c \sum_{i=0}^{l} \|\nabla_i u; \mathcal{B}^{(j)} \cap G\|_{L_{q_i}} \|\nabla_{l-i}\gamma; \mathcal{B}^{(j)} \cap G\|_{L_{q_i p/(q_i-p)}},\end{aligned}$$

where $q_i = pn/[n - p(m - i)]$ if $n > p(m - i)$, $q_i = \infty$ if $n < p(m - i)$ and q_i is an arbitrary positive number in the case $n = p(m - i)$. According to the Sobolev imbedding theorem,

$$\|\nabla_i u; \mathcal{B}^{(j)} \cap G\|_{L_{q_i}} \leq c\, \|u; \mathcal{B}^{(j)} \cap G\|_{W_p^m},$$

$$\|\nabla_{l-i}\gamma; \mathcal{B}^{(j)} \cap G\|_{L_{q_i p/(q_i-p)}} \leq c\, \|\gamma; \mathcal{B}^{(j)} \cap G\|_{W_p^l}.$$

Consequently,

$$\|\gamma u; \mathcal{B}^{(j)} \cap G\|^p_{W_p^l} \leq c\, \|\gamma; \mathcal{B}^{(j)} \cap G\|^p_{W_p^l} \|u; \mathcal{B}^{(j)} \cap G\|^p_{W_p^m}. \qquad (9.1.18)$$

Summing over j and applying the inequality

$$\sum a_j^\alpha \leq \Big(\sum a_j\Big)^\alpha, \quad \text{where } a_j \geq 0, \quad \alpha \geq 1,$$

we complete the proof. □

9.2 Extension of Multipliers to the Complement of a Special Lipschitz Domain

Theorem 9.2.1. *Let $\gamma \in M(W_p^m(G) \to W_p^l(G))$, where m and l are integers, $0 \le l \le m$ and $1 \le p < \infty$. Then $\mathfrak{C}\gamma \in M(W_p^m(\mathbb{R}^n) \to W_p^l(\mathbb{R}^n))$ and*

$$\|\mathfrak{C}\gamma; \mathbb{R}^n\|_{M(W_p^m \to W_p^l)} \le c\, \|\gamma; G\|_{M(W_p^m \to W_p^l)} \,. \tag{9.2.1}$$

Proof. Let $u \in C_0^\infty(\mathbb{R}^n)$. We have

$$\|u\mathfrak{C}\gamma; \mathbb{R}^n\|_{W_p^l} \le \|u\mathfrak{C}\gamma; \mathbb{R}^n\backslash G\|_{L_p} + \|\nabla_l(u\mathfrak{C}\gamma); \mathbb{R}^n\backslash G\|_{L_p} + \|u\gamma; G\|_{W_p^l}. \tag{9.2.2}$$

Let us estimate the first term on the right-hand side. It is clear that

$$|(\mathfrak{C}\gamma)(z)| \le c \int_1^2 |\gamma(x, y + \lambda\delta^*(z))|\, d\lambda = c(\delta^*)^{-1} \int_{\delta^*}^{2\delta^*} |\gamma(x, y+s)|\, ds \,.$$

Using property (ii) of δ^*, we obtain for $z \in \mathbb{R}^n\backslash G$

$$\begin{aligned} |(\mathfrak{C}\gamma)(z)| &\le c\,(\varphi(x) - y)^{-1} \int_{2(\varphi(x)-y)}^{a(\varphi(x)-y)} |\gamma(x, y+s)|\, ds \\ &= c \int_2^a |\gamma(x, y + t(\varphi(x) - y))|\, dt \,. \end{aligned} \tag{9.2.3}$$

This and the Minkowski inequality imply that

$$\begin{aligned} \|u\mathfrak{C}\gamma; \mathbb{R}^n\backslash G\|_{L_p} &\le c\Big[\int_{\mathbb{R}^n\backslash G} |u(z)|^p \Big(\int_2^a |\gamma(x, y + t(\varphi(x) - y))|\, dt\Big)^p dz\Big]^{1/p} \\ &\le c \int_2^a \Big(\int_{\mathbb{R}^n\backslash G} |u(z)\gamma(x, y + t(\varphi(x) - y))|^p\, dz\Big)^{1/p} dt \,. \end{aligned}$$

Let $p > 1$. In view of Lemma 9.1.1, we have

$$\begin{aligned} &\|u\mathfrak{C}\gamma; \mathbb{R}^n\backslash G\|_{L_p} \\ &\le C \sup_{2<t<a} \sup_{E \subset \mathbb{R}^n\backslash G} \frac{\Big(\int_E |\gamma(x, y + t(\varphi(x) - y))|^p\, dz\Big)^{1/p}}{[C_{p,m-l}(E)]^{1/p}} \|u; \mathbb{R}^n\backslash G\|_{W_p^{m-l}} \,. \end{aligned} \tag{9.2.4}$$

By $e(t)$ we denote the image of a compact set E under the mapping

$$z \to \zeta = (\xi, \eta), \qquad \text{where } \xi = x, \ \ \eta = y + t(\varphi(x) - y). \tag{9.2.5}$$

Since

$$\eta \ge y + 2(\varphi(x) - y) > \varphi(x) = \varphi(\xi)$$

for $y < \varphi(x)$, (9.2.5) maps $\mathbb{R}^n \backslash G$ into G. It is clear that $z \to \zeta$ is Lipschitz uniformly with respect to t and that the inverse mapping has the same property. From this fact and the definition (1.2.6) it follows that

$$C_{p,m-l}(E) \sim C_{p,m-l}(e(t)) .$$

Therefore the right-hand side of (9.2.4) does not exceed

$$c \sup_{2<t<a} \sup_{e(t)\subset G} f_0(\gamma; e(t)) \|u; \mathbb{R}^n \backslash G\|_{W_p^{m-l}} .$$

Thus, for $p > 1$,

$$\|u\mathfrak{C}\gamma; \mathbb{R}^n \backslash G\|_{L_p} \leq c\, s_0(\gamma) \|u; \mathbb{R}^n \backslash G\|_{W_p^{m-l}} . \tag{9.2.6}$$

In the case $p = 1$, using Lemma 9.1.3 in place of Lemma 9.1.1, we get the following analogue of the inequality (9.2.4):

$$\|u\mathfrak{C}\gamma; \mathbb{R}^n \backslash G\|_{L_1}$$
$$\leq c \sup_{2<t<a} \sup_{\sigma\in\mathbb{R}^n\backslash G, \rho\in(0,1)} \rho^{m-l-n} \int_{\mathcal{B}_\rho(\sigma)\backslash G} |\gamma(x, y+t(\varphi(x)-y))| dz \|u; \mathbb{R}^n \backslash G\|_{W_p^{m-l}} .$$

Applying the properties of the mapping (9.2.5) which were used earlier in this proof, we arrive at

$$\|u\mathfrak{C}\gamma; \mathbb{R}^n \backslash G\|_{L_1}$$
$$\leq c \sup_{\sigma\in G, \rho\in(0,1)} \rho^{m-l-n} \int_{\mathcal{B}_{c\rho}(\sigma)\cap G} |\gamma(z)|\, dz\, \|u; \mathbb{R}^n \backslash G\|_{W_1^{m-l}} . \tag{9.2.7}$$

Clearly, we may assume the last integral to be taken over $\mathcal{B}_\rho(\sigma) \cap G$ by means of an appropriate change of the constant factor before the supremum. For simplicity of notation in the case $p = 1$ we shall denote the value

$$\sup_{\sigma\in G, \rho\in(0,1)} \rho^{m-l+j-n} \int_{\mathcal{B}_\rho(\sigma)\cap G} |\nabla_j \gamma(z)|\, dz$$

by $s_j(\gamma)$ for the rest of the proof.

Obviously, the second term on the right-hand side of (9.2.2) is not greater than

$$c \sum_{j=0}^{l} \| \, |\nabla_{l-j} u| \, |\nabla_j \mathfrak{C}\gamma)|; \mathbb{R}^n \backslash G\|_{L_p} . \tag{9.2.8}$$

By (9.2.6) and (9.2.7) the term corresponding to $j = 0$ does not exceed

$$c\, s_0\, (\gamma) \|u; \mathbb{R}^n \backslash G\|_{W_p^m} .$$

Next we estimate the other terms in (9.2.8). Applying the operator $D_x^{\varkappa} D_y^k$, $|\varkappa|+k = j \geq 1$, to $\mathfrak{C}\gamma$ and formally differentiating under the integral, we obtain a linear combination of expressions of the form

$$\int_1^2 (D_x^\rho D_y^r \gamma)(x, \varphi(\lambda, z)) \prod_{\nu=1}^r D_x^{\alpha_\nu} D_y^{a_\nu} [\varphi(\lambda, z)] \, \psi(\lambda) \, d\lambda \,, \tag{9.2.9}$$

where

$$0 \leq \rho \leq \varkappa \,, \quad |\rho| + r \leq |\varkappa| + k \,, \quad \sum_{\nu=1}^r \alpha_\nu = \varkappa - \rho \,, \quad \sum_{\nu=1}^r a_\nu = k \,,$$

and $\varphi(\lambda, z) = y + \lambda\delta^*(z)$. For each derivative $(D_x^\rho D_y^r \gamma)(x, \varphi(\lambda, z))$ with $|\rho|+r < |\varkappa| + k$, we take the Taylor expansion

$$(D_x^\rho D_y^r \gamma)(x, y + \lambda\delta^*)(z)$$

$$= \sum_{j=0}^{|\varkappa|+k-|\rho|-r-1} (j!)^{-1} [(\lambda - 1)\delta^*(z)]^j (D_x^\rho D_y^{r+j} \gamma)(x, y + \delta^*(z)) \tag{9.2.10}$$

$$+ \frac{1}{(|\varkappa|+k-|\rho|-r)!} \int_{\delta^*(z)}^{\lambda\delta^*(z)} [\lambda\delta^*(z) - t]^{|\varkappa|+k-|\rho|-r-1} (D_x^\rho D_y^{|\varkappa|+k-|\rho|} \gamma)(x, y+t) dt.$$

In view of (9.1.3),

$$\int_1^2 (\lambda - 1)^j \prod_{\nu=1}^r D_x^{\alpha_\nu} D_y^{a_\nu} [y + \lambda\delta^*(z)] \, \psi(\lambda) \, d\lambda = 0 \,.$$

The absolute value of the integral on the right-hand side of (9.2.10) is majorized by

$$c \, [\delta^*(z)]^{|\varkappa|+k-|\rho|-r-1} \int_{\delta^*(z)}^{\lambda\delta^*(z)} |(D_x^\rho D_y^{|\varkappa|+k-|\rho|} \gamma)(x, y + t)| \, dt \,.$$

Moreover, by the inequality

$$|D_z^\tau \delta^*| \leq c_{|\tau|} (\delta^*)^{1-|\tau|}$$

we have

$$\left| \prod_{\nu=1}^r D_x^{\alpha_\nu} D_y^{a_\nu} [\varphi(\lambda, z)] \right| \leq c \prod_{\nu=1}^r (\delta^*(z))^{1-|\alpha_\nu|-a_\nu}$$

$$= c \, (\delta^*(z))^{r+|\rho|-|\varkappa|-k} \,. \tag{9.2.11}$$

Consequently, for $|\rho| + r < |\varkappa| + k$ the absolute value of integral (9.2.9) does not exceed

$$c\,(\delta^*(z))^{-1}\int_1^2 d\lambda\int_{\delta^*(z)}^{\lambda\delta^*(z)}|\nabla_{|\varkappa|+k}\gamma(x,y+t)|\,dt$$
$$\leq c\,(\delta^*)^{-1}\int_{\delta^*}^{2\delta^*}|(\nabla_{|\varkappa|+k}\gamma)(x,y+s)|\,ds\,. \tag{9.2.12}$$

It follows directly from (9.2.11) that the integral (9.2.9) is also majorized by (9.2.12) in the case $|\rho|+r=|\varkappa|+k$.

Thus, for $j\geq 1$,

$$|\nabla_j\mathfrak{C}\gamma)(z)|\leq c\int_2^a|\nabla_j\gamma(x,y+t(\varphi(x)-y))|\,dt$$

(cf. (9.2.3)). Hence, duplicating the arguments used in the proof of (9.2.6) and (9.2.7) and applying (9.2.3), we find that

$$\|\,|\nabla_{l-j}u|\,|\nabla_j\mathfrak{C}\gamma|;\mathbb{R}^n\backslash G\|_{L_p}\leq c\,s_j(\gamma)\|u;\mathbb{R}^n\backslash G\|_{W_p^m}\,.$$

Hence the second term in (9.2.2) is not greater than

$$c\sum_{j=0}^{l}s_j(\gamma)\|u;\mathbb{R}^n\backslash G\|_{W_p^m}$$

which, together with (9.2.2), (9.2.6) and (9.2.7), gives

$$\|u\mathfrak{C}\gamma;\mathbb{R}^n\|_{W_p^l}\leq c\sum_{j=0}^{l}s_j(\gamma)\|u;\mathbb{R}^n\backslash G\|_{W_p^m}+\|\gamma;G\|_{M(W_p^m\to W_p^l)}\|u;G\|_{W_p^m}\,.$$

By (9.1.15) this completes the proof. □

The next assertion complements Theorem 9.1.1, providing another description of the space $M(W_p^m(G)\to W_p^l(G))$.

Corollary 9.2.1. *Let m and l be integers with $m\geq l\geq 0$ and $p\in(1,\infty)$. Then the space $M(W_p^m(G)\to W_p^l(G))$ consists of the functions γ which are locally integrable with power p, along with their distributional derivatives up to order l, and such that $s_l(\gamma)<\infty$. The equivalence relation*

$$\|\gamma;G\|_{M(W_p^m\to W_p^l)}\sim s_l(\gamma)+\|\gamma;G\|_{L_{1,\mathrm{unif}}} \tag{9.2.13}$$

holds.

Proof. We use Theorem 9.1.1 and show that

$$\|\gamma;G\|_{L_{1,\mathrm{unif}}}\leq c\,s_0(\gamma)\leq s_l(\gamma)+\|\gamma;G\|_{L_{1,\mathrm{unif}}}\,. \tag{9.2.14}$$

The lower estimate for $s_0(\gamma)$ trivially holds. We prove the upper estimate.

By Lemma 2.3.9,

$$s_0(\gamma) \le c \sup_{E\subset\mathbb{R}^n} \frac{\|\mathfrak{C}\gamma; E\|_{L_p}}{[C_{p,m-l}(E)]^{1/p}}$$

$$\le c\Big(\sup_{E\subset\mathbb{R}^n} \frac{\|\nabla_l(\mathfrak{C}\gamma); E\|_{L_p}}{[C_{p,m}(E)]^{1/p}} + \|\mathfrak{C}\gamma; \mathbb{R}^n\|_{L_{1,\mathrm{unif}}}\Big). \tag{9.2.15}$$

Obviously,

$$\sup_{E\subset\mathbb{R}^n} \frac{\|\nabla_l(\mathfrak{C}\gamma); E\|_{L_p}}{[C_{p,m}(E)]^{1/p}} \le c\Big(s_l(\gamma) + \sup_{E\subset\mathbb{R}^n\backslash G} \frac{\|\nabla_l(\mathfrak{C}\gamma); E\|_{L_p}}{[C_{p,m}(E)]^{1/p}}\Big). \tag{9.2.16}$$

By (9.2.6) with γ replaced by $\nabla_l\gamma$ and with $m-l$ replaced by m, we find that

$$\|u\nabla_l(\mathfrak{C}\gamma); \mathbb{R}^n\backslash G\|_{L_p} \le c\, s_l(\gamma)\|u; \mathbb{R}^n\backslash G\|_{W_p^m}, \tag{9.2.17}$$

which, in view of Lemma 9.1.1 with $\mathbb{R}^n\backslash G$ in place of G, implies that

$$\sup_{E\subset\mathbb{R}^n\backslash G} \frac{\|\nabla_l(\mathfrak{C}\gamma); E\|_{L_p}}{[C_{p,m}(E)]^{1/p}} \le c\, s_l(\gamma).$$

Combining the last inequality with (9.2.15) and (9.2.16), and noting that

$$\|\mathfrak{C}\gamma; \mathbb{R}^n\|_{L_{1,\mathrm{unif}}} \le c\,\|\gamma; G\|_{L_{1,\mathrm{unif}}},$$

we arrive at (9.2.14). The proof is complete. □

9.3 Multipliers in a Bounded Domain

9.3.1 Domains with Boundary in the Class $C^{0,1}$

We say that a bounded domain Ω has its boundary in the class $C^{k,1}$, $k = 0, 1, \dots$, if any point of $\partial\Omega$ has a neighborhood in which $\partial\Omega$ can be represented (in a Cartesian coordinate system) by $y = \varphi(x)$, where φ is a function whose derivatives of order k satisfy the Lipschitz condition.

Let $\partial\Omega \in C^{0,1}$ and let ε be a small positive number. We construct a covering of the set

$$\Gamma_\varepsilon = \{z \in \mathbb{R}^n : \mathrm{dist}(z, \partial\Omega) \le \varepsilon\}$$

by domains $U_1, \dots, U_N$ with the following property: for any $i = 1, \dots, N$ there exists a special Lipschitz domain G_i such that $U_i \cap \Omega = U_i \cap G_i$. We add the set $U_0 = \Omega\backslash\Gamma_{\varepsilon/2}$ to the collection $\{U_i\}_{i\ge1}$.

By $\{\varphi_i\}_{i=0}^n$ we denote a set of nonnegative functions such that

$$\text{(i)}\ \ \varphi_i \in C_0^\infty(U_i)\,; \qquad \text{(ii)}\ \ \sum_{0\le i\le N} \varphi_i^2 = 1 \ \text{ on } \ \Omega\,.$$

To define an extension operator from Ω to $\mathbb{R}^n$, we note that for $i = 1, \dots, N$ there exists a linear continuous operator

$$\mathfrak{C}_i\colon W_p^l(U_i \cap \Omega) \to W_p^l(\mathbb{R}^n), \quad p \ge 1.$$

Let $\mathfrak{C}_0$ denote the operator of extension by zero to the exterior of the set U_0. Obviously, the operator

$$\mathfrak{C} = \sum_{0\le i\le N} \varphi_i \mathfrak{C}_i \varphi_i \tag{9.3.1}$$

performs an extension of a function defined on Ω to $\mathbb{R}^n$. We introduce the operator $\mathfrak{K}(h)$, $h > 0$, by

$$\mathfrak{K}(h)\gamma = \sum_{0\le i\le N} \varphi_i \mathfrak{K}_i(h)(\varphi_i\gamma)\,, \tag{9.3.2}$$

where $\mathfrak{K}_i(h)$ is the mollification operator defined by (9.1.4) for a special Lipschitz domain G_i.

9.3.2 Auxiliary Assertions

Lemmas 9.1.1 and 9.1.3 hold for any bounded domain with boundary in the class $C^{0,1}$ except that in their proof we mean by $\mathfrak{C}$ the operator given by (9.3.1).

To restate Theorem 9.1.1 for bounded domains with boundary in the class $C^{0,1}$ we need the following two lemmas.

Lemma 9.3.1. *The estimate*

$$\sup_{e\subset\Omega} \frac{\|\nabla_k\gamma; e\|_{L_p}}{[C_{p,m}(e)]^{1/p}} \le c \sup_{\substack{z\in\Omega\\ \rho\in(0,1)}} \rho^{m-n/p}\big(\|\nabla_l\gamma; \mathcal{B}_\rho(z)\|_{L_p} + \rho^{-l}\|\gamma; \mathcal{B}_\rho(z)\|_{L_p}\big) \tag{9.3.3}$$

holds, where $p > 1$, $k = 0, 1, \dots, l-1$.

Proof. If $mp > n$, then the capacity $C_{p,m}(e)$ of any non-empty compact set e in Ω is separated from zero. Hence the left-hand side of (9.3.3) is equivalent to the norm $\|\nabla_k\gamma; \Omega\|_{L_p}$. Further,

$$\|\nabla_k\gamma; \Omega\|_{L_p} \le c\,(\|\nabla_l\gamma; \Omega\|_{L_p} + \|\gamma; \Omega\|_{L_p})$$

which implies (9.3.3).

For $mp \le n$ the left-hand side of (9.3.3) does not exceed

$$c \sup_{e\subset\Omega} \frac{\|\nabla_k\gamma; e\|_{L_q}}{[C_{p,s}(e)]^{1/p}}\,, \tag{9.3.4}$$

where $s < m$, $q > p$ and the numbers s and q are sufficiently close to m and p respectively. From the inequality

$$\Big(\int_{\Omega} |u|^q \, d\mu\Big)^{p/q} \leq c \sup_{z\in\Omega, \rho\in(0,1)} \rho^{ps-n} \, [\mu(\mathcal{B}_\rho(z)\cap\Omega]^{p/q} \, \|u; \mathbb{R}^n\|^p_{W^s_p} \, ,$$

where μ is a measure in Ω and $u \in C_0^\infty(\mathbb{R}^n)$ (see Lemma 1.3.1), it follows that (9.3.4) is not greater than

$$c \sup_{z\in\Omega, \rho\in(0,1)} \rho^{s-n/p} \|\nabla_k \gamma; \mathcal{B}_\rho(z)\|_{L_q} \, .$$

Since q is close to p, it follows that

$$\rho^{k-l+n(1/p-1/q)} \|\nabla_k \gamma; \mathcal{B}_\rho(z)\|_{L_q} \leq c\,(\|\nabla_l \gamma; \mathcal{B}_\rho(z)\|_{L_p} + \rho^{-l} \|\gamma; \mathcal{B}_\rho(z)\|_{L_p}).$$

Consequently, (9.3.4) is majorized by

$$c \sup_{z\in\Omega, \rho\in(0,1)} \rho^{\mu} (\rho^{m-n/p} \|\nabla_l \gamma; \mathcal{B}_\rho(z)\|_{L_p} + \rho^{m-l-n/p} \|\gamma; \mathcal{B}_\rho(z)\|_{L_p}),$$

where $\mu = l - k + n(1/q - 1/p) + s - m$. Since $\mu > 0$, the result follows. □

Lemma 9.3.2. *The following inequalities hold:*

$$c\,\|\mathfrak{K}(h)\gamma; \Omega\|_{M(W^m_p \to W^l_p)} \leq \|\gamma; \Omega\|_{M(W^m_p \to W^l_p)}$$
$$\leq \liminf_{h\to 0} \|\mathfrak{K}(h)\gamma; \Omega\|_{M(W^m_p \to W^l_p)} \, , \tag{9.3.5}$$

$$\liminf_{h\to 0} s_k(\mathfrak{K}(h)\gamma) \geq s_k(\gamma) \, , \tag{9.3.6}$$

$$s_k(\mathfrak{K}(h)\gamma) \leq c\,[s_k(\gamma) + s_0(\gamma)] \, , \tag{9.3.7}$$

where $s_k(\gamma)$ is the same as in (9.1.10) *with G replaced by Ω.*

Proof. It is clear that

$$\|\mathfrak{K}(h)\gamma; \Omega\|_{M(W^m_p \to W^l_p)} \leq c \sum_{0\leq i\leq N} \|\mathfrak{K}_i(h)(\varphi_i\gamma); G_i\|_{M(W^m_p \to W^l_p)} \, .$$

This and Lemma 9.1.2 imply that

$$\|\mathfrak{K}(h)\gamma; \Omega\|_{M(W^m_p \to W^l_p)} \leq c \sum_{0\leq i\leq N} \|\varphi_i\gamma; G_i\|_{M(W^m_p \to W^l_p)}$$
$$= c \sum_{0\leq i\leq N} \|\varphi_i\gamma; \Omega\|_{M(W^m_p \to W^l_p)} \, .$$

Since $\varphi \in C_0^\infty(\mathbb{R}^n)$, the left inequality in (9.3.5) is proved.

The right inequality in (9.3.5) follows from

$$\|u\gamma;\Omega\|_{W_p^l} = \lim_{h\to 0} \|u\mathfrak{K}(h)\gamma;\Omega\|_{W_p^l} \le \liminf_{h\to 0} \|\mathfrak{K}(h)\gamma;\Omega\|_{M(W_p^m\to W_p^l)} \|u;\Omega\|_{W_p^m}\,.$$

Obviously, for any compact set $E \subset \Omega$

$$\liminf_{h\to 0} s_k(\mathfrak{K}(h)\gamma) \ge \frac{\lim\limits_{h\to 0}\Big(\displaystyle\int_E |\nabla_k(\mathfrak{K}(h)\gamma)|^p\,dx\Big)^{1/p}}{[C_{p,m-l+k}(E)]^{1/p}}\,. \tag{9.3.8}$$

Since

$$\varphi_i\mathfrak{K}_i(h)(\varphi_i\gamma) \to \varphi_i^2\gamma \quad\text{in}\quad W_p^k(\Omega) \quad\text{and}\quad \sum_i \varphi_i^2 = 1\,,$$

the right-hand side of (9.3.8) is equal to $f_k(\gamma;E)$. Thus, (9.3.6) is proved.

Next we turn to the estimate (9.3.7). Since $\varphi \in C_0^\infty(\mathbb{R}^n)$, it follows for small enough h that

$$\int_E \big|\nabla_k[\varphi_i\mathfrak{K}_i(h)(\varphi_i\gamma)]\big|^p\,dx \le c\sum_{0\le j\le k}\int_{E\cap\overline{G}_i} \big|\nabla_j[\mathfrak{K}_i(h)(\varphi_i\gamma)]\big|^p\,dx\,.$$

By Lemma 9.1.2,

$$\sup_{e\subset\overline{G}_i}\frac{\displaystyle\int_e \big|\nabla_j[\mathfrak{K}_i(h)(\varphi_i\gamma)]\big|^p\,dx}{C_{p,m-l+k}(e)} \le \sup_{e\subset\overline{G}_i}\frac{\displaystyle\int_e |\nabla_j(\varphi_i\gamma)|^p\,dx}{C_{p,m-l+k}(e)}\,.$$

Therefore,

$$s_k(\mathfrak{K}(h)\gamma) \le c\sum_{0\le i\le n}\sum_{0\le j\le k}\sup_{e\subset\overline{G}_i}\frac{\displaystyle\int_e |\nabla_j\gamma|^p\,dx}{C_{p,m-l+k}(e)}\,.$$

It remains to use Lemma 9.3.1, noting first that the right-hand side of (9.3.3) does not exceed $c(s_l(\gamma)+s_0(\gamma))$. □

9.3.3 Description of Spaces of Multipliers in a Bounded Domain with Boundary in the Class $C^{0,1}$

Theorem 9.3.1. *Let m and l be integers with $m \ge l \ge 0$ and $p \in (1,\infty)$. The space $M(W_p^m(\Omega)\to W_p^l(\Omega))$ consists of functions $\gamma \in W_p^l(\Omega)$ such that*

$$\sup_{e\subset\Omega}\frac{\|\nabla_l\gamma;e\|_{L_p}}{[C_{p,m}(e)]^{1/p}} < \infty\,. \tag{9.3.9}$$

The following inequalities hold:

$$c\sum_{j=0}^{l}\sup_{e\subset\Omega}\frac{\|\nabla_j\gamma;e\|_{L_p}}{[C_{p,m-l+j}(e)]^{1/p}}\le\|\gamma;\Omega\|_{M(W_p^m\to W_p^l)}$$

$$\le c\Big(\sup_{e\subset\Omega}\frac{\|\nabla_l\gamma;e\|_{L_p}}{[C_{p,m}(e)]^{1/p}}+\|\gamma;\Omega\|_{L_1}\Big)\,. \tag{9.3.10}$$

The proof follows the same lines as that of Corollary 2.3.4 and the theorems preceding it, except that in place of the usual mollification operator we use $\mathfrak{K}(h)$ given by (9.3.2) and Lemma 9.3.2.

The next assertion is proved in the same way as Theorem 9.1.2.

Theorem 9.3.2. *Let m and l be integers with $m \ge l \ge 0$. The space $M(W_1^m(\Omega)\to W_1^l(\Omega))$ consists of functions $\gamma\in W_1^l(\Omega)$ such that*

$$\|\nabla_l\gamma;\mathcal{B}_\rho(z)\cap\Omega\|_{L_1}+\rho^{-l}\|\gamma;\mathcal{B}_\rho(z)\cap\Omega\|_{L_1}\le c\,\rho^{n-m}$$

for all $z\in\Omega$ and $\rho\in(0,1)$. The relation

$$\|\gamma;\Omega\|_{M(W_1^m\to W_1^l)}\sim\sup_{z\in\Omega,\rho\in(0,1)}\rho^{m-n}\sum_{j=0}^{l}\rho^{j-l}\|\nabla_j\gamma;\mathcal{B}_\rho(z)\cap\Omega\|_{L_1}$$

holds.

The following theorem results directly from Theorem 9.1.3.

Theorem 9.3.3. *Let either $mp>n$ and $p\in(1,\infty)$, or $m\ge n$ and $p=1$. Then the space $M(W_p^m(\Omega)\to W_p^l(\Omega))$ coincides with $W_p^l(\Omega)$.*

A corollary of Theorem 9.2.1 and formula (9.3.1) is:

Theorem 9.3.4. *Let $\mathfrak{C}$ be the extension operator defined by (9.3.1) and let $\gamma\in(W_p^m(\Omega)\to W_p^l(\Omega))$, $p\ge 1$. Then $\mathfrak{C}\gamma\in M(W_p^m(\mathbb{R}^n)\to W_p^l(\mathbb{R}^n))$ and*

$$\|\mathfrak{C}\gamma;\mathbb{R}^n\|_{M(W_p^m\to W_p^l)}\le c\,\|\gamma;\Omega\|_{M(W_p^m\to W_p^l)}\,.$$

9.3.4 Essential Norm and Compact Multipliers in a Bounded Lipschitz Domain

Let Ω be a bounded domain with $\partial\Omega\in C^{0,1}$. As in the case of the whole space $\mathbb{R}^n$, we associate with any element $\gamma\in M(W_p^m(\Omega)\to W_p^l(\Omega))$ the essential norm

$$\text{ess}\,\|\gamma;\Omega\|_{M(W_p^m\to W_p^l)}=\inf_{\{T\}}\|\gamma-T;\Omega\|_{W_p^m\to W_p^l}$$

where $\{T\}$ is the collection of all compact linear operators: $W_p^m(\Omega)\to W_p^l(\Omega)$.

To derive two-sided estimates for the essential norm in $M(W_p^m(\Omega) \to W_p^l(\Omega))$ we do not need new arguments beyond those given in Chap. 7. This is even simpler, since m and l are integers and the domain Ω is bounded now. The role of the operator T_* used in Chap. 7 is played by the mapping T_*, defined by

$$(T_* u)(x) = \gamma(x) \sum_j \varphi^{(j)}(x) P^{(j)}(u; x)$$

(cf. the proof of the second part of Theorem 7.2.1). Here we use the following notation: $\{\varphi^{(j)}\}$ is a smooth finite partition of unity subordinate to the covering of $\overline{\Omega}$ by open balls $K_\delta^{(j)}$ with radius δ and with centers $x_j \in \Omega$; $P^{(j)}$ are polynomials of the form

$$\sum_{|\beta| \le m-1} \Big(\frac{x - x_j}{\delta}\Big)^\beta \delta^{-n} \int_{K_\delta^{(j)} \cap \Omega} \psi_\beta\Big(\frac{y - x_j}{\delta}\Big) u(y)\, dy$$

where $\psi_\beta \in C_0^\infty(\mathcal{B}_1)$.

In the same way as in Chap. 7, majorants for $\operatorname{ess} \|\gamma; \Omega\|_{W_p^m \to W_p^l}$ can be obtained from upper bounds for the norms $\|(\gamma - T_*)u; \Omega\|_{W_p^l}$ which are collected in the next assertion (cf. Remark 7.2.3).

Lemma 9.3.3. *Let $\gamma \in M(W_p^m(\Omega) \to W_p^l(\Omega))$ where m and l are integers with $m \ge l > 0$.*

(i) If $p > 1$ and $mp < n$, then

$$\|\gamma - T_*; \Omega\|_{W_p^m \to W_p^l} \le c\Big(\sup_{\{e \subset \Omega:\, d(e) \le \delta\}} \frac{\|\nabla_l \gamma; e\|}{[C_{p,m}(e)]^{1/p}} + \sup_{x \in \Omega} \|\gamma; \mathcal{B}_\delta(x) \cap \Omega\|_{L_p} \Big).$$

In particular,

$$\|\gamma - T_*; \Omega\|_{W_p^l \to W_p^l} \le c\Big(\sup_{\{e \subset \Omega:\, d(e) \le \delta\}} \frac{\|\nabla_l \gamma; e\|_{L_p}}{[C_{p,l}(e)]^{1/p}} + \|\gamma; \Omega\|_{L_\infty} \Big).$$

If $p > 1$ and $mp = n$, then δ is replaced by $\delta^{1/2}$ on the right-hand sides of these inequalities.

(ii) If $m \le n$, then

$$\|\gamma - T_*; \Omega\|_{W_1^m \to W_1^l} \le c\, \delta^{m-n} \sup_{x \in \Omega} \big(\|\nabla_l \gamma; \mathcal{B}_\delta(x) \cap \Omega\|_{L_1} + \delta^{-l} \|\gamma; \mathcal{B}_\delta(x) \cap \Omega\|_{L_1} \big).$$

In particular,

$$\|\gamma - T_*; \Omega\|_{W_1^l \to W_1^l} \le c\Big(\delta^{l-n} \sup_{x \in \Omega} \|\nabla_l \gamma; \mathcal{B}_\delta(x) \cap \Omega\|_{L_1} + \|\gamma; \Omega\|_{L_\infty} \Big).$$

Now we state a theorem on two-sided estimates for the essential norm.

Theorem 9.3.5. *Let $\gamma \in M(W_p^m(\Omega) \to W_p^l(\Omega))$, where m and l are integers with $m \geq l \geq 0$.*

(i) If $p > 1$ and $mp \leq n$, then

$$\begin{aligned}&\operatorname{ess} \|\gamma; \Omega\|_{M(W_p^m \to W_p^l)} \\ &\sim \lim_{\delta \to 0} \sup_{\{e \subset \Omega:\, d(e) \leq \delta\}} \Big(\frac{\|\nabla_l \gamma; e\|_{L_p}}{[C_{p,m}(e)]^{1/p}} + \frac{\|\gamma; e\|_{L_p}}{[C_{p,m-l}(e)]^{1/p}} \Big) \\ &\sim \lim_{\delta \to 0} \Big(\sup_{\{e \subset \Omega:\, d(e) \leq \delta\}} \frac{\|\nabla_l \gamma; e\|_{L_p}}{[C_{p,m}(e)]^{1/p}} + \sup_{\{x \in \Omega, \rho \leq \delta\}} \rho^{m-l-\frac{n}{p}} \|\gamma; \mathcal{B}_\rho(x) \cap \Omega\|_{L_p} \Big).\end{aligned}$$

In particular,

$$\operatorname{ess} \|\gamma; \Omega\|_{MW_p^l} \sim \lim_{\delta \to 0} \sup_{\{e \subset \Omega:\, d(e) \leq \delta\}} \frac{\|\nabla_l \gamma; e\|_{L_p}}{[C_{p,l}(e)]^{1/p}} + \|\gamma; \Omega\|_{L_\infty}.$$

(ii) If $m < n$, then

$$\begin{aligned}\operatorname{ess} \|\gamma; \Omega\|_{M(W_1^m \to W_1^l)} \sim \limsup_{\delta \to 0} \sup_{\substack{x \in \Omega \\ \rho \leq \delta}} \rho^{m-n} \big(&\|\nabla_l \gamma; \mathcal{B}_\rho(x) \cap \Omega\|_{L_1} \\ &+ \rho^{-l} \|\gamma; \mathcal{B}_\rho(x) \cap \Omega\|_{L_1} \big).\end{aligned}$$

In particular,

$$\operatorname{ess} \|\gamma; \Omega\|_{MW_1^l} \sim \limsup_{\delta \to 0} \sup_{\substack{x \in \Omega \\ \rho \leq \delta}} \rho^{l-n} \|\nabla_l \gamma; \mathcal{B}_\rho(x) \cap \Omega\|_{L_1} + \|\gamma; \Omega\|_{L_\infty}.$$

(iii) If $mp > n$ and $p \in (1, \infty)$, or $m \geq n$ and $p = 1$, then

$$\operatorname{ess} \|\gamma; \Omega\|_{M(W_p^m \to W_p^l)} = 0 \quad \text{for } \; m > l$$

and

$$\operatorname{ess} \|\gamma; \Omega\|_{MW_p^l} \sim \|\gamma; \Omega\|_{L_\infty} \quad \text{for } \; m = l\,.$$

This immediately implies:

Proposition 9.3.1. *A function $\gamma \in M(W_p^m(\Omega) \to W_p^l(\Omega))$, where m and l are integers with $m > l \geq 0$, belongs to the space $\overset{\circ}{M}(W_p^m(\Omega) \to W_p^l(\Omega))$ of compact multipliers if and only if*

$$\lim_{\delta \to 0} \sup_{\{e \subset \Omega:\, d(e) \leq \delta\}} \Big(\frac{\|\gamma; e\|_{L_p}}{[C_{p,m-l}(e)]^{1/p}} + \frac{\|\nabla_l \gamma; e\|_{L_p}}{[C_{p,m}(e)]^{1/p}} \Big) = 0$$

or, equivalently,

$$\lim_{\delta \to 0} \Big(\sup_{\{e \subset \Omega:\, d(e) \leq \delta\}} \frac{\|\nabla_l \gamma; e\|_{L_p}}{[C_{p,m}(e)]^{1/p}} + \sup_{\{x \in \Omega, \rho \leq \delta\}} \rho^{m-l-\frac{n}{p}} \|\gamma; \mathcal{B}_\rho(x) \cap \Omega\|_{L_p} \Big) = 0$$

for $p \in (1, \infty)$ *and* $mp \le n$;

$$\lim_{\delta \to 0} \delta^{m-n} \sup_{x \in \Omega} \big(\|\nabla_l \gamma; \mathcal{B}_\delta(x) \cap \Omega\|_{L_1} + \delta^{-l} \|\gamma; \mathcal{B}_\delta(x) \cap \Omega\|_{L_1} \big) = 0$$

for $m < n$. *Finally,*

$$\overset{\circ}{M}(W_p^m(\Omega) \to W_p^l(\Omega)) = W_p^l(\Omega)$$

if $mp > n$ *and* $p \in (1, \infty)$, *or* $m \ge n$ *and* $p = 1$.

Similarly to Theorem 7.2.11, we can obtain the following description of the space of compact multipliers.

Proposition 9.3.2. $\overset{\circ}{M}(W_p^m(\Omega) \to W_p^l(\Omega))$, *where* m *and* l *are integers with* $m > l \ge 0$, *is the completion of* $C^\infty(\overline{\Omega})$ *with respect to the norm in* $M(W_p^m(\Omega) \to W_p^l(\Omega))$.

In concert with this assertion, by $\overset{\circ}{M}W_p^l(\Omega)$ we denote the completion of $C^\infty(\overline{\Omega})$ with respect to the norm of $MW_p^l(\Omega)$.

The following proposition is an analogue of Theorem 7.3.10.

Proposition 9.3.3. *A function* γ *belongs to* $\overset{\circ}{M}W_p^l(\Omega)$, *where* l *is a positive integer, if and only if* $\gamma \in C(\overline{\Omega})$ *and one of the following conditions is satisfied:*

(*i*) *If* $pl \le n$ *and* $p > 1$, *then*

$$\sup_{\{e \subset \Omega:\, d(e) \le \delta\}} \frac{\|\nabla_l \gamma; e\|_{L_p}}{[C_{p,l}(e)]^{1/p}} = o(1) \quad as \quad \delta \to 0\,. \tag{9.3.11}$$

(*ii*) *If* $l < n$, *then*

$$\delta^{l-n} \sup_{x \in \Omega} \|\nabla_l \gamma; \mathcal{B}_\delta(x)\|_{L_1} = o(1) \quad as \quad \delta \to 0\,. \tag{9.3.12}$$

(*iii*) *If* $pl > n$ *and* $p > 1$, *or* $l \ge n$ *and* $p = 1$, *then* $\overset{\circ}{M}W_p^l(\Omega) = W_p^l(\Omega)$.

To conclude this section, we consider a relation between the essential norm and the constant K in the inequality

$$\|\gamma u; \Omega\|_{W_p^l} \le K \|u; \Omega\|_{W_p^m} + C(\gamma) \|u; \Omega\|_{L_p}\,. \tag{9.3.13}$$

Theorem 9.3.6. *Let* $\gamma \in M(W_p^m(\Omega) \to W_p^l(\Omega))$, *where* m *and* l *are integers with* $m \ge l \ge 0$, *and let* $\inf K$ *be the infimum of those* K *for which there exists a constant* $C(\gamma)$ *such that* (9.3.13) *holds for all* $u \in W_p^m(\Omega)$. *Then*

$$\inf K \le \operatorname{ess} \|\gamma; \Omega\|_{M(W_p^m \to W_p^l)} \le c \inf K \tag{9.3.14}$$

where $c = c(\Omega, n, p, l, m)$.

Proof. We set

$$u - \sum_j \varphi^{(j)} P^{(j)}$$

in place of u in (9.3.13). Then

$$\|\gamma u - T_* u; \Omega\|_{W_p^l} \le K \Big\|u - \sum_j \varphi^{(j)} P^{(j)}; \Omega\Big\|_{W_p^m} + C(\gamma) \Big\|u - \sum_j \varphi^{(j)} P^{(j)}; \Omega\Big\|_{L_p}.$$

We have

$$\Big\|u - \sum_j \varphi^{(j)} P^{(j)}; \Omega\Big\|^p_{W_p^m} = \Big\|\sum_j \varphi^{(j)} (u - P^{(j)}); \Omega\Big\|^p_{W_p^m}$$
$$\le c \sum_j \sum_{k=0}^{m} \delta^{-kp} \|u - P^{(j)}; K_\delta^{(j)} \cap \Omega\|^p_{W_p^{m-k}} .$$

Since Ω is Lipschitz, it follows that, for some $c \ge 1$,

$$\|u - P^{(j)}; K_\delta^{(j)} \cap \Omega\|_{W_p^{m-k}} \le c\, \delta^k \|u; K_{c\delta}^{(j)} \cap \Omega\|_{W_p^m} .$$

Therefore,

$$\|(\gamma - T_*)u; \Omega\|_{W_p^l} \le c\,(K + C(\gamma)\delta^m) \|u; \Omega\|_{W_p^m}$$

and hence

$$\mathrm{ess}\, \|\gamma; \Omega\|_{M(W_p^m \to W_p^l)} \le c\,(K + C(\gamma)\delta^m)$$

for any small enough δ. The right estimate in (9.3.14) follows.

Now we turn to the left estimate of (9.3.14). According to the definition of the essential norm, for some compact operator $T\colon W_p^m(\Omega) \to W_p^l(\Omega)$ and for all $u \in W_p^m(\Omega)$, we have

$$\|\gamma u; \Omega\|_{W_p^l} \le (\,\mathrm{ess}\, \|\gamma; \Omega\|_{W_p^m \to W_p^l} + \varepsilon) \|u; \Omega\|_{W_p^m} + \|Tu; \Omega\|_{W_p^l} .$$

We need to show that for any $\varepsilon > 0$ one can find a constant C_ε such that

$$\|Tu; \Omega\|_{W_p^l} \le \varepsilon\, \|u; \Omega\|_{W_p^m} + C_\varepsilon\, \|u; \Omega\|_{L_p} . \tag{9.3.15}$$

We assume that this is not the case. Then for some $\varepsilon > 0$ there exist a function sequence $\{u_j\}$ with $\|u_j; \Omega\|_{W_p^m} = 1$ and a number sequence $\{k_j\}$, $k_j \to +\infty$, such that

$$\|Tu_j; \Omega\|_{W_p^l} > \varepsilon + k_j \|u_j; \Omega\|_{L_p} . \tag{9.3.16}$$

Since the operator $T\colon W_p^m(\Omega) \to W_p^l(\Omega)$ is bounded and the norm of u_j in $W_p^m(\Omega)$ is equal to one, we see by (9.3.16) that $u_j \to 0$ in $L_p(\Omega)$. We select a subsequence from $\{u_j\}$ weakly convergent in $W_p^m(\Omega)$ for which we retain the notation $\{u_j\}$. Let v be its weak limit. Then, for any $g \in L_{p'}(\Omega)$, where $p + p' = pp'$, we have

$$\int_\Omega g u_j \, dx \to \int_\Omega g v \, dx$$

and hence $v = 0$ because $u_j \to 0$ in $L_p(\Omega)$. Since T transforms a sequence weakly convergent in $W_p^m(\Omega)$ into a sequence strongly convergent in $W_p^l(\Omega)$ we have

$$\|Tu_j; \Omega\|_{W_p^l} \to 0,$$

contrary to (9.3.16). □

Proposition 9.3.4. *If $m = l$ and $\gamma \in \mathring{M}(W_p^l(\Omega))$, then*

$$\inf K = \|\gamma; \Omega\|_{L_\infty}.$$

Proof. Let γ_1 be a function in $C^\infty(\overline{\Omega})$ such that $\|\gamma - \gamma_1; \Omega\|_{MW_p^l} < \varepsilon$. Clearly,

$$\|\gamma_1 u; \Omega\|_{W_p^l} \le (\|\gamma; \Omega\|_{L_\infty} \|u; \Omega\|_{W_p^l} + c \sum_{j=1}^{l} \| \, |\nabla_j \gamma_1| \, |\nabla_{l-j} u|; \Omega\|_{L_p}.$$

Hence

$$\|\gamma u; \Omega\|_{W_p^l} \le \big(\|\gamma; \Omega\|_{L_\infty} + 2\varepsilon\big) \|u; \Omega\|_{W_p^l} + c \, \|\gamma_1; \Omega\|_{C^l} \|u; \Omega\|_{W_p^{l-1}}.$$

Since

$$\|u; \Omega\|_{W_p^{l-1}} \le \varepsilon \, \|u; \Omega\|_{W_p^l} + c(\varepsilon) \, \|u; \Omega\|_{L_p},$$

it follows that $\inf K \le \|\gamma; \Omega\|_{L_\infty}$.

Let us estimate $\inf K$ from below. We set

$$u_\delta(x) = \delta^{l-n/p} \eta((x-y)/\delta)$$

into (9.3.13), where $m = l$. We take $y \in \overline{\Omega}$, $\delta > 0$, $\eta \in C_0^\infty(\mathcal{B}_1)$, $\eta(0) = 1$. By definition of $\mathring{M}W_p^l(\Omega)$, the function γ can be assumed to be smooth in $\overline{\Omega}$. One can easily see that

$$\|\gamma u_\delta; \Omega\|_{L_p} = |\gamma(y)| \, \|u_\delta; \Omega\|_{W_p^l} + o(1)$$

and

$$\|u_\delta; \Omega\|_{L_p} = o(1) \quad \text{as } \delta \to 0.$$

This, together with (9.3.13) and the inequality

$$\liminf_{\delta \to 0} \|u_\delta; \Omega\|_{W_p^l} > 0,$$

gives $|\gamma(y)| \le K$. □

Remark 9.3.1. The inequality

$$\|\gamma u; \Omega\|_{W_p^l} \le \left(\|\gamma; \Omega\|_{L_\infty} + \varepsilon\right)\|u; \Omega\|_{W_p^l} + C(\gamma; \varepsilon)\|u; \Omega\|_{L_p}$$

with an arbitrarily small $\varepsilon > 0$ is used in the L_p-theory of elliptic boundary value problems (see, for example, [Tr3], Sect. 5.3.4).

In conclusion, we state an obvious corollary of Theorem 9.3.6 and Proposition 9.3.4.

Corollary 9.3.1. *If* $\gamma \in \overset{\circ}{M}(W_p^l(\Omega))$ *then*

$$\|\gamma; \Omega\|_{L_\infty} \le \text{ess}\, \|\gamma; \Omega\|_{MW_p^l} \le c\, \|\gamma; \Omega\|_{L_\infty} . \tag{9.3.17}$$

Similarly to Theorem 7.3.1, we can prove that the left estimate in (9.3.17) holds for any $\gamma \in MW_p^l(\Omega)$.

9.3.5 The Space $ML_p^1(\Omega)$ for an Arbitrary Bounded Domain

Let Ω be a domain in $\mathbb{R}^n$ with compact closure. By $L_p^1(\Omega)$ we denote the space of functions in $L_{p,\text{loc}}(\Omega)$ with the first distributional derivatives in $L_p(\Omega)$, $p \in [1, \infty)$. We supply $L_p^1(\Omega)$ with the norm

$$\|u; \Omega\|_{L_p^1} = \|\nabla u; \Omega\|_{L_p} + \|u; \omega\|_{L_p} ,$$

where ω is a nonempty domain contained in Ω along with its closure. We can check that the change of ω leads to an equivalent norm.

If Ω is a domain with boundary in the class $C^{0,1}$, then $L_p^1(\Omega) \subset L_p(\Omega)$ and therefore $L_p^1(\Omega) = W_p^1(\Omega)$. Comparing Theorem 2.1.1 with Theorems 9.3.1, 9.3.2 and using Theorem 9.3.4, we obtain that for Lipschitz domains the space $ML_p^1(\Omega)$ coincides with the space of restrictions to Ω of multipliers in $W_p^1(\mathbb{R}^n)$. The following example shows that this fails for arbitrary domains.

Example 9.3.1. Let Ω be the union of the rectangles

$$A_m = \{x\colon 2^{1-m} - \delta_m < x_1 < 2^{1-m},\, 2/3 < x_2 < 1\} ,$$
$$B_m = \{x\colon 2^{1-m} - \varepsilon_m < x_1 < 2^{1-m},\, 1/3 \le x_2 \le 2/3\} ,$$
$$C = \{x\colon 0 < x_1 < 1,\, 0 < x_2 < 1/3\} ,$$

where $\delta_m = 2^{-m-1}$, $\varepsilon_m = 2^{-(m+1)\beta}$, $\beta \ge 1$, $m = 1, 2, \ldots$ (see Fig. 9.1). This domain was proposed by Nikodym in 1933 [Nik] as an example of the failure of the Poincaré inequality.

We show that the function $\gamma(x) = x_1^\lambda$ is a multiplier in $L_p^1(\Omega)$ if and only if $\lambda \ge (\beta + p - 1)/p$. It is clear that this function is a multiplier in $W_{p,\text{loc}}^1(\mathbb{R}^n)$

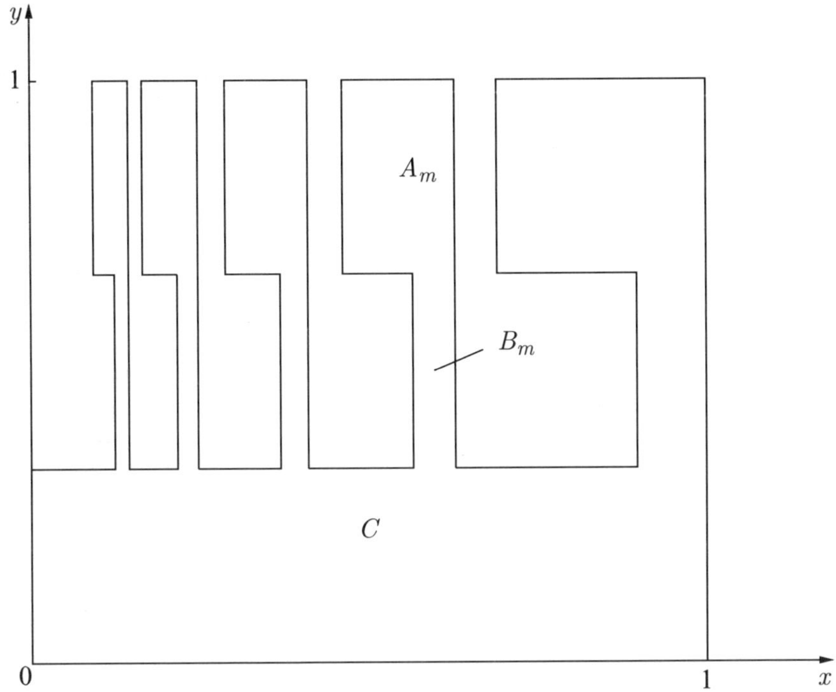

Fig. 9.1. Domain in Example 9.3.1

even for $\lambda > (p-1)/p$. Thus, in the case $(p-1)/p < \lambda < (\beta + p - 1)/p$, the restriction to Ω of the multiplier x_1^λ in $W^1_{p,\mathrm{loc}}(\mathbb{R}^n)$ is not a multiplier in $L^1_p(\Omega)$.

The necessity of the condition $\lambda \geq (\beta + p - 1)/p$ can be checked easily. In fact, let u be a continuous function equal to $(2^{m\beta} m^{-2})^{1/p}$ in A_m and to zero in C, and let u be linear to B_m. Clearly,

$$\|\nabla u; \Omega\|^p_{L_p} = \sum_{m=1}^{\infty} 2^{m\beta}\, m^{-2}\, \mathrm{mes}_2 B_m < \infty\,,$$

$$\|\nabla(\gamma u); \Omega\|^p_{L_p} \geq c \sum_{m=1}^{\infty} 2^{m\beta}\, m^{-2}\, 2^{-(\lambda-1)pm}\, \mathrm{mes}_2 A_m\,.$$

The last series diverges if $\lambda < (\beta + p - 1)/p$.

Now let $\lambda \geq (\beta + p - 1)/p$. By B_m^+ and B_m^- we denote the rectangle B_m raised and lowered by one-third, respectively. We have

$$\delta_m^{-1} \int_{A_m} |u|^p dx - \varepsilon_m^{-1} \int_{B_m^+} |u|^p dx \leq c\, \delta_m^{p-1} \int_{A_m} |\partial u / \partial x_2|^p dx\,.$$

Moreover,

$$\frac{1}{2}\int_{B_m\cup B_m^+}|u|^p dx - \int_{B_m^-}|u|^p dx \le c\int_{B_m\cup B_m^+\cup B_m^-}|\partial u/\partial x_1|^p dx\,.$$

Therefore,

$$\int_{A_m\cup B_m}|u|^p dx \le c\,\frac{\delta_m}{\varepsilon_m}\Big(\int_{A_m\cup B_m\cup B_m^-}|\nabla u|^p dx + \int_{B_m^-}|u|^p dx\Big)\,. \tag{9.3.18}$$

Obviously,

$$\|u\nabla\gamma;\Omega\|^p_{L_p} \le c\Big(\int_C |u|^p dx + \sum_{m=1}^{\infty}\delta_m^{(\lambda-1)p}\int_{A_m\cup B_m}|u|^p dx\Big)\,.$$

Taking into account (9.3.18), we conclude that the sum on the right-hand side does not exceed

$$\begin{aligned} &c\sum_{m=1}^{\infty}\delta_m^{(\lambda-1)p+1-\beta}\Big(\int_{A_m\cup B_m\cup B_m^-}|\nabla u|^p dx + \int_{B_m^-}|u|^p dx\Big)\\ &\qquad\le c\sup_m \delta_m^{(\lambda-1)p+1-\beta}\Big(\int_\Omega|\nabla u|^p dx + \int_C|u|^p dx\Big)\,. \end{aligned}$$

Consequently,

$$\|\nabla(\gamma u);\Omega\|_{L_p} \le c\left(\|\nabla u;\Omega\|_{L_p} + \|u;C\|_{L_p}\right),$$

i.e., $\gamma\in ML^1_p(\Omega)$.

It can be easily shown that $x_1^\lambda\in MW^1_p(\Omega)$ if and only if $\lambda\ge 1$. Therefore, for $(p-1)/p<\lambda<1$ the restriction to Ω of $x_1^\lambda\in MW^1_{p,\mathrm{loc}}$ does not belong to $MW^1_p(\Omega)$.

Below we describe the space of multipliers in $L^1_p(\Omega)$, where Ω is an arbitrary bounded domain. This description is obtained as a corollary of theorems in [Maz15] on necessary and sufficient conditions for the validity of imbeddings of spaces of functions with first derivatives in $L_p(\Omega)$ (for earlier publications see, for example, [Maz1]–[Maz3]).

In what follows, by g and G we denote the so-called admissible subsets of Ω, i.e., bounded sets such that $\Omega\cap\partial g$ and $\Omega\cap\partial G$ are manifolds of the class C^∞. Further, let $\mathrm{clos}_\Omega g$ be the closure of g with respect to Ω. We need the following capacity of the pair of sets g and G:

$$p\text{-}\mathrm{cap}_\Omega(g,G) = \inf\{\|\nabla u;\Omega\|^p_{L_p} : u\in C^\infty(\Omega),\quad u=0 \text{ on } G,\quad u=1 \text{ on } g\}\,.$$

Theorem 9.3.7. ([Maz15]). *Let u be an arbitrary function in $C^\infty(\Omega)$ with $u=0$ on G.*

(i) For $p > 1$*, the inequality*

$$\|\gamma u; \Omega\|_{L_p} \le C\, \|\nabla u; \Omega\|_{L_p} \tag{9.3.19}$$

holds if and only if

$$\int_g |\gamma|^p dx \le \text{const}\ p\text{-cap}_\Omega(g, G),$$

where g *is an admissible set with*

$$clos_\Omega \subset \Omega \backslash clos_\Omega G.$$

The best constant C *in* (9.3.19) *for* $p > 1$ *satisfies the inequality*

$$\frac{(p-1)^{p-1}}{p^p} C^p \le \sup_g \frac{\|\gamma; g\|^p_{L_p}}{p\text{-cap}_\Omega(g, G)} \le C^p\,.$$

(ii) For $p = 1$*, the inequality* (9.3.19) *holds if and only if*

$$\int_g |\gamma|\, dx \le const\, s(\Omega \cap \partial g),$$

where g *is any admissible set with* $clos_\Omega g \subset \Omega \backslash clos_\Omega G$ *and* s *is the* $(n-1)$*-dimensional area.*

The best constant C *in* (9.3.19) *for* $p = 1$ *is given by*

$$C = \sup_g \frac{\|\gamma; g\|_{L_1}}{s(\Omega \cap \partial g)}.$$

Using Theorem 9.3.7, one easily obtains a description of the space $ML^1_p(\Omega)$.

Theorem 9.3.8. *A function* γ *belongs to* $ML^1_p(\Omega)$ *if and only if* $\gamma \in L_\infty(\Omega) \cap MW^1_{p,\text{loc}}(\Omega)$ *and, for some admissible set* G *with* $\overline{G} \subset \Omega$,

$$\sup \frac{\|\nabla\gamma; g\|^p_{L_p}}{p\text{-cap}_\Omega(g, G)} < \infty \quad for \quad p > 1\,,$$

$$\sup \frac{\|\nabla\gamma; g\|_{L_1}}{s(\Omega \cap \partial g)} < \infty \quad for \quad p = 1\,,$$

where the suprema are taken over all admissible sets g *with* $clos_\Omega g \subset \Omega \backslash \overline{G}$.

Proof. The necessity of the condition $\gamma \in MW^1_{p,\text{loc}}(\Omega)$ is obvious and the necessity of the boundedness of γ follows from the inequality

$$\|\gamma^N u; \Omega\|^{1/N}_{L^1_p} \le \|\gamma; \Omega\|_{ML^1_p} \|u; \Omega\|^{1/N}_{L^1_p}\,, \qquad N = 1, 2, \dots\,.$$

Other assertions result from Theorem 9.3.7 and the estimate

$$\|u \nabla \Gamma - \nabla(\Gamma u); \Omega\|_{L_p} \le \|\gamma; \Omega\|_{L_\infty} \|\nabla u; \Omega\|_{L_p}\,,$$

where $\Gamma = \eta\gamma$, $\eta \in C^\infty(\mathbb{R}^n)$, $\eta = 0$ on G, $\eta = 1$ in a neighborhood of $\partial\Omega$, and $0 \le \eta \le 1$. □

Remark 9.3.2. By Theorem 9.3.1 for $p > 1$, the condition $\gamma \in MW^1_{p,\mathrm{loc}}(\Omega)$ can be replaced by

$$\sup \frac{\|\nabla\gamma; g\|^p_{L_p}}{C_{p,1}(e)} < \infty .$$

Here the supremum is taken over all admissible sets g placed at an arbitrary fixed positive distance from $\partial\Omega$.

According to Theorem 9.3.2, $\gamma \in MW^1_{1,\mathrm{loc}}(\Omega)$ if and only if

$$\sup r^{1-n} \|\nabla\gamma; \mathcal{B}_r(x)\|_{L_1} < \infty ,$$

where the supremum is taken over all balls $\mathcal{B}_r(x)$, $x \in \Omega$, placed at an arbitrary fixed positive distance from $\partial\Omega$.

9.4 Change of Variables in Norms of Sobolev Spaces

In this section we introduce and study certain classes of differentiable mappings which are considered as operators in pairs of Sobolev spaces. In Sect. 9.4.1, using spaces of multipliers, we define the so-called (p,l)-diffeomorphisms. We show that these mappings preserve the space W^l_p. With the help of (p,l)-diffeomorphisms we introduce in Sect. 9.4.4 the class of (p,l)-manifolds on which W^l_p can be properly defined. Sect. 9.4.5 concerns mappings of the class $T^{m,l}_p$, i.e., the mappings $U \to V$ which generate continuous operators: $W^m_p(V) \to W^l_p(U)$, where $p \geq 1$, m and l are integers, U and V are open sets in $\mathbb{R}^n$.

9.4.1 (p,l)-Diffeomorphisms

Let V be an open subset of $\mathbb{R}^n$ and let l be a noninteger. By $W^l_p(V)$ we mean the space of functions with the finite norm

$$\|u; V\|_{W^l_p} = \|\nabla_{[l]} u; V\|_{L_p} + \Big(\int_V \int_V |\nabla_{[l]} u(x) - \nabla_{[l]} u(y)|^p |x-y|^{-n-p\{l\}} dx dy\Big)^{1/p}.$$

Let $p \in (1,\infty)$, $l \geq 1$, and let U and V be open subsets of $\mathbb{R}^n$. A quasi-isometric mapping $\varkappa : U \to V$ is called a (p,l)-diffeomorphism if all elements of its Jacobi matrix $\partial\varkappa$ belong to the space of multipliers $MW^l_p(U)$.

The following four lemmas contain some properties of (p,l)-diffeomorphisms. By $\|\partial\varkappa; U\|_{MW^{l-1}_p}$ we denote the sum of the norms of elements of $\partial\varkappa$ in $MW^{l-1}_p(U)$.

Lemma 9.4.1. *Let $u \in W^l_p(V)$, $l \in [1,\infty)$, and let $\varkappa : U \to V$ be a (p,l)-diffeomorphism. Then $u \circ \varkappa \in W^l_p(U)$ and*

$$\|u \circ \varkappa; U\|_{W^l_p} \leq c\, \|u; V\|_{W^l_p}. \tag{9.4.1}$$

Proof. Set $\lambda = \inf \det \partial \varkappa$. Clearly,

$$\|u \circ \varkappa; U\|_{W_p^1} \le \|(\partial\varkappa)^*(\nabla u) \circ \varkappa; U\|_{L_p} + \|u \circ \varkappa; U\|_{L_p}$$

$$\le \lambda^{-1/p}\big(\|\partial\varkappa; U\|_{L_\infty}\|\nabla u; V\|_{L_p} + \|u; V\|_{L_p}\big).$$

In the case $\{l\} > 0$ we have

$$\| u \circ \varkappa; U\|_{W_p^{\{l\}}} = \Big(\int_U \int_U |u(\varkappa(x)) - u(\varkappa(y))|^p |x-y|^{-n-p\{l\}} dxdy\Big)^{1/p}$$

$$+\|u \circ \varkappa; U\|_{L_p} \le \big(\lambda^{-2/p}\|\partial\varkappa; U\|_{L_\infty}^{n/p+\{l\}} + \lambda^{-1/p}\big)\|u; V\|_{W_p^{\{l\}}}.$$

Suppose that (9.4.1) holds for all l with $[l] = 1, \dots, k-1$. Then, for $[l] = k$,

$$\| u \circ \varkappa; U\|_{W_p^l} = \|\nabla(u \circ \varkappa); U\|_{W_p^{l-1}} + \|u \circ \varkappa; U\|_{L_p}$$

$$= \|(\partial\varkappa)^*(\nabla u) \circ \varkappa); U\|_{W_p^{l-1}} + \|u \circ \varkappa; U\|_{L_p}$$

$$\le \|\partial\varkappa; U\|_{MW_p^{l-1}}\|(\nabla u) \circ \varkappa); U\|_{W_p^{l-1}} + \|u \circ \varkappa; U\|_{L_p}.$$

Using the induction assumption, we complete the proof. □

Lemma 9.4.2. *If $\varkappa$ is a (p,l)-diffeomorphism, then $\varkappa^{-1}$ is also a (p,l)-diffeomorphism.*

Proof. For $l = 1$ this assertion is contained in the definition of the (p,l)-diffeomorphism.

Let $1 < l < 2$ and let u be an arbitrary function in $W_p^{l-1}(V)$. Since $\varkappa$ is a bi-Lipschitz mapping, it follows that

$$\|u\, \partial(\varkappa^{-1}); V\|_{W_p^{l-1}} = \|\big(\,(u \circ \varkappa)(\partial\varkappa)^{-1}\big) \circ \varkappa^{-1}; V\|_{W_p^{l-1}}$$

$$\le c\, \|(u \circ \varkappa)(\partial\varkappa)^{-1}; U\|_{W_p^{l-1}}.$$

Hence

$$\|u\, \partial(\varkappa^{-1}); V\|_{W_p^{l-1}} \le c\, \|(\partial\varkappa)^{-1}; U\|_{MW_p^{l-1}}\|u \circ \varkappa; U\|_{W_p^{l-1}}.$$

It remains to use Lemma 9.4.1 and the condition $\partial\varkappa \in MW_p^{l-1}(U)$.

We proceed by induction. Suppose that the lemma is proved for $[l] = 1, \dots, k-1$. Let $[l] = k$. The condition $\partial\varkappa \in MW_p^{l-1}(U)$ implies the inclusion $\partial\varkappa \in MW_p^{l-2}(U)$ and hence, by the induction assumption, $\partial(\varkappa^{-1}) \in MW_p^{l-2}(V)$. This together with Lemma 9.4.1 implies that the matrix $(u \circ \varkappa)(\partial\varkappa)^{-1} \circ \varkappa^{-1}$ belongs to the space $W_p^{l-1}(V)$ provided that $(u\circ\varkappa)(\partial\varkappa)^{-1} \in W_p^{l-1}(U)$. The last inclusion holds since $(\partial\varkappa)^{-1} \in MW_p^{l-1}(U)$ and $u \circ \varkappa \in W_p^{l-1}(U)$ by Lemma 9.4.1. □

Lemma 9.4.3. *Let $\gamma \in MW_p^l(V)$ and let $\varkappa$ be a (p,l)-diffeomorphism. Then*

$$\|\gamma \circ \varkappa; U\|_{MW_p^l} \le c\, \|\gamma; V\|_{MW_p^l}. \tag{9.4.2}$$

Proof. By Lemma 9.4.1, for all $u \in W_p^{l-1}(U)$

$$\|(\gamma \circ \varkappa)u; U\|_{W_p^l} = \|\big(u \circ \varkappa^{-1})\gamma\big) \circ \varkappa; U\|_{W_p^l}$$

$$\le c\, \|\gamma; V\|_{MW_p^l} \|u \circ \varkappa^{-1}; V\|_{W_p^l}. \tag{9.4.3}$$

Since by Lemma 9.4.2 $\varkappa^{-1}$ is a (p,l)-diffeomorphism, Lemma 9.4.1 implies the estimate

$$\|u \circ \varkappa^{-1}; V\|_{W_p^l} \le c\, \|u; U\|_{W_p^l}.$$

Combined with (9.4.3), this completes the proof. □

Lemma 9.4.4. *Let $\varkappa_1 : U \to V$ and $\varkappa_2 : V \to W$ be (p,l)-diffeomorphisms. Then their composition $\varkappa_2 \circ \varkappa_1 : U \to W$ is a (p,l)-diffeomorphism.*

Proof. Since the matrix $\partial(\varkappa_2 \circ \varkappa_1)$ is equal to the product of matrices $(\partial\varkappa_2 \circ \varkappa_1)$ and $\partial\varkappa_1$, it follows that

$$\|\partial(\varkappa_2 \circ \varkappa_1); U\|_{MW_p^{l-1}} \le \|\partial\varkappa_2 \circ \varkappa_1; U\|_{MW_p^{l-1}} \|\partial\varkappa_1; U\|_{MW_p^{l-1}}.$$

We estimate the first factor on the right by (9.4.2) replacing l by $l-1$, γ by $\partial\varkappa_2$ and $\varkappa$ by $\varkappa_1$. □

Remark 9.4.1. The above definition of (p,l)-diffeomorphisms can be generalized replacing MW_p^l by one of the multiplier algebras $A_p^{m,l}$ and $\mathcal{A}_p^{m,l}$ dealt with in Sect. 6.3 and Sect. 6.4.2. Obviously, all results in this subsection remain true.

Remark 9.4.2. Runst and Youssfi [RY] used (p,l)-diffeomorphisms to prove an existence theorem for an equation involving the Jacobian. (For other results in the same area see Dacorogna and Moser [DM1], Ye [Ye], and Sickel and Youssfi [SY].)

9.4.2 More on (p,l)-Diffeomorphisms

In the sequel we sometimes use the norm $|||\cdot|||$ in $W_p^l(V)$, invariant with respect to dilations, which is defined for integer l by

$$|||u; V|||_{W_p^l} = \sum_{j=0}^{l} d^{j-l} \|\nabla_j u; V\|_{L_p}, \tag{9.4.4}$$

where d is the diameter of V. In the case of noninteger l we put

$$\|u; V\|_{W_p^l} = \sum_{j=0}^{[l]} d^{j-l} \|\nabla_j u; V\|_{L_p}$$

$$+ \sum_{j=0}^{[l]} d^{j-[l]} \Big(\int_V \int_V |\nabla_j u(x) - \nabla_j u(y)|^p \frac{dxdy}{|x-y|^{n+p\{l\}}} \Big)^{1/p}. \tag{9.4.5}$$

The norm in $M(W_p^m(V) \to W_p^l(V))$ generated by the norm $\|\cdot\|$ in $W_p^l(V)$ will be denoted by

$$\|\gamma; \Omega\|_{M(W_p^m \to W_p^l)} \,. \tag{9.4.6}$$

In this subsection we collect some properties of (p,l)-diffeomorphisms related to the norm $\|\cdot\|$.

(i) Let $u \in W_p^l(V)$, $l \in [1,\infty)$, and let $\varkappa : U \to V$ be a (p,l)-diffeomorphism. Then $u \circ \varkappa \in W_p^l(U)$ and

$$\|u \circ \varkappa; U\|_{W_p^l} \le c\, \|u; V\|_{W_p^l}. \tag{9.4.7}$$

(ii) If $\varkappa$ is a (p,l)-diffeomorphism, then $\varkappa^{-1}$ is also a (p,l)-diffeomorphism, that is,

$$\|\partial(\varkappa^{-1}); U\|_{MW_p^{l-1}} \le c. \tag{9.4.8}$$

(iii) Let $\gamma \in MW_p^l(V)$ and let $\varkappa$ be a (p,l)-diffeomorphism. Then

$$\|\gamma \circ \varkappa; U\|_{MW_p^l} \le c\, \|\gamma; V\|_{MW_p^l}. \tag{9.4.9}$$

The constants c in *(i)*–*(iii)* depend on $\inf \det \partial\varkappa$, p, l, n, and the norm $\|\partial\varkappa; U\|_{MW_p^{l-1}}$. A similar remark concerns the next property.

(iv) Let $\varkappa_1 : U \to V$ and $\varkappa_2 : V \to W$ be (p,l)-diffeomorphisms. Then their composition $\varkappa_2 \circ \varkappa_1 : U \to W$ is a (p,l)-diffeomorphism, i.e.

$$\|\partial(\varkappa_2 \circ \varkappa_1); U\|_{MW_p^{l-1}} \le c. \tag{9.4.10}$$

Assertions *(i)*–*(iv)* are obtained in Sect. 9.4.1 for the usual norm in W_p^l. The passage to the norm (9.4.6) does not change the proof.

9.4.3 A Particular (p,l)-Diffeomorphism

Let $\varphi : \mathbb{R}^n \to \mathbb{R}$ be a Lipschitz function with the Lipschitz constant L. We introduce the operator $\mathcal{T}$ by the formula

$$(\mathcal{T} f)(x,y) = \int_{\mathbb{R}^n} \zeta(t)\varphi(x+ty)dt, \quad y > 0, \tag{9.4.11}$$

where $\zeta \in C_0^\infty(\mathbb{R}^n)$. By Theorem 8.7.2, if l is an integer, $p \in [1, \infty)$ and $\nabla_s \varphi \in MW_p^{l-1/p}(\mathbb{R}^n)$, $s = 0, 1, \dots, l$, then $\nabla_s(\mathcal{T}\varphi) \in MW_p^l(\mathbb{R}_+^{n+1})$ and

$$\|\nabla_s(\mathcal{T}\varphi); \mathbb{R}_+^{n+1}\|_{MW_p^l} \le c\,\|\nabla_s\varphi; \mathbb{R}^n\|_{MW_p^{l-1/p}}. \tag{9.4.12}$$

We assume further that $\zeta \ge 0$ and

$$\int_{\mathbb{R}^n} \zeta(t)dt = 1.$$

Let, as before,

$$G = \{(x, y) : x \in \mathbb{R}^n, y > \varphi(x)\}$$

and let N be a sufficiently large constant depending on L. We introduce the mapping

$$\lambda : \mathbb{R}_+^{n+1} \ni (\xi, \eta) \to (x, y) \in G \tag{9.4.13}$$

by the equalities

$$x = \xi, \quad y = N\eta + (\mathcal{T}\varphi)(\xi, \eta). \tag{9.4.14}$$

Lemma 9.4.5. *For any $\xi \in \mathbb{R}^n$ the mapping*

$$\alpha_\xi : \mathbb{R}_+ \ni \eta \to y = N\eta + (\mathcal{T}\varphi)(\xi, \eta)$$

is one to one, and the inverse is Lipschitz. Moreover,

$$\Big|\frac{\partial}{\partial y}(\alpha_\xi^{-1}(y))\Big| \le (N - L)^{-1}. \tag{9.4.15}$$

and

$$\big|\alpha_{\xi_1}^{-1}(y) - \alpha_{\xi_2}^{-1}(y)\big| \le c\,L(N - cL)^{-1}\|\xi_1 - \xi_2\|_{\mathbb{R}^n}, \tag{9.4.16}$$

where c is a constant depending on n.

Proof. We fix points $x \in \mathbb{R}^n$ and $y \in \mathbb{R}_+$. The operator

$$\beta : \eta \to N^{-1}\big(y - (\mathcal{T}\varphi)(\xi, \eta)\big)$$

maps the segment

$$\{\eta : |\eta| \le |y - \varphi(x)|\}$$

into itself, because

$$\begin{aligned}|\beta(\eta)| &\le N^{-1}\big(|y - \varphi(x)| + |(\mathcal{T}\varphi)(\xi, \eta) - (\mathcal{T}\varphi)(\xi, 0)|\big)\\ &\le N^{-1}\big(|y - \varphi(x)| + L\,|\eta|\big) \le (1 + L)N^{-1}|y - \varphi(x)|.\end{aligned}$$

Also, β is a contraction operator, since

$$|\beta(\eta_1) - \beta(\eta_2)| \le LN^{-1}|\eta_1 - \eta_2|.$$

Therefore, there exists a unique solution η of the equation

$$N^{-1}\big(y - (\mathcal{T}\varphi)(\xi,\eta)\big) = \eta,$$

or, equivalently, of the equation $\alpha_\xi(\eta) = y$.

Let y_1 and y_2 be arbitrary points in $\mathbb{R}_+$ and let $\eta_j = \alpha_\xi^{-1}(y_j)$, $j = 1, 2$. We have

$$\eta_1 - \eta_2 = N^{-1}\big(y_1 - y_2 - (\mathcal{T}\varphi)(x,\eta_1) + (\mathcal{T}\varphi)(x,\eta_2)\big).$$

Hence

$$|\eta_1 - \eta_2| \le N^{-1}\big(|y_1 - y_2| + L|\eta_1 - \eta_2|\big).$$

Since $N > L$, we arrive at (9.4.15).

The equalities

$$y = N\alpha_{\xi_j}^{-1}(y) + (\mathcal{T}\varphi)(\xi, \alpha_{\xi_j}^{-1}(y)), \qquad j = 1, 2,$$

imply that

$$|\alpha_{\xi_1}^{-1}(y) - \alpha_{\xi_2}^{-1}(y)| \le c\,N^{-1}L\big(\|\xi_1 - \xi_2\|_{\mathbb{R}^n} + |\alpha_{\xi_1}^{-1}(y) - \alpha_{\xi_2}^{-1}(y)|\big).$$

Hence (9.4.16) follows. □

Lemma 9.4.6. *Let l be an integer, $l > 1$, and $p \in [1,\infty)$. Further, let $\nabla\varphi \in MW^{p,l-1-1/p}$. Then the mapping λ defined by (9.4.14) is a (p,l)-diffeomorphism.*

Proof. By Lemma 9.4.5, the inverse mapping λ^{-1} exists, is defined by

$$\xi = x, \quad \eta = \alpha_x^{-1}(y),$$

and satisfies the Lipschitz condition. Its Jakobi matrix $\partial\lambda$ is given by

$$\begin{pmatrix} I & 0 \\ \nabla_\xi(\mathcal{T}f) & N + \partial(\mathcal{T}f)/\partial\eta \end{pmatrix} \tag{9.4.17}$$

where I is the identity $(n-1)\times(n-1)$-matrix. Since $|\partial(\mathcal{T}\varphi)/\partial\eta| \le L$, it follows that

$$\det\partial\lambda = N + \partial(\mathcal{T}\varphi)/\partial\eta \ge N - L > 0.$$

In view of (9.4.12), the elements of $\partial\lambda$ belong to the space $MW_p^{l-1}(\mathbb{R}^n_+)$. □

Remark 9.4.3. By Lemma 9.4.2, the mapping $\varkappa = \lambda^{-1}$ is a (p,l)-diffeomorphism of the domain

$$G = \{(x,y) : x \in \mathbb{R}^{n-1}, y > \varphi(x)\}, \tag{9.4.18}$$

where φ is a Lipschitz function, onto $\mathbb{R}^n_+$. The mapping $\varkappa$ is given by

$$\xi = x, \quad \eta = u(x, y),$$

where u is the unique solution of the equation

$$y = Nu + (\mathcal{T}\varphi)(x, u). \tag{9.4.19}$$

The restrictions of the mappings λ and $\varkappa$ to $\mathbb{R}^{n-1} = \partial\mathbb{R}^n_+$ and ∂G will be also denoted by λ and $\varkappa$, respectively.

Let G be a special Lipschitz domain. By $W_p^{l-1/p}(\partial G)$ we denote the space of traces on ∂G of functions in $W_p^l(G)$. In a similar way, we define the space $W_p^{l-1/p}(\Gamma)$, where Γ is a subset of ∂G.

Since G can be mapped onto $\mathbb{R}^n_+$ by a (p,l)-diffeomorphism, it follows by Theorem 8.7.2 and Lemma 9.4.3 that $MW_p^{l-1/p}(\partial G)$ is the space of traces of functions in $MW_p^l(G)$.

9.4.4 (p, l)-Manifolds

In terms of the (p,l)-diffeomorphisms, we can define in a standard manner (see, for instance, de Rham [dR], Hörmander [H1]) a class of non-smooth n-dimensional manifolds on which Sobolev spaces can be properly defined.

We recall that a topological space $\mathfrak{M}$ is called an n-dimensional manifold, if there exists a collection of homeomorphisms $\{\varphi\}$ of open sets $U_\varphi \in \mathfrak{M}$ onto open subsets of $\mathbb{R}^n$ with $\mathfrak{M} = \cup U_\varphi$.

The pair (φ, U_φ) is called a map (or the coordinate system) and the set of maps is called the atlas.

We say that two maps (φ, U_φ) and (ψ, U_ψ) have a (p,l)-overlapping, if the mapping

$$\varphi\psi^{-1} : \varphi(U_\varphi \cap U_\psi) \to \psi(U_\varphi \cap U_\psi)$$

is a (p,l)-diffeomorphism. By Lemma 9.4.2, the same is true for the inverse mapping.

If any two maps have a (p,l)-overlapping, then we have a (p,l)-atlas. The maximal (p,l)-atlas on $\mathfrak{M}$ is called a (p,l)-structure. By a (p,l)-manifold we mean a manifold with a (p,l)-structure.

Since

$$MW_p^{l-1}(\mathbb{R}^n) = W_{p,\text{unit}}^{l-1}(\mathbb{R}^n) \qquad \text{for} \quad p(l-1) > n,$$

it follows for these values of p and l that the structure on a (p,l)-manifold belongs to the class C^1, whereas for $p(l-1) \le n$ such manifolds are Lipschitz.

For functions defined on a (p,l)-manifold Ω we introduce the space $W_{p,\text{loc}}^l(\Omega)$. Namely, $u \in W_{p,\text{loc}}^l(\Omega)$ if $u \circ \varphi^{-1}$ belongs to $W_{p,\text{loc}}^l(\varphi(U_\varphi))$ for each map (φ, U_φ).

With the help of Lemmas 9.4.1 and 9.4.2 we can prove in a standard way (see [H1], Theorem 2.6.2) that to define the space $W_{p,\text{loc}}^l(\Omega)$ it suffices to use only one arbitrary (p,l)-atlas; i.e., the following assertion holds.

Theorem 9.4.1. *If a function u defined on a (p,l)-manifold Ω is such that*

$$u \circ \varphi^{-1} \in W^l_{p,\mathrm{loc}}(\varphi(U_\varphi))$$

for any map of some atlas, then $u \in W^l_{p,\mathrm{loc}}(\Omega)$. If $\eta_\varphi \in C_0^\infty(\varphi(U_\varphi))$ and the open sets

$$V_\varphi = \{x \in U_\varphi : \eta_\varphi(\varphi(x)) \neq 0\}$$

cover Ω, then, in order to define a topology in the space $W^l_{p,\mathrm{loc}}(\Omega)$, it suffices to introduce the seminorms

$$u \to \|\eta_\varphi(u \circ \varphi^{-1}\|_{W^l_p}.$$

If a manifold Ω is compact, then the topology in $W^l_{p,\mathrm{loc}}(\Omega)$ can be induced by the norm

$$\sum_\varphi \|\eta_\varphi(u \circ \varphi^{-1})\|_{W^l_p},$$

where the sum is taken over all maps of a certain atlas.

Replacing the space $\mathbb{R}^n$ by the closed half-space $\overline{\mathbb{R}^n_+} = \{\zeta \in \mathbb{R}^n : \zeta_n \geq 0\}$ in the definition of a (p,l)-manifold $\mathfrak{M}$, we obtain the definition of a (p,l)-manifold $\mathfrak{M}$ with the boundary $\partial\mathfrak{M}$.

Let l be an integer, $l \geq 2$, and let $\mathfrak{M}$ be a (p,l)-manifold. If $p(l-1) \leq n$, we additionally assume that the (p,l)-structure on $\mathfrak{M}$ belongs to the class C^1. Then the implicit function theorem 9.5.2, which will be proved in Sect. 9.5, implies that the (p,l)-structure on $\partial\mathfrak{M}$ induces the $(p, l-1/p)$-structure on $\partial\mathfrak{M}$.

9.4.5 Mappings $T^{m,l}_p$ of One Sobolev Space into Another

Let U and V be domains in $\mathbb{R}^n$. We say that a mapping $\varkappa : U \to V$ belongs to the class $T^{m,l}_p$ if $u \circ \varkappa \in W^l_p(U)$ for any $u \in W^m_p(V)$ and

$$\|u \circ \varkappa; U\|_{W^l_p} \leq c\, \|u; V\|_{W^m_p}\,. \tag{9.4.20}$$

We limit consideration to integer m and l, $m \geq l \geq 1$. For $m = l$ we write T^l_p instead of $T^{l,l}_p$.

In this subsection we give sufficient and, for some values of p, m, l, necessary and sufficient conditions for a mapping to belong to the class $T^{m,l}_p$. In particular, for $m = l$ we obtain a wider set of mappings than the class of (p,l)-diffeomorphisms.

In what follows, $\varkappa = (\varkappa_1, \ldots, \varkappa_n)$ is a one-to-one mapping with $\partial\varkappa \in W^{l-1}_1(U)$ such that $\det \partial\varkappa$ does not change its sign and

$$\int_U u(\varkappa(z))\, |\det \partial\varkappa(z)|\, dz = \int_V u(\zeta)\, d\zeta \tag{9.4.21}$$

for any $u \in L_1(V)$. (For sufficient conditions ensuring (9.4.21) see, for instance, Vodop'yanov, Gol'dshtein, Reshetnyak [VGR], and Malý [Mal]).

Since $\varkappa$ is a mapping of the class $W_1^l(U)$, it follows that, for any $u \in C^l(V)$ and multi-index α with $|\alpha| \leq l$,

$$D^\alpha[u(\varkappa(z))] = \sum_{1\leq|\beta|\leq|\alpha|} \varphi_\beta^\alpha(z)(D^\beta u)(\varkappa(z)) \qquad (9.4.22)$$

a.e. in U. Here and henceforth,

$$\varphi_\beta^\alpha = \sum_s c_s \prod_{i=1}^n \prod_j D^{s_{ij}} \varkappa_i\,,$$

where the sum is taken over all collections of multi-indices $s = (s_{ij})$ satisfying

$$\sum_{i,j} s_{ij} = \alpha, \quad |s_{ij}| \geq 1, \quad \sum_{i,j}(|s_{ij}| - 1) = |\alpha| - |\beta|.$$

We note that another and more explicit expression for the functions φ_β^α was found by Fraenkel [Fra].

Proposition 9.4.1. *If*

$$|\det(\partial\varkappa \circ \varkappa^{-1})|^{-1/p} \in M(W_p^m(V) \to L_p(V))$$

and

$$(\varphi_\beta^\alpha(|\det \partial\varkappa|^{-1/p})) \circ \varkappa^{-1} \in M(W_p^{m-|\beta|}(V) \to L_p(V))$$

for all multi-indices α and β with $l \geq |\alpha| \geq |\beta| \geq 1$, then the mapping $\varkappa$ belongs to the class $T_p^{m,l}$.

Proof. Inequality (9.4.20) for any $u \in C^l(V) \cap W_p^m(V)$ follows directly from (9.4.21) and (9.4.22). The additional assumption $u \in C^l(V)$ can be removed by approximation. □

We give an example which shows that the conditions of Proposition 9.4.1 are sharp.

Example 9.4.1. We consider the domain

$$U = \{z\colon z_1^2 + \cdots + z_{n-1}^2 < z_n^{2\gamma},\, 0 < z_n < 1\}, \quad \gamma > 0,$$

and the mapping

$$\varkappa\colon z \to \zeta \quad \text{with } \zeta_i = z_i,\ 1 \leq i \leq n-1, \text{ and } \zeta_n = z_n^\gamma.$$

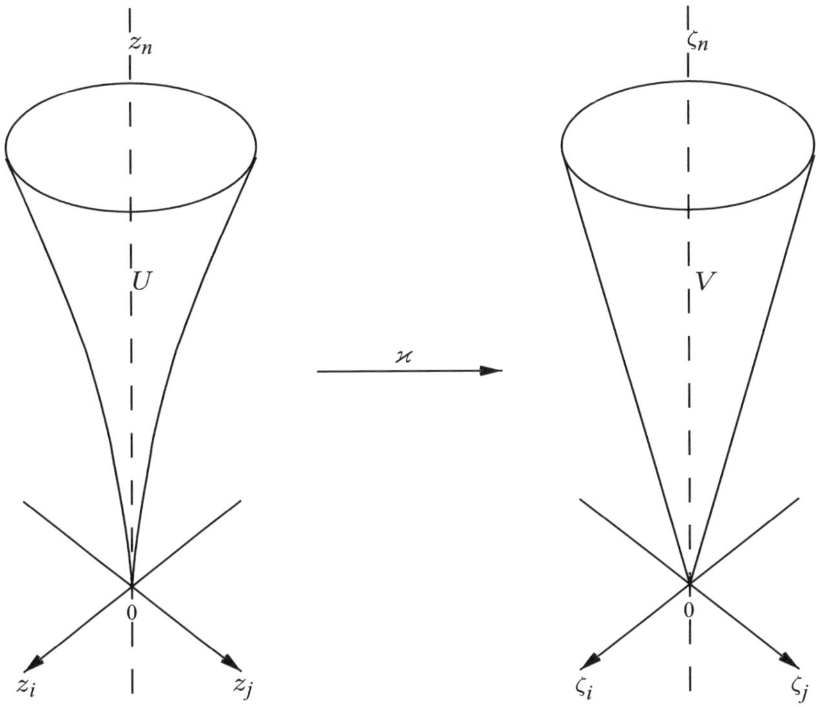

Fig. 9.2. Mapping in the class $T_p^{m,l}$ of a cusp to a cone

It is clear that $\varkappa$ transforms U into the cone

$$V = \{\zeta\colon \zeta_1^2 + \cdots + \zeta_{n-1}^2 < \zeta_n^2, \quad 0 < \zeta_n < 1\}.$$

(See Fig. 9.2) We show that $\varkappa \in T_p^{m,l}$ if and only if

$$\text{either} \quad p(m-1) < n, \quad \gamma \geq \frac{pl-1}{pm-1}, \tag{9.4.23}$$

$$\text{or} \qquad p(m-1) \geq n, \quad \gamma > \frac{pl-1}{p+n-1}. \tag{9.4.24}$$

Let $u(\zeta) = \zeta_n^\sigma$, where $\sigma = 1$ in the case $p(m-1) \geq n$ and σ is a noninteger in the interval

$$(m - n/p, (pl - 1 + \gamma - \gamma n)/\gamma p]$$

for $p(m-1) < n$. Clearly, $u \in W_p^m(V)$. On the other hand, $u(\varkappa(z)) = z_n^{\gamma\sigma}$ and therefore

$$\|\nabla_l(u \circ \varkappa^{-1}); U\|_{L_p}^p \geq c \int_0^1 z_n^{p(\gamma\sigma - l)} \int_0^{z_n^\gamma} r^{n-2}\, dr\, dz_n = \infty\,.$$

Thus, conditions (9.4.23), (9.4.24) are necessary.

Now we turn to the proof of sufficiency. By straightforward computation we have $\varphi^\alpha_\beta(z) = c\, z_n^{|\beta|\gamma-|\alpha|}$ and $\det \partial\varkappa(z) = \gamma\, z_n^{\gamma-1}$. Consequently,

$$(\varphi^\alpha_\beta \circ \varkappa^{-1})(\zeta) = c\,\zeta_n^{|\beta|-|\alpha|/\gamma}$$

and

$$\det(\partial\varkappa \circ \varkappa^{-1})(\zeta) = \gamma z_n^{1-1/\gamma}.$$

To check the conditions of Proposition 9.4.1, it suffices to verify the inequalities

$$\int_V |\zeta|^{(1-\gamma)/\gamma} |w|^p\, d\zeta \le c\,\|w; V\|^p_{W^m_p}$$

and

$$\int_V |\zeta|^{p(|\beta|-|\alpha|/\gamma)+(1-\gamma)/\gamma} |w|^p\, d\zeta \le c\,\|w; V\|^p_{W^{m-|\beta|}_p}$$

for any $w \in W^m_p(V)$. It is known that the Hardy inequality

$$\int_V \frac{|w|^p}{|\zeta|^{pk}}\, d\zeta \le c\,\|w; V\|^p_{W^s_p}\,, \qquad pk < n\,,$$

holds if and only if either $ps \ge n$ or $ps < n$ and $k \le s$. It remains to note that

$$(1-\gamma)/\gamma > -n\,, \qquad (1-\gamma)/\gamma p \ge -m$$

and that, by (9.4.23),

$$p(|\beta| - |\alpha|/\gamma) + (1-\gamma)/\gamma \ge p(1 - l/\gamma) + (1-\gamma)/\gamma > -n$$

and

$$|\beta| - |\alpha|/\gamma + (1-\gamma)/\gamma p \ge |\beta| - m$$

for $p(m-1) < n$. □

Remark 9.4.4. For $|\beta| = |\alpha| = l = m$, one of the conditions of Proposition 9.4.1 takes the form

$$\varphi^\alpha_\beta\, |\det \partial\varkappa|^{-1/p} \in L_\infty(U)\,.$$

We have

$$\frac{\partial^l}{\partial z_{i_1} \cdots \partial z_{i_l}} u(\varkappa(z))$$
$$= \sum_{k_1,\dots,k_l} \Big(\frac{\partial^l u}{\partial\zeta_{k_1} \cdots \partial\zeta_{k_l}}\Big)(\varkappa(z)) \prod_{\nu=1}^{l} \partial\varkappa_{k_\nu}/\partial z_{i_\nu} + \dots\,, \tag{9.4.25}$$

where the terms involving differentiation with respect to ζ of order less than l are omitted. So the condition mentioned above is equivalent to

$$\frac{\partial \varkappa_{k_1}}{\partial z_{i_1}} \cdots \frac{\partial \varkappa_{k_l}}{\partial z_{i_l}} |\det \partial \varkappa|^{-1/p} \in L_\infty(U) . \tag{9.4.26}$$

Here $k_1, \ldots, k_l$, $i_1, \ldots, i_l$ are numbers with values $1, \ldots, n$. The inclusion (9.4.26) can be rewritten in the form

$$\frac{|\partial \varkappa|^{pl}}{|\det \partial \varkappa|} \leq C , \tag{9.4.27}$$

where $C = \text{const}$ and

$$|\partial \varkappa| = \Big(\sum_{i,j=1}^{n} \Big(\frac{\partial \varkappa_i}{\partial z_j} \Big)^2 \Big)^{1/2} .$$

According to the Hadamard determinant inequality,

$$|\det \partial \varkappa| \leq \prod_{i=1}^{n} \Big(\sum_{j=1}^{n} \Big(\frac{\partial \varkappa_i}{\partial z_j} \Big)^2 \Big)^{1/2} .$$

Hence

$$|\det \partial \varkappa| \leq n^{-n/2} |\partial \varkappa|^n .$$

Consequently, for $pl > n$,

$$|\partial \varkappa| \leq (C n^{-n/2})^{1/(pl-n)},$$

i.e. $\varkappa$ is a Lipschitz mapping. In the case $pl < n$ we obtain

$$|\det \partial \varkappa| \geq (n^{pl/2} C^{-1})^{n/(n-pl)} . \tag{9.4.28}$$

Suppose that the mapping $\varkappa$ satisfies (9.4.27) together with $\varkappa^{-1}$. Then $\varkappa$ is Lipschitz for $pl < n$ also. Indeed, replacing $\varkappa$ by $\varkappa^{-1}$ in (9.4.28), we get

$$|\det \partial \varkappa| \leq (n^{pl/2} C^{-1})^{n/(pl-n)} .$$

This estimate and (9.4.27) imply that

$$|\partial \varkappa| \leq n^{n/2(pl-n)} C^{(2n-pl)/(n-pl)pl} .$$

Thus, for $pl \neq n$, the mapping $\varkappa$, which belongs to T_p^l together with $\varkappa^{-1}$, is bi-Lipschitz.

The class of mappings which perform the isomorphism $W_p^1(U) \approx W_p^1(V)$ was studied in [VG], where it is shown that such mappings are bi-Lipschitz for $p \geq n$.

For $pl = n$, (9.4.27) means that κ is a mapping with bounded distortion (see Reshetnyak [Re]). Mappings subjected to (9.4.27) with $p = l = 1$ are called subareal since they either decrease the area of $(n-1)$-dimensional surfaces or increase it with a finite coefficient (see [Maz4], [Maz15], Sect. 3.3.1).

Proposition 9.4.2. *Inequality* (9.4.27) *is necessary for* $\varkappa \in T_p^l$. *The same inequality is equivalent to* $\varkappa \in T_p^1$. *(Hence, by interpolation,* $T_p^l \subset T_p^k$, $k = 1, \dots, l-1$.)

Proof. We put

$$u(\zeta) = \eta(\zeta)|\lambda|^{-l} \exp(i(\lambda, \zeta)),$$

where $\eta \in C_0^\infty(V)$ and $\lambda \in \mathbb{C}^n$, into (9.4.20). Applying the Cauchy formula

$$D_\lambda^\gamma P(0; z) = \gamma!(2\pi i)^{-n} \int_{|\lambda_1|=1} \cdots \int_{|\lambda_n|=1} P(\lambda; z)\lambda^{-\gamma} \frac{d\lambda_1}{\lambda_1} \cdots \frac{d\lambda_n}{\lambda_n}$$

to the polynomial

$$\lambda \to P(\lambda; z) = |\lambda|^l D^\alpha[u(\varkappa(z))],$$

we find that its coefficients belong to $L_p(U)$. Therefore, we may pass to the limit as $|\lambda| \to \infty$ in (9.4.20). As a result, for any unit vector $\theta = (\theta_1, \dots, \theta_n)$ we obtain

$$\Big\| \eta \, |\det(\partial\varkappa \circ \varkappa^{-1})|^{-1/p} \sum_{|\gamma|=l} \theta^\gamma D^\gamma P(0; \cdot) \circ \varkappa^{-1}; V \Big\|_{L_p} \le c \, \|\eta; V\|_{L_p}$$

which by (9.4.25) can be written as

$$\Big\| \eta \, |\det(\partial\varkappa \circ \varkappa^{-1})|^{-1/p} \sum_{k_1, \dots, k_l} \theta_{k_1} \cdots \theta_{k_l} \Big(\frac{\partial \varkappa_{k_1}}{\partial z_{i_1}} \cdots \frac{\partial \varkappa_{k_l}}{\partial z_{i_l}} \Big) \circ \varkappa^{-1}; V \Big\|_{L_p} \le c\|\eta; V\|_{L_p}.$$

Since η is arbitrary, we conclude that the functions

$$|\det(\partial\varkappa \circ \varkappa^{-1})|^{-1/p} \frac{\partial(\theta, \varkappa)}{\partial z_{i_1}} \cdots \frac{\partial(\theta, \varkappa)}{\partial z_{i_l}}$$

are bounded. Hence condition (9.4.27) holds. □

The next assertion concerning conditions for a mapping to belong to $T_p^{m,l}$ follows directly from Proposition 9.4.1 and Theorem 9.3.1.

Proposition 9.4.3. *Let* V *be a bounded domain with boundary in the class* $C^{0,1}$ *and let* $p \in (1, \infty)$. *If, for any compact set* $e \subset \overline{V}$,

$$\text{mes}_n \varkappa^{-1}(e) \le c \, C_{p,m}(e)$$

and, for all multi-indices α, β *with* $l \ge |\alpha| \ge |\beta| \ge 1$,

$$\sup_{e \subset \overline{V}} \frac{\|\varphi_\beta^\alpha; \varkappa^{-1}(e)\|_{L_p}}{[C_{p,m-|\beta|}(e)]^{1/p}} < \infty,$$

then $\varkappa \in T_p^{m,l}$.

Now we present two propositions on necessary and sufficient conditions for a mapping to belong to the class $T_p^{m,l}$.

Proposition 9.4.4. *Let V be a bounded domain with boundary in the class $C^{0,1}$. The mapping κ belongs to $T_1^{m,l}$ if and only if, for any ball $\mathcal{B}_r(\zeta)$ with $\zeta \in \overline{V}$ and $r \in (0,1)$,*

$$\mathrm{mes}_n \varkappa^{-1}(\mathcal{B}_r(\zeta) \cap V) \leq c\, r^{n-m}$$

and, for all multi-indices α and β with $l \geq |\alpha| \geq |\beta| \geq 1$,

$$\sup_{\zeta \in \overline{V}, r \in (0,1)} r^{m-|\beta|-n} \|\varphi_\beta^\alpha; \varkappa^{-1}(\mathcal{B}_r(\zeta) \cap V)\|_{L_1} < \infty .$$

Proof. Sufficiency is an immediate corollary of Proposition 9.4.1 and Theorem 9.3.2.

On the other hand, the inclusion $\varkappa \in T_1^{m,l}$ is equivalent to

$$\Big\| \frac{u}{\det(\partial\varkappa \circ \varkappa^{-1})}; V \Big\|_{L_1} + \sum_{1 \leq |\alpha| \leq l} \Big\| \sum_{1 \leq |\beta| \leq |\alpha|} \Big(\frac{\varphi_\beta^\alpha}{\det \partial\varkappa} \Big) \circ \varkappa^{-1} D^\beta u; V \Big\|_{L_1}$$

$$\leq c\, \|u; V\|_{W_1^m} .$$

It remains to use Theorem 10.1.1 to be proved in the sequel. □

The following two assertions can be proved with the same arguments.

Proposition 9.4.5. *Let V be a bounded Lipschitz domain and let $(m-l)p > n$ and $p \in (1, \infty)$. A mapping $\varkappa$ is an element of $T_p^{m,l}$ if and only if $\mathrm{mes}_n U < \infty$ and $\varphi_\beta^\alpha \in L_p(U)$ for all multi-indices α, β with $l \geq |\alpha| \geq |\beta| \geq 1$.*

Proposition 9.4.6. *Let V be a bounded domain with boundary in the class $C^{0,1}$ and let $p > n$. A mapping $\varkappa$ belongs to T_p^l if and only if (9.4.27) holds and $\varphi_\beta^\alpha \in L_p(U)$ for $l \geq |\alpha| \geq |\beta| \geq 1$.*

For instance, for $p > n$ the inclusion $\varkappa \in T_p^2$ is equivalent to the conditions

$$|\partial\varkappa|^{2p} (|\det \partial\varkappa|)^{-1} \in L_\infty(U) \quad \text{and} \quad \nabla\partial\varkappa \in L_p(U) .$$

Similarly, $\varkappa \in T_p^3$ for $p > n$ if and only if

$$|\partial\varkappa|^{3p} (|\det \partial\varkappa|)^{-1} \in L_\infty(U) , \qquad \nabla_2 \partial\varkappa \in L_p(U)$$

and

$$\sum_{1 \leq \rho, \sigma, \tau \leq n} \frac{\partial \varkappa_r}{\partial z_\rho} \frac{\partial^2 \varkappa_s}{\partial z_\sigma \partial z_\tau} \in L_p(U) , \qquad r, s = 1, \ldots, n .$$

If both U and V are bounded and belong to $C^{0,1}$, then the conditions of Proposition 9.4.5 can be simplified. Namely, the following assertion holds.

Proposition 9.4.7. *Let U and V be bounded domains with boundaries in $C^{0,1}$ and let $p > 1$ and $(m-l)p > n$. A mapping $\varkappa$ belongs to $T_p^{m,l}$ if and only if $\varkappa \in W_p^l(U)$.*

Proof. Since

$$\{\varphi_\beta^\alpha\}|_{|\alpha|=l}^{|\beta|=1} = \{D^\alpha \varkappa_i\}_{|\alpha|=l}^{1\le i\le n}$$

and

$$\{\varphi_\beta^\alpha\}_{|\alpha|=l}^{|\beta|=l} = \Big\{\prod_{\nu=1}^{l} \frac{\partial \varkappa_{k_\nu}}{\partial z_{i_\nu}}\Big\}_{1\le k_\nu\le n}^{1\le i_\nu\le n},$$

the necessity follows from Proposition 9.4.5.

By definition of φ_β^α we obtain

$$\|\varphi_\beta^\alpha; U\|_{L_p} \le c \sum_s \prod_{i=1}^n \prod_j \|D^{s_{ij}} \varkappa_i; U\|_{L_{p|\alpha|/|s_{ij}|}}$$

which, together with the Gagliardo-Nirenberg inequality

$$\|D^{s_{ij}}\varkappa_i; U\|_{L_{p|\alpha|/|s_{ij}|}} \le c\,\|\varkappa_i; U\|_{W_p^{|\alpha|}}^{|s_{ij}|/|\alpha|}\,\|\varkappa_i; U\|_{L_\infty}^{1-|s_{ij}|/|\alpha|}$$

(see [Gag2], [Nir]), completes the proof. □

9.5 Implicit Function Theorems

Using properties of the (p,l)-diffeomorphisms we are in a position to prove the following assertion concerning regularity properties of a function defined implicitly. For the existence of such a function without differentiability assumptions, one may consult Sect. 5.2 in [KP].

Theorem 9.5.1. *Let G be the domain* (9.4.18) *where $\varphi \in C^{0,1}$ and let u be a function in G satisfying*

(*i*) $\nabla u \in MW_p^{l-1}(G)$, $\quad l \ge 2$,

(*ii*) $\mathrm{tr}\, u = 0$, *where* tr *stands for the trace on ∂G,*

(*iii*) $\inf \mathrm{tr}(\partial u/\partial y) > 0$,

where l is an integer and $p \in [1,\infty)$. Then

$$\nabla\varphi \in MW_p^{l-1-1/p}(\mathbb{R}^{n-1}).$$

Proof. We introduce the bi-Lipschitz mapping $\tau : G \ni (x,y) \to (\xi,\eta) \in \mathbb{R}_+^n$ by

$$\xi = x, \quad \eta = y - \varphi(x)$$

and put $v(\xi,\eta) = u(\xi, \eta + \varphi(\xi))$. Since $\partial_z u \in L_\infty(G)$, it follows that $\partial_\xi v \in L_\infty(\mathbb{R}_+^n)$ and almost everywhere in $\mathbb{R}_+^n$

$$\partial_\xi v = (\partial_x u + \partial_y u \nabla\varphi) \circ \tau^{-1}$$

(see, for example, [GR], p.244). Hence, for any $g \in C_0^\infty(\mathbb{R}^n)$ and almost all $\eta \in \mathbb{R}_+$ we have

$$\int_{\mathbb{R}^{n-1}} g(\xi)\partial_\xi v(\xi,\eta)d\xi = \int_{\mathbb{R}^{n-1}} g(\xi)\big((\partial_x u + \partial_y u \nabla\varphi) \circ \tau^{-1}\big)(\xi,\eta)d\xi. \quad (9.5.1)$$

Since $\partial_z u(x,\cdot) \in W_p^{l-1}(\mathbb{R}^1_+)$ and $l \geq 2$, the function $y \to \partial_z u(x,y)$ is continuous for almost all x in $\mathbb{R}^n$. Hence, for almost all $\xi \in \mathbb{R}^{n-1}$ the right-hand side of (9.5.1) is continuous in η, and, in particular, the limit

$$\lim_{\eta\to 0} \int_{\mathbb{R}^{n-1}} g(\xi)\partial_\xi v(\xi,\eta)d\xi = \int_{\mathbb{R}^{n-1}} g(\xi)\big((\partial_x u + \partial_y u \nabla\varphi) \circ \tau^{-1}\big)(\xi,0)d\xi \quad (9.5.2)$$

exists. The function $(\xi,\eta) \to v(\xi,\eta)$ is Lipschitz and $v(\xi,0) = 0$. Therefore,

$$\lim_{\eta\to 0} \int_{\mathbb{R}^{n-1}} g(\xi)\partial_\xi v(\xi,\eta)d\xi = -\lim_{\eta\to 0} \int_{\mathbb{R}^{n-1}} \partial g(\xi) v(\xi,\eta)d\xi = 0.$$

This and (9.5.2) imply that

$$(\partial_x u + \partial_y u \nabla\varphi) \circ \tau^{-1}(\xi,0) = 0 \quad \text{for almost all } \xi \in \mathbb{R}^{n-1},$$

or, equivalently,

$$\partial_x u(x,\varphi(x)) + \partial_y u(x,\varphi(x))\nabla\varphi(x) = 0 \quad \text{for almost all } x \in \mathbb{R}^{n-1}.$$

Thus, the identity

$$\nabla\varphi(x) = -\big(\partial_y u(x,\varphi(x))\big)^{-1}\partial_x u(x,\varphi(x)) \quad (9.5.3)$$

holds for almost all $x \in \mathbb{R}^{n-1}$.

Since $\nabla u \in MW_p^1(G)$, we have by Theorem 8.7.2 that $\operatorname{tr} u_{x_i}$ and $\operatorname{tr} u_y$ belong to $MW_p^{1-1/p}(\partial G)$ or, equivalently, that the functions

$$x \to u_{x_i}(x,\varphi(x)+0) \quad \text{and } x \to u_y(x,\varphi(x)+0)$$

are in $MW_p^{1-1/p}(\mathbb{R}^{n-1})$. The inequality $\inf u_y(x,\varphi(x)+0) > 0$ and the inclusion $u_y(\cdot,\varphi(\cdot)) \in MW_p^{1-1/p}(\mathbb{R}^{n-1})$ imply that $1/u_y(\cdot,\varphi(\cdot)) \in MW_p^{1-1/p}$ $(\mathbb{R}^{n-1})$. Hence the function

$$x \to u_{x_i}(x,\varphi(x)+0)/u_y(x,\varphi(x)+0)$$

belongs to the space $MW_p^{1-1/p}(\mathbb{R}^{n-1})$. Thus,

$$\varphi_{x_i} \in MW_p^{1-1/p}(\mathbb{R}^{n-1}).$$

We proceed by induction. Let k be a positive integer such that $2 \le k < l$. Suppose that $\varphi_{x_i} \in MW_p^{k-1-1/p}(\mathbb{R}^{n-1})$. By Lemma 9.4.6 the mapping $\lambda : \mathbb{R}^n_+ \to G$ defined by (9.4.14) is a (p,k)-diffeomorphism. This and the inclusion $u_{x_i},\ u_y \in MW_p^k(G)$ imply that

$$u_{x_i} \circ \lambda,\ u_y \circ \lambda \in MW_p^k(\mathbb{R}^n_+).$$

By Theorem 8.7.2,

$$\mathrm{tr}(u_{x_i} \circ \lambda),\ \mathrm{tr}(u_y \circ \lambda) \in MW_p^{k-1/p}(\mathbb{R}^{n-1}).$$

Since the function $\mathrm{tr}(u_y \circ \lambda)$ is separated from zero, it follows that the ratio $\mathrm{tr}(u_{x_i} \circ \lambda)/\mathrm{tr}(u_y \circ \lambda)$ belongs to the space $MW_p^{k-1/p}(\mathbb{R}^{n-1})$. It remains to note that (9.5.3) can be written as

$$\nabla\varphi(x) = -\frac{\mathrm{tr}(\nabla_x u \circ \lambda)}{\mathrm{tr}(u_y \circ \lambda)}.$$

□

Now we prove a local variant of Theorem 9.5.1.

Theorem 9.5.2. *Let G be the same domain as in Theorem* 9.5.1. *Further, let ω be an $(n-1)$-dimensional domain and let U be the cylinder*

$$\{(x,y) : x \in \omega, y \in \mathbb{R}\}.$$

Suppose that the function u, defined on $G \cap U$, satisfies the conditions:

(i) $\nabla u \in (MW_{\mathrm{loc}}^{p,l-1}(U \cap \overline{G}))^n$, *where l is an integer, $l \ge 2$, and $p \in [1,\infty)$.*

(ii) $\mathrm{tr}\, u = 0$ on $\omega \cap \partial G$,

(iii) the function $\mathrm{tr}\,\frac{\partial u}{\partial y}$ *is separated from zero on any compact subset of $\omega \cap \partial G$.*

Then

$$\nabla\varphi \in MW_{p,\mathrm{loc}}^{l-1-1/p}(\omega).$$

Proof. Duplicating the beginning of the proof of Theorem 9.5.1 we arrive at (9.5.3) for almost all $x \in \omega$. In the rest of the proof we need only to replace the spaces

$$W_p^s(G),\ W_p^{s-1/p}(\partial G),\ W_p^s(\mathbb{R}^n_+),\ W_p^{s-1/p}(\mathbb{R}^{n-1})$$

by the spaces

$$W_{p,\mathrm{loc}}^s(U \cap G),\ W_{p,\mathrm{loc}}^{s-1/p}(U \cap \partial G),\ W_{p,\mathrm{loc}}^{p,s}(\tau(U \cap \overline{G})),\ W_{p,\mathrm{loc}}^{ps-1/p}(\omega),$$

where $\tau = \lambda^{-1}$.

Remark 9.5.1. Theorems 9.5.1 and 9.5.2 are sharp in the following sense. If $\nabla\varphi \in MW_p^{l-1-1/p}(\mathbb{R}^{n-1})$, then there exists a function u in G satisfying the conditions (i)–(iii). The role of such a function can be played by a solution of equation (9.4.19). In fact, (9.4.19) implies that $u(x, \varphi(x)) = 0$ and

$$\frac{\partial u}{\partial y} = \Big(N + \frac{\partial(\mathcal{T}\varphi)}{\partial u}\Big)^{-1} \geq (N+L)^{-1}.$$

Since τ is a (p,l)-diffeomorphism, it follows that $\nabla u \in MW_p^{l-1}(G)$.

To conclude this section we formulate the implicit mapping theorem analogous to Theorem 9.5.1 and which can be proved in the same way.

Theorem 9.5.3. *Let l and s be integers, $n > s > n-(l-1)p \geq 0$, $z = (x,y)$, $x \in \mathbb{R}^s$, $y \in \mathbb{R}^{n-s}$, and let u φ be the mappings $\mathbb{R}^n \to \mathbb{R}^{n-s}$ and $\mathbb{R}^s \to \mathbb{R}^{n-s}$, respectively, with the properties:*
(i) $u_z \in MW_p^{l-1}(\mathbb{R}^n)$,
(ii) $\operatorname{tr} u = 0$, where tr is the trace on the surface $\{z : x \in \mathbb{R}^s,\, y = \varphi(x)\}$,
(iii) the inverse matrix $(\operatorname{tr} u_y)^{-1}$ exists and its norm is uniformly bounded on the surface $\{z : x \in \mathbb{R}^s,\, y = \varphi(x)\}$.
Then $\varphi_x \in MW_p^{l-1-(n-s)/p}(\mathbb{R}^s)$.

9.6 The Space $M(\mathring{W}_p^m(\Omega) \to W_p^l(\Omega))$

9.6.1 Auxiliary Results

In the present section m and l are integers, Ω is a domain in $\mathbb{R}^n$, $p \in (1,\infty)$, and $\mathring{W}_p^m(\Omega)$ is the completion of $C_0^\infty(\Omega)$ in the norm $W_p^m(\Omega)$.

We define two capacities of a compact set $e \subset \Omega$ by

$$C_{p,l}(e,\Omega) = \inf\{\|u;\Omega\|_{W_p^l}^p : u \in C_0^\infty(\Omega), u \geq 1 \ \text{ on } \ e\}$$

and

$$c_{p,l}(e,\Omega) = \inf\{\|\nabla_l u;\Omega\|_{L_p}^p : u \in C_0^\infty(\Omega), u \geq 1 \ \text{ on } \ e\}.$$

Obviously, if $e_1 \subset e_2$ and $\Omega_1 \supset \Omega_2$, then

$$C_{p,l}(e_1,\Omega_1) \leq C_{p,l}(e_2,\Omega_2)\,.$$

The capacity $c_{p,l}(e,\mathbb{R}^n)$ has the same property of monotonicity. It is also clear that the capacity $c_{p,l}(e,\mathbb{R}^n)$ acquires the factor d^{n-pl} under the similarity transform with coefficient d. The capacity $c_{p,l}(e,\mathbb{R}^n)$ vanishes for any compact set e if $n \leq lp$, $p > 1$.

The Sobolev theorem on the imbedding of $W_p^l(\mathbb{R}^n)$ into $L_\infty(\mathbb{R}^n)$ for $lp > n$, $p > 1$, implies that the capacity $C_{p,l}(e,\mathbb{R}^n)$ is separated from zero.

We present some other known properties of capacity (see [Maz15], Ch. 9):
(*i*) Let $lp < n$ and let e be a compact subset of the ball $\mathcal{B}_\rho$. Then

$$c_{p,l}(e, \mathcal{B}_\rho) \le c\, c_{p,l}(e, \mathbb{R}^n) \tag{9.6.1}$$

where c does not depend on ρ.
(*ii*) For all compact subsets e of the ball $\mathcal{B}_1$,

$$C_{p,l}(e, \mathcal{B}_2) \sim C_{p,l}(e, \mathbb{R}^n)\,. \tag{9.6.2}$$

We recall certain properties of capacity discussed in Sect. 3.1.2:
(*iii*) If $\rho \le 1$, then

$$C_{p,l}(\overline{\mathcal{B}}_\rho) \sim \begin{cases} \rho^{n-pl} & \text{if} \quad n > pl,\ p > 1; \\ (\log 2/\rho)^{1-p} & \text{if} \quad n = pl,\ p > 1\,. \end{cases}$$

(*iv*) If $\rho > 1$, then $C_{p,l}(\overline{\mathcal{B}}_\rho) \sim \rho^n$.
(*v*) If $n > pl$, then $C_{p,l}(e) \ge c\,(\text{mes}_n e)^{(n-pl)/n}$.
(*vi*) If $n = pl$ and $d(e) \le 1$, then

$$C_{p,l}(e) \ge c\,(\log(2^n/\text{mes}_n e))^{1-p}\,.$$

To reveal the dependence of certain constants upon the diameter of the domain, we use the norms (9.4.4) and (9.4.6) in $W_p^k(\Omega)$ and $M(W_p^m(\Omega) \to W_p^l(\Omega))$, respectively.

In the following theorem we present the norms equivalent to the norm (9.4.6). The equivalence means that their ratios are bounded and separated from zero by constants independent of d.

Theorem 9.6.1. *Let Ω be a domain with $\partial\Omega \in C^{0,1}$ and finite diameter $d < \infty$.*

(*i*) *If $p \in (1, \infty)$, then*

$$\|\gamma; \Omega\|_{M(W_p^m \to W_p^l)} \sim \sup_{e \subset \Omega} \frac{\|\nabla_l \gamma; e\|_{L_p}}{[c_{p,m}(e, \mathcal{B}_{ad})]^{1/p}} + \|\gamma; \Omega\|_{L_1}, \tag{9.6.3}$$

where $a > 1$ and $\mathcal{B}_{ad}$ is a ball with center $0 \in \overline{\Omega}$. In the case $mp < n$ we can replace $\mathcal{B}_{ad}$ by $\mathbb{R}^n$. If $m = l$, then the second term is equal to $\|\gamma; \Omega\|_{L_\infty}$.

(*ii*) *If either $pm > n$ and $p > 1$, or $m \ge n$ and $p = 1$, then the relation*

$$\|\gamma; \Omega\|_{M(W_p^m \to W_p^l)} \sim d^{m-n/p} \|\gamma; \Omega\|_{W_p^l} \tag{9.6.4}$$

holds.

(*iii*) *If $m < n$, then*

$$\|\gamma; \Omega\|_{M(W_1^m \to W_1^l)} \sim \sup_{\substack{x \in \Omega, \\ 2\rho < dist(x, \partial\Omega)}} \rho^{m-n}(\|\nabla_l \gamma; \mathcal{B}_\rho(x)\|_{L_1} + \rho^{-l}\|\gamma; \mathcal{B}_\rho(x)\|_{L_1})\,.$$

Proof. For $d = 1$ the assertions formulated above are contained in Theorems 9.3.1-9.3.3. (To obtain (*i*) one must use in addition (9.6.1) and (9.6.2).) The passage from $d = 1$ to $d \in (0, \infty)$ is performed by dilation. □

9.6.2 Description of the Space $M(\mathring{W}_p^m(\Omega) \to W_p^l(\Omega))$

Let Ω be a domain in $\mathbb{R}^n$ with compact closure and $\partial\Omega \in C^{0,1}$. In the next theorem, Q_j are cubes with edge-length d_j forming a Whitney covering of Ω (see [St2] §1, ch. 6). Furthermore, let Q_j^* be a cube homothetic to Q_j with edge length $9d_j/8$. The cubes Q_j^* form a covering of Ω with a finite multiplicity which depends only on n (see [St2]).

Theorem 9.6.2. *Let $1 < p < \infty$ and $mp < n$. The relation*

$$\|\gamma; \Omega\|_{M(\mathring{W}_p^m \to W_p^l)} \sim \sup_j \sup_{e \subset Q_j} \frac{\|\nabla_l \gamma; e\|_{L_p}}{[C_{p,m}(e, Q_j^*)]^{1/p}} + \|\gamma; \Omega\|_{L_1} \tag{9.6.5}$$

holds.

Proof. By $r(x)$ we denote the regularized distance from $x \in \Omega$ to $\partial\Omega$ (see [St2], §2, Ch. 6). It follows from the Hardy inequality

$$\|r^{-m+j}\nabla_j u; \Omega\|_{L_p} \le c\,\|u; \Omega\|_{\mathring{W}_p^m}, \qquad j = 0, 1, \dots, m-1, \tag{9.6.6}$$

that

$$\|u; \Omega\|^p_{\mathring{W}_p^m} \sim \sum_j \|\!|u; Q_j|\!\|^p_{W_p^m} \sim \sum_j \|\!|u; Q_j^*|\!\|^p_{W_p^m} \tag{9.6.7}$$

where $\|\!|\cdot|\!\|$ is the norm defined by (9.4.6). By (9.6.7) we obtain a norm in $M(\mathring{W}_p^m(\Omega) \to W_p^l(\Omega))$ described in terms of the space $M(W_p^m(Q_j) \to W_p^l(Q_j))$. Namely,

$$\|\gamma; \Omega\|_{M(\mathring{W}_p^m \to W_p^l)} \sim \sup_j \|\!|\gamma; Q_j|\!\|_{M(W_p^m \to W_p^l)}\,. \tag{9.6.8}$$

In fact,

$$\|\gamma u; \Omega\|^p_{W_p^l} \le \sum_j \|\!|\gamma u; Q_j|\!\|^p_{W_p^l} \le \sup_j \|\!|\gamma; \Omega|\!\|^p_{M(W_p^m \to W_p^l)} \sum_j \|\!|u; Q_j|\!\|^p_{W_p^m}$$

which, together with (9.6.7), gives the required upper bound for the norm of γ in $M(\mathring{W}_p^m(\Omega) \to W_p^l(\Omega))$.

To justify the lower bound, let $u \in W_p^m(Q_j)$ and let v be an extension of u onto Q_j^* satisfying

$$\|\!|v; Q_j^*|\!\|_{W_p^m} \le c\,\|\!|u; Q_j|\!\|_{W_p^m}\,. \tag{9.6.9}$$

By φ we denote a function in $C_0^\infty(Q_j^*)$, equal to 1 on Q_j and such that

$$|\nabla_k \varphi| = o(d_j^{-k}).$$

We have

$$\begin{aligned}\|\gamma u; Q_j\|_{W_p^l} &\le \|\gamma \varphi v; Q_j^*\|_{W_p^l} \le c\,\|\gamma\varphi v; \Omega\|_{W_p^l}\\ &\le c\,\|\gamma; \Omega|_{M(\mathring{W}_p^m \to W_p^l)}\|\varphi v; \Omega\|_{W_p^m}\\ &\le c\,\|\gamma; \Omega\|_{M(\mathring{W}_p^m \to W_p^l)}\|v; Q_j^*\|_{W_p^m}. \end{aligned} \tag{9.6.10}$$

By (9.6.9) and (9.6.10) we find the lower bound for the norm of γ in $M(\mathring{W}_p^m(\Omega) \to W_p^l(\Omega))$. Relation (9.6.8) is proved.

It remains to use Theorem 9.6.1. The proof is complete. □

Corollary 9.6.1. *Let $1 < p < \infty$. The relations*

$$\|\gamma; \Omega\|_{M(\mathring{W}_p^m \to W_p^l)} \sim \sup_{e \subset \Omega} \Big(\frac{\|\nabla_l \gamma; e\|_{L_p}}{[C_{p,m}(e, \Omega)]^{1/p}} + \frac{\|\gamma; e\|_{L_p}}{[C_{p,m-l}(e, \Omega)]^{1/p}} \Big) \tag{9.6.11}$$

$$\sim \sup_{e \subset \Omega} \frac{\|\nabla_l \gamma; e\|_{L_p}}{[C_{p,m}(e, \Omega)]^{1/p}} + \|\gamma; \Omega\|_{L_1} \tag{9.6.12}$$

hold. These relations still hold if we add the restriction $d(e) \le \delta(e)$, where $d(e)$ is the diameter of e and $\delta(e)$ is the distance from e to $\partial\Omega$.

Proof. In view of (9.6.7),

$$C_{p,m}(e, \Omega) \ge c \sum_j c_{p,m}(e \cap Q_j, Q_j^*)\,, \qquad e \subset \Omega\,,$$

which leads to the required lower bound for the norm of γ in $M(\mathring{W}_p^m(\Omega) \to W_p^l(\Omega))$. The upper bound is obtained via the inequalities

$$c_{p,m}(e, Q_j^*) \ge C_{p,m}(e, Q_j^*) \ge C_{p.m}(e, \Omega)\,.$$

□

The following two statements do not contain capacities.

Corollary 9.6.2. *Let $pm > n$ and $p > 1$, or $m \ge n$ and $p = 1$. Then*

$$\|\gamma; \Omega\|_{M(\mathring{W}_p^m \to W_p^l)} \sim \sup_j d_j^{m-n/p} \|\gamma; Q_j\|_{W_p^l}\,.$$

Corollary 9.6.3. *The relation*

$$\|\gamma; \Omega\|_{M(\mathring{W}_1^m \to W_1^l)} \sim \sup_{\substack{x \in \Omega\,,\\ 2\rho < \delta(x)}} \rho^{m-n} \big(\|\nabla_l \gamma; \mathcal{B}_\rho(x)\|_{L_1} + \rho^{-l}\|\gamma; \mathcal{B}_\rho(x)\|_{L_1}\big)$$

holds.

Corollaries 9.6.2 and 9.6.3 result from (9.6.8) and Theorem 9.6.1.

Part II

Applications of Multipliers to Differential and Integral Operators

10

Differential Operators in Pairs of Sobolev Spaces

The most natural connection of the theory of multipliers with differential operators arises when one is looking for the bounds of the norms and essential norms of these operators mapping one Sobolev space into another.

In Sect. 10.1 we give estimates for the norms of general differential operators performing a mapping between two Sobolev spaces, formulated in terms of their coefficients as multipliers. These estimates involve multiplier norms of the coefficients, and for some values of integrability and smoothness parameters they are two-sided. We also describe a class of differential operators for which their continuity in pairs of Sobolev spaces is equivalent to the inclusion of the coefficients into classes of multipliers without any additional conditions on indices. We give a counterexample showing that in general the inclusion of the coefficients into the natural classes of multipliers is not necessary for the continuity of differential operators.

Estimates for the essential norms of general differential operators is the topic of Sect. 10.2. By the example of a Schrödinger operator in $\mathbb{R}^n$ considered in Sect. 10.3, we outline the role of the essential norm of a multiplier in the Fredholm theory of elliptic differential operators. The last Sect. 10.4 deals with a characterization of pairs of differential operators with constant coefficients which obey the dominance property between L_2 and its weighted counterpart.

10.1 The Norm of a Differential Operator: $W_p^h \to W_p^{h-k}$

In this section we discuss some simple applications of the space $M(W_p^m \to W_p^l)$ to the theory of differential operators, namely, to the question of the continuity of such operators in pairs of Sobolev spaces. As usual, we use the notation $\mathcal{B}_r(x) = \{y \in \mathbb{R}^n : |y - x| < r\}$ and omit $\mathbb{R}^n$ in notations of the norms.

V.G. Maz'ya, T.O. Shaposhnikova, *Theory of Sobolev Multipliers*,
Grundlehren der mathematischen Wissenschaften 337,

10.1.1 Coefficients of Operators Mapping W_p^h into W_p^{h-k} as Multipliers

Lemma 10.1.1. *(i) The operator*

$$P(x, D_x)u = \sum_{|\alpha|\le k} a_\alpha(x) D_x^\alpha u, \qquad x \in \mathbb{R}^n\,, \tag{10.1.1}$$

is a continuous mapping $W_p^h \to W_p^{h-k}$, where $h \ge k$, provided that

$$a_\alpha \in M(W_p^{h-|\alpha|} \to W_p^{h-k})$$

for any multi-index α. The estimate

$$\|P\|_{W_p^h \to W_p^{h-k}} \le c \sum_{|\alpha|\le k} \|a_\alpha\|_{M(W_p^{h-|\alpha|} \to W_p^{h-k})} \tag{10.1.2}$$

holds.

(ii) If $p = 1$ or if $p(h-k) > n$ and $p > 1$, then the relation

$$\|P\|_{W_p^h \to W_p^{h-k}} \sim \sum_{|\alpha|\le k} \|a_\alpha\|_{M(W_p^{h-|\alpha|} \to W_p^{h-k})} \tag{10.1.3}$$

holds.

Proof. The estimate (10.1.2) is obvious, so we need to prove only *(ii)*. Let $x \in \mathbb{R}^n$ and let $\eta \in C_0^\infty(\mathcal{B}_2)$ with $\eta = 1$ on $\mathcal{B}_1$. Further, let $u(y) = \eta((x-y)/\delta)$, where $\delta \in (0, 1]$. By substituting the function u into the inequality

$$\Big\| \sum_{|\alpha|\le k} a_\alpha D^\alpha u \Big\|_{W_p^{h-k}} \le c\, \|u\|_{W_p^h}\,,$$

we obtain

$$\sup_{x\in\mathbb{R}^n} \|a_0; \mathcal{B}_\delta(x)\|_{W_p^{h-k}} \le c\, \delta^{n/p-h}$$

which, together with Theorem 2.1.1, Corollary 4.3.8, and Theorem 5.3.1, shows that $a_0 \in M(W_p^h \to W_p^{h-k})$.

Suppose that $a_\alpha \in M(W_p^{h-|\alpha|} \to W_p^{h-k})$ for $|\alpha| \le \nu$, $\nu \le k-1$, and

$$\sum_{|\alpha|\le\nu} \|a_\alpha\|_{M(W_p^{h-|\alpha|} \to W_p^{h-k})} \le c\, \|P\|_{W_p^h \to W_p^{h-k}}\,.$$

We show that the same holds with ν is replaced by $\nu + 1$. Clearly,

$$\begin{aligned}\Big\| \sum_{|\alpha|\ge\nu+1} a_\alpha D^\alpha u \|_{W_p^{h-k}} &\le \Big\| Pu - \sum_{|\alpha|\le\nu} a_\alpha D^\alpha u \Big\|_{W_p^{h-k}} \\ &\le c\, \|P\|_{W_p^h \to W_p^{h-k}} \|u\|_{W_p^h}\end{aligned}$$

for all $u \in W_p^h$. Putting here

$$u(y) = (x-y)^\alpha \eta((x-y)/\delta), \qquad |\alpha| = \nu + 1,$$

we obtain

$$\sup_{x\in\mathbb{R}^n} \|a_\alpha; \mathcal{B}_\delta(x)\|_{W_p^{h-k}} \le c\,\|P\|_{W_p^h \to W_p^{h-k}}\, \delta^{n/p-h+|\alpha|} .$$

This, together with Theorem 2.1.1, Corollary 4.3.8, and Theorem 5.3.1 implies that $a_\alpha \in M(W_p^{h-|\alpha|} \to W_p^{h-k})$. □

Now we present an analogous result for matrix operators. Let $u(x) = \{u^1(x), u^2(x), \dots, u^N(x)\}$ be an N-tuple vector-valued function. Consider the operator

$$Pu = \Big\{ \sum_{k=1}^N P_{jk}(x, D_x) u^k \Big\}_{j=1}^M , \qquad x \in \mathbb{R}^n , \tag{10.1.4}$$

where

$$P_{jk}(x, D_x) u^k = \sum_{|\alpha| \le s_j + t_k} a_{jk}^{(\alpha)}(x) D_x^\alpha u^k$$

and s_j, t_k are integers.

Theorem 10.1.1. *Let $h \ge s = \max s_j$, $j = 1, \dots, M$.*

(i) The operator P is a continuous mapping

$$P : \prod_{k=1}^N W_p^{t_k+h} \to \prod_{j=1}^M W_p^{h-s_j} , \tag{10.1.5}$$

if

$$a_{jk}^{(\alpha)} \in M(W_p^{t_k+h-|\alpha|} \to W_p^{h-s_j}).$$

The estimate

$$\|P\| \le c \sum_{k=1}^N \sum_{j=1}^M \sum_{|\alpha| \le s_j + t_k} \|a_{jk}^{(\alpha)}\|_{M(W_p^{t_k+h-|\alpha|} \to W_p^{h-s_j})} \tag{10.1.6}$$

holds.

(ii) If $p(h-s) \ge n$ and $p > 1$, or $p = 1$, then the relation

$$\|P\| \sim \sum_{k=1}^N \sum_{j=1}^M \sum_{|\alpha| \le s_j + t_k} \|a_{jk}^{(\alpha)}\|_{M(W_p^{t_k+h-|\alpha|} \to W_p^{h-s_j})}$$

holds.

Proof. Inequality (10.1.6) follows from Lemma 10.1.1. Let

$$\sum_{j=1}^{M} \Big\| \sum_{k=1}^{N} P_{jk}(x, D_x) u^k \Big\|_{W_p^{h-s_j}} \le c\,\|P\| \sum_{k=1}^{N} \|u^k\|_{W_p^{h+t_k}} .$$

We fix i and set $u^k = 0$ provided that $k \ne i$. Then

$$\sum_{j=1}^{M} \|P_{ji}(x, D_x) u^i\|_{W_p^{h-s_j}} \le c\,\|P\| \|u^i\|_{W_p^{h+t_i}}$$

and in the case $p(h-s) > n$, $p > 1$, as well as in the case $p = 1$, Lemma 10.1.1 gives the estimate

$$\sum_{j=1}^{M} \sum_{|\alpha| \le s_j + t_i} \|a_{ji}^{(\alpha)}\|_{M(W_p^{h+t_i-|\alpha|} \to W_p^{h-s_j})} \le c\,\|P\| ,$$

where $i = 1, 2, \dots, N$. □

Next we show that an equivalence relation similar to (10.1.3) with an arbitrary $p \in (1, \infty)$ can be obtained for partial differential operators of a special form.

Theorem 10.1.2. *Let h and s be positive integers, $h \ge 2s$, $1 < p < \infty$ and*

$$P(x, D_x) u = \sum_{j=0}^{s} b_j(x) \Delta^j u,$$

where Δ is the Laplace operator. Then P is a continuous mapping: $W_p^h \to W_p^{h-2s}$ if and only if

$$b_j \in M(W_p^{h-2j} \to W_p^{h-2s}), \quad j = 0, \dots, s.$$

Moreover, the relation

$$\|P\|_{W_p^h \to W_p^{h-2s}} \sim \sum_{j=0}^{s} \|b_j\|_{M(W_p^{h-2j} \to W_p^{h-2s})} \tag{10.1.7}$$

holds.

Proof. The sufficiency as well as the upper bound for the norm of P follows from Lemma 10.1.1.

Suppose that, for all $u \in W_p^h$,

$$\|Pu\|_{W_p^{h-2s}} \le c\,\|u\|_{W_p^h}.$$

Let $u = 1$ in a neighborhood of a compact set e with $d(e) \leq 1$. Then

$$\|\nabla_{h-2s} b_0; e\|_{L_p} + \|b_0; e\|_{L_p} \leq c\, \|u\|_{W_p^h} \, .$$

Consequently,

$$\|\nabla_{h-2s} b_0; e\|_{L_p} \leq c\, [C_{p,h}(e)]^{1/p}$$

and $\|b_0; \mathcal{B}_1(x)\|_{L_p} \leq c$ for all $x \in \mathbb{R}^n$. By Theorem 2.3.3 this means that

$$b_0 \in M(W_p^h \to W_p^{h-2s}).$$

Therefore, the operator $Q\Delta$ with

$$Q = \sum_{j=0}^{s-1} b_{j+1}(x) \Delta^j$$

satisfies the inequality

$$\|Q\Delta u\|_{W_p^{h-2s}} \leq c\|u\|_{W_p^h} \, . \tag{10.1.8}$$

Let $\zeta \in W_p^{h-2}$ and let supp $\zeta \subset \{x = (x_1, \dots, x_n) : 0 < x_i < 1\}$. We put $w(x) = \zeta(x) - \zeta(-x)$ and $w_i(x) = w(x_1, \dots, x_{i-1}, x_i/2, x_{i+1}, \dots, x_n)$ for $i = 1, \dots, n$. Further, let

$$v(x) = w(x) + \sum_{i=1}^{n} \alpha_i w_i(x - a_i) \tag{10.1.9}$$

where a_i are fixed points with $\text{dist}(a_i, a_j) > (8n)^{1/2}$ and α_i are arbitrary constants. It is clear that all the functions on the right-hand side of (10.1.9) are orthogonal to one and have disjoint supports. We show that the coefficients α_i can be selected so that

$$\int x_j v(x)\, dx = 0 \, , \qquad j = 1, \dots, n,$$

which is equivalent to the algebraic system with respect to $\alpha_1, \dots, \alpha_n$:

$$2 \int x_j \zeta(x)\, dx \Big(1 + 2 \sum_{i=1}^{n} \alpha_i (1 + \delta_i^j)\Big) = 0 \, , \qquad j = 1, \dots, n \, ,$$

where δ_i^j is the Kronecker delta. The system is solvable because

$$\det \|1 + \delta_i^j\|_{i,j=1}^n = n + 1.$$

Let u be the harmonic (Newtonian for $n > 2$ and logarithmic for $n = 2$) potential with density v. Since v is orthogonal to $1, x_1, \dots, x_n$ and the diameter of its support is bounded by a constant depending only on n, then

$$\|u\|_{W_p^h} \le c\,\|v\|_{W_p^{h-2}} \le c_1\|\zeta\|_{W_p^{h-2}}\,.$$

This and (10.1.8) imply that

$$\|Q\zeta\|_{W_p^{h-2s}} \le c\,\|\zeta\|_{W_p^{h-2}}\,.$$

By the arbitrariness of the origin, the last inequality holds for all functions $\zeta \in W_p^{h-2}$ supported by any cube of the coordinate grid. Hence it is valid for all $\zeta \in W_p^{h-2}$, i.e. Q is a continuous operator: $W_p^{h-2} \to W_p^{h-2s}$. Now the result follows by successive reduction of order of the operator. □

10.1.2 A Counterexample

Theorem 10.1.2 and part (ii) of Lemma 10.1.1 suggest the hypothesis: relation (10.1.3) holds for all p, h, k, i.e. the coefficients of any differential operator (10.1.1), mapping W_p^h into W_p^{h-k}, are necessarily multipliers in corresponding Sobolev spaces. The next example disproves this conjecture.

Example 10.1.1. Let $x = (x', x_n)$, $x' = (x_1, \dots, x_{n-1})$, $n \ge 3$. We show that the coefficient a of the continuous operator

$$a(x')\frac{\partial}{\partial x_n} : W_2^2 \to L_2 \tag{10.1.10}$$

need not be an element of $M(W_2^1 \to L_2)$.

Suppose that

$$A = \sup_{y\in\mathbb{R}^{n-1}, r\in(0,1)} r^{3-n}\int_{\mathcal{B}_r^{(n-1)}(y)} |a(x')|^2\,dx' < \infty,$$

where $\mathcal{B}_r^{(n-1)}(y)$ is the $(n-1)$-dimensional ball with centre y and radius r. By $\hat{u}(x', \lambda)$ we denote the Fourier transform of the function u with respect to x_n. According to Theorem 1.1.2,

$$\int_{\mathbb{R}^{n-1}} |a(x')\hat{u}(x',\lambda)|^2 dx' \le A\,\|[\hat{u}(\cdot,\lambda)]^2; \mathbb{R}^{n-1}\|_{W_1^2}$$

$$\le c\,A\int_{\mathbb{R}^{n-1}} \big(|\hat{u}(x',\lambda)|\,|\nabla_{2,x'}\hat{u}(x',\lambda)| + |\nabla_{x'}\hat{u}(x',\lambda)|^2 + |\hat{u}(x',\lambda)|^2\big)\,dx'.$$

Consequently,

$$\int_{\mathbb{R}^n} \Big|a(x')\frac{\partial u}{\partial x_n}\Big|^2 dx \le c\,A\int_{\mathbb{R}^1} d\lambda \int_{\mathbb{R}^{n-1}} \big(|\lambda^2\hat{u}(x',\lambda)|\,|\nabla_{2,x'}\hat{u}(x',\lambda)|$$

$$+|\lambda\nabla_{x'}\hat{u}(x',\lambda)|^2 + |\lambda\hat{u}(x',\lambda)|^2\big)\,dx' \le c_1 A\big(\|\nabla_2 u\|_{L_2}^2 + \|\nabla u\|_{L_2}^2\big).$$

Thus the finiteness of the value A is sufficient for the continuity of operator (10.1.10). The necessity of the same condition results from the estimate

$$\|P\|_{W_p^h \to W_p^{h-k}} \geq c \sum_{|\alpha| \leq k} \sup_{x;r\in(0,1)} \left(r^{h-|\alpha|-n/p}\|\nabla_{h-k} a_\alpha; \mathcal{B}_r(x)\|_{L_p}\right.$$

$$\left. + r^{k-|\alpha|-n/p}\|a_\alpha; \mathcal{B}_r(x)\|_{L_p}\right) \tag{10.1.11}$$

derived in the proof of the second part of Lemma 10.1.1 for all $p \in [1, \infty)$.

If (10.1.3) is valid for all p, h, and k, then the continuity of the operator (10.1.10) implies that $a \in M(W_2^1 \to L_2)$. We choose the coefficient a as follows:

$$a(x') = \rho^{-1}|\log \rho|^{\varepsilon-1}\, \eta(x_1, x_2)\, \zeta(x_3, \dots, x_{n-1})$$

where

$$\rho^2 = x_1^2 + x_2^2, \quad 0 < \varepsilon < 1/2, \quad \eta \in C_0^\infty(\mathcal{B}_1^{(2)}), \quad \zeta \in C_0^\infty(\mathcal{B}_1^{(n-3)}).$$

It is clear that for any $y \in \mathbb{R}^{n-1}$ and for all $r \in (0, 1/2)$ we have

$$\int_{\mathcal{B}_r^{(n-1)}(y)} |a(x')|^2\, dx' \leq c\, r^{n-3} \int_{\mathcal{B}_r^{(2)}} \rho^{-2}|\log \rho|^{2(\varepsilon-1)}\, dx_1 dx_2$$

$$= c\, r^{n-3}|\log r|^{2\varepsilon-1}\,.$$

Therefore $A < \infty$. Suppose that

$$\|au; \mathbb{R}^n\|_{L_2} \leq c\, \|u; \mathbb{R}^n\|_{W_2^1}$$

for any $u \in C_0^\infty(\mathbb{R}^n)$. Then, for all $v \in C_0^\infty(\mathbb{R}^2)$,

$$\|\eta\rho^{-1}|\log \rho|^{\varepsilon-1}v; \mathbb{R}^2\|_{L_2} \leq c\, \|v; \mathbb{R}^2\|_{W_2^1}\,.$$

This estimate implies that

$$\int_{\rho<r} \rho^{-2}|\log \rho|^{2(\varepsilon-1)}\, dx_1 dx_2 \leq c\, C_{2,1}(\mathcal{B}_r^{(2)})$$

for $r \in (0, 1/2)$, which contradicts the relation

$$C_{2,1}(\mathcal{B}_r^{(2)}) \sim |\log r|^{-1}.$$

Thus, the operator (10.1.10) with the function a under consideration is continuous, although $a \notin M(W_2^1(\mathbb{R}^n) \to L_2(\mathbb{R}^n))$.

10.1.3 Operators with Coefficients Independent of Some Variables

A necessary and sufficient condition for the continuity of operator (10.1.10) derived in the last example can be generalized to operators with coefficients depending on only some of the variables.

Theorem 10.1.3. *Let $y \in \mathbb{R}^s$, $z \in \mathbb{R}^{n-s}$, where $s \le n$. Further, let h and k be integers with $h \ge k$, and*

$$P(y, D_y, D_z)u = \sum_{0 \le |\beta|+|\gamma| \le k} a_{\beta\gamma}(y) D_y^\beta D_z^\gamma u \, .$$

Then the operator $P(y, D_y, D_z)$ is a continuous mapping: $W_2^h(\mathbb{R}^n) \to W_2^{h-k}(\mathbb{R}^n)$ if and only if the operator

$$P(y, D_y, 0) : W_2^h(\mathbb{R}^s) \to W_2^{h-k}(\mathbb{R}^s)$$

is continuous and, for all multi-indices β and γ,

$$\sup_{y \in \mathbb{R}^s, r \in (0,1)} r^{h-|\beta|-|\gamma|-s/2} \big(\| \nabla_{h-k} a_{\beta\gamma}; \mathcal{B}_r^{(s)}(y) \|_{L_2}$$

$$+ r^{k-h} \| a_{\beta\gamma}; \mathcal{B}_r^{(s)}(y) \|_{L_2} \big) < \infty \tag{10.1.12}$$

where $\mathcal{B}_r^{(s)}(y)$ is the s-dimensional ball with center y and radius r.

The proof of this theorem is based on the following assertion (see [Maz15], Ch. 1).

Lemma 10.1.2. *Let μ be a measure in $\mathbb{R}^s$ such that*

$$K = \sup_{y \in \mathbb{R}^s, r \in (0,1)} r^{-\sigma} \mu(\mathcal{B}_r^{(s)}(y)) < \infty \tag{10.1.13}$$

for a certain $\sigma \in [0, s]$. Further, let l and m be integers, $0 \le l < m$, $\sigma > s - 2(m-l)$. Then, for all $u \in C_0^\infty$,

$$\Big(\int_{\mathbb{R}^s} |\nabla_l u|^2 \, d\mu \Big)^{1/2} \le c \, K \| u; \mathbb{R}^s \|_{W_2^m}^\tau \, \| u; \mathbb{R}^s \|_{L_2}^{1-\tau}, \tag{10.1.14}$$

where $\tau = (2l + s - \sigma)/2m$ and c is a constant independent of u and μ. Moreover, condition (10.1.13) is necessary for the validity of (10.1.14).

Proof of Theorem 10.1.3. *Sufficiency.* First we verify that for $|\gamma| > 0$ the operator

$$a_{\beta\gamma}(y) D_y^\beta D_z^\gamma : W_2^h(\mathbb{R}^n) \to W_2^{h-k}(\mathbb{R}^n)$$

is continuous. With this aim in view, we show the continuity of the operator

$$(D_y^\mu a_{\beta\gamma}(y)) D_y^{\rho+\beta} D_z^{\gamma+\theta} : \; W_2^h(\mathbb{R}^n) \to L_2(\mathbb{R}^n),$$

where $0 \le |\mu| + |\rho| + |\theta| \le h - k$. By $\hat{u}(y, \lambda)$ we denote the Fourier transform of the function u with respect to z. Putting

$$l = |\rho| + |\beta|, \quad \sigma = 2|\beta| + 2|\gamma| + s - 2k - 2|\mu|, \quad m = k + |\mu| + |\rho| + |\theta|$$

in Lemma 10.1.2, we conclude that

$$\int_{\mathbb{R}^s} |(D_y^\mu a_{\beta\gamma}(y)) D_y^{\rho+\beta} \hat{u}(y,\lambda)|^2 \, dy$$
$$\leq c\, A^2_{\beta,\gamma,\mu} \, \|\hat{u}(\cdot,\lambda); \mathbb{R}^s\|^{2\tau}_{W_2^m} \, \|\hat{u}(\cdot,\lambda); \mathbb{R}^s\|^{2(1-\tau)}_{L_2},$$

where $\tau = (|\mu| + |\rho| + k - |\gamma|)/(|\mu| + |\rho| + |\theta| + k)$ and

$$A_{\beta,\gamma,\mu} = \sup_{y \in \mathbb{R}^s, r \in (0,1)} r^{-\sigma/2} \|D_y^\mu a_{\beta\gamma}; \mathcal{B}_r^{(s)}(y)\|_{L_2} .$$

(We note that the condition $\sigma > s - 2(m-l)$ in Lemma 10.1.2 is equivalent to $|\gamma| > 0$.) Multiplying the last inequality by $|\lambda|^{2(|\gamma|+|\theta|)}$ and integrating over λ, we get

$$\|(D_y^\mu a_{\beta\gamma}) D_y^{\rho+\beta} D_z^{\gamma+\theta} u; \mathbb{R}^n\|^2_{L_2}$$
$$\leq c\, A^2_{\beta,\gamma,\mu} \int_{\mathbb{R}^{n-s}} \|\hat{u}(\cdot,\lambda); \mathbb{R}^s\|^{2\tau}_{W_2^m} \, \||\lambda|^m \hat{u}(\cdot,\lambda); \mathbb{R}^s\|^{2(1-\tau)}_{L_2} \, d\lambda$$
$$\leq c\, A^2_{\beta,\gamma,\mu} \|u; \mathbb{R}^n\|^2_{W_2^m} .$$

Since $m \leq h$ and $A_{\beta,\gamma,\mu}$ does not exceed the value (10.1.12), the operator

$$P(y, D_y, D_z) - P(y, D_y, 0) : W_2^h(\mathbb{R}^n) \to W_2^{h-k}(\mathbb{R}^n) \tag{10.1.15}$$

is continuous.

In order to derive the continuity of $P(y, D_y, 0)$ in the same pair of spaces we prove the inequality

$$\|P(y, D_y, 0)v; \mathbb{R}^s\|_{L_2} \leq c\, \|v; \mathbb{R}^s\|^{k/h}_{W_2^h} \, \|v; \mathbb{R}^s\|^{1-k/h}_{L_2} . \tag{10.1.16}$$

Applying Lemma 10.1.2 with $l = |\beta|$ and $m = h$, for any multi-index β with $|\beta| \leq k$, we obtain

$$\|a_{\beta 0} D_y^\beta v; \mathbb{R}^s\|_{L_2}$$
$$\leq c \sup_{y \in \mathbb{R}^n, r \in (0,1)} r^{k-|\beta|-s/2} \|a_{\beta 0}; \mathcal{B}_r^{(s)}(y)\|_{L_2} \, \|v; \mathbb{R}^s\|^{k/h}_{W_2^h} \, \|v; \mathbb{R}^s\|^{1-k/h}_{L_2}$$

which entails (10.1.16). Using (10.1.16) together with (10.1.12), we find that

$$\|P(\cdot, D_y, 0)u; \mathbb{R}^n\|^2_{W_2^{h-k}}$$
$$\leq c \int_{\mathbb{R}^{n-s}} (\|P(\cdot, D_y, 0)\hat{u}(\cdot,\lambda); \mathbb{R}^s\|^2_{W_2^{h-k}} + |\lambda|^{2(h-k)} \|P(\cdot, D_y, 0)\hat{u}(\cdot,\lambda); \mathbb{R}^s\|^2_{L_2}) d\lambda$$

$$\leq c \int_{\mathbb{R}^{n-s}} \left(\|\hat{u}(\cdot,\lambda);\mathbb{R}^s\|^2_{W_2^h} + \|\hat{u}(\cdot,\lambda);\mathbb{R}^s\|^{2k/h}_{W_2^h} \||\lambda|^h \hat{u}(\cdot,\lambda);\mathbb{R}^s\|^{2(1-k/h)}_{L_2} \right) d\lambda$$

$$\leq c \, \|u;\mathbb{R}^n\|^2_{W_2^h}$$

for all $u \in W_2^h(\mathbb{R}^n)$. The sufficiency follows.

Necessity. We note that by replacing s by n in (10.1.12) we obtain an equivalent condition. Then the finiteness of (10.1.12) for all multi-indices β, γ follows from (10.1.11). Thus the first part of this theorem implies the continuity of the operator (10.1.15). Since the mapping

$$P(y, D_y, D_z) : W_2^h(\mathbb{R}^n) \to W_2^{h-k}(\mathbb{R}^n)$$

is continuous, it follows for all $v \in W_2^h(\mathbb{R}^n)$ that

$$\|P(y, D_y, 0)v, \mathbb{R}^n\|_{W_2^{h-k}} \leq c \, \|v;\mathbb{R}^n\|_{W_2^h} .$$

Substituting here $v(x) = \eta(z)u(y)$, where η is a fixed function in $C_0^\infty(\mathbb{R}^{n-s})$ and u is an arbitrary function in $W_2^h(\mathbb{R}^s)$, we complete the proof of Theorem 10.1.3. □

10.1.4 Differential Operators on a Domain

Let U and V be open sets in $\mathbb{R}^n$ and let

$$P(z, D_z)u = \sum_{|\alpha| \leq k} p_\alpha(z) D_z^\alpha u \tag{10.1.17}$$

be a differential operator on U. Given any (p,l)-diffeomorphism $\varkappa : U \to V$ with $l \geq k$, we introduce the differential operator Q on V, defined by

$$Q(u \circ \varkappa^{-1}) = (Pu) \circ \varkappa^{-1}.$$

In view of Lemmas 9.4.1 and 9.4.2, Q maps $W_p^l(V)$ continuously into $W_p^{l-k}(V)$ if and only if P maps $W_p^l(U)$ continuously into $W_p^{l-k}(U)$.

By $O_{p,\mathrm{loc}}^{l,k}(U)$ we denote the class of operators of the form (10.1.17) such that

$$p_\alpha \in M(W_{p,\mathrm{loc}}^{l-|\alpha|}(U) \to W_{p,\mathrm{loc}}^{l-k}(U))$$

for any multi-index α with $|\alpha| \leq k$.

Lemma 10.1.3. *The operator P belongs to the class $O_{p,loc}^{l,k}(U)$ if and only if $Q \in O_{p,loc}^{l,k}(V)$.*

Proof. Let $\zeta = \varkappa(z)$. We have

$$D^\alpha[v(\varkappa(z))] = \sum_{1\le|\beta|\le|\alpha|} (D^\beta v)(\varkappa(z)) \sum_s c_s \prod_{i=1}^n \prod_j D^{s_{ij}} \varkappa_i(z) \,, \qquad (10.1.18)$$

where the sum is taken over all multi-indices $s = (s_{ij})$ such that

$$\sum_{i,j} s_{ij} = \alpha \,, \qquad |s_{ij}| \ge 1, \quad \sum_{i,j}(|s_{ij}| - 1) = |\alpha| - |\beta| \,. \qquad (10.1.19)$$

Let

$$Q(\zeta, D_\zeta) = \sum_{|\beta|\le k} q_\beta(\zeta) D_\zeta^\beta \,.$$

By (10.1.18),

$$q_\beta = \sum_{|\beta|\le|\alpha|\le k} (p_\alpha \circ \varkappa^{-1}) \sum_s c_s \prod_{i=1}^n \prod_j (D^{s_{ij}} \varkappa_i) \circ \varkappa^{-1} \,. \qquad (10.1.20)$$

Since

$$\nabla \varkappa_i \in MW_{p,\mathrm{loc}}^{l-1}(U) \subset MW_{p,\mathrm{loc}}^{l-r}(U), \quad 1 \le r \le l,$$

it follows that

$$D^{s_{ij}} \varkappa_i \in M(W_{p,\mathrm{loc}}^{l-r}(U) \to W_{p,\mathrm{loc}}^{l-r-|s_{ij}|+1}(U)) \,.$$

Therefore,

$$\prod_{i,j} D^{s_{ij}} \varkappa_i \in M(W_{p,\mathrm{loc}}^{l-|\beta|}(U) \to W_{p,\mathrm{loc}}^{l-|\beta|-\sum_{i,j}(|s_{ij}|-1)}(U)),$$

which is the same as

$$\prod_{i,j} D^{s_{ij}} \varkappa_i \in M(W_{p,\mathrm{loc}}^{l-|\beta|}(U) \to W_{p,\mathrm{loc}}^{l-|\alpha|}(U)) \,.$$

It remains to use the condition $p_\alpha \in M(W_{p,\mathrm{loc}}^{l-|\alpha|}(U) \to W_{p,\mathrm{loc}}^{l-k}(U))$ together with Lemmas 9.4.1 and 9.4.2. □

By Lemma 10.1.1, the inclusion $P \in O_{p,\mathrm{loc}}^{l,k}(U)$ is sufficient for the operator P to map $W_{p,\mathrm{loc}}^{l}(U)$ into $W_{p,\mathrm{loc}}^{l-k}(U)$. The inclusion $P \in O_{p,\mathrm{loc}}^{l,k}(U)$ is also necessary for $p = 1$ or $p(l-k) > n$ (see Lemma 10.1.1).

Remark 10.1.1. Let Ω be a (p, l)-manifold. Such manifolds were introduced in Sect. 9.4.4. We define a differential operator of order k, $k \le l$, as a linear mapping P of the space $W_{p,\mathrm{loc}}^{l}(\Omega)$ into the space $W_{p,\mathrm{loc}}^{l-k}(\Omega)$ if, for any map (φ, U_φ), there exists a differential operator P_φ in the class $O_{p,\mathrm{loc}}^{l,k}(\varphi(U_\varphi))$ such that

$$(Pu) \circ \varphi^{-1} = P_\varphi(u \circ \varphi^{-1}) \quad \text{on } \varphi(U_\varphi).$$

By Lemma 10.1.3, it suffices to restrict oneself to the maps of a certain atlas.

10.2 Essential Norm of a Differential Operator

In this section we obtain bounds for the essential norm of a differential operator.

Let $P(x, D_x)$ be a differential operator of order k, defined by (10.1.1), let P_0 be its principal homogeneous part, and let $\operatorname{ess}\|P\|_{W_p^h \to W_p^{h-k}}$ be the essential norm of the mapping

$$P\colon W_p^h \to W_p^{h-k}, \qquad h \geq k.$$

Lemma 10.2.1. *For all $\theta \in \partial\mathcal{B}_1$,*

$$\|P_0(\cdot, \theta)\|_{L_\infty} \leq \operatorname{ess}\|P\|_{W_p^h \to W_p^{h-k}}.$$

The proof is quite similar to that of Theorem 7.3.1, so we only outline it. Let η, φ_k and Q_k be the same functions and cube as in the proof of Theorem 7.3.1. We put

$$v_\xi(y) = |\xi|^{-h}\eta(y)\exp\Big(i\sum_{j=1}^{n}[\xi_j]y_j\Big),$$

where $\xi \in \mathbb{R}^n \backslash \{0\}$ and $[\xi_j]$ is the integer part of ξ_j. By the same argument as in the proof of Theorem 7.3.1, we show that in the first place

$$\lim_{|\xi|\to\infty} \|v_\xi\|_{W_p^h} = A_h\|\eta\|_{L_p}, \qquad A_h = \text{const} > 0,$$

in the second place

$$\lim_{|\xi|\to\infty} \|\varphi_k P v_\xi\|_{W_p^{h-k}} = A_h\|\varphi_k P_0(\cdot, \xi|\xi|^{-1})\eta\|_{L_p},$$

and in the third place

$$\varphi_k T v_\xi \to 0 \quad \text{as} \quad |\xi| \to \infty \quad \text{in} \quad \mathring{W}_p^h(Q_k),$$

where T is a compact operator in W_p^h. Then

$$\limsup_{|\xi|\to\infty} \|\varphi_k P v_\xi\|_{W_p^{h-k}} = \limsup_{|\xi|\to\infty} \|\varphi_k (P - T) v_\xi\|_{W_p^{h-k}}$$

and, by Lemma 7.1.9, for some $\sigma > 0$,

$$\begin{aligned}\limsup_{|\xi|\to\infty} \|\varphi_k P v_\xi|_{W_p^{h-k}} &\leq (1 + O(k^{-\sigma}))\limsup_{|\xi|\to\infty} \|(P - T)v_\xi\|_{W_p^{h-k}} \\ &\leq (1 + O(k^{-\sigma}))\limsup_{|\xi|\to\infty} \|v_\xi\|_{W_p^h}(\operatorname{ess}\|P\|_{W_p^h \to W_p^{h-k}} + \varepsilon).\end{aligned}$$

Consequently,

$$\|\varphi_k P_0(\cdot, \xi|\xi|^{-1})\eta\|_{L_p} \leq (1 + O(k^{-\sigma}))\|\eta\|_{L_p} \operatorname{ess}\|P\|_{W_p^h \to W_p^{h-k}}$$

and finally

$$\|P_0(\cdot, \xi|\xi|^{-1}\eta)\|_{L_p} \leq \|\eta\|_{L_p} \operatorname{ess}\|P\|_{W_p^h \to W_p^{h-k}}.$$

□

Lemma 10.2.2. (*i*) *The estimate*

$$\operatorname{ess} \|P\|_{W_p^h \to W_p^{h-k}} \le c \sum_{|\alpha| \le k} \operatorname{ess} \|a_\alpha\|_{M(W_p^{h-|\alpha|} \to W_p^{h-k})} \tag{10.2.1}$$

holds.

(*ii*) *If $p = 1$ or $p(h-k) > n$, $p > 1$ and P maps continuously W_p^h into W_p^{h-k}, then*

$$\operatorname{ess} \|P\|_{W_p^h \to W_p^{h-k}} \sim \sum_{|\alpha| \le k} \operatorname{ess} \|a_\alpha\|_{M(W_p^{h-|\alpha|} \to W_p^{h-k})} \,. \tag{10.2.2}$$

Proof. (*i*) Let $\varepsilon > 0$ and let T_α be a compact operator which maps $W_p^{h-|\alpha|}$ into W_p^{h-k} and satisfies

$$\|a_\alpha - T_\alpha\|_{W_p^{h-|\alpha|} \to W_p^{h-k}} \le \operatorname{ess} \|a_\alpha\|_{M(W_p^{h-|\alpha|} \to W_p^{h-k})} + \varepsilon \,.$$

Since the operator

$$T = \sum_{|\alpha| \le k} T_\alpha D^\alpha : W_p^h \to W_p^{h-k}$$

is compact, we arrive at (10.2.1).

(*ii*) We begin with the case $p = 1$, $h - k \le n$. Let $\varepsilon > 0$ and let T be a compact operator such that

$$\|P - T\|_{W_1^h \to W_1^{h-k}} \le \operatorname{ess} \|P\|_{W_1^h \to W_1^{h-k}} + \varepsilon \,.$$

Further, let $\eta_{\delta,x}$ and ζ_r be the cutoff functions introduced at the beginning of Sect. 7.1. Duplicating the proof of the upper estimate in Theorem 7.2.1 with obvious changes, we obtain the estimate

$$\|\eta_{\delta,x} P\|_{W_1^h \to W_1^{h-k}} + \|\zeta_r P\|_{W_1^h \to W_1^{h-k}} \le c\,(\operatorname{ess} \|P\|_{W_1^h \to W_1^{h-k}} + \varepsilon) \tag{10.2.3}$$

which, by (10.1.3), is equivalent to

$$\sum_{|\alpha| \le k} \left(\|\eta_{\delta,x} a_\alpha\|_{M(W_1^{h-|\alpha|} \to W_1^{h-k})} + \|\zeta_r a_\alpha\|_{M(W_1^{h-|\alpha|} \to W_1^{h-k})} \right)$$
$$\le c\,(\operatorname{ess} \|P\|_{W_1^h \to W_1^{h-k}} + \varepsilon) \,.$$

It remains to use Theorem 7.2.1.

Next let $p(h-k) > n$ and $p \ge 1$. According to Lemma 10.1.1, $a_\alpha \in M(W_p^{h-|\alpha|} \to W_p^{h-k})$. The inequality

$$\|\zeta_r P\|_{W_p^h \to W_p^{h-k}} \le c\,(\operatorname{ess} \|P\|_{W_p^h \to W_p^{h-k}} + \varepsilon) \,,$$

where ε is sufficiently small and r is sufficiently large, can be obtained in the same way as (10.2.3). This inequality and (10.1.3) give

$$\sum_{|\alpha|\le k} \|\zeta_r a_\alpha\|_{M(W_p^{h-|\alpha|}\to W_p^{h-k})} \le c\,(\operatorname{ess}\|P\|_{W_p^h\to W_p^{h-k}} + \varepsilon)\,.$$

Therefore,

$$\limsup_{|x|\to\infty} \sum_{|\alpha|\le k} \|a_\alpha; \mathcal{B}_1(x)\|_{W_p^{h-k}} \le c \operatorname{ess}\|P\|_{W_p^h\to W_p^{h-k}} \tag{10.2.4}$$

and, by Theorem 7.2.5,

$$\sum_{|\alpha|<k} \operatorname{ess}\|a_\alpha\|_{M(W_p^{h-|\alpha|}\to W_p^{h-k})} \le c \operatorname{ess}\|P\|_{W_p^h\to W_p^{h-k}}\,. \tag{10.2.5}$$

Applying part (i) of the present lemma to the operator $P - P_0$ and using (10.2.5), we arrive at

$$\operatorname{ess}\|P_0\|_{W_p^h\to W_p^{h-k}} \le c \operatorname{ess}\|P\|_{W_p^h\to W_p^{h-k}}$$

which, together with Lemma 10.2.1, shows that

$$\sum_{|\alpha|=k} \operatorname{ess}\|a_\alpha\|_{L_\infty} \le c \operatorname{ess}\|P\|_{W_p^h\to W_p^{h-k}}\,. \tag{10.2.6}$$

Taking into account Theorems 7.3.6 and 7.3.8, and using (10.2.4) and (10.2.6), we find that

$$\sum_{|\alpha|=k} \operatorname{ess}\|a_\alpha\|_{MW_p^{h-k}} \le c \operatorname{ess}\|P\|_{W_p^h\to W_p^{h-k}}\,.$$

This estimate and (10.2.5) imply (ii). □

10.3 Fredholm Property of the Schrödinger Operator

The usefulness of the essential norm of a Sobolev multiplier, whose two-sided estimates were obtained in Chap. 7, can be illustrated by the following application to the Schrödinger operator

$$S := -\Delta + I - \gamma : W_p^m \to W_p^{m-2}, \tag{10.3.1}$$

where I is the imbedding operator: $W_p^m \to W_p^{m-2}$ and $p \in (1,\infty)$. We notice that (10.3.1) holds if and only if $\gamma \in M(W_p^m \to W_p^{m-2})$.

Proposition 10.3.1. *Let* $c_{p,m} = \|(I-\Delta)^{-1}\|_{W_p^{m-2}\to W_p^m}$. *If*

$$\operatorname{ess}\|\gamma\|_{M(W_p^m\to W_p^{m-2})} < c_{p,m}^{-1},$$

then S *is Fredholm.*

Proof. We start with collecting some well-known definitions and facts (see, for example, [Pa]). The operators $R_\pm : W_p^m \to W_p^{m-2}$ are called the right and left regularizers of S if

$$SR_+ = I_{m-2} + T_{m-2}$$

and

$$SR_- = I_m + T_m,$$

respectively, where I_{m-2} and I_m are identity operators in W_p^{m-2} and W_p^m, whereas T_{m-2} and T_m are compact operators in W_p^{m-2} and W_p^m. It is well known and easily seen that the existence of R_+ implies $\dim\ker S < \infty$ and the existence of R_- guarantees the closedness of the range of S and the finiteness of $\dim\ker S$. Hence the existence of both R_+ and R_- ensures that S is Fredholm.

Let T be a compact operator: $W_p^m \to W_p^{m-2}$ such that

$$\|\gamma - T\|_{W_p^m \to W_p^{m-2}} < c_{p,m}^{-1}.$$

Under this condition the inverse operators of

$$I_{m-2} - (\gamma - T)(I - \Delta)^{-1} : W_p^{m-2} \to W_p^{m-2}$$

and

$$I_m - (I - \Delta)^{-1}(\gamma - T) : W_p^m \to W_p^m$$

exist. Now, it is straightforward that

$$(I - \Delta)^{-1}\big(I_{m-2} - (\gamma - T)(I - \Delta)^{-1}\big)^{-1}$$

and

$$\big(I_m - (I - \Delta)^{-1}(\gamma - T)\big)^{-1}(I - \Delta)^{-1}$$

are the right and left regularizers of S. The proof is complete. □

10.4 Domination of Differential Operators in $\mathbb{R}^n$

The discussion in the present section is related to the problem of the domination relations for differential operators with constant coefficients (see, for example, [H1], Sect. 3.3, and [GM]). Here we consider one of the formulations of this problem.

Let $R(D)$ and $P(D)$ be differential operators in $\mathbb{R}^n$ with constant coefficients and let $\mathcal{S}$ be the Schwartz space of infinitely differentiable functions defined on $\mathbb{R}^n$ and tending to zero at infinity, with all its derivatives, faster than an arbitrary positive power of $|x|^{-1}$ (see [Sch], [GSh]). In the following theorem we use the space $L_2((1 + |x|^2)^{k/2})$, $k \in \mathbb{R}$, with the norm

$$\|u\|_{L_2((1+|x|^2)^{k/2})} = \Big(\int |u|^2(1 + |x|^2)^k\,dx\Big)^{1/2}.$$

Theorem 10.4.1. *The inequality*

$$\|R(D)u\|_{L_2((1+|x|^2)^{-l/2})} \leq C\,\|P(D)u\|_{L_2}\,, \qquad l>0\,, \tag{10.4.1}$$

holds for every $u \in \mathcal{S}$ if and only if $R/P \in M(W_2^l \to L_2)$, which is equivalent to

$$\sup_{\{e:d(e)\leq 1\}} \frac{\|R/P; e\|_{L_2}}{[C_{2,l}(e)]^{1/2]}} < \infty\,. \tag{10.4.2}$$

In particular, in the case $2l > n$ the condition (10.4.2) *means that*

$$\sup_{x\in\mathbb{R}^n} \int_{\mathcal{B}_1(x)} |R(\xi)/P(\xi)|^2\, d\xi < \infty\,. \tag{10.4.3}$$

Proof. Sufficiency. The left-hand side in (10.4.1) is equal to

$$\sup_{\varphi} \frac{|(R(D)u, \varphi)|}{\|\varphi\|_{L_2((1+|x|^2)^{l/2})}} = \sup_{\varPhi} \frac{|(PFu, \overline{(R/P)}\varPhi)|}{\|\varPhi\|_{W_2^l}}\,,$$

where F is the Fourier transform in $\mathbb{R}^n$. The last supremum does not exceed

$$\|PFu\|_{L_2} \sup_{\varPhi} \frac{\|(R/P)\varPhi\|_{L_2}}{\|\varPhi\|_{W_2^l}} = \|R/P\|_{M(W_2^l\to L_2)}\|P(D)u\|_{L_2}\,.$$

Necessity. Let

$$Q_\varepsilon(\xi) = (|P(\xi)|^2 + \varepsilon)^{1/2}, \qquad \text{where} \quad \varepsilon = \text{const} > 0,$$

and let $Q_\varepsilon(D) = F^{-1}Q_\varepsilon F$. Since $F\mathcal{S} = \mathcal{S}$, the operator $Q_\varepsilon^{-1}(D) = F^{-1}Q_\varepsilon^{-1}F$ maps $\mathcal{S}$ into itself. We set $u = Q_\varepsilon^{-1}(D)f$, where f is an arbitrary function in $\mathcal{S}$. By (10.4.1),

$$\|R(D)Q_\varepsilon^{-1}(D)f\|_{L_2((1+|x|^2)^{-l/2})} \leq C\,\|f\|_{L_2}$$

which is the same as

$$|(\varPsi, (\overline{R}/Q_\varepsilon)\varPhi)| \leq C\,\|\varPsi\|_{L_2}\|\varPhi\|_{W_2^l}\,.$$

Consequently, for all $\varPhi \in W_2^l$, we have

$$\|(R/Q_\varepsilon)\varPhi\|_{L_2} \leq C\,\|\varPhi\|_{W_2^l}.$$

Passing to the limit as $\varepsilon \to +0$, we complete the proof. □

A rough corollary of Theorem 10.4.1 is the sufficiency of (10.4.3) for the validity of

$$\|R(D)u; K\|_{L_2} \leq C(K)\|P(D)u\|_{L_2}\,, \tag{10.4.4}$$

where $u \in \mathcal{S}$, K is an arbitrary compact set in $\mathbb{R}^n$ and $C(K)$ is a constant independent of u. Namely, the following theorem holds:

Theorem 10.4.2. [Maz8] *Inequality* (10.4.4) *is true if and only if the functions R and P satisfy* (10.4.3).

Proof. We need to prove only the necessity. Let x be a fixed point in $\mathbb{R}^n$, let χ_x be the characteristic function of the ball $\mathcal{B}_1(x)$, and let K be the cube $\{x \in \mathbb{R}^n : |x_i| \le 1, 1 \le i \le n\}$. We define a family of functions $\{u_{\varepsilon,h}\}$ by

$$Fu_{\varepsilon,h} = \Big(\frac{\overline{R}\chi_x}{|P|^2+\varepsilon}\Big)_h ,$$

where $(\varphi)_h$ is the mollification of φ with radius h and ε is a positive number. Clearly, $u_{\varepsilon,h} \in \mathcal{S}$ and it can be put into (10.4.4). We then have

$$\lim_{h\to 0}\|P(D)u_{\varepsilon,h}\|_{L_2} = \Big\|\frac{PR}{|P|^2+\varepsilon}; \mathcal{B}_1(x)\Big\|_{L_2} \le \Big\|\frac{R}{(|P|^2+\varepsilon)^{1/2}}; \mathcal{B}_1(x)\Big\|_{L_2}. \quad (10.4.5)$$

On the other hand,

$$\|R(D)u_{\varepsilon,h}; K\|_{L_2} = c\,\|\psi * RFu_{\varepsilon,h}\|_{L_2} ,$$

where

$$\psi(\xi) = \prod_{1\le i\le n} \xi_i^{-1}\sin\xi_i .$$

Therefore

$$\|R(D)u_{\varepsilon,h}; K\|^2_{L_2} \ge c\int_{\mathcal{B}_1(x)} \Big|\int \psi(\xi-\eta)R(\eta)(Fu_{\varepsilon,h})(\eta)\,d\eta\Big|^2 d\xi .$$

The right-hand side tends to

$$c\int_{\mathcal{B}_1(x)}\Big|\int_{\mathcal{B}_1(x)}\psi(\xi-\eta)\frac{|R(\eta)|^2\,d\eta}{|P(\eta)|^2+\varepsilon}\Big|^2 d\xi$$

as $h \to 0$. Here $|\xi-\eta| < 2$ and consequently $\psi(\xi-\eta) \ge \text{const} > 0$. We arrive at the inequality

$$\liminf_{h\to 0}\|R(D)u_{\varepsilon,h}; K\|_{L_2} \ge c\int_{\mathcal{B}_1(x)}\frac{|R(\eta)|^2\,d\eta}{|P(\eta)|^2+\varepsilon} . \quad (10.4.6)$$

Now, from (10.4.4) for $u_{\varepsilon,h}$ and from (10.4.5) and (10.4.6) we obtain

$$\|R(|P|^2+\varepsilon)^{-1/2}; \mathcal{B}_1(x)\|_{L_2} \le c .$$

Passing to the limit as $\varepsilon \to 0$, we get (10.4.3). □

11

Schrödinger Operator and $M(w_2^1 \to w_2^{-1})$

11.1 Introduction

The results presented in this chapter were obtained in [MV2]. Here a characterization is given for the class of measurable functions (or, more generally, real- or complex-valued distributions) V such that the Schrödinger operator $H = -\Delta + V$ maps the energy space $w_2^1(\mathbb{R}^n)$ to its dual $w_2^{-1}(\mathbb{R}^n)$. Similar results are obtained for the inhomogeneous Sobolev space $W_2^1(\mathbb{R}^n)$. In other words, a complete solution is given to the problem of the relative form-boundedness of the potential energy operator V with respect to the Laplacian $-\Delta$, which is fundamental to quantum mechanics. Relative compactness criteria for the corresponding quadratic forms are established as well. Analogous boundedness and compactness criteria for Sobolev spaces on domains $\Omega \subset \mathbb{R}^n$ are obtained under mild restrictions on $\partial\Omega$.

The abovementioned mapping property of H is equivalent to the classical inequality

$$\left| \int_{\mathbb{R}^n} |u(x)|^2 \, V(x) \, dx \right| \le C \int_{\mathbb{R}^n} |\nabla u(x)|^2 \, dx, \quad u \in C_0^\infty(\mathbb{R}^n), \tag{11.1.1}$$

holding. Here the "indefinite weight" V may change sign, or even be a complex-valued distribution on $\mathbb{R}^n$, $n \ge 3$. (In the latter case, the left-hand side of (11.1.1) is understood as $| < Vu, u > |$, where $< V \cdot, \cdot >$ is the quadratic form associated with the corresponding multiplication operator V.) We also characterize an analogous inequality for the inhomogeneous Sobolev space $W_2^1(\mathbb{R}^n)$, $n \ge 1$:

$$\left| \int_{\mathbb{R}^n} |u(x)|^2 \, V(x) \, dx \right| \le C \int_{\mathbb{R}^n} [\, |\nabla u(x)|^2 + |u(x)|^2 \,] \, dx, \quad u \in C_0^\infty(\mathbb{R}^n). \tag{11.1.2}$$

Such inequalities are used extensively in the spectral and scattering theory of the Schrödinger operator $H = H_0 + V$, where $H_0 = -\Delta$ is the Laplacian on $\mathbb{R}^n$. In particular, (11.1.2) is equivalent to the concept of the relative boundedness of V (potential energy operator) with respect to H_0 in the sense of quadratic forms.

It follows from the polarization identity

$$\bar{u}v = \frac{1}{4}\big(|u+v|^2 - |u-v|^2 - i|u-iv|^2 + i|u+iv|^2\big) \qquad (11.1.3)$$

that (11.1.1) can be restated equivalently in terms of the sesquilinear form $< Vu, v >$ as

$$|< Vu, v >| \le C\, \|\nabla u\|_{L_2}\, \|\nabla v\|_{L_2}, \qquad (11.1.4)$$

for all $u, v \in C_0^\infty(\mathbb{R}^n)$. In other words, it is equivalent to the boundedness of the operator $H = H_0 + V$,

$$H : w_2^1(\mathbb{R}^n) \to w_2^{-1}(\mathbb{R}^n), \qquad n \ge 3. \qquad (11.1.5)$$

Here the energy space $w_2^1(\mathbb{R}^n)$ is defined as the completion of $C_0^\infty(\mathbb{R}^n)$ with respect to the norm $\|\nabla u\|_{L_2}$, and $w_2^{-1}(\mathbb{R}^n)$ is the dual of $w_2^1(\mathbb{R}^n)$. Similarly, (11.1.2) means that H is a bounded operator which maps $W_2^1(\mathbb{R}^n)$ to $W_2^{-1}(\mathbb{R}^n)$, $n \ge 1$.

Note that (11.1.4) means that the distribution V belongs the space of multipliers $M(w_2^1(\mathbb{R}^n) \to w_2^{-1}(\mathbb{R}^n))$.

Before stating the main results we note that we use some expressions involving pseudodifferential operators, e.g. $\nabla \Delta^{-1}$ or $(-\Delta)^{-1/2}$, which will be defined in the main body of this chapter.

As before, $\mathbb{R}^n$ will be omitted in notations of norms and integrals.

Theorem 11.1.1. *Let V be a complex-valued distribution on $\mathbb{R}^n$, $n \ge 3$. Then $V \in M((w_2^1 \to w_2^{-1})$, i.e. (11.1.1) holds, if and only if V is the divergence of a vector field $\mathbf{\Gamma} : \mathbb{R}^n \to \mathbb{C}^n$ such that*

$$\int |u(x)|^2\, |\mathbf{\Gamma}(x)|^2\, dx \le C \int |\nabla u(x)|^2\, dx, \qquad (11.1.6)$$

where the constant is independent of $u \in C_0^\infty$. The vector field $\mathbf{\Gamma} \in \mathbf{L}_{2,\mathrm{loc}}$ can be chosen as $\mathbf{\Gamma} = \nabla \Delta^{-1} V$.

Equivalently, the Schrödinger operator $H = H_0 + V$ acting from w_2^1 to w_2^{-1} is bounded if and only if $V = \operatorname{div} \mathbf{\Gamma}$ with $\mathbf{\Gamma}$ subject to (11.1.6). Furthermore, the corresponding multiplication operator $V : w_2^1 \to w_2^{-1}$ is compact if and only if $V = \operatorname{div} \mathbf{\Gamma}$, where $\mathbf{\Gamma}$ is such that the embedding

$$w_2^1 \subset L_2(|\mathbf{\Gamma}|^2)$$

is compact.

Obviously, (11.1.6) means that $\mathbf{\Gamma} \in M(w_2^1 \to L_2)$. Recall that the last space was discussed in Sect. 2.8.

We remark that once V is written as $V = \operatorname{div} \mathbf{\Gamma}$, the implication

$$(11.1.6) \Longrightarrow (11.1.1)$$

becomes trivial. It follows using integration by parts and the Schwarz inequality (compare with Sect. 2.5, where a similar argument was used to obtain sufficient conditions for the inclusion into $M(W_p^m \to W_p^{-k})$).

On the other hand, the converse statement (11.1.1)$\Longrightarrow$(11.1.6), where $\mathbf{\Gamma} = \nabla \Delta^{-1} V$, is rather delicate. Its proof is based on a special factorization of functions in w_2^1 involving powers P_K^δ of the equilibrium harmonic potential P_K associated with an arbitrary compact set $K \subset \mathbb{R}^n$ of positive Wiener's capacity. New sharp estimates for P_K^δ, where ultimately δ is picked so that $1 < 2\delta < n/(n-2)$, are established in a series of lemmas and propositions in Sect. 11.1.22. One also makes use of the fact that standard Mikhlin-Calderon-Zygmund operators are bounded on L_2 with a weight P_K^δ, and the corresponding operator norm bounds do not depend on K.

We now briefly outline the contents of this chapter. In Sect. 11.2 we define the Schrödinger operator on the energy space w_2^1, and characterize the basic inequality (11.1.1). The compactness problem is treated in Sect. 11.3. Analogous results for the Sobolev space W_2^1 are obtained in Sect. 11.4, while Sect. 11.5 is devoted to similar problems on a domain $\Omega \subset \mathbb{R}^n$ for a broad class of Ω, including those with Lipschitz boundaries.

11.2 Characterization of $M(w_2^1 \to w_2^{-1})$ and the Schrödinger Operator on w_2^1

In this section, we assume that $n \geq 3$, since for the homogeneous space $w_2^1(\mathbb{R}^n)$ our results become vacuous if $n = 1$ and $n = 2$, (11.1.1) then implying that $V = 0$. (Analogous results for inhomogeneous Sobolev spaces $W_2^1(\mathbb{R}^n)$ are valid for all $n \geq 1$; see Sect. 11.4 below.)

For $V \in (C_0^\infty)'$, consider the multiplication operator on C_0^∞ defined by

$$< V u, v > := < V, \bar{u}\, v >, \qquad u,\, v \in C_0^\infty, \tag{11.2.1}$$

where $< \cdot, \cdot >$ represents the usual pairing between C_0^∞ and its dual $(C_0^\infty)'$.

Elements of $w_2^1(\mathbb{R}^n)$, for $n \geq 3$, are weakly differentiable functions $u \in L_{\frac{2n}{n-2}}(\mathbb{R}^n)$ whose first-order weak derivatives lie in $L_2(\mathbb{R}^n)$. By Hardy's inequality, an equivalent norm on w_2^1 is given by

$$\|u\|_{w_2^1} = \left[\int \left(|x|^{-2}\, |u(x)|^2 + |\nabla u(x)|^2 \right) dx \right]^{\frac{1}{2}}.$$

If the sesquilinear form $< V \cdot, \cdot >$ is bounded on $w_2^1 \times w_2^{-1}$:

$$| < V u,\, v > | \leq c\, \|\nabla u\|_{L_2}\, \|\nabla v\|_{L_2}, \qquad u,\, v \in C_0^\infty, \tag{11.2.2}$$

where the constant c is independent of u and v, then $V u \in w_2^{-1}$, and the multiplication operator can be extended by continuity onto w_2^1. As usual, this extension is also denoted by V.

Note that the least constant c in (11.2.2) is equal to the multiplier norm:

$$\|V\|_{M(w_2^1 \to w_2^{-1})} = \sup \{ \|Vu\|_{w_2^{-1}} : \|u\|_{w_2^1} \le 1, \quad u \in C_0^\infty \}.$$

For $V \in M(w_2^1 \to w_2^{-1})$, we extend the form $< V, \bar{u}v >$ defined by the right-hand side of (11.2.1) to the case where both u and v are in w_2^1. This can be done by letting

$$< V u, v > := \lim_{N\to\infty} < V u_N, v_N >,$$

where

$$u = \lim_{N\to\infty} u_N, \quad \text{and} \quad v = \lim_{N\to\infty} v_N \quad \text{in } w_2^1$$

with $u_N, v_N \in C_0^\infty$.

We now state the main result of this section for arbitrary (complex-valued) distributions V. By $\mathbf{L}_{2,\text{loc}} = L_{2,\text{loc}} \otimes \mathbb{C}^n$ we denote the space of vector-functions $\mathbf{\Gamma} = (\Gamma_1, \ldots, \Gamma_n)$ such that $\Gamma_i \in L_{2,\text{loc}}$, $i = 1, \ldots, n$.

Theorem 11.2.1. *Let $V \in (C_0^\infty)'$. Then $V \in M(w_2^1 \to w_2^{-1})$, i.e., the inequality*

$$| < V u, v > | \le c \, \|u\|_{w_2^1} \, \|v\|_{w_2^1} \tag{11.2.3}$$

holds for all $u, v \in C_0^\infty$, if and only if there is a vector field $\mathbf{\Gamma} \in \mathbf{L}_{2,\text{loc}}$ such that $V = \operatorname{div} \mathbf{\Gamma}$, and (11.1.6) *holds for all $u \in C_0^\infty$. The vector field $\mathbf{\Gamma}$ can be chosen in the form*

$$\mathbf{\Gamma} = \nabla \Delta^{-1} V.$$

Proof. Suppose that $V = \operatorname{div} \mathbf{\Gamma}$, where $\mathbf{\Gamma}$ satisfies (11.1.6). Then using integration by parts and the Schwarz inequality we obtain:

$$| < V u, v > | = | < V, \bar{u} \, v > | = | < \mathbf{\Gamma}, v \, \nabla \bar{u} > + < \mathbf{\Gamma}, \bar{u} \, \nabla v > |$$

$$\le \|\mathbf{\Gamma} \bar{v}\|_{\mathbf{L}_2} \, \|\nabla \bar{u}\|_{L_2} + \|\mathbf{\Gamma} u\|_{\mathbf{L}_2} \, \|\nabla v\|_{L_2} \le 2\sqrt{C} \, \|\nabla u\|_{L_2} \, \|\nabla v\|_{L_2},$$

where C is the constant in (11.1.6). This completes the proof of the "if" part of Theorem 11.2.1.

The proof of the "only if" part of Theorem 11.2.1 is based on several lemmas and propositions.

In the next lemma, we show that $\mathbf{\Gamma} = \nabla \Delta^{-1} V \in \mathbf{L}_{2,\text{loc}}$, and give a crude preliminary estimate of the rate of its decay at infinity.

Lemma 11.2.1. *Suppose that*

$$V \in M(w_2^1 \to w_2^{-1}). \tag{11.2.4}$$

Then

$$\mathbf{\Gamma} = \nabla \Delta^{-1} V \in \mathbf{L}_{2,\mathrm{loc}}$$

and

$$V = \operatorname{div} \mathbf{\Gamma} \quad \text{in} \quad (C_0^\infty)'.$$

Moreover, for any ball $\mathcal{B}_R(x_0)$ $(R > 0)$ *and* $\epsilon > 0$,

$$\int_{\mathcal{B}_R(x_0)} |\mathbf{\Gamma}(x)|^2 \, dx \le C(n, \epsilon) \, R^{n-2+\epsilon} \, \|V\|^2_{M(w_2^1 \to w_2^{-1})}, \tag{11.2.5}$$

where $R \ge \max\{1, \, |x_0|\}$.

Proof. Suppose that $V \in M(w_2^1 \to w_2^{-1})$. Define the vector field $\mathbf{\Gamma} \in (C_0^\infty)'$ by

$$< \mathbf{\Gamma}, \, \boldsymbol{\phi} > = - < V, \, \Delta^{-1} \operatorname{div} \boldsymbol{\phi} >, \tag{11.2.6}$$

for every $\boldsymbol{\phi} \in C_0^\infty \otimes \mathbb{C}^n$. In particular,

$$< \mathbf{\Gamma}, \, \nabla \psi > = - < V, \, \psi >, \qquad \psi \in C_0^\infty, \tag{11.2.7}$$

i.e., $V = \operatorname{div} \mathbf{\Gamma}$ in $(C_0^\infty)'$.

We first have to check that the right-hand side of (11.2.6) is well-defined, which a priori is not obvious. For $\boldsymbol{\phi} \in C_0^\infty \otimes \mathbb{C}^n$, let $w = \Delta^{-1} \operatorname{div} \boldsymbol{\phi}$, where $-\Delta^{-1} f = I_2 f$ is the Newtonian potential of $f \in C_0^\infty$. Clearly,

$$w(x) = O(|x|^{1-n}) \quad \text{and} \quad |\nabla w(x)| = O\left(|x|^{-n}\right) \quad \text{as} \quad |x| \to \infty,$$

and hence

$$w = \Delta^{-1} \operatorname{div} \boldsymbol{\phi} \in w_2^1 \cap C^\infty.$$

We will show below that $w = u \, v$, where u is real-valued, and both u and v are in $w_2^1 \cap C^\infty$. Then, since $V \in M(w_2^1 \to w_2^{-1})$, it follows that $< V, \, w > = < V \, u, \, v >$ is defined through the extension of the multiplication operator V as explained above.

For our purposes, it is important to note that this extension of $< V, \, w >$ to the case where $w = \bar{u} v$, and $u, v \in w_2^1 \cap C^\infty$, is independent of the choice of factors u and v. To demonstrate this, we define a real-valued cutoff function

$$\eta_N(x) = \eta(N^{-1}|x|), \quad \text{where} \quad \eta \in C^\infty(\mathbb{R}_+),$$

so that $\eta(t) = 1$ for $0 \le t \le 1$ and $\eta(t) = 0$ for $t \ge 2$. Note that $\nabla \eta_N$ is supported in the annulus $N \le |x| \le 2N$, and $|\nabla \eta_N(x)| \le c \, |x|^{-1}$. It follows easily (for instance, from Hardy's inequality) that

$$\lim_{N \to \infty} \|\eta_N \, u - u\|_{w_2^1} = 0, \qquad u \in w_2^1.$$

Then letting

$$u_N = \eta_N u \quad \text{and} \quad v_N = \eta_N v,$$

so that $\overline{u_N}\, v_N = \eta_N^2\, w$, we define $< V,\, w >$ explicitly by setting:

$$< V,\, w > := \lim_{N\to\infty} < V\, u_N,\, v_N > = \lim_{N\to\infty} < V,\, \eta_N^2 w > .$$

This definition is independent of the choice of η and the factors $u,\ v$. Moreover,

$$| < V, w > | \le C \inf\{\|u\|_{w_2^1}\, \|v\|_{w_2^1} : \, w = \bar{u}\, v;\ u, v \in w_2^1 \cap C^\infty\},$$

where

$$C = \|V\|_{M(w_2^1 \to w_2^{-1})}.$$

Now we fix $\epsilon > 0$ and factorize:

$$w(x) = \Delta^{-1}\mathrm{div}\, \boldsymbol{\phi}(x) = u(x)\, v(x),$$

where

$$u(x) = (1+|x|^2)^{-\frac{n-2+\epsilon}{4}} \quad \text{and} \quad v(x) = (1+|x|^2)^{\frac{n-2+\epsilon}{4}}\, \Delta^{-1}\mathrm{div}\, \boldsymbol{\phi}(x). \tag{11.2.8}$$

Obviously, $u \in w_2^1 \cap C^\infty$, and

$$\|u\|_{w_2^1} = c(n, \epsilon) < \infty.$$

It is easy to see that $v \in w_2^1 \cap C^\infty$ as well. Furthermore, the following statement holds.

Proposition 11.2.1. *Suppose that* $\boldsymbol{\phi} \in C^\infty$, *and* $\mathrm{supp}\, \boldsymbol{\phi} \subset \mathcal{B}_R(x_0)$. *Let* v *be defined by* (11.2.8), *where* $0 < \epsilon < 2$. *Then*

$$\|v\|_{w_2^1} \le c(n, \epsilon)\, R^{\frac{n-2+\epsilon}{2}}\, \|\boldsymbol{\phi}\|_{L_2}, \tag{11.2.9}$$

for $R \ge \max\{1, |x_0|\}$.

Proof. Since $\boldsymbol{\phi}$ is compactly supported, it follows that

$$|\Delta^{-1}\mathrm{div}\, \boldsymbol{\phi}(x)| \le c(n)\, I_1|\boldsymbol{\phi}|(x), \quad x \in \mathbb{R}^n.$$

Hence

$$c(n, \epsilon)\, \|v\|_{w_2^1} \le \|(1+|x|^2)^{\frac{n-2+\epsilon}{4}}\, \nabla\, \Delta^{-1}\mathrm{div}\, \boldsymbol{\phi}(x)\|_{L_2}$$
$$+ \|(1+|x|^2)^{\frac{n-4+\epsilon}{4}}\, I_1|\boldsymbol{\phi}|(x)\|_{L_2}.$$

Note that $\nabla\, \Delta^{-1}\mathrm{div}$ is a Mikhlin-Calderon-Zygmund operator, and that the weight

$$w(x) = (1+|x|^2)^{\frac{n-2+\epsilon}{2}}$$

belongs to the Muckenhoupt class A_2 if $0 < \epsilon < 2$ (see [CF]). Applying the corresponding weighted norm inequality, we have:

$$\|(1+|x|^2)^{\frac{n-2+\epsilon}{4}}\, \nabla\, \Delta^{-1}\mathrm{div}\, \boldsymbol{\phi}(x)\|_{L_2}$$

$$\leq c(n,\epsilon)\, \|(1+|x|^2)^{\frac{n-2+\epsilon}{4}}\, |\phi(x)|\, \|_{L_2}. \tag{11.2.10}$$

The other term is estimated by the weighted Hardy inequality:

$$\int (I_1|\phi|(x))^2\, (1+|x|^2)^{\frac{n-4+\epsilon}{2}}\, dx$$
$$\leq c(n,\epsilon) \int |\phi(x)|^2\, (1+|x|^2)^{\frac{n-2+\epsilon}{2}}\, dx. \tag{11.2.11}$$

Clearly,

$$\|(1+|x|^2)^{\frac{n-2+\epsilon}{4}}\, \phi(x)\|_{L_2} \leq c(n,\epsilon)\, R^{\frac{n-2+\epsilon}{2}}\, \|\phi\|_{L_2}.$$

Hence, combining (11.2.10), (11.2.11), and the preceding estimate, we obtain the desired inequality (11.2.9). The proof of Proposition 11.2.1 is complete. □

Now let us prove (11.2.5). Suppose that $\phi \in C^\infty \otimes \mathbb{C}^n$, and $\operatorname{supp}\phi \subset \mathcal{B}_R(x_0)$. Then by (11.2.6) and Proposition 11.2.1,

$$|<\mathbf{\Gamma}, \phi>| = |<V, u\, v>| \leq \|V\|_{M(w_2^{\frac{1}{2}} \to w_2^{-1})}\, \|u\|_{w_2^{\frac{1}{2}}}\, \|v\|_{w_2^{\frac{1}{2}}}$$
$$\leq C(n,\epsilon)\, R^{\frac{n-2+\epsilon}{2}}\, \|V\|_{M(w_2^{\frac{1}{2}} \to w_2^{-1})}\, \|\phi\|_{L_2}. \tag{11.2.12}$$

Taking the supremum over all ϕ supported in $\mathcal{B}_R(x_0)$ with the unit L_2-norm, we arrive at (11.2.5). The proof of Lemma 11.2.1 is complete.

It remains to prove the main estimate (11.1.6) of Theorem 11.2.1. For this aim, it suffices to establish the inequality

$$\int_e |\mathbf{\Gamma}(x)|^2\, dx \leq c(n)\, \|V\|^2_{M(w_2^{\frac{1}{2}} \to w_2^{-1})}\, c_{2,1}(e), \tag{11.2.13}$$

for every compact set $e \subset \mathbb{R}^n$. Notice that in the special case $e = \overline{\mathcal{B}_R(x_0)}$, the preceding estimate gives a sharper version of (11.2.5):

$$\int_{\mathcal{B}_R(x_0)} |\mathbf{\Gamma}(x)|^2\, dx \leq C(n)\, R^{n-2}\, \|V\|^2_{M(w_2^{\frac{1}{2}} \to w_2^{-1})}, \qquad x_0 \in \mathbb{R}^n,\ R > 0.$$

Without loss of generality we assume that $c_{2,1}(e) > 0$; otherwise $\mathrm{mes}_n e = 0$, and (11.2.13) holds. Denote by $P(x) = P_e(x)$ the equilibrium potential on e. It is well known that P is the Newtonian potential of a positive measure which gives a solution to several variational problems. This measure ν_e is called the *equilibrium measure* for e.

We list some standard properties of ν_e and its potential $P_e(x) = I_2\nu_e(x)$ which will be used below:

$$\begin{aligned}
&(a) && \operatorname{supp}\nu_e \subset e;\\
&(b) && P_e(x) = 1 \quad d\nu_e - \text{a.e.};\\
&(c) && \nu_e(e) = c_{2,1}(e) > 0;\\
&(d) && \|\nabla P_e\|^2_{L_2} = c_{2,1}(e);\\
&(e) && \sup_{x\in\mathbb{R}^n} P_e(x) \leq 1.
\end{aligned} \tag{11.2.14}$$

The rest of the proof of Theorem 11.2.1 is based on some inequalities involving the powers $P_e(x)^\delta$ which are established below.

Proposition 11.2.2. *Let $\delta > \frac{1}{2}$ and let $P = P_e$ be the equilibrium potential of a compact set e of positive capacity. Then*

$$\|\nabla P^\delta\|_{L_2} = \frac{\delta}{\sqrt{2\delta - 1}} \sqrt{c_{2,1}(e)}. \tag{11.2.15}$$

Proof. Clearly,

$$\int |\nabla P(x)^\Delta|^2 \, dx = \delta^2 \int |\nabla P(x)|^2 \, P(x)^{2\delta-2} \, dx. \tag{11.2.16}$$

Using integration by parts, together with the properties $-\Delta P = \nu_e$ (understood in the distributional sense) and $P(x) = 1$ $d\nu_e$-a.e., we have

$$\int |\nabla P(x)|^2 \, P(x)^{2\delta-2} \, dx = \int \nabla P(x) \cdot \nabla P(x) \, P(x)^{2\delta-2} \, dx$$

$$= \int P(x)^{2\delta-1} \, d\nu_e - (2\delta - 2) \int |\nabla P(x)|^2 \, P(x)^{2\delta-2} \, dx$$

$$= c_{2,1}(e) - (2\delta - 2) \int |\nabla P(x)|^2 \, P(x)^{2\delta-2} \, dx.$$

The integration by parts above is easily justified for $\delta > \frac{1}{2}$ by examining the behavior of the potential and its gradient at infinity

$$c_1 \, |x|^{2-n} \le P(x) \le c_2 \, |x|^{2-n},$$

$$|\nabla P(x)| = O\left(|x|^{1-n}\right), \quad \text{as} \quad |x| \to \infty. \tag{11.2.17}$$

It follows from these calculations that

$$(2\delta - 1) \int |\nabla P(x)|^2 \, P(x)^{2\delta-2} \, dx = c_{2,1}(e).$$

Combining this with (11.2.16) yields (11.2.15). The proof of Proposition 11.2.2 is complete. □

Remark 11.2.1. For $\delta \le \frac{1}{2}$, it is easy to see that $\nabla P^\delta \notin L_2$.

In the next lemma we demonstrate that $\|\nabla v\|_{L_2}$ is equivalent to the weighted norm $\|P^{-\delta} \nabla (v P^\delta)\|_{L_2}$.

Lemma 11.2.2. *Let $\delta > 0$, and let $v \in w_2^1$. Then*

$$\|\nabla v\|_{L_2}^2 \le \int |\nabla (v \, P^\delta)(x)|^2 \, \frac{dx}{P(x)^{2\delta}} \le (\delta + 1)(4\delta + 1) \, \|\nabla v\|_{L_2}^2. \tag{11.2.18}$$

Proof. Without loss of generality we may assume that v is real-valued. We first prove (11.2.18) for $v \in C_0^\infty$. The general case will follow using an approximation argument. Clearly,

$$\int |\nabla(v\,P^\delta)(x)|^2 \frac{dx}{P^{2\delta}(x)} = \int |\nabla v(x) + \delta\, v(x)\, \nabla P(x)\, P(x)^{-1}|^2\, dx$$

$$= \int |\nabla v(x)|^2\, dx + \delta^2 \int v(x)^2 \frac{|\nabla P(x)|^2}{P(x)^2}\, dx + 2\delta \int \nabla v \cdot \nabla P(x) \frac{v(x)}{P(x)}\, dx.$$

Integration by parts and the equation $-\Delta P = \nu_e$ (understood in the distributional sense) give

$$2\int \nabla v \cdot \nabla P(x) \frac{v(x)}{P(x)}\, dx = \int v(x)^2 \frac{d\nu_e(x)}{P(x)}\, dx + \int v(x)^2 \frac{|\nabla P(x)|^2}{P(x)^2}\, dx.$$

Using this identity, we rewrite the preceding equation in the form

$$\int |\nabla(v\,P^\delta)(x)|^2 \frac{dx}{P^{2\delta}(x)} = \int |\nabla v(x)|^2\, dx$$

$$+\delta(\delta+1) \int v(x)^2 \frac{|\nabla P(x)|^2}{P(x)^2}\, dx + \delta \int v(x)^2 \frac{d\nu_e(x)}{P(x)}. \tag{11.2.19}$$

The lower estimate in (11.2.18) is now obvious provided the last two terms on the right-hand side of the preceding equation are finite. They are estimated in the following proposition, which holds for Newtonian potentials of arbitrary (not necessarily equilibrium) positive measures.

Proposition 11.2.3. *Let ω be a positive Borel measure on $\mathbb{R}^n$ such that $P(x) = I_2\omega(x) \not\equiv \infty$. Then the inequalities hold:*

$$\int v(x)^2 \frac{|\nabla P(x)|^2}{P(x)^2}\, dx \le 4\,\|\nabla v\|_{L_2}^2, \qquad v \in C_0^\infty, \tag{11.2.20}$$

and

$$\int v(x)^2 \frac{d\omega(x)}{P(x)} \le \|\nabla v\|_{L_2}^2, \qquad v \in C_0^\infty. \tag{11.2.21}$$

Proof. Suppose that $v \in C_0^\infty$. Then $A = \operatorname{supp} v$ is a compact set, and obviously

$$\inf_{x \in A} P(x) > 0.$$

Without loss of generality we may assume that $\nabla P \in \mathbf{L}_{2,\mathrm{loc}}$, and hence the left-hand side of (11.2.20) is finite. (Otherwise we replace ω by its convolution with a compactly supported mollifier: $\omega_t = \omega * \epsilon_t$, and complete the proof by applying the estimates given below to $P(x) = I_2\omega_t(x)$, and then passing to the limit as $t \to \infty$.)

Using integration by parts together with the equation $-\Delta P = \omega$ as above, and applying the Schwarz inequality, we get

$$\int v(x)^2 \frac{|\nabla P(x)|^2}{P(x)^2} \, dx + \int v(x)^2 \frac{d\omega(x)}{P(x)} = 2 \int \nabla v(x) \cdot \nabla P(x) \frac{v(x)}{P(x)} \, dx$$

$$\leq 2 \left(\int v(x)^2 \frac{|\nabla P(x)|^2}{P(x)^2} \, dx \right)^{\frac{1}{2}} \left(\int |\nabla v(x)|^2 \, dx \right)^{\frac{1}{2}}$$

for all $v \in C_0^\infty$. The preceding inequality obviously yields both (11.2.20) and (11.2.21). This completes the proof of Proposition 11.2.3. □

Remark 11.2.2. The constants 4 and 1 respectively in (11.2.20) and (11.2.21) are sharp. Indeed, if ω is a point mass at $x = 0$, it follows that $P(x) = c(n) \, |x|^{2-n}$. Hence, (11.2.20) boils down to the classical Hardy inequality

$$\int |u(x)|^2 \frac{dx}{|x|^2} \leq \frac{4}{(n-2)^2} \int |\nabla u(x)|^2 \, dx, \quad u \in C_0^\infty, \tag{11.2.22}$$

with the best constant $4/(n-2)^2$. To show that the constant in (11.2.21) is sharp, it suffices to let $\omega = \nu_e$ for a compact set e of positive capacity, so that $P(x) = 1$ $d\omega$-a.e. and $\nu_e(e) = c_{2,1}(e)$, and then minimize the right-hand side over all $v \geq 1$ on e, where $v \in C_0^\infty$.

We now complete the proof of Lemma 11.2.2. Combining (11.2.19) with (11.2.20) and (11.2.21) (with ν_e in place of ω), we arrive at the estimate

$$\|\nabla v\|_{L_2}^2 \leq \int |\nabla(v \, P^\delta)(x)|^2 \, \frac{dx}{P(x)^{2\delta}} \leq (\delta + 1)(4\delta + 1) \, \|\nabla v\|_{L_2}^2,$$

for all $v \in C_0^\infty$.

To verify this inequality for arbitrary v in w_2^1, let

$$v = \lim_{N \to \infty} v_N$$

both in w_2^1 and dx-a.e. for $v_N \in C_0^\infty$. Now put v_N in place of v in (11.2.20) and let $N \to \infty$. Using Fatou's lemma, we see that (11.2.20) holds for all $v \in w_2^1$. Hence

$$\lim_{N \to \infty} \int |v_N(x) - v(x)|^2 \, \frac{|\nabla P(x)|^2}{P(x)^2} \, dx = 0,$$

and consequently

$$\lim_{N \to \infty} \int |\nabla(v_N \, P^\delta)(x)|^2 \frac{dx}{P^{2\delta}(x)} = \lim_{N \to \infty} \int |\nabla v_N(x) + \delta v_N(x) \frac{\nabla P(x)}{P(x)}|^2 dx$$

$$= \int |\nabla v(x) + \delta v(x) \frac{\nabla P(x)}{P(x)}|^2 dx = \int |\nabla(v\, P^\delta)(x)|^2 \, \frac{dx}{P^{2\delta}(x)}.$$

Thus, the proof of the general case is completed by putting v_N in place of v in (11.2.18), and letting $N \to \infty$. The proof of Lemma 11.2.2 is complete. □

Remark 11.2.3. In what follows only the lower estimate in (11.2.18) will be used, together with the fact that

$$\|P^{-\delta} \nabla (v P^\delta)\|_{L_2} < \infty$$

for every $v \in w_2^1$.

In the next proposition, we extend the equation $< V,\, w > = - < \Gamma,\, \nabla w >$ to the case where $w = u\, v$, where both u and v lie in w_2^1, are locally bounded, and have a certain decay at infinity.

Proposition 11.2.4. *Suppose that $V \in M(w_2^1 \to w_2^{-1})$, and $\mathbf{\Gamma} = \nabla \Delta^{-1} V \in \mathbf{L}_{2,\mathrm{loc}}$ is defined as in Lemma* 11.2.1. *Suppose that $w = u\, v$, where $u,\, v \in w_2^1$, and*

$$|u(x)| \le C\,(1 + |x|^2)^{-\beta/2}, \quad |v(x)| \le C\,(1 + |x|^2)^{-\beta/2}, \quad x \in \mathbb{R}^n, \tag{11.2.23}$$

for some $\beta > (n-2)/2$. Then $\mathbf{\Gamma} \cdot \nabla \bar{w}$ is integrable, and

$$< V,\, w > = - \int \mathbf{\Gamma} \cdot \nabla \bar{w}(x)\, dx. \tag{11.2.24}$$

Proof. Clearly,

$$\int |\mathbf{\Gamma} \cdot \nabla \bar{w}(x)|\, dx \le \Big(\int |\mathbf{\Gamma}(x)|^2 |u(x)|^2 \, dx \Big)^{\frac{1}{2}} \Big(\int_{\mathbb{R}^n} |\nabla v(x)|^2 \, dx \Big)^{\frac{1}{2}}$$

$$+ \Big(\int |\mathbf{\Gamma}(x)|^2 |v(x)|^2 \, dx \Big)^{\frac{1}{2}} \Big(\int |\nabla u(x)|^2 \, dx \Big)^{\frac{1}{2}}.$$

To show that the right-hand side is finite, note that, for every $\epsilon > 0$ and $R \ge 1$,

$$\int_{|x| \le R} |\mathbf{\Gamma}(x)|^2 \, dx \le C\, R^{n-2+\epsilon}, \tag{11.2.25}$$

by Lemma 11.2.1. It is easy to see that the preceding estimate yields

$$\int |\mathbf{\Gamma}(x)|^2 \, (1 + |x|^2)^{-\beta} \, dx < \infty, \tag{11.2.26}$$

for $\beta > (n-2)/2$. Indeed, pick $\epsilon \in (0,\, 2\beta - n + 2)$, and estimate

$$\int |\mathbf{\Gamma}(x)|^2 (1 + |x|^2)^{-\beta} \, dx \le \int_{|x| \le 1} |\mathbf{\Gamma}(x)|^2 \, dx + \int_{|x| > 1} |\mathbf{\Gamma}(x)|^2 |x|^{-2\beta} \, dx$$

$$\le c_1 + c_2 \int_1^\infty \left(\int_{|x|\le r} |\mathbf{\Gamma}(x)|^2 \, dx \right) r^{-2\beta-1} \, dx$$

$$\le c_1 + c_2 \int_1^\infty r^{n-3-2\beta} \, dx < \infty.$$

From this and (11.2.23) it follows that

$$\int |\mathbf{\Gamma}(x)|^2 |u(x)|^2 \, dx < \infty, \quad \int |\mathbf{\Gamma}(x)|^2 |v(x)|^2 \, dx < \infty.$$

Thus $\mathbf{\Gamma} \cdot \nabla \bar{w}$ is integrable.

To prove (11.2.24), we first assume that both u and v lie in $w_2^1 \cap C^\infty$, and satisfy (11.2.23). Let $\eta_N(x)$ be a smooth cutoff function as in the proof of Lemma 11.2.1. Let $u_N = \eta_N \, u$ and $v_N = \eta_N \, v$. Then by (11.2.7),

$$< V, \, u_N v_N >= - \int \mathbf{\Gamma} \cdot \nabla (\bar{u}_N \, \bar{v}_N)(x) \, dx$$

$$= - \int \mathbf{\Gamma} \cdot \nabla \bar{u}_N(x) \, \bar{v}_N(x) \, dx - \int \mathbf{\Gamma} \cdot \nabla \bar{v}_N(x) \, \bar{u}_N(x) \, dx.$$

Note that

$$0 \le \eta_N(x) \le 1 \quad \text{and} \quad |\nabla \eta_N(x)| \le C \, |x|^{-1},$$

which gives

$$|\mathbf{\Gamma} \cdot \nabla \bar{u}_N(x) \, \bar{v}_N(x)| + |\mathbf{\Gamma} \cdot \nabla \bar{u}_N(x) \, \bar{v}_N(x)| \le C \, |\mathbf{\Gamma}(x)| \, \big(|u(x)| \, |v(x)| |x|^{-1}$$

$$+ |\nabla u(x)| |v(x)| + |\nabla v(x)| |u(x)|\big).$$

Since $v \in w_2^1$), it follows from Hardy's inequality (or directly from (11.2.23)) that $|v(x)| |x|^{-1} \in L_2$. Applying (11.2.26) and the Schwarz inequality, we conclude that the right-hand side of the preceding inequality is integrable. Thus (11.2.24) follows from the dominated convergence theorem in this case.

It remains to show that the C^∞ restriction on u and v can be dropped. We set

$$u_r = u \star \phi_r, \quad v_r = v \star \phi_r,$$

where $\phi_r(x) = r^{-n} \, \phi(x/r)$. Here $\phi \in C_0^\infty$ is a C^∞-mollifier supported in $\mathcal{B}_1$ such that $0 \le \phi(x) \le 1$. It is not difficult to verify that u_r and v_r satisfy estimates (11.2.23). Obviously,

$$|u_r(x)| = |u * \phi_r(x)| \le \mathcal{M} u(x),$$

where $\mathcal{M}$ is the Hardy–Littlewood maximal operator. We can suppose without loss of generality that $(n-2)/2 < \beta < n$ in (11.2.23). Notice that, for $0 < \beta < n$,

$$\mathcal{M}(1 + |x|^2)^{-\beta/2} \le C (1 + |x|^2)^{-\beta/2}, \qquad x \in \mathbb{R}^n.$$

Hence,

$$|u_r(x)| \le \mathcal{M}u(x) \le C(1+|x|^2)^{-\beta/2}, \qquad x \in \mathbb{R}^n, \tag{11.2.27}$$

where C does not depend on r, and a similar estimate holds for v.

We also need the estimate

$$|\nabla u_r(x)| = |\nabla u \star \phi_r(x)| \le \mathcal{M}|\nabla u|(x). \tag{11.2.28}$$

As was shown above,

$$< V,\, u_r v_r >= -\int \mathbf{\Gamma} \cdot \nabla \bar{u}_r(x)\, \bar{v}_r(x)\, dx - \int \mathbf{\Gamma} \cdot \nabla \bar{v}_r(x)\, \bar{u}_r(x)\, dx.$$

Moreover, by (11.2.27) and (11.2.28) we have

$$|\mathbf{\Gamma} \cdot \nabla \bar{u}_r(x) \bar{v}_r(x)| + |\mathbf{\Gamma} \cdot \nabla \bar{v}_r(x)\, \bar{u}_r(x)|$$
$$\le C|\mathbf{\Gamma}(x)|(1+|x|^2)^{-\beta/2}(\mathcal{M}|\nabla u|(x) + \mathcal{M}|\nabla v|(x)).$$

Since $u,\, v \in w_2^1$, and $\mathcal{M}$ is a bounded operator on L_2, it follows that $\mathcal{M}|\nabla u|$ and $\mathcal{M}|\nabla v|$ lie in L_2. Applying (11.2.26) again, we see that the right-hand side of the preceding inequality is integrable. Thus, letting $r \to 0$, and using the dominated convergence theorem, we obtain

$$< V,\, w >= \lim_{r\to 0} < V,\, u_r v_r >= -\int \mathbf{\Gamma} \cdot \nabla \bar{w}(x)\, dx,$$

which completes the proof of Proposition 11.2.4. □

Now we continue the proof of (11.2.13). Suppose that

$$V \in M(w_2^1 \to w_2^{-1}),$$

i.e., the inequality

$$| < V\, u,\, v > | \le \|V\|_{M(w_2^1 \to w_2^{-1})}\, \|u\|_{w_2^1}\, \|v\|_{w_2^1}$$

holds, where $u, v \in w_2^1$.

Let $\boldsymbol{\phi} = (\phi_1, \ldots, \phi_n)$ be an arbitrary vector field in $C_0^\infty \otimes \mathbb{C}^n$, and let

$$w = \Delta^{-1} \operatorname{div} \boldsymbol{\phi} = -I_2 \operatorname{div} \boldsymbol{\phi}, \tag{11.2.29}$$

so that

$$\boldsymbol{\phi} = \nabla w + \mathbf{s}, \quad \operatorname{div} \mathbf{s} = 0.$$

Note that $w \in w_2^1 \cap C^\infty$, since

$$w(x) = O(|x|^{1-n}) \quad \text{and} \quad |\nabla w(x)| = O\left(|x|^{-n}\right) \quad \text{as} \quad |x| \to \infty. \tag{11.2.30}$$

Now set

$$u(x) = P(x)^\delta \quad \text{and} \quad v(x) = \frac{w(x)}{P(x)^\delta}, \tag{11.2.31}$$

where $P(x)$ is the equilibrium potential of a compact set $e \subset \mathbb{R}^n$, and $1 < 2\delta < n/(n-2)$.

By (11.2.14) and (11.2.17) we have

$$0 \le P(x) \le 1 \quad \text{for all } x \in \mathbb{R}^n$$

and

$$P(x) \le c\,|x|^{2-n} \qquad \text{for } |x| \text{ large.}$$

Hence

$$|P(x)|^\delta \le C\,(1+|x|^2)^{-\delta(n-2)/2}.$$

Since $\beta = \delta(n-2) > (n-2)/2$, it follows that u satisfies (11.2.23).

To verify that (11.2.23) holds for $v = w\,P^{-\delta}$, note that

$$\inf_K P(x) > 0$$

for every compact set K, and hence by (11.2.17)

$$P(x)^{-\delta} \le C\,(1+|x|^2)^{\delta(n-2)/2}.$$

Combining this estimate with (11.2.30), we conclude that

$$|v(x)| \le C\,(1+|x|^2)^{-\beta/2},$$

where $\beta = -\delta(n-2) + n - 1 > (n-2)/2$.

By Proposition 11.2.2 and Lemma 11.2.2 both u and v lie in w_2^1. Now applying Proposition 11.2.4, we obtain

$$< Vu,\, v > = < V,\, w > = -\int \mathbf{\Gamma} \cdot \nabla \bar{w}(x)\, dx.$$

Hence,

$$\left| \int \mathbf{\Gamma} \cdot \nabla \bar{w}(x)\, dx \right| \le \|V\|_{M(w_2^1 \to w_2^{-1})}\, \|\nabla u\|_{L_2}\, \|\nabla v\|_{L_2}.$$

By Lemma 11.2.2,

$$\|\nabla v\|_{L_2}^2 \le \int |\nabla(vP^\delta)(x)|^2 \frac{dx}{P(x)^{2\delta}} = \int |\nabla w(x)|^2 \frac{dx}{P(x)^{2\delta}} < \infty.$$

Applying this together with Proposition 11.2.2, we obtain the estimate

$$\left| \int \mathbf{\Gamma} \cdot \nabla \bar{w}(x)\, dx \right| \le C(\delta)\, \|V\|_{M(w_2^1 \to w_2^{-1}(\mathbb{R}^n))}\, c_{2,1}(e)^{\frac{1}{2}} \times \left(\int |\nabla w(x)|^2 \frac{dx}{P(x)^{2\delta}} \right)^{\frac{1}{2}}. \tag{11.2.32}$$

To complete the proof of Theorem 11.2.1, we need one more estimate which involves powers of equilibrium potentials.

Proposition 11.2.5. *Let w be defined by* (11.2.29) *with $\boldsymbol{\phi} \in C_0^\infty \otimes \mathbb{C}^n$. Suppose that $1 < 2\delta < n/(n-2)$. Then*

$$\int |\nabla w(x)|^2 \frac{dx}{P(x)^{2\delta}} \le C(n,\delta) \int |\boldsymbol{\phi}(x)|^2 \frac{dx}{P(x)^{2\delta}}. \tag{11.2.33}$$

Proof. Note that ∇w is related to $\boldsymbol{\phi}$ through the Riesz transforms R_j, $j = 1, \dots, n$ ([St2]):

$$\nabla w = \Big\{\sum_{k=1}^{n} R_j\, R_k \phi_k\Big\}, \qquad j = 1, \dots, n.$$

Since R_j are bounded operators on $L_2(\rho)$ with a weight ρ in the Muckenhoupt class $A_2(\mathbb{R}^n)$ ([CF], [St3]), we have

$$\|\nabla w\|_{L_2(\rho)} \le C\, \|\boldsymbol{\phi}\|_{L_2(\rho)},$$

where the constant C depends only on the Muckenhoupt constant of the weight.

Let $\rho(x) = P(x)^{-2\delta}$. It is easily seen that

$$\inf_{x \in K} P(x) > 0$$

for every compact set K, and hence $P(x)^{-2\delta} \in L_{1,\,\mathrm{loc}}(\mathbb{R}^n)$. It was proved in [MV1] that $P(x)^{2\delta}$ is an A_2-weight, provided that $1 < 2\delta < n/(n-2)$. Moreover, its Muckenhoupt constant depends only on n and δ, but not on the compact set e. (See [MV1], p. 95, the proof of Lemma 3.1 in the case $p = 2$.) Clearly, the same is true for $\rho(x) = P(x)^{-2\delta}$. This completes the proof of Proposition 11.2.5. □

Now we are in a position to complete the proof of Theorem 11.2.1. Recall that from (11.2.7) and Proposition 11.2.4 it follows that

$$< V, w >= -\int \boldsymbol{\Gamma} \cdot \nabla \bar{w}(x)\, dx = -\int \boldsymbol{\Gamma} \cdot \overline{\boldsymbol{\phi}}(x)\, dx.$$

Using (11.2.32) and Proposition 11.2.5, we obtain

$$\left|\int \boldsymbol{\Gamma} \cdot \overline{\boldsymbol{\phi}}(x)\, dx\right| \le C(n,\delta)\, \|V\|_{M(w_2^1 \to w_2^{-1})}\, c_{2,1}(e)^{\frac{1}{2}} \left(\int \frac{|\boldsymbol{\phi}(x)|^2}{P(x)^{2\delta}}\, dx\right)^{\frac{1}{2}}$$

for all $\boldsymbol{\phi} \in C_0^\infty \otimes \mathbb{C}^n$, and hence for all $\boldsymbol{\phi} \in \mathbf{L}_{2,\mathrm{loc}}$.

Let us pick $R > 0$ so that $e \subset \mathcal{B}_R$. Letting $\boldsymbol{\phi} = \chi_{\mathcal{B}_R}\, P^{2\delta}\, \boldsymbol{\Gamma}$ in the preceding inequality, we conclude that

$$\left(\int_{\mathcal{B}_R} |\mathbf{\Gamma}(x)|^2 \, P(x)^{2\delta}(x) \, dx\right)^{\frac{1}{2}} \le C(n,\delta) \, \|V\|_{M(w_2^1 \to w_2^{-1})} \, c_{2,1}(e)^{\frac{1}{2}}.$$

Since $P(x) \ge 1$ dx-a.e. on e (actually $P(x) = 1$ on $e \backslash E$, where E is a polar set, i.e., $c_{2,1}(E) = 0$), it follows that

$$\int_e |\mathbf{\Gamma}(x)|^2 \, dx \le C(n,\delta)^2 \, \|V\|^2_{M(w_2^1 \to w_2^{-1})} \, c_{2,1}(e).$$

Thus, (11.2.13) holds for every compact set $e \subset \mathbb{R}^n$, and hence this yields (11.1.6). The proof of Theorem 11.2.1 is complete. □

We now prove an analogue of Theorem 11.2.1 formulated in terms of $(-\Delta)^{-1/2}V$.

Theorem 11.2.2. *Under the assumptions of Theorem* 11.2.1, *it follows that*

$$V \in M(w_2^1 \to w_2^{-1})$$

if and only if

$$(-\Delta)^{-1/2} V \in M(w_2^1 \to L_2).$$

Proof. By Theorem 11.2.1, $\nabla \Delta^{-1} V \in \mathbf{L}_{2,\mathrm{loc}}$ is well defined in terms of distributions. We now have to show that $(-\Delta)^{-1/2}V$ is also well defined.

Since $\nabla \Delta^{-1} V$ lies in $M(w_2^1 \to L_2) \otimes \mathbb{C}^n$, it follows from Corollary 3.2 in [MV1] that the Riesz transforms R_j $(j = 1, \ldots, n)$ are bounded operators on $M(w_2^1 \to L_2)$. Hence

$$(-\Delta)^{-1/2} \nabla = \{R_j\}_{1 \le j \le n}$$

is a bounded operator from $M(w_2^1 \to L_2)$ to $M(w_2^1 \to L_2) \otimes \mathbb{C}^n$. Then $(-\Delta)^{-1/2}V$ can be defined by

$$(-\Delta)^{-1/2} V = (-\Delta)^{-1/2} \nabla \cdot \nabla \Delta^{-1} V$$

as an element of $M(w_2^1 \to L_2)$. The proof of Theorem 11.2.2 is complete. □

Remark 11.2.4. It is worthwhile to observe that the "naïve" approach is to decompose V into its positive and negative parts: $V = V_+ - V_-$, and to apply the criteria in Sect. 1.2 to both V_+ and V_-. However, this procedure drastically diminishes the class of admissible weights V by ignoring a possible cancellation between V_+ and V_-. This cancellation phenomenon is evident for strongly oscillating weights considered below.

Example 11.2.1. Let us set

$$V(x) = |x|^{N-2} \sin(|x|^N), \tag{11.2.34}$$

where N is an integer, $N \geq 3$, which may be arbitrarily large. Obviously, both V_+ and V_- fail to satisfy (11.1.1) due to the growth of the amplitude at infinity. However,

$$V(x) = \operatorname{div} \mathbf{\Gamma}(x) + O\,(|x|^{-2}), \quad \text{where} \quad \mathbf{\Gamma}(x) = \frac{-1}{N} \frac{\mathbf{x}}{|x|^2} \cos\,(|x|^N). \tag{11.2.35}$$

By Hardy's inequality (11.2.22) with $n \geq 3$, the term $O\,(|x|^{-2})$ in (11.2.35) is harmless, whereas $\mathbf{\Gamma}$ clearly satisfies (11.1.6) since $|\mathbf{\Gamma}(x)|^2 \leq |x|^{-2}$. This shows that V is admissible for (11.1.1), while $|V|$ is obviously not. Similar examples of weights with strong *local* singularities can easily be constructed.

11.3 A Compactness Criterion

In this section we give a compactness criterion for $V \in M(w_2^1 \to w_2^{-1})$. Denote by $\overset{\circ}{M}(w_2^1 \to w_2^{-1})$ the class of compact multiplication operators acting from w_2^1 to w_2^{-1}. Obviously,

$$\overset{\circ}{M}(w_2^1 \to w_2^{-1}) \subset M(w_2^1 \to w_2^{-1}),$$

where the latter class was characterized in the preceding section.

Theorem 11.3.1. *Let $V \in (C_0^\infty)'$ and $n \geq 3$. Then*

$$V \in \overset{\circ}{M}(w_2^1 \to w_2^{-1})$$

if and only if

$$V = \operatorname{div} \mathbf{\Gamma}, \tag{11.3.1}$$

where $\mathbf{\Gamma} = (\Gamma_1, \ldots, \Gamma_n)$ is a vector field such that

$$\Gamma_i \in \overset{\circ}{M}(w_2^1 \to L_2), \quad i = 1, \ldots, n.$$

Moreover, $\mathbf{\Gamma}$ can be represented in the form $\nabla \Delta^{-1} V$, as in Theorem 11.2.1.

Proof. Let V be given by (11.3.1), and let u belong to the unit ball $\mathcal{B}$ in w_2^1. Then

$$V\,u = \operatorname{div}\,(u\,\mathbf{\Gamma}) - \mathbf{\Gamma} \cdot \nabla u. \tag{11.3.2}$$

The set

$$\{\operatorname{div}\,(u\,\mathbf{\Gamma}) \,:\, u \in \mathcal{B}\}$$

is compact in w_2^{-1} because the set

$$\{u\,\mathbf{\Gamma} \,:\, u \in \mathcal{B}\}$$

is compact in $\mathbf{w}_2^{-1}$. The set $\{\mathbf{\Gamma} \cdot \nabla u : u \in \mathcal{B}\}$ is also compact in w_2^{-1} since the set $\{|\nabla u| : u \in \mathcal{B}\}$ is bounded in L_2, and the multiplier operators $\bar{\Gamma}_i$, being adjoint to Γ_i $(i = 1, \dots, n)$, are compact from L_2 to w_2^{-1}. This completes the proof of sufficiency of (11.3.1).

We now prove the necessity. Pick $F \in C^\infty(\mathbb{R}_+)$, where $F(t) = 1$ for $t \le 1$ and $F(t) = 0$ for $t \ge 2$. For $x_0 \in \mathbb{R}^n$, $\delta > 0$, and $R > 0$, define the cutoff functions

$$\varkappa_{\delta,x_0}(x) = F(\delta^{-1}|x - x_0|), \quad \text{and} \quad \xi_R(x) = 1 - F(R^{-1}|x|).$$

Lemma 11.3.1. *If $f \in w_2^{-1}$, then*

$$\lim_{\delta \to 0} \sup_{x_0 \in \mathbb{R}^n} \|\varkappa_{\delta,x_0} f\|_{w_2^{-1}} = 0, \tag{11.3.3}$$

and

$$\lim_{R \to \infty} \|\xi_R f\|_{w_2^{-1}} = 0. \tag{11.3.4}$$

Proof. Let us prove (11.3.3). The distribution f has the form $f = \operatorname{div} \boldsymbol{\phi}$, where $\boldsymbol{\phi} = (\phi_1, \dots, \phi_n) \in \mathbf{L}_2$. Hence,

$$\varkappa_{\delta,x_0} f = \operatorname{div}(\varkappa_{\delta,x_0} \boldsymbol{\phi}) - \boldsymbol{\phi} \nabla \varkappa_{\delta,x_0}.$$

Clearly,

$$\begin{aligned} \|\varkappa_{\delta,x_0} f\|_{w_2^{-1}} &\le \|\varkappa_{\delta,x_0} |\boldsymbol{\phi}| \,\|_{L_2} + c\,\delta \,\|\nabla \varkappa_{\delta,x_0} \cdot \boldsymbol{\phi}\|_{L_2} \\ &\le c \|\, |\boldsymbol{\phi}|; \mathcal{B}_{2\delta}(x_0)\|_{L_2}. \end{aligned}$$

This proves (11.3.3). Since (11.3.4) is derived in a similar way, the proof of Lemma 11.3.1 is complete. □

Lemma 11.3.2. *If $V \in \overset{\circ}{M}(w_2^1 \to w_2^{-1})$, then*

$$\lim_{\delta \to 0} \sup_{x_0 \in \mathbb{R}^n} \|\varkappa_{\delta,x_0} V\|_{\overset{\circ}{M}(w_2^1 \to w_2^{-1})} = 0, \tag{11.3.5}$$

and

$$\lim_{R \to \infty} \|\xi_R V\|_{\overset{\circ}{M}(w_2^1 \to w_2^{-1})} = 0. \tag{11.3.6}$$

Proof. Fix $\epsilon > 0$, and pick a finite number of $f_k \in w_2^{-1}$ such that

$$\|V u - f_k\|_{w_2^{-1}} < \epsilon$$

for $k = 1, \dots, N(\epsilon)$, and for all $u \in \mathcal{B}$, where $\mathcal{B}$ is the unit ball in w_2^1. Note that by Hardy's inequality

$$\sup_{x_0 \in \mathbb{R}^n,\, \delta > 0} \|\varkappa_{\delta,x_0}\|_{M(w_2^1 \to w_2^{-1})} \le c < \infty.$$

Next,

$$\|\varkappa_{\delta,x_0} V u\|_{w_2^{-1}} \le \|\varkappa_{\delta,x_0} (V u - f_k)\|_{w_2^{-1}} + \|\varkappa_{\delta,x_0} f_k\|_{w_2^{-1}}$$
$$\le c\,\epsilon + \|\varkappa_{\delta,x_0} f_k\|_{w_2^{-1}}.$$

Hence,

$$\|\varkappa_{\delta,x_0}\|_{M(w_2^1 \to w_2^{-1})} \le c\,\epsilon + \|\varkappa_{\delta,x_0} f_k\|_{w_2^{-1}}.$$

By Lemma 11.3.1, this gives (11.3.5), and the proof of (11.3.6) is quite similar. The proof of Lemma 11.3.2 is complete. □

We can now complete the proof of the necessity part of Theorem 11.3.1. Suppose that

$$V \in \overset{\circ}{M}\,(w_2^1 \to w_2^{-1}).$$

By Theorem 11.2.1,

$$\|\nabla\,\Delta^{-1}(\xi_R\,V)\|_{M(w_2^1 \to L_2)} \le c\,\|\xi_R\,V\|_{M(w_2^1 \to L_2)}.$$

By the preceding estimate and (11.3.6),

$$\lim_{R\to\infty} \|\nabla\,\Delta^{-1}(\xi_R\,V)\|_{M(w_2^1 \to L_2)} = 0.$$

Hence we can assume without loss of generality that V is compactly supported, e.g., $\operatorname{supp} V \subset \mathcal{B}_1$. To show that

$$\mathbf{\Gamma} = \nabla\,\Delta^{-1} V \in \overset{\circ}{M}\,(w_2^1 \to L_2),$$

consider a covering of the closed unit ball $\overline{\mathcal{B}_1}$ by open balls $\mathcal{B}_k$ $(k = 1, \ldots, n)$ of radius $\sqrt{n}\,\delta$ centered at the nodes x_k of the lattice with mesh size δ. We introduce a partition of unity ϕ_k subordinate to this covering and satisfying the estimate $|\nabla\,\phi_k| \le c\,\delta^{-1}$, so that $\operatorname{supp}\phi_k \subset \mathcal{B}_k^*$, where $\mathcal{B}_k^*$ is the ball of radius $2\sqrt{n}\,\delta$ concentric to $\mathcal{B}_k$. Also, pick $\psi_k \in C_0^\infty(\mathcal{B}_k^*)$, where

$$\phi_k\,\psi_k = \phi_k, \quad \text{and} \quad |\nabla\,\psi_k| \le c\,\delta^{-1}.$$

We have

$$\nabla\,\Delta V = \sum_{k=1}^{N(\delta)} \nabla\,\Delta(\phi_k\,V) = \sum_{k=1}^{N(\delta)} \nabla\,\Delta(\phi_k\,\psi_k\,V)$$
$$= \sum_{k=1}^{N(\delta)} \psi_k\,\nabla\,\Delta(\phi_k\,V) + \sum_{k=1}^{N(\delta)} [\,\nabla\,\Delta,\,\psi_k\,]\,\phi_k\,V,$$

where $[\,A,\,B\,] = A\,B - B\,A$ is the commutator of the operators A and B. Since the multiplicity of the covering

$$\bigcup_{k=1}^{N(\delta)} \mathcal{B}_k$$

depends only on n, it follows that

$$\Big\| \sum_{k=1}^{N(\delta)} \psi_k \, \nabla \, \Delta(\phi_k \, V) \Big\|_{M(w_2^1 \to L_2)} \leq c(n) \sup_{1 \leq k \leq N(\delta)} \| \nabla \, \Delta(\phi_k \, V) \|_{\mathring{M}\,(w_2^1 \to L_2)}.$$

The last supremum is bounded by

$$c \, \|\phi_k \, V\|_{\mathring{M}\,(w_2^1 \to w_2^{-1})},$$

which is made smaller than any $\epsilon > 0$ by choosing $\delta = \delta(\epsilon)$ small enough.

It remains to check that each function

$$\Phi_k := [\, \nabla \, \Delta, \, \psi_k \,] \, \phi_k \, V$$

is a compact multiplier from w_2^1 to L_2, $k = 1, \ldots, n$. Indeed, the kernel of the operator $V \to [\, \nabla \, \Delta, \, \psi_k \,] \, \phi_k \, V$ is smooth, and hence

$$|\Phi_k(x)| = |(\, [\, \nabla \, \Delta, \, \psi_k \,] \, \phi_k \, V)(x)| \leq \, c_k \, (1 + |x|)^{1-n} \, \|\phi_k \, V\|_{w_2^{-1}}$$

$$\leq c_k \, (1 + |x|)^{1-n} \, \|V\|_{\mathring{M}\,(w_2^1 \to w_2^{-1})} \, \|\phi_k\|_{w_2^1} \leq C_k \, (1 + |x|)^{1-n},$$

where the constant C_k does not depend on x. Since $n > 2$, this means that the multiplier operator $\Phi_k \, : \, w_2^1 \, \to \, L_2$ is compact. The proof of Theorem 11.3.1 is complete. □

Remark 11.3.1. The compactness of the multipliers $\Gamma_i \, : \, w_2^1 \, \to \, L_2$, where $i = 1, \ldots, n$, is obviously equivalent to the compactness of the embedding

$$w_2^1 \, \subset \, L_2(|\mathbf{\Gamma}|^2). \tag{11.3.7}$$

An analytic characterization of this property is equivalent to the inequalities

$$\lim_{\delta \to 0} \sup_{\{e : \text{diam}(e) \leq \delta\}} \frac{\displaystyle\int_e |\mathbf{\Gamma}|^2 dx}{c_{2,1}(e)} = 0,$$

$$\lim_{\rho \to \infty} \sup_{e \subset \mathbb{R}^n \setminus \mathcal{B}_\rho} \frac{\displaystyle\int_e |\mathbf{\Gamma}|^2 dx}{c_{2,1}(e)} = 0$$

(see [Maz2] and [Maz15], § 2.5).

11.4 Characterization of $M(W_2^1 \to W_2^{-1})$

In this section, we characterize the class of multipliers $V : W_2^1 \to W_2^{-1}$ for $n \geq 1$. Here $W_2^{-1} = (W_2^1)'$, the dual of W_2^1. Let J_α $(0 < \alpha < +\infty)$ denote the Bessel potential of order α. Every $u \in W_2^1$ can be represented in the form $u = J_1 g$, where

$$c_1 \|g\|_{L_2} \leq \|u\|_{W_2^1} \leq c_2 \|g\|_{L_2}.$$

(See [St2].)

Let $\mathcal{S}$ denote the Schwarz space of infinitely differentiable functions on $\mathbb{R}^n$ and $\mathcal{S}'$ its dual. We say that $V \in \mathcal{S}'$ is a multiplier from W_2^1 to W_2^{-1} if the sesquilinear form defined by

$$< V u, \, v > := < V, \bar{u} v >$$

is bounded on $W_2^1 \times W_2^1$:

$$| < V u, \, v > | \leq c \, \|u\|_{W_2^1} \, \|v\|_{W_2^1}, \quad u, v \in \mathcal{S}, \tag{11.4.1}$$

where the constant c is independent of u and v in Schwartz space $\mathcal{S}$. As in the case of homogeneous spaces, the preceding inequality is equivalent to the boundedness of the corresponding quadratic form; i.e., it suffices to verify (11.4.1) for $u = v$.

If (11.4.1) holds, then V defines a bounded multiplier operator from W_2^1 to W_2^{-1}. (Originally, it is defined on $\mathcal{S}$, but by continuity is extended to W_2^1.) The corresponding class of multipliers is denoted by $M(W_2^1 \to W_2^{-1})$.

Since $I - \Delta : W_2^1 \to W_2^{-1}$ is a bounded operator, it follows that

$$V \in M(W_2^1 \to W_2^{-1})$$

if and only if the operator

$$(I - \Delta) + V : W_2^1 \to W_2^{-1}$$

is bounded.

If V is a locally finite complex-valued measure on $\mathbb{R}^n$, then (11.4.1) can be rewritten in the form

$$\left| \int u(x) \, \overline{v(x)} \, dV(x) \right| \leq c \, \|u\|_{W_2^1} \, \|v\|_{W_2^1}, \tag{11.4.2}$$

where $u, \, v \in \mathcal{S}$.

We now characterize (11.4.2) in the general case of distributions V

Theorem 11.4.1. *Let $V \in \mathcal{S}'$. Then $V \in M(W_2^1 \to W_2^{-1})$ if and only if there exist a vector field $\mathbf{\Gamma} = \{\Gamma_1, \ldots, \Gamma_n\} \in \mathbf{L}_{2,\text{loc}}$ and $\Gamma_0 \in L_{2.\text{loc}}$ such that*

$$V = \operatorname{div} \mathbf{\Gamma} + \Gamma_0, \tag{11.4.3}$$

and

$$\int |u(x)|^2 \, |\Gamma_i(x)|^2 \, dx \le C \, \|u\|^2_{W_2^1}, \qquad i = 0, 1, \ldots, n, \tag{11.4.4}$$

where C does not depend on $u \in \mathcal{S}$.

In (11.4.3), *one can set*

$$\mathbf{\Gamma} = -\nabla \, (I - \Delta)^{-1} V, \quad \text{and} \quad \Gamma_0 = (I - \Delta)^{-1} V. \tag{11.4.5}$$

Proof. Suppose that V is represented in the form (11.4.3), and (11.4.4) holds. Then using integration by parts and the Schwarz inequality, we have

$$\begin{aligned} | < V, \, \bar{u} \, v > | &= | < \mathbf{\Gamma}, \, v \, \nabla \bar{u} > + < \mathbf{\Gamma}, \, \bar{u} \, \nabla v > + < \Gamma_0, \, \bar{u} \, v > | \\ &\le \|\mathbf{\Gamma} v\|_{\mathbf{L}_2} \, \|\nabla u\|_{L_2} + \|\mathbf{\Gamma} u\|_{\mathbf{L}_2} \, \|\nabla v\|_{L_2} + \|\Gamma_0 u\|_{L_2} \, \|v\|_{L_2} \\ &\le 3\sqrt{C} \, \|u\|_{W_2^1} \, \|v\|_{W_2^1}, \end{aligned}$$

where C is the constant in (11.4.4). This proves the "if" part of Theorem 11.4.1.

To prove the "only if" part, define $\mathbf{\Gamma} = \{\Gamma_1, \ldots, \Gamma_n\}$ and Γ_0 by (11.4.5). Then, for every $j = 0, 1, \ldots, n$, we have $\Gamma_j \in L_{2,\text{loc}}$, and the following crude estimate holds:

$$\int_{\mathcal{B}_R(x_0)} |\Gamma_j(x)|^2 \, dx \le C(n, \epsilon) \, R^{n-2+\epsilon} \, \|V\|^2_{M(W_2^1 \to W_2^{-1})}, \tag{11.4.6}$$

where $R \ge \max\{1, |x_0|\}$. The proof is based on the same argument as the proof of Lemma 11.2.1 in the homogeneous case.

Now fix a compact set $e \subset \mathbb{R}^n$ such that $\text{diam}\,(e) \le 1$, and $C_{2,1}(e) > 0$. Denote by $P(x) = P_e(x)$ the equilibrium potential of e which corresponds to the capacity $C_{2,1}$. Letting

$$u(x) = P(x)^\delta \quad \text{and} \quad v(x) = \frac{w(x)}{P(x)^\delta},$$

where $1 < 2\delta < n/(n-2)$ and $w \in \mathcal{S}$, we have

$$| < V, \, w > | \le \|V\|_{M(W_2^1 \to W_2^{-1})} \, \|P^\delta\|_{W_2^1} \, \|\nabla v\|_{W_2^1}.$$

Calculations analogous to those of Propositions 11.2.2 - 11.2.5 yield

$$\|P^\delta\|_{W_2^1} \le C(n, \delta) \, C_{2,1}(e)^{\frac{1}{2}},$$

and

$$\|\nabla v\|_{W_2^1} \le C(n, \delta) \left[\int (|w(x)|^2 + |\nabla w(x)|^2) \frac{dx}{P(x)^{2\delta}} \right]^{\frac{1}{2}}.$$

Combining the preceding inequalities, we obtain

$$| < V, w > | \le C(n,\delta) \, \|V\|_{M(W_2^1 \to W_2^{-1})} \, C_{2,1}(e)^{\frac{1}{2}}$$

$$\times \left[\int (|w(x)|^2 + |\nabla w(x)|^2) \frac{dx}{P(x)^{2\delta}} \right]^{\frac{1}{2}}.$$

Set $w = (1 - \Delta)^{-1} \operatorname{div} \boldsymbol{\phi}$, where $\boldsymbol{\phi}$ is an arbitrary vector-field with components in $\mathcal{S}$. Then the preceding estimate can be restated in the form

$$| < \boldsymbol{\Gamma}, \boldsymbol{\phi} > | \le C(n,\delta) C_{2,1}(e)^{\frac{1}{2}} \left[\int (|w(x)|^2 + |\nabla w(x)|^2) \frac{dx}{P(x)^{2\delta}} \right]^{\frac{1}{2}}. \tag{11.4.7}$$

Unlike in the homogeneous case, $P(x)^{-2\delta}$ is not a Muckenhoupt weight for Bessel potentials. To proceed, we need a localized version of the estimates used in Sect. 11.3.

Lemma 11.4.1. *Let $P(x) = P_e(x)$ be the equilibrium potential of a compact set e of positive capacity $C_{2,1}$ and such that $e \subset \mathcal{B}$, where $\mathcal{B} = \mathcal{B}_1(x_0)$ is the ball of radius 1 centered at $x_0 \in \mathbb{R}^n$. Let $w = (I - \Delta)^{-1} \nabla \psi$, where $\psi \in C^\infty$ and $\operatorname{supp} \psi \subset \mathcal{B}$. Suppose that $1 < 2\delta < n/(n-2)$. Then*

$$\int (|w(x)|^2 + |\nabla w(x)|^2) \frac{dx}{P(x)^{2\delta}} \le C(n,\delta) \int |\psi(x)|^2 \frac{dx}{P(x)^{2\delta}}. \tag{11.4.8}$$

Proof. Let $\nu = \nu_e$ be the equilibrium measure of the compact set e in the sense of the capacitiy $C_{2,1}$, so that $P(x) = J_2\nu(x)$ (see [AH]). Suppose first that $n \ge 3$. Since both $\operatorname{supp} \nu$ and $\operatorname{supp} \psi$ are contained in $\mathcal{B}$, it follows that

$$P(x) = J_2\nu(x) \sim I_2\nu(x) = c(n) \int_{\mathcal{B}} \frac{d\nu(y)}{|x-y|^{n-2}}, \qquad x \in 2\mathcal{B}, \tag{11.4.9}$$

where $2\mathcal{B}$ is a concentric ball of radius 2.

We set $\rho(x) = I_2\nu(x)^{-2\delta}$. Then $\rho(x) \sim P(x)^{-2\delta}$ on $2\mathcal{B}$, and $\rho(x)$ is an A_2-weight (see the proof of Proposition 11.2.4). Note that

$$\nabla w = \nabla^2 (I - \Delta)^{-1} \psi,$$

where

$$\nabla^2 (I - \Delta)^{-1} = \{-R_j \, R_k \, \Delta (I - \Delta)^{-1}\}, \quad j, \, k = 1, \ldots, n.$$

Here R_j, $j = 1, \ldots, n$, are the Riesz transforms which are bounded operators on $L_2(\rho)$ (see [St3]).

Since

$$\Delta (I - \Delta)^{-1} = I - (I - \Delta)^{-1},$$

we have to show that $J_2 = (I - \Delta)^{-1}$ is a bounded operator on $L_2(\rho)$, and its norm is bounded by a constant which depends only on the Muckenhoupt constant of ρ. It is not difficult to see that the same is true for more general operators $J_\alpha = (I - \Delta)^{-\frac{\alpha}{2}}$, where $\alpha > 0$.

Indeed, denote by $G_\alpha(x)$ the kernel of the Bessel potential J_α. Then clearly,

$$|J_\alpha f(x)| = |G_\alpha \star f(x)| \leq c(n, \alpha)\, \mathcal{M} f(x) \sum_{k=-\infty}^{\infty} 2^{kn} \max_{2^k \leq |t| \leq 2^{k+1}} G_\alpha(t),$$

where $\mathcal{M} f(x)$ is the Hardy–Littlewood maximal function. Standard estimates of Bessel kernels $G_\alpha(x)$ (see Sect. 3.2.5), show that

$$\sum_{k=-\infty}^{\infty} 2^{kn} \max_{2^k \leq |t| \leq 2^{k+1}} G_\alpha(t) < \infty,$$

for every $\alpha > 0$. Since $\mathcal{M}$ is bounded on $L_2(\rho)$ (see [St3]), it follows that

$$\|J_\alpha f\|_{L_2(\rho)} \leq C\, \|f\|_{L_2(\rho)}, \tag{11.4.10}$$

where C depends only on n, α, and the Muckenhoupt constant of ρ.

Applying (11.4.10) with $\alpha = 2$, we get

$$\int_{2\mathcal{B}} |\nabla w(x)|^2 \frac{dx}{P(x)^{2\delta}} \leq C(n, \delta) \int |\psi(x)|^2\, \rho(x)\, dx$$
$$\leq C(n, \delta) \int |\psi(x)|^2\, \frac{dx}{P(x)^{2\delta}}.$$

Similarly,

$$|w(x)| = |\nabla\, (I - \Delta)^{-1} \psi(x)| \leq C\, J_1 |\psi|(x)$$

and, by (11.4.10) with $\alpha = 1$,

$$\int_{2\mathcal{B}} |w(x)|^2 \frac{dx}{P(x)^{2\delta}} \leq C \int_{2\mathcal{B}} (J_1 |\psi|(x))^2\, \rho(x)\, dx$$
$$\leq C(n, \delta) \int |\psi(x)|^2\, \rho(x)\, dx \leq C(n, \delta) \int |\psi(x)|^2\, \frac{dx}{P(x)^{2\delta}}.$$

Now suppose that $x \in (2\mathcal{B})^c$. Then, by standard estimates of the Bessel kernel as $|x| \to \infty$ (see Sect. 3.2.5),

$$|\nabla w(x)| = |\nabla^2 J_2 \psi(x)| \leq C(n)\, |x|^{\frac{1-n}{2}}\, e^{-|x|} \int_{\mathcal{B}} |\psi(y)|\, dy$$

and

$$|w(x)| \leq C(n) |\nabla J_2 \psi(x)| \leq C\, |x|^{\frac{-n}{2}}\, e^{-|x|} \int_{\mathcal{B}} |\psi(y)|\, dy.$$

Also, for $x \in (2\mathcal{B})^c$,

$$P(x) = J_2 \nu(x) \sim |x|^{\frac{1-n}{2}}\, e^{-|x|}\, \nu(e), \quad |x| \to \infty,$$

where $\nu(e) = C_{2,1}(e) > 0$.

Now pick δ so that $1 < 2\delta < \min\{2, n/(n-2)\}$. Using the above estimates of $w(x)$, $\nabla w(x)$, and $P(x)$, and the inequality $2\delta < 2$, we get

$$\int_{(2\mathcal{B})^c} (|w(x)|^2 + |\nabla w(x)|^2) \frac{dx}{P(x)^{2\delta}} \leq C(n,\delta)\, \nu(e)^{-2\delta} \Big(\int_{\mathcal{B}} |\psi(y)|\, dy \Big)^2 .$$

By the Schwarz inequality,

$$\Big(\int_{\mathcal{B}} |\psi(y)|\, dy \Big)^2 \leq \int_{\mathcal{B}} |\psi(y)|^2 \frac{dy}{P(y)^{2\delta}} \int_{\mathcal{B}} P(x)^{2\delta}\, dx.$$

Applying Minkowski's inequality and the fact that $2\delta < n/(n-2)$, we obtain

$$\int_{\mathcal{B}} P(x)^{2\delta}\, dx \leq \int_{\mathcal{B}} (I_2\nu)^{2\delta}\, dx \leq C(n,\delta)\, \nu(e)^{2\delta}.$$

Thus,

$$\int_{(2\mathcal{B})^c} (|w(x)|^2 + |\nabla w(x)|^2) \frac{dx}{P(x)^{2\delta}} \leq C(n,\delta) \int |\psi(x)|^2 \frac{dx}{P(x)^{2\delta}}.$$

This completes the proof of (11.4.8) for $n \geq 3$. The cases $n = 1$ and $n = 2$ are treated in a similar way with obvious modifications. The proof of Lemma 11.4.1 is complete. □

Let

$$w = (I - \Delta)^{-1} \operatorname{div} \boldsymbol{\phi}, \quad \text{where} \quad \boldsymbol{\phi} = \{\phi_k\} \in \mathcal{S}.$$

Applying Lemma 11.4.1 with $\psi = \phi_k$, $k = 1, \dots, n$, we obtain

$$\int (|w(x)|^2 + |\nabla w(x)|^2) \frac{dx}{P(x)^{2\delta}} \leq C(n,\delta) \int |\boldsymbol{\phi}(x)|^2 \frac{dx}{P(x)^{2\delta}}.$$

This and (11.4.7) yield

$$| < \boldsymbol{\Gamma}, \boldsymbol{\phi} > | \leq C(n,\delta)\, C_{2,1}(e)^{\frac{1}{2}} \Big[\int |\boldsymbol{\phi}(x)|^2 \frac{dx}{P(x)^{2\delta}} \Big]^{\frac{1}{2}} .$$

By duality, the preceding inequality is equivalent to

$$\int |\boldsymbol{\Gamma}(x)|^2\, P(x)^{2\delta}\, dx \leq C(n,\delta)\, \|V\|^2_{M(W_2^1 \to W_2^{-1})}\, C_{2,1}(e).$$

Since $P(x) \geq 1$ a.e. on e, we obtain the desired estimate

$$\int_e |\boldsymbol{\Gamma}(x)|^2\, dx \leq C(n,\delta)\, \|V\|^2_{M(W_2^1 \to W_2^{-1})}\, C_{2,1}(e).$$

The corresponding inequality with Γ_0 in place of $\boldsymbol{\Gamma}$ is verified in a similar way. By (2.3.3) with $p = 2$ these inequalities are equivalent to (11.4.4). The proof of Theorem 11.4.1 is complete. □

Remark 11.4.1. It is easy to see that in the sufficiency part of Theorem 11.4.1 the restriction on the "lower order" term Γ_0 in (11.4.4) can be relaxed. It is enough to assume that $\Gamma_0 \in L_{1,\mathrm{loc}}$ is such that

$$\int |u(x)|^2 \, |\Gamma_0(x)| \, dx \le C \, \|u\|^2_{W_2^1}. \tag{11.4.11}$$

Finally, we state a compactness criterion in the case of the space W_2^1 analogous to that of Theorem 11.3.1.

Theorem 11.4.2. *Let $V \in S'(\mathbb{R}^n)$, $n \ge 1$. Then*

$$V \in \mathring{M}(W_2^1 \to W_2^{-1})$$

if and only if

$$V = \operatorname{div} \mathbf{\Gamma} + \Gamma_0,$$

where $\mathbf{\Gamma} = (\Gamma_1, \ldots, \Gamma_n)$, and

$$\Gamma_i \in \mathring{M}(W_2^1 \to L_2), \quad i = 0, \ldots, n.$$

Moreover, one can set

$$\mathbf{\Gamma} = -\nabla (I - \Delta)^{-1} V \quad \text{and} \quad \Gamma_0 = (I - \Delta)^{-1} V,$$

as in Theorem 11.4.1.

The proof of Theorem 11.4.2 requires only minor modifications outlined in the proof of Theorem 11.4.1, and is omitted here.

11.5 Characterization of the Space $M(\mathring{w}_2^1(\Omega) \to w_2^{-1}(\Omega))$

Using dilation and the description of the space $M(W_2^1 \to W_2^{-1})$ given in the preceding section, we arrive at the following auxiliary statement.

Corollary 11.5.1. *Let $V \in M(W_2^1 \to W_2^{-1})$. Suppose that there exists a number $d > 0$ such that*

$$|< V, |u|^2 >| \le c \, (\|\nabla u\|^2_{L_2} + d^{-2} \, \|u\|^2_{L_2}), \tag{11.5.1}$$

where c does not depend on $u \in C_0^\infty$. Then V can be represented as

$$V = \operatorname{div} \mathbf{\Gamma} + d^{-1} \, \Gamma_0, \tag{11.5.2}$$

where Γ_0 and $\mathbf{\Gamma} = (\Gamma_1, \ldots, \Gamma_n)$ are in $M(W_2^1 \to L_2)$, and

$$\int |\Gamma_i u(x)|^2 \, dx \le C \, (\|\nabla u\|^2_{L_2} + d^{-2} \, \|u\|^2_{L_2}), \tag{11.5.3}$$

for all $i = 0, 1, \ldots, n$.

Now let Ω be an open set in $\mathbb{R}^n$ such that, for all $u \in C_0^\infty(\Omega)$, Hardy's inequality holds:

$$\int_\Omega |u(x)|^2 \frac{dx}{d_{\partial\Omega}(x)^2} \le \text{const} \int_\Omega |\nabla u(x)|^2 \, dx. \tag{11.5.4}$$

Here $d_{\partial\Omega}(x) = \text{dist}\,(x, \partial\Omega)$. It is well-known that (11.5.4) holds for a wide class of domains including those with Lipschitz boundaries. (See [Dav], [Lew], [MMP] for a discussion of Hardy's inequality and related questions, including best constants, on domains Ω in $\mathbb{R}^n$.)

Let Q_j be the cubes with side-length d_j forming Whitney's covering of Ω (see [St2], Sec. 5.1). Denote by Q_j^* the open cube obtained from Q by dilation with coefficient $\frac{9}{8} d_j$. The cubes Q_j^* form an open covering of Ω of finite multiplicity which depends only on n. By $\{\eta_j\}$ ($\eta_j \in C_0^\infty(Q_j^*)$) we denote a smooth partition of unity subordinate to the covering $\{Q_j\}$ and such that $|\nabla \eta_j(x)| \le c\, d_j^{-1}$. In the proof of the following theorem we also need the functions $\zeta_j \in C_0^\infty(Q_j^*)$ such that

$$\zeta_j(x)\, \eta_j(x) = \eta_j(x), \quad \text{and} \quad |\nabla \zeta_j(x)| \le c\, d_j^{-1}. \tag{11.5.5}$$

In this section we deal with the space $\mathring{w}_2^1(\Omega)$ defined as the completion of $C_0^\infty(\Omega)$ in the norm $\|\nabla u; \Omega\|_{L_2}$. By $w_2^{-1}(\Omega)$ the dual space $(\mathring{w}_2^1(\Omega))'$ is denoted.

The next result is a characterization of the space $M(\mathring{w}_2^1(\Omega) \to w_2^{-1}(\Omega))$.

Theorem 11.5.1. *(i) Let* $d_{\partial\Omega}\,(x) = \text{dist}\,(x, \partial\Omega)$*, and let*

$$V = \text{div}\,\mathbf{\Gamma} + d_{\partial\Omega}^{-1}\, \Gamma_0,$$

where $\mathbf{\Gamma} = \{\Gamma_1, \ldots, \Gamma_n\}$ *and*

$$\Gamma_i \in M(\mathring{w}_2^1(\Omega) \to L_2(\Omega)), \quad i = 0, 1, \ldots, n.$$

Suppose that (11.5.4) *holds. Then* $V \in M(\mathring{w}_2^1(\Omega) \to w_2^{-1}(\Omega))$*, and*

$$\|V; \Omega\|_{M(\mathring{w}_2^1 \to w_2^{-1})} \le c \sum_{0 \le i \le n} \|\Gamma_i; \Omega\|_{M(\mathring{w}_2^1 \to L_2)}. \tag{11.5.6}$$

(ii) Conversely, if

$$V \in M(\mathring{w}_2^1(\Omega) \to w_2^{-1}(\Omega)),$$

then there exist $\mathbf{\Gamma} = (\Gamma_1, \ldots, \Gamma_n)$ *and* Γ_0 *such that*

$$\Gamma_i \in M(\mathring{w}_2^1(\Omega) \to L_2(\Omega)), \quad i = 0, 1, \ldots, n,$$

and

$$V = \text{div}\,\mathbf{\Gamma} + d_{\partial\Omega}^{-1}\, \Gamma_0.$$

Moreover,

$$\sum_{0 \le i \le n} \|\Gamma_i; \Omega\|_{M(\mathring{w}_2^1 \to L_2)} \le C\, \|V; \Omega\|_{M(\mathring{w}_2^1 \to w_2^{-1})}. \tag{11.5.7}$$

Proof. The proof of statement (i) is straightforward (see, e.g., the proof of Theorem 11.4.1 above). To prove (ii), note that, for all $u, v \in C_0^\infty(\Omega)$, and the functions ζ_j with the properties (11.5.5), we have

$$| < V\eta_j,\, u\, v > | = | < V\eta_j,\, \zeta_j u\, \zeta_j\, v > |$$

$$\leq \|V\eta_j; \Omega\|_{M(\mathring{w}_2^1 \to w_2^{-1})} (\|\nabla u\|_{L_2} + d_j^{-1}\, \|u\|_{L_2})\, (\|\nabla v\|_{L_2} + d_j^{-1}\, \|v\|_{L_2}).$$

Hence by Corollary 11.5.1,

$$V\eta_j = \operatorname{div} \mathbf{\Gamma}^{(j)} + d_j^{-1}\, \Gamma_0^{(j)}, \tag{11.5.8}$$

where $\mathbf{\Gamma}^{(j)}$ and $\Gamma_0^{(j)}$ satisfy the inequality

$$\int |\Gamma_i^{(j)} u(x)|^2\, dx$$

$$\leq C\, \|V\eta_j; \Omega\|^2_{M(\mathring{w}_2^1 \to w_2^{-1})} (\|\nabla u\|^2_{L_2} + d_j^{-2}\, \|u\|^2_{L_2}), \tag{11.5.9}$$

for all $i = 0, 1, \dots, n$. Multiplying (11.5.8) by ζ_j, we obtain

$$V\eta_j = \operatorname{div}\, (\zeta_j\, \mathbf{\Gamma}^{(j)}) + d_j^{-1}\, \Gamma_0^{(j)} - \mathbf{\Gamma}^{(j)}\, \nabla \zeta_j.$$

We set

$$\mathbf{\Gamma} = \sum_j \zeta_j\, \mathbf{\Gamma}^{(j)} \quad \text{and} \quad \Gamma_0 = \sum_j (d_j\, \Gamma_0^{(j)} - \mathbf{\Gamma}^{(j)}\, \nabla \zeta_j).$$

If $u \in C_0^\infty(\Omega)$, then

$$\int_\Omega |(|\mathbf{\Gamma}| + |\Gamma_0|)\, u|^2\, dx$$

$$\leq c \sum_j \left(\int_\Omega |\mathbf{\Gamma}^{(j)} \zeta_j\, u|^2\, dx + d_j^{-2} \int_\Omega |(d_j\, \Gamma_0^{(j)} \zeta_j - \mathbf{\Gamma}^{(j)} \nabla \zeta_j)\, \varkappa_j\, u|^2\, dx \right),$$

where $\varkappa_j \in C_0^\infty(Q_j^*)$, and $\varkappa_j = 1$ on $\operatorname{supp} \zeta_j$. By (11.5.9), the last sum does not exceed

$$\sup_j \|V\eta_j; \Omega\|^2_{M(\mathring{w}_2^1 \to w_2^{-1})} \sum_j \int_\Omega (|\nabla(\varkappa_j u)|^2 + d_j^{-2}\, |\varkappa_j u|^2)\, dx.$$

By Hardy's inequality (11.5.4), this is bounded by

$$c\, \|V; \Omega\|^2_{M(\mathring{w}_2^1 \to w_2^{-1})} \int_\Omega |\nabla u|^2\, dx.$$

The proof of Theorem 11.5.1 is complete. □

Remark 11.5.1. In Theorem 11.5.1, one can replace

$$\sum_{0 \le i \le n} \|\Gamma_i; \Omega\|_{M(\mathring{w}_2^1 \to L_2)}$$

by the equivalent norm

$$\sup_j \sup_{e \subset Q_j} \frac{\|(|\mathbf{\Gamma}| + |\Gamma_0|); e\|_{L_2}}{C_{2,1}(e, Q_j^*)^{\frac{1}{2}}}. \tag{11.5.10}$$

In the case $n > 2$, one can use Wiener's capacity $c_{2,1}$ in place of $C_{2,1}(\cdot, Q_j^*)$ (see Sect. 13.2.2).

We now characterize the class of compact multipliers, $\mathring{M}(\mathring{w}_2^1(\Omega) \to w_2^{-1}(\Omega))$. We use the same notation as in the previous section.

Theorem 11.5.2. *Under the assumptions of Theorem* 11.5.1, *a distribution* V *is in* $\mathring{M}(\mathring{w}_2^1(\Omega) \to w_2^{-1}(\Omega))$ *if and only if*

$$V = \operatorname{div} \mathbf{\Gamma} + d_{\partial\Omega}^{-1} \Gamma_0, \tag{11.5.11}$$

where $\Gamma_i \in \mathring{M}(\mathring{w}_2^1(\Omega) \to L_2(\Omega))$ *for* $i = 0, 1, \dots, n$.

Proof. Suppose that V is given by (5.11). Let u be an arbitrary function in the unit ball $\mathcal{B}$ of $w_2^1(\Omega)$. Then

$$V u = \operatorname{div}(u \mathbf{\Gamma}) - \mathbf{\Gamma} + d_{\partial\Omega}^{-1} u \Gamma_0.$$

The set $\{\operatorname{div}(u \mathbf{\Gamma}) : u \in \mathcal{B}\}$ is compact in $w_2^{-1}(\Omega)$ since the set $\{u \mathbf{\Gamma} : u \in \mathcal{B}\}$ is compact in $\mathbf{L}_2(\Omega)$. The sets

$$\{\nabla u \cdot \mathbf{\Gamma} : u \in \mathcal{B}\} \quad \text{and} \quad \{d_{\partial\Omega}^{-1} \Gamma_0 u : u \in \mathcal{B}\}$$

are also compact in $w_2^{-1}(\Omega)$ since the sets

$$\{|\nabla u| : u \in \mathcal{B}\} \quad \text{and} \quad \{d_{\partial\Omega}^{-1} u : u \in \mathcal{B}\}$$

are bounded in $L_2(\Omega)$, and the multiplier operators $\bar{\Gamma}_i : L_2(\Omega) \to w_2^{-1}(\Omega)$, $i = 1, \dots, n$ are compact, being adjoint to Γ_i. This completes the proof of the "if" part of Theorem 11.5.2.

To prove the "only if" part let us assume that $O \in \mathbb{R}^n \backslash \Omega$. Then, for any $x \in \Omega$, it follows that $|x| \ge d_{\partial\Omega}(x)$, and the inequality

$$\int_\Omega \frac{|u(x)|^2}{|x|^2}\, dx \le c \int_\Omega |\nabla u(x)|^2\, dx \tag{11.5.12}$$

follows from (11.5.4).

As in the previous section, we introduce the cutoff functions

$$\varkappa_\delta(x) = F\left(\frac{d_{\partial\Omega}}{\delta}\right),$$

and

$$\xi_R(x) = 1 - F\left(\frac{|x|}{R}\right),$$

where $F \in C^\infty(\mathbb{R}_+)$ such that $F(t) = 1$ for $t \le 1$ and $F(t) = 0$ for $t \ge 2$.

The proofs of the following two lemmas are similar to those of Lemma 11.3.1 and Lemma 11.3.2.

Lemma 11.5.1. *If $f \in w_2^{-1}(\Omega)$, then*

$$\lim_{\delta\to 0} \|\varkappa_\delta f; \Omega\|_{w_2^{-1}} = 0 \tag{11.5.13}$$

and

$$\lim_{R\to\infty} \|\xi_R f; \Omega\|_{w_2^{-1}} = 0. \tag{11.5.14}$$

Lemma 11.5.2. *If $V \in \overset{\circ}{M}(\mathring{w}_2^1(\Omega) \to w_2^{-1}(\Omega))$, then*

$$\lim_{\delta\to 0} \|\varkappa_\delta V; \Omega\|_{\overset{\circ}{M}(\mathring{w}_2^1 \to w_2^{-1})} = 0, \tag{11.5.15}$$

and

$$\lim_{R\to\infty} \|\xi_R V; \Omega\|_{\overset{\circ}{M}(\mathring{w}_2^1 \to w_2^{-1})} = 0. \tag{11.5.16}$$

We now complete the proof of the "only if" part of Theorem 11.5.2. Write V in the form

$$V = \varkappa_\delta V + \xi_R V + (1 - \varkappa_\delta - \xi_R) V.$$

By Theorem 11.5.1 (ii), there exist $\mathbf{\Gamma}_\delta$ and $\Gamma^{(0)}$ such that

$$\varkappa_\delta V = \operatorname{div} \mathbf{\Gamma}_\delta + d_{\partial\Omega}^{-1} \Gamma_\delta^{(0)},$$

where

$$\sum_{0\le i\le n} \|\Gamma_\delta^{(i)}; \Omega\|_{M(\mathring{w}_2^1 \to L_2)} \le C\, \|\varkappa_\delta V; \Omega\|_{M(\mathring{w}_2^1 \to w_2^{-1})}.$$

Analogously,

$$\xi_R V = \operatorname{div} \mathbf{\Gamma}_{(R)} + |x|^{-1} \Gamma_{(R)}^{(0)},$$

where

$$\sum_{0\le i\le n} \|\Gamma_{(R)}^{(i)}; \Omega\|_{M(\mathring{w}_2^1 \to L_2)} \le C\, \|\xi_R V; \Omega\|_{M(\mathring{w}_2^1 \to w_2^{-1})}.$$

Hence, by Lemma 11.5.2,

$$\lim_{\delta\to 0} \sum_{0\le i\le n} \|\Gamma_\delta^{(i)}; \Omega\|_{M(\mathring{w}_2^1 \to L_2)} = 0,$$

and

$$\lim_{R\to\infty} \sum_{0\le i\le n} \|\Gamma_{(R)}^{(i)}; \Omega\|_{M(\mathring{w}_2^1 \to L_2)} = 0.$$

Now we estimate the multiplier

$$V_{\delta,R} := (1 - \varkappa_\delta - \xi_R)\, V.$$

Note that

$$V_{\delta,R} \in \overset{\circ}{M}\,(\overset{\circ}{w}{}_2^1(\Omega) \to w_2^{-1}(\Omega)).$$

Since its support is separated from ∞ and from $\partial\Omega$, it follows that

$$V_{\delta,R} \in \overset{\circ}{M}\,(W_2^1(\mathbb{R}^n) \to W_2^{-1}(\mathbb{R}^n)).$$

By Theorem 11.4.2,

$$V_{\delta,R} = \operatorname{div} \mathbf{\Gamma}_{\delta,R} + \Psi_{\delta,R}, \tag{11.5.17}$$

where all components of $\mathbf{\Gamma}_{\delta,R}$, together with $\Psi_{\delta,R}$, are in $\overset{\circ}{M}\,(W_2^1(\mathbb{R}^n) \to L_2(\mathbb{R}^n))$.

Multiplying, if necessary, both sides of (11.5.17) by a cutoff function as before, we may assume that the supports of $|\mathbf{\Gamma}_{\delta,R}|$ and $\Psi_{\delta,R}$ are in Ω, and are both separated from ∞, and from $\partial\Omega$. Hence, the components of $\mathbf{\Gamma}_{\delta,R}$, as well as $d_{\partial\Omega}\Psi_{\delta,R}$, are in $\overset{\circ}{M}\,(\overset{\circ}{w}{}_2^1(\Omega) \to L_2(\Omega))$. Finally,

$$V = \operatorname{div} \mathbf{\Gamma} + d_{\partial\Omega}^{-1}\, \Gamma^{(0)},$$

where

$$\mathbf{\Gamma} = \mathbf{\Gamma}_\delta + \mathbf{\Gamma}_{(R)} + \mathbf{\Gamma}_{\delta,R},$$

and

$$\Gamma^{(0)} = \Gamma_\delta^{(0)} + |x|^{-1}\, d_{\partial\Omega}\, \Gamma_{(R)}^{(0)} + d_{\partial\Omega}\, \Gamma_{\delta,R}^{(0)}.$$

It remains to note that

$$\mathbf{\Gamma}_\delta, \quad \mathbf{\Gamma}_{(R)}, \quad \Gamma_\delta^{(0)}, \quad \text{and } |x|^{-1}\, d_{\partial\Omega}\, \Gamma_{(R)}^{(0)}$$

are small in the corresponding operator norms, while $\mathbf{\Gamma}_{\delta,R}$ and $\Gamma_{\delta,R}^{(0)}$ are compact. This completes the proof of Theorem 11.5.2. □

11.6 Second-Order Differential Operators Acting from w_2^1 to w_2^{-1}

Further development of the topic of the present section can be found in [MV4]. Here we survey some results of this article, where explicit necessary and sufficient conditions for the boundedness of the general second-order differential operator

$$\mathcal{L} = \sum_{i,\,j=1}^{n} a_{ij}\, \partial_i \partial_j + \sum_{j=1}^{n} b_j\, \partial_j + c$$

with real- or complex-valued distributional coefficients a_{ij}, b_j, and c, acting from the Sobolev space w_2^1 to its dual w_2^{-1}, are found.

For the sake of convenience, let us assume that the principal part of $\mathcal{L}$ is in the divergence form, i.e.,

$$\mathcal{L}\,u = \operatorname{div}(A\,\nabla u) + \mathbf{b}\cdot\nabla u + q\,u, \qquad u \in C_0^\infty, \tag{11.6.1}$$

where

$$A = (a_{ij})_{i,\,j=1}^n \in (C_0^\infty)', \qquad \mathbf{b} = (b_j)_{j=1}^n \in (C_0^\infty)', \qquad \text{and} \quad q \in (C_0^\infty)'.$$

We present necessary and sufficient conditions on A, $\mathbf{b}$, and q which guarantee the boundedness of the sesquilinear form associated with $\mathcal{L}$:

$$|\langle \mathcal{L}\,u,\, v\rangle| \le C\,\|u\|_{w_2^1}\,\|v\|_{w_2^1}, \tag{11.6.2}$$

where the constant C does not depend on $u,\, v \in C_0^\infty$. Equivalently, we characterize all A, $\mathbf{b}$, and q such that

$$\mathcal{L}:\; w_2^1 \to w_2^{-1} \tag{11.6.3}$$

is a bounded operator. We state this boundedness criterion.

For $A = (a_{ij})$, let $A^t = (a_{ji})$ denote the transposed matrix, and let $\operatorname{Div}\colon (C_0^\infty)' \to (C_0^\infty)'$ be the row divergence operator defined by

$$\operatorname{Div}(a_{ij}) = \Big(\sum_{j=1}^n \partial_j\, a_{ij}\Big)_{i=1}^n. \tag{11.6.4}$$

We do not differ in notations between spaces of scalar, vector-, and matrix-valued functions.

Theorem 11.6.1. [MV4] *Let*

$$\mathcal{L} = \operatorname{div}(A\,\nabla\cdot) + \mathbf{b}\cdot\nabla + q,$$

where $A \in (C_0^\infty)'$, $\mathbf{b} \in (C_0^\infty)'$ *and* $q \in (C_0^\infty)'$, $n \ge 2$. *Then the following statements hold.*

(*i*) *The sesquilinear form of* $\mathcal{L}$ *is bounded, i.e.,* (11.6.2) *holds if and only if*

$$\tfrac{1}{2}\,(A + A^t) \in L_\infty,$$

and $\mathbf{b}$ *and* q *can be represented respectively in the form*

$$\mathbf{b} = \mathbf{c} + \operatorname{Div} F, \qquad q = \operatorname{div}\mathbf{h}, \tag{11.6.5}$$

where F *is a skew-symmetric matrix field such that*

$$F - \tfrac{1}{2}\,(A - A^t) \in \mathrm{BMO}, \tag{11.6.6}$$

whereas $\mathbf{c}$ *and* $\mathbf{h}$ *belong to* $M(w_2^1 \to L_2)$.

(ii) If the sesquilinear form of $\mathcal{L}$ is bounded, then $\mathbf{c}$, F, *and* $\mathbf{h}$ *in the decomposition* (11.6.5) *can be determined explicitly by*

$$\mathbf{c} = \nabla(\Delta^{-1}\mathrm{div}\,\mathbf{b}), \qquad \mathbf{h} = \nabla(\Delta^{-1}\,q), \tag{11.6.7}$$

$$F = \Delta^{-1}\mathrm{curl}\,[\mathbf{b} - \tfrac{1}{2}\,\mathrm{Div}\,(A - A^t)] + \tfrac{1}{2}(A - A^t). \tag{11.6.8}$$

where

$$\Delta^{-1}\mathrm{curl}\,[\mathbf{b} - \tfrac{1}{2}\,\mathrm{Div}\,(A - A^t)] \in \mathrm{BMO} \tag{11.6.9}$$

and

$$\nabla(\Delta^{-1}\mathrm{div}\,\mathbf{b}) \quad \text{and} \quad \nabla(\Delta^{-1}\,q)$$

belong to $M(w_2^1 \to L_2)$.

Remark 11.6.1. Condition (11.6.9) in statement *(ii)* of Theorem 11.6.1 may be replaced by

$$\mathbf{b} - \mathrm{Div}\,\tfrac{1}{2}(A - A^t) \in \mathrm{BMO}^{-1}, \tag{11.6.10}$$

which ensures that the decomposition (11.6.5) holds. Here BMO^{-1} stands for the well-known space of distributions that can be represented in the form $f = \mathrm{div}\,\mathbf{g}$ where $\mathbf{g} \in \mathrm{BMO}$.

Remark 11.6.2. In the case $n = 2$, it is shown in [MV4] that (11.6.2) holds if and only if

$$\tfrac{1}{2}\,(A + A^t) \in L_\infty(\mathbb{R}^2),$$
$$\mathbf{b} - \tfrac{1}{2}\,\mathrm{Div}\,(A - A^t) \in \mathrm{BMO}^{-1}(\mathbb{R}^2),$$

and

$$q = \mathrm{div}\,\mathbf{b} = 0.$$

Remark 11.6.3. Expressions like

$$\nabla(\Delta^{-1}\mathrm{div}\,\mathbf{b}), \quad \mathrm{Div}(\Delta^{-1}\mathrm{curl}\,\mathbf{b}), \quad \text{and} \quad \nabla(\Delta^{-1}\,q)$$

used above, which involve nonlocal operators, are defined in the sense of distributions. This is possible since

$$\Delta^{-1}\mathrm{div}\,\mathbf{b}, \qquad \Delta^{-1}\mathrm{curl}\,\mathbf{b}, \qquad \text{and} \quad \Delta^{-1}q$$

can be understood in terms of convergence in the weak-$*$ topology of BMO of, respectively,

$$\Delta^{-1}\,\mathrm{div}\,(\psi_N\,\mathbf{b}), \quad \Delta^{-1}\,\mathrm{curl}\,(\psi_N\,\mathbf{b}), \quad \text{and} \quad \Delta^{-1}\,(\psi_N\,q)$$

as $N \to +\infty$. Here ψ_N is a smooth cutoff function supported on $\{x : |x| < N\}$, and the limits above do not depend on the choice of ψ_N.

It follows from Theorem 11.6.1 that $\mathcal{L} : w_2^1 \to w_2^{-1}$ is bounded if and only if the symmetric part of A is essentially bounded, i.e.,

$$\frac{1}{2}(A + A^t) \in L_\infty$$

and

$$\mathbf{b_1} \cdot \nabla + q : w_2^1 \to w_2^{-1}$$

is bounded, where

$$\mathbf{b_1} = \mathbf{b} - \tfrac{1}{2}\,\mathrm{Div}(A - A^t). \tag{11.6.11}$$

In particular, the principal part

$$\mathcal{P}u = \mathrm{div}(A\,\nabla u) : w_2^1 \to w_2^{-1}$$

is bounded if and only if

$$\tfrac{1}{2}(A + A^t) \in L_\infty \tag{11.6.12}$$

and

$$\mathrm{Div}\,\tfrac{1}{2}(A - A^t) \in \mathrm{BMO}^{-1}. \tag{11.6.13}$$

A simpler condition with

$$\tfrac{1}{2}(A - A^t) \in \mathrm{BMO}$$

in place of (11.6.13) is sufficient, but generally not necessary, unless $n \le 2$.

Thus, the boundedness problem for the general second order differential operator in the divergence form (11.6.1) is reduced to the special case

$$\mathcal{L} = \mathbf{b} \cdot \nabla + q, \qquad \mathbf{b} \in (C_0^\infty)', \quad q \in (C_0^\infty)'. \tag{11.6.14}$$

As a corollary of Theorem 11.6.1, one obtains that, if

$$\mathbf{b} \cdot \nabla + q \ : w_2^1 \to w_2^{-1} \tag{11.6.15}$$

is bounded, then the Hodge decomposition

$$\mathbf{b} = \nabla(\Delta^{-1}\mathrm{div}\,\mathbf{b}) + \mathrm{Div}\,(\Delta^{-1}\mathrm{curl}\,\mathbf{b}) \tag{11.6.16}$$

holds, where

$$\Delta^{-1}\mathrm{curl}\,\mathbf{b} \in \mathrm{BMO}$$

and

$$\int_{|x-y|<r} [\,|\nabla(\Delta^{-1}\mathrm{div}\,\mathbf{b})|^2 + |\nabla(\Delta^{-1}\,q)|^2\,]\,dy \le \mathrm{const}\; r^{n-2} \tag{11.6.17}$$

for all $r > 0$, $x \in \mathbb{R}^n$, in the case $n \ge 3$; in two dimensions $\mathrm{div}\,\mathbf{b} = q = 0$.

The condition (11.6.17) is generally stronger than

$$\Delta^{-1}\mathrm{div}\,\mathbf{b} \in \mathrm{BMO} \qquad \text{and} \qquad \Delta^{-1}\,q \in \mathrm{BMO},$$

while the divergence-free part of $\mathbf{b}$ is characterized by $\Delta^{-1}\mathrm{curl}\,\mathbf{b} \in \mathrm{BMO}$, for all $n \geq 2$.

A close sufficient condition of the Fefferman–Phong type can be stated in the following form:

$$\int_{|x-y|<r} [\,|\nabla(\Delta^{-1}\mathrm{div}\,\mathbf{b})|^2 + |\nabla(\Delta^{-1}\,q)|^2\,]^{1+\epsilon}\,dy \leq \mathrm{const}\, r^{n-2(1+\epsilon)}, \quad (11.6.18)$$

for some $\epsilon > 0$ and all $r > 0$, $x \in \mathbb{R}^n$. This is a consequence of Theorem 11.6.1 coupled with (1.2.50), where

$$|(\Delta^{-1}\mathrm{div}\,\mathbf{b})|^2 + |\nabla(\Delta^{-1}\,q)|^2$$

is used in place of g, $p = 2$, $m = 1$, and $t = 1 + \epsilon$.

It is worth mentioning that the class of potentials obeying (11.6.18) is substantially broader than its subclass

$$\int_{|x-y|<r} (|\mathbf{b}|^2 + |q|)^{1+\epsilon}\,dy \leq \mathrm{const}\, r^{n-2(1+\epsilon)}. \quad (11.6.19)$$

The sufficiency of the preceding condition for (11.6.15) is deduced by a direct application of the original Fefferman–Phong condition and Schwarz's inequality.

More generally, (11.6.15) clearly follows from a cruder estimate,

$$\int_{\mathbb{R}^n} |u|^2\,(|\mathbf{b}|^2 + |q|)\,dx \leq \mathrm{const}\,\|u\|^2_{w_2^1}, \quad u \in C_0^\infty, \quad (11.6.20)$$

which is equivalent to

$$\mathbf{b} \in M(w_2^1 \to L_2) \qquad \text{and} \qquad |q|^{1/2} \in M(w_2^1 \to L_2).$$

However, by replacing (11.6.15) with (11.6.20), one strongly reduces the class of admissible vector fields $\mathbf{b}$ and potentials q. An instructive example for $\mathbf{b}\cdot\nabla$ in the case $q = 0$ is provided by the vector field

$$\mathbf{b}(x) = \left(x_2(x_1^2 + x_2^2)^{-1}, -x_1(x_1^2 + x_2^2)^{-1}, 0, \dots 0\right), \quad x \in \mathbb{R}^n,$$

where $n \geq 2$. An elementary argument involving polar coordinates and a Fourier series expansion shows that this vector field obeys (11.6.15). On the other hand, (11.6.20) fails since $\mathbf{b} \notin L^2_{\mathrm{loc}}$.

We note that, for $q = 0$, (11.6.20) is equivalent to the boundedness of the nonlinear operator

$$u \to |\mathbf{b}\cdot\nabla u| \,:\, w_2^1 \to w_2^{-1}.$$

However, dealing with the linear version $\langle \mathbf{b}\cdot\nabla u,\, u\rangle$ is more difficult.

The main obstacle in the proof of Theorem 11.6.1 is the interaction between the quadratic forms associated with $q - \frac{1}{2}\,\mathrm{div}\,\mathbf{b}$ and the divergence free part of $\mathbf{b}$. To overcome this difficulty, one needs to distinguish the class of vector fields $\mathbf{b}$ such that the commutator inequality

$$\left| \int \mathbf{b} \cdot (u \, \nabla \bar{v} - \bar{v} \, \nabla u) \, dx \right| \leq \text{const} \, \|u\|_{w_1^1} \, \|v\|_{w_2^1} \tag{11.6.21}$$

holds for all $u, v \in C_0^\infty$. In the important special case of irrotational fields where $\mathbf{b} = \nabla f$, the preceding inequality is equivalent to the boundedness of the commutator $[f, \Delta]$ acting from w_2^1 to w_2^{-1}.

We state a complete characterization of those $\mathbf{b}$ which obey (11.6.21) in [MV4]. First, the irrotational part $\mathbf{c} = \nabla(\Delta^{-1}\text{div}\,\mathbf{b})$ of $\mathbf{b}$ belongs to $M(w_2^1 \to L_2)$ and second, $F = \Delta^{-1}\text{curl}\,\mathbf{b}$ belongs to BMO, and $\mathbf{b} = \mathbf{c} + \text{Div}\,F$. These conditions combined turn out to be necessary and sufficient for (11.6.21).

12

Relativistic Schrödinger Operator and $M(W_2^{1/2} \to W_2^{-1/2})$

The material of this chapter is taken from the article [MV3]. The goal is to give necessary and sufficient conditions for the boundedness of the relativistic Schrödinger operator $\mathcal{H} = \sqrt{-\Delta} + Q$ from the Sobolev space $W_2^{1/2}(\mathbb{R}^n)$ to its dual $W_2^{-1/2}(\mathbb{R}^n)$, for an arbitrary real- or complex-valued potential Q on $\mathbb{R}^n$. In other words, a complete characterization of the space $M(W_2^{1/2}(\mathbb{R}^n) \to W_2^{-1/2}(\mathbb{R}^n))$ is obtained.

12.1 Auxiliary Assertions

As before, we omit $\mathbb{R}^n$ in notations of norms and integrals.

Let $\gamma \in (C_0^\infty)'$ be a complex-valued distribution on $\mathbb{R}^n$. As always, we use the same notation for the corresponding multiplication operator $\gamma : C_0^\infty \to (C_0^\infty)'$ defined by

$$\langle \gamma u, v \rangle = \langle \gamma, \bar{u}\, v \rangle \qquad u, v \in C_0^\infty .$$

Given $m, l \in \mathbb{R}$, the inclusion $\gamma \in M\,(W_2^m \to W_2^l)$ means that the sesquilinear form $\langle \gamma \,\cdot, \cdot \rangle$ is bounded:

$$|\langle \gamma\, u,\, v \rangle| = |\langle \gamma,\, \bar{u}\, v \rangle| \le C\, \|u\|_{W_2^m}\, \|v\|_{W_2^{-l}}, \qquad \forall u, v \in C_0^\infty, \qquad (12.1.1)$$

where C does not depend on u, v. The multiplier norm $\|\gamma\|_{M(W_2^m \to W_2^l)}$ is equal to the least bound C in the preceding inequality.

It is easy to see that, in the case $l = -m$, (12.1.1) is equivalent to the quadratic form inequality

$$|\langle \gamma\, u,\, u \rangle| = |\langle \gamma,\, |u|^2 \rangle| \le C'\, \|u\|^2_{W_2^m}, \qquad \forall u \in C_0^\infty . \qquad (12.1.2)$$

To verify this, suppose that

$$\|u\|_{W_2^m} \le 1, \quad \|v\|_{W_2^m} \le 1, \text{ where } u,\, v \in C_0^\infty .$$

V.G. Maz'ya, T.O. Shaposhnikova, *Theory of Sobolev Multipliers*,
Grundlehren der mathematischen Wissenschaften 337,

Applying (12.1.2) together with the polarization identity (11.1.3), we get

$$|\langle \gamma, \bar{u}\, v\rangle| \leq \frac{C'}{4}\left(\|u+v\|^2_{W_2^m} + \|u-v\|^2_{W_2^m} + \|u+iv\|^2_{W_2^m} + \|u-iv\|^2_{W_2^m}\right) \leq 2C'.$$

Hence, (12.1.1) holds for $l = -m$ with $C = 2C'$. Moreover, the least bound C' in (12.1.2) satisfies the inequalities

$$C' \leq \|\gamma\|_{M(W_2^m \to W_2^{-m})} \leq 2C'.$$

Let $|D| = (-\Delta)^{1/2}$. We define the relativistic Schrödinger operator as

$$\mathcal{H} = |D| + Q : C_0^\infty \to (C_0^\infty)',$$

(see [LL], Sect. 7.15), where $Q : C_0^\infty \to (C_0^\infty)'$ is a multiplication operator defined by $Q \in (C_0^\infty)'$. It is well known that actually $|D|$ is a bounded operator from $W_2^{1/2}$ to $W_2^{-1/2}$. Thus, $\mathcal{H}$ can be extended to a bounded operator:

$$\mathcal{H} : W_2^{1/2} \to W_2^{-1/2},$$

if and only if $Q \in M\,(W_2^{1/2} \to W_2^{-1/2})$ or, equivalently, if the quadratic form inequality (12.1.2) holds for $\gamma = Q$ and $m = 1/2$.

From the preceding discussion it follows that $\mathcal{H} : W_2^{1/2} \to W_2^{-1/2}$ is bounded if and only if

$$|\langle Q\, u,\, u\rangle| \leq a\, \langle |D|\, u,\, u\rangle + b\, \langle u,\, u\rangle, \qquad \forall u \in C_0^\infty, \tag{12.1.3}$$

for some $a,\, b > 0$. By definition this means that Q is *relatively form bounded* with respect to $|D|$.

In particular, if Q is real-valued, and $0 < a < 1$ in the preceding inequality, then by the so-called KLMN Theorem ([RS1], Theorem X.17), $\mathcal{H} = |D| + Q$ is defined as a unique self-adjoint operator such that

$$\langle \mathcal{H}u,\, v\rangle = \langle |D|\, u,\, v\rangle + \langle Q\, u, v\rangle, \qquad \forall u \in C_0^\infty.$$

For the complex-valued Q such that (12.1.3) holds with $0 < a < 1/2$, it follows that $\mathcal{H} = |D| + Q$, understood in a similar sense, is an m-sectorial operator ([EE], Theorem IV.4.2).

In the case of $Q \in L_{1,\,\text{loc}}$, (12.1.3) is equivalent to the inequality

$$\left| \int |u(x)|^2\, Q(x)\, dx \right| \leq \text{const}\, \|u\|^2_{W_2^{1/2}}, \qquad \forall u \in C_0^\infty, \tag{12.1.4}$$

and hence to the boundedness of the corresponding sesquilinear form:

$$\left| \int u(x)\, \overline{v(x)}\; Q(x)\, dx \right| \leq \text{const}\, \|u\|_{W_2^{1/2}}\, \|v\|_{W_2^{1/2}},$$

where the constant is independent of $u, v \in C_0^\infty$.

The characterization of potentials Q such that $\mathcal{H} : W_2^{1/2} \to W_2^{-1/2}$ is based on a series of lemmas and propositions presented below, and the results for the nonrelativistic Schrödinger operator obtained in the previous chapter.

By $L_{2,\mathrm{unif}}$, we denote the class of $f \in L_{2,\mathrm{loc}}$ such that

$$\|f\|_{L_{2,\mathrm{unif}}} = \sup_{x\in\mathbb{R}^n} \|\chi_{\mathcal{B}_1(x)} f\|_{L_2} < \infty. \tag{12.1.5}$$

Lemma 12.1.1. *Let* $0 < l < 1$, *and* $m > l$. *Then* $\gamma \in M\,(W_2^m \to W_2^l)$ *if and only if*

$$\gamma \in W_2^{m-l} \to L_2 \qquad \text{and} \qquad |D|^l\gamma \in M\,(W_2^m \to L_2).$$

Moreover,

$$\|\gamma\|_{M(W_2^m\to W_2^l)} \sim \||D|^l\gamma\|_{M(W_2^m\to L_2)} + \|\gamma\|_{M(W_2^{m-l}\to L_2)}\,. \tag{12.1.6}$$

Proof. We first prove the lower estimate for $\|\gamma\|_{M(W_2^m\to W_2^l)}$

$$\||D|^l\gamma\|_{M(W_2^m\to L_2)} + \|\gamma\|_{M(W_2^{m-l}\to L_2)} \le c\,\|\gamma\|_{M(W_2^m\to W_2^l)}\,. \tag{12.1.7}$$

Here and below c denotes a constant which depends only on l, m, and n.

Let $u \in C_0^\infty$. Using the integral representation

$$|D|^l u(x) = c(n,l)\int \frac{u(x)-u(y)}{|x-y|^{n+l}}\,dy \tag{12.1.8}$$

which follows by inspecting the Fourier transforms of both sides, we obtain

$$\begin{aligned}&|D|^l\,(\gamma\, u)(x) - \gamma(x)\,|D|^l\,u(x) - u(x)\,|D|^l\,\gamma(x)\\ &= -c(n,l)\int \frac{(u(x)-u(y))(\gamma(x)-\gamma(y))}{|x-y|^{n+l}}\,dy.\end{aligned}$$

Hence,

$$\big|\,|D|^l\,(\gamma\, u) - \gamma\,|D|^l\,u - u\,|D|^l\gamma\big| \le c\,D_{2,l/2}\,u\cdot D_{2,l/2}\,\gamma, \tag{12.1.9}$$

where, as in Chap. 4,

$$D_{2,s}\,u(x) = \left(\int \frac{|u(x)-u(y)|^2}{|x-y|^{n+2s}}\,dy\right)^{\frac{1}{2}}, \quad s > 0.$$

Next, we estimate

$$\begin{aligned}\|u\cdot|D|^l\gamma\|_{L_2} &\le \big\|\,|D|^l(\gamma\, u)\big\|_{L_2} + \big\|\gamma\,|D|^l u\big\|_{L_2} + c\,\|D_{2,l/2}u\cdot D_{2,l/2}\gamma\|_{L_2}\\ &\le \|\gamma u\|_{W_2^l} + \|\gamma\|_{M(W_2^{m-l}\to L_2)}\,\big\|\,|D|^l u\big\|_{W_2^{m-l}} + c\,\|D_{2,l/2}u\cdot D_{2,l/2}\gamma\|_{L_2}\\ &\le \|\gamma\|_{M(W_2^m\to W_2^l)}\,\|u\|_{W_2^m} + \|\gamma\|_{M(W_2^{m-l}\to L_2)}\,\|u\|_{W_2^m} + c\,\|D_{2,l/2}u\cdot D_{2,l/2}\gamma\|_{L_2}\\ &\le c\,\|\gamma\|_{M(W_2^m\to W_2^l)}\,\|u\|_{W_2^m} + c\,\|D_{2,l/2}u\cdot D_{2,l/2}\gamma\|_{L_2}.\end{aligned} \tag{12.1.10}$$

In the last line we have used the inequality

$$\|\gamma\|_{M(W_2^{m-l} \to L_2)} \le c\,\|\gamma\|_{M(W_2^m \to W_2^l)} \tag{12.1.11}$$

(see (4.3.29)).

To estimate the term

$$\|D_{2,l/2}u \cdot D_{2,l/2}\gamma\|_{L_2},$$

we apply the pointwise estimate (4.2.10)

$$D_{2,l/2}u \le J_s D_{2,l/2}((-\Delta+1)^{s/2}\,u),$$

with $s = m - l/2$, where $J_s = (-\Delta+1)^{-s/2}$ is the Bessel potential of order s. Hence

$$\begin{aligned} \|D_{2,l/2}u \cdot D_{2,l/2}\gamma\|_{L_2} &\le \|J_{m-l/2}D_{2,l/2}((-\Delta+1)^{m/2-l/4}\,u)) \cdot D_{2,l/2}\gamma\|_{L_2} \\ &\le c\,\|D_{2,l/2}\gamma\|_{M(W_2^{m-l/2} \to L_2)}\,\|J_{m-l/2}D_{2,l/2}((-\Delta+1)^{m/2-l/4}\,u))\|_{W_2^{m-l/2}} \\ &\le c\,\|D_{2,l/2}\gamma\|_{M(W_2^{m-l/2} \to L_2)}\,\|D_{2,l/2}(-\Delta+1)^{m/2-l/4}u\|_{L_2} \\ &\le c\,\|D_{2,l/2}\gamma\|_{M(W_2^{m-l/2} \to L_2)}\,\|u\|_{W_2^m}. \end{aligned} \tag{12.1.12}$$

We next notice that, by Corollary 4.3.6,

$$\|D_{2,l/2}\gamma\|_{M(W_2^{m-l/2} \to L_2)} \le c\,\|\gamma\|_{M(W_2^m \to W_2^l)}. \tag{12.1.13}$$

Combining the estimates (12.1.10)–(12.1.13), we obtain

$$\big\|u \cdot |D|^l\gamma\big\|_{L_2} \le c\,\|\gamma\|_{M(W_2^m \to W_2^l)}\,\|u\|_{W_2^m},$$

which is equivalent to the inequality

$$\big\||D|^l\gamma\big\|_{M(W_2^m \to L_2)} \le c\,\|\gamma\|_{M(W_2^m \to W_2^l)}.$$

This, together with (12.1.11), completes the proof of (12.1.7).

We now prove the upper estimate

$$\|\gamma\|_{M(W_2^m \to W_2^l)} \le c\left(\big\||D|^l\gamma\big\|_{M(W_2^m \to L_2)} + \|\gamma\|_{M(W_2^{m-l} \to L_2)}\right). \tag{12.1.14}$$

By (12.1.9),

$$\big\||D|^l(\gamma u)\big\|_{L_2} \le \big\|\gamma|D|^l u\big\|_{L_2} + \big\||D|^l\gamma \cdot u\big\|_{L_2} + c\,\|D_{2,l/2}u \cdot D_{2,l/2}\gamma\|_{L_2}.$$

Using the elementary estimate

$$\|u\|_{W_2^{m-l}} \le c\,\|u\|_{W_2^m},$$

we have

$$\|\gamma u\|_{L_2} \le \|\gamma\|_{M(W_2^{m-l}\to L_2)}\, \|u\|_{W_2^{m-l}} \le c\,\|\gamma\|_{M(W_2^{m-l}\to L_2)}\, \|u\|_{W_2^m}.$$

From these inequalities, combined with the estimate

$$\|D_{2,l/2}u \cdot D_{2,l/2}\gamma\|_{L_2} \le c\,\|\gamma\|_{M(W_2^{m-l/2}\to W_2^{l/2})}\, \|u\|_{W_2^m}$$

established above, it follows that

$$\|\gamma u\|_{W_2^l} \le c\left(\|\gamma\|_{M(W_2^{m-l}\to L_2)}\, \|u\|_{W_2^m} + \||D|^l\gamma\|_{M(W_2^m\to L_2)}\, \|u\|_{W_2^m}\right)$$
$$+c\,\|\gamma\|_{M(W_2^{m-l/2}\to W_2^{l/2})}\, \|u\|_{W_2^m}.$$

Thus,

$$\|\gamma\|_{M(W_2^m\to W_2^l)} \le c\big(\||D|^l\gamma\|_{M(W_2^m\to L_2)} + \|\gamma\|_{M(W_2^{m-l}\to L_2)}$$
$$+\|\gamma\|^{1/2}_{M(W_2^{m-l}\to L_2)}\, \|\gamma\|^{1/2}_{M(W_2^m\to W_2^l)}\big).$$

Combining this with (12.1.11), we find that

$$\|\gamma\|_{M(W_2^m\to W_2^l)} \le c\left(\||D|^l\gamma\|_{M(W_2^m\to L_2)} + \|\gamma\|_{M(W_2^{m-l}\to L_2)}\right).$$

This completes the proof of Lemma 12.1.1. □

Lemma 12.1.2. *Let $0 < l < 1$ and $\frac{n}{2} \ge m > l$. Then $\gamma \in M\,(W_2^m \to W_2^l)$ if and only if*

$$(-\Delta + 1)^{l/2}\gamma \in M\,(W_2^m \to L_2)$$

and

$$\|\gamma\|_{M(W_2^m\to W_2^l)} \sim \|(-\Delta+1)^{l/2}\gamma\|_{M(W_2^m\to L_2)}. \tag{12.1.15}$$

Proof. Recall that a nonnegative weight $w \in L_{1,\mathrm{loc}}$ is said to be in the Muckenhoupt class A_1 if

$$\mathcal{M}w(x) \le c\,w(x) \qquad \text{for almost all } x \in \mathbb{R}^n,$$

where $\mathcal{M}$ is the Hardy–Littlewood maximal operator. The least constant on the right-hand side of the preceding inequality is called the A_1-bound of w.

We need the following statement established in [MV1], Lemma 3.1 (see also [MSh16], Sec. 2.6.3) for the homogeneous Sobolev space $w_p^m(\mathbb{R}^n)$.

Lemma 12.1.3. *Let $\gamma \in M\,(w_p^m \to L_p)$, where $1 < p < \infty$ and $0 < m < \frac{n}{p}$. Suppose that T is a bounded operator on the weighted space $L_p(w)$ for every $w \in A_1$. Suppose additionally that, for all $f \in L_p(w)$, the inequality*

$$\|Tf\|_{L_p(w)} \le C\,\|f\|_{L_p(w)}$$

holds with a constant C which depends only on the A_1-bound of the weight w. Then $T\gamma \in M\,(w_p^m \to L_p)$, and

$$\|T\gamma\|_{M(w_p^m \to L_p)} \le C_1 \|\gamma\|_{M(w_p^m \to L_p)},$$

where the constant C_1 does not depend on γ.

We also need a Fourier multiplier theorem of Mikhlin type for L_p spaces with weights. Let $m \in L_\infty$. Then the Fourier multiplier operator with symbol m is defined on L_2 by $T_m = \mathcal{F}^{-1}\, m\, \mathcal{F}$, where $\mathcal{F}$ and $\mathcal{F}^{-1}$ are respectively the direct and inverse Fourier transforms.

The next lemma follows from the results of Kurtz and Wheeden [KWh], Theorem 1.

Lemma 12.1.4. *Suppose that $1 < p < \infty$ and $w \in A_1$. Suppose also that $m \in C^\infty(\mathbb{R}^n \setminus \{0\})$ satisfies the Mikhlin multiplier condition:*

$$|D^\alpha m(x)| \le C_\alpha\, |x|^{-|\alpha|}, \qquad x \in \mathbb{R}^n \setminus \{0\}, \tag{12.1.16}$$

for every multi-index α such that $0 \le |\alpha| \le n$. Then the inequality

$$\|T_m\, f\|_{L_p(w)} \le C\,\|f\|_{L_p(w)}, \qquad f \in L_p(w) \cap L_2,$$

holds, where C depends only on p, n, the A_1-bound of w, and the constant C_α in (12.1.16).

Corollary 12.1.1. *Suppose that $1 < p < \infty$ and $w \in A_1$. Suppose also that $0 < l \le 2$. Define*

$$m_l(x) = (1 + |x|^2)^{l/2} - |x|^l. \tag{12.1.17}$$

Then

$$\|T_{m_l}\, f\|_{L_p(w)} \le C\,\|f\|_{L_p(w)}, \qquad f \in L_p(w) \cap L_2, \tag{12.1.18}$$

where the constant C depends only on l, p, n, and the A_1-constant of w.

Proof. Clearly,

$$0 \le m_l(x) \le C\,(1 + |x|)^{l-2}, \qquad x \in \mathbb{R}^n.$$

Furthermore, it is easy to see by induction that, for any multi-index α with $|\alpha| \ge 1$, we have the following estimates:

$$|D^\alpha m_l(x)| \le C_{\alpha,l}\,|x|^{l-2-|\alpha|}, \qquad |x| \to \infty,$$

and

$$|D^\alpha m_l(x)| \le C_{\alpha,l}\,|x|^{l-|\alpha|}, \qquad |x| \to 0.$$

Since $0 < l \le 2$, it follows from this that m_l satisfies (12.1.16), and hence by Lemma 12.1.4 the inequality

$$\|T_{m_l} f\|_{L_p(w)} \le C\,\|f\|_{L_p(w)}$$

holds with a constant that depends only on l, p, and the A_1-bound of w. □

Now we are in a position to complete the proof of Lemma 12.1.2. Suppose that $\gamma \in M\,(W_2^m \to W_2^l)$, where $\frac{n}{2} \ge m > l$ and $0 < l < 1$. By Corollary 12.1.1, the operator $T_{m_l} = (1-\Delta)^{l/2} - |D|^l$ is bounded on $L_2(w)$ for every $w \in A_1$, and its norm is bounded by a constant which depends only on l, n, and the A_1-bound of w. Hence by Lemma 12.1.3 it follows that $\gamma \in M\,(w_2^m \to L_2)$ yields

$$T_{m_l}\gamma = \Big((1-\Delta)^{l/2} - |D|^l\Big)\,\gamma \in M\,(w_2^m \to L_2),$$

and

$$\|T_{m_l}\gamma\|_{M(w_2^m \to L_2)} \le c\,\|\gamma\|_{M(w_2^m \to L_2)},$$

where c depends only on l, m, and n.

We need to replace w_2^m in the preceding inequality by W_2^m. To this end, let $\mathcal{B} = \mathcal{B}_1(x_0)$ denote a ball of radius 1 in $\mathbb{R}^n$, and $2\mathcal{B} = \mathcal{B}_2(x_0)$. Suppose that $m < \frac{n}{2}$ (the case $m = \frac{n}{2}$ requires usual modifications). Using Theorem 3.1.2, we obtain that $\gamma \in M\,(W_2^m \to L_2)$ if and only if

$$\sup_{\mathcal{B}}\, \|\chi_{\mathcal{B}}\,\gamma\|_{M(w_2^m \to L_2)} < +\infty,$$

and

$$\|\gamma\|_{M(W_2^m \to L_2)} \sim \sup_{\mathcal{B}}\, \|\chi_{\mathcal{B}}\,\gamma\|_{M(w_2^m \to L_2)}.$$

Hence,

$$\|T_{m_l}\gamma\|_{M(W_2^m \to L_2)} \le c\,\sup_{\mathcal{B}}\, \|\chi_{\mathcal{B}}\,T_{m_l}\gamma\|_{M(w_2^m \to L_2)}.$$

We set

$$\gamma = \chi_{2\mathcal{B}}\,\gamma + \chi_{(2\mathcal{B})^c}\,\gamma$$

and estimate each term separately. By Lemma 12.1.3,

$$\|\chi_{\mathcal{B}}\,T_{m_l}(\chi_{2\mathcal{B}}\,\gamma)\|_{M(w_2^m \to L_2)} \le c\,\sup_{\mathcal{B}}\, \|\chi_{2\mathcal{B}}\,\gamma\|_{M(w_2^m \to L_2)} \le c\,\|\gamma\|_{M(W_2^m \to L_2)}.$$

To estimate the second term, notice that $T_{m_l}(\chi_{(2\mathcal{B})^c}\gamma) \in L_\infty(\mathcal{B})$, and hence

$$\|\chi_{\mathcal{B}}\,T_{m_l}(\chi_{(2\mathcal{B})^c}\gamma)\|_{M(w_2^m \to L_2)} \le c\,\|T_{m_l}(\chi_{(2\mathcal{B})^c}\gamma)(x); \mathcal{B}\|_{L_\infty} \le c\,\|\gamma\|_{M(W_2^m \to L_2)}.$$

Indeed, for $x \in \mathcal{B}$,

$$|T_{m_l}(\chi_{(2\mathcal{B})^c}\gamma)(x)| \le c \int_{|x-y|\ge 1} \frac{|\gamma(y)|}{|x-y|^{n+l}}\,dy \le c \int_1^{+\infty} \frac{\int_{\mathcal{B}_r(x)} |\gamma(y)|dy}{r^{n+l+1}} dr.$$

Since $\gamma \in M\,(W_2^m \to L_2)$, it follows that $\gamma \in L_{2,\mathrm{unif}}$, and hence

$$\int_{\mathcal{B}_r(x)} |\gamma(y)|^2\,dy \le c\,r^n\,\|\gamma\|^2_{M(W_2^m \to L_2)}, \qquad r \ge 1.$$

Consequently,

$$\int_{\mathcal{B}_r(x)} |\gamma(y)|\,dy \le c\,r^{n/2} \|\gamma; \mathcal{B}_r(x)\|_{L_2} \le c\,r^n\,\|\gamma\|_{M(W_2^m \to L_2)}, \qquad r \ge 1.$$

Hence,

$$\|T_{m_l}(\chi_{(2\mathcal{B})^c}\gamma); \mathcal{B}\|_{L_\infty} \le c\,\|\gamma\|_{M(W_2^m \to L_2)}.$$

Thus, we have proved the inequality

$$\left\|\left((1-\Delta)^{l/2} - |D|^l\right)\gamma\right\|_{M(W_2^m \to L_2)} \le c\,\|\gamma\|_{M(W_2^m \to L_2)}.$$

Using this estimate, inequality (12.1.11), and Lemma 12.1.1, we arrive at

$$\|(1-\Delta)^{l/2}\,\gamma\|_{M(W_2^m \to L_2)} \le c\left(\||D|^l\gamma\|_{M(W_2^m \to L_)} + \|\gamma\|_{M(W_2^m \to L_2)}\right)$$

$$\le c\,\|\gamma\|_{M(W_2^m \to W_2^l)}.$$

Conversely, suppose that

$$(1-\Delta)^{l/2}\,\gamma \in M\,(W_2^m \to L_2).$$

It follows from the above estimate of

$$\left\|\left((1-\Delta)^{l/2} - |D|^l\right)\gamma\right\|_{M(W_2^m \to L_2)}$$

that

$$\||D|^l\gamma\|_{M(W_2^m \to L_2)} \le c\left(\|(1-\Delta)^{l/2}\,\gamma\|_{M(W_2^m \to L_2)} + \|\gamma\|_{M(W_2^m \to L_2)}\right).$$

Obviously,

$$\|\gamma\|_{M(W_2^m \to L_2)} \le c\,\|\gamma\|_{M(W_2^{m-l} \to L_2)}.$$

Applying again Lemma 12.1.1 together with the preceding estimates, we have

$$\|\gamma\|_{M(W_2^m \to W_2^l)} \le c\left(\|\,|D|^l\gamma\|_{M(W_2^m \to L_2)} + \|\gamma\|_{M(W_2^{m-l} \to L_2)}\right)$$

$$\leq c\left(\|(1-\Delta)^{l/2}\,\gamma\|_{M(W_2^m\to L_2)}+\|\gamma\|_{M(W_2^{m-l}\to L_2)}\right).$$

It remains to obtain the estimate

$$\|\gamma\|_{M(W_2^{m-l}\to L_2)}\leq c\,\|(1-\Delta)^{l/2}\,\gamma\|_{M(W_2^m\to L_2)}.$$

Since $(1-\Delta)^{l/2}\,\gamma\in M\,(W_2^m\to L_2)$, it follows that

$$\int_e |(1-\Delta)^{l/2}\,\gamma|^2\,dx\leq\ \|(1-\Delta)^{l/2}\,\gamma\|^2_{M(W_2^m\to L_2)}\,C_{2,m}\,(e),$$

for every compact set $e\subset\mathbb{R}^n$. Hence, for every ball $\mathcal{B}_r(a)$,

$$\int_{\mathcal{B}_r(a)}|(1-\Delta)^{l/2}\,\gamma|^2\,dx\leq c\,\|(1-\Delta)^{l/2}\,\gamma\|^2_{M(W_2^m\to L_2)}\,r^{n-2m},\quad 0<r\leq 1,$$

and in particular

$$\|(1-\Delta)^{l/2}\,\gamma\|_{L_2,\mathrm{unif}}\leq c\,\|(1-\Delta)^{l/2}\,\gamma\|_{M(W_2^m\to L_2)}.$$

In view of the properties (1.2.4), (1.2.5) of the Bessel function G_l, it is easy to derive the pointwise estimate

$$|\gamma(x)|\leq\int G_l(x-t)\,|(1-\Delta)^{l/2}\,\gamma(t)|\,dt$$

$$\leq c\left(\int_{|z|\leq 1}\frac{|(1-\Delta)^{l/2}\,\gamma(x+z)|}{|z|^{n-l}}\,dz+\|(1-\Delta)^{l/2}\,\gamma\|_{L_2,\mathrm{unif}}\right).$$

Using Lemma 2.3.7 together with the preceding pointwise estimate, we deduce that

$$|\gamma(x)|\leq c\,(\mathcal{M}\,(1-\Delta)^{l/2}\,\gamma(x))^{1-\frac{l}{m}}\left(\sup_{0<r\leq 1,\,a\in\mathbb{R}^n}\frac{\int_{\mathcal{B}_r(a)}|(1-\Delta)^{l/2}\,\gamma|^2dy}{r^{n-2m}}\right)^{\frac{l}{2m}}$$

$$+c\,\|(1-\Delta)^{l/2}\gamma\|_{L_2,\mathrm{unif}}\leq c\,(\mathcal{M}\,(1-\Delta)^{l/2}\,\gamma(x))^{1-\frac{l}{m}}\,\|(1-\Delta)^{l/2}\gamma\|^{\frac{l}{m}}_{M(W_2^m\to L_2)}$$

$$+\,c\,\|(1-\Delta)^{l/2}\gamma\|_{M(W_2^m\to L_2)},$$

where $\mathcal{M}$ is the Hardy–Littlewood maximal operator. Using the preceding estimates, together with the boundedness of $\mathcal{M}$ on the space $M\,(W_2^m\to L_2)$ (see (2.3.22)), we obtain

$$\left\||\gamma|^{\frac{m}{m-l}}\right\|^{1-\frac{l}{m}}_{M(W_2^m\to L_2)}\leq c\,\|(1-\Delta)^{l/2}\,\gamma\|_{M(W_2^m\to L_2)}.$$

By Lemma 2.3.6 it follows that

$$\|\gamma\|_{M(W_2^{m-l}\to L_2)}\leq c\,\left\||\gamma|^{\frac{m}{m-l}}\right\|^{1-\frac{l}{m}}_{M(W_2^m\to L_2)}\leq c\,\|(1-\Delta)^{l/2}\,\gamma\|_{M(W_2^m\to L_2)}.$$

The proof of Lemma 12.1.2 is complete. □

12.1.1 Main Result

Now we are in a position to obtain an analytic characterization of the space of multipliers $M(W_2^{1/2} \to W_2^{-1/2})$.

Theorem 12.1.1. *Let $\gamma \in (C_0^\infty)'$. Then*

$$\gamma \in M\,(W_2^{1/2} \to W_2^{-1/2})$$

if and only if

$$\varPhi = (-\varDelta + 1)^{-1/4}\gamma \in M\,(W_2^{1/2} \to L_2).$$

Furthermore,

$$\|\gamma\|_{M(W_2^{1/2} \to W_2^{-1/2})} \sim \|\varPhi\|_{M(W_2^{1/2} \to L_2)}.$$

Proof. To prove the "if" part, it suffices to verify that, for every $u \in C_0^\infty$ and $\varPhi = (-\varDelta + 1)^{-1/4}\gamma \in M\,(W_2^{1/2} \to L_2)$, the inequality

$$\left| \int |u|^2 \gamma\, dx \right| \le C\, \|\varPhi\|_{M(W_2^{1/2} \to L_2)}\, \|u\|^2_{W_2^{1/2}} \tag{12.1.19}$$

holds. Here the integral on the left-hand side is understood in the sense of quadratic forms:

$$\int |u|^2 \gamma\, dx = \langle\, \gamma u,\, u \rangle,$$

where $\langle\, \gamma\cdot,\, \cdot \rangle$ is the quadratic form associated with the multiplier operator γ, as explained in detail in [MV2].

Since $\gamma = (-\varDelta + 1)^{1/4}\varPhi$, we have

$$\left| \int |u|^2\, \gamma\, dx \right| = \left| \int (-\varDelta + 1)^{1/4}\varPhi \cdot\, |u|^2\, dx \right|$$

$$\le \left| \int \left((-\varDelta + 1)^{1/4} - |D|^{1/2} \right) \varPhi \cdot\, |u|^2\, dx \right| + \left| \int |D|^{1/2}\, \varPhi \cdot\, |u|^2\, dx \right|.$$

Note that

$$(-\varDelta + 1)^{1/4} - |D|^{1/2} = T_{m_{1/2}},$$

where T_{m_l} is the Fourier multiplier operator defined by (12.1.17). By Corollary 12.1.1, $T_{m_{1/2}}$ is a bounded operator on $L_2(w)$ for any A_1-weight w, and its norm depends only on the A_1-bound of w. Hence by Lemma 12.1.3 it follows that

$$\left((-\varDelta + 1)^{1/4} - |D|^{1/2} \right) \varPhi \in M\,(W_2^{1/2} \to L_2)$$

and

$$\left\| \left((-\varDelta + 1)^{1/4} - |D|^{1/2} \right) \varPhi \right\|_{M(W_2^{1/2} \to L_2)} \le C\, \|\varPhi\|_{M(W_2^{1/2} \to L_2)}.$$

Using this estimate and the Schwarz inequality, we get

$$\left|\int \left((-\Delta+1)^{1/4} - |D|^{1/2}\right) \Phi \cdot |u|^2\, dx\right|$$

$$\le C\, \|((-\Delta+1)^{1/4} - |D|^{1/2})\Phi \cdot u\|_{L_2}\, \|u\|_{L_2}$$
$$\le C\, \|\Phi\|_{M(W_2^{1/2}\to L_2)} \|u\|^2_{W_2^{1/2}}.$$

Hence, in order to prove (12.1.19), it suffices to establish the inequality

$$\left|\int |D|^{1/2}\, \Phi \cdot |u|^2\, dx\right| \le C\, \|\Phi\|_{M(W_2^{1/2}\to L_2)}\, \|u\|^2_{W_2^{1/2}}. \tag{12.1.20}$$

By duality,

$$\left|\int |D|^{1/2}\, \Phi \cdot |u|^2\, dx\right| = \left|\int \Phi\, (|D|^{1/2}\, |u|^2)\, dx\right|,$$

where $\Phi \in L_{2,\mathrm{loc}}$, and the integral on the right-hand side is well-defined.

Notice that for $u \in C_0^\infty$,

$$|D|^{1/2}\, |u|^2(x) = c \int \frac{|u(x)|^2 - |u(y)|^2}{|x-y|^{n+1/2}}\, dy.$$

Using the identity

$$|a|^2 - |b|^2 = |a-b|^2 - 2\mathrm{Re}\, [\bar{b}\, (b-a)]$$

with $b = u(x)$ and $a = u(y)$, and integrating against $dy/|x-y|^{n+1/2}$, we get

$$\int \frac{|u(x)|^2 - |u(y)|^2}{|x-y|^{n+1/2}}\, dy = \int \frac{|u(x)-u(y)|^2}{|x-y|^{n+1/2}}\, dy$$
$$-2\, \mathrm{Re} \left[\overline{u(x)} \int \frac{u(x)-u(y)}{|x-y|^{n+1/2}}\, dy\right].$$

Hence,

$$\left||D|^{1/2}\, |u|^2(x)\right| \le c \left(2\, |u(x)| \left|\int \frac{u(x)-u(y)}{|x-y|^{n+1/2}}\, dy\right| + \int \frac{|u(x)-u(y)|^2}{|x-y|^{n+1/2}}\, dy\right)$$
$$= 2c\, |u(x)| \left||D|^{1/2}\, u(x)\right| + c\, |D_{2,1/4} u(x)|^2.$$

Using the preceding inequality, we estimate

$$\left|\int \Phi\, |D|^{1/2}\, |u|^2\, dx\right|$$

$$\le c\, \|\Phi\, u\|_{L_2} \left\||D|^{1/2}\, u\right\|_{L_2} + c \int |\Phi|\, |D_{2,1/4} u|^2\, dx$$

$$\le c\,\|\Phi\|_{W_2^{1/2}\to L_2}\,\|u\|^2_{W_2^{1/2}} + c\int |\Phi|\,|D_{2,1/4}J_{1/2}f|^2\,dx,$$

where $f = (-1+\Delta)^{1/4}\,u$. The last integral is bounded by

$$\int |\Phi|\,|J_{1/4}\,D_{2,1/4}\,J_{1/4}\,f|^2\,dx$$

$$\le c\int |\Phi|\,M(D_{2,1/4}\,J_{1/4}\,f)\,|J_{1/2}\,D_{2,1/4}\,J_{1/4}\,f|\,dx$$

$$\le c\,\|\mathcal{M}(D_{2,1/4}\,J_{1/4}\,f)\|_{L_2}\,\|\Phi\,J_{1/2}\,D_{2,1/4}\,J_{1/4}\,f\|_{L_2}$$

$$\le c\,\|D_{2,1/4}\,J_{1/4}\,f\|_{L_2}\,\|\Phi\|_{M(W_2^{1/2}\to L_2)}\,\|J_{1/2}\,D_{2,1/4}\,J_{1/4}\,f\|_{W_2^{1/2}}$$

$$\le c\,\|\Phi\|_{M(W_2^{1/2}\to L_2)}\,\|f\|^2_{L_2} = c\,\|\Phi\|_{W_2^{1/2}\to L_2}\,\|u\|^2_{W_2^{1/2}}.$$

In the preceding chain of inequalities we first applied the inequality (1.2.28)

$$J_{1/4}\,g \le c\,(\mathcal{M}g)^{1/2}\,(J_{1/2}\,g)^{1/2},$$

with $g = |D_{2,1/4}\,J_{1/4}\,f|$, and then the Hardy–Littlewood maximal inequality for the operator $\mathcal{M}$. This completes the proof of (12.1.19).

To prove the "only if" part of the Theorem, we show that

$$\|\Phi\|_{M(W_2^{1/2}\to L_2)} \le c\,\|\gamma\|_{M(W_2^{1/2}\to W_2^{-1/2})}.$$

The proof of this estimate is based on the extension of the distribution $\gamma \in M\,(W_2^{1/2} \to W_2^{-1/2}))$ to the higher-dimensional Euclidean space, and subsequent application of the characterization of the class of multipliers $M\,(W_2^1(\mathbb{R}^{n+1}) \to W_2^{-1}(\mathbb{R}^{n+1}))$ obtained in Sect. 11.4.

We denote by $\gamma\otimes\delta$ the distribution on $\mathbb{R}^{n+1}$ defined by

$$\langle \gamma\otimes\delta,\, u(x, x_{n+1})\rangle = \langle \gamma,\, u(x,0)\rangle,$$

where $x = (x_1,\dots,x_n) \in \mathbb{R}^n$, and $\delta = \delta(x_{n+1})$ is the delta-function supported by $x_{n+1} = 0$. It is not difficult to see that

$$\|\gamma\otimes\delta; \mathbb{R}^{n+1}\|_{M(W_2^1\to W_2^{-1})} \sim \|\gamma; \mathbb{R}^n\|_{M(W_2^{1/2}\to W_2^{-1/2})}.$$

This follows from the well-known fact that the space of traces on $\mathbb{R}^n$ of functions in $W_2^1(\mathbb{R}^{n+1})$ coincides with $W_2^{1/2}(\mathbb{R}^n)$, with the equivalence of norms (see Lemma 8.7.1). Indeed, for any $U, V \in C_0^\infty(\mathbb{R}^{n+1})$ let

$$u(x) = U(x,0) \qquad \text{and} \qquad v(x) = V(x,0).$$

Then by the trace estimate mentioned above

$$\|u; \mathbb{R}^n\|_{W_2^{1/2}} \le c\,\|U; \mathbb{R}^{n+1}\|_{W_2^1}$$

and hence

$$|\langle \gamma \otimes \delta, \overline{U}\, V\rangle| = |\langle \gamma, \overline{u}\, v\rangle| \le \|\gamma; \mathbb{R}^n\|_{M(W_2^{1/2}\to W_2^{-1/2})} \|u; \mathbb{R}^n\|_{W_2^{1/2}} \|v; \mathbb{R}^n\|_{W_2^{1/2}}$$

$$\le c^2 \|\gamma; \mathbb{R}^n\|_{M(W_2^{1/2}\to W_2^{-1/2})} \|U; \mathbb{R}^{n+1}\|_{W_2^1} \|V; \mathbb{R}^{n+1}\|_{W_2^1}.$$

This gives the estimate

$$\|\gamma \otimes \delta; \mathbb{R}^{n+1}\|_{M(W_2^1\to W_2^{-1})} \le c^2 \|\gamma; \mathbb{R}^n\|_{M(W_2^{1/2}\to W_2^{-1/2})}.$$

The converse inequality (which is not used below) follows similarly by extending $u, v \in C_0^\infty(\mathbb{R}^n)$ to $U, V \in W_2^1(\mathbb{R}^{n+1})$ with the corresponding estimates of norms.

For the rest of the proof, it will be convenient to introduce the notation

$$J_s^{(n+1)} = (-\Delta_{n+1} + 1)^{-s/2}, \quad s > 0,$$

for the Bessel potential of order s on $\mathbb{R}^{n+1}$; here Δ_{n+1} denotes the Laplacian on $\mathbb{R}^{n+1}$.

Now by Theorem 11.4.1 we obtain that

$$\gamma \otimes \delta \in M\,(W_2^1(\mathbb{R}^{n+1}) \to W_2^{-1}(\mathbb{R}^{n+1}))$$

if and only if

$$J_1^{(n+1)}(\gamma \otimes \delta) \in M\,(W_2^1(\mathbb{R}^{n+1}) \to L_2(\mathbb{R}^{n+1})),$$

and

$$\|J_1^{(n+1)}(\gamma \otimes \delta); \mathbb{R}^{n+1}\|_{M(W_2^1\to L_2)} \le c\, \|\gamma \otimes \delta; \mathbb{R}^{n+1}\|_{M(W_2^1\to W_2^{-1})}$$

$$\le c_1 \|\gamma; \mathbb{R}^n\|_{M(W_2^{1/2}\to W_2^{-1/2})}.$$

Next, pick $0 < \epsilon < 1/2$ and observe that

$$J_1^{(n+1)} = (-1 + \Delta_{n+1})^{1/4+\epsilon/2}\, J_{\epsilon+3/2}^{(n+1)}.$$

Using Lemma 12.1.2 with $l = 1/2 + \epsilon$, $m = 1$, and $J_{\epsilon+3/2}^{(n+1)}(\gamma \otimes \delta)$ in place of γ, we deduce that

$$\|J_1^{(n+1)}(\gamma \otimes \delta); \mathbb{R}^{n+1}\|_{M(W_2^1\to L_2)} \sim \|J_{\epsilon+3/2}^{(n+1)}(\gamma \otimes \delta); \mathbb{R}^{n+1}\|_{M(W_2^1\to W_2^{1/2+\epsilon})}.$$

As was proved above, the left-hand side of the preceding relation is bounded by a constant multiple of $\|\gamma; \mathbb{R}^n\|_{M(W_2^{1/2}\to W_2^{-1/2})}$.

Thus,

$$\|J_{\epsilon+3/2}^{(n+1)}(\gamma \otimes \delta); \mathbb{R}^{n+1}\|_{M(W_2^1\to W_2^{1/2+\epsilon})} \le c\, \|\gamma; \mathbb{R}^n\|_{M(W_2^{1/2}\to W_2^{-1/2})}.$$

Passing to the trace on $\mathbb{R}^n = \{x_{n+1} = 0\}$ in the multiplier norm on the left-hand side, we obtain

$$\|\mathrm{Trace}\, J^{(n+1)}_{\epsilon+3/2}(\gamma \otimes \delta); \mathbb{R}^n\|_{M(W_2^{1/2} \to W_2^{\epsilon})} \le c\, \|\gamma; \mathbb{R}^n\|_{M(W_2^{1/2} \to W_2^{-1/2})}.$$

We now observe that

$$\mathrm{Trace}\, J^{(n+1)}_{\epsilon+3/2}(\gamma \otimes \delta) = \mathrm{const}\, J^{(n)}_{\epsilon+1/2}(\gamma),$$

which follows immediately by inspecting the corresponding Fourier transforms.

In other words,

$$\|J^{(n)}_{\epsilon+1/2}\, \gamma; \mathbb{R}^n\|_{M(W_2^{1/2} \to W_2^{\epsilon})} \le c\, \|\gamma; \mathbb{R}^n\|_{M(W_2^{1/2} \to W_2^{-1/2})}. \tag{12.1.21}$$

From this estimate and Lemma 12.1.2 with $l = \epsilon$, $m = 1/2$, and with γ replaced by $J^{(n)}_{\epsilon+1/2}\gamma$, it follows that

$$\|J^{(n)}_{1/2}\, \gamma\|_{M(W_2^{1/2} \to L_2)} = \|(-\Delta + 1)^{\epsilon/2} J^{(n)}_{\epsilon+1/2}\gamma\|_{M(W_2^{1/2} \to L_2)}$$

$$\le c\, \|J^{(n)}_{\epsilon+1/2}\gamma\|_{M(W_2^{1/2} \to W_2^{\epsilon})} \le C\, \|\gamma\|_{M(W_2^{1/2} \to W_2^{-1/2})}.$$

Thus,

$$\Phi = J^{(n)}_{1/2}\, \gamma \in M\, (W_2^{1/2} \to L_2)$$

and

$$\|\Phi\|_{M(W_2^{1/2} \to L_2)} \le C\, \|\gamma\|_{M(W_2^{1/2} \to W_2^{-1/2})}.$$

The proof of Theorem 12.1.1 is complete. □

The theorem just proved can be reformulated as follows.

Theorem 12.1.2. *Let $Q \in (C_0^\infty)'$, $n \ge 1$. The following statements are equivalent:*

(i) The relativistic Schrödinger operator $\mathcal{H} = \sqrt{-\Delta} + Q$ is bounded from $W_2^{1/2}$ to $W_2^{-1/2}$.

(ii) The inequality

$$|\langle Qu,\, u\rangle| \le C\, \|u\|^2_{W_2^{1/2}} \tag{12.1.22}$$

holds for all $u \in C_0^\infty$.

(iii) $\Phi = (-\Delta + 1)^{-1/4} Q \in L_{2,\,\mathrm{loc}}$, and the inequality

$$\int |u(x)|^2\, |\Phi(x)|^2\, dx \le C\, \|u\|^2_{W_2^{1/2}} \tag{12.1.23}$$

holds for all $u \in C_0^\infty$.

12.2 Corollaries of the Form Boundedness Criterion and Related Results

Theorem 12.1.1 combined with the criteria for the trace inequality with a nonnegative measure (see Theorem 3.1.4 and Remark 3.1.3) implies

Theorem 12.2.1. *Let $Q \in (C_0^\infty)'$, $n \geq 1$, and let $\mathcal{H} = \sqrt{-\Delta} + Q$. Then $\mathcal{H} : W_2^{1/2} \to W_2^{-1/2}$ is bounded if and only if*

$$\Phi = (-\Delta + 1)^{-1/4} Q \in L_{2,\,\mathrm{loc}}$$

and any one of the following equivalent conditions holds:

(i) For every compact set $e \subset \mathbb{R}^n$,

$$\int_e |\Phi(x)|^2\, dx \leq C\, C_{2,1/2}(e), \tag{12.2.1}$$

where the constant does not depend on e.

(ii) The function $J_{1/2}\, |\Phi|^2$ is finite a.e., *and*

$$J_{1/2}\, \left(J_{1/2}\, |\Phi|^2\right)^2 (x) \leq C\, J_{1/2}\, |\Phi|^2(x) \qquad \text{a.e.} \tag{12.2.2}$$

Here $J_{1/2} = (-\Delta + 1)^{-1/4}$ is the Bessel potential of order $1/2$.

(iii) For every dyadic cube P_0 in $\mathbb{R}^n$ of side length $\ell(P_0) \leq 1$,

$$\sum_{P \subseteq P_0} \left[\frac{\int_P |\Phi(x)|^2\, dx}{(\mathrm{mes}_n P)^{1-1/2n}} \right]^2 \mathrm{mes}_n P \leq C \int_{P_0} |\Phi(x)|^2\, dx, \tag{12.2.3}$$

where the sum is taken over all dyadic cubes P contained in P_0, and the constant does not depend on P_0.

Some simpler either necessary or sufficient conditions which do not involve capacities are discussed in this section.

The following *necessary* condition is immediate from (12.2.1) and the known estimates of the capacity of the ball in $\mathbb{R}^n$ (see Sect. 1.2.2).

Corollary 12.2.1. *Suppose that $Q \in (C_0^\infty)'$, $n \geq 1$. Suppose also that $\mathcal{H} = \sqrt{-\Delta} + Q : W_2^{1/2} \to W_2^{-1/2}$ is a bounded operator. Then, for every ball $\mathcal{B}_r(a)$ in $\mathbb{R}^n$,*

$$\int_{\mathcal{B}_r(a)} |\Phi(x)|^2\, dx \leq c\, r^{n-1}, \quad 0 < r \leq 1, \qquad n \geq 2, \tag{12.2.4}$$

and

$$\int_{\mathcal{B}_r(a)} |\Phi(x)|^2\, dx \leq \frac{c}{\log \frac{2}{r}}, \quad 0 < r \leq 1, \qquad n = 1, \tag{12.2.5}$$

where the constant c does not depend on $a \in \mathbb{R}^n$ and r.

We notice that the class of distributions Q such that $\Phi = (-\Delta+1)^{-1/4}Q$ satisfies (12.2.4) can be regarded as a Morrey space of order $-1/2$.

Combining Theorem 12.2.1 with the Fefferman-Phong condition (see Sect. 1.2.6) applied to $|\Phi|^2$, we arrive at *sufficient* conditions involving Morrey spaces of negative order. (Strictly speaking, the Fefferman-Phong condition [F2] was originally established for estimates in the homogeneous Sobolev space w_2^1 of order $m = 1$. However, it can be carried over to Sobolev spaces W_2^m for all $0 < m \leq n/2$. See, e.g., [KeS] or [MV1], p. 98.)

Corollary 12.2.2. *Suppose that $Q \in (C_0^\infty)'$, $n \geq 2$. Suppose also that $\Phi = (-\Delta+1)^{-1/4}Q$, and $t > 1$. Then $\mathcal{H}$ is a bounded operator from $W_2^{1/2}$ to $W_2^{-1/2}$ if*

$$\int_{\mathcal{B}_r(a)} |\Phi(x)|^{2t}\, dx \leq C\, r^{n-t}, \quad 0 < r \leq 1, \tag{12.2.6}$$

where the constant does not depend on $a \in \mathbb{R}^n$ and r.

Remark 12.2.1. It is worth mentioning that the condition (12.2.6) defines a class of potentials which is strictly broader than the (relativistic) Fefferman-Phong class of Q such that

$$\int_{\mathcal{B}_r(a)} |Q(x)|^t\, dx \leq \text{const}\, r^{n-t}, \quad 0 < r \leq 1, \qquad n \geq 2, \tag{12.2.7}$$

for some $t > 1$.

This follows from the observation that if one replaces Q by $|Q|$ in (12.2.6), then obviously the resulting class defined by

$$\int_{\mathcal{B}_r(a)} (J_{1/2}|Q|)^{2t}\, dx \leq \text{const}\, r^{n-t}, \quad 0 < r \leq 1, \qquad n \geq 2, \tag{12.2.8}$$

becomes smaller, but still contains some singular measures, together with all functions in the Fefferman-Phong class (12.2.7). (This was noticed earlier in [MV1], Proposition 3.5.)

A smaller but more conventional class of admissible potentials appears when one replaces $C_{2,1/2}(e)$ on the right-hand side of (12.2.1) by its lower bound involving the Lebesgue measure of $e \subset \mathbb{R}^n$, as shown by the following result.

Corollary 12.2.3. *Suppose that $Q \in (C_0^\infty)'$, $n \geq 1$. Suppose also that $\Phi = (-\Delta+1)^{-1/4}Q$. Then $\mathcal{H} = \sqrt{-\Delta} + Q$ is a bounded operator from $W_2^{1/2}$ to $W_2^{-1/2}$ if, for every measurable set $e \subset \mathbb{R}^n$,*

$$\int_e |\Phi(x)|^2\, dx \leq c\, (\text{mes}_n e)^{(n-1)/n}, \quad \text{diam}\,(e) \leq 1, \qquad n \geq 2, \tag{12.2.9}$$

or

$$\int_e |\Phi(x)|^2\, dx \le \frac{c}{\log \frac{2}{\text{mes}_n e}}, \qquad \text{diam}\,(e) \le 1, \qquad n = 1, \tag{12.2.10}$$

where the constant c does not depend on e.

We remark that (12.2.9), without the extra assumption diam $(e) \le 1$, is equivalent to $\Phi \in L_{2n,\infty}$, where $L_{p,\infty}$ is the Lorentz (weak L_p) space of functions f such that

$$|\{x \in \mathbb{R}^n : |f(x)| > t\}| \le \frac{C}{t^p}, \qquad t > 0.$$

In particular, (12.2.9) holds if $\Phi \in L_{2n}$ or, equivalently, $Q \in W_{2n}^{-1/2}$.

Furthermore, if $\Phi \in L_\infty$, then obviously (12.2.9) holds as well, since

$$C_{2,1/2}(e) \ge C\, \text{mes}_n e,$$

if diam $(e) \le 1$. This leads to the sufficient condition $\Phi \in L_{2n} + L_\infty$, $n \ge 2$.

It is worth noting that (12.2.9) defines a substantially broader class of admissible potentials than the standard (in the relativistic case) class $Q \in L_n + L_\infty$, $n \ge 2$ ([LL], Sec. 11.3). This is a consequence of the imbedding

$$L_n \subset W_{2n}^{-1/2}, \qquad n \ge 2,$$

which follows from the classical Sobolev imbedding $W_p^{1/2} \subset L_r$ for $p = 2n/(2n-1)$ and $r = n/(n-1)$, $n \ge 2$. Indeed, by duality, the latter is equivalent to

$$L_n = (L_r)' \subset (W_p^{1/2})' = W_{2n}^{-1/2}.$$

Similarly, in the one-dimensional case, the class of potentials defined by (12.2.10) is wider than the standard class $L_{1+\epsilon}(\mathbb{R}^1) + L_\infty(\mathbb{R}^1)$, $\epsilon > 0$.

It is easy to see that actually $Q \in L_n(\mathbb{R}^n) + L_\infty(\mathbb{R}^n)$ if $n \ge 2$, or $Q \in L_{1+\epsilon}(\mathbb{R}^1) + L_\infty(\mathbb{R}^1)$ if $n = 1$, is sufficient for the inequality

$$\int |u(x)|^2\, |Q(x)|\, dx \le C\, \|u\|^2_{W_2^{1/2}}, \quad u \in C_0^\infty,$$

which is a "naïve" version of (12.1.4), where Q is replaced by $|Q|$.

We conclude this chapter with mentioning the article by Frank and Seiringer [FrS] which is mostly devoted to sharp Hardy type inequalities involving Besov type seminorms and their generalizations. In particular, the authors found the optimal value of the constant $\mathcal{C}_{n,s,p}$ in the inequality

$$\int_{\mathbb{R}^n}\int_{\mathbb{R}^n} \frac{|u(x) - u(y)|^p}{|x-y|^{n+ps}}\, dx\, dy \ge \mathcal{C}_{n,s,p} \int_{\mathbb{R}^n} \frac{|u(x)|^p}{|x|^{ps}}\, dx, \tag{12.2.11}$$

where $0 < s < 1$ and $1 \le p < n/s$. Moreover, they extend this result to functionals on the left-hand side of (12.2.11) with $|x-y|^{-n-ps}$ replaced by an arbitrary symmetric and nonnegative, but not necessarily translation invariant, kernel $k(x,y)$:

$$E[u] := \int_{\mathbb{R}^n}\int_{\mathbb{R}^n} |u(x) - u(y)|^p k(x,y)\, dx\, dy.$$

In [FrS], a sufficient condition for the following general version of Hardy's inequality

$$\int_{\mathbb{R}^n} Q(x)\, |u(x)|^p dx \le E[u], \quad u \in C_0^\infty(\mathbb{R}^n),$$

is found. The function Q is assumed to be of the form

$$Q(x) = 2\,\omega(x)^{1-p} \int_{\mathbb{R}^n} (\omega(x) - \omega(y))|\omega(x) - \omega(y)|^{p-2}\, k(x,y)\, dy \qquad (12.2.12)$$

with a certain positive function ω. The integral on the right-hand side of (12.2.12) might be divergent and some regularization of the principal value type is needed for its definition. In particular, the representation

$$Q(x) = 2\int_{\mathbb{R}^n} \Big(1 - \frac{\omega(x)}{\omega(y)}\Big) \frac{dy}{|x-y|^{n+1}} \qquad (12.2.13)$$

with $\omega > 0$ proves to be sufficient for the inequality

$$\int_{\mathbb{R}^n} Q(x)|u(x)|^2\, dx \le \int_{\mathbb{R}^n}\int_{\mathbb{R}^n} \frac{|u(x) - u(y)|^2}{|x-y|^{n+1}} dx\, dy, \quad u \in C_0^\infty(\mathbb{R}^n). \quad (12.2.14)$$

Note that (12.2.14) does not imply that $Q \in M(w_2^{1/2} \to w_2^{-1/2})$ if Q is not nonnegative. However, for $Q \ge 0$ the representation (12.2.13) with a positive factor in place of 2 is sufficient for the inclusion $Q \in M(w_2^{1/2} \to w_2^{-1/2})$ (and hence for $Q \in M(W_2^{1/2} \to W_2^{-1/2})$). It is of interest to investigate the question of necessity.

13

Multipliers as Solutions to Elliptic Equations

In Sects. 13.1–13.3 of this chapter, solutions of second-order linear and quasi-linear elliptic differential equations and systems are considered as multipliers in certain spaces of differentiable functions in a domain Ω. On one hand, this can be of interest for the theory of functions, since it leads to new characterizations of multipliers and, on the other hand, for the theory of partial differential equations, since it allows us to obtain a priori information about the solutions in spaces different from the usual ones.

In Sect. 13.4 we obtain coercive estimates in multiplier spaces for solutions of linear elliptic systems in a half-space. The last Sect. 13.5 is devoted to regularity of solutions to higher order semilinear elliptic equations.

13.1 The Dirichlet Problem for the Linear Second-Order Elliptic Equation in the Space of Multipliers

Let us start with a multiplier analogue of the classical unique solvability of a linear second-order equation in the variational sense.

By Ω we denote a bounded domain in $\mathbb{R}^n$ with $\partial\Omega \in C^{0,1}$. Let

$$\mathcal{L}u = -\sum_{i,j=1}^{n} \frac{\partial}{\partial x_i}\Big(a_{ij}(x)\frac{\partial u}{\partial x_j}\Big). \tag{13.1.1}$$

Suppose that the coefficients a_{ij} are real measurable bounded functions on Ω and that the matrix $\|a_{ij}\|$ is symmetric and uniformly positive definite.

We consider the Dirichlet problem

$$\mathcal{L}u = 0 \text{ in } \Omega, \quad u - g \in \mathring{W}_2^1(\Omega),$$

where $g \in W_2^1(\Omega)$. This problem is uniquely solvable.

Grundlehren der mathematischen Wissenschaften 337,

Theorem 13.1.1. *If $g \in MW_2^1(\Omega)$, then $u \in MW_2^1(\Omega)$. Moreover,*

$$u - g \in M(W_2^1(\Omega) \to \mathring{W}_2^1(\Omega))$$

and

$$\|u;\Omega\|_{MW_2^1} \le c\,\|g;\Omega\|_{MW_2^1}\,. \tag{13.1.2}$$

Proof. By the maximum principle for variational solutions of the equation $\mathcal{L}u = 0$ we have

$$\|u;\Omega\|_{L_\infty} \le \|g;\Omega\|_{L_\infty}\,. \tag{13.1.3}$$

Hence the function $\gamma = u - g$ belongs to $\mathring{W}_2^1(\Omega) \cap L_\infty(\Omega)$. The definition of a variational solution yields

$$\int_\Omega a_{ij}\frac{\partial\varphi}{\partial x_i}\frac{\partial\gamma}{\partial x_j}\,dx = -\int_\Omega a_{ij}\frac{\partial\varphi}{\partial x_i}\frac{\partial g}{\partial x_j}\,dx \tag{13.1.4}$$

for any $\varphi \in \mathring{W}_2^1(\Omega)$. Let v be an arbitrary function in $W_2^1(\Omega) \cap L_\infty(\Omega)$. It is easily seen that γv and γv^2 belong to $\mathring{W}_2^1(\Omega)$. We set $\varphi = \gamma v^2$ in (13.1.4). Then

$$\int_\Omega a_{ij}\frac{\partial(\gamma v)}{\partial x_i}\frac{\partial(\gamma v)}{\partial x_j}\,dx = \int_\Omega a_{ij}\frac{\partial v}{\partial x_i}\frac{\partial v}{\partial x_j}\gamma^2\,dx - \int_\Omega a_{ij}\frac{\partial(\gamma v)}{\partial x_i}v\frac{\partial g}{\partial x_j}\,dx - \int_\Omega \gamma\, v\, a_{ij}\frac{\partial v}{\partial x_i}\frac{\partial g}{\partial x_j}\,dx\,.$$

Consequently,

$$c\,\|\nabla(\gamma v);\Omega\|_{L_2}^2 \le \|\gamma;\Omega|_{L_\infty}^2\|\nabla v;\Omega\|_{L_2}^2 + \|\nabla(\gamma v);\Omega\|_{L_2}\|v\nabla g;\Omega\|_{L_2} + \|\gamma;\Omega\|_{L_\infty}\|\nabla v;\Omega\|_{L_2}\|v\nabla g;\Omega\|_{L_2}\,. \tag{13.1.5}$$

Clearly,

$$\|v\nabla g;\Omega\|_{L_2} \le \|vg;\Omega\|_{W_2^1} + \|g;\Omega\|_{L_\infty}\|v;\Omega\|_{W_2^1}\,.$$

Since

$$\|g;\Omega\|_{L_\infty} \le \|g;\Omega\|_{MW_2^1},$$

it follows that

$$\|v\nabla g;\Omega\|_{L_2} \le 2\,\|g;\Omega\|_{MW_2^1}\|v;\Omega\|_{W_2^1}\,.$$

This inequality, together with (13.1.3) and (13.1.5), implies that

$$c\,\|\nabla(\gamma v);\Omega\|_{L_2}^2 \le 8\,\|g;\Omega\|_{MW_2^1}^2\|v;\Omega\|_{W_2^1}^2 + 2\,\|\nabla(\gamma v);\Omega\|_{L_2}\|g;\Omega\|_{MW_2^1}\|v;\Omega\|_{W_2^1}\,.$$

Hence

$$\|\gamma v;\Omega\|_{W_2^1} \le c\,\|g;\Omega\|_{MW_2^1}\|v;\Omega\|_{W_2^1}$$

or, which is the same,

$$\|uv;\Omega\|_{W_2^1} \le c\,\|g;\Omega\|_{MW_2^1}\|v;\Omega\|_{W_2^1}\,. \tag{13.1.6}$$

Since $W_2^1(\Omega) \cap L_\infty(\Omega)$ is dense in $W_2^1(\Omega)$ and $\gamma v \in \mathring{W}_2^1(\Omega)$ for all $v \in W_2^1(\Omega) \cap L_\infty(\Omega)$, we have (13.1.6) and $\gamma v \in \mathring{W}_2^1(\Omega)$ for any $v \in W_2^1(\Omega)$. □

Remark 13.1.1. Let Ω be a bounded domain in $\mathbb{R}^n$ with $\partial\Omega \in C^{0,1}$. By Theorem 8.7.2, $MW_2^{1/2}(\mathbb{R}^{n-1})$ is the space of traces on $\mathbb{R}^{n-1}$ of functions in $MW_2^1(\mathbb{R}^n_+)$. This clearly implies that $\varphi \in MW_2^{1/2}(\partial\Omega)$ has an extension $g \in MW_2^1(\Omega)$ for which

$$\|g;\Omega\|_{MW_2^1} \sim \|\varphi;\partial\Omega\|_{MW_2^{1/2}}\,.$$

This, together with Theorem 13.1.1, proves the unique solvability of the Dirichlet problem

$$\mathcal{L}u = 0 \quad \text{in} \quad \Omega\,, \qquad u|_{\partial\Omega} = \varphi \in MW_2^{1/2}(\partial\Omega)$$

in $MW_2^1(\Omega)$.

Remark 13.1.2. Let Ω be a bounded domain in $\mathbb{R}^n$ with $\partial\Omega \in C^{0,1}$. In Theorem 13.1.1 we used the space $M(W_p^m(\Omega) \to \mathring{W}_p^l(\Omega))$. Let us show that

$$M(W_p^m(\Omega) \to \mathring{W}_p^l(\Omega)) = \mathring{W}_p^l(\Omega) \cap M(W_p^m(\Omega) \to W_p^l(\Omega))\,.$$

We denote the left-hand side of this equality by A and the right-hand side by B. Since $1 \in W_p^m(\Omega)$, it follows that $A \subset B$. Let $u \in W_p^m(\Omega)$, $\gamma \in B$ and let $\{u_\nu\}_{\nu\ge1}$ be a sequence of functions in $C^\infty(\overline{\Omega})$ such that $u_\nu \to u$ in $W_p^m(\Omega)$. Then $\gamma u_\nu \in \mathring{W}_p^l(\Omega)$ and

$$\|\gamma u - \gamma u_\nu;\Omega\|_{W_p^l} \le \|\gamma;\Omega\|_{M(W_p^m \to W_p^l)}\|u - u_\nu;\Omega\|_{W_p^m} = o(1)\,.$$

Consequently $\gamma u \in \mathring{W}_p^l(\Omega)$, that is, $\gamma \in A$.

13.2 Bounded Solutions of Linear Elliptic Equations as Multipliers

13.2.1 Introduction

In this section we study bounded solutions of a linear elliptic second-order equation without any requirements on their boundary values. It is shown that, under some conditions on the right-hand side of the equation, such solutions are multipliers in certain function spaces.

Let Ω be a domain in $\mathbb{R}^n$ with compact closure and sufficiently smooth boundary $\partial\Omega$. By $W^1_{2,\beta}(\Omega), \beta \in \mathbb{R}^1$ we denote the space of functions $u \in W^1_{2,\mathrm{loc}}(\Omega)$, having the finite norm

$$\|u;\Omega\|_{W^1_{2,\beta}} = \left(\int_\Omega \rho(x)^\beta |\nabla u|^2 dx \; + \|u;\Omega\|^2_{L_2}\right)^{1/2},$$

where $\rho(x) = \text{ dist } (x, \partial\Omega)$.

Theorem 13.2.1 below states that bounded solutions of the above mentioned equations are multipliers in the space $W^1_{2,\beta}(\Omega)$ for $\beta > 1$. For $\beta < 1$ this fact is not true since the space of traces of functions from $W^1_{2,\beta}(\Omega)$ with $\beta < 1$ is $W_2^{(1-\beta)/2}(\partial\Omega)$, which does not contain all bounded functions on $\partial\Omega$. The case $\beta = 1$ is special. It is considered in Theorems 13.2.2-13.2.4, in which solutions from $L_\infty(\Omega)$ appear as multipliers acting into the space $W^1_{2,1}(\Omega)$ from some function spaces more narrow than $W^1_{2,1}(\Omega)$. In Theorems 13.2.3 and 13.2.4 we deal with the weighted Hilbert space $W^1_{2,w(\rho)}(\Omega)$ similar to $W^1_{2,\beta}(\Omega)$, where the role of ρ^β is played by a weight $w(\rho)$. Here it is shown that all bounded solutions belong to the class $M(W^1_{2,w(\rho)}(\Omega) \to W^1_{2,1}(\Omega))$ if and only if $1/w \in L(0,1)$.

13.2.2 The Case $\beta > 1$

Let $\mathcal{L}$ be the uniformly elliptic operator (13.1.1) with sufficiently smooth coefficients in $\bar{\Omega}$ and let $a_{ij} = a_{ji}$. We consider the equation

$$\mathcal{L}\gamma = f + \text{div } g, \tag{13.2.1}$$

where f is a scalar function while g is a vector-valued function from $L_{2,\mathrm{loc}}(\Omega)$. By a variational solution of (13.2.1) we mean a function $\gamma \in W^1_{2,\mathrm{loc}}(\Omega)$, satisfying

$$\int_\Omega a_{ij} \frac{\partial \gamma}{\partial x_j}\frac{\partial \eta}{\partial x_i} dx = \int_\Omega (f\eta - g \; \nabla\eta) dx, \tag{13.2.2}$$

where η is an arbitrary function from $W^1_2(\Omega)$ with compact support in Ω.

Theorem 13.2.1. *Let $\beta > 1$ and let the functions f and g satisfy the condition*

$$\rho^{\beta/2}(|f|^{1/2} + |g|) \in M(W^1_{2,\beta}(\Omega) \to L_2(\Omega)).$$

Then any variational solution $\gamma \in L_\infty(\Omega)$ of equation (13.2.1) *belongs to the space $MW^1_{2,\beta}(\Omega)$, and the estimate*

$$\|\gamma;\Omega\|_{MW^1_{2,\beta}} \le c\,\big(\|\rho^{\beta/2}|f|^{1/2};\Omega\|^2_{M(W^1_{2,\beta}\to L_2)}$$

$$+ \|\rho^{\beta/2} g;\Omega\|_{M(W^1_{2,\beta}\to L_2)} + \|\gamma;\Omega\|_{L_\infty}\big) \tag{13.2.3}$$

holds with a constant c independent of f, g and γ.

Proof. Let R denote a solution of the Dirichlet problem

$$\mathcal{L}R = 1 \quad \text{in } \Omega, \qquad R = 0 \text{ on } \partial\Omega.$$

From Giraud's theorem on the sign of the normal derivative (see[Mir], Sect. 3.5) and the boundedness of the gradient of R it follows that $c_1\rho \le R \le c_2\rho$ on Ω. Choosing a sufficiently small $\delta > 0$, we introduce a family of functions $\{\zeta_\tau\}$ from $C_0^\infty(\Omega)$, $0 < \tau < \delta$, such that $\zeta_\tau = 1$ on the set

$$\Omega_\tau = \{x \in \Omega : \rho(x) > \tau\},$$

and $0 \le \zeta_\tau \le 1$, $|D^\alpha \zeta_\tau| \le c\,\tau^{-|\alpha|}$ for all multi-indices α.

We set

$$\eta = R^\beta \zeta_\tau^2 u^2 \gamma$$

in (13.2.2), where u is an arbitrary function from $W^1_{2,\beta}(\Omega)$. Then the left-hand side of (13.2.2) can be written as

$$\int_\Omega a_{ij} \frac{\partial \gamma}{\partial x_j} \frac{\partial \eta}{\partial x_i} dx = \int_\Omega a_{ij} \frac{\partial(\gamma u)}{\partial x_j} \frac{\partial(\gamma u)}{\partial x_i} R^\beta \zeta_\tau^2 dx - \int_\Omega a_{ij} \frac{\partial u}{\partial x_j} \frac{\partial u}{\partial x_i} \gamma^2 R^\beta \zeta_\tau^2 dx$$

$$+ \int_\Omega a_{ij} \frac{\partial(\gamma u)}{\partial x_j} \frac{\partial(R^\beta \zeta_\tau^2)}{\partial x_i} \gamma u\, dx - \int_\Omega a_{ij} \frac{\partial u}{\partial x_j} \frac{\partial(R^\beta \zeta_\tau^2)}{\partial x_i} \gamma^2 u\, dx. \tag{13.2.4}$$

Since

$$|\nabla(R^\beta \zeta_\tau^2)| \le c\, R^{\beta-1} \zeta_\tau,$$

the third term on the right-hand side of (13.2.4) does not exceed

$$c\, \|\gamma; \Omega\|_{L_\infty} \int_\Omega \zeta_\tau |\nabla(\gamma u)|\, |u|\, \rho^{\beta-1} dx$$

$$\le c\, \|\gamma; \Omega\|_{L_\infty} \|\zeta_\tau |\nabla(\gamma u)| \rho^{\beta/2}; \Omega\|_{L_2} \|u\, \rho^{(\beta-2)/2}; \Omega\|_{L_2}. \tag{13.2.5}$$

By the Hardy inequality,

$$\|u\, \rho^{(\beta-2)/2}; \Omega\|_{L_2} \le c\, \|u; \Omega\|_{W^1_{2,\beta}}. \tag{13.2.6}$$

We estimate the fourth term on the right-hand side of (13.2.4) using (13.2.6):

$$\left| \int_\Omega a_{ij} \frac{\partial u}{\partial x_j} \frac{\partial(R^\beta \zeta_\tau^2)}{\partial x_i} \gamma^2 u\, dx \right| \le c\, \|\gamma; \Omega\|^2_{L_\infty} \int_\Omega |\nabla u|\, |u|\, \rho^{\beta-1} dx$$

$$\le c\, \|\gamma; \Omega\|^2_{L_\infty} \|u; \Omega\|^2_{W^1_{2,\beta}}$$

Consequently, (13.2.4) implies that

$$\int_\Omega a_{ij}\frac{\partial(\gamma u)}{\partial x_j}\frac{\partial(\gamma u)}{\partial x_i}R^\beta\zeta_\tau^2\,dx \le c\,\|\gamma;\Omega\|^2_{L_\infty}\|u;\Omega\|^2_{W^1_{2,\beta}}$$

$$+\Big|\int_\Omega \big(fR^\beta\zeta_\tau^2u^2\gamma - g\,\nabla(R^\beta\zeta_\tau^2u^2\gamma)\big)dx\Big|. \tag{13.2.7}$$

We have

$$\Big|\int_\Omega fR^\beta\zeta_\tau^2u^2\gamma\,dx\Big|$$

$$\le c\,\|\gamma;\Omega\|_{L_\infty}\|\rho^{\beta/2}|f|^{1/2};\Omega\|^2_{M(W^1_{2,\beta}\to L_2)}\|u;\Omega\|^2_{W^1_{2,\beta}}. \tag{13.2.8}$$

Also,

$$\Big|\int_\Omega g\nabla(R^\beta\zeta_\tau^2u^2\gamma)dx\Big| \le c\int_\Omega |g|\,\rho^\beta\zeta_\tau|u|\,|\nabla(\gamma u)|\,dx$$

$$+c\,\|\gamma;\Omega\|_{L_\infty}\int_\Omega |gu|\,|\nabla(R^\beta\zeta_\tau u)|dx \le$$

$$\le c\,\|\rho^{\beta/2}\,g;\Omega\|_{M(W^1_{2,\beta}\to L_2)}\,\|\zeta_\tau\,\rho^{\beta/2}\,\nabla(\gamma u);\Omega\|_{L_2}\,\|u;\Omega\|_{W^1_{2,\beta}}$$

$$+c\,\|\gamma;\Omega\|_{L_\infty}\|\rho^{\beta/2}\,g\,u;\Omega\|_{L_2}\,\big\|\rho^{-\beta/2}\,\nabla(R^\beta\zeta_\tau^2u);\Omega\big\|_{L_2}.$$

Since by (13.2.6) the last norm does not exceed $c\,\|u;\Omega\|_{W^1_{2,\beta}}$, it follows that

$$\Big|\int_\Omega g\;\nabla(R^\beta\zeta_\tau^2u^2\gamma)dx\Big| \le c\,\|\rho^{\beta/2}g;\Omega\|_{M(W^1_{2,\beta}\to L_2)}\big(\|\zeta_\tau\,\rho^{\beta/2}\,\nabla(\gamma u);\Omega\|_{L_2}$$

$$+\,\|\gamma;\Omega\|_{L_\infty}\|u;\Omega\|_{W^1_{2,\beta}}\big)\|u;\Omega\|_{W^1_{2,\beta}}.$$

Combining this estimate with (13.2.8) and (13.2.7), we find that

$$\|\zeta_\tau\,\rho^{\beta/2}\,\nabla(\gamma u);\Omega\|_{L_2} \le c\,\big(\|\gamma;\Omega\|_{L_\infty} + \|\rho^{\beta/2}\,|f|^{1/2};\Omega\|^2_{M(W^1_{2,\beta}\to L_2)}$$

$$+\|\rho^{\beta/2}\,g;\Omega\|_{M(W^1_{2,\beta}\to L_2)}\big)\,\|u;\Omega\|_{W^1_{2,\beta}}.$$

Passing to the limit as $\tau\to 0$, we complete the proof. □

Remark 13.2.1. The same proof shows that for $0\le\beta<1$ the assertion of Theorem 13.2.1 remains valid for bounded solutions $\gamma\in W^1_2(\Omega)$ of equation (13.2.1) satisfying the Dirichlet condition $\gamma=0$ on $\partial\Omega$.

We next describe the space $MW^1_{2,\beta}(\Omega)$ for $\beta>1$. By Q_j we denote the cubes with edge lengths d_j forming a Whitney covering of Ω. Let Q_j^* be the cube in Ω, concentric to Q_j and with edge length $9d_j/8$. The cubes Q_j^* form a covering of Ω of finite multiplicity, depending only on n.

We introduce the relative capacity of a compact set $e\subset\Omega$,

$$\text{cap}(e,\Omega) = \inf\{\|\nabla u;\Omega\|^2_{L_2} : u\in C_0^\infty(\Omega),\ u=1\ \text{ on }\ e\}.$$

Proposition 13.2.1. *For $\beta > 1$*

$$\|\gamma; \Omega\|_{MW^1_{2,\beta}} \sim \sup_j \sup_{e \subset Q_j} \frac{\|\nabla \gamma; e\|_{L_2}}{[\mathrm{cap}(e, Q_j^*)]^{1/2}} + \|\gamma; \Omega\|_{L_\infty}.$$

Proof. First of all we note that for $\beta > 1$ the Hardy inequality (13.2.6) leads to the relation

$$\|u; \Omega\|^2_{W^1_{2,\beta}} \sim \int_\Omega \rho(x)^\beta \left(|\nabla u|^2 + \rho(x)^{-2} u^2\right) dx.$$

By $\{\eta_j\}$ we denote a partition of unity, subordinate to the covering $\{Q_j\}$ and such that $|\nabla \eta_j| \leq c\, d_j^{-1}$. Assume also that $\{\zeta_j\}$ is a sequence of smooth functions with

$$\mathrm{supp}\, \zeta_j \subset Q_j^*, \quad \zeta_j = 1 \quad \text{on } Q_j \quad \text{and } |\nabla \zeta_j| \leq c\, d_j^{-1}.$$

By the last equivalence relation and the Poincaré inequality, we have

$$\|u; \Omega\|^2_{W^1_{2,\beta}} \sim \sum_j d_j^\beta\, \|\nabla(\eta_j u); \Omega\|^2_{L_2}.$$

The same holds if we replace η_j by ζ_j on the right-hand side. This relation and Theorem 1.2.2 give

$$\|\gamma u; \Omega\|^2_{W^1_{2,\beta}} \leq c \sum_j d_j^\beta\, \|\nabla(\gamma \eta_j \zeta_j u); \Omega\|^2_{L_2}$$

$$\leq c \sup_j \sup_{e \subset Q_j} \frac{\|\nabla(\gamma \eta_j); e\|^2_{L_2}}{\mathrm{cap}(e, Q_j^*)} \sum_j d_j^\beta\, \|\nabla(\zeta_j u); \Omega\|^2_{L_2}$$

$$\leq c \Big(\sup_j \sup_{e \subset Q_j} \frac{\|\nabla \gamma; e\|^2_{L_2}}{\mathrm{cap}(e, Q_j)} + \|\gamma; \Omega\|^2_{L_\infty} d_j^{-2} \frac{\mathrm{mes}_n e}{\mathrm{cap}(e, Q_j)} \Big) \|u; \Omega\|^2_{W^1_{2,\beta}}.$$

Since the Poincaré inequality implies the estimate

$$\mathrm{cap}(e, Q_j^*) \geq c\, d_j^2\, \mathrm{mes}_n e,$$

we obtain the required upper estimate for the norm of γ in $MW^1_{2,\beta}(\Omega)$.

The inequality

$$\|\gamma; \Omega\|_{L_\infty} \leq \|\gamma; \Omega\|_{MW^1_{2,\beta}}$$

is obtained by standard arguments (see Proposition 2.7.4). Setting $u \in C_0^\infty(Q_j^*)$, $u = 1$ on a compact $e \subset Q_j$, in

$$\int_\Omega \rho(x)^\beta |\nabla u|^2 dx \leq \|\gamma; \Omega\|^2_{MW^1_{2,\rho}} \|u; \Omega\|^2_{W^1_{2,\beta}},$$

we obtain

$$\|\nabla \gamma; e\|^2_{L_2} \leq c\ \mathrm{cap}(e, Q_j^*) \|\gamma; \Omega\|^2_{MW^1_{2,\beta}}.$$

The proposition is proved. □

In a similar manner we can derive the relation

$$\|h\,\rho^{\beta/2};\Omega\|_{M(W^1_{2,\beta}\to L_2)} \sim \sup_j \sup_{e\subset Q_j} \frac{\|h;e\|_{L_2}}{[\mathrm{cap}(e,Q_j^*)]^{1/2}},$$

where $\beta > 1$. Therefore, the estimate of the solution γ obtained in Theorem 13.2.1 can be written in the form

$$\sup_j \sup_{e\subset Q_j} \frac{\|\nabla\gamma;e\|_{L_2}}{[\mathrm{cap}(e,Q_j^*)]^{1/2}} \le c\Big(\|\gamma;\Omega\|_{L_\infty} + \sup_j \sup_{e\subseteq Q_j} \frac{\|f;e\|_{L_1}+\|g;e\|_{L_2}}{[\mathrm{cap}(e,Q_j^*)]^{1/2}}\Big).$$

In the case $n > 2$ we can replace $\mathrm{cap}(e,Q_j^*)$ by the Wiener capacity $c_{2,1}$, and the supremum with respect to j and with respect to $e \subset Q_j$, by the supremum over compact sets e with $d(e) \le \rho(e)$, where $d(e)$ is the diameter of e and $\rho(e)$ is the distance from e to $\partial\Omega$.

It follows from Proposition 13.2.1 that the spaces of multipliers in $W^1_{2,\beta}(\Omega)$ are isomorphic for all $\beta > 1$. Since the proof of the lower estimate for $\|\gamma\|_{MW^1_{2,\beta}(\Omega)}$ remains valid for $\beta = 1$ as well, it follows that in the case $\beta > 1$ we have the imbedding $MW^1_{2,\beta}(\Omega) \subset MW^1_{2,1}(\Omega)$. We show that the last imbedding is strict. Indeed, let

$$\gamma(x) = \sin\log\rho(x) \quad\text{and}\quad u(x) = |\log\rho(x)|^{1/2-\varepsilon},\ \ \varepsilon > 0.$$

One checks directly that $u \in W^1_{2,1}(\Omega)$ whereas $\gamma u \notin W^1_{2,1}(\Omega)$.

13.2.3 The Case $\beta = 1$

By $S(\Omega)$ we mean the space of functions from $W^1_{2,\mathrm{loc}}(\Omega)$ with the finite norm

$$\|u;\Omega\|_S = \Big(\int_\Omega \rho(x)|\nabla u|^2dx + \|u;\Omega\|^2_{L_2}\Big)^{1/2} + \int_0^\delta \|\nabla u;\Gamma_\tau\|_{L_2}d\tau,$$

where δ is a small positive number and Γ_τ is the boundary of the domain Ω_τ.

Theorem 13.2.2. *Suppose that the functions f and g are subject to*

$$\rho^{1/2}(|f|^{1/2}+|g|)+|g|^{1/2} \in M(S(\Omega)\to L_2(\Omega)).$$

Then any variational solution $\gamma \in L_\infty(\Omega)$ of (13.2.1) belongs to the space $M(S(\Omega)\to W^1_{2,1}(\Omega))$ and

$$\|\gamma;\Omega\|_{M(S\to W^1_{2,1})} \le c\,\big(\|\gamma;\Omega\|_{L_\infty} + \|(\rho\,|f|+|g|)^{1/2};\Omega\|^2_{M(S\to L_2)} + \|\rho^{1/2}g;\Omega\|_{M(S\to L_2)}\big) \tag{13.2.9}$$

with a constant c, independent of f, g and γ.

Proof. We use the notations R and ζ_τ, introduced in the proof of Theorem 13.2.1. We set $\eta = R\zeta_\tau^2 u^2 \gamma$ in (13.2.2), where u is an arbitrary function from $S(\Omega)$. Then the left-hand side in (13.2.2) takes the form

$$\int_\Omega a_{ij} \frac{\partial(\gamma u)}{\partial x_j} \frac{\partial(\gamma u)}{\partial x_i} R\zeta_\tau^2\, dx - \int_\Omega a_{ij} \frac{\partial u}{\partial x_j} \frac{\partial u}{\partial x_i} \gamma^2 R\zeta_\tau^2\, dx$$

$$+ \int_\Omega a_{ij} \frac{\partial(\gamma u)}{\partial x_j} \frac{\partial(R\zeta_\tau^2)}{\partial x_i} \gamma u\, dx - \int_\Omega a_{ij} \frac{\partial u}{\partial x_j} \frac{\partial (R\zeta_\tau^2)}{\partial x_i} \gamma^2 u\, dx. \tag{13.2.10}$$

The third term on the right-hand side of (13.2.10) is equal to

$$\frac{1}{2}\int_\Omega (\gamma u)^2 \mathcal{L}(R\zeta_\tau^2)\, dx = \frac{1}{2}\int_\Omega (\gamma u)^2 \zeta_\tau^2 dx + \int_\Omega (\gamma u)^2 \Big(a_{ij} \frac{\partial R}{\partial x_i} \frac{\partial \zeta_\tau^2}{\partial x_j} + \frac{1}{2} R\mathcal{L}(\zeta_\tau^2)\Big) dx.$$

The absolute value of the last integral does not exceed

$$c\,\|\gamma;\Omega\|^2_{L_\infty} \tau^{-1} \int_{\Omega\backslash\Omega_\tau} u^2 dx \le c_1 \|\gamma;\Omega\|^2_{L_\infty} \sup_{0<\tau<\delta} \|u;\Gamma_\tau\|^2_{L_2}.$$

Clearly,

$$\|u;\Gamma_\tau\|_{L_2} \le \int_0^\delta \|\nabla u;\Gamma_\sigma\|_{L_2} d\sigma + c\,\delta^{-1}\|u;\Omega_{\delta/2}\backslash\Omega_\delta\|_{L_2}. \tag{13.2.11}$$

Consequently, the absolute value of the third term in (13.2.10) is not greater than

$$c\,\|\gamma;\Omega\|^2_{L_2}\|u;\Omega\|^2_S. \tag{13.2.12}$$

Next we estimate the fourth term on the right-hand side of (13.2.10). We have

$$\Big|\int_\Omega a_{ij}\frac{\partial u}{\partial x_j}\frac{\partial(R\zeta_\tau^2)}{\partial x_i}\gamma^2 u\, dx\Big| \le c\,\|\gamma;\Omega\|^2_{L_\infty}\big(\|u;\Omega_\delta\|^2_{W_2^1} + \int_{\Omega\backslash\Omega_\delta} |\nabla u|\,|u| dx\big).$$

Using (13.2.11), we find that the last integral does not exceed

$$c\int_0^\delta \|\nabla u;\Gamma_\tau\|_{L_2}\|u;\Gamma_\tau\|_{L_2} d\tau \le c \sup_{0<\tau<\delta} \|u;\Gamma_\tau\|_{L_2}\|u;\Omega\|_S \le c_1\|u;\Omega\|^2_S$$

and, consequently, the absolute value of the fourth term on the right-hand side of (13.2.10), as well as the third one, is majorized by the product (13.2.12). Combining this fact with (13.2.10) and (13.2.2), we arrive at the estimate

$$\int_\Omega a_{ij}\frac{\partial(\gamma u)}{\partial x_j}\frac{\partial(\gamma u)}{\partial x_i} R\zeta_\tau^2\, dx \le c\,\|\gamma;\Omega\|^2_{L_\infty}\|u;\Omega\|^2_S$$

$$+ \Big|\int_\Omega (fR\zeta_\tau^2 u^2\gamma - g\,\nabla(R\zeta_\tau^2 u^2 \gamma)) dx\Big|. \tag{13.2.13}$$

The second term on the right-hand side of this inequality does not exceed

$$c\,\|\gamma;\Omega\|_{L_\infty}\Big(\|(\rho\,|f|)^{1/2}u;\Omega\|^2_{L_2}+\int_\Omega |gu\nabla(R\zeta_\tau^2 u)|dx\Big)$$
$$+\int_\Omega |g|\,|\nabla(\gamma u)|R\zeta_\tau^2\;|u|\,dx. \tag{13.2.14}$$

Obviously, the first term in brackets has the majorant

$$\|(\rho\,|f|)^{1/2};\Omega\|^2_{M(S\to L_2)}\|u;\Omega\|^2_S,$$

while the third one is not greater than

$$c\,\|\zeta_\tau\rho^{1/2}\,\nabla(\gamma u);\Omega\|_{L_2}\,\|\rho^{1/2}g\;u;\Omega\|_{L_2}$$
$$\leq c\,\|\zeta_\tau\rho^{1/2}\,\nabla(\gamma u);\Omega\|_{L_2}\,\|\rho^{1/2}g;\Omega\|_{M(S\to L_2)}\,\|u;\Omega\|_S.$$

We estimate the second term in the brackets in (13.2.14):

$$\int_\Omega |gu\nabla(R\zeta_\tau^2 u)|\,dx\leq c\,\|\,|g|^{1/2}u;\Omega\|^2_{L_2}+c\int_\Omega \rho\,|g|\,|u\nabla u|\,dx$$
$$\leq c\big(\||g|^{1/2};\Omega\|^2_{M(S\to L_2)}+\|\rho^{1/2}g;\Omega\|_{M(S\to L_2)}\big)\|u;\Omega\|^2_S.$$

Combining the above estimates with (13.2.13) and noting that $R\geq c\rho$, we find that the norm $\|\rho^{1/2}\nabla(\gamma u);\Omega\|_{L_2}$ does not exceed the right-hand side of (13.2.9) multiplied by $\|u;\Omega\|_S$. The theorem is proved. □

13.2.4 Solutions as Multipliers from $W^1_{2,w(\rho)}(\Omega)$ into $W^1_{2,1}(\Omega)$

Using Theorem 13.2.2, we show that bounded solutions of (13.2.1) are multipliers acting from a weighted Hilbert space, intermediate between $W^1_{2,1}(\Omega)$ and $W^1_{2,\beta}(\Omega)$ with $\beta<1$, into $W^1_{2,1}(\Omega)$.

Let w be a continuous function on $[0,\infty)$ such that $w(\rho)\geq c\rho$ for $\rho>0$. We say that a function $u\in W^1_{2,\mathrm{loc}}(\Omega)$ belongs to $W^1_{2,w(\rho)}(\Omega)$ if it has the finite norm

$$\|u;\Omega\|_{W^1_{2,w(\rho)}}=\Big(\int_\Omega w(\rho(x))|\nabla\;u|^2dx+\|u;\Omega\|^2_{L_2}\Big)^{1/2}.$$

In particular, for $w(\rho)=\rho$ we have $W^1_{2,w(\rho)}(\Omega)=W^1_{2,\beta}(\Omega)$.

Theorem 13.2.3. *Let*

$$\int_0^\delta\frac{d\tau}{w(\tau)}<\infty. \tag{13.2.15}$$

Suppose also that

$$\rho^{1/2}(|f|+|g|)+|g|^{1/2}\in M(W^1_{2,w(\rho)}(\Omega)\to L_2(\Omega)).$$

Then any variational solution $\gamma \in L_\infty(\Omega)$ of (13.2.1) *belongs to the space*

$$M(W^1_{2,w(\rho)}(\Omega) \to W^1_{2,1}(\Omega))$$

and

$$\|\gamma; \Omega\|_{M(W^1_{2,w(\rho)} \to W^1_{2,1})} \le c\big(\|\gamma; \Omega\|_{L_\infty} + \|\rho\,|f| + |g|; \Omega\|^2_{M(W^1_{2,w(\rho)} \to L_2)} + \|\rho^{1/2} g; \Omega\|_{M(W^1_{2,w(\rho)} \to L_2)}\big) \quad (13.2.16)$$

with a constant c independent of f, g and γ.

Proof. Since

$$\int_0^\delta \|\nabla u; \Gamma_\tau\|_{L_2} d\tau \le c\Big(\int_0^\delta \frac{d\tau}{w(\tau)}\Big)^{1/2} \Big(\int_{\Omega\setminus\Omega_\delta} w(\rho(x))|\nabla u|^2 dx\Big)^{1/2},$$

it follows that $W^1_{2,w(\rho)}(\Omega) \subset S(\Omega)$ and

$$\|u; \Omega\|_S \le c\|u; \Omega\|_{W^1_{2,w(\rho)}}.$$

It remains to make use of Theorem 13.2.2. □

The next theorem shows that the condition (13.2.15) is also necessary.

Theorem 13.2.4. *If any variational solution $\gamma \in L_\infty(\Omega)$ of the equation $\mathcal{L}\gamma = 0$ belongs to $M(W^1_{2,w(\rho)}(\Omega) \to W^1_{2,1}(\Omega))$, then the condition* (13.2.15) *is satisfied.*

Proof. Suppose (13.2.15) does not hold. We introduce a positive Lipschitz function u in Ω, coinciding near $\partial\Omega$ with the function

$$x \to v(\rho(x)) = \Big(\int_{\rho(x)}^{2\delta} \frac{d\tau}{w(\tau)}\Big)^{1/2-\sigma},$$

where σ is an arbitrarily small positive number. Clearly, $u \in W^1_{2,w(\rho)}(\Omega)$. Therefore, for any $\gamma \in L_\infty(\Omega)$, the norm $\|\rho^{1/2}\nabla(\gamma u); \Omega\|_{L_2}$ is finite and, consequently,

$$\|\, \rho^{1/2}\, u \nabla\gamma; \Omega\|_{L_2} < \infty.$$

Let $\mathcal{A}$ and $\mathcal{B}$ denote the Banach spaces of solutions of the equation $\mathcal{L}\gamma = 0$ with the finite norms $\|\gamma; \Omega\|_{L_\infty}$ and

$$\|\gamma; \Omega\|_{L_\infty} + \|\rho^{1/2} u\ \nabla\gamma; \Omega\|_{L_2},$$

respectively. Let I be the identity mapping of $\mathcal{A}$ into $\mathcal{B}$ which is, obviously, linear, continuous, and one-to-one. By what we have just proved, I maps $\mathcal{B}$

onto $\mathcal{A}$ and, consequently, by Banach's theorem, it is an isomorphism. Thus, we have the estimate

$$\|\rho^{1/2}u\ \nabla\gamma; \Omega\|_{L_2} \le c\,\|\gamma; \Omega\|_{L_\infty}$$

with a constant c independent of γ.

For any sufficiently small ε we have

$$\Big(\int_\varepsilon^{2\delta} \frac{d\tau}{w(\tau)}\Big)^{1/2-\sigma} \|\rho^{1/2}\nabla\gamma; \Omega \setminus \Omega_\varepsilon\|_{L_2} \le c\,\|\gamma; \Omega\|_{L_\infty}.$$

This inequality and the maximum principle imply that

$$\|\rho^{1/2}\nabla\gamma; \Omega\|_{L_2} \le 2c\Big(\int_\varepsilon^{2\delta} \frac{d\tau}{w(\tau)}\Big)^{\sigma-1/2} \|\gamma; \partial\Omega\|_{L_\infty} + c(\varepsilon)\|\gamma; \Omega_{\varepsilon/2}\|_{L_2}.$$

Since solutions of $\mathcal{L}\gamma = 0$ obey the estimate

$$\|\gamma; \partial\Omega\|_{L_2} \le c\left(\|\rho^{1/2}\nabla\gamma; \Omega\|_{L_2} + \|\gamma; \Omega_\delta\|_{L_2}\right)$$

(see for instance, [Maz5], Sect.2), it follows that

$$\|\gamma; \partial\Omega\|_{L_2} \le c_2\Big(\int_\varepsilon^{2\delta} \frac{d\tau}{w(\tau)}\Big)^{\sigma-1/2} \|\gamma; \partial\Omega\|_{L_\infty} + c(\varepsilon)\|\gamma; \Omega_{\varepsilon/2}\|_{L_2}.$$

Given an arbitrary sequence of solutions γ, whose traces on $\partial\Omega$ belong to the unit ball in $L_\infty(\partial\Omega)$, we select a subsequence $\{\gamma_m\}_{m\ge 0}$ convergent in $L_2(\Omega_{\varepsilon/2})$. Then

$$\limsup_{m,k\to\infty} \|\gamma_m - \gamma_k; \partial\Omega\|_{L_2} \le 2\,c_2\Big(\int_\varepsilon^{2\delta} \frac{d\tau}{w(\tau)}\Big)^{\sigma-1/2}.$$

By assumption, the right-hand side tends to zero as $\varepsilon \to 0$. Hence $\{\gamma_m\}_{m\ge 0}$ converges in $L_2(\partial\Omega)$. Thus, the unit ball in $L_\infty(\partial\Omega)$ is compact in $L_2(\partial\Omega)$ which, of course, is not true. The theorem is proved. □

13.3 Solvability of Quasilinear Elliptic Equations in Spaces of Multipliers

In the present section we collect theorems on the solvability of boundary value problems for quasilinear second-order elliptic equations and systems in spaces of multipliers. Sects. 13.3.1 and 13.3.3 concern the case of a single equation in divergent form, while in Sect. 13.3.2 we examine a system of the same type. It is shown that bounded variational solutions of these equations and systems belong to the space $MW_p^1(\Omega)$ of multipliers in $W_p^1(\Omega)$ under certain conditions. A similar assertion concerning solutions of nondivergence equations in the space $MW_2^2(\Omega)$ is contained in Sect. 13.3.4.

13.3.1 Scalar Equations in Divergence Form

Here we deal with a variational formulation of the mixed boundary value problem

$$\frac{\partial a_i(x, \gamma, \nabla\gamma)}{\partial x_i} = \varphi(x, \gamma, \nabla\gamma), \quad x \in \Omega, \tag{13.3.1}$$

$$\gamma|_{\Gamma} = 0, \quad a_i(x, \gamma, \nabla\gamma)\cos(\nu, x_i)\big|_{\partial\Omega\backslash\Gamma} = 0, \tag{13.3.2}$$

where Γ is a subset of the boundary $\partial\Omega$ and ν is a normal to $\partial\Omega$.

Let Ω be an arbitrary bounded domain in $\mathbb{R}^n$. We assume that the functions $a_i(x, \xi_0, \xi)$ are measurable with respect to x for all $\xi_0, \xi = (\xi_1, ..., \xi_n)$, continuous with respect (ξ_0, ξ) for almost all $x \in \Omega$, and, for $x \in \Omega, |\xi_0| \le Q$, satisfy the inequalities

$$a_i(x, \xi_0, \xi)\xi_i \ge c_1|\xi|^p - f_1(x) \tag{13.3.3}$$

and

$$\sum_{i=1}^{n} |a_i(x, \xi_0, \xi)| \le c_2|\xi|^{p-1} + f_2(x), \tag{13.3.4}$$

where Q, c_1, c_2 are positive constants, f_1 and f_2 are nonnegative functions, $\xi \ne 0$ and

$$f_1^{1/p} \in M(W_p^1(\Omega) \to L_p(\Omega)), \; f_2^{1/(p-1)} \in M(W_p^1(\Omega) \to L_p(\Omega)), \; p > 1.$$

By $\mathring{W}_p^1(\Omega, \Gamma)$ we denote the completion, in the norm of space $W_p^1(\Omega)$, of the set of functions in $W_p^1(\Omega)$ with compact supports in $\overline{\Omega}\backslash\Gamma$. It is standard that the intersection $\mathring{W}_p^1(\Omega, \Gamma) \cap L_\infty(\Omega)$ is dense in $\mathring{W}_p^1(\Omega, \Gamma)$.

Let $|\gamma(x)| \le Q$ for almost all $x \in \Omega$. By a variational solution of problem (13.3.1)–(13.3.2), we mean a function $\gamma \in \mathring{W}_p^1(\Omega, \Gamma)$ satisfying the identity

$$\int_\Omega a_i(x, \gamma, \nabla\gamma)\frac{\partial v}{\partial x_i}dx = \int_\Omega \varphi(x, \gamma, \nabla\gamma)v\,dx \tag{13.3.5}$$

for any function $v \in \mathring{W}_p^1(\Omega, \Gamma)$.

Theorem 13.3.1. *Let γ be a variational solution of problem* (13.3.1)–(13.3.2) *with $|\gamma(x)| \le Q$ for almost all $x \in \Omega$. Further, let*

$$|\varphi(x, \xi_0, \xi)| \le k\,|\xi|^p + g(x), \tag{13.3.6}$$

where $k = \text{const} > 0$ and

$$g^{1/p} \in M(W_p^1(\Omega) \to L_p(\Omega)).$$

Then

$$\gamma \in MW_p^1(\Omega).$$

Proof. Let u be an arbitrary element of $(W_p^1 \cap L_\infty)(\Omega)$ and let γ be a variational solution of problem (13.3.1)–(13.3.2). We set

$$v = (e^{\lambda\gamma} - 1)|u|^p$$

in (13.3.5), where λ is a positive constant whose value will be chosen later. It is easily seen that this function belongs to the space $\mathring{W}_p^1(\Omega, \Gamma)$. We have

$$\lambda \int_\Omega a_i(x, \gamma, \nabla\gamma) \frac{\partial \gamma}{\partial x_i} e^{\lambda\gamma} |u|^p dx = -p \int_\Omega a_i(x, \gamma, \nabla\gamma) \frac{\partial u}{\partial x_i} |u|^{p-1} \mathrm{sgn}\, u (e^{\lambda\gamma} - 1) dx$$
$$+ \int_\Omega \varphi(x, \gamma, \nabla\gamma)(e^{\lambda\gamma} - 1)|u|^p dx.$$

Using (13.3.3) and (13.3.4), we obtain

$$\lambda\, c_1 \int_\Omega e^{\lambda\gamma} |u\nabla\gamma|^p dx \le \lambda \int_\Omega e^{\lambda\gamma} f_1 |u|^p dx + p\, c_2 \int_\Omega |u\nabla\gamma|^{p-1} |\nabla u| (e^{\lambda\gamma} - 1) dx$$
$$+ \int_\Omega f_2 |\nabla u|\, |u|^{p-1} (e^{\lambda\gamma} - 1)\, dx + \int_\Omega \varphi\, |u|^p (e^{\lambda\gamma} - 1)\, dx.$$

By (13.3.6) the last integral does not exceed

$$k \int_\Omega |\nabla\gamma|^p\, |u|^p (e^{\lambda\gamma} - 1)\, dx + \int_\Omega g\, |u|^p (e^{\lambda\gamma} - 1)\, dx.$$

Setting $\lambda = 2k/c_1$ and $\varepsilon = e^{\lambda Q}$, we arrive at the inequality

$$\frac{k}{2\varepsilon} \int_\Omega |u\nabla\gamma|^p dx \le \frac{2k}{c_1} \varepsilon \int_\Omega f_1 |u|^p dx + (\varepsilon - 1) \int_\Omega g |u|^p dx$$
$$+ p\, c_2\, (\varepsilon - 1) \int_\Omega |u\nabla\gamma|^{p-1} |\nabla u|\, dx + (\varepsilon - 1) \int_\Omega f_2 |\nabla u|\, |u|^{p-1} dx.$$

Hence

$$\|\nabla(u\gamma); \Omega\|_{L_p} \le c\Big[\Big(\int_\Omega (g + f_1 + f_2^{p/(p-1)}) |u|^p dx\Big)^{1/p} + \|\nabla u; \Omega\|_{L_p}\Big].$$

Since the functions $g^{1/p}, f_1^{1/p}, f_2^{1/(p-1)}$ belong to $M(W_p^1(\Omega) \to L_p(\Omega))$, the result follows. □

13.3.2 Systems in Divergence Form

Theorem 13.3.1 has a partial generalization to systems of equations of the type (13.3.1). In the theorem below we assume λ, a_i, and φ to be vector functions with values in $\mathbb{R}^m$. In (13.3.5), by v we mean an m-dimensional vector function.

Theorem 13.3.2. *Suppose that vector-valued functions a_i and φ obey conditions* (13.3.3), (13.3.4), (13.3.6), *where f_1, f_2, g are the same as above. Let the constant k in* (13.3.6) *be sufficiently small. If γ is a variational solution of the system* (13.3.1), *satisfying the boundary conditions* (13.3.2) *and $|\gamma(x)| \le Q$ for almost all $x \in \Omega$, then $\gamma \in MW_p^1(\Omega)$.*

Proof. We set $v = \gamma\,|u|^p$ in (13.3.5), where u is an arbitrary function in $(W_p^l \cap L_\infty)(\Omega)$. It is easily seen that the vector function $\gamma|u|^p$ belongs to the space $\mathring{W}_p^1(\Omega, \Gamma)$. We have

$$\int_\Omega a_i(x,\gamma,\nabla\gamma)\frac{\partial\gamma}{\partial x_i}|u|^p dx = -p\int_\Omega \gamma a_i(x,\gamma,\nabla\gamma)\frac{\partial u}{\partial x_i}|u|^{p-2}u\,dx$$
$$+\int_\Omega \varphi(x,\gamma,\nabla\gamma)\gamma\,|u|^p dx.$$

From this inequality and condition (13.3.3), we obtain

$$\int_\Omega |\nabla(\gamma u)|^p dx \le c\Big(\int_\Omega |a_i(x,\gamma,\nabla\gamma)|\,|\nabla u|\,|u|^{p-1}|\gamma| dx$$
$$+\int_\Omega |\varphi(x,\gamma,\nabla\gamma)|\,|\gamma|\,|u|^p dx + \int_\Omega f_1|u|^p dx + \int_\Omega |\gamma|^p|\nabla u|^p dx\Big). \tag{13.3.7}$$

Using condition (13.3.4) and the Hölder inequality to estimate the first integral on the right-hand side of (13.3.7), we find that it is dominated by

$$c\,\|\gamma;\Omega\|_{L_\infty}\big(\|\nabla(\gamma u);\Omega\|_{L_p}^{p-1}\|\nabla u;\Omega\|_{L_p} + \|\gamma;\Omega\|_{L_\infty}^{p-1}\|\nabla u;\Omega\|_{L_p}^p$$
$$+\|f_2^{1/(p-1)};\Omega\|_{M(W_p^1\to L_p)}^{p-1}\|\nabla u;\Omega\|_{L_p}^p\big). \tag{13.3.8}$$

By (13.3.6), the second integral on the right-hand side of (13.3.7) does not exceed

$$\|\gamma;\Omega\|_{L_\infty}\int_\Omega |u|^p(k\,|\nabla\gamma|^p + g)dx.$$

Consequently,

$$\|\nabla(\gamma u);\Omega\|_{L_p} \le c_0\|\gamma;\Omega\|_{L_\infty}\big(\|\nabla u;\Omega\|_{L_p} + \|f_2^{1/(p-1)};\Omega\|_{M(W_p^1\to L_p)}^{1-1/p}\,\|\nabla u;\Omega\|_{L_p}\big)$$
$$+ k^{1/p}\|\gamma;\Omega\|_{L_\infty}^{1/p}\,\|\nabla(\gamma u);\Omega\|_{L_p} + k^{1/p}\|\gamma;\Omega\|_{L_\infty}^{1+1/p}\|\nabla u;\Omega\|_{L_p}$$
$$+ \|\gamma;\Omega\|_{L_\infty}^{1/p}\,\|g^{1/p};\Omega\|_{M(W_p^1\to L_p)}\|\nabla u;\Omega\|_{L_p}\big).$$

In view of the smallness of k we may assume that

$$c_0\,k^{1/p}\,\|\gamma;\Omega\|_{L_\infty}^{1/p} < 1/2. \tag{13.3.9}$$

Then

$$\|\nabla(\gamma u);\Omega\|_{L_p} \le c\,\|\nabla u;\Omega\|_{L_p}.$$

The theorem is proved. □

Comparison with the case of a single equation, considered in Theorem 13.3.1, may give the impression that the result obtained in Theorem 13.3.2 for a system is weaker because of the condition that the constant k in (13.3.6) is small (see (13.3.9)). However, we show by the following example that this restriction is not only dictated by the proof, but really is necessary.

Example. Let Ω be the disk

$$\Omega = \{x + iy = re^{i\theta} : \ 0 < r < e^{-1}, \, 0 \le \theta < 2\pi\}.$$

Consider the quasilinear elliptic system

$$\Delta\gamma_1 = \frac{-2\gamma_1 - 2 - \big((\log\log r^{-1})^{-2} + (\log\log r^{-1})^{-1}\big)\gamma_2}{|\gamma|^2 + 2\gamma_1 + 2}|\nabla\gamma|^2,$$

$$\Delta\gamma_2 = \frac{-2\gamma_2 + \big((\log\log r^{-1})^{-2} + (\log\log r^{-1})^{-1}\big)(\gamma_1 + 1)}{|\gamma|^2 + 2\gamma_1 + 2}|\nabla\gamma|^2, \quad (13.3.10)$$

where

$$\gamma = (\gamma_1, \gamma_2) \qquad \text{and} \qquad |\nabla\gamma|^2 = |\nabla\gamma_1|^2 + |\nabla\gamma_2|^2.$$

We subject the vector function γ to the homogeneous Dirichlet condition on $\partial\Omega$. It can be checked directly that the boundary value problem just formulated has the solution

$$\gamma_1 = \cos\,(\log\log r^{-1})^2 - 1,$$

$$\gamma_2 = \sin\,(\log\log r^{-1})^2 \qquad (13.3.11)$$

in the space $\mathring{W}_2^1(\Omega) \cap L_\infty(\Omega)$. In fact, the substitution $v = (\gamma_1 + 1) + i\gamma_2$ reduces the system (13.3.10) to the scalar equation

$$\Delta v = \frac{-2v + i(\log\log r^{-1})^{-2}v + i(\log\log r^{-1})^{-1}v}{|v|^2 + 1}|\nabla v|^2,$$

After writing the Laplacian in polar coordinates one readily checks that

$$v = e^{i(\log\log r^{-1})^2}$$

satisfies the last equation.

It is clear that $|\gamma(x)| \le 2$ on $\bar{\Omega}$. Since

$$|\nabla\gamma(x)| = 2\,\frac{\log\log r^{-1}}{r\log r^{-1}},$$

it follows that

$$\|\nabla\gamma; \Omega\|_{L_2} = c\int_1^\infty \frac{(\log t)^2}{t^2}dt < \infty.$$

If the vector function γ is a multiplier in $W_2^1(\Omega)$, then the inequality

$$\int_\Omega |\nabla\gamma|^2 u^2 dx \le c\,\|u;\Omega\|^2_{W_2^1}$$

holds for all $u \in W_2^1(\Omega)$, which is equivalent to the estimate

$$\int_\Omega \Big(\frac{\log\log r^{-1}}{r\log r^{-1}}\Big)^2 u^2 dx \le c\,\|u;\Omega\|^2_{W_2^1}. \tag{13.3.12}$$

We set here

$$u(x) = \begin{cases} 1 & \text{for } r \le \varepsilon^2; \\ 0 & \text{for } r > \varepsilon; \\ \dfrac{\log\varepsilon r^{-1}}{\log\varepsilon^{-1}} & \text{for } \varepsilon^2 \le r \le \varepsilon, \end{cases}$$

where ε is a small positive number. Then the right-hand side of (13.3.12) is $O((\log\varepsilon^{-1})^{-1})$, while the left-hand side majorizes the expression

$$c\int_0^{\varepsilon^2} \Big(\frac{\log\log r^{-1}}{r\log r^{-1}}\Big)^2 r dr \ge c\int_{\log\varepsilon^{-2}}^{\infty} \frac{(\log t)^2}{t^2} dt \ge c\,\frac{(\log\log\varepsilon^{-1})^2}{\log\varepsilon^{-1}}.$$

Thus, (13.3.12) is false and $\gamma \notin MW_2^1(\Omega)$. It remains to note that the vector-valued function $\gamma = (\gamma_1, \gamma_2)$ given by (13.3.11) satisfies a system for which (13.3.6) is valid, but that k is not sufficiently small.

13.3.3 Dirichlet Problem for Quasilinear Equations in Divergence Form

In this subsection we extend Theorem 13.1.1 to a class of quasilinear equations. As above, we assume Ω to be an open bounded subset of $\mathbb{R}^n$. Let the functions $A_i(x,\xi)$ be measurable with respect to x for all $\xi = (\xi_1, ..., \xi_n)$, and continuous for almost all $x \in \Omega$, and let the inequalities

$$A_i(x,\xi)\xi_i \ge c_1|\xi|^p, \quad \sum_{i=1}^n |A_i(x,\xi)| \le c_2\,|\xi|^{p-1} \tag{13.3.13}$$

be satisfied for any ξ, where c_1, c_2 are positive constants and $p > 1$. Further suppose that the monotonicity condition

$$\big(A_i(x,v) - A_i(x,w)\big)(v_i - w_i) > 0$$

is satisfied for $v \ne w$.

By the solution to the Dirichlet problem for the equation

$$\frac{\partial A_i(x,\nabla u)}{\partial x_i} = 0$$

we mean a function $u \in W_p^1(\Omega)$, satisfying

$$\int_\Omega A_i(x, \nabla u) \frac{\partial v}{\partial x_i} dx = 0, \quad u - g \in \mathring{W}_p^1(\Omega), \tag{13.3.14}$$

where g is a given function from $W_p^1(\Omega)$, and v is any function from $\mathring{W}_p^1(\Omega)$.

It is known (see [LeL])that, under the above conditions on A_i, the problem (13.3.14) has a unique solution from the space $W_p^1(\Omega)$. If g is bounded, then the solution u is also bounded by the maximum principle.

The following theorem on the unique solvability of the problem (13.3.14) in a multiplier space basically follows from Theorem 13.3.1.

Theorem 13.3.3. *If $g \in MW_p^1(\Omega)$, then the solution u of* (13.3.14) *belongs to the space $MW_p^1(\Omega)$.*

Proof. Set $\gamma = u - g$ and introduce the notation

$$a_i(x, \nabla\gamma(x)) := A_i(x, \nabla\gamma(x) + \nabla g(x)).$$

By (13.3.13) we have

$$a_i(x, \xi)\xi_i = A_i(x, \xi + \nabla g(x))\xi_i \geq c_1|\xi + \nabla g(x)|^p \geq c_1\big(2^{1-p}|\xi|^p - |\nabla g(x)|^p\big),$$

$$\sum_{i=1}^n |a_i(x, \xi)| = \sum_{i=1}^n |A_i(x, \xi + \nabla g(x))| \leq c_2\,\big(|\xi| + |\nabla g(x)|\big)^{p-1}.$$

Since

$$\nabla g \in M(W_p^1(\Omega) \to L_p(\Omega)),$$

the functions a_i satisfy (13.3.3) and (13.3.4). It remains to use Theorem 13.3.1. $\square$

Let Ω be a bounded domain in $\mathbb{R}^n$ with $\partial\Omega \in C^{0,1}$ (that is, a Lipschitz graph domain). In Sect. 8.7.4 it is shown that $MW_p^{1-1/p}(\mathbb{R}^{n-1})$ is the space of traces on $\mathbb{R}^{n-1}$ of functions from $MW_p^1(\mathbb{R}^n_+)$. It readily follows that any function φ from the space $MW_p^{1-1/p}(\partial\Omega)$ has an extension g onto Ω such that

$$c^{-1}\|g; \Omega\|_{MW_p^1} \leq \|\varphi; \partial\Omega\|_{MW_p^{1-1/p}} \leq c\,\|g; \Omega\|_{MW_p^1},$$

where c is a constant depending only on Ω and $c > 1$. The next assertion follows from Theorem 13.3.3.

Corollary 13.3.1. *The Dirichlet problem*

$$\frac{\partial A_i(x, \nabla u)}{\partial x_i} = 0, \quad u \in W_p^1(\Omega),\ u\big|_{\partial\Omega} = \varphi,$$

where $\varphi \in MW_p^{1-1/p}(\partial\Omega)$, has a unique solution in the space $MW_p^1(\Omega)$.

13.3.4 Dirichlet Problem for Quasilinear Equations in Nondivergence Form

We next prove a theorem showing that solutions to the Dirichlet problem for nondivergence quasilinear elliptic equations belong to the space $MW_2^2(\Omega)$.

Let Ω be a domain with C^2 boundary and let Q be a nonnegative constant. Suppose that the functions $F(x,\xi_0,\xi)$ and $a_{ij}(x,\xi_0,\xi), i,j = 1,...,n$, are measurable with respect to x for all $\xi_0 \in (-Q,Q)$, $\xi = (\xi_1,...,\xi_n) \in \mathbb{R}^n$, and continuous with respect to ξ_0, ξ for almost all $x \in \Omega$. We assume that the functions a_{ij} are bounded and that there exists a positive constant c such that

$$a_{ij}\xi_i\xi_j \geq c\,|\xi|^2$$

for the same ξ_0, ξ. Further, for some $\varepsilon > 0$, let the coefficients a_{ij} satisfy the Cordes condition

$$\Big(2n-4+\frac{3}{n+1}\Big)\Big[n\sum_{i,j=1}^{n} a_{ij}^2 - \Big(\sum_{i=1}^{n} a_{ii}\Big)^2\Big] \leq (1-\varepsilon)\Big(\sum_{i=1}^{n} a_{ii}\Big)^2$$

(see [Cor]), which restricts the dispersion of eigenvalues of the matrix $\|a_{ij}\|_{i,j=1}^n$. Suppose that

$$|F(x,\xi_0,\xi)| \leq k\,|\xi|^2 + g(x)$$

for $|\xi_0| \leq Q$, where

$$g \in M(W_2^2(\Omega) \to L_2(\Omega))$$

and k is a small constant.

Consider the Dirichlet problem

$$a_{ij}(x,u,\nabla u)\frac{\partial^2 u}{\partial x_i \partial x_j} = F(x,u,\nabla u) \ \text{ in } \Omega, \quad u\big|_{\partial\Omega} = \varphi, \tag{13.3.15}$$

where

$$u \in W_2^2(\Omega), \quad |u(x)| \leq Q \text{ almost everywhere in } \Omega,$$

and $\varphi \in MW_2^{3/2}(\partial\Omega)$. We set

$$v = u - \Phi, \tag{13.3.16}$$

where Φ is an extension of φ onto $\mathbb{R}^n$ belonging to the class $MW_2^2(\mathbb{R}^n)$ (see Theorem 8.6.1). Then the problem (13.3.15) is equivalent to the problem

$$b_{ij}(x,v,\nabla v)\frac{\partial^2 v}{\partial x_i \partial x_j} = G(x,v,\nabla v) \text{ in } \Omega, \quad v|_{\partial\Omega} = 0,$$

where

$$b_{ij}(x,\xi_0,\xi) = a_{ij}(x,\xi_0+\Phi(x),\xi+\nabla\Phi(x))$$

and

$$G(x, \xi_0, \xi) = F(x, \xi_0 + \Phi(x), \xi + \nabla\Phi(x)) + b_{ij}(x, \xi_0, \xi)\frac{\partial^2 \Phi}{\partial x_i \partial x_j}.$$

It is clear that the coefficients b_{ij} satisfy the same conditions as a_{ij}. Moreover,

$$|G(x, \xi_0, \xi)| \le 2\,k\,|\xi|^2 + h(x), \tag{13.3.17}$$

where

$$h = 2k|\nabla\Phi|^2 + c\,|\nabla_2\Phi|.$$

Lemma 13.3.1. *The function h belongs to the space $M(W_2^2(\Omega) \to L_2(\Omega))$.*

Proof. For any function $w \in C_0^\infty(\mathbb{R}^n)$ we have

$$\int_{\mathbb{R}^n} |\nabla\Phi|^4 w^2 dx = -\int_{\mathbb{R}^n} \Phi \operatorname{div}\big(|\nabla\Phi|^2(\nabla\Phi)w^2\big)dx$$

$$\le c\,\|\Phi;\mathbb{R}^n\|_{L_\infty}\Big(\int_{\mathbb{R}^n} |\nabla\Phi|^2|\nabla_2\Phi|w^2 dx + \int_{\mathbb{R}^n} |\nabla\Phi|^3\,|w|\,|\nabla w|\,dx\Big)$$

$$\le c\,\|\Phi;\mathbb{R}^n\|_{L_\infty}\|\,|\nabla\Phi|^2 w;\mathbb{R}^n\|_{L_2}\Big(\|\,|\nabla_2\Phi|w;\mathbb{R}^n\|_{L_2} + \|\,|\nabla\Phi|\,|\nabla w|\,;\mathbb{R}^n\|_{L_2}\Big).$$

Since

$$\|\nabla_2\Phi;\mathbb{R}^n\|_{M(W_2^2\to L_2)} + \|\nabla\Phi;\mathbb{R}^n\|_{M(W_2^1\to L_2)} \le c\,\|\Phi;\mathbb{R}^n\|_{MW_2^2}, \tag{13.3.18}$$

it follows that

$$\|\,|\nabla\Phi|^2 w;\mathbb{R}^n\|_{L_2} \le c\,\|\Phi;\mathbb{R}^n\|_{L_\infty}\|\Phi;\mathbb{R}^n\|_{MW_2^2}\|w;\mathbb{R}^n\|_{W_2^2}.$$

It is clear that the last inequality is valid for all $w \in W_2^2(\mathbb{R}^n)$. Let w be any function in $W_2^2(\Omega)$, extended onto $\mathbb{R}^n$ and such that

$$\|w;\mathbb{R}^n\|_{W_2^2} \le c\,\|w;\Omega\|_{W_2^2},$$

where c is independent of w. Then

$$\|\,|\nabla\Phi|^2 w;\Omega\|_{L_2} \le c\,\|w;\Omega\|_{W_2^2},$$

which means that

$$|\nabla\Phi|^2 \in M(W_2^2(\Omega) \to L_2(\Omega)).$$

Using the inclusion

$$g \in M(W_2^2(\Omega) \to L_2(\Omega)),$$

by (13.3.18) we find that the function $|\nabla_2\Phi|$ belongs to class $M(W_2^2(\Omega) \to L_2(\Omega))$. The proof is complete. □

Theorem 13.3.4. *If* $\varphi \in MW_2^{3/2}(\partial\Omega)$ *and* u *is a solution of problem* (13.3.15) *in the space* $W_2^2(\Omega)$ *such that* $|u| \le Q$, *then* u *belongs to the space* $MW_2^2(\Omega)$.

Proof. It is sufficient to assume that $n \ge 4$, since otherwise $MW_2^2(\Omega) = W_2^2(\Omega)$. Let e be an arbitrary compact subset of $\bar{\Omega}$ and let σ_e be the $C_{2,2}$-capacitary Bessel potential of e (see Sect. 3.6.2). We show that the function v, given by (13.3.16), belongs to the space $MW_2^2(\Omega)$. Since $\sigma_e \in L_\infty(\mathbb{R}^n)$ and $W_2^2(\Omega) \cap L_\infty(\Omega)$ is an algebra with respect to multiplication, the product $\sigma_e v$ is a function from $W_2^2(\Omega)$. We have

$$b_{ij}\,\frac{\partial^2(\sigma_e v)}{\partial x_i \partial x_j} = \sigma_e\, G - 2b_{ij}\,\frac{\partial \sigma_e}{\partial x_i}\,\frac{\partial v}{\partial x_j} - v\, b_{ij}\,\frac{\partial^2 \sigma_e}{\partial x_i \partial x_j}.$$

According to [Cor], the estimate

$$\|\sigma_e v; \Omega\|_{W_2^2} \le c\left(\|\sigma_e G; \Omega\|_{L_2} + \|\, |\nabla \sigma_e|\, |\nabla v|\, ; \Omega\|_{L_2} + \|v \nabla_2 \sigma_e; \Omega\|_{L_2}\right)$$

is valid. Hence we find from (13.3.17) and Lemma 13.3.1 that

$$\|\sigma_e v; \Omega\|_{W_2^2} \le c\Big(k\, \|\sigma_e |\nabla v|^2; \Omega\|_{L_2} + \|\, |\nabla \sigma_e| |\nabla v|\, ; \Omega\|_{L_2}$$
$$+ \left(\|h; \Omega\|_{M(W_2^2 \to L_2)} + \|v; \Omega\|_{L_\infty}\right) \|\sigma_e; \mathbb{R}^n\|_{W_2^2}\Big). \tag{13.3.19}$$

We have

$$\sigma_e = (1-\Delta)^{-2}\mu \quad \text{and} \quad (1-\Delta)\sigma_e = (1-\Delta)^{-1}\mu,$$

where μ is a nonnegative measure. By G_4 and G_2 we denote the kernels of the integral operators $(1-\Delta)^{-2}$ and $(1-\Delta)^{-1}$. In view of the relations given in Subsect. 13.3.4,

$$|\nabla G_4| = \begin{cases} O(|x|^{3-n}) & \text{for } |x| \le 1, \\ O(|x|^{(3-n)/2} e^{-|x|} & \text{for } |x| > 1. \end{cases}$$

Also,

$$|G_2(x)| \ge c \begin{cases} |x|^{2-n} & \text{for } |x| \le 1, \\ |x|^{(1-n)/2} e^{-|x|} & \text{for } \ |x| > 1 \end{cases}$$

and

$$|G_4(x)| \ge c \begin{cases} |x|^{4-n} & \text{for } |x| \le 1,\ n > 4, \\ \log|x|^{-1} & \text{for } |x| < 1/2,\ n = 4, \\ |x|^{(3-n)/2} e^{-|x|} & \text{for } \ |x| > 1. \end{cases}$$

Therefore,

$$|\nabla G_4| = O(G_4^{1/2} G_2^{1/2})$$

and

$$|\nabla \sigma_e(x)|^2 \le c\, \sigma_e(x)(1-\Delta)\sigma_e(x). \tag{13.3.20}$$

We use this inequality to estimate the second norm on the right-hand side of (13.3.19):

$$\begin{aligned}\| \,|\nabla \sigma_e|\,|\nabla v|\,; \Omega\|_{L_2} &\leq c\,\|\sigma_e|\nabla v|^2; \Omega\|_{L_2}^{1/2}\|(1-\Delta)\sigma_e; \Omega\|_{L_2}^{1/2} \\ &\leq k\,\|\sigma_e|\nabla v|^2; \Omega\|_{L_2} + c\,k^{-1}\|\sigma_e; \mathbb{R}^n\|_{W_2^2}. \qquad (13.3.21)\end{aligned}$$

Now, (13.3.21) and (13.3.19) imply that

$$\begin{aligned}\|\,\sigma_e v; \Omega\|_{W_2^2} &\leq c\Big(2k\|\sigma_e|\nabla v|^2; \Omega\|_{L_2} \\ &+ (\|h; \Omega\|_{M(W_2^2\to L_2)} + \|v; \Omega\|_{L_\infty} + k^{-1})\|\sigma_e; \mathbb{R}^n\|_{W_2^2}\Big). \qquad (13.3.22)\end{aligned}$$

We estimate the first norm on the right-hand side. Since $v \in \mathring{W}_2^2(\Omega) \cap W_2^1(\Omega)$, integration by parts yields

$$\begin{aligned}I := &\int_\Omega \sigma_e^2|\nabla v|^4 dx \\ &\leq c\,\|v; \Omega\|_{L_\infty}\Big(\int_\Omega \sigma_e|\nabla \sigma_e|\,|\nabla v|^3 dx + \int_\Omega \sigma_e^2|\nabla v|^2|\nabla_2 v| dx\Big),\end{aligned}$$

which together with (13.3.20) gives the estimate

$$\begin{aligned}I \leq c\,\|v; \Omega\|_{L_\infty}\Big(&\int_\Omega (\sigma_e|\nabla v|^2)^{3/2}((1-\Delta)\sigma_e)^{1/2} dx \\ &+ I^{1/2}\|\sigma_e|\nabla_2 v|\,; \Omega\|_{L_2}\Big).\end{aligned}$$

Consequently,

$$I \leq c\,\|v; \Omega\|_{L_\infty}\Big(I^{3/4}\|\sigma_e; \mathbb{R}^n\|_{W_2^2}^{1/2} + I^{1/2}\|\sigma_e|\nabla_2 v|\,; \Omega\|_{L_\infty}\Big)$$

and hence

$$\begin{aligned}I^{1/2} \leq c\,\|v; \Omega\|_{L_\infty}\big(&I^{1/4}\|\sigma_e; \mathbb{R}^n\|_{W_2^2}^{1/2} + \|\sigma_e v; \Omega\|_{W_2^2} \\ &+\|\,|\nabla \sigma_e|\,|\Delta v|\,; \Omega\|_{L_2} + \|v\nabla_2 \sigma_e; \Omega\|_{L_2}\big).\end{aligned}$$

Applying (13.3.20), we conclude that

$$\begin{aligned}I^{1/2} \leq c\,\|v; \Omega\|_{L_\infty}\Big(&\|\sigma_e v; \Omega\|_{W_2^2} + \|v; \Omega\|_{L_\infty}\|\sigma_e; \mathbb{R}^n\|_{W_2^2} \\ &+ kI^{1/2} + c\,k^{-1}\|\sigma_e; \mathbb{R}^n\|_{W_2^2}\Big).\end{aligned}$$

Since $k\,\|v; \mathbb{R}^n\|_{L_\infty}$ is small, it follows that

$$\|\sigma_e|\nabla v|^2; \Omega\|_{L_2} = I^{1/2} \leq c\,\|v; \Omega\|_{L_\infty}\Big(\|\sigma_e v; \mathbb{R}^n\|_{W_2^2} + c\,k^{-1}\|\sigma_e; \mathbb{R}^n\|_{W_2^2}\Big).$$

The required estimate is obtained for the first norm on the right-hand side of (13.3.22). We find from (13.3.22) that

$$\|\sigma_e v; \Omega\|_{W_2^2} \leq c\left(\|h; \Omega\|_{M(W_2^2 \to L_2)} + k^{-1}\right)\|\sigma_e; \mathbb{R}^n\|_{W_2^2}.$$

Therefore,

$$\int_e |\nabla_2 v|^2 dx \leq c\left(\|h; \Omega\|_{M(W_2^2 \to L_2)} + k^{-1}\right)^2 C_{2,2}(e).$$

Using the equivalent norm for $\|v; \Omega\|_{MW_2^2}$, we complete the proof. □

13.4 Coercive Estimates for Solutions of Elliptic equations in Spaces of Multipliers

It is well known that solutions of elliptic boundary value problems satisfy coercive estimates in Sobolev spaces (see [ADN2]). The purpose of this section is to show that similar estimates are valid for norms in classes of multipliers acting in a Sobolev space or in a pair of Sobolev spaces.

13.4.1 The Case of Operators in $\mathbb{R}^n$

Theorem 13.4.1. *Let P be an elliptic (in the sense of Douglis and Nirenberg) operator* (10.1.4), *where $M = N$. Let the coefficients of P be constant. Further let $\gamma = \{\gamma^1, \dots, \gamma^N\}$ be a vector-valued function in the space*

$$\prod_k W_{p,\mathrm{loc}}^{h+t_k} \cap \prod_k M(W_p^{r-h-t_k} \to L_p)$$

and let

$$P\gamma \in \prod_j M(W_p^r \to W_p^{h-s_j}),$$

where $r \geq h + t_k \geq 0$, $r \geq h - s_j \geq 0$, $1 \leq j, k \leq N$. Then

$$\gamma \in \prod_k M(W_p^r \to W_p^{h+t_k})$$

and the estimate

$$\|\gamma\|_{\prod_k M(W_p^r \to W_p^{h+t_k})}$$
$$\leq C\left(\|P\gamma\|_{\prod_j M(W_p^r \to W_p^{h-s_j})} + \|\gamma\|_{\prod_k M(W^{r-h+t_k} \to L_p)}\right) \tag{13.4.1}$$

holds.

Proof. It is known that for all $u \in \prod_k W_p^{t_k+h}$

$$\|u\|_{\prod_k W_p^{t_k+h}} \leq C_1\big(\|Pu\|_{\prod_j W_p^{h-s_j}} + \|u\|_{L_p}\big)\,.$$

Consequently, for all $\varphi \in C_0^\infty$,

$$\|\gamma_\rho \varphi\|_{\prod_k W_p^{t_k+h}}$$
$$\leq C_1\big(\|\varphi P\gamma_\rho\|_{\prod_j W_p^{h-s_j}} + \|\gamma_\rho\varphi\|_{L_p} + \|[\varphi, P]\gamma_\rho\|_{\prod_j W_p^{h-s_j}}\big)\,, \tag{13.4.2}$$

where $[\varphi, P]$ is the commutator of P and the operator of multiplication by φ. As usual, by γ_ρ we denote a mollification of γ with radius ρ.

It is clear that

$$\|\varphi P\gamma_\rho\|_{\prod_j W_p^{h-s_j}} \leq \|P\gamma_\rho\|_{\prod_j M(W_p^r \to W_p^{h-s_j})}\|\varphi\|_{W_p^r} \tag{13.4.3}$$

and

$$\|\gamma_\rho\varphi\|_{L_p} \leq \|\gamma_\rho\|_{\prod_k M(W_p^{r-h-t_k} \to L_p)}\|\varphi\|_{W_p^{r-h-\min_k t_k}}\,. \tag{13.4.4}$$

It remains to estimate the third term in (13.4.2). For any multi-index α, $|\alpha| \leq s_j + t_k$, we have

$$\|[\varphi, D^\alpha]\gamma_\rho\|_{\prod_j W_p^{h-s_j}} \leq c \sum_{0<\beta\leq\alpha} \|D^\beta\varphi\, D^{\alpha-\beta}\gamma_\rho\|_{\prod_k W_p^{h-|\alpha|+t_k}}$$
$$\leq c_1 \sum_{0<\beta\leq\alpha} \|D^{\alpha-\beta}\gamma_\rho\|_{\prod_k M(W_p^{r-|\beta|} \to W_p^{h-|\alpha|+t_k})}\|D^\beta\varphi\|_{W_p^{r-|\beta|}}$$

which, together with Lemma 2.3.3, gives

$$\|[\varphi, D^\alpha]\gamma_\rho\|_{\prod_j W_p^{h-s_j}} \leq c \sum_{1\leq\nu\leq|\alpha|} \|\gamma_\rho\|_{\prod_k M(W_p^{r-\nu} \to W_p^{h-\nu+t_k})}\|\varphi\|_{W_p^r}$$
$$\leq \big(\varepsilon\,\|\gamma_\rho\|_{\prod_k M(W_p^r \to W_p^{h+t_k})} + c(\varepsilon)\|\gamma_\rho\|_{\prod_k M(W_p^{r-h-t_k} \to L_p)}\|\varphi\|_{W_p^r}\big). \tag{13.4.5}$$

Now, (13.4.5) and (13.4.2)–(13.4.4) imply that

$$\|\gamma_\rho\varphi\|_{\prod_k W_p^{h+t_k}} \leq C_2\big(\|P\gamma_\rho\|_{\prod_j M(W_p^r \to W_p^{h-s_j})}$$
$$+\varepsilon\,\|\gamma_\rho\|_{\prod_k M(W_p^r \to W_p^{h+t_k})} + c(\varepsilon)\|\gamma_\rho\|_{\prod_k M(W_p^{r-h-t_k} \to L_p)}\big)\|\varphi\|_{W_p^r}.$$

Consequently, γ_ρ satisfies (13.4.1). Since $P\gamma_\rho = (P\gamma)_\rho$, the result follows by Lemma 2.3.1. □

Remark 13.4.1. If $r - h = t_k$ for all $k = 1, \ldots, N$, then the additional assumption

$$\gamma \in \prod_k M(W_p^{r-h-t_k} \to L_p)$$

is equivalent to $\gamma \in L_\infty$ and the estimate (13.4.1) takes the form

$$\|\gamma\|_{MW_p^r} \le C(\|P\gamma\|_{\prod_j M(W_p^r \to W_p^{h-s_j})} + \|\gamma\|_{L_\infty}) .$$

In particular, the inequality

$$\|\gamma\|_{MW_p^r} \le C(\|P\gamma\|_{M(W_p^r \to W_p^{r-2\sigma})} + \|\gamma\|_{L_\infty}) \tag{13.4.6}$$

holds for a scalar elliptic operator P of order 2σ with constant coefficients. Here we *a priori* assume that

$$\gamma \in W_{p,\mathrm{loc}}^r \cap L_\infty \qquad \text{and} \qquad P\gamma \in M(W_p^r \to W_p^{r-2\sigma}),\ \ r \ge 2\sigma.$$

Remark 13.4.2. We show that the norm $\|\gamma\|_{L_\infty}$ on the right-hand side of (13.4.6) cannot be omitted. Let $P = -\Delta + 1$, $r = 2$, $2p < n$, and let $\gamma_0(x) = \eta(x)\log|x|$, where $\eta \in C_0^\infty(\mathcal{B}_1)$ with $\eta(0) = 1$. Then

$$P\gamma_0 = O(|x|^{-2}) \qquad \text{and} \qquad \mathrm{supp} P\gamma_0 \subset \mathcal{B}_1 .$$

Therefore, $P\gamma_0 \in M(W_p^2 \to L_p)$. Suppose that

$$\|\gamma\|_{MW_p^2} \le c\,\|P\gamma\|_{M(W_p^2 \to L_p)}$$

for all $\gamma \in W_{p,\mathrm{loc}}^2 \cap L_\infty$. Let us substitute a mollification $(\gamma_0)_\rho$ of the function γ_0 into the last estimate. Then, according to Corollary 2.3.1 and Lemma 2.3.1,

$$\|(\gamma_0)_\rho\|_{L_\infty} \le C\|(P\gamma_0)_\rho\|_{M(W_p^2 \to L_p)} \le C\|P\gamma_0\|_{M(W_p^2 \to L_p)} .$$

Passing to the limit as $\rho \to 0$, we obtain a contradiction.

13.4.2 Boundary Value Problem in a Half-Space

Let us consider an operator $\{P, \mathrm{tr}P_1, \ldots, \mathrm{tr}P_\sigma\}$ of the boundary value problem in $\mathbb{R}_+^{n+1}$, where P and P_j are differential operators of orders 2σ and σ_j, respectively, and tr stands for the trace at the boundary. Suppose that the coefficients of operators P and P_j are constant and the operators $\{P; \mathrm{tr}P_j\}$ form an elliptic boundary value problem in $\mathbb{R}_+^{n+1}$ (see, for instance, [H1], Sect. 10.1).

We need the following two auxiliary assertions.

Lemma 13.4.1. *Let $[\gamma]_\rho$ be a mollification of γ in $\mathbb{R}_+^{n+1}$ with respect to variables $x \in \mathbb{R}^n$ with nonnegative kernel and radius ρ. Then*

$$\begin{aligned}\|[\gamma]_\rho; \mathbb{R}_+^{n+1}\|_{M(W_p^m \to W_p^l)} &\le \|\gamma; \mathbb{R}_+^{n+1}\|_{M(W_p^m \to W_p^l)} \\ &\le \liminf_{\rho \to 0} \|[\gamma]_\rho; \mathbb{R}_+^{n+1}\|_{M(W_p^m \to W_p^l)}\end{aligned} \tag{13.4.7}$$

and

$$\begin{aligned}\|[\gamma]_\rho; \mathbb{R}^n\|_{M(W_p^{m-1/p} \to W_p^{l-1/p})} &\le \|\gamma; \mathbb{R}^n\|_{M(W_p^{m-1/p} \to W_p^{l-1/p})} \\ &\le \liminf_{\rho \to 0} \|[\gamma]_\rho; \mathbb{R}^n\|_{M(W_p^{m-1/p} \to W_p^{l-1/p})} .\end{aligned} \tag{13.4.8}$$

The proof of (13.4.7) is analogous to that of Lemma 2.3.1, and (13.4.8) was proved in Lemma 4.3.3.

Lemma 13.4.2. *If $\gamma \in W^l_{p,\mathrm{unif}}(\mathbb{R}^{n+1}_+)$ and $[\gamma]_\rho$ is a mollification of γ in $\mathbb{R}^{n+1}$ with respect to variables $x \in \mathbb{R}^n$, then $[\gamma]_\rho \in MW^l_p(\mathbb{R}^{n+1}_+)$.*

Proof. Obviously all derivatives with respect to x of the function $\psi = [\gamma]_\rho$ belong to $W^l_{p,\mathrm{unif}}(\mathbb{R}^{n+1}_+)$. Therefore all derivatives of ψ up to order l, except for $\partial^l \psi/\partial y^l$, are bounded. What is more, for all $y \in \mathbb{R}^1_+$,

$$\sup_x \int_y^{y+1} |\partial^l \psi(x,t)/\partial t^l|^p \, dt < \infty \, .$$

Using these properties and the estimate

$$|u(x,y)|^p \le c \sum_{j=0}^{l} \int_y^{y+1} |\partial^j u(x,t)/\partial t^j|^p \, dt \, ,$$

we arrive at $\|\psi u\|_{W^l_p} \le C \|u\|_{W^l_p}$. □

Theorem 13.4.2. *Let*

$$\gamma \in W^h_{p,\mathrm{loc}}(\mathbb{R}^{n+1}_+) \cap M(W^{r-h}_p(\mathbb{R}^{n+1}_+) \to L_p(\mathbb{R}^{n+1}_+)),$$

where r and h are integers, $r \ge h - \sigma_j > 0$, and $r \ge h - 2\sigma$. Further, let

$$P\gamma \in M(W^r_p(\mathbb{R}^{n+1}_+) \to W^{h-2\sigma}_p(\mathbb{R}^{n+1}_+))$$

and

$$\mathrm{tr} P_j \gamma \in M(W^{r-1/p}_p(\mathbb{R}^n) \to W^{h-\sigma_j-1/p}_p(\mathbb{R}^n)).$$

Then

$$\gamma \in M(W^r_p(\mathbb{R}^{n+1}_+) \to W^h_p(\mathbb{R}^{n+1}_+))$$

and

$$\|\gamma; \mathbb{R}^{n+1}_+\|_{M(W^r_p \to W^h_p)} \le C\Big(\|P\gamma; \mathbb{R}^{n+1}_+\|_{M(W^r_p \to W^{h-2\sigma}_p)}$$
$$+ \sum_{j=1}^{\sigma} \|\mathrm{tr} P_j \gamma; \mathbb{R}^n\|_{M(W^{r-1/p}_p \to W^{h-\sigma_j-1/p}_p)} + \|\gamma; \mathbb{R}^{n+1}_+\|_{M(W^{r-h}_p \to L_p)}\Big). \quad (13.4.9)$$

Proof. It is known (see, for instance, [Tr3], Sect. 5.3.3) that, for all $u \in W^h_p(\mathbb{R}^{n+1}_+)$,

$$\|u; \mathbb{R}^{n+1}_+\|_{W^h_p} \le C_1\Big(\|Pu; \mathbb{R}^{n+1}_+\|_{W^{h-2\sigma}_p}$$
$$+ \sum_{j=1}^{\sigma} \|\mathrm{tr} P_j u; \mathbb{R}^n\|_{W^{h-\sigma_j-1/p}_p} + \|u; \mathbb{R}^{n+1}_+\|_{L_p}\Big). \quad (13.4.10)$$

First we assume that γ belongs to $M(W_p^r(\mathbb{R}_+^{n+1}) \to W_p^h(\mathbb{R}_+^{n+1}))$. By virtue of (13.4.10), for all $\varphi \in C_0^\infty(\overline{\mathbb{R}_+^{n+1}})$,

$$\|\gamma\varphi; \mathbb{R}_+^{n+1}\|_{W_p^h} \leq C\Big(\|\varphi P\gamma; \mathbb{R}_+^{n+1}\|_{W_p^{h-2\sigma}} + \sum_{j=1}^{\sigma} \|\mathrm{tr}(\varphi P_j\gamma); \mathbb{R}^n\|_{W_p^{h-\sigma_j-1/p}}$$
$$+\|\varphi\gamma; \mathbb{R}_+^{n+1}\|_{L_p} + \|[\varphi, P]\gamma; \mathbb{R}_+^{n+1}\|_{W_p^{h-2\sigma}}$$
$$+\sum_{j=1}^{\sigma} \|\mathrm{tr}([\varphi, P_j]\gamma); \mathbb{R}^n\|_{W_p^{h-\sigma_j-1/p}}\Big), \tag{13.4.11}$$

where $[\varphi, P]$, $[\varphi, P_j]$ are commutators of P, P_j and the operator of multiplication by φ. It is clear that

$$\|\varphi P\gamma; \mathbb{R}_+^{n+1}\|_{W_p^{h-2\sigma}} \leq \|P\gamma; \mathbb{R}_+^{n+1}\|_{M(W_p^r \to W_p^{h-2\sigma})} \|\varphi; \mathbb{R}_+^{n+1}\|_{W_p^r},$$
$$\|\varphi\gamma; \mathbb{R}_+^{n+1}\|_{L_p} \leq \|\gamma; \mathbb{R}_+^{n+1}\|_{M(W_p^{r-h} \to L_p)} \|\varphi; \mathbb{R}_+^{n+1}\|_{W_p^{r-h}}, \tag{13.4.12}$$
$$\|\mathrm{tr}(\varphi P_j\gamma); \mathbb{R}^n\|_{W_p^{h-\sigma_j-1/p}} \leq \|\mathrm{tr} P_j\gamma; \mathbb{R}^n\|_{M(W_p^{r-1/p} \to W_p^{h-\sigma_j-1/p})} \|\mathrm{tr}\varphi; \mathbb{R}^n\|_{W_p^{r-1/p}}.$$

Let us estimate the norm $\|[\varphi, P]\gamma; \mathbb{R}_+^{n+1}\|_{W_p^{h-2\sigma}}$. For any multi-index α with $|\alpha| \leq 2\sigma$ we have

$$\|[\varphi, D^\alpha]\gamma; \mathbb{R}_+^{n+1}\|_{W_p^{h-2\sigma}} \leq c \sum_{0<\beta\leq\alpha} \|D^\beta\varphi D^{\alpha-\beta}\gamma; \mathbb{R}_+^{n+1}\|_{W_p^{h-|\alpha|}}$$
$$\leq c_1 \sum_{0<\beta\leq\alpha} \|D^{\alpha-\beta}\gamma; \mathbb{R}_+^{n+1}\|_{M(W_p^{r-|\beta|} \to W_p^{h-|\alpha|})} \|D^\beta\varphi; \mathbb{R}_+^{n+1}\|_{W_p^{r-|\beta|}}.$$

This and Corollary 2.4.1 yield

$$\|[\varphi, D^\alpha]\gamma; \mathbb{R}_+^{n+1}\|_{W_p^{h-2\sigma}} \leq c \sum_{0<\beta\leq\alpha} \|\gamma; \mathbb{R}_+^{n+1}\|_{M(W_p^{r-|\beta|} \to W_p^{h-|\beta|})} \|\varphi; \mathbb{R}_+^{n+1}\|_{W_p^r}.$$

By (2.3.13) the expression on the right-hand side does not exceed

$$c \sum_{0<\beta\leq\alpha} \|\gamma; \mathbb{R}_+^{n+1}\|_{M(W_p^r \to W_p^h)}^{1-|\beta|/h} \|\gamma; \mathbb{R}_+^{n+1}\|_{M(W_p^{r-h} \to L_p)}^{|\beta|/h} \|\varphi; \mathbb{R}_+^{n+1}\|_{W_p^r}.$$

Thus for any $\varepsilon > 0$,

$$\|[\varphi, P]\gamma; \mathbb{R}_+^{n+1}\|_{W_p^{h-2\sigma}} \leq \big(\varepsilon \|\gamma; \mathbb{R}_+^{n+1}\|_{M(W_p^r \to W_p^h)}$$
$$+c(\varepsilon) \|\gamma; \mathbb{R}_+^{n+1}\|_{M(W_p^{r-h} \to L_p)}\big) \|\varphi; \mathbb{R}_+^{n+1}\|_{W_p^r}. \tag{13.4.13}$$

We estimate the norm

$$\|\mathrm{tr}([\varphi, P_j]\gamma); \mathbb{R}^n\|_{W_p^{h-\sigma_j-1/p}}.$$

For any multi-index α, $|\alpha| \leq \sigma_j$, we have

$$\|\mathrm{tr}([\varphi, D^\alpha]\gamma); \mathbb{R}^n\|_{W_p^{h-\sigma_j-1/p}} \leq c \sum_{0<\beta\leq\alpha} \|\mathrm{tr}(D^\beta \varphi D^{\alpha-\beta}\gamma); \mathbb{R}^n\|_{W_p^{h-|\alpha|-1/p}}$$

$$\leq c \sum_{0<\beta\leq\alpha} \|\mathrm{tr} D^{\alpha-\beta}\gamma; \mathbb{R}^n\|_{M(W_p^{r-|\beta|-1/p} \to W_p^{h-|\alpha|-1/p})} \|\mathrm{tr} D^\beta \varphi; \mathbb{R}^n\|_{W_p^{r-|\beta|-1/p}}.$$

It is clear that

$$\|\mathrm{tr} D^{\alpha-\beta}\gamma; \mathbb{R}^n\|_{M(W_p^{r-|\beta|-1/p} \to W_p^{h-|\alpha|-1/p})} \leq c \|D^{\alpha-\beta}\gamma; \mathbb{R}^{n+1}_+\|_{M(W_p^{r-|\beta|} \to W_p^{h-|\alpha|})}.$$

This inequality and Corollary 2.4.1 imply that

$$\|\mathrm{tr}([\varphi, D^\alpha]\gamma); \mathbb{R}^n\|_{W_p^{h-\sigma_j-1/p}} \leq c \sum_{0<\beta\leq\alpha} \|\gamma; \mathbb{R}^{n+1}_+\|_{M(W_p^{r-|\beta|} \to W_p^{h-|\beta|})} \|\varphi; \mathbb{R}^{n+1}_+\|_{W_p^r}.$$

The right-hand side of this inequality was estimated earlier. Therefore for any $\varepsilon > 0$ the norm $\|\mathrm{tr}([\varphi, P_j]\gamma); \mathbb{R}^n\|_{W_p^{h-\sigma_j-1/p}}$ is majorized by the right-hand side of (13.4.13). Therefore, using (13.4.11)–(13.4.13), we arrive at

$$\|\gamma\varphi; \mathbb{R}^{n+1}_+\|_{W_p^h} \leq C_1 \Big(\|P\gamma; \mathbb{R}^{n+1}_+\|_{M(W_p^r \to W_p^{h-2\sigma})}$$
$$+ \sum_{j=1}^{\sigma} \|\mathrm{tr} P_j \gamma; \mathbb{R}^n\|_{M(W_p^{r-1/p} \to W_p^{h-\sigma_j-1/p})} + \varepsilon \|\gamma; \mathbb{R}^{n+1}_+\|_{M(W_p^r \to W_p^h)}$$
$$+ c(\varepsilon) \|\gamma; \mathbb{R}^{n+1}_+\|_{M(W_p^{r-h} \to L_p)} \Big) \|\varphi; \mathbb{R}^{n+1}_+\|_{W_p^r}.$$

Consequently (13.4.9) holds.

Now we get rid of the assumption

$$\gamma \in M(W_p^r(\mathbb{R}^{n+1}_+) \to W_p^h(\mathbb{R}^{n+1}_+)).$$

Let η_y be a function in $C_0^\infty(\mathbb{R}^{n+1})$ defined by $\eta_y(z) = \eta(z - y)$, $y \in \mathbb{R}^{n+1}$. Since

$$\|\Gamma\eta_y; \mathbb{R}^{n+1}_+\|_{W_p^l} \leq \|\Gamma; \mathbb{R}^{n+1}_+\|_{M(W_p^m \to W_p^l)} \|\eta_y; \mathbb{R}^{n+1}_+\|_{W_p^m}$$
$$\leq c \|\Gamma; \mathbb{R}^{n+1}_+\|_{M(W_p^m \to W_p^l)}$$

for all $y \in \overline{\mathbb{R}^{n+1}_+}$, it follows that

$$M(W_p^m(\mathbb{R}^{n+1}_+) \to W_p^l(\mathbb{R}^{n+1}_+)) \subset W^l_{p,\mathrm{unif}}(\mathbb{R}^{n+1}_+).$$

The inclusion

$$M(W_p^{m-1/p}(\mathbb{R}^n) \to W_p^{l-1/p}(\mathbb{R}^n)) \subset W^{l-1/p}_{p,\mathrm{unif}}(\mathbb{R}^n)$$

can be derived in a similar way. This inclusion and the conditions of Theorem imply that

$$P\gamma \in W^{h-2\sigma}_{p,\mathrm{unif}}(\mathbb{R}^{n+1}_+) \qquad \text{and} \qquad \mathrm{tr}P_j\gamma \in W^{h-\sigma_j-1/p}_{p,\mathrm{unif}}(\mathbb{R}^n).$$

The last fact, together with a local coercive estimate (see [ADN1], Ch. 15), leads to $\gamma \in W^h_{p,\mathrm{unif}}(\mathbb{R}^{n+1}_+))$. It remains to substitute the mollification of γ with respect to the variables $x \in \mathbb{R}^n$ into (13.4.9) and to use Lemmas 13.4.1 and 13.4.2. □

Remark 13.4.3. In the same way one can prove a generalization of Theorem 13.4.2 to elliptic boundary value problems for a system, elliptic in the sense of Douglis-Nirenberg (cf. Theorem 13.4.1).

Remark 13.4.4. If $r = h$, then (13.4.9) takes the form

$$\|\gamma; \mathbb{R}^{n+1}_+\|_{MW^r_p} \le C\Big(\|P\gamma; \mathbb{R}^{n+1}_+\|_{M(W^r_p \to W^{r-2\sigma}_p)}$$
$$+ \sum_{j=1}^{\sigma} \|\mathrm{tr}P_j\gamma; \mathbb{R}^n\|_{M(W^{r-1/p}_p \to W^{r-\sigma_j-1/p}_p)} + \|\gamma; \mathbb{R}^{n+1}_+\|_{L_\infty}\Big). \tag{13.4.14}$$

Here we *a priori* assume that

$$\gamma \in W^r_{p,\mathrm{loc}}(\mathbb{R}^{n+1}_+) \cap L_\infty(\mathbb{R}^{n+1}_+)$$

and

$$P\gamma \in M(W^r_p(\mathbb{R}^{n+1}_+) \to W^{r-2\sigma}_p(\mathbb{R}^{n+1}_+)), \quad r \ge 2\sigma.$$

13.4.3 On the L_∞-Norm in the Coercive Estimate

We show by an example that the norm $\|\gamma; \mathbb{R}^{n+1}_+\|_{L_\infty}$ on the right-hand side of (13.4.14) cannot be omitted even if the operator $\{P, P_1, \ldots, P_\sigma\}$ satisfies (13.4.10) without the norm $\|u; \mathbb{R}^{n+1}_+\|_{L_p}$ (cf. Remark 13.4.2).

Let

$$P = -\Delta + 1, \quad \mathrm{tr}P_1 = \partial/\partial x_{n+1}|_{\mathbb{R}^n}, \quad r = 2, \ 2p < n.$$

By η we denote a function in $C_0^\infty(\mathbb{R}^{n+1})$ with support in the unit ball centered at the origin. Let

$$\eta(0) = 1 \qquad \text{and} \quad \partial\eta/\partial x_{n+1} = 0 \quad \text{for} \quad x_{n+1} = 0.$$

We put

$$\Gamma(z) = \eta(z) \log|z|.$$

It is clear that

$$\Gamma \in W^2_p(\mathbb{R}^{n+1}_+), \quad P\Gamma = O(|z|^{-2}) \text{ for } |z| < 1, \quad P\Gamma = 0 \text{ for } |z| > 1.$$

Therefore,

$$P\Gamma \in M(W_p^2(\mathbb{R}_+^{n+1}) \to L_p(\mathbb{R}_+^{n+1})).$$

Further we notice that $\mathrm{tr}P_1\Gamma = 0$. We suppose that

$$\|\gamma; \mathbb{R}_+^{n+1}\|_{MW_p^2} \le C\big(\|P\gamma; \mathbb{R}_+^{n+1}\|_{M(W_p^2 \to L_p)} + \|\mathrm{tr}P_1\gamma; \mathbb{R}^n\|_{M(W_p^{2-1/p} \to W_p^{1-1/p})}\big)$$

for all $\gamma \in W_p^2(\mathbb{R}_+^{n+1}) \cap L_\infty(\mathbb{R}_+^{n+1})$, and we substitute a mollification $[\Gamma]_\rho$ of Γ with respect to variables $x \in \mathbb{R}^n$ into the last inequality. Since $MW_p^r(\mathbb{R}_+^{n+1}) \subset L_\infty(\mathbb{R}_+^{n+1})$ and $P[\Gamma]_\rho \in M(W_p^2 \to L_p)$ by Lemma 13.4.1, it follows that

$$\|[\Gamma]_\rho; \mathbb{R}_+^{n+1}\|_{L_\infty} \le C\,\|P[\Gamma]_\rho; \mathbb{R}_+^{n+1}\|_{M(W_p^2 \to L_p)}\,.$$

The right-hand side of this inequality is uniformly bounded with respect to ρ by Lemma 13.4.1, although the left-hand side tends to infinity as $\rho \to 0$, and we have the required contradiction.

13.5 Smoothness of Solutions to Higher Order Elliptic Semilinear Systems

13.5.1 Composition Operator in Classes of Multipliers

Here we study a composition operator in the space $M(W_p^m(\Omega) \to W_p^l(\Omega))$ with integer m and l, $m \ge l \ge 0$ and $p \in (1, \infty)$. One can consult general properties of such operators in the book [KZPS].

Let a function $(x, \xi) \to f$ be defined on $\Omega \times \mathcal{B}_R$, where $\mathcal{B}_R = \{\xi \in \mathbb{R}^s : |\xi| < R\}$ and Ω is a domain in $\mathbb{R}^n$ with compact closure and the boundary $\partial\Omega$ of the class $C^{0,1}$. We assume that, for almost all $x \in \Omega$ and all $\xi \in \mathcal{B}_R$, there exist partial derivatives of f up to order l independent of the order of differentiation and satisfying the Carathéodory conditions, that is, they are measurable in x for all $\xi \in \mathcal{B}_R$ and continuous in ξ for almost all $x \in \Omega$.

Consider the non-linear mapping $\gamma \to F(\gamma)$ defined by

$$F(\gamma)(x) = f(x, \gamma(x)), \qquad x \in \Omega,$$

where $\gamma(x) = (\gamma_1(x), \dots, \gamma_s(x))$ is the mapping $\Omega \to \mathcal{B}_R$. We additionally assume that f and all its partial derivatives up to order $l-1$ obey the rule of differentiation of a composition function (see [MaMi]) if $\gamma \in C^\infty(\bar\Omega)$.

The next assertion contains the main result concerning the operator $F(\gamma)$.

Theorem 13.5.1. *Let $\gamma \in MW_p^m(\Omega)$, $m \ge l$, and let $|\gamma(x)| \le R$ for almost all $x \in \Omega$. Further, let*

$$|D_\xi^\beta D_x^\alpha f(x, \xi)| \le G_{|\beta|}(x), \qquad |\beta| + |\alpha| \le l, \tag{13.5.1}$$

where

$$G_{|\beta|} \in M(W_p^{m-|\beta|}(\Omega) \to L_p(\Omega)).$$

Then

$$F(\gamma) \in M(W_p^m(\Omega) \to W_p^l(\Omega)). \tag{13.5.2}$$

Proof. Let $\mathfrak{K}(h)\gamma$ stand for the mollification of $\gamma \in MW_p^m(\Omega)$ defined with the help of a partition of unity on Ω used in Sect. 9.1.2. For brevity we put $\gamma_h = \mathfrak{K}(h)\gamma$. The mollification has the following properties: $\gamma_h \in C^\infty(\bar{\Omega})$, $|\gamma_h(x)| \le R$, if $|\gamma(x)| \le R$, and $\gamma_h \in MW_p^m(\Omega)$ (see Sects. 9.1.2, 9.1.3).

Suppose that the theorem is proved for the pairs $(l-1, m)$ and $(l-1, m-1)$ and that

$$\|uF(\gamma_h); \Omega\|_{MW_p^{l-1}} \le C\,\|u; \Omega\|_{W_p^r}, \quad r = m-1, m,$$

with a constant independent of h. We have

$$\frac{\partial}{\partial x_j}\left[F(\gamma_h(x))\right] = \frac{\partial f}{\partial x_j}(x, \gamma_h(x)) + \sum_{i=1}^{s} \frac{\partial f}{\partial \xi_j}(x, \gamma_h(x)) \frac{\partial(\gamma_h)_i(x)}{\partial x_j}. \tag{13.5.3}$$

By (13.5.1),

$$\left|D_\xi^\beta D_x^\alpha \frac{\partial f}{\partial \xi_i}\right| \le G_{|\beta|+1}, \quad |\beta| + |\alpha| \le l-1,$$

where

$$G_{|\beta|+1} \in M(W_p^{m-1-|\beta|}(\Omega) \to L_p(\Omega)).$$

Hence, by the induction hypothesis for the pair $(l-1, m-1)$ it follows that the functions $\partial f(\cdot, \gamma_h(\cdot))/\partial \xi_i$ belong to $M(W_p^{m-1}(\Omega) \to W_p^{l-1}(\Omega))$. Also, the inequality

$$\left\|u\frac{\partial f}{\partial \xi_i}(\cdot, \gamma_h(\cdot)); \Omega\right\|_{W_p^{l-1}} \le C\,\|u; \Omega\|_{W_p^{m-1}} \tag{13.5.4}$$

holds with a constant C independent of h. Since

$$\begin{aligned}\left\|u\frac{\partial(\gamma_i)_h}{\partial x_j}; \Omega\right\|_{W_p^{m-1}} &\le \left\|\frac{\partial(\gamma_i)_h}{\partial x_j}; \Omega\right\|_{M(W_p^m \to W_p^{m-1})} \|u; \Omega\|_{W_p^m} \\ &\le \left\|\frac{\partial \gamma_i}{\partial x_j}; \Omega\right\|_{M(W_p^m \to W_p^{m-1})} \|u; \Omega\|_{W_p^m}\end{aligned} \tag{13.5.5}$$

(see Lemma 9.3.2), it follows that the sum over i in (13.5.3) belongs to the space $M(W_p^m(\Omega) \to W_p^{l-1}(\Omega))$.

Similarly, by (13.5.1)

$$\left|D_\xi^\beta D_x^\alpha \frac{\partial f}{\partial x_j}\right| \le G_{|\beta|}(x), \qquad |\beta| + |\alpha| \le l-1,$$

where

$$G_{|\beta|} \in M(W_p^{m-|\beta|}(\Omega) \to L_p(\Omega)).$$

Therefore, by the induction hypothesis for the pair $(l-1, m)$ it follows that the derivatives $\partial f(\cdot, \gamma_h(\cdot))/\partial x_j$ belong to $M(W_p^m(\Omega) \to W_p^{l-1}(\Omega))$ and that

$$\left\| u \frac{\partial f}{\partial x_j}(\cdot, \gamma_h(\cdot)); \Omega \right\|_{W_p^{l-1}} \leq C \, \|u; \Omega\|_{W_p^m} \tag{13.5.6}$$

with a constant C independent of h. Thus,

$$\nabla_x F(\gamma_h) \in M(W_p^m(\Omega) \to W_p^{l-1}(\Omega)).$$

Now, (13.5.4)–(13.5.6) imply that

$$\|u \nabla_x F(\gamma_h); \Omega\|_{W_p^{l-1}} \leq C \, \|u; \Omega\|_{W_p^m} \tag{13.5.7}$$

with a constant C independent of h.

Using (13.5.1) with $|\beta| = |\alpha| = 0$, we find that the norms of the functions $F(\gamma_h)$ are uniformly bounded in $L_p(\Omega)$. Moreover, (13.5.1) with $m = l$ implies that the derivatives

$$D_x^\beta f(x, \gamma_h(x)), \qquad |\beta| = m,$$

are bounded on $\Omega \times \mathcal{B}_R$ uniformly with respect to h. Hence $F(\gamma_h)$ is uniformly bounded on $\Omega \times \mathcal{B}_R$. This fact and Theorem 9.1.1 imply the inequality

$$\|u \, F(\gamma_h); \Omega\|_{W_p^l} \leq C \, \|u; \Omega\|_{W_p^m} \tag{13.5.8}$$

with a constant C independent of h.

The induction basis, that is (13.5.8) with $l = 0$ follows directly from the inequality $|f(x, \xi)| \leq G_0(x)$, where

$$G_0 \in M(W_p^m(\Omega) \to L_p(\Omega)).$$

To complete the proof, it remains to show that $D^\alpha[uF(\gamma_h)]$ with $u \in C^\infty(\bar{\Omega})$ tends to $D^\alpha[uF(\gamma)]$, $|\alpha| = l$, for almost all $x \in \Omega$, and then to use estimate (13.5.8). The result will follow by passage to the limit in (13.5.8) and the Fatou theorem.

Since $f(x, \xi)$ and all its partial derivatives up to order l are continuous in ξ for almost all $x \in \Omega$, and

$$\lim_{h \to 0} D^\alpha \gamma_h = \lim_{h \to 0} (D^\alpha \gamma)_h = D^\alpha \gamma$$

for almost all $x \in \Omega$, it follows that for such x the limit

$$\lim_{h \to 0} D^\alpha [uF(\gamma_h)] = g(x)$$

exists. Let us show that g coincides with $D^\alpha[uF(\gamma)]$. Let η be an arbitrary function in $C_0^\infty(\Omega)$ and let $(\, , \,)$ stand for the scalar product in $L_2(\Omega)$. We have

$$(D^\alpha[u \, F(\gamma_h)], \eta) = (-1)^{|\alpha|}(u \, F(\gamma_h), D^\alpha \eta). \tag{13.5.9}$$

Since $F(\gamma_h)$ tends to $F(\gamma)$ for almost all $x \in \Omega$ and $|F(\gamma_h)(x)| \le G_0(x)$, it follows by Lebesgue's theorem that the right-hand side of (13.5.9) has the limit

$$(-1)^{|\alpha|}(u\,F(\gamma), D^\alpha \eta) = (D^\alpha[u\,F(\gamma)], \eta).$$

Hence

$$\lim_{h\to 0}(D^\alpha[u\,F(\gamma_h)], \eta) = (g, \eta) = (D^\alpha[u\,F(\gamma)], \eta)$$

and $g = D^\alpha[u\,F(\gamma)]$. Thus, $D^\alpha[u\,F(\gamma_h)]$ tends to $D^\alpha[u\,F(\gamma)]$ almost everywhere in Ω. The proof is complete. □

Remark 13.5.1. When proving the equality

$$\lim_{h\to 0}(D^\alpha[u\,F(\gamma_h)], \eta) = (D^\alpha[u\,F(\gamma)], \eta) \ \text{ for a.e. } x \in \Omega,$$

we showed in passing that the chain rule of differentiation of $f(x, \gamma(x))$ holds for all its partial derivatives of order $l-1$, provided that all partial derivatives of f up to order l satisfy the Carathéodory conditions and $\gamma \in MW_p^l(\Omega)$.

Remark 13.5.2. According to Theorem 9.1.1, the condition

$$G_k \in M(W_p^{m-k}(\Omega) \to L_p(\Omega)), \quad p \in (1, \infty),$$

is equivalent to

$$\sup_e \frac{\displaystyle\int_e |G_k(x)|^p dx}{C_{p,m-k}(e)} < \infty, \tag{13.5.10}$$

where e is an arbitrary compact subset of $\bar{\Omega}$. In the case $m = k$, (13.5.10) should be replaced by the inclusion $G_k \in L_\infty(\Omega)$. For $p = 1$ the role of the condition (13.5.10) is played by

$$\sup_{x\in\bar{\Omega}, 0<r<1} r^{m-k-n} \int_{\mathcal{B}_r(x)\cap\bar{\Omega}} |G_k(y)|dy < \infty.$$

13.5.2 Improvement of Smoothness of Solutions to Elliptic Semilinear Systems

Let Ω be a bounded domain with smooth boundary and let $P(x, D_x)$ be a $(s \times s)$-matrix-valued elliptic operator of order $2k$. Further, let

$$P_1(x, D_x), \dots, P_k(x, D_x)$$

be differential operators of orders $m_1, \dots, m_k \le 2k - 1$. We assume that the coefficients of P and P_j are smooth and that the system of boundary operators $\{P_1, \cdots, P_k\}$ is normal (see [LiM2], vol. 1, Sect. 1.4).

The next auxiliary assertion follows directly from Remark 13.4.4 by a standard localization argument.

Lemma 13.5.1. *If*

$$\gamma \in W^{2k}_{p,\mathrm{loc}}(\Omega) \cap L_\infty(\Omega)$$

and

$$P\gamma \in M(W^s_p(\Omega) \to W^{s-2k}_p(\Omega)), \ \ s > 2k, \quad \mathrm{tr} P_j\gamma = 0 \ \text{ on } \partial\Omega,$$

then $\gamma \in MW^s_p(\Omega)$.

Consider the boundary value problem

$$\begin{cases} P(x, D_x)\gamma(x) = f(x, \gamma(x)) & \text{in } \Omega, \\ \mathrm{tr} P_j(x, D_x)\gamma = 0 & \text{on } \partial\Omega, \ \ j = 1, \dots, k, \end{cases}$$

where $\gamma \in W^{2k}_{p,\mathrm{loc}}(\Omega) \cap L_\infty(\Omega)$ and $f(x, \gamma(x))$ satisfies the conditions immediately preceding Theorem 13.5.1 with l replaced by $m - 2k$, $m \geq 2k + 1$.

Theorem 13.5.2. *Let* $|\gamma(x)| \leq R$ *for almost all* $x \in \Omega$. *Assume that* (13.5.2) *holds with* $l = m - 2k$ *for almost all* $x \in \Omega$ *and for all* $\xi \in \mathcal{B}_R$, *and that*

$$G_{|\beta|} \in M(W^{m-1-|\beta|}_p(\Omega) \to L_p(\Omega)).$$

Then $\gamma \in MW^m_p(\Omega)$.

Proof. Since

$$G_0 \in M(W^{m-1}_p(\Omega) \to L_p(\Omega)) \subset M(W^{2k}_p(\Omega) \to L_p(\Omega)),$$

it follows by Lemma 13.5.1 that $\gamma \in MW^{2k}_p(\Omega)$. For all multi-indices α and β such that $|\alpha| + |\beta| \leq m - 2k$, we have

$$|D^\beta_\xi D^\alpha_x f| \leq G_{|\beta|}(x), \qquad \text{where } \ G_{|\beta|} \in M(W^{m-1-|\beta|}_p(\Omega) \to L_p(\Omega)).$$

Therefore, by Theorem 13.5.1,

$$f(\cdot, \gamma(\cdot)) \in M(W^{m-1}_p(\Omega) \to W^{m-2k}_p(\Omega)).$$

Hence

$$f(\cdot, \gamma(\cdot)) \in M(W^m_p(\Omega) \to W^{m-2k}_p(\Omega)).$$

This inclusion and Lemma 13.5.1 imply that $\gamma \in MW^m_p(\Omega)$. The result follows. □

Remark 13.5.3. For $pl > n$, when the space $M(W^m_p(\Omega) \to W^l_p(\Omega))$ coincides with $W^l_p(\Omega)$ (see Theorem 9.3.3), Theorem 13.5.2 contains the same result as Theorem 1 in [Poh2].

Remark 13.5.4. We finish this chapter by noting that during recent years multipliers in spaces of differentiable functions turned out to be useful in the study of nonlinear evolution equations. The first application of such a kind seems to be made by Lasiecka who proved the existence of compact local attractors for a von Karman dissipative system with nonlinear dissipation [Las].

Lemarié-Rieusset applied the space $M(h^m_p \to h^l_p)$ to the problem of uniqueness for the Navier-Stokes system [LR] (see also [Ger], [MaPa], and [LRM]).

14

Regularity of the Boundary in L_p-Theory of Elliptic Boundary Value Problems

14.1 Description of Results

The purpose of this chapter is to give applications of the theory of multipliers, developed earlier, to elliptic boundary value problems in domains with 'non-regular' boundaries.

We consider an operator $\{P, \text{tr } P_1, \dots, \text{tr } P_h\}$ of the general elliptic boundary value problem with smooth coefficients in a bounded domain $\Omega \subset \mathbb{R}^n$. We assume that ord $P = 2h \le l$ and ord $P_j = k_j < l$, where l is an an integer. The trace operator on the boundary $\partial\Omega$ is denoted by tr.

It is well known that the mapping

$$\{P; \text{tr}P_j\} : W_p^l(\Omega) \to W_p^{l-2h}(\Omega) \times \prod_{j=1}^{h} W_p^{l-k_j-1/p}(\partial\Omega), \tag{14.1.1}$$

where $1 < p < \infty$, is Fredholm, i.e., it has a finite index and a closed range, provided that the boundary is sufficiently smooth. In particular, for all $u \in W_p^l$ the *a priori* estimate

$$\|u; \Omega\|_{W_p^l} \le c\Big(\|Pu; \Omega\|_{W_p^{l-2h}} + \sum_{j=1}^{h} \|\text{tr}P_j u; \partial\Omega\|_{W_p^{l-k_j-1/p}} + \|u; \Omega\|_{L_1}\Big) \tag{14.1.2}$$

holds; the last norm on the right-hand side can be omitted in the case of a unique solution (see [ADN1], [H1] *et al.*).

The analytic background to these fundamental assertions of elliptic L_p-theory is the study of the boundary value problem with constant coefficients in $\mathbb{R}^n_+$ and subsequent localization of the original problem, with the help of a partition of unity together with a local mapping of the domain onto a half-space. The smoothness of the coefficients, and hence that of the solution of the obtained boundary value problem in $\mathbb{R}^n_+$, depends on the smoothness of the surface $\partial\Omega$. It is well known that the above mentioned properties of the operator (14.1.1) fail where the boundary has singularities.

V.G. Maz'ya, T.O. Shaposhnikova, *Theory of Sobolev Multipliers*, Grundlehren der mathematischen Wissenschaften 337,

In this chapter $\partial\Omega$ is characterized either in terms of spaces of multipliers for $p(l-1) \le n$ or in terms of fractional Sobolev spaces for $p(l-1) > n$. We use the usual procedure of localization of the boundary value problem. The novel aspect is the application of properties of multipliers and, in particular, theorems on their traces on the boundary. This makes less stringent the conditions on the domain Ω which ensure the main results of elliptic L_p-theory.

Sections 14.2 and 14.4 contain auxiliary results. In 14.3 we show that the mapping (14.1.1) is Fredholm in the case $p(l-1) \le n$ provided that the boundary $\partial\Omega$ belongs to the class $M_p^{l-1/p}(\delta)$. This means that for each point of the boundary there exists a neighborhood U and a Lipschitz function φ such that

$$U \cap \Omega = \{(x, y) \in U : x \in \mathbb{R}^{n-1},\ y > \varphi(x)\} \tag{14.1.3}$$

and

$$\|\nabla\varphi; \mathbb{R}^{n-1}\|_{MW_p^{l-1-1/p}} \le \delta\ .$$

Here δ is a small constant and MW_p^s is the space of multipliers in W_p^s for $s > 0$ and the space L_∞ for $s \le 0$.

For $p(l-1) > n$ the mapping $\{P; \mathrm{tr}\ P_j\}$ is Fredholm provided that $\partial\Omega$ belongs to the class $W_p^{l-1/p}$.

In Sect. 14.5 we consider specifically the first boundary value problem for a strongly elliptic operator P in divergence form. We study two variants of this problem which differ in the description of the boundary data. In the first formulation, called generalized, we look for a solution $u \in W_p^l(\Omega)$ of the equation

$$Pu = f \in W_p^{l-2h}(\Omega), \qquad l \ge h,$$

satisfying the condition

$$u - g \in W_p^l(\Omega) \cap \mathring{W}_p^h(\Omega),$$

where g is a given function in $W_p^l(\Omega)$. It is shown that this problem has a unique solution if $\partial\Omega$ is in the class $M_p^{l+1-h-1/p}(\delta)$ for $p(l-h) \le n$ and $\partial\Omega$ belongs to the class $W_p^{l+1-h-1/p}$ for $p(l-h) > n$. In the second, stronger, formulation, the boundary data are prescribed by means of some differential operators P_j, $1 \le j \le h$. We prove that such a problem is solvable for $h > 1$ if $\partial\Omega$ belongs to the class $M_p^{l-1/p}$ defined by the condition $\nabla\varphi \in MW_p^{l-1-1/p}$, where φ is the same as in (14.1.3), and if the Lipschitz constant of φ is small. In the case $p(l-1) > n$, this condition is equivalent to $\partial\Omega \in W_p^{l-1/p}$.

The inclusion $\partial\Omega \in W_p^{l-1/p}$ for $p(l-1) > n$ is not only sufficient but also necessary for solvability of the Dirichlet problem in the second formulation (see Sect. 14.6).

In Sect. 14.6.3 we give an analytic description of the class $M_p^{l-1/p}(\delta)$ involving a capacity and obtain some simpler conditions for the inclusion of $\partial\Omega$ into $M_p^{l-1/p}(\delta)$. For instance, if the norm $\|\nabla\varphi; \mathbb{R}^{n-1}\|_{L_\infty}$ is small and φ

belongs to the Besov space $B^{l-1/p}_{q,p}(\mathbb{R}^{n-1})$ with $q \in [p(n-1)/(p(l-1)-1), \infty]$ for $p(l-1) < n$ and $q \in (p, \infty]$ for $p(l-1) = n$, then $\partial\Omega$ belongs to $M_p^{l-1/p}(\delta)$. Putting $q = \infty$ we obtain that this inclusion follows from the convergence of the integral

$$\int_0 [\omega_{l-1}(t)/t]^p\, dt,$$

where ω_{l-1} is the modulus of continuity of the vector-function $\nabla_{l-1}\varphi$.

In view of the imbedding $B^{l-1/p}_{\infty,p} \subset W_p^{l-1/p}$, the last condition is also sufficient for $\partial\Omega \in W_p^{l-1/p}$. Using this fact, one can immediately derive the following assertions from our theorems.

The inclusion $\omega_{l-1}(t)/t \in L_p(0,1)$ provides the Fredholm property of the operator $\{P; \mathrm{tr} P_j\}$ as well as the unique solvability of the Dirichlet problem in the second formulation. Moreover, the unique solvability in $W_p^l(\Omega)$ of the Dirichlet problem in the first (generalized) formulation is obtained under the assumption

$$\omega_{l-h}(t)/t \in L_p(0,1).$$

In Sect. 14.6.3 we note that even these, the roughest of our sufficient conditions, are precise in a sense.

The proof of a theorem in 14.6.3, which contains a local characterization of the class $M_p^{l-1/p}(\delta)$, is given in 14.7.

14.2 Change of Variables in Differential Operators

Consider the domain

$$G = \{z = (x,y) \in \mathbb{R}^n : x \in \mathbb{R}^{n-1}, y > \varphi(x)\},$$

where φ is a function satisfying the Lipschitz condition

$$|\varphi(x_1) - \varphi(x_2)| \le L\,|x_1 - x_2|.$$

The following assertion characterizes coefficients of a differential operator under a change of variables.

Proposition 14.2.1. *Let G be a special Lipschitz domain and let λ be an arbitrary (p,l)-diffeomorphism $\mathbb{R}^n_+ \to G$, $\varkappa = \lambda^{-1}$. Further, let*

$$R(z, D_z) = \sum_{0\le|\alpha|\le h} a_\alpha(z) D_z^\alpha\,, \qquad z \in G\,,$$

and

$$S(\zeta, D_\zeta) = \sum_{0\le|\beta|\le h} b_\beta(\zeta) D_\zeta^\beta\,, \qquad \zeta \in \mathbb{R}^n_+\,,$$

be differential operators in G *and* $\mathbb{R}^n_+$ *such that*

$$Sv = [R(v \circ \varkappa)] \circ \lambda \,. \tag{14.2.1}$$

If

$$a_\alpha \in M(W_p^{l-|\alpha|}(G) \to W_p^{l-h}(G))$$

for all multi-indices α*, then*

$$b_\beta \in M(W_p^{l-|\beta|}(\mathbb{R}^n_+) \to W_p^{l-h}(\mathbb{R}^n_+))$$

and

$$\|b_\beta; \mathbb{R}^n_+\|_{M(W_p^{l-|\beta|} \to W_p^{l-h})} \le c \sum_{|\beta| \le |\alpha| \le h} \|a_\alpha; G\|_{M(W_p^{l-|\alpha|} \to W_p^{l-h})} \,. \tag{14.2.2}$$

The proof follows the same lines as that of Lemma 10.1.3.
We note that according to (10.1.20) the equality

$$b_\beta = \sum_{|\beta| \le |\alpha| \le h} (a_\alpha \circ \lambda) \sum c_s \prod_{i,j} (D_z^{s_{ij}} \varkappa_i) \circ \lambda \tag{14.2.3}$$

holds.

Lemma 14.2.1. *Let* G *denote a special Lipschitz domain and let* λ *be an arbitrary* (p, l)*-diffeomorphism:* $\mathbb{R}^n_+ \to G$*. Further, let* R *be a homogeneous differential operator of order* h *with constant coefficients and let* S *be an operator defined by* (14.2.1)*. Then*

$$\|S - R; \mathbb{R}^n_+\|_{M(W_p^l \to W_p^{l-h})} \le c \, \|I - \partial\lambda; \mathbb{R}^n_+\|_{MW_p^{l-1}} \,, \tag{14.2.4}$$

where c *is a continuous function of the norm of* $\partial\lambda$ *in* $MW_p^{l-1}(\mathbb{R}^n_+)$*. (Here and henceforth by the norm of a matrix we mean the sum of the norms of its elements.)*

Proof. We put $\varkappa = \lambda^{-1}$ and

$$a = \|I - \partial\lambda; \mathbb{R}^n_+\|_{MW_p^{l-1}}.$$

Let $S_1(\zeta, D_\zeta)$ denote the principal homogeneous part of the operator S. Since

$$S_1(\zeta, \rho) = S((\partial\varkappa)^* \rho) \circ \lambda$$

for any vector $\rho \in \mathbb{R}^n$, it follows that every coefficient of S_1 differs from the corresponding coefficient of R by $O(a)$ in the norm of $MW_p^{l-h}(\mathbb{R}^n_+)$. Hence,

$$\|S_1 - R; \mathbb{R}^n_+\|_{M(W_p^l \to W_p^{l-h})} \le c\, a.$$

Consider the coefficients of S which multiply the derivatives of order $|\beta| < h$. Let formula (14.2.3) relate the coefficients a_α and b_β of the operators R

and S. Since R is homogeneous, we have $|\alpha| = h$ in (14.2.3). Hence by (14.2.2) every term in (14.2.3) with $|\beta| < h$ contains at least one factor $D^{s_{ij}} \varkappa_i(z)$ for which $|s_{ij}| > 1$. Noting that such a factor is equal to $D^{s_{ij}}[\varkappa_i(z) - z_i]$, we obtain

$$\|b_\beta; \mathbb{R}^n_+\|_{M(W^{l-|\beta|}_p \to W^{l-h}_p)} \le c\,\|I - \partial\varkappa; G\|_{MW^{l-1}_p} \le c\,a$$

(see the proof of Lemma 10.1.3). Therefore,

$$\|S - S_1; \mathbb{R}^n_+\|_{M(W^l_p \to W^{l-h}_p)} \le c\,a.$$

□

Duplicating the proof of Lemma 14.2.1 with obvious changes and using the properties of (p, l)-diffeomorphisms given in Sect. 9.4.1, we obtain the following local variant of Lemma 14.2.1.

Lemma 14.2.2. *Let all conditions of Proposition* 14.2.1 *be satisfied. Then for each* $v \in W^l_p(\mathbb{R}^n_+)$ *with support in* $\mathcal{B}_r \cap \overline{\mathbb{R}^n_+}$,

$$\|(S - R)v; \mathbb{R}^n_+\|_{W^{l-h}_p} \le c\,\|I - \partial\lambda; \mathcal{B}_r \cap \mathbb{R}^n_+\|_{MW^{l-1}_p} \|v; \mathbb{R}^n_+\|_{W^l_p}\,, \qquad (14.2.5)$$

where c *is a constant independent of* $r \in (0, 1)$.

For $p(l-1) > n$ it follows from (9.6.4) that (14.2.5) is equivalent to

$$\|(S - R)v; \mathbb{R}^n_+\|_{W^{l-h}_p} \le c\,r^{l-1-n/p} \|I - \partial\lambda; \mathcal{B}_r \cap \mathbb{R}^n_+\|_{W^{l-1}_p} \|v; \mathbb{R}^n_+\|_{W^l_p}\,. \quad (14.2.6)$$

14.3 Fredholm Property of the Elliptic Boundary Value Problem

14.3.1 Boundaries in the Classes $M^{l-1/p}_p$, $W^{l-1/p}_p$, and $M^{l-1/p}_p(\delta)$

Let Ω be a bounded domain with $\partial\Omega \in C^{0,1}$. We introduce the class $M^{l-1/p}_p$ $(l = 2, 3, \dots)$ of boundaries $\partial\Omega$, satisfying the following condition. For every point of $\partial\Omega$ there exists an n-dimensional neighborhood in which $\partial\Omega$ is specified (in a certain Cartesian coordinate system) by a function φ such that

$$\nabla\varphi \in MW^{l-1-1/p}_p(\mathbb{R}^{n-1}).$$

Furthermore, by definition, $M^{1-1/p}_p = C^{0,1}$.

We say that $\partial\Omega$ belongs to the class $W^{l-1/p}_p$ if $\partial\Omega$ can be locally specified by a function $\varphi \in W^{l-1/p}_p(\mathbb{R}^{n-1})$. Since

$$MW^{l-1-1/p}_p(\mathbb{R}^{n-1}) \subset W^{l-1-1/p}_{p,\mathrm{loc}}(\mathbb{R}^{n-1}), \quad l \ge 2,$$

and $C^{0,1}(\mathbb{R}^{n-1}) \subset W_{p,\mathrm{loc}}^{l-1/p}(\mathbb{R}^{n-1})$, it follows that any bounded domain Ω with $\partial\Omega \in M_p^{l-1/p}$ satisfies $\partial\Omega \in W_p^{l-1/p}$.

According to Corollary 4.3.8, for $p(l-1) > n$ we have

$$\|\nabla\varphi; \mathbb{R}^{n-1}\|_{MW_p^{l-1-1/p}} \sim \sup_{x\in\mathbb{R}^{n-1}} \|\nabla\varphi; \mathcal{B}_1(x)\|_{W_p^{l-1-1/p}} .$$

Therefore, the classes $M_p^{l-1/p}$ and $W_p^{l-1/p}$ coincide for $p(l-1) > n$.

For a bounded domain Ω with $\partial\Omega \in C^{0,1}$, by $W_p^{l-l/p}(\partial\Omega)$ we denote the space of traces on $\partial\Omega$ of functions in $W_p^l(\Omega)$. Taking into account the analogous fact for special Lipschitz domains of the class $M_p^{l-1/p}$ (see Sect. 9.4.3), we obtain that $MW_p^{l-1/p}(\partial\Omega)$ is the space of traces of functions in $MW_p^l(\Omega)$.

Let $P, P_1, \dots, P_k$ be differential operators in $\bar{\Omega}$ of orders $2h, k_1, \dots, k_h$, respectively, where $2h \le l$ and $k_j < l$. Suppose that the coefficients of P and P_j belong to $C^{l-2h}(\bar{\Omega})$ and $C^{l-k_j}(\bar{\Omega})$, respectively. (This restriction can be removed by the use of spaces of multipliers, but we do not want to complicate the formulations.) We assume that the operators P, tr $P_1, \dots,$ tr P_n form an elliptic boundary value problem at every point $O \in \partial\Omega$ with respect to the hyperplane $y = 0$ and that P is an elliptic operator in Ω.

In our subsequent exposition the following additional condition on Ω will play an important role.

The class $M_p^{l-1/p}(\delta)$. We say that $\partial\Omega$ belongs to the class $M_p^{l-1/p}(\delta)$ if for each point $O \in \partial\Omega$ there exists a neighborhood U and a special Lipschitz domain $G = \{z = (x, y) : x \in \mathbb{R}^{n-1}, y > \varphi(x)\}$ such that $U \cap \Omega = U \cap G$ and

$$\|\nabla\varphi; \mathbb{R}^{n-1}\|_{MW_p^{l-1-1/p}} \le \delta .$$

Here $p(l-1) \le n$ and δ is a constant which depends on the coefficients of the principal homogeneous parts of P, $P_1, \dots, P_h$ calculated at the point O in the coordinate system (x, y). For $l = 1$ the role of the last inequality is played by

$$\|\nabla\varphi; \mathbb{R}^{n-1}\|_{L_\infty} \le \delta.$$

Obviously, the boundaries in $M_p^{l-1/p}(\delta)$ belong to the class $M_p^{l-1/p}$ and, therefore, to the class $W_p^{l-1/p}$. In 14.6.3 we give an equivalent description of $M_p^{l-1/p}(\delta)$ and discuss sufficient conditions for the inclusion into this class.

14.3.2 A Priori L_p-Estimate for Solutions and Other Properties of the Elliptic Boundary Value Problem

In the next two theorems we consider separately the cases $p(l-1) \le n$ and $p(l-1) > n$.

Theorem 14.3.1. *If* $p(l-1) \le n$, $1 < p < \infty$, *and if* $\partial\Omega$ *belongs to the class* $M_p^{l-1/p}(\delta)$, *then* (14.1.2) *holds for any* $u \in W_p^l(\Omega)$.

Proof. We retain the notation used in the definition of $M_p^{l-1/p}(\delta)$. Let U be an open ball with a small radius, $\sigma \in C_0^\infty(U)$, and R and R_j be the principal homogeneous parts of the operators P and P_j with "frozen" coefficients at the point O. Clearly,

$$\|(P-R)(\sigma u); U\cap\Omega\|_{W_p^{l-2h}} \le \varepsilon\,\|\sigma u; U\cap\Omega\|_{W_p^l} + c\,\|\sigma u; U\cap\Omega\|_{W_p^{l-1}}\,, \quad (14.3.1)$$

where ε is a small positive number (the required smallness is defined by the coefficients of the operators R, R_1, …, R_h). An analogous estimate holds for the norm of $(P_j - R_j)(\sigma u)$ in $W_p^{l-k_j}(U \cap \Omega)$.

By definition of the class $M_p^{l-1/p}(\delta)$, the Lipschitz constant of φ is small, so we can put $N = 1$ in the definition (9.4.14) of the mapping $\lambda : \mathbb{R}^n_+ \to G$. Then, from (9.4.12), we obtain that $\partial\lambda$ differs from the identity matrix by $O(\delta)$ in the norm of $MW_p^{l-1}(\mathbb{R}^n_+)$. It is well known (see, for instance, [ADN1], [Tr3], Sect. 5.3.3) that, for all $v \in W_p^l(\mathbb{R}^n_+)$ with supports in $\mathcal{B}_1 \cap \overline{\mathbb{R}^n_+}$,

$$\|v; \mathbb{R}^n_+\|_{W_p^l} \le c\Big(\|Rv; \mathbb{R}^n_+\|_{W_p^{l-2h}} + \sum_{j=1}^h \|\mathrm{tr} R_j v; \mathbb{R}^{n-1}\|_{W_p^{l-k_j-1/p}}\Big)\,.$$

Since δ is small, we can replace here R by R_j and S by S_j (see Lemma 14.2.2). From the estimate obtained after this change, it follows that

$$\|\sigma u; U\cap\Omega\|_{W_p^l} \le c\Big(\|R(\sigma u); U\cap\Omega\|_{W_p^{l-2h}} + \sum_{j=1}^h \|\mathrm{tr} R_j(\sigma u); U\cap\partial\Omega\|_{W_p^{l-k_j-1/p}}\Big)\,.$$

This inequality and (14.3.1) entail

$$\|\sigma u; \Omega\|_{W_p^l} \le c\Big(\|P(\sigma u); \Omega\|_{W_p^{l-2h}} + \sum_{j=1}^h \|\mathrm{tr} P_j(\sigma u); \partial\Omega\|_{W_p^{l-k_j-1/p}} + \|\sigma u; \Omega\|_{W_p^{l-1}}\Big)\,.$$

Summing over all sufficiently small neighborhoods U which generate a covering of $\bar{\Omega}$, we arrive at

$$\|u; \Omega\|_{W_p^l} \le c\Big(\|Pu; \Omega\|_{W_p^{l-2h}} + \sum_{j=1}^h \|\mathrm{tr} P_j u; \partial\Omega\|_{W_p^{l-k_j-1/p}} + \|u; \Omega\|_{W_p^{l-1}}\Big)\,.$$

It remains to use the known inequality

$$\|u; \Omega\|_{W_p^{l-1}} \le \varepsilon\,\|u; \Omega\|_{W_p^l} + c(\varepsilon)\|u; \Omega\|_{L_1},$$

where ε is any positive number. Estimate (14.1.2) is proved. □

Theorem 14.3.2. *If* $p(l-1) > n$, $1 < p < \infty$, *and if* $\partial\Omega \in W_p^{l-1/p}$, *then the conclusion of Theorem* 14.3.1 *holds.*

Proof. From the condition $\partial\Omega \in W_p^{l-1/p}$ and the Sobolev embedding theorem, it follows that $\partial\Omega \in C^1$. We place the origin at the point $O \in \partial\Omega$ and direct the axis Oy along the interior normal to $\partial\Omega$. Let U be the neighborhood of O in the definition of the class $W_p^{l-1/p}$, i.e. $U \cap \Omega = U \cap G$, where $G = \{z : x \in \mathbb{R}^{n-1}, y > \varphi(x)\}$ and $\varphi \in W_p^{l-1/p}(\mathbb{R}^{n-1})$. Let ε be a small positive number, which will be specified later, and let $\mathcal{B}_\rho = \{z \in \mathbb{R}^n : |z| < \rho\}$.

We choose a small number ρ such that

$$\|\nabla\varphi; \mathcal{B}_\rho \cap \mathbb{R}^{n-1}\|_{L_\infty} < \varepsilon$$

and $\overline{\mathcal{B}_{2\rho}} \subset U$. Let $\tau \in C_0^\infty(\mathcal{B}_2)$, $\tau = 1$ on $\mathcal{B}_1$, and $\tau_\rho(z) = \tau(z/\rho)$. We introduce the function $\varphi^* = \varphi\tau_\rho$ on $\mathbb{R}^{n-1}$ and note that

$$\|\nabla\varphi^*; \mathbb{R}^{n-1}\|_{L_\infty} < c\varepsilon.$$

We also define the extension Φ of φ^* onto $\mathbb{R}^n_+$ by $\Phi = \mathcal{T}\varphi^*$. According to (9.4.12), where φ is replaced by φ^*,

$$\|\nabla\Phi; \mathbb{R}^n_+\|_{L_\infty} \le c\varepsilon\,. \tag{14.3.2}$$

Since

$$\|\varphi^*; \mathbb{R}^{n-1}\|_{W_p^{l-1/p}} \le c(\rho)\|\varphi; \mathbb{R}^{n-1}\|_{W_p^{l-1/p}}\,,$$

we have

$$\|\Phi; \mathbb{R}^n_+\|_{W_p^l} \le c(\rho)\|\varphi; \mathbb{R}^{n-1}\|_{W_p^{l-1/p}}\,.$$

Now let r be a small positive number such that $r < \rho$ and

$$r^{l-1-n/p}\|\Phi; \mathbb{R}^n_+\|_{W_p^l} < \varepsilon\,. \tag{14.3.3}$$

It follows from the inequality

$$|\!|\!|\nabla\Phi; \mathcal{B}_r \cap \mathbb{R}^n_+|\!|\!|_{W_p^{l-1}} \le c\left(\|\nabla_l\Phi; \mathcal{B}_r \cap \mathbb{R}^n_+\|_{L_p} + r^{1-l+n/p}\|\nabla\Phi; \mathcal{B}_r \cap \mathbb{R}^n_+\|_{L_\infty}\right)$$

and the estimates (14.3.2) and (14.3.3) that

$$r^{l-1-n/p}|\!|\!|\nabla\Phi; \mathcal{B}_r \cap \mathbb{R}^n_+|\!|\!|_{W_p^{l-1}} \le c\varepsilon\,.$$

According to (9.6.4), this means that

$$|\!|\!|\nabla\Phi; \mathcal{B}_r \cap \mathbb{R}^n_+|\!|\!|_{MW_p^{l-1}} \le c\varepsilon\,.$$

Using the function Φ, we define the mapping λ by (9.4.14) with $N = 1$. By the last inequality,

$$|\!|\!|I - \partial\lambda; \mathcal{B}_r \cap \mathbb{R}^n_+|\!|\!|_{MW_p^{l-1}} \le c\varepsilon\,.$$

Now it suffices to duplicate the arguments we have already used in Theorem 14.3.1, with $\mathcal{B}_r$ in place of the ball U, and estimate (14.2.6) instead of (14.2.5).

The following assertion can be deduced in a standard way from the *a priori* estimate (14.1.2) (see, for instance [H1], §10.5; [Tr3], Sect. 5.4.3).

Proposition 14.3.1. *Let the domain Ω satisfy the conditions of either Theorem* 14.3.1 *or Theorem* 14.3.2.

(i) If the kernel of the operator (14.1.1) *is trivial, then the norm* $\|u;\Omega\|_{L_1}$ *in* (14.1.2) *can be omitted.*

(ii) The kernel of the operator (14.1.1) *is finite-dimensional.*

(iii) The range of the operator (14.1.1) *is closed.*

Proof. (*i*) Suppose that the assertion is not true. Then there exists a sequence of functions $\{v_m\}_{m\geq 1}$ in $W_p^l(\Omega)$ such that

$$\|v_m;\Omega\|_{W_p^l} = 1, \tag{14.3.4}$$

$$\|Pv_m;\Omega\|_{W_p^{l-2h}} + \sum_{j=1}^{h} \|\mathrm{tr}P_j v_m;\partial\Omega\|_{W_p^{l-k_j-1/p}} \to 0\,. \tag{14.3.5}$$

We can select a subsequence of $\{v_m\}$, also denoted by $\{v_m\}$, which weakly converges in $W_p^l(\Omega)$ to a function $v \in W_p^l(\Omega)$. Since the imbedding operator $W_p^l(\Omega) \to L_1(\Omega)$ is compact, we can assume that $v_m \to v$ in $L_1(\Omega)$. Substituting $v_m - v_k$ into (14.1.2), we have $v_m \to v$ in $W_p^l(\Omega)$. This and (14.3.5) imply that $v \in \ker\{P;\mathrm{tr}P_j\}$, i.e. $v = 0$, which contradicts (14.3.4).

(*ii*) It follows from (14.1.2) that, for all $v \in \ker\{P;\mathrm{tr}P_j\}$,

$$\|v;\Omega\|_{W_p^l} \leq c\,\|v;\Omega\|_{L_1}\,.$$

Therefore a unit sphere in $\ker\{P;\mathrm{tr}P_j\}$, considered as a subspace of $W_p^l(\Omega)$, is compact and the dimension of the kernel is finite.

(*iii*) Since $\dim\ker\{P;\mathrm{tr}P_j\} < \infty$, there exists a projection operator Π which acts parallel to $\ker\{P;\mathrm{tr}P_j\}$.

Duplicating the arguments used in part (*i*) of the present proof, we obtain

$$\|v;\Omega\|_{W_p^l} \leq c\Big(\|Pv;\Omega\|_{W_p^{l-2h}} + \sum_{j=1}^{h} \|\mathrm{tr}P_j v;\partial\Omega\|_{W_p^{l-k_j-1/p}}\Big)$$

for all $v \in W_p^l(\Omega)$, which implies that the range of the operator $\{P;\mathrm{tr}\,P_j\}$ is closed. □

Next we derive a local *a priori* estimate for solutions of the elliptic boundary value problem (cf. [ADN1], Sect. 15).

Proposition 14.3.2. *Let the domain Ω satisfy the condition of either Theorem* 14.3.1 *or Theorem* 14.3.2. *Further, let U and V be open subsets of $\mathbb{R}^n$ with $\overline{U} \subset V$, and let $u \in W_p^l(V\cap\Omega)$. Then*

$$\|u; U \cap \Omega\|_{W_p^l} \leq c\Big(\|Pu; V \cap \Omega\|_{W_p^{l-2h}} + \sum_{j=1}^{h} \|\mathrm{tr} P_j u; V \cap \partial\Omega\|_{W_p^{l-k_j-1/p}} + \|u; V \cap \Omega\|_{L_1}\Big) . \tag{14.3.6}$$

Proof. Let U and V be concentric balls with radii r and ρ, $r < \rho$. Further, let either $\bar{V} \subset \Omega$ or the centre of the balls be placed on $\partial\Omega$. It suffices to prove the proposition under this additional assumption. We introduce the sets

$$C_0 = U \cap \Omega, \quad C_k = \{x \in \Omega : \delta_k < \rho - |x| < \delta_{k-1}\},$$

where $k = 1, 2, \dots$ and $\delta_k = (\rho - r)2^{-k}$. Let

$$D_0 = C_0 \cup C_l, \quad D_k = C_{k-1} \cup C_k \cup C_{k+1}, \;\; k = 1, 2, \dots .$$

We construct a C^∞-partition of unity $\{\sigma_k\}_{k\geq 0}$ subordinate to the covering $\{D_k\}_{k\geq 0}$ of $V \cap \Omega$ and satisfying

$$|D^\alpha \sigma_k| = O(\delta_k^{-|\alpha|})$$

for any multi-index α.

Applying the *a priori* estimate (14.1.2) to $\sigma_k v$, we obtain

$$\|\sigma_k v; \Omega\|_{W_p^l} \leq c\Big(\|P(\sigma_k v); \Omega\|_{W_p^{l-2h}} + \sum_{j=1}^{h} \|\mathrm{tr} P_j(\sigma_k v); \partial\Omega\|_{W_p^{l-k_j-1/p}} + \|\sigma_k v; \Omega\|_{L_1}\Big) ,$$

which implies that

$$\|\sigma_k v; \Omega\|_{W_p^l} \leq c\Big(\|\sigma_k P v; \Omega\|_{W_p^{l-2h}} + \sum_{j=1}^{h} \|\sigma_k \mathrm{tr} P_j v; \partial\Omega\|_{W_p^{l-k_j-1/p}} + \delta_k^{-l}\|v; D_k\|_{W_p^{l-1}}\Big) .$$

Let M be a sufficiently large positive number. We have

$$\begin{aligned}\sum_{k=0}^{\infty} \delta_k^M \|v; C_k\|_{W_p^l} &\leq c\Big(\sum_{k=0}^{\infty} \delta_k^M \|\sigma_k P v; \Omega\|_{W_p^{l-2h}} \\ &+ \sum_{k=0}^{\infty} \delta_k^M \sum_{j=1}^{h} \|\sigma_k \,\mathrm{tr} P_j v; \partial\Omega\|_{W_p^{l-k_j-1/p}} + \sum_{k=0}^{\infty} \delta_k^{M-l} \|v; D_k\|_{W_p^{l-1}}\Big) \\ &\leq c\Big(\|Pv; V \cap \partial\Omega\|_{W_p^{l-2h}} + \sum_{j=1}^{h} \|\mathrm{tr} P_j v; V \cap \partial\Omega\|_{W_p^{l-k_j-1/p}} \\ &+ \sum_{k=0}^{\infty} \delta_k^{M-l} \|v; D_k\|_{W_p^{l-1}}\Big) .\end{aligned}$$

Note that, for any $\varepsilon > 0$ and for some positive N,

$$\|v; D_k\|_{W_p^{l-1}} \le \varepsilon\, \delta_k^l \|v; D_k\|_{W_p^l} + c(\varepsilon)\delta_k^{-N}\|v; D_k\|_{L_1}\,.$$

Consequently,

$$\sum_{k=0}^{\infty} \delta_k^M \|v; C_k\|_{W_p^l} \le c\Big(\|Pv; V\cap\Omega\|_{W_p^{l-2h}} + \sum_{j=1}^{h} \|\mathrm{tr} P_j v; V\cap\partial\Omega\|_{W_p^{l-k_j-1/p}}$$

$$+c(\varepsilon)\|v; V\cap\Omega\|_{L_1} + \varepsilon \sum_{k=0}^{\infty} \delta_k^M \|v; D_k\|_{W_p^l}\Big)\,.$$

Clearly, the last sum can be removed by changing c. The result follows. □

Using the same properties of P, P_j, λ, $\varkappa$ as those used in the proof of Theorems 14.3.1 and 14.3.2, one can establish the existence of a right regularizer by the same argument as in, for instance, [Wl], Sect. 13.

Proposition 14.3.3. *Let the domain Ω satisfy the condition of either Theorem 14.3.1 or Theorem 14.3.2. Then there exists a linear bounded operator*

$$R : W_p^{l-2h}(\Omega) \times \prod_{j=1}^{h} W_p^{l-k_j-1/p}(\partial\Omega) \to W_p^l(\Omega)$$

such that $\{P; \mathrm{tr}P_j\}R = I + K$. Here I and K are the identity and compact operators respectively.

A direct corollary of Proposition 14.3.1 and 14.3.3 is:

Theorem 14.3.3. *Let $\partial\Omega$ belong to $M_p^{l-1/p}(\delta)$ for $p(l-1) \le n$ and to $W_p^{l-1/p}$ for $p(l-1) > n$. Then the operator (14.1.1) is Fredholm, that is, its null space is finite-dimensional and its range is closed and has a finite codimension.*

In the following sections we consider the Dirichlet problem in more detail.

14.4 Auxiliary Assertions

14.4.1 Some Properties of the Operator $\mathcal{T}$

In this subsection, $\mathcal{T}$ is the operator defined by (9.4.11).

Lemma 14.4.1. *Let α be an n-tuple multi-index and let k, r be nonnegative integers with $k \ge |\alpha| - r \ge 0$. Then the operator*

$$M(W_p^{k-1/p}(\mathbb{R}^{n-1})) \ni \gamma \to \eta^r(D^\alpha \mathcal{T}\gamma)(\zeta) \in M(W_p^k(\mathbb{R}_+^n) \to W_p^{k-|\alpha|+r}(\mathbb{R}_+^n))$$

is continuous.

Proof. Clearly,

$$\eta^r (D^\alpha \mathcal{T}\gamma)(\zeta) = D^\beta \sum_{0\leq|\nu|\leq r} c_\nu \eta^{|\nu|} (D^\nu \mathcal{T}\gamma)(\zeta) ,$$

where β is a multi-index of order $|\alpha| - r$, $c_\nu = \text{const}$. The operator

$$\gamma \xrightarrow{\mathcal{T}_\nu} \eta^{|\nu|}(D^\nu \mathcal{T}\gamma)(\zeta)$$

has the same form as $\mathcal{T}$ and therefore it maps $MW_p^{k-1/p}(\mathbb{R}^{n-1})$ into $M(W_p^k(\mathbb{R}^n_+))$. Hence, the continuity of the operator

$$M(W_p^{k-1/p}(\mathbb{R}^{n-1})) \ni \gamma \to D^\beta \mathcal{T}_\nu \gamma \in M(W_p^k(\mathbb{R}^n_+) \to W_p^{k-|\alpha|+r}(\mathbb{R}^n_+))$$

follows from Corollary 2.4.1. □

The next assertion follows directly from the lemma just proved.

Corollary 14.4.1. *Let G be a special Lipschitz domain, let α be a positive n-tuple multi-index and let r be a nonnegative integer with $l \geq |\alpha| - r > 0$. Then the function $\zeta \to \eta^r(D^\alpha \mathcal{T}\varphi)(\zeta)$ belongs to the space $M(W_p^{l-1}(\mathbb{R}^n_+) \to W_p^{l-|\alpha|+r}(\mathbb{R}^n_+))$ and*

$$\|\eta^r D^\alpha \mathcal{T}\varphi; \mathbb{R}^n_+\|_{M(W_p^{l-1}\to W_p^{l-|\alpha|+r})} \leq c\,\|\nabla\varphi; \mathbb{R}^{n-1}\|_{MW_p^{l-1-1/p}} .$$

14.4.2 Properties of the Mappings λ and $\varkappa$

Let G be a special Lipschitz domain and let λ be the mapping (9.4.13) defined by (9.4.14). As in Sect. 9.4.3, we denote by $\varkappa$ the inverse mapping to λ. We shall assume that $L < 1$ and $N = 1$.

From Corollary 14.4.1 we deduce

Corollary 14.4.2. *Let α be a multi-index. Let r be a nonnegative integer with $l \geq |\alpha| - r + 1 > 0$, and write $\lambda(\zeta) = \{\lambda_1(\zeta), \dots, \lambda_n(\zeta)\}$. Then the function*

$$\zeta \to \eta^r(D^\alpha \partial\lambda_i)(\zeta)$$

belongs to $M(W_p^{l-1}(\mathbb{R}^n_+) \to W_p^{l-1-|\alpha|+r}(\mathbb{R}^n_+))$ and

$$\|\eta^r D^\alpha(\partial\lambda - I); \mathbb{R}^n_+\|_{M(W_p^{l-1}\to W_p^{l-1-|\alpha|+r})} \leq c\,\|\nabla\varphi; \mathbb{R}^{n-1}\|_{MW_p^{l-1-1/p}} .$$

A similar assertion concerning the mapping $\varkappa$ needs a separate proof.

Lemma 14.4.2. *Let α be a multi-index, let r be a nonnegative integer with $l \geq |\alpha| - r + 1 > 0$, and write $\varkappa(z) = \{\varkappa_1(z), \dots, \varkappa_n(z)\}$. Then the function*

$$z \to (\eta^r D^\alpha \partial \varkappa_i)(z)$$

belongs to $M(W_p^{l-1}(G) \to W_p^{l-1-|\alpha|+r}(G))$ and

$$\|\eta^r D^\alpha (\partial\varkappa - I); G\|_{M(W_p^{l-1} \to W_p^{l-|\alpha|+r-1})} \leq c \, \|\nabla \varphi; \mathbb{R}^{n-1}\|_{MW_p^{l-1-1/p}} .$$

Proof. For $|\alpha| = 0$ the result follows from the definition of a (p,l)-diffeomorphism. Suppose that the lemma is proved for $|\alpha| < N$. Let $|\alpha| = N$, $r < N$. For any multi-index δ of order $N-1$,

$$\begin{aligned}(D^\delta \partial\varkappa)(z) &= D^\delta [\partial\lambda(\varkappa(z))]^{-1} \\ &= \sum_{1 \leq |\beta| \leq |\delta|} [D^\beta (\partial\lambda)^{-1}](\varkappa(z)) \sum c_s \prod_{i=1}^{n} \prod_j D^{s_{ij}} \varkappa_i(z) ,\end{aligned}$$

where the summation is taken over all collections of multi-indices $s = (s_{ij})$ such that

$$\sum s_{ij} = \delta, \quad |s_{ij}| \geq 1, \quad \sum (|s_{ij}| - 1) = |\delta| - |\beta|.$$

Therefore, the expression $(\eta^r D^\delta \partial\varkappa)(z)$ is the sum of the products of two factors

$$\pi_1(z) = c_\beta \left[\eta^{|\beta|} D^\beta (\partial\lambda)^{-1}\right](\varkappa(z))$$

and

$$\pi_2(z) = \prod_{i=1}^{n} \prod_j (\eta^{r-|\beta|} D^{s_{ij}} \varkappa_i)(z) .$$

Corollary 14.4.2 implies that the function

$$\zeta \to \eta^{|\beta|} D^\beta (\partial\lambda)^{-1}$$

belongs to the space $MW_p^{l-1}(\mathbb{R}^n_+)$, and since $\varkappa$ is a (p,l)-diffeomorphism, it follows that $\pi_1 \in MW_p^{l-1}(G)$.

We introduce positive integers σ_{ij} such that

$$\sigma_{ij} \leq |s_{ij}|, \qquad \sum (\sigma_{ij} - 1) = r - |\beta|.$$

Then

$$\pi_2(z) = \prod_{i=1}^{n} \prod_j (\eta^{\sigma_{ij}-1} D^{s_{ij}} \varkappa_i)(z) .$$

Since $|\beta| \geq 1$, we have $|s_{ij}| \leq N-1$ and by the induction hypothesis the function

$$z \rightarrow (\eta^{\sigma_{ij}-1} D^{s_{ij}} \varkappa_i)(z)$$

belongs to

$$M(W_p^q(G) \rightarrow W_p^{q-1-|s_{ij}|+\sigma_{ij}}(G)), \quad q = |s_{ij}| - \sigma_{ij}, \dots, l-1.$$

Hence

$$\pi_2 \in M(W_p^{l-1}(G) \rightarrow W_p^{l-2-\sum(|s_{ij}|-\sigma_{ij})}(G)) = M(W_p^{l-1}(G) \rightarrow W_p^{l-2-|\delta|+r}(G)).$$

Noting that $|\alpha| = 1 + |\delta|$ and $\pi_1 \in MW_p^{l-1}(G)$, we obtain

$$\pi_1 \pi_2 \in M(W_p^{l-1}(G) \rightarrow W_p^{l-1-|\alpha|-r}(G)).$$

Since the space W_p^l is invariant under the (p,l)-diffeomorphisms, a function u on ∂G belongs to $W_p^{l-1/p}(\partial G)$ if and only if $u \circ \mathrm{tr}\ \lambda \in W_p^{l-1/p}(\mathbb{R}^{n-1})$. We put

$$\|u; \partial G\|_{W_p^{l-1/p}} = \|u \circ \mathrm{tr}\lambda; \mathbb{R}^{n-1}\|_{W_p^{l-1/p}}.$$

Taking here an arbitrary (p,l)-diffeomorphism: $\mathbb{R}^n_+ \rightarrow G$ instead of λ, we obtain an equivalent norm (see Lemmas 9.4.1, 9.4.2 and 9.4.4).

14.4.3 Invariance of the Space $W_p^l \cap \mathring{W}_p^h$ Under a Change of Variables

In this subsection we present auxiliary assertions which will be used later in the study of conditions for solvability of the Dirichlet problem in $W_p^l(\Omega)$.

As before, let $G = \{z = (x,y) : x \in \mathbb{R}^{n-1}, y > \varphi(x)\}$ and let ∂G belong to the class $M_p^{l+1-h-1/p}$, where l and h are integers with $l \geq h \geq 1$. In other words,

$$\nabla \varphi \in MW_p^{l-h-1/p}(\mathbb{R}^n) \qquad \text{if} \quad l > h$$

and

$$\varphi \in C^{0,1}(\mathbb{R}^{n-1}) \qquad \text{if} \quad l = h.$$

Let λ be the mapping $\mathbb{R}^n_+ \ni (\xi, \eta) \rightarrow (x,y) \in G$, defined by (9.4.13), and let $\varkappa = \lambda^{-1}$.

Lemma 14.4.3. *Let* $v \in (\mathring{W}_p^k \cap W_p^{t+k})(G)$, *where* $0 \leq t \leq l-h$, *and let* $\partial G \in M_p^{l+1-h-1/p}$. *Then*

$$\|\eta^{-k} v; G\|_{W_p^t} \leq c\, \|v; G\|_{W_p^{t+k}}. \tag{14.4.1}$$

Proof. Since $\eta(z)$ is equivalent to $y-\varphi(x)$, the inequality (14.4.1) with $t=0$ follows from the Hardy inequality,

$$\int_{\varphi(x)}^{\infty} |v(x,y)|^p \frac{dy}{(y-\varphi(x))^{pk}} \le c \int_{\varphi(x)}^{\infty} \left|\frac{\partial^k v}{\partial y^k}(x,y)\right|^p dy \tag{14.4.2}$$

a.e. in $\mathbb{R}^{n-1}$. Let the lemma be proved for all $t<T$ and $k>K$. We have

$$\|\eta^{-K} v; G\|_{W_p^T} \le \|\nabla(\eta^{-K} v); G\|_{W_p^{T-1}} + \|\eta^{-K} v; G\|_{L_p} .$$

The second term on the right-hand side is estimated by (14.4.2) and the first one does not exceed

$$\|\eta^{-K} \nabla v; G\|_{W_p^{T-1}} + K\|\eta^{-K-1} v \nabla\eta; G\|_{W_p^{T-1}} . \tag{14.4.3}$$

Since $\partial G \in M_p^{l+1-h-1/p}$, it follows that

$$\nabla\eta \in MW_p^{l-h}(G) \subset MW_p^{T-1}(G).$$

Hence the sum (14.4.3) is dominated by

$$\|\eta^{-K} \nabla v; G\|_{W_p^{T-1}} + c\,\|\eta^{-K-1} v; G\|_{W_p^{T-1}} .$$

Using the induction hypothesis, we complete the proof. □

Lemma 14.4.4. *For all $u \in W_p^l(G) \cap \mathring{W}_p^h(G)$, the inequality*

$$\|u\circ\lambda; \mathbb{R}_+^n\|_{W_p^l} \le c\,\|u; G\|_{W_p^l} \tag{14.4.4}$$

holds.

Proof. We have

$$\int_{\mathbb{R}_+^n} |D_\zeta^\alpha[u(\lambda(\zeta))]|^p \, d\zeta$$
$$\le c \sum_{1\le|\beta|\le l} \int_{\mathbb{R}_+^n} \left|(D^\beta u)(\lambda(\zeta)) \sum_s c_s \prod_{i,j} D_\zeta^{s_{ij}} \lambda_i(\zeta)\right|^p d\zeta \,,$$

where α is an arbitrary positive multi-index of order l and $s=(s_{ij})$ is the set of multi-indices satisfying (10.1.19). Hence,

$$\int_{\mathbb{R}_+^n} |D_\zeta^\alpha[u(\lambda(\zeta))]|^p \, d\zeta$$

$$\le c \sum_{1\le|\beta|\le l} \int_G \left|(D^\beta u)(z) \sum_s c_s \prod_{i,j} (D^{s_{ij}}\lambda_i)(\varkappa(z))\right|^p dz \,\|\partial\varkappa; G\|_{L_\infty}^p . \tag{14.4.5}$$

Let $|\beta| \geq h$ or, what is the same, $l - |\beta| \leq l - h$. Since $\nabla \lambda_i \in MW_p^k(\mathbb{R}_+^n)$ for $k \leq l - h$, we have

$$D^{s_{ij}} \lambda_i \in M(W_p^k(\mathbb{R}_+^n) \to W_p^{k-|s_{ij}|+1}(\mathbb{R}_+^n))$$

for $|s_{ij}| - 1 \leq k \leq l - h$. Consequently,

$$\prod_{i,j} D^{s_{ij}} \lambda_i \in M\big(W_p^{\sum_{i,j}(|s_{ij}|-1)}(\mathbb{R}_+^n) \to L_p(\mathbb{R}_+^n)\big) = M(W_p^{l-|\beta|}(\mathbb{R}_+^n) \to L_p(\mathbb{R}_+^n)).$$

Since $\varkappa$ is a $(p, l-h)$-diffeomorphism, we obtain

$$\prod_{i,j} D^{s_{ij}} \lambda_i \circ \varkappa \in M(W_p^{l-|\beta|}(G) \to L_p(G)) .$$

Therefore, the terms on the right-hand side of (14.4.5) which correspond to multi-indices β of order $|\beta| \geq h$ are majorized by $c \, \|u; G\|^p_{W_p^l}$.

Now suppose that $|\beta| \leq h - 1$. By (14.4.1) the function

$$z \to \eta(z)^{|\beta|-h} (D^\beta u)(z)$$

belongs to the space $W_p^{l-h}(G)$. By σ_{ij} we denote integers subject to

$$1 \leq \sigma_{ij} \leq |s_{ij}|, \quad \sum_{i,j} (\sigma_{ij} - 1) = h - |\beta|.$$

Such numbers exist, since

$$\sum_{i,j} (|s_{ij}| - 1) = l - |\beta| \qquad \text{and} \quad l \geq h.$$

By Corollary 14.4.2, the function

$$\zeta \to \eta^{\sigma_{ij}-1} (D^{s_{ij}} \lambda_i)(\zeta)$$

belongs to $M(W_p^k(\mathbb{R}_+^n) \to W_p^{k-|s_{ij}|+\sigma_{ij}}(\mathbb{R}_+^n))$. Using the identity

$$\prod_{i,j} (D^{s_{ij}} \lambda_i)(\zeta) = \eta^{|\beta|-h} \prod_{i,j} \eta^{\sigma_{ij}-1} (D^{s_{ij}} \lambda_i)(\zeta) ,$$

we observe that the function

$$\zeta \to \eta^{h-|\beta|} \prod (D^{s_{ij}} \circ \lambda_i)(\zeta)$$

belongs to $M(W_p^{l-h}(\mathbb{R}_+^n) \to L_p(\mathbb{R}_+^n))$. Since $\varkappa$ is a $(p, l-h)$-diffeomorphism, the function

$$z \to \eta(z)^{h-|\beta|} \prod (D^{s_{ij}} \lambda_i)(\varkappa(z))$$

is an element of $M(W_p^{l-h}(G) \to L_p(G))$. Therefore, the terms with $|\beta| \le h-1$ on the right-hand side of (14.4.5) are bounded by $c\,\|u;G\|^p_{W_p^l}$. We also have

$$\|u;G\|_{L_p} \sim \|u\circ\lambda;\mathbb{R}^n_+\|_{L_p}$$

because λ is a bi-Lipschitz mapping. This relation and (14.4.5) give the estimate (14.4.4). □

Now we prove an analogous assertion concerning the mapping $\varkappa$.

Lemma 14.4.5. *For each $v\in W_p^l(\mathbb{R}^n_+)\cap \mathring{W}_p^h(\mathbb{R}^n_+)$, the inequality*

$$\|v\circ\varkappa;G\|_{W_p^l} \le c\,\|v;\mathbb{R}^n_+\|_{W_p^l} \tag{14.4.6}$$

holds.

Proof. It is sufficient to derive the estimate

$$\int_\Omega |D_z^\alpha[v(\varkappa(z))]|^p\,dz \le c\,\|v;\mathbb{R}^n_+\|^p_{W_p^l}.$$

The left-hand side is dominated by

$$c\sum_{1\le|\beta|\le l}\int_{\mathbb{R}^n_+}\Big|(D_\zeta^\beta v)(\zeta)\sum_s c_s\prod_{i,j}(D^{s_{ij}}\varkappa_i)(\lambda(\zeta))\Big|^p\,d\zeta\,\|\partial\lambda;\mathbb{R}^n_+\|^p_{L_\infty} \tag{14.4.7}$$

(cf. (14.4.5)).

Let $|\beta|\ge h$. Repeating the same arguments as in Lemma 14.4.4 with $\mathbb{R}^n_+$ replaced by G, λ by $\varkappa$, u by v and *vice versa*, we obtain that the terms in (14.4.7) with $|\beta|\ge h$ do not exceed $c\,\|v;\mathbb{R}^n_+\|^p_{W_p^l}$.

Now let $|\beta|\le h-1$. Since $v\in W_p^l(\mathbb{R}^n_+)\cap\mathring{W}_p^h(\mathbb{R}^n_+)$, it follows that $\eta^{h-|\beta|}(D^\beta v)(\zeta)$ belongs to $W_p^{l-h}(\mathbb{R}^n_+)$ and its norm is dominated by $c\,\|v;\mathbb{R}^n_+\|_{W_p^l}$. According to Lemma 14.4.2,

$$\eta^{|s_{ij}|-1}(D^{s_{ij}}\varkappa_i)(\lambda(\zeta))$$

is a multiplier in $W_p^{l-h}(\mathbb{R}^n_+)$. Hence

$$\eta^{|\beta|-h}\prod(D^{s_{ij}}\varkappa_i)(\lambda(\zeta))$$

is a multiplier in the same space.

14.4.4 The Space W_p^{-k} for a Special Lipschitz Domain

We assume G to be a special Lipschitz domain. In other words, we put $l = h$ in the conjectures of the preceding subsection. Let us retain the notation of Sect. 14.4.3.

We introduce the space $W_p^{-k}(G)$ of linear functionals on $\mathring{W}_p^k(G)$, where $p + p' = pp'$, $k = 0, 1, \ldots$.

One can immediately check the continuity of the operator $D^\alpha : W_p^s(G) \to W_p^{s-|\alpha|}(G)$ for any $s = 0, \pm 1, \ldots$.

By Lemmas 14.4.4 and 14.4.5, the mapping λ performs an isomorphism between $\mathring{W}_{p'}^k(G)$ and $\mathring{W}_{p'}^k(\mathbb{R}_+^n)$. Therefore, λ maps $W_p^{-k}(G)$ onto $W_p^{-k}(\mathbb{R}_+^n)$ isomorphically.

The following assertion, which is proved in a standard way, gives one of possible realizations of $W_p^{-k}(G)$.

Proposition 14.4.1. *Any linear functional on $\mathring{W}_{p'}^k(G)$ can be identified with a distribution $f \in (C_0^\infty(G))'$ of the form*

$$f(z) = \sum_{|\alpha| \le k} D^\alpha f_\alpha(z)\,, \tag{14.4.8}$$

where f_α is a function such that

$$\eta^{k-|\alpha|} f_\alpha \in L_p(G).$$

The norm of this functional is equivalent to the norm

$$\|f\| = \inf \Big\| \Big(\sum_{|\alpha| \le k} \eta^{2(k-|\alpha|)} f_\alpha^2 \Big)^{1/2}; G \Big\|_{L_p}\,,$$

the infimum being taken over all collections $\{f_\alpha\}_{|\alpha| \le k}$ in (14.4.8).

Proof. By Lemma 14.4.1, the space $\mathring{W}_{p'}^k(G)$ can be supplied with the norm

$$\Big\| \Big(\sum_{|\alpha| \le k} \eta^{2(k-|\alpha|)} (D^\alpha u)^2 \Big)^{1/2}; G \Big\|_{L_{p'}}\,.$$

Therefore the right-hand side of (14.4.8) is a linear functional on $\mathring{W}_{p'}^k(G)$, and

$$\|f\| \le \Big\| \Big(\sum_{|\alpha| \le k} \eta^{2(k-|\alpha|)} f_\alpha^2 \Big)^{1/2}; G \Big\|_{L_p}\,.$$

To express an arbitrary linear functional on $\mathring{W}_{p'}^k$ in the form (14.4.8), we consider the space $\mathbf{L}_{p'}(G)$ of vectors $\mathbf{v} = \{v_\alpha\}_{|\alpha| \le k}$ with components in $L_{p'}(G)$ endowed with the norm

$$\|(\sum_{|\alpha|\le k} v_\alpha^2)^{1/2}; G\|_{L_{p'}}.$$

Further, let

$$\Lambda_k = \{(-1)^{|\alpha|}\eta^{|\alpha|-k}D^\alpha\}_{|\alpha|\le k}\,.$$

Since the space $\overset{\circ}{W}{}^k_{p'}(G)$ is complete, the range of the operator $\Lambda_k : \overset{\circ}{W}{}^k_{p'}(G) \to \mathbf{L}_{p'}(G)$ is a closed subspace of $\mathbf{L}_{p'}(G)$. Let $u \in \overset{\circ}{W}{}^k_{p'}(G)$ and let $f(u)$ be the value of the functional $f \in W_p^{-k}(G)$ on u. We define the functional Φ by $\Phi(\mathbf{v}) = f(u)$ on the set of vectors $\mathbf{v}$ which can be expressed in the form $\Lambda_k u$. Then $\|\Phi\| = \|f\|$ and, by the Hahn-Banach theorem, Φ can be extended to a linear functional on $\mathbf{L}_{p'}(G)$ with the same norm. Consequently,

$$\Phi(\Lambda_k u) = \sum_{|\alpha|\le k}\int_G g_\alpha(-1)^{|\alpha|}\eta^{|\alpha|-k}D^\alpha u\,dz\,,$$

where $g_\alpha \in L_p(G)$. It remains to put $f_\alpha = \eta^{|\alpha|-k}g_\alpha$. □

Lemma 14.4.6. *Let $v \in \overset{\circ}{W}{}^{t+k}_p(G)$, where $t < 0$, k is a nonnegative integer, and $t + k \ge 0$. Then* (14.4.1) *holds.*

Proof. We have

$$\|\eta^{-k}v; G\|_{W_p^t} = \sup_{w\in\overset{\circ}{W}{}^{-t}_{p'}} \frac{(\eta^{-k}v, w)}{\|w\|_{W^{-t}_{p'}}}\,.$$

By Lemma 14.4.1,

$$(\eta^{-k}v, w) \le \|\eta^{-k-t}v; G\|_{L_p}\|\eta^t w\|_{L_{p'}} \le c\,\|v; G\|_{W_p^{t+k}}\|w\|_{W^{-t}_{p'}}\,.$$

□

Corollary 14.4.3. *Let $\eta^k v \in W_p^{t+k}(G)$, where t and k are the same numbers as in Lemma* 14.4.6. *Then*

$$\|v; G\|_{W_p^t} \le c\,\|\eta^k v; G\|_{W_p^{t+k}}.$$

Proof. Let $w \in \overset{\circ}{W}{}^{-t}_{p'}$. We have

$$(v, w) \le \|\eta^k v; G\|_{W_p^{t+k}}\|\eta^{-k}w; G\|_{W^{-t-k}_{p'}}\,.$$

Using Lemma 14.4.6 with p and t replaced by p' and $-t-k$, we obtain

$$\|\eta^{-k}w; G\|_{W^{-t-k}_{p'}} \le c\,\|w; G\|_{W^{-t}_{p'}}\,.$$

The result follows. □

14.4.5 Auxiliary Assertions on Differential Operators in Divergence Form

Lemma 14.4.7. *Let $\partial G \in M_p^{l+1-h-1/p}$, $l \geq h$, and let*

$$Pu = \sum_{|\alpha|,|\beta|\leq h} (-1)^{|\alpha|} D^\alpha (a_{\alpha\beta}(z) D^\beta u) \,. \tag{14.4.9}$$

If

$$\eta^{h-|\beta|} a_{\alpha\beta} \in M(W_p^{l-h}(G) \to W_p^{l-2h+|\alpha|}(G)) \quad \text{for } l-2h+|\alpha| \geq 0$$

and

$$\eta^{2h-|\alpha|-|\beta|} a_{\alpha\beta} \in MW_p^{l-h}(G)$$

for $l-2h+|\alpha|<0$, then P is a continuous operator:

$$(W_p^l \cap \mathring{W}_p^h)(G) \to W_p^{l-2h}(G)$$

and its norm does not exceed $c\,A$, where

$$A = \sum_{|\beta|\leq h} \Big(\sum_{|\alpha|\geq 2h-l} \|\eta^{h-|\beta|} a_{\alpha\beta}; G\|_{M(W_p^{l-h} \to W_p^{l-2h+|\alpha|})}$$
$$+ \sum_{|\alpha|<2h-l} \|\eta^{2h-|\alpha|-|\beta|} a_{\alpha\beta}; G\|_{MW_p^{l-h}} \Big) \,.$$

Proof. Let $u \in (W_p^l \cap \mathring{W}_p^h)(G)$. According to Lemma 14.4.3,

$$\eta^{|\beta|-h} D^\beta u \in W_p^{l-h}(G).$$

Consequently,

$$a_{\alpha\beta} D^\beta u \in W_p^{l-2h+|\alpha|}(G)$$

and

$$D^\alpha (a_{\alpha\beta} D^\beta u) \in W_p^{l-2h}(G)$$

for $|\alpha| \geq 2h-l$.

Now let $|\alpha| < 2h-l$. By Corollary 14.4.3, we have

$$\|D^\alpha (a_{\alpha\beta} D^\beta u); G\|_{W_p^{l-2h}}$$

$$\leq \|a_{\alpha\beta} D^\beta u; G\|_{W_p^{l-2h+|\alpha|}} \leq c\,\|\eta^{h-|\alpha|} a_{\alpha\beta} D^\beta u; G\|_{W_p^{l-h}}$$

$$\leq c\,\|\eta^{2h-|\alpha|-|\beta|} a_{\alpha\beta}; G\|_{MW_p^{l-h}} \|\eta^{|\beta|-h} D^\beta u; G\|_{W_p^{l-h}} \,.$$

Using inequality (14.4.1) to estimate the last norm, we arrive at

$$\|D^\alpha (a_{\alpha\beta} D^\beta u); G\|_{W_p^{l-2h}} \leq c\,A \|u; G\|_{W_p^l} \,.$$

□

Henceforth in this subsection, R is a differential operator of order $2h$ with constant coefficients having the form

$$R(D) = \sum_{|\alpha|,|\beta|\leq h} (-1)^{|\alpha|} D^\alpha (a_{\alpha\beta} D^\beta)$$

and S is the operator defined by (14.2.1).

We retain the notation as well as the conditions imposed on the domain G in Sect. 14.4.3.

Lemma 14.4.8. *For all $v \in (W_p^l \cap \mathring{W}_p^h)(\mathbb{R}_+^n)$ with $l \geq h$,*

$$\|(S-R)v; \mathbb{R}_+^n\|_{W_p^{l-2h}} \leq c\, \|I - \partial\lambda; \mathbb{R}_+^n\|_{MW_p^{l-h}}\, \|v; \mathbb{R}_+^n\|_{W_p^l} ,$$

where c is a continuous function of the norm of $\partial\lambda$ in $MW_p^{l-h}(\mathbb{R}_+^n)$ independent of v.

Proof. Changing variables in the bilinear form $(R\varphi, \psi)$, where $\varphi, \psi \in C_0^\infty(G)$, we obtain

$$S(\zeta, D_\zeta) = \frac{1}{\det \partial\lambda(\zeta)} \sum_{|\mu|,|\delta|\leq h} (-1)^{|\mu|} D_\zeta^\mu [\det \partial\lambda(\zeta) f_{\mu\delta}(\zeta) D_\zeta^\delta] ,$$

where

$$f_{\mu\delta}(\zeta) = \sum_{|\alpha|=|\beta|=h} (a_{\alpha\beta} \circ \lambda) \sum_{\{s\}} c_s \prod_{i,j} (D^{s_{ij}} \varkappa_i)(\lambda(\zeta)) \sum_{\{t\}} c_t \prod_{i,j} (D^{t_{ij}} \varkappa_i)(\lambda(\zeta)) .$$

By $\{s\}$ and $\{t\}$ we denote collections of multi-indices s_{ij} and t_{ij} such that

$$\sum_{i,j} s_{ij} = \alpha , \qquad |s_{ij}| \geq 1 , \qquad \sum_{i,j} (|s_{ij}| - 1) = h - |\mu| ,$$

$$\sum_{i,j} t_{ij} = \beta , \qquad |t_{ij}| \geq 1 , \qquad \sum_{i,j} (|t_{ij}| - 1) = h - |\delta| .$$

Since

$$uD^\mu u = \sum_{\mu\geq\gamma>0} c_{\mu\gamma} D^\gamma (v D^{\mu-\gamma} u),$$

it follows that

$$S(\zeta, D_\zeta) = \sum_{|\gamma|,|\delta|\leq h} (-1)^{|\gamma|} D_\zeta^\gamma (b_{\gamma\delta}(\zeta) D_\zeta^\delta),$$

where

$$b_{\gamma\delta}(\zeta) = \sum_{\mu\geq\gamma>0} (-1)^{|\mu|-|\gamma|} c_{\mu\gamma} \Big[D_\zeta^{\mu-\gamma} \Big(\frac{1}{\det \partial\lambda(\zeta)} \Big) \Big] \det \partial\lambda(\zeta) f_{\mu\delta}(\zeta) .$$

In particular, if $|\gamma| = |\delta| = h$, we have

$$b_{\gamma\delta}(\zeta) = \sum_{|\alpha|=|\beta|=h} a_{\alpha\beta} P_{\alpha\beta\gamma\delta}(\partial\varkappa \circ \lambda) \,,$$

where $P_{\alpha\beta\gamma\delta}$ is a polynomial of elements of the matrix $\partial\varkappa$. Also $P_{\alpha\beta\gamma\delta}(I) = 1$ if $\alpha = \gamma$, $\beta = \delta$, and $P_{\alpha\beta\gamma\delta}(I) = 0$ if $\alpha \neq \gamma$ or $\beta \neq \delta$.

Let

$$S_0(\zeta, D_\zeta) = \sum_{|\gamma|=|\delta|=h} (-1)^h D_\zeta^\gamma (b_{\gamma\delta}(\zeta) D_\zeta^\delta) \,, \qquad S_1 = S - S_0 \,.$$

It is clear that

$$\begin{aligned}\|(S_0 - R)v; \mathbb{R}_+^n\|_{W_p^{l-2h}} &\leq \sum_{\alpha,\beta,\gamma,\delta} \|(P_{\alpha\beta\gamma\delta}(\partial\varkappa \circ \lambda) - P_{\alpha\beta\gamma\delta}(I)) D_\zeta^\delta v; \mathbb{R}_+^n\|_{W_p^{l-h}} \\ &\leq c\, \|I - \partial\lambda; \mathbb{R}_+^n\|_{MW_p^{l-h}} \|v; \mathbb{R}_+^n\|_{W_p^l} \\ &\leq c\, \|\nabla\varphi; \mathbb{R}^{n-1}\|_{MW_p^{l+1-h-1/p}} \|v; \mathbb{R}_+^n\|_{W_p^l} \,.\end{aligned}$$

Next we derive an analogous estimate for the norm

$$\|(S - S_0)v; \mathbb{R}_+^n\|_{W_p^{l-2h}} .$$

According to Lemma 14.4.7, is suffices to prove the two inequalities

$$\|\eta^{h-|\delta|} b_{\gamma\delta}; \mathbb{R}_+^n\|_{M(W_p^{l-h} \to W_p^{l-2h+|\gamma|})} \leq c\, \|\nabla\varphi; \mathbb{R}^{n-1}\|_{MW_p^{l+1-h-1/p}} \tag{14.4.10}$$

for $|\gamma| \leq h$, $|\delta| \leq h$, $|\gamma| + |\delta| < 2h$, $l - 2h + |\gamma| \geq 0$, and

$$\|\eta^{2h-|\gamma|-|\delta|} b_{\gamma\delta}; \mathbb{R}_+^n\|_{MW_p^{l-h}} \leq c\, \|\nabla\varphi; \mathbb{R}^{n-1}\|_{MW_p^{l+1-h-1/p}} . \tag{14.4.11}$$

for $|\gamma| \leq h$, $|\delta| \leq h$, $|\gamma| + |\delta| < 2h$, $l - 2h + |\gamma| < 0$.

By Corollary 14.4.2,

$$D_\zeta^{\mu-\gamma}(1/\det \partial\lambda(\zeta)) \in M(W_p^{l-h}(\mathbb{R}_+^n) \to W_p^{l-h-|\mu|+|\gamma|}(\mathbb{R}_+^n)) \tag{14.4.12}$$

and, for $\mu > \gamma$,

$$\begin{aligned}&\|D_\zeta^{\mu-\gamma}(1/\det \partial\lambda(\zeta)); \mathbb{R}_+^n\|_{M(W_p^{l-h} \to W_p^{l-h-|\mu|+|\gamma|})} \\ &\qquad \leq c\, \|\nabla\varphi; \mathbb{R}^{n-1}\|_{MW_p^{l+1-h-1/p}} \,.\end{aligned} \tag{14.4.13}$$

We show that

$$\eta^{h-|\delta|} f_{\mu\delta} \in M(W_p^{l-h-|\mu|+|\gamma|}(\mathbb{R}_+^n) \to W_p^{l-2h+|\gamma|}(\mathbb{R}_+^n)) \,. \tag{14.4.14}$$

Applying Corollary 14.4.2 once more, we obtain

$$\eta^{h-|\delta|} \prod_{i,j} (D^{t_{ij}} \varkappa_i) \circ \lambda \in MW_p^{l-h}(\mathbb{R}^n_+) \,,$$

$$\prod_{i,j} (D^{s_{ij}} \varkappa_i) \circ \lambda \in M(W_p^{l-h}(\mathbb{R}^n_+) \to W_p^{l-2h+|\mu|}(\mathbb{R}^n_+))$$

$$\subset M(W_p^{l-h-|\mu|+|\gamma|}(\mathbb{R}^n_+) \to W_p^{l-2h+|\gamma|}(\mathbb{R}^n_+)) \tag{14.4.15}$$

and therefore the inclusion (14.4.14) holds. Since, for $\mu = \gamma$, at least one of the exponents t_{ij} and s_{ij} is greater than 1, it follows that

$$\|\eta^{h-|\delta|} f_{\gamma\delta}; \mathbb{R}^n_+\|_{M(W_p^{l-h} \to W_p^{l-2h+|\gamma|})} \leq c \, \|\nabla \varphi; \mathbb{R}^{n-1}\|_{MW_p^{l+1-h-1/p}} \,. \tag{14.4.16}$$

Now (14.4.10) results directly from (14.4.12)–(14.4.14), and (14.4.16).

Next we turn to the proof of (14.4.11). By virtue of Corollary 14.4.2, the inclusion (14.4.15) holds. Moreover,

$$\eta^{|\mu|-|\gamma|} D_\zeta^{\mu-\gamma} (1/\det \partial\lambda(\zeta)) \in MW_p^{l-h}(\mathbb{R}^n_+) \,,$$

and

$$\eta^{h-|\mu|} \prod_{i,j} (D^{s_{ij}} \varkappa_i) \circ \lambda \in MW_p^{l-h}(\mathbb{R}^n_+) \,.$$

To obtain (14.4.11), it remains to note that we always have either $\mu > \gamma$ or one of the exponents t_{ij}, s_{ij} is greater than one, and then to apply Corollary 14.4.2 once more. □

With minor modifications in the above proof and using the properties of (p, l)-diffeomorphisms given in Sect. 9.4.2, we arrive at the following local variant of Lemma 14.4.8.

Lemma 14.4.9. *For all $v \in (W_p^l \cap \mathring{W}_p^h)(\mathbb{R}^n_+)$, $l \geq h$, with supports in $\mathcal{B}_r \cap \overline{\mathbb{R}^n_+}$, the inequality*

$$\|(S - R)v; \mathbb{R}^n_+\|_{W_p^{l-2h}} \leq c \, \|\!|I - \partial\lambda; \mathcal{B}_r \cap \mathbb{R}^n_+|\!\|_{MW_p^{l-h}} \|v; \mathbb{R}^n_+\|_{W_p^l} \tag{14.4.17}$$

holds. For $p(l - h) > n$ it follows from (9.6.4) that (14.4.17) is equivalent to

$$\|(S - R)v; \mathbb{R}^n_+\|_{W_p^{l-2h}}$$

$$\leq c r^{l-h-n/p} \|\!|I - \partial\lambda; \mathcal{B}_r \cap \mathbb{R}^n_+|\!\|_{W_p^{l-h}} \, \|v; \mathbb{R}^n_+\|_{W_p^l} \,. \tag{14.4.18}$$

14.5 Solvability of the Dirichlet Problem in $W_p^l(\Omega)$

14.5.1 Generalized Formulation of the Dirichlet Problem

Let Ω be open subset of $\mathbb{R}^n$ and let P be the operator (14.4.9), where $a_{\alpha\beta} \in C^{l-h}(\bar{\Omega})$, $l \geq h$. Further, let the Gårding inequality

$$\operatorname{Re} \int_\Omega \sum_{|\alpha|=|\beta|=h} a_{\alpha\beta} D^\alpha u \overline{D^\beta u}\, dz \geq c\, \|u; \Omega\|^2_{W_2^h} \tag{14.5.1}$$

hold for $u \in C_0^\infty(\Omega)$.

We say that $u \in W_p^l(\Omega)$ is a solution of the generalized Dirichlet problem in $W_p^l(\Omega)$ if

$$Pu = f\,, \qquad u - g \in W_p^l(\Omega) \cap \mathring{W}_p^h(\Omega)\,, \tag{14.5.2}$$

where f and g are given functions in the spaces $W_p^{l-2h}(\Omega)$ and $W_p^l(\Omega)$ respectively.

By $W_p^{-k}(\Omega)$, $k = 1, 2, \dots$, we mean the space of linear continuous functionals in $\mathring{W}_{p'}^k(\Omega)$.

14.5.2 A Priori Estimate for Solutions of the Generalized Dirichlet Problem

Following the proof of Theorem 14.3.1 and using Lemma 14.4.8 in place of Lemma 14.2.1, we arrive at:

Theorem 14.5.1. *If $p(l-h) \leq n$, $1 < p < \infty$, and if $\partial\Omega$ belongs to the class $M_p^{l+1-h-1/p}(\delta)$, then*

$$\|u; \Omega\|_{W_p^l} \leq c\,(\|Pu; \Omega\|_{W_p^{l-2h}} + \|u; \Omega\|_{L_1}) \tag{14.5.3}$$

for all $u \in (W_p^l \cap \mathring{W}_p^h)(\Omega)$.

Duplicating the proof of Theorem 14.3.2 and using (14.4.18) instead of (14.2.6), we obtain:

Theorem 14.5.2. *If $p(l-h) > n$, $1 < p < \infty$, and $\partial\Omega \in W_p^{l+1-h-1/p}$, then Theorem 14.5.1 holds.*

Next we state two corollaries of (14.5.3) which are analogous to Propositions 14.3.1 and 14.3.2.

Proposition 14.5.1. *Let Ω satisfy the conditions of either Theorem 14.5.1 or Theorem 14.5.2.*

(i) If the kernel of the operator

$$P : (W_p^l \cap \mathring{W}_p^h)(\Omega) \to W_p^{l-2h}(\Omega) \tag{14.5.4}$$

is trivial, then the norm $\|u; \Omega\|_{L_1}$ in (14.5.3) can be omitted.

(ii) The kernel of the operator (14.5.4) is finite-dimensional and the range of this operator is closed.

Proposition 14.5.2. *Let Ω satisfy the conditions of either Theorem* 14.5.1 *or Theorem* 14.5.2. *Further, let U and V be open bounded subsets of $\mathbb{R}^n$ with $\bar{U} \subset V$, and let $u \in (W_p^l \cap \mathring{W}_p^h)(\Omega)$. Then*

$$\|u; U \cap \Omega\|_{W_p^l} \le c\,(\|Pu; V \cap \Omega\|_{W_p^{l-2h}} + \|u; V \cap \Omega\|_{L_1})\,.$$

14.5.3 Solvability of the Generalized Dirichlet Problem

Let the Gårding inequality (14.5.1) hold for all $u \in C_0^\infty(\Omega)$. Then, as is well known, the equation $Pu = f$ with $f \in W_2^{-h}(\Omega)$ is uniquely solvable in $\mathring{W}_2^h(\Omega)$.

Theorem 14.5.3. *Let $\partial\Omega \in M_p^{l+1-h-1/p}$ for $p(l-h) \le n$ and let $\partial\Omega$ belong to the class $W_p^{l+1-h-1/p}$ for $p(l-h) > n$.*

(*i*) *If $f \in W_p^{l-2h}(\Omega) \cap W_2^{-h}(\Omega)$ and $g \in W_p^l(\Omega) \cap W_2^h(\Omega)$, $1 < p < \infty$, and if $u \in W_2^h(\Omega)$ is such that $Pu = f$ and $u - g \in \mathring{W}_2^h(\Omega)$, then $u \in W_p^l(\Omega)$ and $u - g \in \mathring{W}_p^h(\Omega)$.*

(*ii*) *The problem* (14.5.2) *has one and only one solution $u \in W_p^l(\Omega)$.*

Proof. It is sufficient to assume that $g = 0$.

(*i*) First let $p(l-h) \le n$. We put $\varphi_\varepsilon(x) = \varepsilon + \Phi(x, \varepsilon)$, where Φ is an extension of φ defined by $\Phi = \mathcal{T}\varphi$ (cf. (9.4.11)). We introduce the domain $G_\varepsilon = \{z = (x, y) : x \in \mathbb{R}^{n-1}, y > \varphi_\varepsilon(x)\}$. Since $1 + \partial\Phi/\partial\eta > 0$, it follows that $\{G_\varepsilon\}$ is an increasing family, $\overline{G_\varepsilon} \subset G$, and $G_\varepsilon \to G$ as $\varepsilon \to +0$.

By part (*i*) of Theorem 8.7.2 we have

$$\|\nabla\varphi_\varepsilon; \mathbb{R}^{n-1}\|_{MW_p^{l+1-h-1/p}} \le c\,\|\nabla\Phi; \mathbb{R}^n_+\|_{MW_p^{l-h+1}}\,.$$

This inequality and (9.4.12) imply that

$$\|\nabla\varphi_\varepsilon; \mathbb{R}^{n-1}\|_{MW_p^{l+1-h-1/p}} \le c\,\|\nabla\varphi; \mathbb{R}^{n-1}\|_{MW_p^{l+1-h-1/p}}\,, \tag{14.5.5}$$

with a constant c independent of ε. Let $\Omega_\varepsilon = \Omega\backslash(\bar{U}\backslash G_\varepsilon)$ and let u_ε be a solution of $Pu = f$ in $\mathring{W}_2^h(\Omega_\varepsilon)$. It is known (see, for instance, Nečas [Ne], 6.6, Ch. 3), that $u_\varepsilon \to u$ in $\mathring{W}_2^h(\Omega)$. By U_1 we denote an open set such that $\overline{U_1} \subset U$. Since $\varphi_\varepsilon \in C^\infty(\mathbb{R}^{n-1})$, we have $u_\varepsilon \in W_p^l(U_1 \cap \Omega_\varepsilon)$ by the known theorem on the regularity of weak solutions of elliptic boundary value problems near a smooth part of a boundary. This and Proposition 14.5.2 give the estimate

$$\|u_\varepsilon; U_2 \cap \Omega_\varepsilon\|_{W_p^l} \le c\,(\|f; U_1 \cap \Omega_\varepsilon\|_{W_p^{l-2h}} + \|u_\varepsilon; U_1 \cap \Omega_\varepsilon\|_{L_1})\,,$$

where U_2 is an open set, $\overline{U_2} \subset U_1$, and c does not depend on ε. Hence the left-hand side is uniformly bounded with respect to ε. Now, if we fix a domain ω such that $\bar{\omega} \subset \Omega$, then the upper limit

$$\limsup_{\varepsilon \to +0} \|u_\varepsilon; \omega\|_{W_p^l}$$

is majorized by a constant independent of ω. From $\{u_\varepsilon\}$ we select a sequence which is weakly convergent in $W_p^l(\omega)$. This sequence converges in $W_2^h(\omega)$, and hence its weak limit in $W_p^l(\omega)$ coincides with u. Therefore, $u \in W_p^l(\omega)$, and the $W_p^l(\omega)$-norm of u is uniformly bounded with respect to ω. Thus $u \in W_p^l(\Omega)$. The identity of the spaces $W_p^h(\Omega) \cap \mathring{W}_2^h(\Omega)$ and $\mathring{W}_p^h(\Omega)$ for Lipschitz domains Ω is known.

The case $p(l-1) > n$ can be treated in the same way, with (14.5.5) replaced by

$$\|\varphi_\varepsilon; \mathbb{R}^{n-1}\|_{W_p^{l+1-h-1/p}} \le c\,\|\varphi; \mathbb{R}^{n-1}\|_{W_p^{l+1-h-1/p}}\ .$$

(*ii*) For $p \ge 2$, the assertion follows from the unique solvability of the problem in $\mathring{W}_2^h(\Omega)$, together with the first part of the theorem.

Consider the case $p < 2$. By P^t we denote the operator formally conjugate to P. The coefficients of P^t belong to $C^{l-2h}(\bar{\Omega})$, and Gårding's inequality (14.5.1) holds for P^t too. Recall that the property $\partial\Omega \in M_p^{1-1/p}$ implies that the Lipschitz constants of the functions φ which locally specify $\partial\Omega$ are small. Hence the equation

$$P^t v = F \in W_{p'}^{-h}(\Omega) \quad \text{with } v \in \mathring{W}_{p'}^h(\Omega)$$

is uniquely solvable in $\mathring{W}_{p'}^h(\Omega)$. Let u be a solution of the homogeneous problem (14.5.2) and let $\{v_m\}_{m\ge 1}$ be a sequence of functions in $C_0^\infty(\Omega)$, which converges to v in $\mathring{W}_{p'}^h(\Omega)$. Then

$$0 = \lim \sum_{|\alpha|,|\beta|\le h} (a_{\alpha\beta} D^\alpha u, D^\beta v_m) = \sum_{|\alpha|,|\beta|\le h} (a_{\alpha\beta} D^\alpha u, D^\beta v) = (u, F)\ ,$$

and the uniqueness property of problem (14.5.2) follows.

Let

$$f_m \in C^l(\bar{\Omega}),\ \ m = 1, 2, \ldots, \qquad f_m \to f \ \text{ in } W_p^{l-2h}(\Omega).$$

By u_m we denote a function in $\mathring{W}_2^h(\Omega)$ satisfying $Pu_m = f_m$. According to the first part of the theorem, $u_m \in W_p^l(\Omega) \cap \mathring{W}_p^h(\Omega)$. By part (*i*) of Proposition 14.5.1,

$$\|u_m - u_k; \Omega\|_{W_p^l} \le c\,\|f_m - f_k; \Omega\|_{W_p^{l-2h}}\ .$$

Thus $\{u_m\}$ converges in $W_p^l(\Omega) \cap \mathring{W}_p^h(\Omega)$ and its limit satisfies $Pu = f$. □

14.5.4 The Dirichlet Problem Formulated in Terms of Traces

The first boundary value problem (14.5.2) is not a particular case of the general boundary value problem formulated in Sect. 14.3.1. In the present subsection we study the Dirichlet problem in another formulation which is analogous to that considered in 14.3.1

Let P be the elliptic operator (14.4.9) with coefficients $a_{\alpha\beta}$ in $C^{l-h}(\bar{\Omega})$, $l \geq h$, for which the inequality (14.5.1) holds. We assume $\partial\Omega$ to be in the class $C^{0,1}$.

We introduce a sufficiently small finite open covering $\{U\}$ of $\bar{\Omega}$ and a corresponding partition of unity $\{\zeta_U\}$. Let

$$P_{jU} = \partial^{j-1}/\partial y^{j-1}, \ j = 1, \dots, h, \quad \text{if } U \cap \partial\Omega \neq \varnothing$$

and

$$P_{jU} = 0 \qquad \text{if } U \cap \partial\Omega = \varnothing.$$

The Dirichlet boundary conditions will be prescribed by the operators

$$P_j = \sum_U \zeta_U P_{jU}.$$

We give a formulation of the Dirichlet problem. Let us look for a function $u \in W_p^l(\Omega)$ such that

$$Pu = f \quad \text{in} \quad \Omega, \qquad \text{tr} P_j u = f_j \quad \text{on} \quad \partial\Omega, \quad j = 1, \dots, h\,, \tag{14.5.6}$$

where f and f_j are functions in $W_p^{l-2h}(\Omega)$ and $W_p^{l+1-j-1/p}(\partial\Omega)$ respectively.

It is clear that any solution of the problem (14.5.2) is a solution of (14.5.6) with $f_j = \text{tr} P_j g$. The following lemma shows that an opposite statement holds if $\partial\Omega \in M_p^{l-1/p}$.

Lemma 14.5.1. *Let $G = \{z = (x, y) : x \in \mathbb{R}^{n-1}, y > \varphi(x)\}$ be a domain with ∂G in the class $M_p^{l-1/p}$ and let $f_1, \dots, f_h$ be arbitrary functions in $W_p^{l-1/p}(\partial G)$, ..., $W_p^{l+1-h-1/p}(\partial G)$. Then there exists a function $g \in W_p^l(G)$ such that*

$$\text{tr}(\partial^{j-1} g/\partial y^{j-1}) = f_j, \quad j = 1, \dots, h.$$

Proof. We use the notation λ, $\varkappa$, and $\mathcal{T}\varphi$ introduced in Sect. 9.4.3. We have

$$[(\partial/\partial y)^{j-1} g] \circ \lambda = [(N + \partial(\mathcal{T}\varphi)/\partial\eta)^{-1}(\partial/\partial\eta)]^{j-1}(g \circ \lambda)\,.$$

Since $\nabla(\mathcal{T}\varphi) \in MW_p^l(\mathbb{R}_+^n)$, it follows that

$$[(\partial/\partial y)^{j-1} g] \circ \lambda = \sum_{\nu=1}^{j} a_{\nu j} (\partial/\partial\eta)^{\nu-1}(g \circ \lambda)\,, \qquad j = 1, \dots, h\,, \tag{14.5.7}$$

where

$$a_{\nu j} \in M(W_p^{l-\nu+1}(\mathbb{R}_+^n) \to W_p^{l-j+1}(\mathbb{R}_+^n)), \quad a_{jj} = (N + \partial(\mathcal{T}\varphi)/\partial\eta)^{1-j}.$$

We note that (14.5.7) is a triangular algebraic system with respect to $(\partial/\partial\eta)^{\nu-1}(u \circ \lambda)$. Hence

$$(\partial/\partial\eta)^{\nu-1}(g\circ\lambda)=\sum_{j}^{\nu} b_{j\nu}[(\partial/\partial y)^{j-1}g]\circ\lambda\,,\quad \nu=1,\ldots,h\,,$$

where

$$b_{j\nu}\in M(W_p^{l-j+1}(\mathbb{R}^n_+)\to W_p^{l-\nu+1}(\mathbb{R}^n_+)).$$

Taking into account that

$$\operatorname{tr} b_{j\nu}\in M(W_p^{l+1-j-1/p}(\mathbb{R}^{n-1})\to W_p^{l+1-\nu-1/p}(\mathbb{R}^{n-1})),$$

we obtain

$$(\operatorname{tr} b_{j\nu})f_j\circ\lambda\in W_p^{l+1-\nu-1/p}(\mathbb{R}^{n-1}).$$

Therefore, there exists a function $H\in W_p^l(\mathbb{R}^n_+)$ such that

$$\operatorname{tr}(\partial/\partial\eta)^{\nu-1}H=\sum_{\nu=1}^{j}(\operatorname{tr} b_{j\nu})f_j\circ\lambda\,.$$

Setting $g=H\circ\varkappa$, we complete the proof. □

Since both formulations (14.5.2) and (14.5.6) of the Dirichlet problem are equivalent for domains with boundary in the class $M_p^{l-1/p}$, then, from Theorem 14.5.3, we obtain:

Theorem 14.5.4. *Let any of the following conditions hold:*

(α) $h=1$, $p(l-1)\le n$; $\partial\Omega\in M_p^{l-1/p}(\delta)$;

(β) $h=1$, $p(l-1)>n$; $\partial\Omega\in W_p^{l-1/p}$;

(γ) $h>1$, $\partial\Omega\in M_p^{l-1/p}$ *and* $\partial\Omega$ *is locally defined by equations of the form* $y=\varphi(x)$, *where* φ *is a function with a small Lipschitz constant (for* $p(l-1)>n$, *this is equivalent to* $\partial\Omega\in W_p^{l-1/p}$).

Then the operator

$$\{P;\operatorname{tr}P_j\}:W_p^l(\Omega)\to W_p^{l-2h}(\Omega)\times\prod_{j=1}^{h}W_p^{l+1-j-1/p}(\partial\Omega)$$

is an isomorphism.

Proof. For $h=1$ the result follows from Theorem 14.5.3.

Let $h>1$. According to (2.3.8),

$$\|\nabla\varphi;\mathbb{R}^{n-1}\|_{MW_p^{l-h-1/p}}\le c\,\|\nabla\varphi;\mathbb{R}^{n-1}\|^{\alpha}_{MW_p^{l-1-1/p}}\|\nabla\varphi;\mathbb{R}^{n-1}\|^{1-\alpha}_{L_\infty}$$

with $\alpha=(p(l-h)-1)/(p(l-1)-1)$. Consequently, it follows from (γ) that

$$\partial\Omega\in M_p^{l+1-h-1/p}(\delta)\qquad\text{if }\ p(l-h)\le n$$

and

$$\partial\Omega\in W_p^{l+1-h-1/p}\qquad\text{if }\ p(l-h)>n.$$

By Theorem 14.5.3 the proof is complete. □

Thus, by changing the formulation of the Dirichlet problem, we have obtained its solvability in $W_p^l(\Omega)$ under stricter assumptions on Ω (compare the last theorem with Theorem 14.5.3). The exception is the second-order operator P, i.e., $h = 1$, when admissible classes of domains coincide for both formulations.

In the following section we discuss the necessity of the conditions in the last theorem.

14.6 Necessity of Assumptions on the Domain

14.6.1 A Domain Whose Boundary is in $M_2^{3/2} \cap C^1$ but does not Belong to $M_2^{3/2}(\delta)$

In this subsection we give an example which shows that for $p(l-1) \le n$ and for $h = 1$ the condition $\partial\Omega \in M_p^{l-1/p}(\delta)$ in part (α) of Theorem 14.5.4 cannot be replaced by the assumption that $\partial\Omega$ belongs to the class $M_p^{l-1/p} \cap C^{l-1}$. To be precise, we construct a domain Ω with $\partial\Omega \in M_2^{3/2} \cap C^1$ for which the problem

$$-\Delta u = f \quad \text{in} \quad \Omega\,, \qquad \text{tr}\, u = 0 \quad \text{on} \quad \partial\Omega \tag{14.6.1}$$

is not solvable in $W_2^2(\Omega)$ for all $f \in L_2(\Omega)$. This means that the smallness of $\|\nabla\varphi; \mathbb{R}^{n-1}\|_{MW_2^{1/2}}$ in the definition of $M_2^{3/2}(\delta)$ is essential for the solvability of the problem (14.6.1) in $W_2^2(\Omega)$.

Let the domain Ω be specified in a neighborhood of O by the inequality $y > C\varphi(x)$, where C is a positive constant and

$$\varphi(x) = \eta(x, 0)|x_1|/\log(1/|x_1|).$$

Here and henceforth η is a function in $C_0^\infty(\mathcal{B}_{1/2})$ with $\eta = 1$ on the ball $\mathcal{B}_{1/4}$.

We introduce the domain

$$\omega = \{\xi : x_1 + ix_2 : |\xi| < 1/2, x_2 > C|x_1|/\log(1/|x_1|)\}$$

and we denote by $\zeta(\xi)$ the conformal mapping of ω onto the half-disc $\{\zeta : \operatorname{Im}\zeta > 0, |\zeta| < 1\}$ with the fixed point $\xi = 0$. Let $\xi = i\rho\exp(i\theta)$ and let ω be given in polar coordinates (ρ, θ) as

$$\omega = \{\xi : \rho < 1/2, |\theta| < \pi/2 + \varphi(\rho)\}.$$

It is easily checked that

$$\varphi(\rho) = C(\log 1/\rho)^{-1} + O((\log 1/\rho)^{-3})\,.$$

According to an asymptotic formula due to Warschawski [Wa],

$$\operatorname{Im}\zeta(\xi) = c\Big(\exp\Big(-\pi\int_\rho^{1/2}\frac{dr}{r(\pi+2\varphi(r))}\Big)\Big)\Big(\cos\frac{\pi\theta}{\pi+2\varphi(\rho)}+o(1)\Big)$$

$$= c\,\rho(\log 1/\rho)^{2C/\pi}\Big(\cos\frac{\pi\theta}{\pi+2\varphi(\rho)}+o(1)\Big) \quad \text{as } \rho\to+0.$$

It is clear that, for $C \ge \pi/4$,

$$\int_\omega \frac{(\operatorname{Im}\zeta(\xi))^2}{\rho^4(\log\rho)^2}\,dx_1dx_2 = \infty\,. \tag{14.6.2}$$

Next we need the following assertion.

Lemma 14.6.1. *If $h \in W_2^2(\omega)\cap \mathring{W}_2^1(\omega)$, then*

$$\int_\omega \frac{h^2 dx_1dx_2}{\rho^4(\log\rho)^2} \le c\,\|h;\omega\|^2_{W_2^2}\,. \tag{14.6.3}$$

Proof. First we show that, for any $g \in W_2^1(\omega)$,

$$\int_\omega g^2\frac{dx_1dx_2}{\rho^2(\log\rho)^2} \le c\,\|g;\omega\|^2_{W_2^1}\,. \tag{14.6.4}$$

Clearly, to prove (14.6.4) it suffices to assume that ω is the half-disc $\{\xi : \rho < 1/2, |\theta| < \pi/2\}$. After integration of the Hardy inequality

$$\int_0^{1/2} g^2\frac{d\rho}{\rho(\log\rho)^2} \le 4\int_0^{1/2}\Big(\frac{\partial g}{\partial\rho}\Big)^2\rho\,d\rho$$

with respect to the variable θ, we arrive at (14.6.4). Putting $g = \partial h/\partial x_1$ and $g = \partial h/\partial x_2$ into (14.6.4), we obtain

$$\int_\omega (\nabla h)^2\frac{dx_1dx_2}{\rho^2(\log\rho)^2} \le c\,\|h;\omega\|^2_{W_2^2}\,.$$

Let $C_\rho = \{\xi : |\xi| = \rho\}$. Since $h = 0$ on $\partial\omega$, it follows that for almost all $\rho > 0$

$$\int_{\omega\cap C_\rho} h^2\,d\theta \le c\int_{\omega\cap C_\rho}(\partial h/\partial\theta)^2\,d\theta \le c\,\rho^2\int_{\omega\cap C_\rho}(\nabla h)^2\,d\theta\,,$$

which leads to

$$\int_\omega h^2\frac{dx_1dx_2}{\rho^4(\log\rho)^2} \le c\int_\omega(\nabla h)^2\frac{dx_1dx_2}{\rho^2(\log\rho)^2}\,.$$

□

By (14.6.2) and (14.6.3), we see that the function u defined on Ω by

$$u(z) = \eta(2z)\mathrm{Im}\ \zeta(x_1 + ix_2)$$

does not belong to $W_2^2(\Omega)$. On the other hand, u is in $\mathring{W}_2^1(\Omega)$ and satisfies the equation

$$-\Delta u = f, \quad f \in L_2(\Omega).$$

Consequently, the boundary value problem (14.6.1) is solvable in $\mathring{W}_2^2(\Omega)$ if and only if $C < \pi/4$.

Obviously, $\partial\Omega$ is in the class C^1. We show that $\partial\Omega \in M_2^{3/2}$, i.e. $\nabla\varphi \in MW_2^{1/2}(\mathbb{R}^{n-1})$. With this aim in view, we verify that the gradient of the function ψ defined as

$$\psi(z) = \eta(z) r \log r\ , \quad \text{where } r = (x_1^2 + y^2)^{1/2}\ ,$$

belongs to the space $MW_2^1(\mathbb{R}^n)$ (cf. Theorem 8.7.1). Clearly, $\nabla\psi \in L_\infty(\mathbb{R}^n)$ and it remains to prove that

$$\nabla_2\psi \in M(W_2^1(\mathbb{R}^n) \to L_2(\mathbb{R}^n)).$$

In fact, for all $u \in W_2^1(\mathbb{R}^n)$,

$$\|u\nabla_2\psi; \mathbb{R}^n\|_{L_2}^2 \le c \int_{\mathcal{B}_{1/2}} \left|\frac{u}{r\log r}\right|^2 dz \le c\,\|u; \mathcal{B}_1\|_{W_p^1}^2\ .$$

Thus, $\partial\Omega \in M_2^{3/2}$.

14.6.2 Necessary Conditions for Solvability of the Dirichlet Problem

The next assertion, which follows directly from the Implicit Function Theorem 9.5.2, shows that the condition $\partial\Omega \in W_p^{l-1/p}$ with $p(l-1) > n$ is necessary for the solvability of the problem (14.5.6) in $W_p^l(\Omega)$ for an operator P of higher than the second order.

Theorem 14.6.1. *Let Ω be a bounded Lipschitz domain and let l be an integer, $l \ge 2h$, $p(l-1) > n$, $1 < p < \infty$, and $h > 1$. If there exists a solution $u \in W_p^l(\Omega)$ of the problem*

$$Pu = 0 \ \text{ in } \Omega, \quad \mathrm{tr}\, u = 0, \quad \mathrm{tr}\, P_2 u = 1, \ \ \mathrm{tr}\, P_j u = 0, \ j = 3, \dots, h\ , \qquad (14.6.5)$$

then $\partial\Omega \in W_p^{l-1/p}$.

Under the additional assumption $\partial\Omega \in C^{l-2,1}$, we can prove the necessity of the inclusion $\partial\Omega \in W_p^{l-1/p}$ for $p(l-1) \le n$.

Theorem 14.6.2. *Let $\partial\Omega$ be in the class $C^{l-2,1}$ and let l be an integer, $l \geq 2h$, $p(l-1) \leq n$, $1 < p < \infty$, and $h > 1$. If there exists a solution $u \in W_p^l(\Omega)$ of problem* (14.6.5)*, then $\partial\Omega \in W_p^{l-1/p}$.*

Proof. We use the same notation as in the formulation of Theorem 9.5.2. Since

$$\nabla_x u, \;\; u_y \in W_{p.\mathrm{loc}}^{l-1}(U \cap \bar{G}) \quad \text{and } \varphi \in C^{l-2,1}(\mathbb{R}^{n-1}),$$

it follows that

$$\nabla_x u \circ \lambda, \quad u_y \circ \lambda \in W_{p.\mathrm{loc}}^{l-1}(\varkappa(U \cap \bar{G})).$$

Therefore, $\mathrm{tr}(\nabla_x u \circ \lambda)$ and $\mathrm{tr}(u_y \circ \lambda)$ belong to $W_{p,\mathrm{loc}}^{l-1-1/p}(\omega)$. Now we note that $\partial\Omega$ is in the class C^h and so, by the known coercive estimate for solutions of the elliptic boundary value problem in the variational form (cf. [ADN1], Sect. 15), we have $u \in W_q^h(\Omega)$ for any $q < \infty$. In particular, $\nabla u \in C(\bar{\Omega})$. Since the space $(W_{p,\mathrm{loc}}^{l-1-1/p} \cap L_\infty)(\omega)$ is a multiplication algebra, the vector function

$$\nabla\varphi = \mathrm{tr}\;(\nabla_x u \circ \lambda)/\mathrm{tr}\;(u_y \circ \lambda)$$

belongs to $W_{p,\mathrm{loc}}^{l-1-1/p}(\omega)$. □

We consider the case of the second-order operator P.

Theorem 14.6.3. *Let l be an integer, $l \geq 2$, $1 < p < \infty$, $h = 1$, and $P(1) \leq 0$. Let Ω be a domain with $\partial\Omega \in C^1$ and let the normal to $\partial\Omega$ satisfy the Dini condition. If, for a nonpositive function $f \in C_0^\infty(\Omega)$, there exists a solution $u \in W_p^l(\Omega)$ of the problem*

$$Pu = f \quad in \quad \Omega, \quad \mathrm{tr}\; u = 0\,, \tag{14.6.6}$$

then $\partial\Omega \in W_p^{l-1/p}$.

Proof. It is sufficient to note that the interior normal derivative at any point of $\partial\Omega$ is positive by Giraud's theorem (see, for example, [Mir], Sect. 3.5) and then to duplicate the proof of Theorem 14.6.2. □

14.6.3 Boundaries of the Class $M_p^{l-1/p}(\delta)$

Let Ω be a bounded Lipschitz domain. We denote by O an arbitrary point of $\partial\Omega$ and introduce a neighborhood U of O such that $\Omega \cap U = G \cap U$ with $G = \{(x, y) : x \in \mathbb{R}^{n-1}, y > \varphi(x)\}$.

By Theorem 4.1.1, the norm of $\nabla\varphi$ in $MW_p^{l-1-1/p}(\mathbb{R}^{n-1})$ is equivalent to

$$\sup_{e \subset \mathbb{R}^{n-1}} \frac{\|D_{p,l-1/p}\varphi; e\|_{L_p}}{[C_{p,l-1-1/p}(e)]^{1/p}} + \|\nabla\varphi; \mathbb{R}^{n-1}\|_{L_\infty}\,. \tag{14.6.7}$$

(Here we can restrict ourselves to compact sets e with $d(e) \le 1$.) Thus the definition of the class $M_p^{l-1/p}(\delta)$ means that the sum (14.6.7) is sufficiently small.

The following assertion, whose proof is postponed to Sect. 14.7, gives a local characterization of the class $M_p^{l-1/p}(\delta)$.

Theorem 14.6.4. *Let $p(l-1) \le n$. The class $M_p^{l-1/p}(\delta)$ has the following equivalent description. For any point $O \in \partial\Omega$ there exists a neighborhood U and a special Lipschitz domain $G = \{z = (x, y) : x \in \mathbb{R}^{n-1}, y > \varphi(x)\}$ such that $U \cap \Omega = U \cap G$ and*

$$\lim_{\varepsilon \to 0} \Big(\sup_{e \subset \mathcal{B}_\varepsilon} \frac{\|D_{l-1/p}(\varphi; \mathcal{B}_\varepsilon); e\|_{L_p}}{[C_{p,l-1-1/p}(e)]^{1/p}} + \|\nabla\varphi; \mathcal{B}_\varepsilon\|_{L_\infty} \Big) \le c\,\delta\,. \tag{14.6.8}$$

Here $\mathcal{B}_\varepsilon$ is the ball with centre at O and radius ε, c is a constant which depends only on l, p, n, and

$$D_{j-1/p}(\varphi; \mathcal{B}_\varepsilon)(x) = \Big(\int_{\mathcal{B}_\varepsilon} |\nabla_{j-1}\varphi(x) - \nabla_{j-1}\varphi(y)|^p |x-y|^{-n+2-p}\, dy \Big)^{1/p}.$$

Theorem 14.6.4 and properties (v), (vi) of the capacity formulated in Sect. 9.6.1 lead to:

Corollary 14.6.1. *(i) If $n > p(l-1)$ and*

$$\lim_{\varepsilon \to 0} \Big(\sup_{e \subset \mathcal{B}_\varepsilon} \frac{\|D_{l-1/p}(\varphi; \mathcal{B}_\varepsilon); e\|_{L_p}}{(\mathrm{mes}_{n-1} e)^{[n-p(l-1)]/(n-1)p}} + \|\nabla\varphi; \mathcal{B}_\varepsilon\|_{L_\infty} \Big) < c\,\delta\,,$$

then $\partial\Omega \in M_p^{l-1/p}(\delta)$.

(ii) If $n = p(l-1)$ and

$$\lim_{\varepsilon \to 0} \Big(\sup_{e \subset \mathcal{B}_\varepsilon} \|D_{l-1/p}(\varphi; \mathcal{B}_\varepsilon); e\|_{L_p} |\log(\mathrm{mes}_{n-1} e)|^{(p-1)/p} + \|\nabla\varphi; \mathcal{B}_\varepsilon\|_{L_\infty} \Big) < c\,\delta\,,$$

then $\partial\Omega \in M_p^{l-1/p}(\delta)$.

Now we derive another test for the inclusion into $M_p^{l-1/p}(\delta)$ involving the Besov space $B_{q,p}^m$ (cf. Corollary 14.6.2 below).

We say that the boundary of the Lipschitz domain Ω belongs to $B_{q,p}^{l-1/p}$ $(l = 1, 2, \dots,)$ if, for any point of $\partial\Omega$, there exists a neighborhood in which $\partial\Omega$ is specified in Cartesian coordinates by a function φ satisfying

$$\int_{\mathbb{R}^{n-1}} \Big(\int_{\mathbb{R}^{n-1}} |\nabla_{l-1}\varphi(x+h) - \nabla_{l-1}\varphi(x)|^q\, dx \Big)^{p/q} |h|^{2-n-p}\, dh < \infty\,.$$

Corollary 14.6.2. *Let $p(l-1) \le n$ and let Ω be a bounded Lipschitz domain with $\partial\Omega \in B^{l-1/p}_{q,p}$ where*

$$q \in [p(n-1)/(p(l-1)-1), \infty] \quad \text{if } p(l-1) < n$$

and

$$q \in (p, \infty] \qquad \text{if } p(l-1) = n.$$

Further, let $\partial\Omega$ be locally defined in Cartesian coordinates by $y = \varphi(x)$, where φ is a function with a Lipschitz constant less than $c\,\delta$. Then $\partial\Omega \in M^{l-1/p}_p(\delta)$.

Proof. We have

$$\begin{aligned}\|D_{l-1/p}(\varphi;\mathcal{B}_\varepsilon);e\|^p_{L_p} &\le \int_{\mathcal{B}_\varepsilon} |h|^{-n+2-p}\,dh \int_e |\nabla_{l-1}\varphi(x+h) - \nabla_{l-1}\varphi(x)|^p\,dx \\ &\le (\mathrm{mes}_{n-1}e)^{1-p/q} \int_{\mathcal{B}_\varepsilon} |h|^{-n+2+p}\,dh \\ &\quad \times \Big(\int_{\mathcal{B}_\varepsilon} |\nabla_{l-1}\varphi(x+h) - \nabla_{l-1}\varphi(x)|^q\,dx\Big)^{p/q}.\end{aligned}$$

Then the result follows from part (i) of Corollary 14.6.1. □

Corollary 14.6.2 can be made sharper in the case $p(l-1) = n$, by virtue of part (ii) of Corollary 14.6.1, if one uses the Orlicz space $L_{t^p(\log_+ t)^{p-1}}$ instead of L_q, but we shall not go into this.

Setting $q = \infty$ in Corollary 14.6.2, we obtain a simple sufficient condition for the inclusion into $M^{l-1/p}_p(\delta)$ formulated in terms of the modulus of continuity $\omega_{l-1}(t)$ of $\nabla_{l-1}\varphi$:

$$\int_0 \Big(\frac{\omega_{l-1}(t)}{t}\Big)^p dt < \infty\,. \tag{14.6.9}$$

Since $B^{l-1/p}_{\infty,p} \subset W^{l-1/p}_p$, it follows that (14.6.9) is sufficient for $\partial\Omega \in W^{l-1/p}_p$.

We show that (14.6.9) is in a sense a sharp condition for solvability of the Dirichlet problems (14.6.5) and (14.6.6) in $W^l_p(\Omega)$.

We have shown in 4.4.3 that, for any increasing function $\omega \in C[0,1]$ satisfying the inequalities (4.4.15) as well as the condition

$$\int_0^1 \Big(\frac{\omega(t)}{t}\Big)^p dt = \infty,$$

one can construct a function φ on $\mathbb{R}^{n-1}$ such that

(i) the continuity modulus of $\nabla_{l-1}\varphi$ does not exceed $c\omega$ with $c = \text{const}$;

(ii) supp $\varphi \subset Q_{2\pi}$, where $Q_d = \{x \in \mathbb{R}^{n-1} : |x_i| < d\}$;

(iii) $\varphi \notin W^{l-1/p}_p(\mathbb{R}^{n-1})$.

By Ω we denote a bounded domain in $\mathbb{R}^n$ such that

$$\Omega \cap \{z : x \in Q_{3\pi}, |y| < 1\} = \{z : x \in Q_{3\pi},\ \varphi(x) < y < 1\}\,.$$

Further, we assume that $\partial\Omega$ is a surface of the class C^∞ in the exterior of the set $\{z : x \in Q_{2\pi}, y = \varphi(x)\}$.

By Theorems 14.6.1–14.6.3 the problems (14.6.5), (14.6.6) for this Ω have no solutions in $W_p^l(\Omega)$.

Remark 14.6.1. Suppose that, for any point $O \in \partial\Omega$, there exists a neighborhood U such that $U \cap \Omega$ is C^l-diffeomorphic to the domain

$$\{(x, y) : y > \varphi(x_1, \ldots, x_{n-s})\}, \quad 2 \le s \le n-1,$$

i.e. 'the dimensions of singularities of $\partial\Omega$ are not less than $s-1$'. Then all properties of domains with boundaries in $M_p^{l-1/p}(\delta)$ remain valid after the change from $n-1$ to $n-s$. This follows from the definition of the class $M_p^{l-1/p}(\delta)$, from Theorem 8.7.1 and from the fact that, if ψ is defined in $\mathbb{R}^n$ and ψ depends on $n-s+1$ variables only, the norms

$$\|\psi; \mathbb{R}^n\|_{MW_p^l}, \quad \|\psi; \mathbb{R}^{n-s+1}\|_{MW_p^l}$$

are equivalent (see Proposition 2.7.2).

14.7 Local Characterization of $M_p^{l-1/p}(\delta)$

In this section we prove Theorem 14.6.4 which gives a local characterization of surfaces in the class $M_p^{l-1/p}(\delta)$. By $\mathcal{B}_\epsilon$ we mean the ball in $\mathbb{R}^{n-1}$ of radius ϵ centered at the origin.

14.7.1 Estimates for a Cutoff Function

Let η be an even function in $C_0^\infty(-1, 1)$ with $\eta = 1$ on $(-1/2, 1/2)$. For $z \in \mathbb{R}^n$ we define

$$\eta_\epsilon(z) = \begin{cases} \eta(|z|/\epsilon), & \text{if } p(l-1) < n, \\ \eta(\log \epsilon / \log |z|), & \text{if } p(l-1) = n. \end{cases}$$

Clearly, supp $\eta_\epsilon \subset \mathcal{B}_\varepsilon$ and

$$|\nabla_j \eta_\epsilon(z)| \le \begin{cases} c\,\epsilon^{-j}, & \text{if } p(l-1) < n, \\ c\,|\log |z||^{-1} |z|^{-j}, & \text{if } p(l-1) = n. \end{cases}$$

Lemma 14.7.1. *Let* $p(l-1)=n$. *Then*

$$\int_{\mathcal{B}_\epsilon} \frac{|\nabla_j \eta_\epsilon(x) - \nabla_j \eta_\epsilon(y)|^p}{|x-y|^{n-2+p}} dy \le c_j |\log|x||^{-p} |x|^{1-p-pj}, \tag{14.7.1}$$

where $x \in \mathcal{B}_\epsilon, j = 0, 1, \dots$.

Proof. Since

$$D^\alpha \eta_\epsilon(x) = |x|^{-|\alpha|} \sum_{k=1}^{|\alpha|} \sigma_k(\log\epsilon/\log|x|)(\log\epsilon)^{-k},$$

where $\sigma_k \in C_0^\infty(-1,1)$, it follows that

$$|\nabla_j \eta_\epsilon(x) - \nabla_j \eta_\epsilon(y)| \le \begin{cases} c_j |\log\epsilon|^{-1} |x-y||x|^{-j-1} \\ \quad \text{if } |x|/2 \le |y| \le 2|x|, \\ \\ c_j |\log\epsilon|^{-1} (\max\{|x|,|y|\})^{-j} \\ \text{if } j>0 \text{ and } |y|<|x|/2 \text{ or } |x|<|y|/2, \\ \\ c_j|\log\epsilon|^{-1}|\log(|x|/|y|)| \\ \text{if } j=0 \text{ and } |y|<|x|/2 \text{ or } |x|<|y|/2. \end{cases}$$

These inequalities imply that

$$\int_{\mathcal{B}_\epsilon} \frac{|\nabla_j \eta_\epsilon(x) - \nabla_j \eta_\epsilon(y)|^p}{|x-y|^{n-2+p}} dy \le c_j |\log\epsilon|^{-p} |x|^{1-p-pj}, \tag{14.7.2}$$

which is equivalent to (14.7.1) for $x \in \mathcal{B}_\epsilon \backslash \mathcal{B}_{\epsilon^3}$. Let $x \in \mathcal{B}_{\epsilon^3}$. We then have

$$\int_{\mathcal{B}_\epsilon \backslash \mathcal{B}_{\epsilon^2}} \frac{|\delta_{0,j} - \nabla_j \eta_\epsilon(y)|^p}{|y|^{n-2+p}} dy \le c \int_{\mathcal{B}_\epsilon} \frac{|\nabla_j \eta_\epsilon(x) - \nabla_j \eta_\epsilon(y)|^p}{|x-y|^{n-2+p}} dy,$$

where $\delta_{0,j}=1$ for $j=0$ and $\delta_{0,j}=0$ for $j>0$. Consequently,

$$\int_{\mathcal{B}_\epsilon \backslash \mathcal{B}_{\epsilon^2}} \frac{|\delta_{0,j} - \nabla_j \eta_\epsilon(y)|^p}{|y|^{n-2+p}} dy \le c|\log\epsilon|^{-p} |x|^{1-p-pj}.$$

Setting $|x|=\epsilon^3$ in the last inequality and observing that

$$t^{3(p+pj-1)} |\log t|^p$$

increases near $t=0$, we obtain (14.7.1). □

14.7.2 Description of $M_p^{l-1/p}(\delta)$ Involving a Cutoff Function

The aim of this subsection is to prove the following assertion on a local characterization of $M_p^{l-1/p}(\delta)$ involving a cutoff function. Hereafter, without loss of generality, we assume that $\varphi(0) = 0$.

Lemma 14.7.2. *A surface $\partial\Omega$ belongs to the class $M_p^{l-1/p}(\delta)$ if and only if for any $O \in \partial\Omega$ there exists a neighborhood U and a special Lipschitz domain $G = \{z = (x, y) : x \in \mathbb{R}^{n-1}, y > \varphi(x)\}$ such that $U \cap \Omega = U \cap G$ and*

$$\limsup_{\epsilon \to 0} \|\nabla(\eta_\epsilon \varphi); \mathbb{R}^{n-1}\|_{MW_p^{l-1-1/p}} \le c\,\delta, \tag{14.7.3}$$

where c is a constant which depends on l, p, n, and η_ϵ is the function introduced in the previous subsection.

Proof. Clearly, (14.7.3) implies that $\partial\Omega \in M_p^{l-1/p}(\delta)$. In order to obtain the converse assertion, it is sufficient to derive the estimate

$$\|\nabla(\eta_\epsilon \varphi); \mathbb{R}^{n-1}\|_{MW_p^{l-1-1/p}} \le c\, \|\nabla\varphi; \mathbb{R}^{n-1}\|_{MW_p^{l-1-1/p}}. \tag{14.7.4}$$

Let $\Phi = \mathcal{T}\varphi$ be an extension of φ, defined in Sect. 8.7.2. By Theorem 8.7.1,

$$\|\nabla(\eta_\epsilon \varphi); \mathbb{R}^{n-1}\|_{MW_p^{l-1-1/p}} \le c\, \|\nabla(\eta_\epsilon \Phi); \mathbb{R}^n_+\|_{MW_p^{l-1}}. \tag{14.7.5}$$

For any function $u \in W_p^{l-1}(\mathbb{R}^n_+)$, we have

$$\|u\nabla(\eta_\epsilon \Phi); \mathbb{R}^n_+\|_{W_p^{l-1}} \le c\Big(\sum_{j=0}^{l-1} \|\Phi\, |\nabla_{j+1}\eta_\epsilon|\, |\nabla_{l-1-j}u|; \mathbb{R}^n_+\|_{L_p}$$

$$+\sum_{j=0}^{l-1}\sum_{k=0}^{j} \||\nabla_{j+1-k}\Phi|\, |\nabla_k \eta_\epsilon|\, |\nabla_{l-1-j}u|; \mathbb{R}^n_+\|_{L_p}\Big). \tag{14.7.6}$$

Let $p(l-1) < n$. The first sum on the right-hand side of (14.7.6) is dominated by

$$c\, \|\nabla\Phi; \mathbb{R}^n_+\|_{L_\infty} \sum_{j=0}^{l-1} \|r^{-j}\nabla_{l-1-j}u; \mathbb{R}^n_+\|_{L_p}$$

and the second one is not greater than

$$\sum_{j=0}^{l-1}\sum_{k=1}^{j} \|\nabla_{j+1-k}\Phi; \mathbb{R}^n_+\|_{M(W_p^{j-k} \to L_p)} \|r^{-k}\nabla_{l-1-j}u; \mathbb{R}^n_+\|_{W_p^{j-k}}.$$

From Corollary 2.4.1 and the inclusion $MW_p^s(\mathbb{R}^n_+) \subset MW_p^t(\mathbb{R}^n_+), s > t$, it follows that

$$\|\nabla_{j+1-k}\Phi; \mathbb{R}^n_+\|_{M(W_p^{j-k}\to L_p)} \le c\,\|\nabla\Phi; \mathbb{R}^n_+\|_{MW_p^{j-k}}$$
$$\le c\,\|\nabla\Phi; \mathbb{R}^n_+\|_{MW_p^{l-1}}. \qquad (14.7.7)$$

Moreover, by Hardy's inequality

$$\|r^{-k}\nabla_{l-1-j}u; \mathbb{R}^n_+\|_{W_p^{j-k}} \le c\,\|u; \mathbb{R}^n_+\|_{W_p^{l-1}}$$

with $p(l-1) < n$ and $l-1 \ge j \ge k$, we obtain

$$\|\nabla(\eta_\epsilon\Phi); \mathbb{R}^n_+\|_{MW_p^{l-1}} \le c\,\|\nabla\Phi; \mathbb{R}^n_+\|_{MW_p^{l-1}}. \qquad (14.7.8)$$

This, together with Theorem 8.7.2 and (14.7.5), implies (14.7.4) for $p(l-1)<n$.

Now let $p(l-1) = n$. The first sum on the right-hand side of (14.7.6) does not exceed

$$c\,\|\nabla\Phi; \mathbb{R}^n_+\|_{L_\infty}\Big(|\log\epsilon|^{-1}\sum_{j=0}^{l-2}\|r^{-j}\nabla_{l-1-j}u; \mathbb{R}^n_+\|_{L_p}$$
$$+\|r^{l-1}(\log r)^{-1}u; \mathcal{B}_{1/2}\cap\mathbb{R}^n_+\|_{L_p}\Big),$$

and the second one is not greater than

$$\sum_{j=0}^{l-1}\sum_{k=0}^{j}\|\nabla_{j+1-k}\Phi; \mathbb{R}^n_+\|_{M(W_p^{j-k}\to L_p)}\|r^{-k}(\log r)^{-1}\eta_{1/2}\nabla_{l-1-j}u; \mathbb{R}^n_+\|_{W_p^{j-k}}.$$

Using Hardy's inequality

$$\|r^{-k}(\log r)^{-1}v; \mathbb{R}^n_+\|_{W_p^{j-k}} \le c\,\|v; \mathbb{R}^n_+\|_{W_p^{j}},$$

where v is a function supported by $\mathcal{B}_{1/2}$, we obtain

$$\|r^{-k}(\log r)^{-1}\eta_{1/2}\nabla_{l-1-j}u; \mathbb{R}^n_+\|_{W_p^{j-k}} \le c\,\|u; \mathbb{R}^n_+\|_{W_p^{l-1}},$$

which, together with (14.7.7), gives (14.7.8). This, in combination with Theorem 8.7.2 and (14.7.5), leads to (14.7.4) for $p(l-1) = n$. The result follows. □

14.7.3 Estimate for s_1

Clearly, if

$$\sup_{e\subset\mathbb{R}^{n-1}}\frac{\|D_{p,l-1/p}\varphi; e\|_{L_p}}{[C_{p,l-1-1/p}(e)]^{1/p}} + \|\nabla\varphi, \mathbb{R}^{n-1}\|_{L_\infty} \le c\,\delta,$$

then (14.6.8) holds. So we must prove the sufficiency of (14.6.8).

According to Theorem 4.1.1, the condition (14.7.3) means that

$$\sup_e \frac{\|D_{p,l-1/p}(\eta_\epsilon\varphi); e\|_{L_p}}{[C_{p,l-1-1/p}(e)]^{1/p}} + \|\nabla(\eta_\epsilon\varphi); \mathcal{B}_\epsilon\|_{L_\infty} < c\,\delta \tag{14.7.9}$$

for sufficiently small $\epsilon > 0$. Hereafter, e is a compact set in $\mathbb{R}^{n-1}$ with $d(e) < 1$. Since $\varphi(0) = 0$, we derive from (14.7.1) that

$$\|\nabla(\eta_\epsilon\varphi); \mathcal{B}_\epsilon\|_{L_\infty} \le c\,\|\nabla\varphi; \mathcal{B}_\epsilon\|_{L_\infty}.$$

The first term in (14.7.9) is majorized by

$$\sup_{e\subset\mathbb{R}^{n-1}} \Big(\frac{\|D_{p,l-1/p}(\eta_\epsilon\varphi); e\backslash\mathcal{B}_\epsilon\|_{L_p}}{[C_{p,l-1-1/p}(e\backslash\mathcal{B}_\epsilon)]^{1/p}} + \frac{\|D_{p,l-1/p}(\eta_\epsilon\varphi); e\cap\mathcal{B}_\epsilon\|_{L_p}}{[C_{p,l-1-1/p}(e\cap\mathcal{B}_\epsilon)]^{1/p}}\Big). \tag{14.7.10}$$

(If either $e\backslash\mathcal{B}_\epsilon$ or $e\cap\mathcal{B}_\epsilon$ has zero capacity, then the corresponding term is equal to zero). Consequently, the supremum in (14.7.9) is not greater than $s_1 + s_2 + s_3$, where

$$s_1 = \sup_{e\subset\mathbb{R}^{n-1}\backslash\mathcal{B}_\epsilon} \frac{\|D_{p,l-1/p}(\eta_\epsilon\varphi); e\|_{L_p}}{[C_{p,l-1-1/p}(e)]^{1/p}},$$

$$s_2 = \sup_{e\subset\mathcal{B}_\epsilon} \frac{\Big(\displaystyle\int_e dx \int_{\mathbb{R}^{n-1}\backslash\mathcal{B}_\epsilon} \frac{|\nabla_{l-1}(\eta_\epsilon\varphi)(x) - \nabla_{l-1}(\eta_\epsilon\varphi)(y)|^p}{|x-y|^{n-2+p}}dy\Big)^{1/p}}{[C_{p,l-1-1/p}(e)]^{1/p}},$$

$$s_3 = \sup_{e\subset\mathcal{B}_\epsilon} \frac{\|D_{l-1/p}(\eta_\epsilon\varphi; \mathcal{B}_\epsilon); e\|_{L_p}}{[C_{p,l-1-1/p}(e)]^{1/p}}.$$

The goal of this subsection is to give an estimate for s_1.

Lemma 14.7.3. *If* (14.6.8) *holds, then* $s_1 \le c\,\delta$ *for sufficiently small* ϵ.

Proof. We have

$$s_1^p = \sup_{e\subset\mathbb{R}^{n-1}\backslash\mathcal{B}_\epsilon} \frac{\displaystyle\int_{\mathcal{B}_\epsilon} |\nabla_{l-1}(\eta_\epsilon\varphi)(y)|^p dy \int_e |x-y|^{2-n-p}dx}{C_{p,l-1-1/p}(e)}.$$

Let $q = (n-1)/(p(l-1)-1)$ if $p(l-1) < n$, and let $q \in [1,\infty)$ if $p(l-1) = n$. Since $y \in \operatorname{supp}\eta_\epsilon \subset \mathcal{B}_{\epsilon/2}$, it follows that

$$\begin{aligned}\int_e |x-y|^{2-n-p}dx &\le (\mathrm{mes}_{n-1}e)^{1-1/q}\Big(\int_{\mathbb{R}^{n-1}\backslash\mathcal{B}_\epsilon} |x-y|^{(2-n-p)q}dx\Big)^{1/q}\\ &\le c\,(\mathrm{mes}_{n-1}e)^{1-1/q}\epsilon^{2-n-p+(n-1)/q}.\end{aligned} \tag{14.7.11}$$

We use the inequalities

$$C_{p,l-1-1/p}(e) \geq \begin{cases} c\,(\text{mes}_{n-1}e)^{1-1/q} & \text{if } p(l-1) < n, \\ c\,(\log(2^n/\text{mes}_{n-1}e))^{1-p} & \text{if } p(l-1) = n. \end{cases} \tag{14.7.12}$$

Then, for $p(l-1) < n$,

$$\int_e |x-y|^{2-n-p}dx \leq c\,\epsilon^{p(l-2)+1-n}(\text{mes}_{n-1}e)^{1-1/q}.$$

In the case $p(l-1) = n$, we have

$$(\log(2^n/\text{mes}_{n-1}e))^{p-1}\int_e |x-y|^{2-n-p}dx$$
$$\leq c\,(\text{mes}_{n-1}e)^{1-1/q}(\log(2^n/\text{mes}_{n-1}e))^{p-1}\epsilon^{2-n-p+(n-1)/q}. \tag{14.7.13}$$

If $\text{mes}_{n-1}e \leq \epsilon^{n-1}$, then the right-hand side is dominated by $c\,\epsilon^{1-p}|\log\epsilon|^{p-1}$. If $\text{mes}_{n-1}e > \epsilon^{n-1}$, then, setting $q = 1$ in (14.7.13), we obtain the same majorant $c\,\epsilon^{1-p}|\log\epsilon|^{p-1}$. Thus

$$s_1^p \leq \begin{cases} c\,\epsilon^{p(l-2)+1-n}\displaystyle\int_{\mathcal{B}_\epsilon} |\nabla_{l-1}(\eta_\epsilon\varphi)|^p dy & \text{if } p(l-1) < n, \\ c\,\epsilon^{1-p}|\log\epsilon|^{p-1}\displaystyle\int_{\mathcal{B}_\epsilon} |\nabla_{l-1}(\eta_\epsilon\varphi)(y)|^p dy & \text{if } p(l-1) = n. \end{cases}$$

In the case $p(l-1) < n$, this implies that

$$s_1^p \leq c\epsilon^{p(l-2)+1-n}\Big(\epsilon^{(1-l)p}\int_{\mathcal{B}_\epsilon} |\varphi|^p dy + \sum_{j=0}^{l-2}\epsilon^{-jp}\int_{\mathcal{B}_\epsilon} |\nabla_{l-1-j}\varphi|^p dy\Big).$$

We introduce the notation

$$\langle v;\mathcal{B}_\epsilon\rangle_{p,l-1-1/p} = \|D_{l-1-1/p}(v;\mathcal{B}_\epsilon);\mathcal{B}_\epsilon\|_{L_p}$$

and use the inequality

$$\int_{\mathcal{B}_\epsilon} |\nabla_{l-2-j}v|^p dy$$
$$\leq c\,\Big(\epsilon^{p(j+1)-1}\langle v;\mathcal{B}_\epsilon\rangle^p_{p,l-1-1/p} + \epsilon^{p(j+2-l)}\int_{\mathcal{B}_\epsilon} |v|^p dy\Big). \tag{14.7.14}$$

Then

$$s_1^p \leq c\Big(\epsilon^{-p+1-n}\int_{\mathcal{B}_\epsilon} |\varphi|^p dy + \epsilon^{1-n}\int_{\mathcal{B}_\epsilon} |\nabla\varphi|^p dy + \epsilon^{p(l-1)-n}\langle\varphi;\mathcal{B}_\epsilon\rangle^p_{p,l-1/p}\Big)$$
$$\leq c\,\big(\epsilon^{p(l-1)-n}\langle\varphi;\mathcal{B}_\epsilon\rangle^p_{p,l-1/p} + \|\nabla\varphi;\mathcal{B}_\epsilon\|^p_{L_\infty}\big),$$

which, together with (14.6.8), gives $s_1 \leq c\delta$.

Now let $p(l-1)=n$. In the case $l=2$, we have

$$s_1^p \le c\,\epsilon^{1-p}|\log\epsilon|^{p-1}\|\nabla\varphi;\mathcal{B}_\epsilon\|_{L_\infty}^p\Big(\int_{\mathcal{B}_\epsilon}|\eta_\epsilon|^p dy+\int_{\mathcal{B}_\epsilon}|y|^p|\nabla\eta_\epsilon|^p dy\Big).$$

Since

$$\int_{\mathcal{B}_\epsilon}|\eta_\epsilon|^p dy \le c\int_{\mathcal{B}_\epsilon}|y|^p|\nabla\eta_\epsilon|^p dy \le c\,\epsilon^{p-1}|\log\epsilon|^{-p},$$

we obtain

$$s_1^p \le c\,|\log\epsilon|^{-1}\|\nabla\varphi;\mathcal{B}_\epsilon\|_{L_\infty}^p.$$

Suppose that $l>2$ and $p(l-1)=n$. We write

$$s_1^p \le c\,\epsilon^{1-p}|\log\epsilon|^{p-1}\Big(|\log\epsilon|^{-p}\int_{\mathcal{B}_\epsilon}|y|^{(1-l)p}|\varphi|^p dy$$

$$+\sum_{j=1}^{l-2}|\log\epsilon|^{-p}\int_{\mathcal{B}_\epsilon}|y|^{-jp}|\nabla_{l-1-j}\varphi|^p dy+\int_{\mathcal{B}_\epsilon}|\eta_\epsilon\nabla_{l-1}\varphi|^p dy\Big). \qquad (14.7.15)$$

The first term in brackets does not exceed

$$c\,\epsilon^{p-1}|\log\epsilon|^{-p}\|\nabla\varphi;\mathcal{B}_\epsilon\|_{L_\infty}^p. \qquad (14.7.16)$$

Using the inequality

$$\int_{\mathcal{B}_\epsilon}|y|^{-jp}|\nabla_{l-2-j}v|^p dy \le c\Big(\int_{\mathcal{B}_\epsilon}|\nabla_{l-2}v|^p dy+\epsilon^{p(2-l)}\int_{\mathcal{B}_\epsilon}|v|^p dy\Big)$$

with $v=\partial\varphi/\partial y_i$, we conclude that the sum with respect to j in (14.7.15) is majorized by

$$c\,\epsilon^{p-1}|\log\epsilon|^{-p}\Big(\epsilon^{1-p}\int_{\mathcal{B}_\epsilon}|\nabla_{l-1}\varphi|^p dy+\|\nabla\varphi;\mathcal{B}_\epsilon\|_{L_\infty}^p\Big). \qquad (14.7.17)$$

We apply (14.7.14) with $j=0$ to the vector function $v=\nabla\varphi$. Then (14.7.17) is not greater than

$$c\,\epsilon^{p-1}|\log\epsilon|^{-p}\big(\langle\varphi:\mathcal{B}_\epsilon\rangle_{p,l-1/p}^p+\|\nabla\varphi;\mathcal{B}_\epsilon\|_{L_\infty}^p\big). \qquad (14.7.18)$$

Now we turn to an estimate for the last integral in (14.7.15). The inequality

$$\|w;\mathcal{B}_\epsilon\|_{L_p} \le c\,\epsilon^{1-1/p}\langle w;\mathcal{B}_\epsilon\rangle_{p,1-1/p}$$

holds for all w defined on $\mathcal{B}_\epsilon$ and vanishing outside $\mathcal{B}_{(1-c)\epsilon}$, $c\in(0,1)$. Hence,

$$\epsilon^{1-p}\int_{\mathcal{B}_\epsilon}|\eta_\epsilon\nabla_{l-1}\varphi|^p dy \le c\,\langle\nabla_{l-1}(\eta_\epsilon\varphi);\mathcal{B}_\epsilon\rangle_{p,1-1/p}^p$$

$$\leq c\Big(\langle\nabla_{l-1}\varphi;\mathcal{B}_\epsilon\rangle^p_{p,1-1/p}+\int_{\mathcal{B}_\epsilon}|\nabla_{l-1}\varphi(x)|^p dx\int\frac{|\eta_\epsilon(x)-\eta_\epsilon(y)|^p}{|x-y|^{n-2+p}}dy\Big). \quad (14.7.19)$$

According to Lemma 14.7.1,

$$\int\frac{|\eta_\epsilon(x)-\eta_\epsilon(y)|^p}{|x-y|^{n-2+p}}dy\leq c\,|\log\epsilon|^{-p}|x|^{1-p}.$$

Therefore,

$$\langle\nabla_{l-1}(\eta_\epsilon\varphi);\mathcal{B}_\epsilon\rangle^p_{p,1-1/p}$$
$$\leq c\Big(\langle\nabla_{l-1}\varphi;\mathcal{B}_\epsilon\rangle^p_{p,1-1/p}+|\log\epsilon|^{-p}\int_{\mathcal{B}_\epsilon}|\nabla_{l-1}\varphi(y)|^p|y|^{1-p}dy\Big).$$

Since $p(l-1)=n$ and $l>2$, we see that $p<n$ and Hardy's inequality

$$\int_{\mathcal{B}_\epsilon}|v|^p|y|^{1-p}dy\leq c\Big(\langle v;\mathcal{B}_\epsilon\rangle^p_{p,1-1/p}+\epsilon^{1-p}\int_{\mathcal{B}_\epsilon}|v|^p dy\Big)$$

holds. Setting here $\nabla_{l-1}\varphi$ as v, we get

$$\langle\nabla_{l-1}(\eta_\epsilon\varphi);\mathcal{B}_\epsilon\rangle^p_{p,1-1/p}$$
$$\leq c\Big(\langle\varphi;\mathcal{B}_\epsilon\rangle^p_{p,l-1/p}+\epsilon^{1-p}|\log\epsilon|^{-p}\int_{\mathcal{B}_\epsilon}|\nabla_{l-1}\varphi(y)|^p dy\Big).$$

By (14.7.14), the last integral does not exceed

$$\epsilon^{p-1}(\langle\varphi;\mathcal{B}_\epsilon\rangle^p_{p,l-1/p}+\|\nabla\varphi;\mathcal{B}_\epsilon\|^p_{L_\infty})$$

and, therefore,

$$\langle\nabla_{l-1}(\eta_\epsilon\varphi);\mathcal{B}_\epsilon\rangle^p_{p,1-1/p}\leq c\left(\langle\varphi;\mathcal{B}_\epsilon\rangle^p_{p,l-1/p}+|\log\epsilon|^{-p}\|\varphi;\mathcal{B}_\epsilon\|^p_{L_\infty}\right). \quad (14.7.20)$$

Consequently,

$$\int_{\mathcal{B}_\epsilon}|\eta_\epsilon\nabla_{l-1}\varphi|^p dy\leq c\,\epsilon^{p-1}\left(\langle\varphi;\mathcal{B}_\epsilon\rangle^p_{p,l-1/p}+|\log\epsilon|^{-p}\|\nabla\varphi;\mathcal{B}_\epsilon\|^p_{L_\infty}\right). \quad (14.7.21)$$

Substituting (14.7.14), (14.7.18) and (14.7.21) into (14.7.15), we derive

$$s_1^p\leq c\left(|\log\epsilon|^{p-1}\langle\varphi;\mathcal{B}_\epsilon\rangle^p_{p,l-1/p}+\|\nabla\varphi;\mathcal{B}_\epsilon\|^p_{L_\infty}\right)$$

which, together with (14.6.8), gives $s_1\leq c\delta$.

14.7.4 Estimate for s_2

Lemma 14.7.4. *If* (14.6.8) *holds, then* $s_2\leq c\,\delta$ *for sufficiently small* ϵ.

Proof. Clearly,

$$s_2^p = \sup_{e \subset \mathcal{B}_\epsilon} \frac{\displaystyle\int_e |\nabla_{l-1}(\eta_\epsilon \varphi)|^p dx \int_{\mathbb{R}^{n-1}\setminus \mathcal{B}_\epsilon} |x-y|^{2-n-p} dy}{C_{p,l-1-1/p}(e)}.$$

Let $p(l-1) < n$. By q we denote any number sufficiently close to p, such that $q > p$. We have

$$s_2^p \le c \sup_{e \subset \mathcal{B}_\epsilon} \sum_{j=0}^{l-1} \epsilon^{-(j+1)p+1} \frac{\displaystyle\int_e |\nabla_{l-1-j}\varphi|^p dx}{C_{p,l-1-1/p}(e)}$$

$$\le c \sum_{j=0}^{l-1} \epsilon^{-(j+1)p+1} \sup_{e \subset \mathcal{B}_\epsilon} (\mathrm{mes}_{n-1} e)^{1-p/q} \frac{\displaystyle\Big(\int_e |\nabla_{l-1-j}\varphi|^q dx\Big)^{p/q}}{C_{p,l-1-1/p}(e)}. \tag{14.7.22}$$

From Lemma 1.3.1 and the equality

$$W_p^{l-1}(\mathbb{R}^n)|_{\mathbb{R}^{n-1}} = W_p^{l-1-1/p}(\mathbb{R}^{n-1}),$$

it follows that

$$\Big(\int_{\mathbb{R}^{n-1}} |u|^q d\mu\Big)^{p/q} \le c \sup_{x\in \mathcal{B}_\epsilon, \rho \in (0,\epsilon)} \frac{[\mu(\mathcal{B}_\rho(x))]^{p/q}}{C_{p,l-1-1/p}(\mathcal{B}_\rho)} \|u; \mathbb{R}^{n-1}\|^p_{W_p^{l-1-1/p}},$$

where μ is a measure with supp $\mu \subset \mathcal{B}_\epsilon$. Therefore

$$\sup_{e \subset \mathcal{B}_\epsilon} \frac{[\mu(e)]^{p/q}}{C_{p,l-1-1/p}(e)} \le c \sup_{x\in \mathcal{B}_\epsilon, \rho \in (0,\epsilon)} \frac{[\mu(\mathcal{B}_\rho(x))]^{p/q}}{C_{p,l-1-1/p}(\mathcal{B}_\rho)}. \tag{14.7.23}$$

This estimate and (14.7.22) imply that

$$s_2^p \le c \sum_{j=0}^{l-1} \epsilon^{-(j+1)p+1+(n-1)(1-p/q)}$$
$$\times \sup_{x\in \mathcal{B}_\epsilon, \rho \in (0,\epsilon)} \Big(\Big(\int_{\mathcal{B}_\rho(x)} |\nabla_{l-1-j}\varphi|^q dy\Big)^{p/q} \rho^{p(l-1)-n}\Big).$$

Since $-(j+1)p+1+(n-1)(1-p/q) \le (n-1)(1-p/q)+1-p < 0$ for any q sufficiently close to p, we obtain

$$s_2^p \le c \sum_{j=0}^{l-2} \sup_{x\in \mathcal{B}_\epsilon, \rho \in (0,\epsilon)} \Big(\int_{\mathcal{B}_\rho(x)} |\nabla_{l-1-j}\varphi|^q dy \rho^{q(l-j-2)+1-n}\Big)^{p/q} + c\,\|\nabla\varphi; \mathcal{B}_\epsilon\|^p_{L_\infty}.$$

We use the inequality

$$\rho^{l-j-2-(n-1)/q}\Big(\int_{\mathcal{B}_\rho}|\nabla_{l-2-j}v|^q dy\Big)^{1/q}$$
$$\leq c\,\rho^{l-1-1/p-(n-1)/p}\langle v;\mathcal{B}_\rho\rangle_{p,l-1-1/p}+c\Big(\rho^{1-n}\int_{\mathcal{B}_\rho}|v|^p dy\Big)^{1/p}, \quad (14.7.24)$$

$j=0,\cdots,l-2$, which follows by dilation from the continuity of the embedding of $W_p^{l-1-1/p}(\mathcal{B}_1)$ into $W_q^{l-2-j}(\mathcal{B}_1)$. Then

$$s_2^p\leq c\sup_{x\in\mathcal{B}_\epsilon,\rho\in(0,\epsilon)}\frac{\langle\nabla\varphi;\mathcal{B}_\rho(x)\rangle^p_{p,l-1-1/p}}{\rho^{n-1-p(l-1-1/p)}}+c\,\|\nabla\varphi;\mathcal{B}_\epsilon\|^p_{L_\infty}. \quad (14.7.25)$$

Let $p(l-1)=n$. By Hölder's inequality and (14.7.23) we have

$$s_2^p\leq c\,\epsilon^{1-p}\sup_{e\subset\mathcal{B}_\epsilon}(\mathrm{mes}_{n-1}e)^{1-p/q}\frac{\Big(\int_e|\nabla_{l-1}(\eta_\epsilon\varphi)|^q dx\Big)^{p/q}}{C_{p,l-1-1/p}(e)}$$
$$\leq c\,\epsilon^{1-p+(n-1)(1-p/q)}\sup_{x\in\mathcal{B}_\epsilon,\rho\in(0,\epsilon)}|\log\rho|^{p-1}\Big(\int_{\mathcal{B}_\rho(x)}|\nabla_{l-1}(\eta_\epsilon\varphi)|^q dy\Big)^{p/q}.$$

Hence

$$s_2^p\leq c\sup_{x\in\mathcal{B}_\epsilon,\rho\in(0,\epsilon)}\Big[\rho^{1-p+(n-1)(1-p/q)}|\log\rho|^{-1}$$
$$\times\sum_{j=1}^{l-2}\Big(\int_{\mathcal{B}_\rho(x)}|y|^{-jq}|\nabla_{l-1-j}\varphi(y)|^q dy\Big)^{p/q}$$
$$+\rho^{1-p+(n-1)(1-p/q)}|\log\rho|^{-1}\Big(\int_{\mathcal{B}_\rho(x)}|y|^{-(l-1)q}|\varphi(y)|^q dy\Big)^{p/q}$$
$$+\epsilon^{1-p+(n-1)(1-p/q)}|\log\rho|^{p-1}\Big(\int_{\mathcal{B}_\rho(x)}|\eta_\epsilon\nabla_{l-1}\varphi|^q dy\Big)^{p/q}\Big]. \quad (14.7.26)$$

We apply the following variant of Hardy's inequality:

$$\Big(\int_{\mathcal{B}_\rho}|y|^{-jq}|\nabla_{l-2-j}v|^q dy\Big)^{1/q}$$
$$\leq c\,\rho^{1-1/p+(n-1)(1/q-1/p)}\langle v;\mathcal{B}_\rho\rangle_{p,l-1-1/p}$$
$$+c\,\rho^{2-l+(n-1)(1/q-1/p)}\Big(\int_{\mathcal{B}_\rho}|v|^p dy\Big)^{1/p}, \quad j=1,\cdots,l-2,$$

with $\nabla\varphi$ instead of v. Then the first term on the right-hand side of (14.7.26) is dominated by

$$c\,|\log\rho|^{-1}\big(\langle\varphi;\mathcal{B}_\rho(x)\rangle_{p,l-1/p}^p+\|\nabla\varphi;\mathcal{B}_\rho(x)\|_{L_\infty}^p\big)$$

$$\le c\,\big(\langle\varphi;\mathcal{B}_{2\epsilon}\rangle_{p,l-1/p}^p+\|\nabla\varphi;B_{2\epsilon}\|_{L_\infty}^p\big).$$

Obviously, the second term on the right-hand side of (14.7.26) does not exceed

$$c\,\rho^{1-p+(n-1)(1-p/q)}|\log\rho|^{-1}\Big(\int_{\mathcal{B}_\rho(x)}|y|^{-(l-2)q}dy\Big)^{p/q}\|\nabla\varphi;\mathcal{B}_\rho(x)\|_{L_\infty}^p$$
$$\le c\,|\log\rho|^{-1}\|\nabla\varphi;\mathcal{B}_\rho(x)\|_{L_\infty}^p.$$

We turn to an estimate of the third term on the right-hand side of (14.7.26). Let s be a number sufficiently close to q with $s>q$. By Hölder's inequality this term is majorized by

$$\epsilon^{1-p+(n-1)(1-p/q)}\rho^{(n-1)p(1/q-1/s)}|\log\rho|^{p-1}\Big(\int_{\mathcal{B}_\rho(x)}|\eta_\epsilon\nabla_{l-1}\varphi|^s dy\Big)^{p/s}$$
$$\le \epsilon^{1-p+(n-1)(1-p/s)}|\log\epsilon|^{p-1}\Big(\int_{\mathcal{B}_\epsilon}|\eta_\epsilon\nabla_{l-1}\varphi|^s dy\Big)^{p/s}. \tag{14.7.27}$$

The inequality

$$\|w;\mathcal{B}_\epsilon\|_{L_s}\le c\,\epsilon^{(n-1)(1/s-1/p)+1-1/p}\langle w;\mathcal{B}_\epsilon\rangle_{p,1-1/p}$$

holds for all w, defined on $\mathcal{B}_\epsilon$ and vanishing outside $\mathcal{B}_{(1-c)\epsilon}$, $c\in(0,1)$. Hence the right-hand side of (14.7.27) is not greater than

$$c\,|\log\epsilon|^{p-1}\langle\eta_\epsilon\nabla_{l-1}\varphi;\mathcal{B}_\epsilon\rangle_{p,1-1/p}^p.$$

By (14.7.21) this last term does not exceed

$$c\,|\log\epsilon|^{p-1}\big(\langle\varphi;\mathcal{B}_\epsilon\rangle_{p,1-1/p}^p+|\log\epsilon|^{-p}\|\nabla\varphi;\mathcal{B}_\epsilon\|_{L_\infty}^p\big).$$

Using the estimates which were obtained for three terms in (14.7.26), we arrive at

$$s_2^p\le c\Big(\sup_{x\in\mathcal{B}_\epsilon,\rho\in(0,\epsilon)}|\log\rho|^{p-1}\langle\varphi;\mathcal{B}_\rho(x)\rangle_{p,l-1/p}^p+\|\nabla\varphi;\mathcal{B}_{2\epsilon}\|_{L_\infty}^p\Big).$$

The result follows from (14.6.8). □

14.7.5 Estimate for s_3

Lemma 14.7.5. *If* (14.6.8) *holds, then* $s_3\le c\,\delta$ *for sufficiently small* ϵ.

Proof. We have

$$\|D_{l-1/p}(\eta_\epsilon\varphi;\mathcal{B}_\epsilon);e\|^p_{L_p} \le c\Big(\sum_{\substack{|\alpha|+|\beta|=l-1\\ |\alpha|>0}} \|D_{1-1/p}(D^\alpha\eta_\epsilon D^\beta\varphi;\mathcal{B}_\epsilon);e\|^p_{L_p}$$

$$+\|D_{l-1/p}(\varphi;\mathcal{B}_\epsilon);e\|^p_{L_\rho} + \int_e |\nabla_{l-1}\varphi(x)|^p dx \int_{\mathcal{B}_\epsilon} \frac{|\eta_\epsilon(x)-\eta_\epsilon(y)|^p}{|x-y|^{n-2+p}}dy\Big). \quad (14.7.28)$$

Clearly,

$$\|D_{1-1/p}(D^\alpha\eta_\epsilon D^\beta\varphi;\mathcal{B}_\epsilon);e\|^p_{L_p}$$

$$\le c\Big(\int_e |D^\beta\varphi(x)|^p dx \int_{\mathcal{B}_\epsilon} \frac{|D^\alpha\eta_\epsilon(x)-D^\alpha\eta_\epsilon(y)|^p}{|x-y|^{n-2+p}}dy$$

$$+\int_{\mathcal{B}_\epsilon} |D^\alpha\eta_\epsilon(y)|^p dy \int_e \frac{|D^\beta\varphi(x)-D^\beta\varphi(y)|^p}{|x-y|^{n-2+p}}dx\Big). \quad (14.7.29)$$

Let $p(l-1)<n$. Using (14.7.2), we obtain

$$s_3^p \le c \sup_{e\subset\mathcal{B}_\epsilon} \sum_{j=0}^{l-1} \epsilon^{-(j+1)p+1} \frac{\int_e |\nabla_{l-1-j}\varphi(x)|^p dx}{C_{p,l-1-1/p}(e)}$$

$$+c\sup_{e\subset\mathcal{B}_\epsilon}\sum_{j=0}^{l-1}\epsilon^{-jp}\frac{\int_e dx\int_{\mathcal{B}_\epsilon}\frac{|\nabla_{l-1-j}\varphi(x)-\nabla_{l-1-j}\varphi(y)|^p}{|x-y|^{n-2+p}}dy}{C_{p,l-1-1/p}(e)}. \quad (14.7.30)$$

The first supremum is bounded by the right-hand side of (14.7.25) (see the beginning of the proof of Lemma 14.7.4).

Let q denote a number sufficiently close to p and such that $q>p$. By Hölder's inequality and (14.7.23), the second supremum on the right-hand side of (14.7.30) does not exceed

$$c \sup_{\xi\in\mathcal{B}_\epsilon,\rho\in(0,\epsilon)} \sum_{j=1}^{l-1} \epsilon^{-jp+n(1-p/q)}\rho^{p(l-1)-n}$$

$$\times\Big[\epsilon^{(p-q)/p}\int_{\mathcal{B}_\rho(\xi)}\Big(\int_{\mathcal{B}_\epsilon}\frac{|\nabla_{l-1-j}\varphi(x)-\nabla_{l-1-j}\varphi(y)|^p}{|x-y|^{n-2+p}}dy\Big)^{q/p}dx\Big]^{p/q}$$

$$+c\sup_{e\subset\mathcal{B}_\epsilon}\frac{\|D_{l-1/p}(\varphi;\mathcal{B}_\epsilon);e\|^p_{L_p}}{C_{p,l-1-1/p}(e)}. \quad (14.7.31)$$

We note that for $j=1,\cdots,l-2$

$$\epsilon^{(p-q)/p}\int_{\mathcal{B}_\rho(\xi)}dx\Big(\int_{\mathcal{B}_\epsilon}\frac{|\nabla_{l-1-j}\varphi(x)-\nabla_{l-1-j}\varphi(y)|^p}{|x-y|^{n-2+p}}dy\Big)^{q/p}$$

$$\le \epsilon^{(p-q)/p} \int_{\mathcal{B}_\rho(\xi)} dx \int_{\mathcal{B}_\epsilon} \frac{|\nabla_{l-1-j}\varphi(x) - \nabla_{l-1-j}\varphi(y)|^q}{|x-y|^{n-2+q}} dy \Big(\int_{\mathcal{B}_\epsilon} \frac{dy}{|x-y|^{n-2}} \Big)^{(q-p)/q}$$

$$\le c \Big(\int_{\mathcal{B}_\rho(\xi)} dx \int_{\mathcal{B}_{2\rho}(\xi)} \frac{|\nabla_{l-1-j}\varphi(x) - \nabla_{l-1-j}\varphi(y)|^q}{|x-y|^{n-2+q}} dy$$

$$+ \int_{\mathcal{B}_\epsilon \backslash \mathcal{B}_{2\rho}(\xi)} \int_{\mathcal{B}_\rho(\xi)} \frac{|\nabla_{l-1-j}\varphi(x) - \nabla_{l-1-j}\varphi(y)|^q}{|x-y|^{n-2+q}} dx\, dy \Big). \qquad (14.7.32)$$

The first term on the right-hand side of (14.7.32) is dominated by

$$c \, \langle \nabla\varphi; \mathcal{B}_{2\rho}(\xi) \rangle^q_{q,l-1-j-1/q}$$

which does not exceed

$$c \left(\rho^{(\frac{1}{q}-\frac{1}{p})n+j} \langle \nabla\varphi; \mathcal{B}_{2\rho}(\xi) \rangle_{p,l-1-1/p} + \rho^{-l+1+j+\frac{n}{q}} \| \nabla\varphi; \mathcal{B}_{2\rho}(\xi) \|_{L_\infty} \right)^q. \quad (14.7.33)$$

Since $|x-y| > \rho$ in the second term on the right-hand side of (14.7.32), this term is majorized by

$$c \int_{\mathcal{B}_{2\epsilon} \backslash \mathcal{B}_\rho} \frac{dz}{|z|^{n-2+q}} \int_{\mathcal{B}_\rho(\xi)} \left(|\nabla_{l-1-j}\varphi(x+z)|^q + |\nabla_{l-1-j}\varphi(x)|^q \right) dx$$

$$\le c \int_{\mathcal{B}_{2\epsilon} \backslash \mathcal{B}_\rho} \Big(\frac{\|\nabla_{l-1-j}; \mathcal{B}_\rho(\xi+z)\|^q_{L_q}}{|z|^{n-2+q}} + c\,\rho^{1-q} \|\nabla_{l-1-j}; \mathcal{B}_\rho(\xi+z)\|^q_{L_q} \Big) dz$$

which does not exceed

$$c \sup_{\xi \in \mathcal{B}_{3\epsilon}} \left(\rho^{q[n(1/q-1/p)+j]} \langle \varphi; \mathcal{B}_\rho(\xi) \rangle^q_{p,l-1/p} + \rho^{(-l+1+j)q+n} \| \nabla\varphi; \mathcal{B}_\rho(\xi) \|^q_{L_\infty} \right).$$

Note that $n(1/q - 1/p) + j > 0$ for $j = 1, \ldots, l-2$, because q is close to p. Hence, the terms in the second sum in (14.7.30) with $j = 1, \ldots, l-2$ are estimated by

$$c \sup_{\xi \in \mathcal{B}_{3\epsilon}, \rho \in (0,\epsilon)} \epsilon^{-jp+n(1-p/q)} \big(\rho^{q[n(1/q-1/p)+j]} \rho^{p(l-1)-n} \langle \varphi; \mathcal{B}_\rho(\xi) \rangle^p_{p,l-1/p}$$

$$+ \rho^{jp+n(p/q-1)} \| \nabla\varphi; \mathcal{B}_\rho(\xi) \|^p_{L_\infty} \big)$$

$$\le c \sup_{\xi \in \mathcal{B}_{3\epsilon}, \rho \in (0,\epsilon)} \left(\rho^{p(l-1)-n} \langle \varphi; \mathcal{B}_\rho(\xi) \rangle^p_{p,l-1/p} + \| \nabla\varphi; \mathcal{B}_\rho(\xi) \|^p_{L_\infty} \right).$$

The term with $j = l-1$ in (14.7.31) has the majorant

$$\epsilon^{-(l-1)p+(n-1)(1-p/q)} \rho^{p(l-1)-n} \| \nabla\varphi; \mathcal{B}_{2\epsilon} \|^p_{L_\infty} \Big[\int_{\mathcal{B}_\rho(\xi)} dx \Big(\int_{\mathcal{B}_\epsilon} \frac{dy}{|x-y|^{n-2}} \Big)^{q/p} \Big]^{p/q}$$

$$\le c\,\epsilon^{1-(l-1)p+(n-1)(1-p/q)} \rho^{p(l-1)-n+(n-1)p/q} \| \nabla\varphi; \mathcal{B}_{2\epsilon} \|^p_{L_\infty} \le c \, \| \nabla\varphi; B_{2\epsilon} \|^p_{L_\infty}.$$

Finally, the second supremum in (14.7.30) is dominated by

$$c \sup_{\xi\in\mathcal{B}_\epsilon,\rho\in(0,\epsilon)} ((\rho^{p(l-1)-n}\langle\varphi;\mathcal{B}_{2\rho}(\xi)\rangle^p_{p,l-1/p} + \|\nabla\varphi;\mathcal{B}_{2\rho}(\xi)\|^p_{L_\infty})$$
$$+ c \sup_{e\subset\mathcal{B}_\epsilon} \frac{\|D_{l-1/p}(\varphi;\mathcal{B}_\epsilon);e\|^p_{L_p}}{C_{p,l-1-1/p}(e)}.$$

Taking into account the estimate for the first supremum in (14.7.30), obtained previously, we complete the proof for the case $p(l-1)<n$.

Let $p(l-1)=n$. We have

$$\|D_{l-1/p}(\eta_\epsilon\varphi;\mathcal{B}_\epsilon);e\|^p_{L_p} \le \sigma_1+\sigma_2+\sigma_3 + c\,\|D_{l-1/p}(\varphi;\mathcal{B}_\epsilon);e\|^p_{L_p}, \qquad (14.7.34)$$

where

$$\sigma_1 = c\sum_{j=0}^{l-2}\int_e |\nabla_{l-1-j}\varphi(x)|^p dx \int_{\mathcal{B}_\epsilon} \frac{|\nabla_j\eta_\epsilon(x)-\nabla_j\eta_\epsilon(y)|^p}{|x-y|^{n-2+p}}dy,$$

$$\sigma_2 = c\sum_{j=1}^{l-2}\int_{\mathcal{B}_\epsilon} |\nabla_j\eta_\epsilon(y)|^p dy \int_e \frac{|\nabla_{l-1-j}\varphi(x)-\nabla_{l-1-j}\varphi(y)|^p}{|x-y|^{n-2+p}}dx,$$

$$\sigma_3 = c\int_e dx\int_{\mathcal{B}_\epsilon} \frac{|\varphi(x)\nabla_{l-1}\eta_\epsilon(x)-\varphi(y)\nabla_{l-1}\eta_\epsilon(y)|^p}{|x-y|^{n-2+p}}dy.$$

By (14.7.2),

$$\int_{\mathcal{B}_\epsilon} \frac{|\nabla_j\eta_\epsilon(x)-\nabla_j\eta_\epsilon(y)|^p}{|x-y|^{n-2+p}}dy \le c\,|\log|x||^{-p}|x|^{1-p(j+1)}.$$

Therefore,

$$\sigma_1 \le c\sum_{j=0}^{l-2}\int_e |\nabla_{l-1-j}\varphi(x)|^p|x|^{1-p(j+1)}|\log|x||^{-p}dx$$
$$\le c\sum_{j=0}^{l-3}\Big(\int_{\mathcal{B}_\epsilon}|\nabla_{l-1-j}\varphi(x)|^{p(n-1)/(n-p(j+1))}dx\Big)^{(n-p(j+1))/(n-1)}$$
$$\times\Big(\int_e \frac{dx}{|x|^{n-1}|\log|x||^{p(n-1)/(p(j+1)-1)}}\Big)^{(p(j+1)-1)/(n-1)}$$
$$+ c\,\|\nabla\varphi;\mathcal{B}_\epsilon\|^p_{L_\infty}\int_e |x|^{1-n}|\log|x||^{-p}dx.$$

The function $t^{n-1}|\log t|^\alpha$ increases near $t=0$. Hence, among all sets e with a fixed mes_{n-1}, the integral

$$\int_e |x|^{1-n}|\log||^{-\alpha}dx$$

attains its maximum at a ball with centre $x = 0$. Consequently, for $\alpha > 1$, we have

$$\int_e |x|^{1-n} |\log |x||^{-\alpha} dx \le c\, |\log \mathrm{mes}_{n-1} e|^{1-\alpha} \tag{14.7.35}$$

and hence,

$$\sigma_1 \le c \sum_{j=0}^{l-3} |\log \mathrm{mes}_{n-1} e|^{1-p-(n-p(j+1))/(n-1)} \|\nabla_{l-1-j}\varphi; \mathcal{B}_\epsilon\|^p_{L_{p(n-1)/(n-p(j+1))}}$$
$$+ c\, |\log \mathrm{mes}_{n-1} e|^{1-p} \|\nabla\varphi; \mathcal{B}_\epsilon\|^p_{L_\infty}. \tag{14.7.36}$$

This and (14.7.24) with $q = p(n-1)/[n-p(j+1)]$, $\rho = \epsilon$ and $\nabla\varphi$ as v yield

$$\sigma_1 \le c\, |\log \mathrm{mes}_{n-1} e|^{1-p} \big(\langle \varphi; \mathcal{B}_\epsilon \rangle^p_{p,l-1/p} + \|\nabla\varphi; \mathcal{B}_\epsilon\|^p_{L_\infty} \big).$$

We now estimate the sum σ_2. Note that

$$\int_e \frac{|\nabla_{l-1-j}\varphi(x) - \nabla_{l-1-j}\varphi(y)|^p}{|x-y|^{n-2+p}} dx$$
$$\le c \Big(\int_{\{x\in e: |x| > 2|y|\}} |\nabla_{l-1-j}\varphi(x)|^p \frac{dx}{|x|^{n-2+p}}$$
$$+ |\nabla_{l-1-j}\varphi(y)|^p \int_{\{x\in e: |x| > 2|y|\}} \frac{dx}{|x|^{n-2+p}}$$
$$+ \int_{\{x\in e: |x| \le 2|y|\}} \frac{|\nabla_{l-1-j}\varphi(x) - \nabla_{l-1-j}\varphi(y)|^p}{|x-y|^{n-2+p}} dx \Big).$$

Therefore, $\sigma_2 \le \sigma_2^{(1)} + \sigma_2^{(2)} + \sigma_2^{(3)}$, where

$$\sigma_2^{(1)} = c \sum_{j=1}^{l-2} \int_e |\nabla_{l-1-j}\varphi(x)|^p \frac{dx}{|x|^{n-2+p}} \int_{\mathcal{B}_{|x|/2}} |\nabla_j \eta_\epsilon(y)|^p dy,$$
$$\sigma_2^{(2)} = c \sum_{j=1}^{l-2} \int_e \frac{dx}{|x|^{n-2+p}} \int_{\mathcal{B}_{|x|/2}} |\nabla_{l-1-j}\varphi(y)|^p |\nabla_j \eta_\epsilon(y)|^p dy,$$
$$\sigma_2^{(3)} = c \sum_{j=1}^{l-2} \int_{\mathcal{B}_\epsilon} |\nabla_j \eta_\epsilon(y)|^p dy \int_{\{x\in e: |x|\le 2|y|\}} \frac{|\nabla_{l-1-j}\varphi(x) - \nabla_{l-1-j}\varphi(y)|^p}{|x-y|^{n-2+p}} dx.$$

Using the definition of η_ϵ given in Sect. 14.7.1, we obtain

$$\sigma_2^{(1)} \le c \sum_{j=1}^{l-2} \int_e \frac{|\nabla_{l-1-j}\varphi(x)|^p}{|x|^{n-2+p}} dx \int_{\mathcal{B}_{|x|/2}} |\log |y||^{-p} |y|^{-jp} dy$$
$$\le c \sum_{j=1}^{l-2} \int_e \frac{|\nabla_{l-1-j}\varphi(x)|^p dx}{|x|^{(j+1)p-1} |\log |x||^p}.$$

A majorant for this sum was found when estimating σ_1. It is equal to

$$c\,|\log \mathrm{mes}_{n-1} e|^{1-p}\big(\langle\varphi;\mathcal{B}_{2\epsilon}\rangle^p_{p,l-1/p}+\|\nabla\varphi;\mathcal{B}_{2\epsilon}\|^p_{L_\infty}\big). \tag{14.7.37}$$

Clearly,

$$\sigma_2^{(2)}\le c\sum_{j=1}^{l-1}\int_e\frac{dx}{|x|^{n-2+p}}\int_{\mathcal{B}_{|x|/2}}|\nabla_{l-1-j}\varphi(y)|^p|y|^{-jp}|\log|y||^{-p}dy$$

$$\le c\sum_{j=1}^{l-2}\int_e\frac{dx}{|x|^{n-2+p}|\log|x||^p}\int_{\mathcal{B}_{|x|/2}}|\nabla_{l-1-j}\varphi(y)|^p|y|^{-jp}dy.$$

In view of the inequality

$$\int_{\mathcal{B}_{|x|/2}}|\nabla_{l-1-j}\varphi(y)|^p|y|^{-jp}dy\le c\,|x|^{p-1}\big(\langle\varphi;\mathcal{B}_{|x|/2}\rangle^p_{p,l-1/p}+\|\nabla\varphi;\mathcal{B}_{|x|/2}\|^p_{L_\infty}\big)$$

we have

$$\sigma_2^{(2)}\le c\int_e\frac{dx}{|x|^{n-1}|\log|y||^p}\big(\langle\varphi;\mathcal{B}_\epsilon\rangle^p_{p,l-1/p}+\|\nabla\varphi;\mathcal{B}_\epsilon\|^p_{L_\infty}\big).$$

This inequality, together with (14.7.35), gives (14.7.37) as a majorant of $\sigma_2^{(2)}$.

The value $\sigma_2^{(3)}$ does not exceed

$$c\sum_{j=1}^{l-2}\int_{\mathcal{B}_\epsilon}\frac{dx}{|y|^{jp}|\log|y||^p}\int_{\{x\in e:|x|\le 2|y|\}}\frac{|\nabla_{l-1-j}\varphi(x)-\nabla_{l-1-j}\varphi(y)|^p}{|x-y|^{n-2+p}}dx.$$

Changing the order of integration and using the monotonicity of $t^{jp}|\log t|^p$ near $t=0$, we find that

$$\sigma_2^{(3)}\le c\sum_{j=1}^{l-1}\int_e\frac{dx}{|x|^{jp}|\log|x||^p}\int_{\mathcal{B}_\epsilon}\frac{|\nabla_{l-1-j}\varphi(x)-\nabla_{l-1-j}\varphi(y)|^p}{|x-y|^{n-2+p}}dy.$$

We apply Hölder's inequality and (14.7.35). Then, for $q_j=(n-1)/(n-1-jp)$,

$$\sigma_2^{(3)}\le c\sum_{j=1}^{l-2}\Big(\int_e|x|^{1-n}|\log|x||^{(1-n)/j}dx\Big)^{jp/(n-1)}$$

$$\times\Big[\int_{\mathcal{B}_\epsilon}\Big(\int_{\mathcal{B}_\epsilon}\frac{|\nabla_{l-1-j}\varphi(x)-\nabla_{l-1-j}\varphi(y)|^p}{|x-y|^{n-2+p}}dy\Big)^{q_j}dx\Big]^{1/q_j}$$

$$\le c\sum_{j=1}^{l-2}|\log \mathrm{mes}_{n-1}e|^{1-p-1/q_j}$$

$$\times\Big[\int_{\mathcal{B}_\epsilon}\Big(\int_{\mathcal{B}_\epsilon}|\nabla_{l-1-j}\varphi(x+h)-\nabla_{l-1-j}\varphi(x)|^p\frac{dh}{|h|^{n-2+p}}\Big)^{q_j}dx\Big]^{1/q_j}.$$

Now, by Minkowski's inequality,

$$\sigma_2^{(3)} \le c\,|\log \text{mes}_{n-1} e|^{1-p}$$

$$\times \sum_{j=1}^{l-1} \int_{\mathcal{B}_\epsilon} \Big(\int_{\mathcal{B}_\epsilon} |\nabla_{l-1-j}\varphi(x+h) - \nabla_{l-1-j}\varphi(x)|^{pq_j}\, dx \Big)^{1/q_j} \frac{dh}{|h|^{n-2+p}}.$$

Since

$$W_p^{l-1-1/p}(\mathbb{R}^{n-1}) \subset B_{pq_j,p}^{l-j-1-1/p}(\mathbb{R}^{n-1}) \quad \text{with} \quad j = 1, \cdots, l-2,$$

(see [Bes] or [Tr3], Sect. 2.8.1), it follows that

$$\int_{\mathcal{B}_\epsilon} \Big(\int_{\mathcal{B}_\epsilon} |\nabla_{l-2-j} v(x+h) - \nabla_{l-2-j} v(x)|^{pq_j}\, dx \Big)^{1/q_j} \frac{dh}{|h|^{n-2-p}}$$

$$\le c\,(\langle v; \mathcal{B}_{3\epsilon}\rangle_{p,l-1-1/p}^p + \epsilon^{1-n} \|v; \mathcal{B}_{3\epsilon}\|_{L_\infty}^p).$$

Therefore,

$$\sigma_2^{(3)} \le c\,|\log \text{mes}_{n-1} e|^{1-p} \big(\langle \varphi; \mathcal{B}_{3\epsilon}\rangle_{p,l-1/p}^p + \|\nabla\varphi; \mathcal{B}_{3\epsilon}\|_{L_\infty}^p \big).$$

Taking into account the estimates for $\sigma_2^{(1)}$ and $\sigma_2^{(2)}$ which were obtained above, we find that σ_2 is bounded from above by the right-hand side of the last inequality.

To obtain an estimate for σ_3 we note that

$$\int_e dx \int_{\mathcal{B}_{2|x|}\setminus \mathcal{B}_{|x|/2}} |\varphi(x)\nabla_{l-1}\eta_\epsilon(x) - \varphi(y)\nabla_{l-1}\eta_\epsilon(y)|^p \frac{dy}{|x-y|^{n-2+p}}$$

$$\le c\,\|\nabla\varphi; \mathcal{B}_\epsilon\|_{L_\infty}^p \int_e \frac{dx}{|x|^{(l-1)p} |\log|x||^p} \int_{\mathcal{B}_{2|x|}} \frac{dy}{|x-y|^{n-2}}$$

$$\le c\,\|\nabla\varphi; \mathcal{B}_\epsilon\|_{L_\infty}^p \int_e \frac{dx}{|x|^{n-1} |\log|x||^p}. \tag{14.7.38}$$

Moreover,

$$\int_e dx \int_{\mathcal{B}_{|x|/2}} |\varphi(x)\nabla_{l-1}\eta_\epsilon(x) - \varphi(y)\nabla_{l-1}\eta_\epsilon(y)|^p \frac{dy}{|x-y|^{n-2+p}}$$

$$\le c\,\|\nabla\varphi; \mathcal{B}_\epsilon\|_{L_\infty}^p \int_e dx \int_{\mathcal{B}_{|x|/2}} \frac{(|x|^{2-l}|\log|x||^{-1} + |y|^{2-l}|\log|y||^{-1})^p}{|x-y|^{n-2+p}} dy$$

$$\le c\,\|\nabla\varphi; \mathcal{B}_\epsilon\|_{L_\infty}^p \int_e \frac{dx}{|x|^{n-2+p}} \int_{\mathcal{B}_{|x|/2}} \frac{dy}{|y|^{n-p} |\log|y||^p}$$

$$\le c\,\|\nabla\varphi; \mathcal{B}_\epsilon\|_{L_\infty}^p \int_e \frac{dx}{|x|^{n-1} |\log|x||^p}. \tag{14.7.39}$$

In the same way we obtain

$$\int_e dx \int_{\mathbb{R}^{n-1}\setminus\mathcal{B}_{2|x|}} |\varphi(x)\nabla_{l-1}\eta_\epsilon(x) - \varphi(y)\nabla_{l-1}\eta_\epsilon(y)|^p \frac{dy}{|x-y|^{n-2+p}}$$

$$\le c\,\|\nabla\varphi;\mathcal{B}_\epsilon\|^p_{L_\infty} \int_e dx \int_{\mathbb{R}^{n-1}\setminus\mathcal{B}_{2|x|}} \frac{(|x|^{2-l}|\log|x||^{-1} + |y|^{2-l}|\log|y||^{-1})^p}{|x-y|^{n-2+p}} dy$$

$$\le c\,\|\nabla\varphi;\mathcal{B}_\epsilon\|^p_{L_\infty} \int_e \frac{dx}{|x|^{(l-2)p}|\log|x||^p} \int_{\mathbb{R}^{n-1}\setminus\mathcal{B}_{2|x|}} \frac{dy}{|y|^{n-2+p}}$$

$$\le c\,\|\nabla\varphi;\mathcal{B}_\epsilon\|^p_{L_\infty} \int_e \frac{dx}{|x|^{n-1}|\log|x||^p}. \tag{14.7.40}$$

Adding the estimates (14.7.38) – (14.7.40) and applying (14.7.35), we arrive at

$$\sigma_3 \le c\,|\log \mathrm{mes}_{n-1} e|^{1-p} \|\nabla\varphi;\mathcal{B}_\epsilon\|^p_{L_\infty}.$$

Now from (14.7.34) and the estimates for $\sigma_1, \sigma_2, \sigma_3$, it follows that

$$\|D_{l-1/p}(\eta_\epsilon\varphi;\mathcal{B}_\epsilon); e\|^p_{L_p} \le c\,|\log \mathrm{mes}_{n-1} e|^{1-p} \big(\langle\varphi;\mathcal{B}_{3\epsilon}\rangle^p_{p,l-1/p}$$
$$+ \|\nabla\varphi;\mathcal{B}_{3\epsilon}\|^p_{L_\infty}\big) + c\,\|D_{l-1/p}(\varphi;\mathcal{B}_\epsilon); e\|^p_{L_p}.$$

This, together with (14.7.31), gives the estimate

$$s_3^p \le c\Big(\langle\varphi;\mathcal{B}_{3\epsilon}\rangle^p_{p,l-1/p} + \|\nabla\varphi;\mathcal{B}_{3\epsilon}\|^p_{L_\infty} + \sup_{e\subset\mathcal{B}_\epsilon} \frac{\|D_{l-1/p}(\varphi;\mathcal{B}_\epsilon); e\|^p_{L_p}}{C_{p,l-1-1/p}(e)}\Big)$$

which together with (14.6.8) gives $s_3 \le c\,\delta$. □

Proof of Theorem 14.6.4. Lemmas 14.7.3–14.7.5 and the inequality (14.7.10) imply that (14.7.9) holds for sufficiently small ϵ. As was pointed out at the beginning of Sect. 14.7.3, the estimate (14.7.9) is equivalent to the inclusion $\partial\Omega \in M_p^{l-1/p}(\delta)$. The result follows.

Remark 14.7.1. The results of the present chapter were obtained in [MSh10]. Filonov applied these results to the study of the Maxwell operator in Lipschitz domains [Fil].

15

Multipliers in the Classical Layer Potential Theory for Lipschitz Domains

In this chapter we give applications of Sobolev multipliers to the question of higher regularity in fractional Sobolev spaces of solutions to boundary integral equations generated by the classical boundary value problems for the Laplace equation in and outside a Lipschitz domain. Since the sole Lipschitz graph property of $\partial\Omega$ does not guarantee higher regularity of solutions, we are forced to select an appropriate subclass of Lipschitz domains whose description involves a space of multipliers. For domains of this subclass we develop a solvability and regularity theory analogous to the classical one for smooth domains. We also show that the chosen subclass of Lipschitz domains proves to be the best possible in a certain sense. We end the chapter with a brief discussion of boundary integral equations of linear elastostatics.

15.1 Introduction

We study the internal and external Dirichlet problems

$$\Delta u_+ = 0 \text{ in } \Omega, \quad \operatorname{tr} u_+ = \Phi_+ \text{ on } \partial\Omega, \tag{$\mathcal{D}_+$}$$

and

$$\Delta u_- = 0 \text{ in } \mathbb{R}^n \backslash \overline{\Omega}, \quad \operatorname{tr} u_- = \Phi_- \text{ on } \partial\Omega,$$
$$u_-(x) = O(|x|^{2-n}) \quad \text{as } |x| \to \infty, \tag{$\mathcal{D}_-$}$$

where the boundary trace is denoted by tr, as well as the internal and external Neumann problems

$$\Delta v_+ = 0 \text{ in } \Omega, \quad \frac{\partial v_+}{\partial \nu} = \Psi_+ \text{ on } \partial\Omega, \tag{$\mathcal{N}_+$}$$

and

$$\Delta v_- = 0 \text{ in } \mathbb{R}^n \backslash \overline{\Omega}, \quad \frac{\partial v_-}{\partial \nu} = \Psi_- \text{ on } \partial\Omega,$$

V.G. Maz'ya, T.O. Shaposhnikova, *Theory of Sobolev Multipliers*,
Grundlehren der mathematischen Wissenschaften 337,

$$v_-(x) = O(|x|^{2-n}) \text{ as } |x| \to \infty, \qquad (\mathcal{N}_-)$$

where ν stands for the outer normal with respect to Ω.

In what follows, we exclude the case $n = 2$, which will simplify the presentation. The changes required in formulations, in comparison with dimensions $n > 2$, are the same as in the logarithmic potential theory for smooth contours. Our proofs, given for $n > 2$, apply to the two-dimensional case after minor changes.

A classical method for solving the problems $(\mathcal{D}_\pm)$ and $(\mathcal{N}_\pm)$ is representation of their solutions using the double layer potential

$$D\sigma(z) = \int_{\partial\Omega} \frac{\partial}{\partial \nu_\zeta} \Gamma(\zeta - z)\, \sigma(\zeta) ds_\zeta, \quad z \in \mathbb{R}^n \backslash \partial\Omega,$$

and the single layer potential

$$S\rho(z) = \int_{\partial\Omega} \Gamma(\zeta - z)\, \rho(\zeta) ds_\zeta, \quad z \in \mathbb{R}^n \backslash \partial\Omega,$$

where Γ is the fundamental solution of Δ with singularity at the origin. Putting $u_\pm = D\sigma_\pm$ and $v_\pm = S\rho_\pm$, one arrives at the boundary integral equations

$$\left(\pm\tfrac{1}{2} I + D\right) \sigma_\pm = \Phi_\pm, \qquad (2_\pm)$$

and

$$\left(\mp\tfrac{1}{2} I + D^*\right) \rho_\pm = \Psi_\pm, \qquad (3_\pm)$$

where D^* is the adjoint of D given by

$$D^*\rho(z) = \int_{\partial\Omega} \frac{\partial}{\partial \nu_z} \Gamma(\zeta - z)\, \rho(\zeta) ds_\zeta.$$

Looking for solutions of the problems $(\mathcal{D}_\pm)$ and $(\mathcal{N}_\pm)$ with boundary data $\Phi_\pm = \Phi$ and $\Psi_\pm = \Psi$ in the form $u_\pm = S\rho$ and $v_\pm = D\sigma$, one obtains the integral equations on $\partial\Omega$

$$S\rho = \Phi, \qquad (15.1.1)$$

and

$$\frac{\partial}{\partial \nu} D\sigma = \Psi. \qquad (15.1.2)$$

Let ℓ be *positive and noninteger*. In the case $p(\ell - 1) > n - 1$, our sole restriction on Ω is the inclusion of its boundary in the class W_p^ℓ, which means that *every function* φ *in* (14.1.3) *belongs to* $W_p^\ell(\mathbb{R}^{n-1})$.

In the opposite case $p(\ell - 1) \le n - 1$, we assume that $\partial\Omega$ *belongs to the class* M_p^ℓ *if every point* $O \in \partial\Omega$ *has a neighborhood* U *such that* $\Omega \cap U$ *is given by* (14.1.3) *with* $\varphi \in C^{0,1}(\mathbb{R}^{n-1})$ *subject to*

$$\nabla\varphi \in MW_p^{\ell-1}(\mathbb{R}^{n-1})$$

(here and elsewhere we do not differ between spaces of scalar and vector-valued functions in our notation). Furthermore, *the surface* $\partial\Omega$ *is said to be in the class* $M_p^\ell(\delta)$ *if*

$$\|\nabla\varphi;\mathbb{R}^{n-1}\|_{MW_p^{\ell-1}} \le \delta, \tag{15.1.3}$$

where δ is a positive number. Obviously,

$$M_p^\ell = \bigcup_{\delta>0} M_p^\ell(\delta).$$

These definitions are in accordance with those in Sect. 14.3.1, where we dealt with the particular case $\ell = l - 1/p$ with integer l.

Several conditions, either necessary or sufficient for $\partial\Omega \in M_p^\ell(\delta)$, will be discussed in Sect. 15.5. In particular, the inclusion

$$\partial\Omega \in M_p^\ell(0) := \bigcap_{\delta>0} M_p^\ell(\delta)$$

is guaranteed by the condition

$$\int_0^1 \left(\frac{\omega_q(\nabla_{[\ell]}\varphi, t)}{t\{\ell\}}\right)^p \frac{dt}{t} < \infty, \tag{15.1.4}$$

where $\omega_q(\nabla_k\varphi, t)$ is the L_q continuity modulus of the vector

$$\nabla_k\varphi = \{\partial^\alpha\varphi/\partial x_1^{\alpha_1}\dots, \partial x_{n-1}^{\alpha_{n-1}}\}$$

with $\alpha_1+\dots+\alpha_{n-1} = |\alpha| = k$, and q is any number satisfying $(n-1)/(\ell-1) \le q \le \infty$ for $p(\ell-1) < n-1$ and $p < q \le \infty$ for $p(\ell-1) = n-1$.

Clearly, any surface in the class $C^{\ell+\epsilon}$, $\epsilon > 0$, belongs to $M_p^\ell(0)$. However, there are surfaces in C^ℓ which are not in M_p^ℓ. Note that $\partial\Omega \in M_p^\ell$ may have vertices and edges on $\partial\Omega$ in the case $p(\ell-1) < n-1$.

We formulate our main result concerning the boundary integral equations $(2_\pm)$-(15.1.2). In the statement of this result and in the sequel, the notation $W_p^s(\partial\Omega) \ominus g$ with $g \in (W_p^s(\partial\Omega))^*$ stands for the subspace of functions $\psi \in W_p^s(\partial\Omega)$ such that

$$\int_{\partial\Omega} \psi g ds = 0.$$

Theorem 15.1.1. *Let* $n > 2$, $p \in (1,\infty)$, *and let* ℓ *be a noninteger with* $\ell > 1$. *Suppose that* $\partial\Omega$ *is connected,* $\partial\Omega \in W_p^\ell$ *for* $p(\ell-1) > n-1$ *and* $\partial\Omega \in M_p^\ell(\delta)$ *with some* $\delta = \delta(n,p,\ell) > 0$ *for* $p(\ell-1) \le n-1$. *Then the following assertions hold.*

(*i*) *The operator* $\frac{1}{2}I + D$ *is an isomorphism of* $W_p^\ell(\partial\Omega)$.
(*ii*) *The operator* $\frac{1}{2}I + D^*$ *is an isomorphism of* $W_p^{\ell-1}(\partial\Omega)$.
(*iii*) *The operator* S *maps* $W_p^{\ell-1}(\partial\Omega)$ *isomorphically onto* $W_p^\ell(\partial\Omega)$.
(*iv*) *The operator* $(\partial/\partial\nu)D$ *maps* $W_p^\ell(\partial\Omega)$ *continuously into* $W_p^\ell(\partial\Omega) \ominus 1$.

There exists a continuous inverse

$$\left(\frac{\partial}{\partial\nu}D\right)^{-1} : W_p^{\ell-1}(\partial\Omega)\ominus 1 \to W_p^{\ell}(\partial\Omega)\ominus 1.$$

(*v*) *There exists a continuous inverse*

$$\left(-\tfrac{1}{2}I + D\right)^{-1} : W_p^{\ell}(\partial\Omega)\ominus 1 \to W_p^{\ell}(\partial\Omega)\ominus \frac{\partial P}{\partial\nu},$$

where P *is the harmonic capacitary potential of* Ω *and*

$$\partial P/\partial\nu \in W_p^{\ell-1}(\partial\Omega)\cap(W_p^{\ell}(\partial\Omega))^*.$$

The equality $\left(-\tfrac{1}{2}I + D\right)1 = 0$ *holds.*
(*vi*) *There exists a continuous inverse*

$$\left(-\tfrac{1}{2}I + D^*\right)^{-1} : W_p^{\ell-1}(\partial\Omega)\ominus 1 \to W_p^{\ell-1}(\partial\Omega)\ominus 1.$$

The equality

$$\left(-\frac{1}{2}I + D^*\right)\partial P/\partial\nu = 0$$

holds.

Counterexamples in Sect. 15.6 show that Theorem 15.1.1 fails if $M_p^{\ell}(\delta)$ is replaced by M_p^{ℓ}.

The invertibility properties of the operators

$$\pm\frac{1}{2}I + D, \qquad \pm\frac{1}{2}I + D^*, \qquad S, \qquad \text{and} \quad (\partial/\partial\nu)D$$

in Theorem 15.1.1 result from solvability properties of the problems $(\mathcal{D}_\pm)$ and $(\mathcal{N}_\pm)$ collected in the next Theorem 15.1.2 which is of independent interest. The continuity properties of D, D^*, S, and $(\partial/\partial\nu)D$ stated in Theorem 15.1.1 are deduced from the part of Theorem 15.1.2 concerning the transmission problem

$$\Delta w_+ = 0 \text{ in } \Omega, \quad \Delta w_- = 0 \text{ in } \mathbb{R}^n\backslash\overline{\Omega},$$
$$\text{tr } w_+ - \text{tr } w_- = \Phi, \quad \frac{\partial w_+}{\partial\nu} - \frac{\partial w_-}{\partial\nu} = \Psi \text{ on } \partial\Omega,$$
$$w_-(x) = O(|x|^{2-n}) \text{ as } |x|\to\infty. \qquad (\mathcal{T})$$

In the formulation of Theorem 15.1.2 and in the sequel, we use the weighted Sobolev space $W_p^{k,\alpha}(\Omega)$ endowed with the norm

$$\|u;\Omega\|_{W_p^{k,\alpha}} = \left(\int_\Omega (\text{dist}(z,\partial\Omega))^{p\alpha}|\nabla_k u(z)|^p dz\right)^{1/p} + \|u,\Omega\|_{L_p}. \qquad (15.1.5)$$

Also, $W^{k,\alpha}_{p,\mathrm{loc}}(\mathbb{R}^n\backslash\overline{\Omega})$ stands for the space of functions subject to

$$\|u;\mathcal{B}\backslash\overline{\Omega}\|_{W^{k,\alpha}_p}<\infty$$

for an arbitrary open ball $\mathcal{B}$ containing $\overline{\Omega}$.

Theorem 15.1.2. *Let $n>2$, $p\in(1,\infty)$, and $\alpha=1-\{\ell\}-1/p$, where ℓ is a noninteger with $\ell>1$. Suppose that $\partial\Omega\in W^\ell_p$ for $p(\ell-1)>n-1$, and $\partial\Omega\in M^\ell_p(\delta)$ with some $\delta=\delta(n,p,\ell)$ for $p(\ell-1)\le n-1$. Then*

(i) For every $\Phi_+\in W^\ell_p(\partial\Omega)$ there exists a unique solution of the problem $(\mathcal{D}_+)$, $u_+\in W^{[\ell]+1,\alpha}_p(\Omega)$, and

$$\|u_+;\Omega\|_{W^{[\ell]+1,\alpha}_p}\le c\,\|\Phi_+;\partial\Omega\|_{W^\ell_p}. \tag{15.1.6}$$

This solution is represented uniquely as $(D\sigma_+)_+$ with $\sigma_+\in W^\ell_p(\partial\Omega)$ subject to equation (2_+). Moreover, u_+ can be represented uniquely in the form $S\rho$ with $\rho\in W^{\ell-1}_p(\partial\Omega)$ subject to equation (15.1.1).

(ii) For every $\Phi_-\in W^\ell_p(\partial\Omega)$ there exists a unique solution of the problem $(\mathcal{D}_-)$, $u_-\in W^{[\ell]+1,\alpha}_{p,loc}(\mathbb{R}^n\backslash\overline{\Omega})$ and, for every ball $\mathcal{B}$ with $B\supset\overline{\Omega}$,

$$\|u_-;\mathcal{B}\backslash\overline{\Omega}\|_{W^{[\ell]+1,\alpha}_p}\le c(\mathcal{B})\,\|\Phi_-;\partial\Omega\|_{W^\ell_p}. \tag{15.1.7}$$

This solution is represented uniquely in the form

$$u_-(z)=(D\sigma_-)(z)+C\,\Gamma(z),\quad z\in\mathbb{R}^n\backslash\overline{\Omega},$$

where C is a constant, the singularity of the fundamental solution Γ is situated in Ω, and $\sigma_-\in W^\ell_p(\partial\Omega)\ominus 1$ is a solution of the equation

$$\left(-\tfrac{1}{2}I+D\right)\sigma_-=\Phi_--C\Gamma\ \text{ on }\ \partial\Omega. \tag{15.1.8}$$

Moreover, u_- can be represented uniquely in the form $S\rho$ with $\rho\in W^{\ell-1}_p(\partial\Omega)$ subject to equation (15.1.1).

(iii) For every $\Psi_+\in W^{\ell-1}_p(\partial\Omega)\ominus 1$ there exists a unique solution of the problem $(\mathcal{N}_+)$, $v_+\in W^{[\ell]+1,\alpha}_p(\Omega)$, subject to $v_+\perp 1$ on Ω and satisfying

$$\|v_+;\Omega\|_{W^{[\ell]+1,\alpha}_p}\le c\,\|\Psi_+;\partial\Omega\|_{W^{\ell-1}_p}. \tag{15.1.9}$$

This solution is represented uniquely in the form

$$v_+(z)=(S\rho_+)(z)+C,\quad z\in\Omega,$$

where C is a constant, $\rho_+\in W^{\ell-1}_p(\partial\Omega)\ominus 1$ and ρ_+ satisfies (3_+). Moreover, v_+ can be represented uniquely as

$$v_+(z)=(D\sigma)(z)+C,\quad z\in\Omega,$$

where C is a constant and $\sigma\in W^\ell_p(\partial\Omega)\ominus 1$ satisfies (15.1.2).

(iv) For every $\Psi_- \in W_p^{\ell-1}(\partial\Omega)$ *there exists a unique solution of the problem* $(\mathcal{N}_-)$, $v_- \in W_{p,loc}^{[\ell]+1,\alpha}(\mathbb{R}^n\backslash\overline{\Omega})$ *and, for every ball* $\mathcal{B}$ *with* $\mathcal{B}\supset\overline{\Omega}$,

$$\|v_-;\mathcal{B}\backslash\overline{\Omega}\|_{W_p^{[\ell]+1,\alpha}} \le c(\mathcal{B})\|\Psi_-;\partial\Omega\|_{W_p^{\ell-1}}. \tag{15.1.10}$$

The solution is represented uniquely in the form $(S\rho_-)_-$ *with* $\rho_- \in W_p^{\ell-1}(\partial\Omega)$ *subject to equation* (3_-). *Moreover,* v_- *can be represented uniquely as*

$$v_-(z) = (D\sigma)(z) + C\Gamma(z), \quad z\in\mathbb{R}^n\backslash\overline{\Omega},$$

where

$$C = -\int_{\partial\Omega}\Psi_- ds, \qquad \sigma\in W_p^{\ell}(\partial\Omega)\ominus 1$$

and σ *satisfies the equation*

$$\frac{\partial}{\partial\nu}(D\sigma)_- = \Psi_- \; - \; C\frac{\partial}{\partial\nu}\Gamma_-. \tag{15.1.11}$$

(v) For every $(\Phi,\Psi)\in W_p^{\ell}(\partial\Omega)\times W_p^{\ell-1}(\partial\Omega)$ *there exists a unique solution of the problem* $(\mathcal{T})$

$$(w_+,w_-)\in W_p^{[\ell]+1,\alpha}(\Omega)\times W_{p,\mathrm{loc}}^{[\ell]+1,\alpha}(\mathbb{R}^n\backslash\overline{\Omega})$$

and, for every ball $\mathcal{B}$ *with* $\mathcal{B}\supset\overline{\Omega}$,

$$\|w_+;\Omega\|_{W_p^{[\ell]+1,\alpha}} + \|w_-;\mathcal{B}\backslash\overline{\Omega}\|_{W_p^{[\ell]+1,\alpha}}$$

$$\le c(\mathcal{B})\big(\|\Phi;\partial\Omega\|_{W_p^{\ell}} + \|\Psi;\partial\Omega\|_{W_p^{\ell-1}}\big). \tag{15.1.12}$$

This solution is given explicitly by

$$w_\pm = (S\Psi)_\pm + (D\Phi)_\pm \quad \text{on } \mathbb{R}^n\backslash\partial\Omega. \tag{15.1.13}$$

This theorem follows essentially from Theorem 15.2.1 in Sect. 15.2 concerning the $W_p^{[\ell]+1,\alpha}$-solvability of the Dirichlet, Neumann, and transmission problems for equations with nonzero right-hand sides. A typical statement, contained in Theorem 15.2.1, runs as follows.

Let $n\ge 2$, $1<p<\infty$, $\ell>1$, *and* $\{\ell\}>0$. *If* $\partial\Omega\in W_p^{\ell}$ *for* $p(\ell-1)>n-1$ *and* $\partial\Omega\in M_p^{\ell}(\delta)$ *with some* $\delta=\delta(n,p,\ell)$, *for* $p(\ell-1)\le n-1$, *then the mapping*

$$W_p^{[\ell]+1,\alpha}(\Omega)\ni u\to\{\Delta u,\ \operatorname{tr}\ u\}\in W_p^{[\ell]-1,\alpha}(\Omega)\times W_p^{\ell}(\partial\Omega) \tag{15.1.14}$$

is isomorphic.

In the case $p(\ell-1)>n-1$ this last assertion can be inverted for a subclass of Lipschitz domains: the isomorphism property of the mapping (15.1.14) implies that $\partial\Omega\in W_p^{\ell}$ (Theorem 15.6.1).

Note that this implication fails for the whole class of Lipschitz domains. As for the case $p(\ell-1) \le n-1$, several examples in Sects. 15.5 and 15.6 illustrate the sharpness of the condition $\partial\Omega \in M_p^\ell(\delta)$ in formulations of Theorems 15.1.1-15.2.1. In particular, Example 15.6.6 shows that in general the condition $\partial\Omega \in M_p^\ell(\delta)$ in Theorem 15.2.1 cannot be improved by $\partial\Omega \in M_p^\ell \cap C^{[\ell]}$.

We outline the structure of this chapter. In Sect. 15.2.1 we introduce and study a class of mappings, the so-called (p,k,α)-diffeomorphisms, preserving $W_p^{k,\alpha}$ and which play a crucial role in the subsequent study of the boundary value problems.

Properties of the problems $(\mathcal{D}_\pm)$, $(\mathcal{N}_\pm)$, and $(\mathcal{T})$ are obtained in Sect. 15.2 (Proposition 15.3.1). The next section deals with continuity properties of the potentials and their normal derivatives. Here, in particular, definitions of all integral operators involved in Theorem 15.1.1 are given. Proof of Theorems 15.1.1 and 15.1.2 can be found in Sect. 15.4.

The short Sect. 15.5 is devoted to a discussion of the class $M_p^\ell(\delta)$. In Sect. 15.6 we give a number of examples of domains which demonstrate the sharpness of our solvability results for the Dirichlet and Neumann problems as well as for the corresponding integral equations.

15.2 Solvability of Boundary Value Problems in Weighted Sobolev Spaces

15.2.1 (p,k,α)-Diffeomorphisms

In this section U and V are open subsets of $\mathbb{R}^n_+ = \{z=(x,y) : x \in \mathbb{R}^{n-1}, y > 0\}$. By $W_p^{k,\alpha}(V)$ we denote the space of functions with the finite norm

$$\Big(\int_V (\min\{1,y\})^{p\alpha} (|\nabla_k v(x,y)|^p + |v(x,y)|^p) dz \Big)^{1/p},$$

where k is a positive integer, $-1 < p\alpha < p-1$, and $1 \le p \le \infty$. By analogy with the definition of the (p,l)-diffeomorphism given in Sect. 9.4.1, a bi-Lipschitz homeomorphism $\varkappa : U \to V$ will be called a (p,k,α)-diffeomorphism if the elements of its Jacobi matrix $\partial\varkappa$ belong to the space of multipliers $MW_p^{k-1,\alpha}(U)$.

The next two propositions contain basic properties of (p,k,α)-diffeomorphisms, verified in the same way as the corresponding properties of (p,l)-diffeomorphisms in Chap. 9. By $\|\partial\varkappa, U\|_{MW_p^{k-1,\alpha}}$ we denote the sum of the norms of the elements of $\partial\varkappa$ in the space $MW_p^{k-1,\alpha}(U)$.

Proposition 15.2.1. *(i) If $u \in W_p^{k,\alpha}(V)$ and $\varkappa$ is a (p,k,α)-diffeomorphism: $U \to V$, then $u \circ \varkappa \in W_p^{k,\alpha}(U)$ and*

$$\|u \circ \varkappa; U\|_{W_p^{k,\alpha}} \le c\, \|u; V\|_{W_p^{k,\alpha}}.$$

(ii) If $\varkappa$ is a (p,k,α)-diffeomorphism, then $\varkappa^{-1}$ is also a (p,k,α)-diffeomorphism.

(iii) If $\gamma \in MW_p^{k,\alpha}(V)$ and $\varkappa$ is a (p,k,α)-diffeomorphism, then $\gamma \circ \varkappa \in MW_p^{k,\alpha}(U)$ and

$$\|\gamma \circ \varkappa; U\|_{MW_p^{k,\alpha}} \le c\,\|\gamma; V\|_{MW_p^{k,\alpha}}.$$

(iv) If $\varkappa_1 : U \to V$ and $\varkappa_2 : V \to W$ are (p,k,α)-diffeomorphisms then their composition $\varkappa_2 \circ \varkappa_1 : U \to W$ is a (p,k,α)-diffeomorphism.

Let $\mathcal{T}$ denote the extension operator defined by (9.4.11), where $\zeta(\tau) = 0$ for $|\tau| \ge 1$. Consider a domain

$$G = \{(x,y) : x \in \mathbb{R}^{n-1},\ y > \varphi(x)\}, \tag{15.2.1}$$

where φ is a Lipschitz function such that $\varphi(0) = 0$ and $|\nabla\varphi(x)| \le L$ for almost all $x \in \mathbb{R}^{n-1}$. We introduce the mapping

$$\varkappa :\ \mathbb{R}^n_+ \ni (\xi,\eta) \to (x,y) \in G$$

by the equalities

$$x = \xi,\ \ y = N\eta + (\mathcal{T} f)(\xi,\eta), \tag{15.2.2}$$

where N is a sufficiently large constant depending on L.

Proposition 15.2.2. *Let ℓ be a noninteger with $\ell > 1$, and let $p \in (1,\infty)$. If $\nabla\varphi \in MW_p^{\ell-1}(\mathbb{R}^{n-1})$, then $\varkappa$ is a $(p,[\ell]+1,\alpha)$-diffeomorphism.*

We say that a function f defined on ∂G belongs to the space $W_p^\ell(\partial G)$ if the function $\mathbb{R}^{n-1} \ni x \to f(x,\varphi(x))$ belongs to $W_p^\ell(\mathbb{R}^{n-1})$. This can be written as

$$f \in W_p^\ell(\partial G) \Leftrightarrow f \circ \varkappa|_{\mathbb{R}^{n-1}} \in W_p^\ell(\mathbb{R}^{n-1}). \tag{15.2.3}$$

By (15.2.3) and Proposition 15.2.1 (i), the inclusion $u \in W_p^{[\ell]+1,\alpha}(G)$ implies that $\operatorname{tr} u \in W_p^\ell(\partial G)$ and there exists a linear extension operator: $W_p^\ell(\partial G) \to W_p^{[\ell]+1,\alpha}(G)$.

Note that (15.2.2) gives an extension of $\varkappa$ to the lower half-space $\mathbb{R}^n_- = \{(x,y) : x \in \mathbb{R}^{n-1},\ y < 0\}$:

$$\mathbb{R}^n_- \ni (\xi,\eta) \to (x,y) \in \mathbb{R}^n \backslash \overline{G}$$

and this extension has the same properties as the original mapping $\varkappa$. We preserve the same notation $\varkappa$ for the extended mapping so that, now, $\varkappa$ is a quasi-isometric mapping of $\mathbb{R}^n$ onto $\mathbb{R}^n$ and a $(p,[\ell]+1,\alpha)$-diffeomorphic mapping of $\mathbb{R}^n_+$ and $\mathbb{R}^n_-$ onto G and $\mathbb{R}^n\backslash\overline{G}$, respectively.

15.2.2 Weak Solvability of the Dirichlet Problem

We need the following assertion which is similar in flavor to [GG] and Sect. 5.7.2 in [Tr4].

Lemma 15.2.1. *Let $p \in (1,\infty)$, and let $0 < \alpha + 1/p$. Suppose that the Lipschitz constant of the function φ in* (15.1.3) *does not exceed a sufficiently small constant depending on n, p, and α. Then the mapping*

$$W_p^{1,\alpha}(\Omega) \ni u \to \{\Delta u, \operatorname{tr} u\} \in W_p^{-1,\alpha}(\Omega) \times W_p^{1-\alpha-1/p}(\partial\Omega)$$

is an isomorphism.

Proof. It is well known that the fractional Sobolev space $W_p^{k-\alpha-1/p}(\mathbb{R}^{n-1})$ is the space of traces on $\mathbb{R}^{n-1}$ of functions in the space $W_p^{k,\alpha}(\mathbb{R}^n_+)$, where $p \in (1,\infty)$ (see [Usp]). Since $\partial\Omega \in C^{0,1}$, it follows from this result that the Dirichlet problem

$$\Delta u = F \text{ in } \Omega, \quad u = \Phi \text{ on } \partial\Omega \tag{15.2.4}$$

with $F \in W_p^{[\ell]-1,\alpha}(\Omega)$ and $\Phi \in W_p^{\ell}(\partial\Omega)$ can be reduced to the case $\Phi = 0$.

Let $\mathring{W}_q^1(\Omega)$ be the completion of $C_0^\infty(\Omega)$ in the norm of $W_q^1(\Omega)$ and let $W_{q'}^{-1}(\Omega)$ stand for the dual of $W_q^1(\Omega)$, where $q+q' = qq'$. We choose $s = s(p,\alpha)$ so that the imbeddings

$$\mathring{W}_p^{1,\alpha}(\Omega) \subset \mathring{W}_s^1(\Omega), \tag{15.2.5}$$

$$\mathring{W}_{s'}^1(\Omega) \subset \mathring{W}_p^{1,\alpha}(\Omega), \tag{15.2.6}$$

$$W_p^{-1,\alpha}(\Omega) \subset W_s^{-1}(\Omega), \tag{15.2.7}$$

hold. By Hölder's inequality these imbeddings follow from

$$s' \ge p, \quad s < \frac{p}{1+\alpha p}, \qquad \text{for } \alpha > 0,$$
$$s' \ge p, \quad s \le p, \qquad \text{for } \alpha = 0,$$
$$s' > \frac{p}{1+\alpha p}, \qquad \text{for } \alpha < 0.$$

We can put, for example,

$$s = \frac{1}{2}\Big(1 + \min\{\frac{p}{p-1-\alpha p}, p\}\Big), \quad \text{for } \alpha \le 0$$

and

$$s = \frac{1}{2}\Big(1 + \min\{\frac{p}{1+\alpha p}, p\}\Big), \qquad \text{for } \alpha > 0.$$

Since $s' > 2$, the operator

$$\mathring{W}_{s'}^1(\Omega) \ni u \to \Delta u \in W_{s'}^{-1}(\Omega) \tag{15.2.8}$$

is a monomorphism. We now show the existence of a bounded inverse to (15.2.8) defined on $W^{-1}_{s'}(\Omega)$.

Let $F \in W^{-1}_{s'}(\Omega)$, and let $u \in W^1_2(\Omega)$ be a solution to the problem (15.2.4) with $\Phi = 0$. We denote by U a small coordinate neighborhood of a point $O \in \partial\Omega$ and by V an open set such that $O \in V$ and $\overline{V} \subset U$. We take a function $\chi \in C_0^\infty(U)$ with $\chi = 1$ on V. Then

$$\Delta(\chi u) = [\Delta, \chi]u + \chi F.$$

Let $\varkappa$ be the bi-Lipschitz diffeomorphism: $\mathbb{R}^n_+ \ni (\xi, \eta) \to (x, y) \in G$ defined by

$$x = \xi, \qquad y = \eta + \varphi(\xi), \tag{15.2.9}$$

and let σ denote its inverse. Clearly, σ maps $U \cap \partial\Omega$ onto an open subset of the hyperplane $\eta = 0$. Now, $(\chi u) \circ \varkappa$ satisfies the boundary value problem

$$\operatorname{div}(A\nabla((\chi u) \circ \varkappa)) = (\chi F) \circ \varkappa + ([\Delta, \chi]u) \circ \varkappa \quad \text{on } \mathbb{R}^n_+, \tag{15.2.10}$$

$$(\chi u) \circ \varkappa\big|_{\mathbb{R}^{n-1}} = (\chi\Phi) \circ (\varkappa\big|_{\mathbb{R}^{n-1}}), \tag{15.2.11}$$

where

$$A = (\partial\sigma \circ \varkappa)^*(\partial\sigma \circ \varkappa). \tag{15.2.12}$$

Obviously, the right-hand side of (15.2.10) belongs to $W^{-1}_{s'}(\Omega)$. Therefore, the function

$$v := (\chi u) \circ \varkappa \in W^1_{s'}(\mathbb{R}^n_+)$$

is a solution of the problem

$$\operatorname{div}(A\nabla v) - v = H \text{ on } \mathbb{R}^n_+, \quad v\big|_{\mathbb{R}^{n-1}} = 0, \tag{15.2.13}$$

where

$$H = (\chi F) \circ \varkappa + ([\Delta, \chi]u) \circ \varkappa - (\chi u) \circ \varkappa. \tag{15.2.14}$$

Clearly,

$$\|I - \partial\varkappa; \mathbb{R}^n_+\|_{L_\infty} \le c\,\|\nabla\varphi; \mathbb{R}^{n-1}\|_{L_\infty},$$

which implies that

$$\|I - A; \mathbb{R}^n_+\|_{L_\infty} < \varepsilon, \tag{15.2.15}$$

where ε is sufficiently small.

It is a classical fact that the Dirichlet problem

$$-\Delta w + w = g_0 + \operatorname{div}\mathbf{g} \quad \text{on } \mathbb{R}^v_+, \;\; w|_{\mathbb{R}^{n-1}} = 0, \tag{15.2.16}$$

with $g_0 \in L_q(\mathbb{R}^n_+)$ and $\mathbf{g} \in (L_q(\mathbb{R}^n_+))^n$, $1 < q < \infty$, is uniquely solvable in $\mathring{W}^1_q(\mathbb{R}^n_+)$. (This follows from the explicit representation of w by Green's function and the continuity of a singular integral operator in $L_q(\mathbb{R}^n)$.) Let

$(I - \Delta)^{-1}$ stand for the inverse operator of the problem (15.2.16). We write (15.2.13) in the form

$$v - (I - \Delta)^{-1} S v = (\Delta - I)^{-1} H, \tag{15.2.17}$$

with H given by (15.2.14), and

$$S v = \operatorname{div}((A - I)\nabla v). \tag{15.2.18}$$

This leads to the Neumann series

$$v = \sum_{j=0}^{\infty} ((I - \Delta)^{-1} S)^j (\Delta - I)^{-1} H, \tag{15.2.19}$$

where the operator $(I - \Delta)^{-1} S$ has a small norm in $W_q^1(\mathbb{R}_+^n)$ for every $q \in (2, s']$, by (15.2.15). Hence,

$$\|v; \mathbb{R}_+^n\|_{W_q^1} \leq c\, \|(\Delta - I)^{-1} H; \mathbb{R}_+^n\|_{W_q^1}.$$

Using (15.2.14) and the arbitrariness of the point $O \in \partial\Omega$, we obtain

$$\|u; \Omega\|_{W_q^1} \leq c\, (\|F; \Omega\|_{W_q^{-1}} + \|u; \Omega\|_{L_q}). \tag{15.2.20}$$

By Sobolev's imbedding theorem, $u \in L_{2n/(n-2)}(\Omega)$ if $n > 2$. Thus,

$$u \in W^1_{2n/(n-2)}(\Omega)$$

by (15.2.20). Using Sobolev's theorem again, we see that

$$u \in L_{2n/(n-4)}(\Omega) \qquad \text{if} \quad n > 4$$

and

$$u \in W^1_{s'}(\Omega) \qquad \text{if} \quad n \leq 4.$$

Repeating this argument m times, where $m > n(s' - 2)/2s'$, we conclude that $u \in W^1_{s'}(\Omega)$ and arrive at the estimate

$$\|u; \Omega\|_{W^1_{s'}} \leq c\, (\|F; \Omega\|_{W^{-1}_{s'}} + \|u; \Omega\|_{L_{2n/(n-2)}}).$$

This implies that

$$\begin{aligned} \|u; \Omega\|_{W^1_{s'}} &\leq c\, (\|F; \Omega\|_{W^{-1}_{s'}} + \|u; \Omega\|_{W^1_2}) \\ &\leq c\, (\|F; \Omega\|_{W^{-1}_{s'}} + \|F; \Omega\|_{W^{-1}_2}) \leq c\, \|F; \Omega\|_{W^{-1}_{s'}}. \end{aligned}$$

Hence, the operator (15.2.8) is isomorphic. By duality, the operator

$$\mathring{W}^1_s(\Omega) \ni u \to \Delta u \in W^{-1}_s(\Omega) \tag{15.2.21}$$

is isomorphic as well. This fact, combined with (15.2.5), shows that the operator

$$\Delta : \mathring{W}_p^{1,\alpha}(\Omega) \to W_p^{-1,\alpha}(\Omega)$$

is a monomorphism.

Let $F \in W_{s'}^{-1}(\Omega)$, and let $u \in W_{s'}^{1}(\Omega)$ be a solution of (15.2.4) with $\Phi = 0$. By (15.2.6), we see that $u \in \mathring{W}_p^{1,\alpha}(\Omega)$. It remains to prove the estimate

$$\|u; \Omega\|_{W_p^{1,\alpha}} \leq c\,\|F; \Omega\|_{W_p^{-1,\alpha}}. \tag{15.2.22}$$

It is well known that there exists a bounded inverse $(I - \Delta)^{-1}$ of the operator $I - \Delta$ in $\mathbb{R}^n_+$ with zero Dirichlet data on $\mathbb{R}^{n-1}$, acting from $W_p^{k-2,\alpha}(\mathbb{R}^n_+)$ into $W_p^{k,\alpha}(\mathbb{R}^n_+)$, $k = 1, 2, \dots$ (cfr. [GG], [Tr4], Sect. 5.3.2). Using this inverse, we write (15.2.13) in the form (15.2.17) and arrive at the Neumann series (15.2.19), where the operator $(I - \Delta)^{-1}S$ has a small norm in $W_p^{1,\alpha}(\mathbb{R}^n_+)$, by (15.2.15). Hence,

$$\|v; \mathbb{R}^n_+\|_{W_p^{1,\alpha}} \leq c\,\|(\Delta - I)^{-1}H; \mathbb{R}^n_+\|_{W_p^{1,\alpha}}.$$

Using the arbitrariness of the point $O \in \partial\Omega$ and (15.2.14), we obtain

$$\|u; \Omega\|_{W_p^{1,\alpha}} \leq c\,(\|F; \Omega\|_{W_p^{-1,\alpha}} + \|u; \Omega\|_{W_p^{0,\alpha}}). \tag{15.2.23}$$

It follows from the one-dimensional Hardy inequality that

$$\|u; \Omega\|_{W_p^{0,\alpha}} \leq \varepsilon_0\|u; \Omega\|_{W_p^{1,\alpha}} + C(\varepsilon_0)\|u; \Omega\|_{L_1}. \tag{15.2.24}$$

for any sufficiently small $\varepsilon_0 > 0$. Since the operator (15.2.21) is isomorphic and the imbedding (15.2.7) holds, we have

$$\|u; \Omega\|_{L_1} \leq c_1\,\|u; \Omega\|_{W_s^1} \leq c_2\,\|F; \Omega\|_{W_s^{-1}} \leq c_3\,\|F; \Omega\|_{W_p^{-1,\alpha}},$$

which together with (15.2.23) and (15.2.24) completes the proof of Lemma 15.2.1. □

15.2.3 Main Result

Let $W_p^{k,\alpha}(\Omega)$ be the weighted Sobolev space endowed with the norm (8.1.1). We also need the weighted Sobolev space $W_p^{k,\alpha}(\mathbb{R}^n\backslash\overline{\Omega})$ supplied with the norm

$$\|v; \mathbb{R}^n\backslash\overline{\Omega}\|_{W_p^{k,\alpha}} = \Big(\int\limits_{\mathbb{R}^n\backslash\overline{\Omega}} (\min\{\mathrm{dist}(x, \partial\Omega), 1\})^{p\alpha}(|\nabla_k v(x)|^p + |v(x)|^p)dx\Big)^{1/p}.$$

Using a partition of unity and properties of the special Lipschitz domain (15.2.1) mentioned at the end of Sect. 15.2.1, we can introduce the space $W_p^\ell(\partial\Omega)$ and show that it is the trace space for both $W_p^{[\ell]+1,\alpha}(\Omega)$ and $W_p^{[\ell]+1,\alpha}(\mathbb{R}^n\backslash\overline{\Omega})$. We also need the space $\mathring{W}_p^{1,\alpha}(\Omega)$ obtained by completion of $C_0^\infty(\Omega)$ in the norm of $W_p^{1,\alpha}(\Omega)$.

By $W_p^{-1,\alpha}(\Omega)$ we denote the space of distributions $F = g_0 + \operatorname{div}\mathbf{g}$, with $g_0 \in W_p^{0,\alpha}(\Omega)$ and $\mathbf{g} \in (W_p^{0,\alpha}(\Omega))^n$. We supply $W_p^{-1,\alpha}(\Omega)$ with the norm

$$\|F;\Omega\|_{W_p^{-1,\alpha}} = \inf\big(\|g_0;\Omega\|_{W_p^{0,\alpha}} + \|\mathbf{g};\Omega\|_{(W_p^{0,\alpha})^n}\big),$$

where the infimum is taken over all representations $F = g_0 + \operatorname{div}\mathbf{g}$.

The next theorem contains all the information on auxiliary boundary value problems $(\mathcal{D}_\pm)$, $(\mathcal{N}_\pm)$, and $(\mathcal{I})$ to be used in the sequel.

Theorem 15.2.1. *Let $p \in (1,\infty)$, and let $\alpha = 1 - \{\ell\} - 1/p$, where ℓ is a noninteger with $\ell > 1$. Suppose that $\partial\Omega \in W_p^\ell$ for $p(\ell-1) > n-1$ and $\partial\Omega \in M_p^\ell(\delta)$ with some $\delta = \delta(n,p,\ell)$ for $p(\ell-1) \le n-1$.*

The five mappings

$$W_p^{[\ell]+1,\alpha}(\Omega) \ni u \to \{\Delta u, \operatorname{tr} u\} \in W_p^{[\ell]-1,\alpha}(\Omega) \times W_p^\ell(\partial\Omega), \tag{15.2.25}$$

$$W_p^{[\ell]+1,\alpha}(\mathbb{R}^n\backslash\overline{\Omega}) \ni u \to \{\Delta u - u, \operatorname{tr} u\} \in W_p^{[\ell]-1,\alpha}(\mathbb{R}^n\backslash\overline{\Omega}) \times W_p^\ell(\partial\Omega), \tag{15.2.26}$$

$$W_p^{[\ell]+1,\alpha}(\Omega) \ni u \to \{\Delta u - u,\ \partial u/\partial\nu\} \in W_p^{[\ell]-1,\alpha}(\Omega) \times W_p^{\ell-1}(\partial\Omega), \tag{15.2.27}$$

$$W_p^{[\ell]+1,\alpha}(\mathbb{R}^n\backslash\overline{\Omega}) \ni u \to \{\Delta u - u,\ \partial u/\partial\nu\} \in W_p^{[\ell]-1,\alpha}(\mathbb{R}^n\backslash\overline{\Omega}) \times W_p^{\ell-1}(\partial\Omega), \tag{15.2.28}$$

$$\begin{aligned} &W_p^{[\ell]+1,\alpha}(\Omega) \times W_p^{[\ell]+1,\alpha}(\mathbb{R}^n\backslash\overline{\Omega}) \ni (u_+, u_-) \\ &\quad \to \Big\{\Delta u_+,\ \Delta u_- - u_-,\ \operatorname{tr}(u_+ - u_-),\ \frac{\partial u_+}{\partial\nu} - \frac{\partial u_-}{\partial\nu}\Big\} \\ &\quad \in \{W_p^{[\ell]-1,\alpha}(\Omega) \times W_p^{[\ell]-1,\alpha}(\mathbb{R}^n\backslash\overline{\Omega}) \times W_p^\ell(\partial\Omega) \times W_p^{\ell-1}(\partial\Omega)\} \end{aligned} \tag{15.2.29}$$

are all isomorphisms.

Proof. The continuity of the mappings (15.2.25)–(15.2.29) is obvious. Dealing with their invertibility, we restrict ourselves to a detailed treatment of (15.2.25), since the analysis of (15.2.26)–(15.2.29) is essentially the same.

Let us show that the Dirichlet problem (15.2.4) with $F \in W_p^{[\ell]-1,\alpha}(\Omega)$ and $\Phi \in W_p^\ell(\partial\Omega)$ is uniquely solvable in $W_p^{[\ell]+1,\alpha}(\Omega)$, and that

$$\|u; \Omega\|_{W_p^{[\ell]+1,\alpha}} \leq c\left(\|F; \Omega\|_{W_p^{[\ell]-1,\alpha}} + \|\Phi; \partial\Omega\|_{W_p^\ell}\right).$$

By Lemma 15.2.1, the problem (15.2.4) has a unique solution $u \in W_p^{1,\alpha}(\Omega)$. Therefore, in order to prove Theorem 15.2.1 we only need to show that the solution u belongs to $W_p^{[\ell]+1,\alpha}(\Omega)$ and to estimate its norm in this space.

Let U be a coordinate neighborhood of a point $O \in \partial\Omega$ and let V denote an open set such that $O \in V$ and $\overline{V} \subset U$. We take a function $\chi \in C_0^\infty(U)$ with $\chi = 1$ on V. Then

$$\Delta(\chi u) = [\Delta, \chi]u + \chi F.$$

Let $\varkappa$ be the $(p, [\ell]+1, \alpha)$-diffeomorphism defined by (15.2.2), where $N = 1$, and let σ denote its inverse. Clearly, σ maps $U \cap \partial\Omega$ onto an open subset of the hyperplane $\eta = 0$. Now, $(\chi u) \circ \varkappa$ satisfies the boundary value problem

$$\operatorname{div}(A\nabla((\chi u) \circ \varkappa)) = \frac{(\chi F) \circ \varkappa + ([\Delta, \chi]u) \circ \varkappa}{\det(\partial\sigma \circ \varkappa)} \quad \text{on } \mathbb{R}_+^n, \tag{15.2.30}$$

$$(\chi u) \circ \varkappa\big|_{\mathbb{R}^{n-1}} = (\chi\Phi) \circ (\varkappa\big|_{\mathbb{R}^{n-1}}), \tag{15.2.31}$$

where

$$A = \frac{(\partial\sigma \circ \varkappa)^*(\partial\sigma \circ \varkappa)}{\det(\partial\sigma \circ \varkappa)}. \tag{15.2.32}$$

By Proposition 15.2.1 (*i*), (*iii*), the right-hand side of (15.2.30) belongs to $W_p^{[\ell]-1,\alpha}(\mathbb{R}_+^n)$ and the Dirichlet data (15.2.31) are in $W_p^\ell(\mathbb{R}^{n-1})$. These data have an extension $\Theta \in W_p^{[\ell]+1,\alpha}(\mathbb{R}_+^n)$. Therefore, the function

$$v := (\chi u) \circ \varkappa - \Theta \in W_p^{1,\alpha}(\mathbb{R}_+^n)$$

is a solution of the problem

$$\operatorname{div}(A\nabla v) - v = H \text{ on } \mathbb{R}_+^n, \quad v\big|_{\mathbb{R}^{n-1}} = 0, \tag{15.2.33}$$

where

$$H = \frac{(\chi F) \circ \varkappa + ([\Delta, \chi]u) \circ \varkappa}{\det(\partial\sigma \circ \varkappa)} - \operatorname{div}(A\nabla\Theta) + \Theta - (\chi u) \circ \varkappa. \tag{15.2.34}$$

We shall consider the cases $p(\ell-1) \leq n-1$ and $p(\ell-1) > n-1$ separately.

The case $p(\ell-1) \leq n-1$. Let $\partial\Omega \in M_p^\ell(\delta)$. By (9.4.17) and Theorem 8.7.1,

$$\|I - \partial\varkappa; \mathbb{R}_+^n\|_{MW_p^{[\ell],\alpha}} \leq c\,\|\nabla\varphi; \mathbb{R}^{n-1}\|_{MW_p^{\ell-1}}.$$

This along with (15.1.3) implies that

$$\|I - A; \mathbb{R}_+^n\|_{MW_p^{[\ell],\alpha}} \leq c\,\|\nabla\varphi; \mathbb{R}^{n-1}\|_{MW_p^{\ell-1}} \leq c\,\delta. \tag{15.2.35}$$

We can replace $[\ell]$ on the left-hand side of (15.2.35) by any $k = 0, 1, \dots [\ell]$ because of the imbedding $MW_p^{[\ell],\alpha}(\mathbb{R}_+^n) \subset MW_p^{k,\alpha}(\mathbb{R}_+^n)$. This imbedding follows from

$$MW_p^{0,\alpha}(\mathbb{R}_+^n) = L_\infty(\mathbb{R}_+^n) \supset MW_p^{k,\alpha}(\mathbb{R}_+^n)$$

by interpolation between $W_p^{[\ell],\alpha}(\mathbb{R}_+^n)$ and $W_p^{0,\alpha}(\mathbb{R}_+^n)$ (see [Tr4], Sect. 3.4.2).

It is standard that there exists a bounded inverse $(I-\Delta)^{-1}$ to the operator $I - \Delta$ in $\mathbb{R}_+^n$ with zero Dirichlet data on $\mathbb{R}^{n-1}$, acting from $W_p^{k,\alpha}(\mathbb{R}_+^n)$ into $W_p^{k-2,\alpha}(\mathbb{R}_+^n)$, $k = 0, 1, \dots$ (see [Tr4], Sect. 5.3.2).

We write (15.2.33) in the form

$$v - (I - \Delta)^{-1} S v = (\Delta - I)^{-1} H \tag{15.2.36}$$

with H given by (15.2.34) and

$$Sv = \operatorname{div}((A - I)\nabla v).$$

This leads to the Neumann series

$$v = \sum_{j=0}^{\infty} ((I - \Delta)^{-1} S)^j (\Delta - I)^{-1} H$$

where the operator $(I-\Delta)^{-1}S$ has a small norm in $W_p^{k+1,\alpha}(\mathbb{R}_+^n)$, $k = 0, 1, \dots,$ by (15.2.35).

Since $H \in W_p^{0,\alpha}(\mathbb{R}_+^n)$ and $(\Delta - I)^{-1}H \in W_p^{2,\alpha}(\mathbb{R}_+^n)$, it follows that $v \in W_p^{2,\alpha}(\mathbb{R}_+^n)$ and therefore, $\chi u \in W_p^{2,\alpha}(\Omega)$. Using the arbitrariness of the point $O \in \partial\Omega$ we derive that $u \in W_p^{2,\alpha}(\Omega)$ which completes the proof for $\ell < 2$.

Let $\ell > 2$. Using Proposition 15.2.1 and $u \in W_p^{2,\alpha}(\Omega)$, we obtain $H \in W_p^{1,\alpha}(\mathbb{R}_+^n)$ which implies that $v \in W_p^{3,\alpha}(\mathbb{R}_+^n)$ by (15.2.36). Repeating this argument several times if necessary, we conclude that $u \in W_p^{[\ell]+1,\alpha}(\Omega)$. This is the required result for $p(\ell - 1) \le n - 1$.

The case $p(\ell - 1) > n - 1$. We have

$$\|A; \mathbb{R}_+^n\|_{W_p^{[\ell],\alpha}} \le c\, \|\varphi; \mathbb{R}^{n-1}\|_{W_p^\ell}. \tag{15.2.37}$$

Without loss of generality we may assume that $\|\nabla\varphi; \mathbb{R}^{n-1}\|_{L_\infty} < \delta$, where δ is sufficiently small. Then

$$\|I - A; \mathbb{R}_+^n\|_{L_\infty} \le c\delta. \tag{15.2.38}$$

We introduce a cutoff function $\zeta \in C_0^\infty(\mathcal{B}_2)$ with $\zeta = 1$ on $\mathcal{B}_1$, and set $\zeta_\varepsilon(\xi, \eta) = \zeta(\xi/\varepsilon, \eta/\varepsilon)$, where ε is a small positive number. By (15.2.33)

$$\operatorname{div}(A\nabla(\zeta_\varepsilon v)) - \varepsilon^{-2}\zeta_\varepsilon v = K \text{ on } \mathbb{R}_+^n, \quad \zeta_\varepsilon v\big|_{\mathbb{R}^{n-1}} = 0, \tag{15.2.39}$$

where

$$K = \zeta_\varepsilon H + \nabla\zeta_\varepsilon A \nabla v + \operatorname{div}(v A \nabla \zeta_\varepsilon) - \varepsilon^{-2}\zeta_\varepsilon v$$

with H and v defined as in the case $p(\ell-1) \le n-1$.

We know that $u \in W_p^{1,\alpha}(\Omega)$. Let us suppose that $u \in W_p^{k,\alpha}(\Omega)$, $1 < k \le [\ell]$. Then $v \in W_p^{k,\alpha}(\mathbb{R}_+^n)$ and $H \in W_p^{k-1,\alpha}(\mathbb{R}_+^n)$, which implies $K \in W_p^{k-1,\alpha}(\mathbb{R}_+^n)$. We introduce the new coordinates $(\xi/\varepsilon, \eta/\varepsilon)$ and use the notations $\tilde{A}$, $\tilde{v}$ and $\tilde{K}$ for A, v, and K as functions of $(\xi/\varepsilon, \eta/\varepsilon)$. Written in these dilated variables, the problem (15.2.39) becomes

$$(I-\Delta)(\zeta\tilde{v}) - \operatorname{div}((\tilde{A}-I)\nabla(\zeta\tilde{v})) = \varepsilon^2\tilde{K} \text{ on } \mathbb{R}_+^n, \quad \zeta\tilde{v}\big|_{\mathbb{R}^{n-1}} = 0.$$

By (15.2.37)

$$\|\nabla_{[\ell]}\tilde{A}; \mathbb{R}_+^n\|_{W_p^{0,\alpha}} \le c\,\varepsilon^{\ell-1-(n-1)/p}\|\varphi; \mathbb{R}^{n-1}\|_{W_p^{\ell}}.$$

Also, (15.2.38) holds with A replaced by $\tilde{A}$. Therefore, $\|\tilde{A}-I; \mathbb{R}_+^n\|_{MW_p^{[\ell],\alpha}}$ is sufficiently small. This implies that the operator P given by

$$Pw = (I-\Delta)^{-1}\operatorname{div}((\tilde{A}-I)\nabla w)$$

is contractive in $W_p^{k+1,\alpha}(\mathbb{R}_+^n)$. Hence, $\zeta\tilde{v} \in W_p^{k+1,\alpha}(\mathbb{R}_+^n)$ which implies that $u \in W_p^{k+1,\alpha}(\Omega)$. This gives the result for the mapping (15.2.25).

Only trivial changes in the above argument are needed in order to treat the mappings (15.2.26)–(15.2.29). □

Now we deduce certain properties of the problems $(\mathcal{D}_\pm)$, $(\mathcal{N}_\pm)$ and $(\mathcal{I})$ from Theorem 15.2.1.

Proposition 15.2.3. *Let Ω satisfy the conditions in Theorem* 15.2.1. *Then:*

(i) For every $\Phi_+ \in W_p^\ell(\partial\Omega)$ there exists a unique solution $u_+ \in W_p^{[\ell]+1,\alpha}(\Omega)$ of $(\mathcal{D}_+)$ subject to (15.1.6).

(ii) For every $\Phi_- \in W_p^\ell(\partial\Omega)$ there exists a unique solution $u_- \in W_{p,loc}^{[\ell]+1,\alpha}(\mathbb{R}^n\backslash\overline{\Omega})$ of $(\mathcal{D}_-)$ subject to (15.1.7).

(iii) For every $\Psi_+ \in W_p^{\ell-1}(\partial\Omega) \ominus 1$ there exists a unique solution $v_+ \in W_p^{[\ell]+1,\alpha}(\Omega)$ of $(\mathcal{N}_+)$ subject to $v_+ \perp 1$ on Ω and (15.1.9).

(iv) For every $\Psi_- \in W_p^{\ell-1}(\partial\Omega)$ there exists a unique solution $v_- \in W_{p,loc}^{[\ell]+1,\alpha}(\mathbb{R}^n\backslash\overline{\Omega})$ of $(\mathcal{N}_-)$ subject to (15.1.10).

(v) For every $(\Phi, \Psi) \in W_p^\ell(\partial\Omega) \times W_p^{[\ell]+1,\alpha}(\partial\Omega)$ there exists a unique solution $(w_+, w_-) \in W_p^{[\ell]+1,\alpha}(\Omega) \times W_{p,loc}^{[\ell]+1,\alpha}(\mathbb{R}^n\backslash\overline{\Omega})$ of $(\mathcal{I})$ subject to (15.1.12).

Proof. Assertion (i) was justified in Theorem 15.2.1.

Let us prove (ii). Since the local Lipschitz constant of $\partial\Omega$ is small, the unique solvability of $(\mathcal{N}_-)$ in $W_{p,\text{loc}}^{1,\alpha}(\mathbb{R}^n\backslash\overline{\Omega})$ is standard. It suffices to prove that the solution $u \in W_{p,\text{loc}}^{1,\alpha}(\mathbb{R}^n\backslash\overline{\Omega})$ belongs to $W_{p,\text{loc}}^{[\ell]+1,\alpha}(\mathbb{R}^n\backslash\Omega)$. Let $\chi \in C_0^\infty(\mathbb{R}^n)$, $\chi = 1$ on $\overline{\Omega}$. Clearly,

$$(I - \Delta)(\chi u) = -\chi u - [\Delta, \chi]u \text{ on } \mathbb{R}^n \backslash \overline{\Omega}, \quad \text{tr}(\chi u) = 0 \text{ on } \partial\Omega.$$

Since

$$\chi u + [\Delta, \chi]u \in W_p^{k-1,\alpha}(\mathbb{R}^n \backslash \overline{\Omega}) \qquad \text{for} \quad u \in W_p^{k,\alpha}(\mathbb{R}^n \backslash \overline{\Omega}),$$

it follows from Lemma 15.2.1 with $[\ell]$ replaced by k that $u \in W_p^{k+1,\alpha}(\mathbb{R}^n \backslash \overline{\Omega})$. Letting $k = 1, \dots, [\ell]$, we arrive at (ii).

Proofs of (iii)–(v) require only obvious changes in the argument just used.

□

15.3 Continuity Properties of Boundary Integral Operators

We collect basic properties of the potentials $D\sigma$ and $S\rho$ with $\sigma \in W_p^\ell(\partial\Omega)$ and $\rho \in W_p^{\ell-1}(\partial\Omega)$ where, as usual, $p \in (1, \infty)$, $\ell > 1$, and $\{\ell\} > 0$.

Proposition 15.3.1. *Let the notations $D\sigma$ and $S\rho$ refer to the double and single layer potentials defined on $\mathbb{R}^n \backslash \partial\Omega$. For almost all $Q \in \partial\Omega$ there exist the seven limits*

$$(D\sigma)(Q) := \lim_{\varepsilon \to 0} \frac{1}{|\partial \mathcal{B}_1|} \int\limits_{\partial\Omega \backslash \mathcal{B}_\varepsilon(Q)} \frac{(\zeta - Q, \nu(\zeta))}{|\zeta - Q|^n} \sigma(\zeta) ds_\zeta,$$

$$(D^*\sigma)(Q) := \lim_{\varepsilon \to 0} \frac{1}{|\partial \mathcal{B}_1|} \int\limits_{\partial\Omega \backslash \mathcal{B}_\varepsilon(Q)} \frac{(\zeta - Q, \nu(Q))}{|\zeta - Q|^n} \sigma(\zeta) ds_\zeta,$$

$$\lim_{\substack{z \to Q \\ z \in \Omega}} (D\sigma)(z) = \big(\tfrac{1}{2} I + D\big)\sigma(Q), \tag{15.3.1}$$

$$\lim_{\substack{z \to Q \\ z \in \mathbb{R}^n \backslash \overline{\Omega}}} (D\sigma)(z) = \big(-\tfrac{1}{2} I + D\big)\sigma(Q), \tag{15.3.2}$$

$$(S\rho)(Q) := \lim_{\substack{z \to Q \\ z \in \mathbb{R}^n \backslash \partial\Omega}} (S\rho)(z) = \frac{-1}{|\partial \mathcal{B}_1|(n-2)} \int\limits_{\partial\Omega} \frac{\rho(\zeta) ds_\zeta}{|\zeta - Q|^{n-2}}, \tag{15.3.3}$$

$$\frac{\partial}{\partial \nu}(S\rho)_+(Q) := \lim_{\substack{z \to Q \\ z \in \Omega}} (\nu(Q), (\nabla S\rho)(z)) = \big(-\tfrac{1}{2} I + D^*\big)\rho(Q), \tag{15.3.4}$$

$$\frac{\partial}{\partial \nu}(S\rho)_-(Q) := \lim_{\substack{z \to Q \\ z \in \mathbb{R}^n \backslash \overline{\Omega}}} (\nu(Q), (\nabla S\rho)(z)) = \big(\tfrac{1}{2} I + D^*\big)\rho(Q), \tag{15.3.5}$$

where $(S\rho)_+$ and $(S\rho)_-$ are the restrictions of $S\rho$ to Ω and $\mathbb{R}^n \backslash \overline{\Omega}$.

These classical properties of the layer potentials can be found in [Verc] for σ and ρ in $L_p(\partial\Omega)$, where $z \to Q$ means a nontangential approach. As a justification, a reference is given in [Verc] to the methods developed in [CMM], [Ca3], and [FJR]. However, for our more regular σ and ρ, the above identities can be deduced directly by using the convergence of the integral

$$\int\limits_{\partial\Omega} \frac{|\sigma(\zeta)-\sigma(z)|^p + |\rho(\zeta)-\rho(z)|^p}{|\zeta - z|^{n-1+p\{\ell\}}} ds_\zeta$$

for almost every $z \in \partial\Omega$.

Proposition 15.3.2. *The operators D, D^*, and S satisfy*

$$\| \, D\sigma; \partial\Omega\|_{W_p^\ell} \le c\,\|\sigma, \partial\Omega\|_{W_p^\ell} \tag{15.3.6}$$

$$\| \, (D\sigma)_+; \Omega\|_{W_p^{[\ell]+1,\alpha}} \le c\,\|\sigma; \partial\Omega\|_{W_p^\ell} \tag{15.3.7}$$

$$\| \, (D\sigma)_-; \mathcal{B}\backslash\overline{\Omega}\|_{W_p^{[\ell]+1,\alpha}} \le c(\mathcal{B})\,\|\sigma, \partial\Omega\|_{W_p^\ell} \tag{15.3.8}$$

$$\| \, S\rho; \partial\Omega\|_{W_p^\ell} \le c\,\|\rho; \partial\Omega\|_{W_p^{\ell-1}} \tag{15.3.9}$$

$$\| \, (S\rho)_+; \Omega\|_{W_p^{[\ell]+1,\alpha}} \le c\,\|\rho; \partial\Omega\|_{W_p^{\ell-1}} \tag{15.3.10}$$

$$\| \, (S\rho)_-; \mathcal{B}\backslash\overline{\Omega}\|_{W_p^{[\ell]+1,\alpha}} \le c(\mathcal{B})\,\|\rho, \partial\Omega\|_{W_p^{\ell-1}} \tag{15.3.11}$$

$$\| \, D^*\rho; \partial\Omega\|_{W_p^{\ell-1}} \le c\,\|\rho; \partial\Omega\|_{W_p^{\ell-1}}, \tag{15.3.12}$$

where $(D\rho)_\pm$ and $(S\rho)_\pm$ are the restrictions of $D\sigma$ and $S\rho$ to Ω and $\mathbb{R}^n\backslash\overline{\Omega}$, respectively, and $\mathcal{B}$ is an arbitrary ball containing $\overline{\Omega}$.

Proof. Let us prove (15.3.6)–(15.3.8). Suppose that $\sigma \in W_p^\ell(\partial\Omega)$. By Proposition 15.2.3 (v), the transmission problem ($\mathcal{I}$) with $\Phi = \sigma$ and $\Psi = 0$ has a unique solution

$$(w_+, w_-) \in W_p^{[\ell]+1,\alpha}(\Omega) \times W_{p,\mathrm{loc}}^{[\ell]+1,\alpha}(\mathbb{R}^n\backslash\overline{\Omega})$$

subject to

$$\|w_+; \Omega\|_{W_p^{[\ell]+1,\alpha}} \le c\,\|\sigma; \partial\Omega\|_{W_p^\ell}, \tag{15.3.13}$$

$$\|w_-; \mathcal{B}\backslash\Omega\|_{W_p^{[\ell]+1,\alpha}} \le c(\mathcal{B})\,\|\sigma; \partial\Omega\|_{W_p^\ell}. \tag{15.3.14}$$

By Green's formula, $w_\pm = D(w_+ - w_-) = D\sigma$ on $\mathbb{R}^n\backslash\partial\Omega$ which implies (15.3.7), (15.3.8), and

$$\|\mathrm{tr}\; w_+, \partial\Omega\|_{W_p^\ell} \le c\,\|\sigma, \partial\Omega\|_{W_p^\ell}.$$

Since $D\sigma = \mathrm{tr}\, w_+ - \sigma/2$ by (15.3.1), this last inequality leads to (15.3.6).

Combining (15.3.6) with (15.3.1) and (15.3.2), we see that $\mathrm{tr}\,(D\sigma)_+$ and $\mathrm{tr}\,(D\sigma)_-$ belong to $W_p^\ell(\partial\Omega)$. This together with Theorem 15.1.2 (*i*), (*ii*) lead to (15.3.7) and (15.3.8).

We turn to the proof of (15.3.9)–(15.3.12). Let $\rho \in W_p^{\ell-1}(\partial\Omega)$. By Proposition 15.2.3 (*v*) the transmission problem $(\mathcal{T})$ with $\Phi = 0$ and $\Psi = \rho$ has a unique solution

$$(w_+, w_-) \in W_p^{[\ell]+1,\alpha}(\Omega) \times W_{p,\mathrm{loc}}^{[\ell]+1,\alpha}(\mathbb{R}^n\backslash\overline{\Omega})$$

subject to

$$\|w_+;\Omega\|_{W_p^{[\ell]+1,\alpha}} \le c\,\|\Psi;\partial\Omega\|_{W_p^{\ell-1}}$$

and

$$\|w_-;\mathcal{B}\backslash\Omega\|_{W_p^{[\ell]+1,\alpha}} \le c(\mathcal{B})\,\|\Psi,\partial\Omega\|_{W_p^{\ell-1}}.$$

By Green's formula,

$$w_\pm = S\Big(\frac{\partial w_+}{\partial\nu} - \frac{\partial w_-}{\partial\nu}\Big) = (S\Psi)_\pm \tag{15.3.15}$$

which implies (15.3.10), (15.3.11), and

$$\|\mathrm{tr}\ w_-;\partial\Omega\|_{W_p^\ell} + \|\frac{\partial w_-}{\partial\nu};\partial\Omega\|_{W_p^{\ell-1}} \le c\,\|\rho;\partial\Omega\|_{W_p^{\ell-1}}. \tag{15.3.16}$$

Since

$$D^*\rho = \partial w_-/\partial\nu - \frac{1}{2}\rho$$

by (15.3.5), we arrive at (15.3.12). Finally, (15.3.9) follows from (15.3.15) and (15.3.16). □

We finish this section with a discussion of properties of the normal derivatives of the double layer potential with density in $W_p^\ell(\partial\Omega)$. By (15.3.7), the trace of $\nabla(D\sigma)_+$ belongs to $W_p^{\ell-1}(\partial\Omega)$ and defines a continuous operator: $W_p^\ell(\partial\Omega) \to W_p^{\ell-1}(\partial\Omega)$.

We need the following weighted extension of Proposition 2.7.5 which is proved in the same way.

Proposition 15.3.3. *Let $\Gamma \in MW_p^{k,\alpha}(\mathbb{R}^n_+)$ and let $\Gamma_0 := \|\Gamma;\mathbb{R}^n_+\|_{L_\infty}$. If $g \in C^{k-1}([-\Gamma_0,\Gamma_0])$, then $g(\Gamma) \in MW_p^{k,\alpha}(\mathbb{R}^n_+)$ and*

$$\|g(\Gamma);\mathbb{R}^n_+\|_{MW_p^{k,\alpha}} \le c\sum_{j=0}^k \|g^{(j)};[-\Gamma_0,\Gamma_0]\|_{L_\infty}\|\Gamma;\mathbb{R}^n_+\|^j_{MW_p^{k,\alpha}}.$$

Corollary 15.3.1. *Let* $\gamma \in W_p^\ell(\mathbb{R}^{n-1})$ *and let*

$$\gamma_0 := \|\gamma; \mathbb{R}^{n-1}\|_{L_\infty}.$$

Suppose that $g \in C^{[\ell],1}([-\gamma_0, \gamma_0])$. *Then* $g(\gamma) \in MW_p^\ell(\mathbb{R}^{n-1})$ *and*

$$\|g(\gamma); \mathbb{R}^{n-1}\|_{MW_p^\ell} \leq c \sum_{j=0}^{[\ell]+1} \|g^{(j)}; [-\gamma_0, \gamma_0]\|_{L_\infty} \|\gamma; \mathbb{R}^{n-1}\|^j_{MW_p^\ell}.$$

Proof. The result follows from Proposition 15.3.3 by putting $\Gamma = \mathcal{T}\gamma$, where $\mathcal{T}$ is defined by (9.4.11), and using Theorem 8.7.1.

Proposition 15.3.4. *Let* $\sigma \in W_p^\ell(\partial\Omega)$. *The operator defined by*

$$\frac{\partial}{\partial\nu}(D\sigma)_+(P) := \big(\nu(P), \text{tr } \nabla(D\sigma)_+\big) \tag{15.3.17}$$

maps $W_p^\ell(\partial\Omega)$ *into* $W_p^{\ell-1}(\partial\Omega) \ominus 1$ *continuously, and*

$$\frac{\partial}{\partial\nu}(D\sigma)_+ = \frac{\partial}{\partial\nu}(D\sigma)_- \quad a.e. \text{ on } \partial\Omega. \tag{15.3.18}$$

Proof. The components of ν, expressed in a local cartesian system (x, y), depend smoothly on $\nabla\varphi$, where φ is the function in (14.1.3). Since $\nabla\varphi \in MW_p^{\ell-1}(\mathbb{R}^{n-1})$, we conclude by Proposition 15.3.3 that

$$\nu \in MW_p^{\ell-1}(\partial\Omega). \tag{15.3.19}$$

Hence the operator

$$W_p^\ell(\partial\Omega) \ni \sigma \to \frac{\partial}{\partial\nu}(D\sigma)_+(P) \in W_p^{\ell-1}(\partial\Omega)$$

is continuous.

Let us consider the solution (w_+, w_-) of problem $(\mathcal{T})$ with the boundary conditions

$$\text{tr}\, w_+ - \text{tr}\, w_- = \sigma \text{ and } \frac{\partial w_+}{\partial\nu} - \frac{\partial w_-}{\partial\nu} = 0 \text{ a.e. on } \partial\Omega. \tag{15.3.20}$$

By Green's formula,

$$w_+ = D\,\text{tr}\, w_+ - S\frac{\partial w_+}{\partial\nu} \text{ and } S\frac{\partial w_-}{\partial\nu} = D\,\text{tr}\, w_- \text{ on } \Omega. \tag{15.3.21}$$

Analogously,

$$w_- = S\frac{\partial w_-}{\partial\nu} - D\,\text{tr}\, w_- \text{ and } S\frac{\partial w_+}{\partial\nu} = D\,\text{tr}\, w_+ \text{ on } \mathbb{R}^n\backslash\overline{\Omega}. \tag{15.3.22}$$

Hence,

$$w_+ = D(\operatorname{tr} w_+ - \operatorname{tr} w_-) = D\sigma \text{ on } \Omega \tag{15.3.23}$$

and

$$w_- = D(\operatorname{tr} w_+ - \operatorname{tr} w_-) = D\sigma \text{ on } \mathbb{R}^n\backslash\overline{\Omega}.$$

Now, equality (15.3.18) is a consequence of (15.3.20), and $\partial(D\sigma)_+/\partial\nu \perp 1$ follows from (15.3.23). □

The proposition just proved enables us to introduce the operator $(\partial/\partial\nu)D$ by

$$\Big(\frac{\partial}{\partial\nu} D\Big)\sigma := \frac{\partial}{\partial\nu}(D\sigma)_\pm \tag{15.3.24}$$

and to conclude that $(\partial/\partial\nu)D$ maps $W_p^\ell(\partial\Omega)$ into $W_p^{\ell-1}(\partial\Omega) \ominus 1$.

15.4 Proof of Theorems 15.1.1 and 15.1.2

15.4.1 Proof of Theorem 15.1.1

The continuity of the operators

$$\begin{aligned} D &: W_p^\ell(\partial\Omega) \to W_p^\ell(\partial\Omega) \\ D^* &: W_p^{\ell-1}(\partial\Omega) \to W_p^{\ell-1}(\partial\Omega) \\ S &: W_p^{\ell-1}(\partial\Omega) \to W_p^\ell(\partial\Omega) \\ \frac{\partial}{\partial\nu} D &: W_p^\ell(\partial\Omega) \to W_p^{\ell-1}(\partial\Omega) \ominus 1 \end{aligned}$$

was established in Propositions 15.3.2 and 15.3.4.

Solvability of equation (2_+). Let $u_+ \in W_p^{[\ell]+1,\alpha}(\Omega)$ solve $(\mathcal{D}_+)$ with $\Phi_+ \in W_p^\ell(\partial\Omega)$. Then $\partial u_+/\partial\nu \in W_p^{\ell-1}(\partial\Omega)$. We find a solution $v_- \in W_{p,\mathrm{loc}}^{[\ell]+1,\alpha}(\mathbb{R}^n\backslash\overline{\Omega})$ of problem $(\mathcal{N}_-)$ with $\Psi_- := \partial u_+/\partial\nu$. By Green's formula,

$$u_+ = D \operatorname{tr} u_+ - S\,\frac{\partial u_+}{\partial\nu} \text{ and } S\frac{\partial v_-}{\partial\nu} = D \operatorname{tr} v_- \text{ on } \Omega.$$

Hence, $u_+ = D(\operatorname{tr} u_+ - \operatorname{tr} v_-)$ on Ω. This together with (15.3.1) shows that

$$\sigma_+ := \operatorname{tr} u_+ - \operatorname{tr} v_- \in W_p^\ell(\partial\Omega)$$

is a solution of (2_+).

We have

$$\begin{aligned} \|\sigma_+;\partial\Omega\|_{W_p^\ell} &\le \|\operatorname{tr} u_+;\partial\Omega\|_{W_p^\ell} + \|\operatorname{tr} v_-;\partial\Omega\|_{W_p^\ell} \\ &\le c\big(\|u_+;\Omega\|_{W_p^{[\ell]+1,\alpha}} + \|v_-;\mathcal{B}\backslash\overline{\Omega}\|_{W_p^{[\ell]+1,\alpha}}\big). \end{aligned} \tag{15.4.1}$$

By Proposition 15.2.3 (*iv*) and (15.3.19)

$$\|v_-;\mathcal{B}\backslash\overline{\Omega}\|_{W_p^{[\ell]+1,\alpha}} \le c\,\|\frac{\partial u_+}{\partial\nu};\partial\Omega\|_{W_p^{\ell-1}} \le c\,\|u_+;\Omega\|_{W_p^{[\ell]+1,\alpha}}.$$

The last norm does not exceed $c\,\|\Phi_+,\partial\Omega\|_{W_p^\ell}$ by Proposition 15.3.1 (*i*) which together with (15.4.1) leads to the estimate

$$\|\sigma_+;\partial\Omega\|_{W_p^\ell} \le c\,\|\Phi_+;\partial\Omega\|_{W_p^\ell}. \tag{15.4.2}$$

Uniqueness for equation (2_+). Let

$$(\frac{1}{2}I + D)\sigma = 0 \quad \text{with} \quad \sigma \in W_p^\ell(\partial\Omega).$$

By Proposition 15.2.3 (*v*) we can find a solution

$$(w_+, w_-) \in W_p^{[\ell]+1,\alpha}(\Omega) \times W_{p,\text{loc}}^{[\ell]+1,\alpha}(\mathbb{R}^n\backslash\overline{\Omega})$$

of the transmission problem for the Laplace equation on $\mathbb{R}^n\backslash\partial\Omega$ with boundary conditions (15.3.20). By (15.3.21), $w_+ = (D\sigma)_+$. It follows from (15.3.1) and the definition of σ that $\operatorname{tr} w_+ = 0$. In view of Proposition 15.2.3 (*i*), $w_+ = 0$ which together with (15.3.20) implies that $\partial w_-/\partial\nu = 0$. Proposition 15.2.3 (*iv*) gives $w_- = 0$ and hence $\sigma = \operatorname{tr} w_+ - \operatorname{tr} w_- = 0$. This completes the proof of (*i*).

Solvability of equation (3_-). Let $v_- \in W_{p,\text{loc}}^{[\ell]+1,\alpha}(\mathbb{R}^n\backslash\overline{\Omega})$ solve $(\mathcal{N}_-)$ with $\Psi_- \in W_p^{\ell-1}(\partial\Omega)$. Then $\operatorname{tr} v_- \in W_p^\ell(\partial\Omega)$. We find a solution $u_+ \in W_p^{[\ell]+1,\alpha}(\Omega)$ of $(\mathcal{D}_+)$ with $\Phi_+ := \operatorname{tr} v_-$. By Green's formula,

$$v_- = S(\partial v_-/\partial\nu - \partial u_+/\partial\nu)$$

which implies that

$$\rho_- = \partial v_-/\partial\nu - \partial u_+/\partial\nu \in W_p^{\ell-1}(\partial\Omega)$$

satisfies (3_-).

By (15.3.19),

$$\|\rho_-;\partial\Omega\|_{W_p^{\ell-1}} \le c\,(\|\operatorname{tr}\nabla v_-;\partial\Omega\|_{W_p^{\ell-1}} + \|\operatorname{tr}\nabla u_+;\partial\Omega\|_{W_p^{\ell-1}})$$

$$\le c\,(\|v_-;\mathcal{B}\backslash\overline{\Omega}\|_{W_p^{[\ell]+1,\alpha}} + \|u_+;\Omega\|_{W_p^{[\ell]+1,\alpha}}).$$

Using Proposition 15.2.3 (*i*), we see that the last norm does not exceed $c\,\|\operatorname{tr} v_-;\partial\Omega\|_{W_p^\ell}$ which is majorized by $\|v_-;\mathcal{B}\backslash\overline{\Omega}\|_{W_p^{[\ell]+1,\alpha}}$. Hence,

$$\|\rho_-;\partial\Omega\|_{W_p^{\ell-1}} \le c\,\|v_-;\mathcal{B}\backslash\overline{\Omega}\|_{W_p^{[\ell]+1,\alpha}}.$$

Reference to Proposition 15.2.3 (*iv*) yields the estimate

$$\|\rho_-, \partial\Omega\|_{W_p^{\ell-1}} \le c\,\|\Psi_-, \partial\Omega\|_{W_p^{\ell-1}}.$$

Uniqueness for equation (3_-). Let

$$(\frac{1}{2}I + D^*)\rho_- = 0, \quad \text{where} \quad \rho_- \in W_p^{\ell-1}(\partial\Omega).$$

By Proposition 15.2.3 (*v*) we can find a solution $(w_+, w_-) \in W_p^{[\ell]+1,\alpha}(\Omega) \times W_{p,\mathrm{loc}}^{[\ell]+1,\alpha}(\mathbb{R}^n\backslash\overline{\Omega})$ of the transmission problem for the Laplace equation on $\mathbb{R}^n\backslash\partial\Omega$ with boundary conditions

$$\operatorname{tr} w_+ - \operatorname{tr} w_- = 0 \;\text{ and }\; \frac{\partial w_-}{\partial\nu} - \frac{\partial w_+}{\partial\nu} = \rho_- \;\text{ on } \partial\Omega. \tag{15.4.3}$$

By Green's formula,

$$w_- = S\,\frac{\partial w_-}{\partial\nu} - D\,w_- \;\text{ and }\; S\frac{\partial w_+}{\partial\nu} - D\,w_+ = 0 \;\text{ on } \mathbb{R}^n\backslash\overline{\Omega}. \tag{15.4.4}$$

Hence

$$w_- = S\,(\frac{\partial w_-}{\partial\nu} - \frac{\partial w_+}{\partial\nu}) = S\rho_- \;\text{ on } \mathbb{R}^n\backslash\overline{\Omega}.$$

By (15.3.5),

$$\frac{\partial w_-}{\partial\nu} = (\tfrac{1}{2}I + D^*)\rho_-$$

which implies that $\partial w_-/\partial\nu = 0$ on $\partial\Omega$. Using Theorem 15.1.2 (*iv*), we see that $w_- = 0$ on $\mathbb{R}^n\backslash\overline{\Omega}$. This and (15.4.3) gives $\operatorname{tr} w_+ = 0$. Proposition 15.2.3 (*i*) shows that $w_+ = 0$. Therefore, $\rho_- = 0$ by (15.4.3). This completes the proof of assertion (*ii*).

We turn to assertion (*iii*).

Solvability of equation (15.1.1). Let $u_+ \in W_p^{[\ell]+1,\alpha}(\Omega)$ be a solution of $(\mathcal{D}_+)$ with $\Phi_+ := \Phi \in W_p^\ell(\partial\Omega)$. By u_- we denote a solution of $(\mathcal{D}_-)$ with $\Phi_- := \Phi$, $u_- \in W_{p,\mathrm{loc}}^{[\ell]+1,\alpha}(\mathbb{R}^n\backslash\overline{\Omega})$. Using Green's formula we obtain

$$u_+ = S(\partial u_-/\partial\nu - \partial u_+/\partial\nu)$$

which together with (15.3.19) implies that

$$\rho = \partial u_-/\partial\nu - \partial u_+/\partial\nu \in W_p^{\ell-1}(\partial\Omega).$$

Hence, ρ is a solution of (15.1.1). We have

$$\|\rho;\partial\Omega\|_{W_p^{\ell-1}} \le \|\frac{\partial u_+}{\partial\nu};\partial\Omega\|_{W_p^{\ell-1}} + \|\frac{\partial u_-}{\partial\nu};\partial\Omega\|_{W_p^{\ell-1}}$$

$$\le c\left(\|u_+;\Omega\|_{W_p^{[\ell]+1,\alpha}} + \|u_-;\mathcal{B}\backslash\overline{\Omega}\|_{W_p^{[\ell]+1,\alpha}}\right)$$

and, in view of Proposition 15.2.3 (i), (ii), we obtain

$$\|\rho; \partial\Omega\|_{W_p^{\ell-1}} \le c\,\|\Phi; \partial\Omega\|_{W_p^\ell}.$$

Uniqueness for equation (15.1.1). Let $\rho \in W_p^{\ell-1}(\partial\Omega)$ and $S\rho = 0$ on $\partial\Omega$. By (15.3.3), $\mathrm{tr}\,(S\rho)_\pm = 0$ which together with Proposition 15.2.3 (i), (ii) implies that $(S\rho)_\pm = 0$. Since

$$\rho = \partial(S\rho)_-/\partial\nu - \partial(S\rho)_+/\partial\nu$$

by (15.3.4) and (15.3.5), it follows that $\rho = 0$.

Our next goal is assertion (iv).

Solvability of equation (15.1.2). Let $\Psi \in W_p^{\ell-1}(\partial\Omega) \ominus 1$. By Proposition 15.3.1 (iii) there exists a solution $v_+ \in W_p^{[\ell]+1,\alpha}(\Omega)$ of $(\mathcal{N}_+)$ with boundary data Ψ, unique up to an arbitrary constant term. By v_- we denote a unique $W_{p,\mathrm{loc}}^{[\ell]+1,\alpha}(\mathbb{R}^n\backslash\overline{\Omega})$-solution of $(\mathcal{N}_-)$ with the same boundary data ψ which exists by Proposition 15.2.3 (iv). Let $\sigma = \mathrm{tr}\,v_+ - \mathrm{tr}\,v_-$. Then (15.3.20) holds and, by (15.3.21), $v_+ = D\,\sigma$. This together with (15.3.24) gives (15.1.2). Choosing the value of an arbitrary constant term in v_+, we obtain $\sigma \perp 1$.

We have

$$\|\sigma; \partial\Omega\|_{W_p^\ell} \le \|\mathrm{tr}\,v_+ - \overline{\mathrm{tr}\,v_+}; \partial\Omega\|_{W_p^\ell} + \|\mathrm{tr}\,v_- - \overline{\mathrm{tr}\,v_-}; \partial\Omega\|_{W_p^\ell},$$

where the bar over a function stands for its mean value. Hence,

$$\|\sigma; \partial\Omega\|_{W_p^\ell} \le \|v_+ - \overline{\mathrm{tr}\,v_+}; \Omega\|_{W_p^{[\ell]+1,\alpha}} + \|v_-; \mathcal{B}\backslash\overline{\Omega}\|_{W_p^{[\ell]+1,\alpha}},$$

where $\mathcal{B}$ is a ball containing $\overline{\Omega}$. Using Proposition 15.2.3 (iii), (iv), we obtain

$$\|\sigma; \partial\Omega\|_{W_p^\ell} \le c\,\|\Psi; \partial\Omega\|_{W_p^{\ell-1}}.$$

Uniqueness for equation (15.1.2). Let $\sigma \in W_p^{\ell-1}(\partial\Omega)$ and let $\partial(D\sigma)/\partial\nu = 0$ on $\partial\Omega$. By (15.3.24), $\partial(D\sigma)_\pm/\partial\nu = 0$ and therefore, by Proposition 15.2.3 (ii), (iv),

$$(D\sigma)_+ = const, \qquad (D\sigma)_- = 0.$$

It follows from

$$\sigma = \mathrm{tr}\,(D\sigma)_+ - \mathrm{tr}\,(D\sigma)_-$$

that $\sigma = const$.

Solvability of equation (2_-). We recall that the capacitary potential P of Ω is a unique solution of $(\mathcal{D}_-)$ with the Dirichlet data 1 and that

$$-\int_{\partial\Omega} \frac{\partial P}{\partial\nu} ds = \mathrm{cap}\,\Omega > 0.$$

Suppose that

$$\Phi_- \in W_p^\ell(\partial\Omega) \ominus \partial P/\partial\nu.$$

Let $u_- \in W^{[\ell]+1,\alpha}_{p,\mathrm{loc}}(\mathbb{R}^n\backslash\overline{\Omega})$ satisfy problem $(\mathcal{D}_-)$. Then

$$\partial u_-/\partial\nu \in W^{\ell-1}_p(\partial\Omega)$$

and

$$\int_{\partial\Omega}\frac{\partial u_-}{\partial\nu}ds = \int_{\partial\Omega}\frac{\partial u_-}{\partial\nu}\,\mathrm{tr}\,P ds = \int_{\partial\Omega}\Phi_-\frac{\partial P}{\partial\nu}ds = 0.$$

By Proposition 15.2.3 (iii), there exists a solution $v_+ \in W^{[\ell]+1,\alpha}_p(\Omega)$ of $(\mathcal{N}_+)$ with $\Psi_+ = \partial u_-/\partial\nu$ and $v_+ \perp 1$ on Ω. By Green's formula,

$$u_- = S\,\frac{\partial u_-}{\partial\nu} - D\,\mathrm{tr}\,u_- \;\text{ and }\; S\,\frac{\partial v_+}{\partial\nu} = D\,\mathrm{tr}\,v_+ \;\text{ on }\; \mathbb{R}^n\backslash\overline{\Omega}.$$

Hence, $u_- = D(\mathrm{tr}\,v_+ - \mathrm{tr}\,u_-)$. This together with (15.3.2) shows that

$$\mathrm{tr}\,v_+ - \mathrm{tr}\,u_- \in W^{\ell}_p(\partial\Omega)$$

is a solution of (2_-).

From $(D1)_- = 0$ and (15.3.2) we find

$$(-\tfrac{1}{2}I + D)1 = 0.$$

Therefore, the function

$$\sigma_- := \mathrm{tr}\,v_+ - \mathrm{tr}\,u_- - \overline{\mathrm{tr}\,v_+} + \overline{\mathrm{tr}\,u_-}$$

satisfies (2_-). Clearly,

$$\begin{aligned}\|\sigma_-;\partial\Omega\|_{W^{\ell}_p} &\le c\left(\|\mathrm{tr}\,v_+;\partial\Omega\|_{W^{\ell}_p} + \|\mathrm{tr}\,u_-;\partial\Omega\|_{W^{\ell}_p}\right)\\ &\le c\big(\|v_+;\Omega\|_{W^{[\ell]+1,\alpha}_p} + \|u_-;\mathcal{B}\backslash\overline{\Omega}\|_{W^{[\ell]+1,\alpha}_p}\big).\end{aligned} \tag{15.4.5}$$

In view of Proposition 15.2.3 (ii) and (15.3.19)

$$\|v_+;\Omega\|_{W^{[\ell]+1,\alpha}_p} \le c\,\Big\|\frac{\partial u_-}{\partial\nu},\partial\Omega\Big\|_{W^{\ell-1}_p} \le c\,\|u_-;\mathcal{B}\backslash\overline{\Omega}\|_{W^{[\ell]+1,\alpha}_p}$$

for an arbitrary ball $\mathcal{B}\supset\overline{\Omega}$. The last norm does not exceed $c\,\|\Phi_-;\partial\Omega\|_{W^{\ell}_p}$, by Proposition 15.2.3 (ii), which together with (15.4.5) leads to

$$\|\sigma_-;\partial\Omega\|_{W^{\ell}_p} \le c\,\|\Phi_-;\partial\Omega\|_{W^{\ell}_p}.$$

Uniqueness for equation (2_-). Suppose that $\sigma \in W^{\ell}_p(\partial\Omega)$ and

$$(-\tfrac{1}{2}I + D)^{-1}\sigma = 0.$$

By Proposition 15.2.3 (v) we can find a solution

$$(w_+, w_-) \in W^{[\ell]+1,\alpha}_p(\Omega)\times W^{[\ell]+1,\alpha}_{p,\mathrm{loc}}(\mathbb{R}^n\backslash\overline{\Omega})$$

of the transmission problem for the Laplace equation on $\mathbb{R}^n \backslash \partial\Omega$ with boundary conditions (15.3.20). In view of (15.3.22), $w_- = (D\sigma)_-$. It follows from (15.3.2) and the definition of σ that $\operatorname{tr} w_- = 0$. By Proposition 15.3.1 (*ii*), $w_- = 0$ which together with (15.3.20) implies that $\partial w_+/\partial\nu = 0$. Proposition 15.2.3 (*iii*) gives $w_+ = const$ and hence

$$\sigma = \operatorname{tr} w_+ - \operatorname{tr} w_- = const.$$

The result follows since $\sigma \perp 1$.

Solvability of equation (3_+). Let $v_+ \in W_p^{[\ell]+1,\alpha}(\Omega)$ solve $(\mathcal{N}_+)$ with $\Psi_+ \in W_p^{\ell-1}(\partial\Omega) \ominus 1$. We assume that $v_+ \perp 1$ on Ω. We find a solution $u_- \in W_{p,\mathrm{loc}}^{[\ell]+1,\alpha}(\mathbb{R}^n \backslash \overline{\Omega})$ of $(\mathcal{D}_-)$ with $\Phi_- := \operatorname{tr} v_+ \in W_p^{\ell}(\partial\Omega)$. By Green's formula,

$$v_+ = D \operatorname{tr} v_+ - S\frac{\partial v_+}{\partial\nu} \quad \text{and} \quad S\frac{\partial u_-}{\partial\nu} = D \operatorname{tr} u_- \quad \text{on } \Omega.$$

Hence,

$$v_+ = S\Big(\frac{\partial u_-}{\partial\nu} - \frac{\partial v_+}{\partial\nu}\Big).$$

This together with (15.3.4) shows that

$$\partial u_-/\partial\nu - \partial v_+/\partial\nu \in W_p^{\ell-1}(\partial\Omega)$$

is a solution of (3_+).

Since $S\partial P/\partial\nu = 1$ on Ω, it follows from (15.3.4) that

$$(-\tfrac{1}{2}I + D^*)\partial P/\partial\nu = 0.$$

Therefore, the function

$$\rho_+ := \frac{\partial u_-}{\partial\nu} - \frac{\partial v_+}{\partial\nu} + C\frac{\partial P}{\partial\nu}, \quad C = const, \tag{15.4.6}$$

satisfies (3_+). The constant C can be chosen so that $\rho_+ \perp 1$ on $\partial\Omega$. By (15.4.6) and (15.3.19),

$$\begin{aligned}\|\rho_+;\partial\Omega\|_{W_p^{\ell-1}} &\le c\,(\|\operatorname{tr}\nabla v_+;\partial\Omega\|_{W_p^{\ell-1}} + \|\operatorname{tr}\nabla u_-;\partial\Omega\|_{W_p^{\ell-1}})\\ &\le c\,\big(\|v_+;\Omega\|_{W_p^{[\ell]+1,\alpha}} + \|u_-;\mathcal{B}\backslash\overline{\Omega}\|_{W_p^{[\ell]+1,\alpha}}\big).\end{aligned}$$

By Proposition 15.2.3 (*iii*), the last norm does not exceed $c\,\|\operatorname{tr} v_+;\partial\Omega\|_{W_p^\ell}$ which is majorized by $c\,\|v_+;\Omega\|_{W_p^{[\ell]+1,\alpha}}$. Hence,

$$\|\rho_+;\partial\Omega\|_{W_p^{\ell-1}} \le c\,\|v_+;\Omega\|_{W_p^{[\ell]+1,\alpha}}.$$

Reference to Proposition 15.2.3 (*iii*) yields in the estimate

$$\|\rho_+;\partial\Omega\|_{W_p^{\ell-1}} \le c\,\|\Psi_+;\partial\Omega\|_{W_p^{\ell-1}}.$$

Uniqueness for equation (3_+). Let

$$(-\tfrac{1}{2}I + D^*)\rho_+ = 0, \quad \text{where} \quad \rho_+ \in W_p^{\ell-1}(\partial\Omega) \ominus 1.$$

By Proposition 15.2.3 (v) we can find a solution

$$(w_+, w_-) \in W_p^{[\ell]+1,\alpha}(\Omega) \times W_{p,\text{loc}}^{[\ell]+1,\alpha}(\mathbb{R}^n \backslash \overline{\Omega})$$

of the transmission problem for the Laplace equation on $\mathbb{R}^n \backslash \partial\Omega$ with the boundary conditions

$$\operatorname{tr} w_+ - \operatorname{tr} w_- = 0 \text{ and } \frac{\partial w_-}{\partial \nu} - \frac{\partial w_+}{\partial \nu} = \rho_+ \text{ on } \partial\Omega. \tag{15.4.7}$$

By Green's formula,

$$w_+ = D \operatorname{tr} w_+ - S\,\frac{\partial w_+}{\partial \nu} \text{ and } S\,\frac{\partial w_-}{\partial \nu} = D \operatorname{tr} w_- \text{ on } \Omega. \tag{15.4.8}$$

Hence,

$$w_+ = S\Big(\frac{\partial w_-}{\partial \nu} - \frac{\partial w_+}{\partial \nu}\Big) = S\,\rho_+ \text{ on } \Omega.$$

Using (15.3.4), we have

$$\frac{\partial w_+}{\partial \nu} = (-\tfrac{1}{2}I + D^*)\rho_+$$

which implies that $\partial w_+/\partial \nu = 0$ on $\partial\Omega$. Using Proposition 15.2.3 (iii), we see that $w_+ = const$ on Ω. This and (15.4.7) gives $\operatorname{tr} w_- = const$ which implies that $w_- = const P$. Using (15.4.7) again, we obtain $\rho_+ = const\, \partial P/\partial \nu$. This together with $\rho_+ \perp 1$ completes the proof of assertion (v). □

15.4.2 Proof of Theorem 15.1.2

Now, we are in a position to prove Theorem 15.1.2 stated in Introduction.

All assertions concerning the solvability of the problems $(\mathcal{D}_\pm)$, $(\mathcal{N}_\pm)$, and $(\mathcal{T})$, as well as the estimates (15.1.6)–(15.1.12) have been proved in Proposition 15.2.3. We need to justify the representations of solutions to these problems by layer potentials.

(i) By Theorem 15.1.1 (i), there exists a unique solution $\sigma_+ \in W_p^\ell(\partial\Omega)$ to equation (2_+). By (15.3.1) and (15.3.7), $(D\sigma_+)_+$ is a solution of $(\mathcal{D}_+)$ in $W_p^{[\ell]+1,\alpha}(\Omega)$. Hence, $u_+ = (D\sigma_+)_+$ by Proposition 15.2.3 (i).

Theorem 15.1.1 (iii) implies the existence of a unique solution $\rho \in W_p^{\ell-1}(\partial\Omega)$ of (15.1.1). From (15.3.3) and (15.3.10) we obtain that $(S\rho)_+$ is a solution of $(\mathcal{D}_+)$ in $W_p^{[\ell]+1,\alpha}(\Omega)$. Hence, $u_+ = (S\rho)_+$ by Proposition 15.2.3 (i).

(ii) By Theorem 15.1.1 (v), (15.1.8) has a solution $\sigma_- \in W_p^\ell(\partial\Omega) \ominus 1$ if and only if

$$\int_{\partial\Omega} (\Phi_- - C\,\Gamma) \frac{\partial P}{\partial\nu} ds = 0,$$

which is equivalent to

$$C = \int_{\partial\Omega} \Phi_- \frac{\partial P}{\partial\nu} ds.$$

By (15.3.2) and (15.3.8), the function $(D\sigma_-)_- + C\,\Gamma_-$ is a solution in $W^{[\ell]+1,\alpha}_{p,\mathrm{loc}}(\mathbb{R}^n \backslash \overline{\Omega})$ to $(\mathcal{D}_-)$. Hence,

$$u_- = (D\sigma_-)_- + C\,\Gamma_-$$

by Proposition 15.2.3 (ii).

According to Theorem 15.1.1 (iii), there exists a unique solution $\rho \in W^{\ell-1}_p(\partial\Omega)$ of (15.1.1). Using (15.3.3) and (15.3.11), we find that $(S\rho)_-$ is a solution of $(\mathcal{D}_-)$ in $W^{[\ell]+1,\alpha}_{p,\mathrm{loc}}(\mathbb{R}^n \backslash \overline{\Omega})$. Hence, $u_- = (S\rho)_-$ by Proposition 15.2.3 (ii).

(iii) Theorem 15.1.1 (vi) implies the existence of a unique solution $\rho_+ \in W^{\ell-1}_p(\partial\Omega) \ominus 1$ of equation (3_+). From (15.3.4) and (15.3.10) we obtain that $(S\rho)_+$ is a solution of $(\mathcal{N}_+)$ in $W^{[\ell]+1,\alpha}_p(\Omega)$. Therefore,

$$v_+ = (S\rho)_+ + C.$$

The constant C can be chosen to ensure that $v_+ \perp 1$ on Ω.

By Theorem 15.1.1 (iv), there exists a unique solution $\sigma \in W^{\ell}_p(\partial\Omega) \ominus 1$ of (15.1.2). From (15.3.17) and (15.3.7) we find that $(D\sigma)_+ + C$ is a solution of $(\mathcal{N}_+)$ in $W^{[\ell]+1,\alpha}_p(\Omega)$. Choosing C to ensure the orthogonality of $(D\sigma)_+ + C$ and 1 on Ω, we conclude that

$$v_+ = (D\sigma)_+ + C.$$

(iv) Theorem 15.1.1 (ii) implies the existence of a unique solution $\rho_- \in W^{\ell-1}_p(\partial\Omega)$ of equation (3_-). It follows from (15.3.5) and (15.3.11) that $(S\rho_-)_-$ is a solution of $(\mathcal{N}_-)$ in $W^{[\ell]+1,\alpha}_{p,\mathrm{loc}}(\mathbb{R}^n \backslash \overline{\Omega})$. Hence, $v_- = (S\rho_-)_-$ by Proposition 15.2.3 (iv).

By Theorem 15.1.1 (iv), there exists a unique solution $\sigma \in W^{\ell}_p(\partial\Omega) \ominus 1$ of (15.1.11) provided that

$$C = -\int_{\partial\Omega} \Psi_- ds.$$

It follows from (15.3.24) and (15.3.8) that $(D\sigma)_- + C\Gamma_-$ is a solution of $(\mathcal{N}_-)$ in the space $W^{[\ell]+1,\alpha}_{p,\mathrm{loc}}(\mathbb{R}^n \backslash \overline{\Omega})$. Therefore,

$$v_- = (D\sigma)_- + C\Gamma_-$$

by Proposition 15.2.3 (iv).

(v) We note that $(S\Psi)_+ + (D\Phi)_+$ belongs to $W^{[\ell]+1,\alpha}_p(\Omega)$ and that $(S\Psi)_- + (D\Phi)_-$ belongs to $W^{[\ell]+1,\alpha}_{p,\mathrm{loc}}(\mathbb{R}^n \backslash \overline{\Omega})$ by (15.3.7), (15.3.10) and (15.3.8),

(15.3.11), respectively. Furthermore, $(S\Psi)_\pm + (D\Phi)_\pm$ satisfies the boundary conditions of problem $(\mathcal{T})$ by (15.3.1), (15.3.2), (15.3.3), and (15.3.5). The equality

$$w_\pm = (S\Psi)_\pm + (D\Phi)_\pm$$

results from Proposition 15.2.3 (v). The proof of Theorem 15.1.2 is complete. □

15.5 Properties of Surfaces in the Class $M_p^\ell(\delta)$

Let $p(\ell-1) \le n-1$. According to Theorem 4.1.1, the condition $\partial\Omega \in M_p^\ell(\delta)$ is equivalent to the inequality

$$\|D_{p,\ell}\varphi; \mathbb{R}^{n-1}\|_{M(W_p^{\ell-1}\to L_p)} + \|\nabla\varphi; \mathbb{R}^{n-1}\|_{L_\infty} < c\,\delta, \tag{15.5.1}$$

where

$$D_{p,\ell}\,\varphi(x) = \Big(\int_{\mathbb{R}^{n-1}} |\nabla_{[\ell]}\varphi(x+h) - \nabla_{[\ell]}\varphi(x)|^p |h|^{1-n-p\{\ell\}} dh\Big)^{1/p}.$$

The following local characterization of $M_p^\ell(\delta)$ is obtained in the same way as Lemma 14.7.2, where $\ell = l - 1/p$ with integer l. Let η be an even function in $C_0^\infty(-1,1)$ with $\eta = 1$ on $(-1/2, 1/2)$. We put

$$\eta_\varepsilon(z) = \begin{cases} \eta(|z|/\varepsilon) & \text{if } p(\ell-1) < n, \\ \eta(\log\varepsilon/\log|z|) & \text{if } p(\ell-1) = n. \end{cases}$$

Lemma 15.5.1. *A surface $\partial\Omega$ belongs to the class $M_p^\ell(\delta)$ if and only if for any $O \in \partial\Omega$ there exists a neighborhood U and a special Lipschitz domain $G = \{z = (x,y) : x \in \mathbb{R}^{n-1}, y > \varphi(x)\}$ such that $U \cap \Omega = U \cap G$ and*

$$\limsup_{\varepsilon\to 0} \|\nabla(\eta_\varepsilon\varphi); \mathbb{R}^{n-1}\|_{MW_p^{\ell-1}} \le c\,\delta, \tag{15.5.2}$$

where c is a constant which depends on ℓ, p, n, and η_ϵ is introduced above.

A proof of the next local condition, equivalent to $\partial\Omega \in M_p^\ell(\delta)$, follows the same lines as that of Theorem 14.6.4.

Theorem 15.5.1. *A surface $\partial\Omega$ belongs to the class $M_p^\ell(\delta)$ if and only if for every point $O \in \partial\Omega$ there exists a neighborhood U such that (14.1.3) holds with φ satisfying*

$$\lim_{\varepsilon\to 0}\Bigg(\sup_{e\subset\mathcal{B}_\varepsilon} \frac{\|D_{p,\ell}(\varphi,\mathcal{B}_\varepsilon); e\|_{L_p}}{\big(C_{\ell-1,p}(e)\big)^{1/p}} + \|\nabla\,\varphi; \mathcal{B}_\varepsilon\|_{L_\infty}\Bigg) \le c\,\delta, \tag{15.5.3}$$

where $\mathcal{B}_\varepsilon = \{\zeta \in \mathbb{R}^{n-1}, |\zeta| < \varepsilon\}$,

$$D_{p,\ell}(\varphi, \mathcal{B}_\varepsilon)(x) = \left(\int_{\mathcal{B}_\varepsilon} \frac{|\nabla_{[\ell]}\varphi(x) - \nabla_{[\ell]}\varphi(\zeta)|^p}{|x-\zeta|^{n-1+p\{\ell\}}} d\zeta \right)^{1/p},$$

and c *is a constant depending on* n, p, *and* ℓ.

Simpler conditions sufficient for $\partial\Omega \in M_p^\ell(\delta)$ can be derived from (15.5.3) combined with the well-known inequalities between the capacity and the Lebesgue measure (see Propositions 3.1.2, 3.1.3):

We have $\partial\Omega \in M_p^\ell(\delta)$ *if either*

(*a*) $p(\ell - 1) < n - 1$ *and*

$$\lim_{\varepsilon \to 0} \left(\sup_{e \subset \mathcal{B}_\varepsilon} \frac{\|D_{p,\ell}(\varphi, \mathcal{B}_\varepsilon); e\|_{L_p}}{(\mathrm{mes}_{n-1}(e))^{\frac{n-1-p(\ell-1)}{(n-1)p}}} + \|\nabla \varphi; \mathcal{B}_\varepsilon\|_{L_\infty} \right) \le c\,\delta,$$

or

$p(\ell - 1) = n - 1$ *and*

$$\lim_{\varepsilon \to 0} \left(\sup_{e \subset \mathcal{B}_\varepsilon} |\log(\mathrm{mes}_{n-1}(e))|^{(p-1)/p} \|D_{p,\ell}(\varphi, \mathcal{B}_\varepsilon); e\|_{L_p} + \|\nabla \varphi; \mathcal{B}_\varepsilon\|_{L_\infty} \right) \le c\,\delta.$$

This leads to the following condition, sufficient for $\partial\Omega \in M_p^\ell(0)$:

$$\partial\Omega \in B_{q,p}^\ell \ \text{ and } \ \|\nabla\varphi; \mathbb{R}^{n-1}\|_{L_\infty} < \delta.$$

The condition $\partial\Omega \in B_{q,p}^\ell$ can be improved for $p(\ell - 1) = n - 1$, if the Orlicz space $L_{t^p(\log_+ t)^{p-1}}$ is used instead of L_q with an arbitrary q, but we shall not go into this. Note that $\partial\Omega \in B_{\infty,p}^\ell$ means that the continuity modulus $\omega_{[\ell]}$ of $\nabla_{[\ell]}\varphi$ satisfies

$$\int_0^1 \left(\frac{\omega_{[\ell]}(t)}{t^{\{\ell\}}} \right)^p \frac{dt}{t} < \infty, \tag{15.5.4}$$

which implies, in particular, that any surface $\partial\Omega$ in the class $C^{[\ell],\{\ell\}+\varepsilon}$ with an arbitrary $\varepsilon > 0$ belongs to $M_p^\ell(0)$. □

The next example shows that the condition (15.5.4), sufficient for $\partial\Omega \in M_p^\ell(0)$, is sharp. It demonstrates, in particular, that there exist surfaces in $C^{[\ell],\{\ell\}}$ which do not belong to M_p^ℓ.

Example 15.5.1. Let T denote a domain in $\mathbb{R}^2$ with compact closure and boundary ∂T. By $\mathcal{B}_r^{(2)}$ we denote the open disk of a sufficiently small radius r centered at an arbitrary point $O \in \partial T$. We assume that

$$\mathcal{B}_r^{(2)} \cap T = \{(x_1, y) \in \mathcal{B}_r^{(2)} : x_1 \in \mathbb{R}^1,\, y > F(x_1)\}.$$

Let $\mathcal{B}_\rho^{(n-2)} = \{x' \in \mathbb{R}^{n-2} : |x'| < \rho\}$, where $x' = (x_2, \dots, x_n)$ and let $\eta \in C_0^\infty(\mathcal{B}_2^{(n-2)})$ with $\eta = 1$ on $\mathcal{B}_1^{(n-2)}$. Also let $\varphi(x) = F(x_1)\,\eta(x')$ and $U = \mathcal{B}_r^{(2)} \times \mathcal{B}_2^{(n-2)}$. We construct a bounded domain $\Omega \subset \mathbb{R}^n$ satisfying (14.1.3) whose boundary is smooth outside U. According to Example 4.4.1, for any increasing function $\omega \in C[0,1]$ satisfying the inequality

$$\delta \int_\delta^1 \omega(t)\frac{dt}{t^2} + \int_0^\delta \omega(t)\frac{dt}{t} \le c\,\omega(\delta),$$

as well as the condition

$$\int_0^1 \left(\frac{\omega(t)}{t\{\ell\}}\right)^p \frac{dt}{t} = \infty, \tag{15.5.5}$$

one can construct a function φ of the above form such that the continuity modulus of $\nabla_{[\ell]}\varphi$ does not exceed $c\omega$ with $c = const$, and

$$\varphi \notin W_p^\ell(\mathbb{R}^{n-1}). \tag{15.5.6}$$

Therefore, $\partial\Omega \notin M_p^\ell$. In the case $\partial\Omega \in C^{[\ell],\{\ell\}}$ we have $\omega(t) = t^{\{\ell\}}$ which implies (15.5.5). Hence, the last inclusion is not sufficient for $\partial\Omega$ to be in M_p^ℓ. □

Now we show that surfaces in the class $M_p^\ell(\delta)$ with $p(\ell-1) < n-1$ may have conic vertices and s-dimensional edges if $s < n-1-p(\ell-1)$.

Example 15.5.2. Let s be an integer, $1 \le s \le n-1$, and let $x = (x_1, \dots, x_{n-1}) \in \mathbb{R}^{n-1}$. We use the notations $\xi = (x_1, \dots, x_s)$ and $\eta = (x_{s+1}, \dots, x_{n-1})$. Consider the domain $G = K_{n-s} \times \mathbb{R}^s$, where K_{n-s} is the $(n-s)$-dimensional cone

$$\{(\eta, y) : y > -A\,|\eta|\}, \qquad A = const > 0. \tag{15.5.7}$$

The well-known equivalence relation

$$\|v; \mathbb{R}^{n-1}\|_{W_p^{\ell-1}} \sim \left(\int_{\mathbb{R}^s} \|v(\xi,\cdot); \mathbb{R}^{n-1-s}\|^p_{W_p^\ell}\, d\xi\right)^{1/p} + \left(\int_{\mathbb{R}^{n-1-s}} \|v(\cdot,\eta); \mathbb{R}^s\|^p_{W_p^\ell} d\eta\right)^{1/p} \tag{15.5.8}$$

(see (4.2.3)) implies that the Hardy-type inequality

$$\int_{\mathbb{R}^{n-1}} \frac{|v|^p dx}{|\eta|^{p(\ell-1)}} \le c\,\|v; \mathbb{R}^{n-1}\|^p_{W_p^{\ell-1}} \tag{15.5.9}$$

holds for all $v \in C_0^\infty(\mathbb{R}^{n-1})$ if and only if

$$\int_{\mathbb{R}^{n-1-s}} \frac{|w|^p d\eta}{|\eta|^{p(\ell-1)}} \le c\,\|w; \mathbb{R}^{n-1-s}\|^p_{W_p^{\ell-1}}$$

holds for all $w \in C_0^\infty(\mathbb{R}^{n-1-s})$. It is standard that the last inequality is valid if and only if $p(\ell-1) < n-1-s$. One can easily check that $D_{p,\ell}|\eta| = c\,|\eta|^{1-\ell}$. Hence, (15.5.9) is equivalent to

$$D_{p,\ell}|\eta| \in M(W_p^{\ell-1}(\mathbb{R}^{n-1}) \to L_p(\mathbb{R}^{n-1})).$$

By (4.3.89), the last inclusion can be written as $\nabla|\eta| \in MW_p^{\ell-1}(\mathbb{R}^{n-1})$. Thus, the domain G belongs to $M_p^\ell \cap C^{0,1}$ if and only if $s < n-1-p(\ell-1)$. Under this restriction on the dimension of the edge, $\partial G \in M_p^\ell(c\,A)$. □

Remark 15.5.1. Suppose that for any point $O \in \partial\Omega$ there exists a neighborhood U such that $U \cap \Omega$ is C^∞-diffeomorphic to the domain

$$\mathbb{R}^s \times \{(x,y) : y > f(x_{s+1}, \dots, x_{n-1})\}, \qquad 0 \le s \le n-2,$$

i.e. the dimensions of boundary singularities are at most $n-1-s$. Then (15.5.8) shows that (15.1.3) is equivalent to

$$\|\nabla\varphi; \mathbb{R}^{n-1-s}\|_{MW_p^{\ell-1}} \le c\,\delta$$

and, in particular, it takes the form

$$\|\nabla\varphi; \mathbb{R}^{n-1-s}\|_{W^{\ell-1}_{p,\mathrm{unif}}} \le c\,\delta,$$

if $n-1-s < p(\ell-1) \le n-1$. In other words, $\partial\Omega \in M_p^\ell(\delta)$ if and only if the $(n-1-s)$-dimensional domain $\{(x,y) : y > \varphi(x_{s+1}, \dots, x_{n-1})\}$ belongs to $M_p^\ell(c\,\delta)$.

15.6 Sharpness of Conditions Imposed on $\partial\Omega$

15.6.1 Necessity of the Inclusion $\partial\Omega \in W_p^\ell$ in Theorem 15.2.1

We start by showing that the condition $\partial\Omega \in W_p^\ell$ is necessary for the solvability in $W_p^{[\ell]+1,\alpha}(\Omega)$ of the Dirichlet problem

$$\Delta u = g \in W_p^{[\ell]-1,\alpha}(\Omega), \quad u|_{\partial\Omega} = \Phi \in W_p^\ell(\partial\Omega) \tag{15.6.1}$$

provided that Ω is subject to some regularity assumptions. It is worth noting that certain additional conditions on $\partial\Omega$ should be imposed to guarantee the above statement. For example, it is well known that the problem

$$\Delta u = g \in L_2(\Omega), \quad u|_{\partial\Omega} = 0$$

is uniquely solvable in $W_2^2(\Omega)$ for any convex domain which is not necessarily in $W_2^{3/2}$ ($p = 2$, $\alpha = 0$, $\ell = 3/2$).

Theorem 15.6.1. *Let one of the following conditions hold:*

Either $\ell \in (1, 2)$, $\partial\Omega \in C^1$, *and the continuity modulus* ω *of the normal to* $\partial\Omega$ *satisfies the Dini condition*

$$\int_0^1 \omega(t)\frac{dt}{t} < \infty, \tag{15.6.2}$$

or $\ell > 2$ *and* $\partial\Omega \in C^{[\ell]-1,1}$.

Then $\partial\Omega \in W_p^\ell$ *if, for every* $\Phi \in W_p^\ell(\partial\Omega)$, *the problem* (15.6.1) *has a solution* $u \in W_p^{[\ell]+1,\alpha}(\Omega)$, *where* $\alpha = 1 - \{\ell\} - 1/p$.

Proof. Let Φ be a nonnegative function vanishing on $U \cap \partial\Omega$, where U is an arbitrary coordinate neighborhood. It is well known that (15.6.2) guarantees that $u \in C^1(\overline{\Omega})$ and the outer normal derivative at any point of $U \cap \partial\Omega$ is positive. Let us use the mapping $\lambda = \varkappa^{-1}$ with $\varkappa$ introduced in Sect. 9.4.3.

Since $\varphi \in C^{[\ell]-1,1}(\mathbb{R}^{n-1})$ and $\nabla_x u$, $u_y \in W^{[\ell],\alpha}(V \cap \Omega)$ for any set V with $\overline{V} \subset U$, it follows that $\nabla_x u \circ \lambda$ and $u_y \circ \lambda$ belong to $W^{[\ell],\alpha}(\varkappa(V \cap \Omega))$. Therefore, $\operatorname{tr}(\nabla_x u \circ \lambda)$ and $\operatorname{tr}(u_y \circ \lambda)$ are in $W_p^\ell(\varkappa(V \cap \partial\Omega))$. Observing that

$$(W_p^\ell \cap L_\infty)(\varkappa(V \cap \partial\Omega))$$

is a multiplication algebra, we conclude that

$$\nabla\varphi = -\operatorname{tr}(\nabla_x u \circ \lambda)/\operatorname{tr}(u_y \circ \lambda) \in W_p^\ell(\varkappa(V \cap \partial\Omega)).$$

The result follows from the arbitrariness of V and U. □

Remark 15.6.1. By the above proof we have shown that the inclusion $\partial\Omega \in W_p^\ell$ is necessary for the solvability of $(\mathcal{D}_+)$ in $W^{[\ell]+1,\alpha}(\Omega)$ for all $\Phi \in W_p^\ell(\partial\Omega)$ under the conditions imposed on $\partial\Omega$ in Theorem 15.6.1. Note that $\partial\Omega \in W_p^\ell$ is also sufficient in the case $p(\ell - 1) > n - 1$ by Theorem 15.1.2.

15.6.2 Sharpness of the Condition $\partial\Omega \in B_{\infty,p}^\ell$

Using Remark 15.6.1 we show in the following example that no condition on $\partial\Omega$ weaker than $\partial\Omega \in B_{\infty,p}^\ell$ (condition (15.5.4)) can give the solvability of problem $(\mathcal{D}_+)$ in $W^{[\ell]+1,\alpha}(\Omega)$ for all $\Phi_+ \in W_p^\ell(\partial\Omega)$. We recall that $\partial\Omega \in B_{\infty,p}^\ell$ is sufficient for $\partial\Omega \in M_p^\ell(\delta)$ and hence for this solvability in the case $p(\ell - 1) \leq n - 1$ (see Theorem 15.1.2 and Sect. 15.4).

Example 15.6.1. Let Ω be the domain described in Example 15.5.1. By (15.5.6) and Theorem 15.6.1, problem $(\mathcal{D}_+)$ for Ω is not generally solvable in $W_p^{[\ell]+1,\alpha}(\Omega)$ if $\Phi_+ \in W_p^\ell(\partial\Omega)$.

The next example of the same nature demonstrates the sharpness of the conditions $\partial\Omega \in W_p^\ell$ and $\partial\Omega \in B_{\infty,p}^\ell$ for the solvability of the Neumann problem.

Example 15.6.2. We use the domains T and Ω from Example 15.5.1. Let ∂T be a simple contour and let (α,β) denote the arc $\mathcal{B}_r^{(2)} \cap \partial T$. We choose an arbitrary point $\tau \in \partial T \backslash (\alpha,\beta)$ and introduce a function $\gamma \in W_p^{\ell-1}(\partial T)$ equal to zero on (α,β) and at the point τ, negative on (τ,α) and positive on (β,τ). We require also that γ is orthogonal to one on ∂T. Since $\partial T \in C^{[\ell]-1,1}$, the problem

$$\Delta h = 0 \text{ in } \mathbb{R}^2 \backslash \overline{T}, \quad \partial h/\partial\nu = \gamma \text{ on } \partial T$$

has a solution $h \in (W_{p,\mathrm{loc}}^{[\ell]} \cap L_\infty)(\mathbb{R}^2 \backslash \overline{T})$.

Let $\zeta \in C_0^\infty(\mathbb{R}^2)$ with $\zeta = 1$ on $\overline{T}$, and let η be the cutoff function from Example 15.5.1. The function $v(x,y) = h(x_1,y)\,\zeta(x_1,y)\,\eta(x')$ satisfies the Neumann problem

$$\Delta v - v = g \text{ in } \mathbb{R}^n \backslash \overline{\Omega}, \quad \partial v/\partial\nu = \Psi \text{ on } \partial\Omega \tag{15.6.3}$$

with

$$g = h\Delta\eta + 2\eta\nabla h \nabla\zeta \in W_p^{[\ell]}(\mathbb{R}^n \backslash \overline{\Omega}) \subset W_p^{[\ell]-1,\alpha}(\mathbb{R}^n \backslash \overline{\Omega})$$

and

$$\psi = \eta \partial h/\partial\nu + h\partial\eta/\partial x \in W_p^{\ell-1}(\partial\Omega).$$

If the problem (15.6.3) is solvable in the space $W_p^{[\ell]+1,\alpha}(\mathbb{R}^n \backslash \overline{\Omega})$ for all

$$g \in W_p^{[\ell]-1,\alpha}(\mathbb{R}^n \backslash \overline{\Omega}) \qquad \text{and} \quad \Psi \in W_p^{\ell-1}(\partial\Omega),$$

then $v \in W_p^{[\ell]+1,\alpha}(\mathbb{R}^n \backslash \overline{\Omega})$ and hence $h \in W_{p,\mathrm{loc}}^{[\ell]+1,\alpha}(\mathbb{R}^2 \backslash \overline{T})$. By χ we denote a conjugate harmonic function of h such that $h(\alpha) = 0$. Clearly, $\chi \in W_p^{[\ell]+1,\alpha}(\mathbb{R}^2 \backslash \overline{T})$. Since the first derivative of $\chi|_{\partial T}$ is equal to γ, it follows that $\chi = 0$ on $\mathcal{B}_r^{(2)} \cap \partial T$ and $\chi \geq 0$ on ∂T.

Repeating the proof of Theorem 15.6.1 with $\mathbb{R}^2 \backslash \overline{T}$ and $\chi|_{\partial T}$ instead of Ω and φ, respectively, we obtain $\partial T \in W_p^\ell$ which implies that $\partial\Omega \in W_p^\ell$. However, this is not true in view of (15.5.6), and therefore (15.6.3) is not solvable in $W_p^{[\ell]+1,\alpha}(\mathbb{R}^n \backslash \overline{\Omega})$, in general. Thus, no matter how weak the violation of the inclusion $\partial\Omega \in B_{\infty,p}^\ell$ be, it may lead to the breakdown of the solvability in $W_p^{[\ell]+1,\alpha}(\Omega)$, $p(\ell-1) \leq n$, for the Neumann problem (15.6.3).

15.6.3 Sharpness of the Condition $\partial\Omega \in M_p^\ell(\delta)$ in Theorem 15.2.1

It was mentioned preceding Theorem 15.6.1, that the inclusion $\partial\Omega \in W_p^\ell$ is not necessary for the solvability of the Dirichlet problem in $W_p^{[\ell]+1,\alpha}(\Omega)$. Hence, there is no necessity of the condition $M_p^\ell(\delta)$. However, we show in this section that the inclusion $\partial\Omega \in M_p^\ell(\delta)$ is best possible in a certain sense. In fact, the following two examples demonstrate that the inequality (15.1.3), where $p(\ell-1) < n-1$ and δ is not small, is not sufficient, in general, for the $W_p^{[\ell]+1,\alpha}$-solvability of the Dirichlet and Neumann problems.

Example 15.6.3. Let a domain Ω coincide with the domain G in Example 15.5.2 in a neighborhood of the origin. We adopt the same notations as in Example 15.5.2.

Let u be a positive harmonic function in Ω, satisfying $\operatorname{tr} u = \Phi_+ \in W_p^l(\partial\Omega)$ with Φ_+ vanishing on $U \cap \partial\Omega$. It is well known that for small $r = (|\eta|^2 + y^2)^{1/2}$

$$u(x) = C(\xi) r^\lambda \Theta(\omega) + O(r^{\lambda_1}), \tag{15.6.4}$$

where $1 > \lambda_1 > \lambda > 0$, $\omega = (\eta/r, y/r)$, Θ is smooth on $\{(\eta, y) \in K_{n-s} : r = 1\}$, and C is smooth and positive near the origin of $\mathbb{R}^s$. Moreover, the asymptotic relation (15.6.4) is infinitely differentiable and therefore $u \in W_p^{[\ell]+1,\alpha}(\Omega)$ if and only if $n-1-s > p(\ell-\lambda)$. If $s < n-2$, λ can be made arbitrarily small by choosing sufficiently large A in (15.5.7). In the case $s = n-2$, we have $\lambda > 1/2$, and $\lambda - 1/2$ can be made arbitrarily small by increasing the value of A.

According to Example 15.5.2, $\partial\Omega \in M_p^\ell$ if and only if $n-1-s > p(\ell-1)$. At the same time, one can choose A to have $u \notin W_p^{[\ell]+1,\alpha}(\Omega)$ if and only if $n-1-s < p\ell$ for $s < n-2$, and $1 < p(\ell - 1/2)$ for $s = n-2$. Thus, the inclusion $\partial\Omega \in M_p^\ell \cap C^{0,1}$ does not imply the solvability of $(\mathcal{D}_+)$ in $W_p^{[\ell]+1,\alpha}(\Omega)$ for all $\Phi_+ \in W_p^\ell(\partial\Omega)$ if

$$p\ell > n-1-s > p(\ell-1) \qquad \text{for} \quad s < n-2$$

and

$$p(\ell - 1/2) > 1 > p(\ell-1) \qquad \text{for} \quad s = n-2.$$

□

In the next example we demonstrate that the inclusion $\partial\Omega \in M_p^\ell(\delta)$ in Theorem 15.2.1 cannot be replaced by $\partial\Omega \in M_p^\ell \cap C^{[\ell]}$, for a particular choice of p and ℓ.

Example 15.6.4. Let the domain Ω be described in a neighborhood of O by the inequality $y > \varphi(x)$, where

$$\varphi(x) = C\,\eta(x,0)|x_1|/\log(1/|x_1|)$$

with $C \ge \pi/4$ and $\eta \in C_0^\infty(\mathcal{B}_{1/2})$, $\eta = 1$ on $\mathcal{B}_{1/4}$. By $\zeta(t)$ we denote the conformal mapping of the domain

$$\{t = x_1 + ix_2 : \ |t| < 1/2, \ x_2 > C|x_1|/\log(1/|x_1|)\}$$

into the half-disk $\{\zeta : \operatorname{Im}\zeta > 0, |\zeta| < 1\}$ with $\zeta(0) = 0$. By Sect. 14.6.1, the function $u(z) = \eta(2z)\operatorname{Im}\zeta(x_1 + ix_2)$ does not belong to $W_2^2(\Omega)$ and satisfies the Dirichlet problem

$$\Delta u = f \in L_2(\Omega), \quad \operatorname{tr} u = \Phi \in W_2^{3/2}(\partial\Omega). \tag{15.6.5}$$

Replacing Ω by $\mathbb{R}^n\backslash\overline{\Omega}$ and using the function $v(z) = \eta(2z)\operatorname{Re}\zeta(x_1 + ix_2)$, we arrive at a solution of the Neumann problem

$$\Delta v - v = f \in L_2(\Omega), \quad \partial v/\partial\nu = \Psi \in W_2^{1/2}(\partial\Omega), \tag{15.6.6}$$

which does not belong to $W_2^2(\Omega)$. Thus, there is no solvability in $W_2^2(\Omega)$ and in $W_2^2(\mathbb{R}^n\backslash\overline{\Omega})$ of problems (15.6.5) and (15.6.6) in spite of the inclusion $\partial\Omega \in M_2^{3/2} \cap C^1$.

The same result can be obtained for the problem $(\mathcal{N}_-)$ by making small changes in the above argument. We require that $\mathbb{R}^n\backslash\overline{\Omega}$ coincides with the domain G from Example 15.5.2 near the origin O. We note that there exists a harmonic function in $\mathbb{R}^n\backslash\overline{\Omega}$ satisfying

$$\partial u/\partial\nu = \Psi_- \in W_p^{\ell-1}(\partial\Omega)$$

with $\Psi_- = 0$ in a neighborhood of O such that the asymptotic representation (15.6.4) holds with $C(0) \neq 0$. The rest of the argument is literally the same as for $(\mathcal{D}_+)$.

15.6.4 Sharpness of the Condition $\partial\Omega \in M_p^\ell(\delta)$ in Theorem 15.1.1

Here we give counterexamples concerning the solutions σ and ρ of the integral equations (2_+) and (3_-). First we show that the solvability properties of (2_+) and (3_-) proved in Theorem 15.1.1 may fail if $\partial\Omega \in M_p^\ell \cap C^{0,1}$ and $\partial\Omega \notin M_p^\ell(\delta)$.

Example 15.6.5. Let us consider the domain Ω at the beginning of Example 15.5.2 with $n = 3$ and $s = 0$. Now we deal with the three-dimensional conic singularity $\{z = (r, \theta, \omega) : r > 0,\ 0 \le \theta < \pi - \varepsilon,\ 0 \le \omega < 2\pi\}$, where $\varepsilon > 0$ and θ is the angle between y-axis and z.

We asume that the functions Φ_+ and Ψ_- in (2_+) and (3_-) vanish near the vertex of this cone. It was proved in [LeM] that solutions of (2_+) and (3_-) have the asymptotic representations

$$\sigma_+(z) = \sigma(0) + c_1\,|z|^\lambda + O(|z|^{1+\varepsilon}),$$
$$\rho_-(z) = c_2\,|z|^{\lambda-1}(1 + O(|z|^\mu))$$

with $\mu > 0$, $0 < \lambda < 1$, and nonzero c_1 and c_2. The exponent λ can be made arbitrarily small by diminishing the value of ε. Also note that these asymptotic formulae can be differentiated. According to Example 15.5.2, $\partial\Omega \in M_p^\ell$ if and only if $p(\ell-1) < 2$. However, for $p\ell > 2$, we can choose A in the cone (15.3.22) so large that $\sigma_+ \notin W_p^\ell(\partial\Omega)$ and $\rho_- \notin W_p^{\ell-1}(\partial\Omega)$.

Now, suppose that $\rho_- \in W_p^{\ell-1}(\partial\Omega)$ is a solution of (15.1.1) with $\Phi = 1$ near O, $0 \le \Phi \le 1$ on $\partial\Omega$. Let us denote the solutions of the interior and exterior Dirichlet problems for the Laplace equation with the same boundary data Φ by u_+ and u_-. It is well known that

$$\nabla_k u_-(z) = o(|z|^{k-1}) \quad \text{as } |z| \to 0 \text{ for } \ k = 1, 2$$

and

$$u_+(z) = c\,|z|^\lambda\, \alpha(\theta)\,(1 + o(|z|^\mu)) \quad \text{as } |z| \to 0,$$

where $\mu > 0$, α is smooth, $\alpha'(\pi - \varepsilon) \neq 0$, and $\lambda > 0$ can be made arbitrarily small by choosing a sufficiently small $\varepsilon > 0$. The above asymptotics of u_+ can be differentiated. Hence,

$$\rho = \partial u_-/\partial\nu - \partial u_+/\partial\nu$$

has the differentiable representation

$$\rho(z) = c_3\,|z|^{\lambda-1}(1 + o(|z|^\mu))$$

which contradicts the inclusion $\rho \in W_p^{\ell-1}(\partial\Omega)$. □

Next, we give an example demonstrating that, in general, the condition $\partial\Omega \in M_p^\ell(\delta)$ in Theorem 15.1.1 *(iii)* cannot be improved by $\partial\Omega \in M_p^\ell \cap C^{[\ell]}$.

Example 15.6.6. Consider the same domain Ω as in Example 15.6.4. Let $\rho \in W_2^{1/2}(\partial\Omega)$ be a solution of (15.1.1) with $\Phi = 1$ near O and $0 \le \Phi \le 1$ on $\partial\Omega$. By u_+ and u_- we mean the solution of the interior and exterior Dirichlet problems for the Laplace equation with $\operatorname{tr} u_\pm = \Phi$. Using the conformal mapping $t \to \zeta(t)$ one can show that u_+ has the differentiable asymptotic representation

$$u_+(z) = H(\xi)\,\operatorname{Im}\zeta(t)\,(1 + |\log|t|\,|^{-1}) \quad \text{as } |t| \to 0,$$

where $\xi = (x_3, \ldots, x_{n-1}, y)$ and H is a smooth function with $H(0) \neq 0$. We also have $\nabla_k u_-(z) = o(|z|^{k-1})$ as $|z| \to 0$ for $k = 1, 2$. Hence $\rho = \partial u_-/\partial\nu - \partial u_+/\partial\nu$ has the differentiable representation

$$\rho(z) = c\,H(\xi)\,|\log|t|\,|^{2C/\pi}\,(1 + |\log|t|\,|^{-1})$$

for sufficiently small $|\xi|$ and $|t| \to 0$. One can check directly that the function on the right-hand side does not belong to $W_2^{1/2}$ in any neighborhood of O for $C \ge \pi/4$. If the condition $\partial\Omega \in W_2^{3/2}(\delta)$ in Theorem 15.1.1 *(iii)* could be replaced by $\partial\Omega \in W_2^{3/2} \cap C^1$, one would have a contradiction.

Remark 15.6.2. For the history of boundary integral equations generated by elliptic boundary value problems in domains with nonsmooth boundaries see [Ke2], [Maz17]. In particular, a comprehensive theory of integral equations on the boundaries of Lipschitz graph domains was developed in [JK1], [JK2], [CMM], [Verc], [Ke1], [Ca4], [Fab], [FKV], [DKV], [Cos], [MT1]–[MT5], and [MM]. All these works concern solvability and regularity properties either in $L_p(\partial\Omega)$ or in fractional Sobolev spaces $W_p^\ell(\partial\Omega)$, $0 < \ell < 1$.

Under the assumption that $\partial\Omega$ is sufficiently smooth, one can apply such powerful tools as pseudodifferential calculus to equations $(2_\pm)$–(15.1.2), which results in a comprehensive theory of their solvability in various spaces of differentiable functions.

The regularity theory for equations $(2_\pm)$–(15.1.2) with respect to the scale of the fractional Sobolev spaces $W_p^\ell(\partial\Omega)$ is developed in this chapter under weak smoothness assumptions on $\partial\Omega$, when the corresponding results in the theory of pseudodifferential operators on $\partial\Omega$ are unavailable at the present time. As a substitute, we rely upon an approach proposed in [Maz13], [Maz14], [Maz16], [Maz17], which reduces the study of boundary integral equations to the study of the inverse operators of auxiliary boundary value problems.

The exposition of Sects. 15.1–15.6 follows the paper [MSh23].

15.7 Extension to Boundary Integral Equations of Elasticity

In principle, the Laplace operator we dealt with in this chapter can be replaced by the operator

$$\sum_{1\le i,j\le n} A_{ij}\frac{\partial^2}{\partial z_i \partial z_j}$$

with constant matrix coefficients $A_{ij} = \|A_{ij}^{rs}\|_{r,s=1}^m$, subject to the symmetry condition $A_{ij}^{rs} = A_{ji}^{sr}$ and the Legendre-Hadamard strong ellipticity condition

$$(A_{ij}\eta,\eta)\xi_i\xi_j \ge c|\xi|^2|\eta|^2, \quad c = \text{const} > 0,$$

for all vectors $\xi \in \mathbb{R}^n$ and $\eta \in \mathbb{R}^m$. The statement of the interior and exterior Dirichlet problems for a bounded Lipschitz domain $\Omega \in \mathbb{R}^n$ does not change, whereas the Neumann condition is replaced by

$$\sum_{1\le i,j\le n} \nu_i A_{ij} \,\mathrm{tr}\, \frac{\partial u_\pm}{\partial z_j} = \Psi_\pm$$

with $\nu = (\nu_1, \dots, \nu_n)$ standing for the outer unit normal with respect to Ω.

In particular, one may include the Dirichlet and traction problems for the Lamé system of linear elastostatics. We preserve the same notations for boundary value problems and elastic potentials as in the harmonic potential theory developed previously. Also, we make no difference in notations of spaces of scalar and vector-valued functions.

Let Ω be a domain in $\mathbb{R}^3$ with compact closure and boundary $\partial\Omega$. We study the internal and external Dirichlet problems for the Lamé system

$$\begin{aligned} \mu\,\Delta u_+ + (\lambda+\mu)\nabla \mathrm{div}\, u_+ &= 0 \text{ in } \Omega, \\ \mathrm{tr}\, u_+ &= \Phi_+ \text{ on } \partial\Omega, \end{aligned} \qquad (\mathcal{D}_+)$$

and

$$\begin{aligned} \mu\,\Delta u_- + (\lambda+\mu)\nabla \mathrm{div}\, u_- &= 0 \text{ in } \mathbb{R}^3\backslash\overline{\Omega}, \\ \mathrm{tr}\, u_- &= \Phi_- \text{ on } \partial\Omega, \\ u_-(x) &= O(|x|^{-1}) \quad \text{as } |x|\to\infty, \end{aligned} \qquad (\mathcal{D}_-)$$

where the boundary trace is denoted by tr and

$$\mu > 0, \qquad 3\lambda + 2\mu > 0,$$

as well as the internal and external Neumann problems

$$\begin{aligned} \mu\,\Delta v_+ + (\lambda+\mu)\nabla \mathrm{div}\, v_+ &= 0 \text{ in } \Omega, \\ \mathcal{J} v_+ &= \Psi_+ \text{ on } \partial\Omega, \end{aligned} \qquad (\mathcal{N}_+)$$

and

$$\begin{aligned} \mu\,\Delta v_- + (\lambda+\mu)\nabla \mathrm{div}\, v_- &= 0 \text{ in } \mathbb{R}^3\backslash\overline{\Omega}, \\ \mathcal{J} v_- &= \Psi_- \text{ on } \partial\Omega, \\ v_-(x) &= O(|x|^{-1}) \text{ as } |x|\to\infty, \end{aligned} \qquad (\mathcal{N}_-)$$

where $\mathcal{J}$ is the traction operator given by

$$\mathcal{J}\Big(\frac{\partial}{\partial x}, \nu_x\Big)u = 2\,\mu\frac{\partial u}{\partial \nu_x} + \lambda\nu_x\cdot \mathrm{div}\, u + \mu\,\nu_x \times \mathrm{rot}\, u.$$

We also need the transmission problem

$$\begin{aligned} \mu\,\Delta w_+ + (\lambda+\mu)\nabla \mathrm{div}\, w_+ &= 0 \text{ in } \Omega, \\ \mu\,\Delta w_- + (\lambda+\mu)\nabla \mathrm{div}\, w_- &= 0 \text{ in } \mathbb{R}^3\backslash\overline{\Omega}, \\ \mathrm{tr}\ w_+ - \mathrm{tr}\ w_- = \Phi \quad \text{and } \mathcal{J}w_+ &- \mathcal{J}w_- = \Psi \text{ on } \partial\Omega, \\ w_-(x) &= O(|x|^{-1}) \text{ as } |x|\to\infty. \end{aligned} \qquad (\mathcal{T})$$

We collect properties of the problems $(\mathcal{D}_\pm)$, $(\mathcal{N}_\pm)$, and $(\mathcal{T})$ in the following statement.

Theorem 15.7.1. *Let $p \in (1,\infty)$ and $\alpha = 1 - \{\ell\} - 1/p$, where ℓ is a non-integer with $\ell > 1$. Suppose that $\partial\Omega \in W_p^\ell$ for $p(\ell-1) > 2$ and $\partial\Omega \in M_p^\ell(\delta)$ with some $\delta = \delta(p,\ell)$ for $p(\ell-1) \le 2$. Then*

(i) For every $\Phi_+ \in W_p^\ell(\partial\Omega)$ there exists a unique solution u_+ of $(\mathcal{D}_+)$ in $W_p^{[\ell]+1,\alpha}(\Omega)$.

(ii) For every $\Phi_- \in W_p^\ell(\partial\Omega)$ there exists a unique solution u_- of $(\mathcal{D}_-)$ in $W_{p,loc}^{[\ell]+1,\alpha}(\mathbb{R}^n\backslash\overline{\Omega})$.

(iii) For every $\Psi_+ \in W_p^{\ell-1}(\partial\Omega) \ominus 1$ there exists a unique solution v_+ of $(\mathcal{N}_+)$ in $W_p^{[\ell]+1,\alpha}(\Omega)$, subject to $v_+ \perp 1$ on Ω.

(iv) For every $\Psi_- \in W_p^{\ell-1}(\partial\Omega)$ there exists a unique solution v_- of $(\mathcal{N}_-)$ in $W_{p,loc}^{[\ell]+1,\alpha}(\mathbb{R}^n\backslash\overline{\Omega})$.

(v) For every $(\Phi,\Psi) \in W_p^\ell(\partial\Omega) \times W_p^{\ell-1}(\partial\Omega)$ there exists a unique solution

$$(w_+, w_-) \in W_p^{[\ell]+1,\alpha}(\Omega) \times W_{p,\mathrm{loc}}^{[\ell]+1,\alpha}(\mathbb{R}^n\backslash\overline{\Omega})$$

of $(\mathcal{T})$.

The proof is essentially the same as that of Theorem 15.1.2.

The following two results in the theory of elastic potentials are parallel to Proposition 15.3.2 and Theorem 15.1.1, and can be proved in a similar way. We restrict ourselves to the solution of $(\mathcal{D}_\pm)$ by means of the single layer potential as well as $(\mathcal{D}_\pm)$ and $(\mathcal{N}_\pm)$ by means of the double layer potential.

We recall that the Kelvin-Somigliana tensor $\Gamma = \|\Gamma_{ij}\|_{i,j=1}^3$, where

$$\Gamma_{ij}(x) = -\frac{\lambda+\mu}{8\pi\mu(\lambda+3\mu)}\Big(\frac{\lambda+3\mu}{\lambda+\mu}\frac{\delta_i^j}{|x|} + \frac{x_i\,x_j}{|x|^3}\Big),$$

is a fundamental solution of the Lamé system, and we introduce the elastic single layer potential

$$(S\rho)(x) = \int_{\partial\Omega} \Gamma(x-\xi)\,\rho(\xi)ds.$$

Theorem 15.7.2. *Let $\partial\Omega$ satisfy the conditions in Theorem 15.7.1. Then*

$$\|\, S\rho; \partial\Omega\|_{W_p^\ell} \le c\,\|\rho; \partial\Omega\|_{W_p^{\ell-1}}, \tag{15.7.1}$$

$$\|\, (S\rho)_+; \Omega\|_{W_p^{[\ell]+1,\alpha}} \le c\,\|\rho; \partial\Omega\|_{W_p^{\ell-1}}, \tag{15.7.2}$$

$$\|\, (S\rho)_-; \mathcal{B}\backslash\overline{\Omega}\|_{W_p^{[\ell]+1,\alpha}} \le c(\mathcal{B})\,\|\rho, \partial\Omega\|_{W_p^{\ell-1}}, \tag{15.7.3}$$

where $(S\rho)_\pm$ are the restrictions of $S\rho$ to Ω and $\mathbb{R}^n\backslash\overline{\Omega}$, respectively, and $\mathcal{B}$ is an arbitrary ball containing $\overline{\Omega}$.

Let D be the elastic double layer potential defined by

$$(D\chi)(x) = \int_{\partial\Omega} \mathcal{J}\Big(\frac{\partial}{\partial\xi}, \nu_\xi\Big)\big(\Gamma(x-\xi)\big)\, \chi(\xi)\, ds, \quad x \notin \partial\Omega.$$

If $u_+ = D\chi_+$, then χ_+ satisfies the integral equation

$$\tfrac{1}{2}\chi_+ + D\chi_+ = \Phi_+ \quad \text{on} \quad \partial\Omega, \tag{15.7.4}$$

which is understood in the same sense as in [Ke1] and [Fab]. The solution of $(\mathcal{D}_-)$ may be represented as the sum

$$(D\chi)(x) + a\,\Gamma(x,0) + b\,\mathrm{rot}\,\Gamma(x,0),$$

where a and b are unknown constant vectors. The triple (χ, a, b) satisfies the equation

$$-\tfrac{1}{2}\chi_- + D\chi_- + a\,\Gamma(\cdot,0) + b\,\mathrm{rot}\,\Gamma(\cdot,0) = -\Phi_-. \tag{15.7.5}$$

Representing solutions of problems $(\mathcal{N}_\pm)$ in the form $S\chi_\pm$, one arrives at the equations

$$-\tfrac{1}{2}\chi_+ + D^*\chi_+ = \Psi_+, \tag{15.7.6}$$

$$\tfrac{1}{2}\chi_- + D^*\chi_- = \Psi_-, \tag{15.7.7}$$

where D^* is the adjoint of D.

Here is the main result concerning the integral equations (15.7.4)–(15.7.7).

Theorem 15.7.3. *Let $\partial\Omega$ satisfy the conditions in Theorem* 15.7.1. *Then*

(i) the operators D and D^* are bounded on $W_p^\ell(\partial\Omega)$ and $W_p^{\ell-1}(\partial\Omega)$, respectively;

(ii) for $\Phi_\pm \in W_p^\ell(\partial\Omega)$ the equations (15.7.4) *and* (15.7.5) *are uniquely solvable in $W_p^\ell(\partial\Omega)$ and $W_p^\ell(\partial\Omega) \times \mathbb{R}^3 \times \mathbb{R}^3$;*

(iii) there exists a continuous inverse of $\frac{1}{2}I + D^*$ on the space $W_p^{\ell-1}(\partial\Omega)$. Equation (15.7.6) *has a solution in $W_p^{\ell-1}(\partial\Omega)$ for an arbitrary Ψ_+ orthogonal to all rigid motions.*

Remark 15.7.1. A straightforward modification of our arguments used in the harmonic potential theory developed in this chapter leads to analogous higher regularity results in the theory of hydrodynamic potentials related to the Stokes system

$$\nu\Delta u - \nabla p = 0, \quad \mathrm{div}\, u = 0,$$

(for background, see [Lad], [Ke1], [Fab], and Sects. 2.2–2.4 in [Maz17]).

16

Applications of Multipliers to the Theory of Integral Operators

In this chapter it is shown that Sobolev multipliers are useful for the study of integral operators. First, in Sect. 16.1 we consider an arbitrary convolution operator acting in a pair of weighted L_2-spaces and collect corollaries of the theory of multipliers providing criteria of boundedness and compactness of the convolutions and a characterization of their spectra. Next we turn to classical singular integral operators acting in Sobolev spaces. In Sect. 16.2 a calculus of these operators is developed under the assumption that their symbols belong to classes of multipliers in Sobolev spaces. Finally, in Sect. 16.3 sharp conditions for continuity of the singular integral operators acting from W_2^m to W_2^l are found. These conditions are formulated in terms of certain classes of multipliers.

16.1 Convolution Operator in Weighted L_2-Spaces

Let

$$K : u \to k * u$$

be a convolution operator with the kernel k. The results of Sects. 3.6 and 4.6 for the case $p = 2$ can be interpreted as theorems on properties of K considered as a mapping

$$K : L_2((1+|x|^2)^{m/2}) \to L_2((1+|x|^2)^{l/2}), \quad m \geq l \geq 0, \tag{16.1.1}$$

where

$$\|u\|_{L_2((1+|x|^2)^{k/2})} = \Big(\int |u|^2 (1+|x|^2)^k \, dx\Big)^{1/2}.$$

For example, the operator K is continuous if and only if its symbol, i.e. the Fourier transform Fk, belongs to $M(W_2^m \to W_2^l)$. By Theorem 4.1.1 this is equivalent to the following properties: $Fk \in W^l_{2,\mathrm{unif}}$ and, for every compact set $e \subset \mathbb{R}^n$,

V.G. Maz'ya, T.O. Shaposhnikova, *Theory of Sobolev Multipliers*,
Grundlehren der mathematischen Wissenschaften 337,

$$\int_e [D_{2,l}(Fk)]^2\, dx \le \text{const}\ C_{2,m}(e)\,,$$

where

$$(D_{2,l}u)(x) = \Big(\int |\nabla_{[l],x}(u(x+h)-u(x))|^2 |h|^{-n-2\{l\}}\, dh\Big)^{1/2}$$

if $\{l\} > 0$, and

$$(D_{2,l}u)(x) = |\nabla_l u(x)|$$

if $\{l\} = 0$. Moreover,

$$\|K\| \sim \sup_e \Big(\|Fk\|_{L_{2,\text{unif}}} + \frac{\|D_{2,l}(Fk); e\|_{L_2}}{[C_{2,m}(e)]^{1/2}}\Big)\,.$$

In the case $2m > n$,

$$\|K\| \sim \|Fk\|_{W^l_{2,\text{unif}}}\,.$$

Theorem 4.6.1 describes properties of a function of the operator K considered as the mapping (16.1.1). Namely, let $0 < l < 1$ and let φ be a complex-valued function of a complex argument with $\varphi(0) = 0$. By $\varphi(K)$ we denote the convolution operator with the symbol $\varphi(Fk)$. If

$$|\varphi(t+\tau) - \varphi(t)| \le A|\tau|^\rho,$$

where $|\tau| < 1$ and $\rho \in (0,1)$, we derive the following assertion from Theorem 4.6.1. If the operator (16.1.1) is continuous, then the operator

$$\varphi(K) : L_2((1+|x|^2)^{(m-l+r)/2}) \to L_2((1+|x|^2)^{r/2})$$

with $r \in (0, l\rho)$ is continuous as well.

Results of Chap. 7 imply two-sided estimates for the essential norm and conditions for compactness of the convolution K:

(*i*) If $m > l$ and $2m \le n$, then

$$\text{ess}\,\|K\| \sim \lim_{\delta\to 0}\Big(\sup_{\{e:d(e)\le\delta\}} \frac{\|D_{2,l}(Fk); e\|_{L_2}}{(C_{2,m}(e))^{\frac12}} + \sup_{\substack{x\in\mathbb{R}^n\\ \rho\le\delta}} \rho^{m-l-\frac{n}{2}} \|Fk; \mathcal{B}_\rho(x)\|_{L_2}\Big)$$

$$+ \lim_{r\to\infty}\Big(\sup_{\substack{e\subset\mathbb{R}^n\setminus\mathcal{B}_r\\ d(e)\le 1}} \frac{\|D_{2,l}(Fk); e\|_{L_2}}{(C_{2,m}(e))^{\frac12}} + \sup_{x\in\mathbb{R}^n\setminus\mathcal{B}_r} \|Fk; \mathcal{B}_1(x)\|_{L_2}\Big), \qquad (16.1.2)$$

where $d(e)$ is the diameter of a compact set $e \subset \mathbb{R}^n$.

(*ii*) If $m > l$ and $2m > n$, then

$$\text{ess}\,\|K\| \sim \limsup_{|x|\to\infty} \|Fk; \mathcal{B}_1(x)\|_{W^l_2}. \qquad (16.1.3)$$

Hence K is compact if and only if either $m > l$, $2m \le n$, and

$$\lim_{\delta\to 0}\Big(\sup_{\{e:d(e)\le\delta\}}\frac{\|D_{2,l}(Fk);e\|_{L_2}}{\big(C_{2,m}(e)\big)^{\frac{1}{2}}}+\sup_{\substack{x\in\mathbb{R}^n\\ \rho\le\delta}}\rho^{m-l-\frac{n}{2}}\|Fk;\mathcal{B}_\rho(x)\|_{L_2}\Big)=0,$$

$$\lim_{r\to\infty}\Big(\sup_{\substack{e\subset\mathbb{R}^n\backslash\mathcal{B}_r\\ d(e)\le 1}}\frac{\|D_{2,l}Fk;e\|_{L_2}}{(C_{2,m}(e))^{\frac{1}{2}}}+\sup_{x\in\mathbb{R}^n\backslash\mathcal{B}_r}\|Fk;\mathcal{B}_1(x)\|_{L_2}\Big)=0;$$

or $m>l$, $2m>n$ and

$$\lim_{|x|\to\infty}\|Fk;\mathcal{B}_1(x)\|_{W_2^l}=0.$$

According to Corollary 3.6.1, a complex number λ belongs to the spectrum $\sigma(K)$ of an operator K, continuous in $L_2((1+|x|^2)^{l/2})$, $l>0$, if and only if

$$(Fk-\lambda)^{-1}\notin L_\infty.$$

Let $\lambda\in\sigma(K)$. By Theorem 3.6.1, λ is an eigenvalue of K if and only if

$$\lim_{\rho\to 0}\rho^{-n}\,C_{2,l}(\mathcal{B}_\rho(x)\backslash Z_\lambda)=0$$

for all x in a set of positive measure, where $Z_\lambda=\{\xi\in\mathbb{R}^n:(Fk)(\xi)=\lambda\}$. This condition is equivalent to

$$C_{2,l}(G\backslash Z_\lambda)<C_{2,l}(G)\qquad\text{for some open set } G. \tag{16.1.4}$$

By the same Theorem 3.6.1, λ belongs to the residual spectrum $\sigma_r(K)$ if and only if (16.1.4) does not hold and $C_{2,l}(Z_\lambda)>0$. Furthermore, λ belongs to the continuous spectrum of K if and only if $C_{2,l}(Z_\lambda)=0$.

If λ is a point of the spectrum of an operator K, continuous in $L_2((1+|x|^2)^{-l/2})$, then Theorem 3.6.1 implies that

$$\lambda\in\sigma_p(K)\Longleftrightarrow C_{2,l}(Z_\lambda)>0$$

and

$$\lambda\in\sigma_c(K)\Longleftrightarrow C_{2,l}(Z_\lambda)=0\,.$$

Consequently, the convolution acting in $L_2((1+|x|^2)^{-l/2})$ has no residual spectrum.

16.2 Calculus of Singular Integral Operators with Symbols in Spaces of Multipliers

In this section we demonstrate that the spaces MW_p^l and $\mathring{M}W_p^l$ are useful in construction of a calculus of singular integral operators acting in W_p^l, $1<p<\infty$, $l=0,1,\dots$.

First we quote basic definitions of the theory of singular integrals (see Mikhlin and Prössdorf [MiP]).

Let α be a bounded measurable function defined on $\mathbb{R}^n \times \partial\mathcal{B}_1$, orthogonal to one on $\partial\mathcal{B}_1$, and let $\alpha_0^{(0)} \in L_\infty(\mathbb{R}^n)$. The singular integral operator is defined by

$$(Au)(x) = \alpha_0^{(0)}(x)u(x) + \int_{\mathbb{R}^n} \frac{\alpha(x,\theta)}{r^n} u(y)\,dy\,, \qquad x \in \mathbb{R}^n\,, \quad (16.2.1)$$

where $r = |y-x|$, $\theta = (y-x)/r$ and the integral is interpreted in the sense of the Cauchy principal value. We express α as a series in spherical harmonics

$$\alpha(x,\theta) = \sum_{m=1}^{\infty}\sum_{k=1}^{k_m} \alpha_m^{(k)}(x) Y_m^{(k)}(\theta)\,, \qquad (16.2.2)$$

where k_m is the number of spherical harmonics $Y_m^{(k)}$ of order m. Then (16.2.1) and (16.2.2) imply the formal expansion

$$(Au)(x) = \alpha_0^{(0)}(x)u(x) + \sum_{m=1}^{\infty}\sum_{k=1}^{k_m} \alpha_m^{(k)}(x) \int_{\mathbb{R}^n} \frac{Y_m^{(k)}(\theta)}{r^n} u(y)\,dy\,. \quad (16.2.3)$$

We put $(S_0^{(0)}u)(x) = u(x)$, $k_0 = 0$ and

$$(S_m^{(k)}u)(x) = \int_{\mathbb{R}^n} \frac{Y_m^{(k)}(\theta)}{r^n} u(y)\,dy\,.$$

It is known that $S_m^{(k)} = \mu_m F^{-1} Y_m^{(k)} F$, where F is the Fourier transform in $\mathbb{R}^n$, $\mu_0 = 1$ and

$$\mu_m = i^{-m}\pi^{n/2}\frac{\Gamma(m/2)}{\Gamma((n+m)/2)}\,, \qquad |\mu_m| \sim m^{-n/2}\,, \qquad (16.2.4)$$

for $m \geq 1$ with Γ standing for the Gamma function. Hence the operator A can be written in the form

$$(Au)(x) = F^{-1}_{\xi\to x}[a(x,\xi/|\xi|)(Fu)(\xi)]\,,$$

where a is defined by

$$a(x,\theta) = \sum_{m=0}^{\infty}\sum_{k=1}^{k_m} \mu_m \alpha_m^{(k)}(x) Y_m^{(k)}(\theta) \qquad (16.2.5)$$

and is called the symbol of the singular integral operator A.

Next we introduce the space $C^\infty(MW_p^l, \partial\mathcal{B}_1)$ of infinitely differentiable functions defined on $\partial\mathcal{B}_1$ with range in MW_p^l. The space $C^\infty(\mathring{M}W_p^l, \partial\mathcal{B}_1)$ is defined in a similar way.

Lemma 16.2.1. *If $a \in C^\infty(MW_p^l, \partial\mathcal{B}_1)$, then the singular integral operator A with the symbol a is continuous in W_p^l and can be expressed as the series*

$$\sum_{m=0}^{\infty}\sum_{k=1}^{k_m} \alpha_m^{(k)} S_m^{(k)}$$

which converges in the operator norm in W_p^l.

Proof. From (16.2.4), (16.2.5), and the definition of the space $C^\infty(MW_p^l, \partial\mathcal{B}_1)$, it follows that for any positive integer N there exists a constant C_N such that

$$\|\alpha_m^{(k)}; \mathbb{R}^n\|_{MW_p^l} \le C_N m^{-N} \qquad m \ge 1\,.$$

It remains to make use of the fact that the singular convolution operator $S_m^{(k)}$ is continuous in W_p^l and its norm increases no faster than a certain degree of m as $m \to \infty$. □

Henceforth, A, B, and C are singular integral operators in $\mathbb{R}^n$ with the symbols $a(x,\theta)$, $b(x,\theta)$, and $c(x,\theta)$, where $x \in \mathbb{R}^n$ and $\theta \in \partial\mathcal{B}_1$.

Theorem 16.2.1. *Let AB be a singular operator with the symbol ab and let $A \circ B$ be the composition of operators A and B.*

If $a \in C^\infty(MW_p^l, \partial\mathcal{B}_1)$ and there exists a function $b_\infty \in C^\infty(\partial\mathcal{B}_1)$ such that $b - b_\infty \in C^\infty(\overset{\circ}{M}W_p^l, \partial\mathcal{B}_1)$, then the operator $AB - A \circ B$ is compact in W_p^l.

Proof. By Lemma 16.2.1, it is sufficient to consider the operators A and B expressed in the form of finite sums

$$\sum_{m,k} \alpha_m^{(k)} S_m^{(k)}\,, \qquad \sum_{q,r} \beta_q^{(r)} S_q^{(r)}\,.$$

It is clear that

$$\begin{aligned}
A \circ B &= F^{-1}\Big(\sum_{m,k,q,r} \mu_m \alpha_m^{(k)} \mu_q \beta_q^{(r)} Y_m^{(k)} Y_q^{(r)}\Big) F \\
&= \sum_{m,k,q,r} \alpha_m^{(k)} \beta_q^{(r)} \mu_m F^{-1} Y_m^{(k)} F \mu_q F^{-1} Y_q^{(r)} F \\
&= \sum_{m,k,q,r} \alpha_m^{(k)} \beta_q^{(r)} S_m^{(k)} S_q^{(r)}\,.
\end{aligned}$$

On the other hand,

$$\begin{aligned}
AB &= \sum_{m,k,q,r} \alpha_m^{(k)} S_m^{(k)} \beta_q^{(r)} S_q^{(r)} \\
&= \sum_{m,k,q,r} \alpha_m^{(k)} \beta_q^{(r)} S_m^{(k)} S_q^{(r)} + \sum_{m,k,q,r} \alpha_m^{(k)} [S_m^{(k)}, \beta_q^{(r)}] S_q^{(r)}\,,
\end{aligned}$$

where $[X, Y] = XY - YX$. Therefore,

$$AB - A \circ B = \sum_{m,k,q,r} \alpha_m^{(k)} [S_m^{(k)}, \beta_q^{(r)}] S_q^{(r)}, \tag{16.2.6}$$

and it remains to show that the commutator $[S_m^{(k)}, \beta_q^{(r)}]$ is compact in W_p^l. This assertion is contained in the following lemma.

Lemma 16.2.2. *Let $\gamma \in \mathring{M}W_p^l$, and let A be a singular integral operator with the symbol $a(\theta)$, where $a \in C^\infty(\partial\mathcal{B}_1)$. Then the commutator $[\gamma, A]$ is compact in W_p^l.*

Proof. Let $\{\gamma_j\}$ be a sequence of functions, $\gamma_j \in C_0^\infty$, and let γ_j converge to γ in MW_p^l. Then the operators $(\gamma - \gamma_j)A$ and $A(\gamma - \gamma_j)$ tend to zero in the operator norm in W_p^l. The compactness of the mapping $[\gamma_j, A]$ in W_p^l is well known and can be easily verified. □

The following theorem contains conditions for the operator $AB - A \circ B$ to be of order -1 in W_p^l (cf. [KN]).

Theorem 16.2.2. *If $a \in C^\infty(MW_p^{l+1}, \partial\mathcal{B}_1)$ and $\nabla_x b \in C^\infty(MW_p^l, \partial\mathcal{B}_1)$, then the operator $AB - A \circ B$ maps W_p^l continuously into W_p^{l+1}. Here AB is a singular operator with the symbol ab, and $A \circ B$ is the composition of operators A and B.*

This assertion follows from (16.2.6) and the next lemma.

Lemma 16.2.3. *Let a function γ satisfy the Lipschitz condition and let $\nabla\gamma \in MW_p^l$. Further, let A be a singular integral operator with the symbol $a(\xi)$, where $a \in C^\infty(\partial\mathcal{B}_1)$. Then the commutator $[\gamma, A]$ satisfies the inequality*

$$\|[\gamma, A]\|_{W_p^l \to W_p^{l+1}} \le c \, \|\nabla\gamma\|_{MW_p^l} .$$

Proof. For $l = 0$ the assertion is a known result due to Calderon [Ca2]. Let the lemma be proved for all $l = 0, 1, \ldots, k-1$. Then, for all $u \in W_p^{k+1}$,

$$\|[\gamma, A]u\|_{W_p^{k+1}} \le \sum_{j=1}^{n} \Big\| \frac{\partial}{\partial x_j} [\gamma, A]u \Big\|_{W_p^k} + \|[\gamma, A]u\|_{W_p^k} . \tag{16.2.7}$$

In view of the imbedding $MW_p^k \subset MW_p^{k-1}$, the last term in (16.2.7) is estimated by the induction hypothesis. Since

$$(\partial/\partial x_j)[\gamma, A] = (\partial\gamma/\partial x_j)A - A(\partial\gamma/\partial x_j) + [\gamma, A](\partial/\partial x_j),$$

it follows that

$$\left\|\frac{\partial}{\partial x_j}[\gamma, A]u\right\|_{W_p^k} \leq 2\|A\|_{W_p^k \to W_p^k}\left\|\frac{\partial\gamma}{\partial x_j}\right\|_{MW_p^k}\|u\|_{W_p^k} + \left\|[\gamma, A]\frac{\partial u}{\partial x_j}\right\|_{W_p^k}.$$

Applying the induction hypothesis to the last norm, we complete the proof. □

To conclude this section we formulate two corollaries on the regularization of a singular integral operator which follow from Theorems 16.2.1 and 16.2.2.

Corollary 16.2.1. *Let there exist a function* $a_\infty \in C^\infty(\partial\mathcal{B}_1)$ *such that*

$$a - a_\infty \in C^\infty(\overset{\circ}{M}W_p^l, \partial\mathcal{B}_1).$$

Further, let $c = 1/a \in L_\infty(\mathbb{R}^n \times \partial\mathcal{B}_1)$. *Then*

$$c \in C^\infty(MW_p^l, \partial\mathcal{B}_1)$$

and

$$c - c_\infty \in C^\infty(\overset{\circ}{M}W_p^l, \partial\mathcal{B}_1),$$

where $c_\infty = 1/\alpha_\infty$. *Moreover, the operators* $A \circ C - I$ *and* $C \circ A - I$ *are compact in* W_p^l.

Corollary 16.2.2. *Let* $a \in L_\infty(\mathbb{R}^n \times \partial\mathcal{B}_1)$ *and let*

$$\nabla_x a \in C^\infty(MW_p^l, \partial\mathcal{B}_1).$$

Further, let $c = 1/a \in L_\infty(\mathbb{R}^n \times \partial\mathcal{B}_1)$. *Then*

$$\nabla_x c \in C^\infty(MW_p^l, \partial\mathcal{B}_1)$$

and the operators $A \circ C - I$ *and* $C \circ A - I$ *map* W_p^l *continuously into* W_p^{l+1}.

Remark 16.2.1. The condition of infinite differentiability of the symbols on $\partial\mathcal{B}_1$ can be replaced everywhere in this section by the condition of their sufficient smoothness.

16.3 Continuity in Sobolev Spaces of Singular Integral Operators with Symbols Depending on x

Here we give conditions for the boundedness of a singular integral operator, acting from the Sobolev class $W_2^m(\mathbb{R}^n)$ into $W_2^l(\mathbb{R}^n)$ with $m \geq l \geq 0$. The symbol may depend not only on the angular variable $\theta \in \partial\mathcal{B}$ but also on the space variable $x \in \mathbb{R}^n$. Here $\partial\mathcal{B}$ stands for the unit sphere in $\mathbb{R}^n$ centered at the origin. It will be shown that the conditions, which are stated in terms of a certain space of multipliers, are sharp.

16.3.1 Function Spaces

Let μ be a measurable function defined on $\mathbb{R}^{n-1}$ and satisfying the inequalities

$$\mu(\xi) \geq c \quad \text{and} \quad \mu(\xi+\eta) \leq \big(1 + c\,|\xi|^{Q}\big)\mu(\eta),$$

where c and Q are positive constants. By $\mathcal{H}_\mu(\mathbb{R}^{n-1})$ we denote the completion of $C_0^\infty(\mathbb{R}^{n-1})$ in the norm

$$\|v;\mathbb{R}^{n-1}\|_{\mathcal{H}_\mu} = \Big(\int_{\mathbb{R}^{n-1}} |\mu(\xi)(Fv)(\xi)|^2 d\xi\Big)^{1/2}, \tag{16.3.1}$$

where F is the Fourier transform in $\mathbb{R}^{n-1}$. We obtain the space of Bessel potentials $H_2^l(\mathbb{R}^{n-1}), l \in \mathbb{R}^1$, by setting $\mu(\xi) = (1+|\xi|^2)^l$. The space $\mathcal{H}_\mu$ was introduced and studied in [H2], [VP]. In particular, it was shown in [H1], [VP] that $\mathcal{H}_\mu(\mathbb{R}^{n-1})$ is embedded into the space $C(\mathbb{R}^{n-1})$ of continuous and bounded functions on $\mathbb{R}^{n-1}$ if and only if

$$\int_{\mathbb{R}^{n-1}} \frac{d\xi}{\mu(\xi)^2} < \infty. \tag{16.3.2}$$

Everywhere in this section we assume that (16.3.2) *holds.*

We suppose that μ is weakly subadditive, that is,

$$\mu(\xi+\eta) \leq c\,(\mu(\xi)+\mu(\eta)), \quad c = \text{const}.$$

An easy modification of the proof of a similar result for H_2^l given in [Pe1] shows that the space $\mathcal{H}_\mu(\mathbb{R}^{n-1})$ is an algebra with respect to pointwise multiplication if μ satisfies (16.3.2). The converse assertion also holds. In fact, since $\mu(\xi) \geq c > 0$, we have

$$c\,\|u^N;\mathbb{R}^{n-1}\|_{L_2} \leq \|u^N;\mathbb{R}^{n-1}\|_{\mathcal{H}_\mu} \leq c_1^N\,\|u;\mathbb{R}^{n-1}\|^N_{\mathcal{H}_\mu}$$

for all $u \in \mathcal{H}_\mu(\mathbb{R}^{n-1})$, where $N = 1, 2, \ldots$ and the constants c and c_1 do not depend on N. Taking the N-th root and passing to the limit as $N \to \infty$, we arrive at

$$\|u;\mathbb{R}^{n-1}\|_{L_\infty} \leq c_1\|u;\mathbb{R}^{n-1}\|_{\mathcal{H}_\mu}.$$

Consequently, $\mathcal{H}_\mu(\mathbb{R}^{n-1}) \subset C(\mathbb{R}^{n-1})$, which is equivalent to (16.3.2).

We supply the sphere $\partial\mathcal{B}$ with a structure of the class C^∞ by introducing a family of coordinate neighborhoods $\{U_k\}$ and a family of diffeomorphisms $\varphi_k : U_k \to \mathbb{R}^{n-1}$. Further, let $\{\nu_k\}$ be a smooth partition of unity on $\partial\mathcal{B}$ subordinate to the covering $\{U_k\}$.

A function σ defined on $\partial\mathcal{B}$ belongs to the space $\mathcal{H}_\mu(\partial\mathcal{B})$ if

$$(\nu_k\sigma)\circ\varphi_k^{-1} \in \mathcal{H}_\mu(\mathbb{R}^{n-1})$$

for all k. The norm in $\mathcal{H}_\mu(\partial\mathcal{B})$ is introduced by

$$\|\sigma;\, \partial\mathcal{B}\|_{\mathcal{H}_\mu} = \Big(\sum_k \|(\nu_k \sigma) \circ \varphi_k^{-1};\, \mathbb{R}^{n-1}\|^2_{\mathcal{H}_\mu} \Big)^{1/2}$$

Similarly to $\mathcal{H}_\mu(\mathbb{R}^{n-1})$, the space $\mathcal{H}_\mu(\partial\mathcal{B})$ is an algebra with respect to multiplication if and only if (16.3.2) holds. The same condition is equivalent to the embedding $\mathcal{H}_\mu(\partial\mathcal{B}) \subset C(\partial\mathcal{B})$.

Let $\mathcal{B}(x)$ denote the unit ball in $\mathbb{R}^n$ centered at x, and let $\mathcal{B} = \mathcal{B}(0)$. We need the space $H^{l,\mu}(\mathcal{B} \times \partial\mathcal{B})$ of functions $\mathcal{B} \times \partial\mathcal{B} \ni (x,\theta) \to u(x,\theta)$ with the finite norm

$$\Big(\int_{\mathcal{B}} (\|\nabla_l u(x,\cdot); \partial\mathcal{B}\|^2_{\mathcal{H}_\mu} + \|u(x,\cdot); \partial\mathcal{B}\|^2_{\mathcal{H}_\mu}) dx \Big)^{1/2}$$

for integer $l \geq 0$, and with the finite norm

$$\Big(\int_{\mathcal{B}} \int_{\mathcal{B}} \|\nabla_{[l],x} u(x,\cdot) - \nabla_{[l],y} u(y,\cdot); \partial\mathcal{B}\|^2_{\mathcal{H}_\mu} \frac{dxdy}{|x-y|^{n+2\{l\}}}$$
$$+ \int_{\mathcal{B}} \|u(y,\cdot); \partial\mathcal{B}\|^2_{\mathcal{H}_\mu} dy \Big)^{1/2}$$

for noninteger $l > 0$. Further, we introduce the space $H^{l,\mu}(\mathbb{R}^n \times \partial\mathcal{B})$ of functions

$$\mathbb{R}^n \times \partial\mathcal{B} \ni (x,\theta) \to u(x,\theta)$$

endowed with the norm

$$\|u; \mathbb{R}^n \times \partial\mathcal{B}\|_{H^{l,\mu}} = \Big(\int_{\mathbb{R}^n} \big((\mathcal{D}_{l,\mu} u(x))^2 + (\mathcal{D}_{0,\mu} u(x))^2 \big) dx \Big)^{1/2},$$

where

$$\mathcal{D}_{l,\mu} u(x) = \|\nabla_{l,x} u(x,\cdot); \partial\mathcal{B}\|_{\mathcal{H}_\mu} \tag{16.3.3}$$

for $\{l\} = 0$, and

$$\mathcal{D}_{l,\mu} u(x) = \Big(\int_{\mathbb{R}^n} \|\nabla_{[l],x} u(x+h,\cdot) - \nabla_{[l],x} u(x,\cdot); \partial\mathcal{B}\|^2_{\mathcal{H}_\mu} \frac{dh}{|h|^{n+2\{l\}}} \Big)^{1/2} \tag{16.3.4}$$

for $\{l\} > 0$.

We say that a function γ defined on $\mathbb{R}^n \times \partial\mathcal{B}$ belongs to the space of multipliers $M(H^{m,\mu} \to H^{l,\mu})$ if $\gamma u \in H^{l,\mu}(\mathbb{R}^n \times \partial\mathcal{B})$ for all $u \in H^{m,\mu}(\mathbb{R}^n \times \partial\mathcal{B})$.

Since the operator

$$H^{m,\mu}(\mathbb{R}^n \times \partial\mathcal{B}) \ni u \to \gamma u \in H^{l,\mu}(\mathbb{R}^n \times \partial\mathcal{B})$$

is closed, it is bounded. As a norm in $M(H^{m,\mu} \to H^{l,\mu})$ we take the norm of the multiplication operator:

$$\|\gamma; \mathbb{R}^n \times \partial\mathcal{B}\|_{M(H^{m,\mu} \to H^{l,\mu})}$$
$$= \sup\{\|\gamma u; \mathbb{R}^n \times \partial\mathcal{B}\|_{H^{l,\mu}} : \|u; \mathbb{R}^n \times \partial\mathcal{B}\|_{H^{m,\mu}} \leq 1\}.$$

We use the notation $MH^{l,\mu}$ instead of $M(H^{l,\mu} \to H^{l,\mu})$.

16.3.2 Description of the Space $M(H^{m,\mu} \to H^{l,\mu})$

Here we characterize the space $M(H^{m,\mu} \to H^{l,\mu})$. Consider first the case $l = 0$.

Lemma 16.3.1. *A function γ defined on $\mathbb{R}^n \times \partial\mathcal{B}$ belongs to the space $M(H^{m,\mu} \to H^{0,\mu})$ if and only if $\gamma \in H^{0,\mu}(\mathcal{B}(x) \times \partial\mathcal{B})$ for an arbitrary unit ball $\mathcal{B}(x)$, and for any compact set $e \subset \mathbb{R}^n$*

$$\|\gamma; e \times \partial\mathcal{B}\|^2_{H^{0,\mu}} \le c\, C_{2,m}(e),$$

where c is a constant which does not depend upon e. Moreover, the equivalence relation

$$\|\gamma; \mathbb{R}^n \times \partial\mathcal{B}\|_{M(H^{m,\mu} \to H^{0,\mu})} \sim \sup_{e \subset \mathbb{R}^n} \left(\frac{\int_e \|\gamma(x,\cdot); \partial\mathcal{B}\|^2_{\mathcal{H}_\mu} dx}{C_{2,m}(e)} \right)^{1/2} \tag{16.3.5}$$

holds.

Proof. Necessity. We substitute $u(x,\theta) := u(x)$, where $u \in W_2^m(\mathbb{R}^n)$, into the inequality

$$\left(\int_{\mathbb{R}^n} \|\gamma(x,\cdot)u(x,\cdot); \partial\mathcal{B}\|^2_{\mathcal{H}_\mu} dx \right)^{1/2} \le c\, \|u; \mathbb{R}^n \times \partial\mathcal{B}\|_{H^{m,\mu}}.$$

Then

$$\left(\int_{\mathbb{R}^n} \|\gamma(x,\cdot); \partial\mathcal{B}\|^2_{\mathcal{H}_\mu} |u(x)|^2 dx \right)^{1/2} \le c\, \|u\|_{W_2^m}.$$

By Theorems 1.2.2 and 3.1.4, the exact constant in this inequality is equivalent to the right-hand side of (16.3.5).

Sufficiency. Since the space $\mathcal{H}_\mu(\partial\mathcal{B})$ is an algebra under the condition (16.3.2), it follows that

$$\|\gamma u; \mathbb{R}^n \times \partial\mathcal{B}\|^2_{H^{0,\mu}} \le c \int_{\mathbb{R}^n} \|\gamma; \partial\mathcal{B}\|^2_{\mathcal{H}_\mu} \|u; \partial\mathcal{B}\|^2_{\mathcal{H}_\mu} dx$$

$$= c \sum_j \int_{\mathbb{R}^{n-1}} |\mu(\xi)|^2 \int_{\mathbb{R}^n} \|\gamma; \partial\mathcal{B}\|^2_{\mathcal{H}_\mu} |F[\nu_j(\phi_j^{-1}(\xi)) u(x, \phi_j^{-1}(\xi))]|^2 dx d\xi.$$

Applying Lemma 16.3.1 to the internal integral, one obtains

$$\|\gamma u; \mathbb{R}^n \times \partial\mathcal{B}\|^2_{H^{0,\mu}} \le c \sup_{e \subset \mathbb{R}^n} \frac{\int_e \|\gamma(x,\cdot); \partial\mathcal{B}\|^2_{\mathcal{H}_\mu} dx}{C_{2,m}(e)}$$

$$\times \Big(\int_{\mathbb{R}^n} \int_{\mathbb{R}^n} \sum_j \int_{\mathbb{R}^{n-1}} |\mu(\xi)|^2 |F \Delta_h \nabla_{[m],x} \big(\nu_j(\phi_j^{-1}(\xi)) u(x, \phi_j^{-1}(\xi))\big)|^2 d\xi dx \frac{dh}{|h|^{n+2\{m\}}}$$

$$+ \int_{\mathbb{R}^n} \sum_j \int_{\mathbb{R}^{n-1}} |\mu(\xi)|^2 |F\big(\nu_j(\phi_j^{-1}(\xi)) u(x, \phi_j^{-1}(\xi))\big)|^2 d\xi dx \Big),$$

where

$$\Delta_h v(x,\theta) = v(x+h,\theta) - v(x,\theta).$$

Hence, using the definition of $\mathcal{H}_\mu(\partial\mathcal{B})$,we arrive at

$$\|\gamma u;\mathbb{R}^n \times \partial\mathcal{B}\|^2_{H^{0,\mu}} \le c \sup_{e\subset\mathbb{R}^n} \frac{\int_e \|\gamma;\partial\mathcal{B}\|^2_{\mathcal{H}_\mu}}{C_{2,m}(e)} \|u;\mathbb{R}^n \times \partial\mathcal{B}\|^2_{H^{m,\mu}}.$$

The proof is complete. □

Remark 16.3.1. According to (3.1.14), in Lemma 16.3.1 we may restrict ourselves to compact sets e satisfying $\operatorname{diam}(e) \le 1$.

In order to obtain sharp two-sided estimates for the norm in $M(H^{m,\mu} \to H^{l,\mu})$ for $m \ge l > 0$ we should prove some auxiliary assertions which are derived in the same way as the corresponding assertions on multipliers in the classes $M(H_2^m(\mathbb{R}^n) \to H_2^l(\mathbb{R}^n))$ (see Sects. 2.3 and 3.2). When doing this we should replace $|\gamma(x)|$ by $\|\gamma(x,\cdot);\partial\mathcal{B}_1\|_{\mathcal{H}_\mu}$ and replace $S_l u(x)$ by $\mathcal{D}_{l,\mu}u(x)$ defined by (16.3.3) and (16.3.4). As a result we arrive at the following description of the class $M(H^{m,\mu} \to H^{l,\mu})$.

Theorem 16.3.1. *The equivalence relation holds:*

$$\|\gamma;\mathbb{R}^n \times \partial\mathcal{B}\|_{M(H^{m,\mu}\to H^{l,\mu})} \sim \sup_{e:d(e)\le 1} \Big(\frac{\int_e (\mathcal{D}_{l,\mu}\gamma(x))^2 dx}{C_{2,m}(e)} \Big)^{1/2}$$
$$+ \begin{cases} \sup\limits_{x\in\mathbb{R}^n} \Big(\int_{\mathcal{B}(x)} \|\gamma(y,\cdot);\partial\mathcal{B}\|^2_{\mathcal{H}_\mu} dy \Big)^{1/2} & m > l, \\ \limsup\limits_{x\in\mathbb{R}^n} \|\gamma(x,\cdot);\partial\mathcal{B}\|_{\mathcal{H}_\mu} & m = l. \end{cases} \tag{16.3.6}$$

The restriction $d(e) \le 1$ can be omitted.

Remark 16.3.2. In the same way as for the space $M(W_2^m \to W_2^l)$ (cf. Sect. 4.3.4) we can check that $M(H^{m,\mu} \to H^{l,\mu})$ is continuously embedded into $M(H^{m-l,\mu} \to H^{0,\mu})$. Since the spaces $H^{m,\mu}(\mathbb{R}^n \times \partial\mathcal{B})$ form an interpolation scale in m (see, for instance, [Tr3], Sec.1.18.5), we have for any $j \in [0,l]$

$$\|\gamma;\mathbb{R}^n \times \partial\mathcal{B}\|_{M(H^{m-j,\mu}\to H^{l-j,\mu})}$$
$$\le c\,\|\gamma;\mathbb{R}^n \times \partial\mathcal{B}\|^{(l-j)/l}_{M(H^{m,\mu}\to H^{l,\mu})} \|\gamma;\mathbb{R}^n \times \partial\mathcal{B}\|^{j/l}_{M(H^{m-l,\mu}\to H^{0,\mu})}. \tag{16.3.7}$$

The embedding $M(H^{m,\mu} \to H^{l,\mu}) \subset M(H^{m-l,\mu} \to H^{0,\mu})$ together with (16.3.7) implies that the space $M(H^{m,\mu} \to H^{l,\mu})$ is continuously embedded into $M(H^{m-j,\mu} \to H^{l-j,\mu})$. From this and Theorem 16.3.1 it follows that (16.3.6) is equivalent to

$$\|\gamma; \mathbb{R}^n \times \partial\mathcal{B}\|_{M(H^{m,\mu}\to H^{l,\mu})}$$

$$\sim \sup_{e:d(e)\le 1} \Big(\sum_{j=0}^{[l]} \frac{\int_e (\mathcal{D}_{l-j,\mu}\gamma(x))^2 dx}{C_{2,m-j}(e)} + \sum_{j=0}^{[l]} \frac{\int_e (\mathcal{D}_{j,\mu}\gamma(x))^2 dx}{C_{2,m-l+j}(e)}\Big)^{1/2}. \tag{16.3.8}$$

For $m = l$ the term corresponding to $j = 0$ in the second sum should be replaced by

$$\operatorname*{ess\,sup}_{x\in\mathbb{R}^n} \|\gamma(x,\cdot); \partial\mathcal{B}\|^2_{\mathcal{H}_\mu}. \tag{16.3.9}$$

Clearly, for integer l both sums in (16.3.8) coincide. The restriction $d(e) \le 1$ can be omitted.

Duplicating the proof of Corollary 4.3.8, we arrive at the following assertion.

Corollary 16.3.1. *For $2m > n$*

$$\|\gamma; \mathbb{R}^n \times \partial\mathcal{B}\|_{M(H^{m,\mu}\to H^{l,\mu})}$$

$$\sim \sup_{x\in\mathbb{R}^n} \Big(\int_{\mathcal{B}(x)} (\mathcal{D}_{l,\mu}\gamma(y))^2 dy + \int_{\mathcal{B}(x)} \|\gamma(y,\cdot); \partial\mathcal{B}\|^2_{\mathcal{H}_\mu} dy\Big)^{1/2}. \tag{16.3.10}$$

One can verify directly that the right-hand side of (16.3.10) is equivalent to the norm $\|\gamma; \mathcal{B} \times \partial\mathcal{B}\|_{H^{l,\mu}}$.

From Theorem 16.3.1 upper estimates for the norm in $M(H^{m,\mu} \to H^{l,\mu})$ can be obtained, using the lower estimates for the capacity of a compact set stated in terms of its Lebesgue measure mes_n (see Propositions 3.1.2 and 3.1.3).

Corollary 16.3.2. *For $2m < n$*

$$c\,\|\gamma; \mathbb{R}^n \times \partial\mathcal{B}\|_{M(H^{m,\mu}\to H^{l,\mu})}$$

$$\le \sup_{e:d(e)\le 1} \frac{\Big(\int_e (\mathcal{D}_{l,\mu}\gamma(x))^2 dx\Big)^{1/2}}{(\mathrm{mes}_n e)^{\frac{1}{2}-\frac{m}{n}}}$$

$$+ \sup_{x\in\mathbb{R}^n} \Big(\int_{\mathcal{B}(x)} \|\gamma(y,\cdot); \partial\mathcal{B}\|^2_{\mathcal{H}_\mu} dy\Big)^{1/2}. \tag{16.3.11}$$

For $2m = n$

$$c\,\|\gamma; \mathbb{R}^n \times \partial\mathcal{B}\|_{M(H^{m,\mu}\to H^{l,\mu})}$$

$$\leq \sup_{e:d(e)\leq 1}\Big(\log\frac{2^n}{\text{mes}_n e}\Big)^{1/2}\Big(\int_e (\mathcal{D}_{l,\mu}\gamma(x))^2 dx\Big)^{1/2}$$

$$+\sup_{x\in\mathbb{R}^n}\Big(\int_{\mathcal{B}(x)}\|\gamma(y,\cdot);\partial\mathcal{B}\|^2_{\mathcal{H}_\mu}dy\Big)^{1/2}. \tag{16.3.12}$$

In the case $m = l$ the second term on the right-hand sides of (16.3.11) *and* (16.3.12) *should be replaced by* (16.3.9).

16.3.3 Main Result

Let σ be a measurable function on $\mathbb{R}^n$ with values in $L_2(\partial\mathcal{B})$. For any $u \in C_0^\infty(\mathbb{R}^n)$ we define the singular integral operator $\mathcal{S}$ with symbol σ by the equality

$$\mathcal{S}u(x) = \mathcal{F}^{-1}_{\xi\to x}[\sigma(x,\xi/|\xi|)(\mathcal{F}u)(\xi)], \tag{16.3.13}$$

where $\mathcal{F}$ is the Fourier transform in $\mathbb{R}^n$ and $\mathcal{F}^{-1}$ is its inverse.

In what follows we use the notation

$$\mathcal{K} = \Big(\int_{\mathbb{R}^{n-1}}\frac{d\tau}{\mu(\tau)^2}\Big)^{1/2}. \tag{16.3.14}$$

As before, we omit $\mathbb{R}^n$ in notations of spaces and norms.

Theorem 16.3.2. *Let $\mathcal{K} < \infty$ and let*

$$\sigma \in M(H^{m,\mu} \to H^{l,\mu}), \quad m \geq l \geq 0. \tag{16.3.15}$$

Then the operator (16.3.13) *maps W_2^m continuously into W_2^l. Moreover, the estimate*

$$\|\mathcal{S}\|_{W_2^m\to W_2^l} \leq c\,\mathcal{K}\|\sigma\|_{M(H^{m,\mu}\to H^{l,\mu})} \tag{16.3.16}$$

holds.

Proof. Let $x,\xi \in \mathbb{R}^n, \theta = \xi/|\xi|$ and let u be an arbitrary function from $C_0^\infty(\mathbb{R}^n)$. We write the operator $\mathcal{S}$ as

$$\mathcal{S}u(x) = \int_{\partial\mathcal{B}}\int_0^\infty e^{2\pi i x\xi}\sigma(x,\theta)\mathcal{F}u(\xi)|\xi|^{n-1}d|\xi|d\theta$$

or, briefly,

$$\mathcal{S}u(x) = \int_{\partial\mathcal{B}}\sigma(x,\theta)v(x,\theta)d\theta, \tag{16.3.17}$$

where

$$v(x,\theta) = \int_0^\infty e^{2\pi i x\xi}\mathcal{F}u(\xi)|\xi|^{n-1}d|\xi| \tag{16.3.18}$$

and

$$\mathcal{F}u(\xi) = \int_{\mathbb{R}^n} e^{-2\pi i y\xi}u(y)dy.$$

Using the C^∞ structure on $\partial\mathcal{B}$ introduced above, we have

$$\mathcal{S}u(x) = \sum_k \int_{\mathbb{R}^{n-1}} \nu_k(\varphi_k^{-1}(t))\sigma(x,\varphi_k^{-1}(t))v(x,\varphi_k^{-1}(t))|J_k(t)|dt,$$

where J_k is the Jacobian of the mapping φ_k^{-1}. Let $\eta_k \in C_0^\infty(U_k)$ be such that $\eta_k\nu_k = \nu_k$. We put

$$\sigma_k(x,t) = \nu_k(\varphi_k^{-1}(t))\sigma(x,\varphi_k^{-1}(t))$$

and

$$v_k(x,t) = \eta_k(\varphi_k^{-1}(t))v(x,\varphi_k^{-1}(t))|J_k(t)|.$$

By Parseval's theorem,

$$\mathcal{S}u(x) = \sum_k \int_{\mathbb{R}^{n-1}} \sigma_k(x,t)v_k(x,t)dt$$

$$= \sum_k \int_{\mathbb{R}^{n-1}} F\sigma_k(x,\tau)\overline{F^{-1}\overline{v_k(x,\tau)}}d\tau, \tag{16.3.19}$$

where F is the Fourier transform in $\mathbb{R}^{n-1}$.

By (16.3.18), we obtain

$$\overline{F^{-1}\overline{v_k(x,\tau)}} = \int_{\mathbb{R}^{n-1}} e^{-2\pi i\tau t}\eta_k(\varphi_k^{-1}(t))v(x,\varphi_k^{-1}(t))|J_k(t)|dt$$

$$= \int_{\partial\mathcal{B}} e^{-2\pi i\tau\varphi_k(\theta)}\eta_k(\theta)v(x,\theta)d\theta$$

$$= \int_{\mathbb{R}^n} e^{2\pi i x\xi}\eta_k(\theta)e^{-2\pi i\tau\varphi_k(\theta)}\mathcal{F}u(\xi)d\xi. \tag{16.3.20}$$

The last integral can be interpreted as a family of singular integral convolution operators $E_k(\tau)$, depending on a parameter $\tau \in \mathbb{R}^{n-1}$, with symbols

$$\eta_k(\theta)e^{-2\pi i\tau\varphi_k(\theta)}, \quad k = 1, 2, \dots$$

Now, it follows from (16.3.19) and (16.3.20) that $\mathcal{S}$ can be represented in the form

$$\mathcal{S}u(x) = \sum_k \int_{\mathbb{R}^{n-1}} F\sigma_k(x,\tau)E_k(\tau)u(x)d\tau. \tag{16.3.21}$$

Let l be a noninteger and let

$$D_l w(x) = \Big(\int_{\mathbb{R}^n} |\Delta_h \nabla_{[l]} w(x)|^2 \frac{dh}{|h|^{n+2\{l\}}}\Big)^{1/2}.$$

We have

$$|D_l \mathcal{S}u(x)|^2$$

$$\leq c\sum_{j=0}^{[l]}\sum_{k}\Big(\int_{\mathbb{R}^n}\Big(\int_{\mathbb{R}^{n-1}}|F\nabla_{j,x}\sigma_k(x+h,\tau)|\,|\Delta_h\nabla_{[l]-j,x}E_k(\tau)u(x)|d\tau\Big)^2\frac{dh}{|h|^{n+2\{l\}}}$$

$$+\int_{\mathbb{R}^n}\Big(\int_{\mathbb{R}^{n-1}}|F\Delta_h\nabla_{[l]-j,x}\sigma_k(x,\tau)|\,|\nabla_{j,x}E_k(\tau)u(x)|d\tau\Big)^2\frac{dh}{|h|^{n+2\{l\}}}\Big).$$

The right-hand side does not exceed

$$c\sum_{j=0}^{[l]}\sum_{k}\Big(\int_{\mathbb{R}^n}\int_{\mathbb{R}^{n-1}}|\mu(\tau)F\nabla_{j,x}\sigma_k(x+h,\tau)|^2 d\tau$$

$$\times\int_{\mathbb{R}^{n-1}}|\Delta_h\nabla_{[l]-j,x}E_k(\lambda)u(x)|^2\frac{d\lambda}{\mu(\lambda)^2}\frac{dh}{|h|^{n+2\{l\}}}$$

$$+\int_{\mathbb{R}^n}\int_{\mathbb{R}^{n-1}}|\mu(\tau)F\Delta_h\nabla_{[l]-j,x}\sigma_k(x,\tau)|^2 d\tau$$

$$\times\int_{\mathbb{R}^{n-1}}|(\nabla_{j,x}E_k(\lambda)u(x)|^2\frac{d\lambda}{\mu(\lambda)^2}\frac{dh}{|h|^{n+2\{l\}}}\Big). \tag{16.3.22}$$

Consequently,

$$\|D_l\mathcal{S}u\|^2_{L_2}$$

$$\leq c\sum_{j=0}^{[l]}\sum_{k}\Big(\int_{\mathbb{R}^n}\|\nabla_{j,x}\sigma_k(x,\cdot);\mathbb{R}^{n-1}\|^2_{\mathcal{H}_\mu}\int_{\mathbb{R}^{n-1}}|(D_{l-j}E_k(\lambda)u)(x)|^2\frac{d\lambda}{\mu(\lambda)^2}dx$$

$$+\int_{\mathbb{R}^n}\Big(\int_{\mathbb{R}^n}\|\Delta_h\nabla_{[l]-j,x}\sigma_k(x,\cdot);\mathbb{R}^{n-1}\|^2_{\mathcal{H}_\mu}\frac{dh}{|h|^{n+2\{l\}}}\Big)$$

$$\times\Big(\int_{\mathbb{R}^{n-1}}|(\nabla_j E_k(\lambda)u)(x)|^2\frac{d\lambda}{\mu(\lambda)^2}\Big)dx\Big).$$

This inequality and Lemma 16.3.1 imply that

$$\|D_l\mathcal{S}u\|^2_{L_2}$$

$$\leq c\sum_{j=0}^{[l]}\sum_{k}\Big(\sup_e\frac{\int_e\|\nabla_{j,x}\sigma(x,\cdot);\mathbb{R}^{n-1}\|^2_{\mathcal{H}_\mu}dx}{C_{2,m-l+j}(e)}\int_{\mathbb{R}^{n-1}}\|E_k(\lambda)u\|^2_{W_2^m}\frac{d\lambda}{\mu(\lambda)^2}$$

$$+\sup_e\frac{\int_e\|D_{l-j}\sigma_k(x,\cdot);\mathbb{R}^{n-1}\|^2_{\mathcal{H}_\mu}dx}{C_{2,m-j}(e)}\int_{\mathbb{R}^{n-1}}\|E_k(\lambda)u\|^2_{W_2^m}\frac{d\lambda}{\mu(\lambda)^2}\Big).$$

Since the operators $E_k(\lambda)$ are uniformly bounded in W_2^m, it follows that $\|D_l\mathcal{S}u\|_{L_2}$ does not exceed

$$\mathcal{K} \sup_e \Big(\sum_{j=0}^{[l]} \frac{\int_e (\mathcal{D}_{j,\mu}\gamma(x))^2 dx}{C_{2,m-l+j}(e)} + \sum_{j=0}^{[l]} \frac{\int_e (\mathcal{D}_{l-j,\mu}\gamma(x))^2 dx}{C_{2,m-j}(e)} \Big)^{1/2} \|u\|_{W_2^m},$$

which together with Remark 16.3.2 gives

$$\|D_l \mathcal{S} u\|_{L_2} \le c\mathcal{K} \|\sigma\|_{M(H^{m,\mu} \to H^{l,\mu})} \|u\|_{W_2^m}. \tag{16.3.23}$$

For integer l the proof is similar and somewhat easier. In particular, the counterpart of (16.3.22) is

$$c \sum_{j=0}^{l} \sum_k \int_{\mathbb{R}^{n-1}} |\mu(\tau) F \nabla_{j,x} \sigma_k(x,\tau)|^2 d\tau \int_{\mathbb{R}^{n-1}} |\nabla_{l-j,x} E_k(\lambda) u(x)|^2 \frac{d\lambda}{\mu(\lambda)^2}.$$

Duplicating the above arguments, we arrive at an analogue of (16.3.23) with D_l replaced by ∇_l on the left-hand side. This together with the inequality

$$\|\mathcal{S} u\|_{L_2} \le c\mathcal{K} \|\sigma\|_{M(H^{m,\mu} \to H^{0,\mu})} \|u\|_{W_2^m},$$

corresponding to $l = 0$, completes the proof.

Remark 16.3.3. We show that Theorem 16.3.2 is sharp in a sense.

Let the symbol of $\mathcal{S}$ have the form $a(x)b(\theta)$, where $x \in \mathbb{R}^n$ and $\theta \in \partial\mathcal{B}$, and let $b \in \mathcal{H}_\mu(\partial\mathcal{B})$ and $|b(\theta)| \ge \text{const} > 0$. Clearly, the mapping

$$\mathcal{S} : W_2^m \to W_2^l$$

is continuous if and only if the operator of multiplication by a is a continuous operator from W_2^m into W_2^l. In other words, (16.3.15) follows from the continuity of $\mathcal{S}$.

Now let $\mathcal{S}$ be the operator (16.3.13) with symbol $b(\theta)$, where $\theta \in \partial\mathcal{B}$. Its continuity from W_2^m into W_2^l is equivalent to the inequality

$$|b(\theta)|(1 + |\xi|^2)^{(l-m)/2} \le \text{const}$$

which gives the boundedness of b. Therefore, if the operator $\mathcal{S} : W_2^m \to W_2^l$ is continuous for any $b \in \mathcal{H}_\mu(\partial\mathcal{B})$, then $\mathcal{H}_\mu(\partial\mathcal{B}) \subset L_\infty(\partial\mathcal{B})$, which is equivalent to $\mathcal{K} < \infty$.

16.3.4 Corollaries

Now, we give sufficient conditions for the continuity of the operator $\mathcal{S} : W_2^m \to W_2^l$, which follow from Theorem 16.3.2 and from either necessary and sufficient or sufficient conditions for a function to belong to $M(H^{m,\mu} \to H^{l,\mu})$ (see Sect. 16.3.2).

The next assertion is a direct corollary of Theorems 16.3.1 and 16.3.2.

Corollary 16.3.3. *The estimate* (16.3.16) *is equivalent to*

$$\|\mathcal{S}\|_{W_2^m \to W_2^l} \le c\,\mathcal{K}\Big(\sup_{\{e\subset\mathbb{R}^n : d(e)\le 1\}} \frac{\displaystyle\int_e \big(\mathcal{D}_{l,\mu}\sigma(x)\big)^2 dx}{C_{2,m}(e)}$$

$$+ \sup_{x\in\mathbb{R}^n} \int_{\mathcal{B}(x)} \|\sigma(y,\cdot);\partial\mathcal{B}\|^2_{\mathcal{H}_\mu} dy\Big)^{1/2} \tag{16.3.24}$$

for $m > l \ge 0$. *For* $m = l$ *the second term on the right-hand side of* (16.3.24) *should be replaced by*

$$\lim \sup_{x\in\mathbb{R}^n} \|\sigma(x,\cdot);\partial\mathcal{B}\|^2_{\mathcal{H}_\mu}. \tag{16.3.25}$$

Theorem 16.3.2 and Corollary 16.3.1 imply the following assertion.

Corollary 16.3.4. *Let* $2m > n$. *The inequality* (16.3.16) *is equivalent to*

$$\|\mathcal{S}\|_{W_2^m \to W_2^l} \le c\,\mathcal{K} \sup_{x\in\mathbb{R}^n} \|\sigma;\mathcal{B}(x)\times\partial\mathcal{B}\|_{H^{l,\mu}}. \tag{16.3.26}$$

Combining Theorem 16.3.2 with Corollary 16.3.2 we can remove the capacity from inequality (16.3.24) as follows.

Corollary 16.3.5. *Let* $2m < n$. *Then*

$$\|\mathcal{S}\|_{W_2^m \to W_2^l} \le c\,\mathcal{K}\Big(\sup_{e\subset\mathbb{R}^n : \mathrm{diam}(e)\le 1} \frac{\displaystyle\int_e \big(\mathcal{D}_{l,\mu}\sigma(x)\big)^2 dx}{(\mathrm{mes}_n e)^{1-2m/n}}$$

$$+ \sup_{x\in\mathbb{R}^n} \int_{\mathcal{B}(x)} \|\sigma(y,\cdot);\partial\mathcal{B}\|^2_{\mathcal{H}_\mu} dy\Big)^{1/2}. \tag{16.3.27}$$

For $2m = n$ *the expression* $(\mathrm{mes}_n e)^{1-2m/n}$ *should be replaced by*

$$\big(\log(2^n/\mathrm{mes}_n e)\big)^{-1}.$$

In the case $m = l$ *the second term on the right-hand side of* (16.3.27) *should be replaced by* (16.3.25).

Remark 16.3.4. Theorem 16.3.2 and its corollaries can be directly extended to classical pseudo-differential operators with symbols of the form

$$\zeta(\xi)\sum_{k=1}^N \sigma_k(x,\xi/|\xi|)|\xi|^{r_k},$$

where $r_1 > \cdots > r_N$ and $\zeta \in C^\infty(\mathbb{R}^{n-1})$ with $\zeta(\xi) = 1$ for $|\xi| > 2$ and $\zeta(\xi) = 0$ for $|\xi| < 1$ (see [KN] for a theory of these operators).

References

[Ad1] D.R. Adams, *A trace inequality for generalized potentials*, Studia Math., **48**:1 (1973), 99-105.

[Ad2] D.R. Adams, *A note on Riesz potentials*, Duke Math. J., **42**:4 (1975), 99-105.

[Ad3] D.R. Adams, *On the existence of capacitary strong type estimates in* $\mathbb{R}^n$, Ark. Mat., **14** (1976), 125-140.

[AF] D.R. Adams and M. Frazier, *Composition operators on potential spaces*, Proc. Amer. Math. Soc., **114** (1992), 155-165.

[AH] D.R. Adams and L.-I. Hedberg, *Function Spaces and Potential Theory*, Springer, 1996.

[AM] D.R. Adams and N.G. Meyers, *Bessel potentials. Inclusion relations among classes of exceptional sets*, Indiana Univ. Math. J., **22**:9 (1973), 873-905.

[APo] D.R. Adams and J.C. Polking, *The equivalence of two definitions of capacity*, Proc. Amer. Math. Soc., **37** (1973), 529-534.

[AX] D.R. Adams and J. Xiao, *Strong type estimates for homogeneous Besov capacities*, Math. Ann., **325** (2003), 695-709.

[ADN1] S. Agmon, A. Douglis, and L. Nirenberg, *Estimates near the boundary for the solutions of elliptic equations satisfying general boundary values, I.*, Comm. Pure Appl. Math., **12** (1959), 623-727.

[ADN2] S. Agmon, A. Douglis, and L. Nirenberg, *Estimates near the boundary for the solutions of elliptic equations satisfying general boundary values, II.*, Comm. Pure Appl. Math., **17** (1964), 35-92.

[AB] L. Ahlfors and A. Beurling, *Conformal invariants and function-theoretic null-sets*, Acta Math. **83** (1950), 623-727.

[AN] F. Ali Mehmeti and S. Nicaise, *Banach algebras of functions on nonsmooth domains*, Oper. Theory Adv. Appl., **102**, Birkhäuser, Basel, 1998, 11-20.

[Am] H. Amann, *Multiplication in Sobolev and Besov spaces*, Nonlinear Analysis, Scuola Normale Superiore, Pisa, 1991, 27-50.

[And] K.F. Andersen, *Weighted inequalities for convolutions*, Proc. AMS, **123**:4 (1995), 1129-1136.

[AMS] N. Aronszajn, F. Mulla, and P. Szeptycki, *On spaces of potentials connected with L^p-spaces*, Ann. Inst. Fourier, **13** (1963), 211-306.

[BB] B.M. Benchekroun, A. Benkirane, *Sur l'algèbre d'Orlicz-Sobolev*, Bull. Belg. Math. Soc., **2**:4 (1995), 463-476.

[BG] C. Bennet and J.E. Gilbert, *Homogeneous algebras on the circle: II. Multipliers, Ditkin conditions*, Ann. Inst. Fourier, **22**:3 (1972), 21-50.

[Bes] O.V. Besov, *Investigation of a family of function spaces in connection with imbedding and extension theorems*, Trudy Mat. Inst. Steklov, **60** (1961), 42-81.

[BIN] O.V. Besov, V.P. Il'in, and S.M. Nikol'skii, *Integral Representations of Functions and Imbedding Theorems*, Vol I, 1978, and Vol. II, 1979, John Wiley & Sons, New York-Toronto-London.

[Beu] A. Beurling, *Construction and analysis of some convolution algebras*, Ann. Inst. Fourier (Grenoble), **14** (1964), 1-32.

[Bl1] N.K. Bliev, *On products of functions in Nikolskii-Besov spaces*, Izv. AN Kazach. SSR, Ser. Phis.-Mat., no. 5 (1979), 69-71.

[Bl2] N.K. Bliev, *Homeomorphisms of Beltrami equation in fractional spaces*, Differential and integral equations. Boundary value problems, Tbilisi, 1979, 33-43.

[Blo] S. Bloom, *Pointwise multipliers of weighted BMO spaces*, Proc. Amer. Math. Soc., **105** (1989), 950-960.

[Bo] G. Bourdaud, *Localizations des espaces de Besov*, Studia Math., **90** (1988), 153-163.

[Bur] V. Burenkov, *Sobolev Spaces on Domains*, Teubner-Texte zur Mathematik, 137. B. G. Teubner, Stuttgart, Leipzig, 1998.

[Ca1] A.P. Calderon, *Lebesgue spaces of differentiable functions and distributions*, Proc. Sympos. Pure Math., **4** (1961), 33-49.

[Ca2] A.P. Calderon, *Commutators of singular integral operators*, Proc. Nat. Acad. Sci. USA, **53** (1965), 1092-1099.

[Ca3] A.P. Calderon, *Algebra of singular integral operators*, Proc. Symp. Pure Math., **10**, AMS, Providence, R.I., 1967.

[Ca4] A.P. Calderon, *Boundary value problems for the Laplace equation in Lipschitz domains*, Recent progress in Fourier Analysis, Sci. Publ., Amsterdam, 1985, 33-48.

[Cam] S. Campanato, *Proprietà di hölderianità di alcune classi di funzioni*, Ann. Scuola Norm. Sup. Pisa, **17** (1963), 175-188.

[Car] L. Carleson, *Interpolation by bounded analytic functions and the corona problem,* Ann. Math., **76** (1962), 547-559.

[COV1] C. Cascante, J.M. Ortega, and I.E. Verbitsky, *Nonlinear potentials and two weight trace inequalities for general dyadic and radial kernels*, Indiana Univ. Math. J., **53** (2004), 845-882.

[COV2] C. Cascante, J.M. Ortega, and I.E. Verbitsky, *On L^p–L^q trace inequalities*, J. London Math. Soc., **74**:2 (2006), 497-511.

[ChWW] S.-Y. A. Chang, J. M. Wilson, and T. H. Wolff, *Some weighted norm inequalities concerning the Schrödinger operators*, Comment. Math. Helv., **60** (1985), 217-246.

[CF] R. R. Coifman and C. Fefferman, *Weighted norm inequalities for maximal functions and singular integrals*, Studia Math., **51** (1974), 241-250.

[CMM] R. R. Coifman, A. McIntosh, and I. Meyer, *L' intégrale de Cauchy définit un opérateur borné sur L^2 pour les courbes Lipschitziennes*, Ann. of Math., **116** (1982), 361-387.

[Cor] H.O. Cordes, *Die erste Randwertaufgabe bei Differentialgleichungen zweiter Ordnung in mehr als zwei Variabeln*, Math. Ann., **131**:3 (1956), 278-312.

[Cos] M. Costabel, *Boundary integral operators on Lipschitz domains: elementary results*, SIAM J. Math. Anal., **19**:3 (1988), 613-623.

[DM1] B. Dacorogna and J. Moser, *On a partial differential equation involving the Jacobian determinant*, Ann. Inst. Henri Poincaré, **7** (1991), 1-26.

[DKV] B.E.J. Dahlberg, C.E. Kenig, and G.C. Verchota, *Boundary value problems for the systems of elastostatics in Lipschitz domains*, Duke Math. J., **57**:3 (1988), 795-818.

[Dav] E.B. Davies, *A review of Hardy inequalities*, The Maz'ya Anniversary Collection, Eds. J. Rossmann, P. Takáč, and G. Wildenhain, Operator Theory: Advances and Applications, Vol. 110, Birkhäuser, 1999, 55-67, Basel–Boston–Berlin.

[dR] G. de Rham, *Variétés Différentiables*, Hermann, Paris, 1960.

[DH] A. Devinatz and I.I. Hirschman, *Multiplier transformations on $l^{2,\alpha}$*, Annals of Math., **69**:3 (1959), 575-587.

[DM2] D. Drihem and M. Moussai, *On the pointwise multiplication in Besov and Lizorkin-Triebel spaces*, Int. J. Math. Math. Sci. Art. ID 76182 (2006), 1-18.

[DS] N. Dunford and J.T. Schwartz, *Linear Operators. Part I: General Theory*, Interscience Publishers, 1967.

[EE] D.E. Edmunds and W.D. Evans, *Spectral Theory and Differential Operators*, Clarendon Press, Oxford, 1987.

[ES] D.E. Edmunds and E. Shargorodsky, *The inner variation of an operator and the essential norm of pointwise multipliers in function spaces*, Houston J. Math., **31**:3 (2005), 841-855.

[Fab] E.B. Fabes, *Boundary value problems of linear elastostatics and hydrostatics on Lipschitz domains*, Proc. Cent. Math. Anal. Aust. Nat. Univ., **9** (1985), 27-45.

[FJR] E.B. Fabes, M. Jodeit, and N.M. Riviere, *Potential techniques for boundary value problems in C^1 domains*, Acta Math., **141**:3-4 (1978), 165-186.

[FKV] E.B. Fabes, C.E. Kenig, and G.C. Verchota, *The Dirichlet problem for the Stokes system on Lipschitz domains*, Duke Math. J., **57**:3 (1988), 769-793.

[Fe1] H. Federer, *Curvature measures*, Trans. AMS, **93**:3 (1959), 418-491.

[Fe2] H. Federer, *The area of nonparametric surface*, Proc. AMS, **11**:3 (1960), 436-439.

[Fe3] H. Federer, *Geometric Measure Theory*, Springer, 1969.

[F1] C. Fefferman, *Characterizations of bounded mean oscillation*, Bull. AMS, **77** (1971), 587-588.

[F2] C. Fefferman, *The uncertainty principle*, Bull. AMS, **9** (1983), 129-206.

[Fil] N. Filonov, *Principal singularities of the magnetic field component in resonators with boundary of a given class of smoothness*, Algebra i Analiz, **9**:2 (1997), 241-255.

[FR] W.H. Fleming and R.W. Rishel, *An integral formula for total gradient variation*, Arch. Math., **11**:3 (1960), 218-222.

[Fra] L.E. Fraenkel, *Formulae for high derivatives of composite functions*, Math. Proc. Camb. Soc., **77** (1971), 587-588.

[FrS] R.L. Frank and R. Seiringer, *Non-linear ground state representations and sharp Hardy inequalities*, arXiv:0803.0503.

[Fr] J. Franke, *On the spaces $F^s_{p,q}$ of Triebel-Lizorkin type: Pointwise multipliers and spaces on domains*, Math. Nachr., **125** (1986), 29-68.

[FrJ] M. Frazier and B. Jawerth, *A discrete transform and decompositions of distribution spaces*, J. Funct. Analysis, **93** (1990), 34-170.

[Gag1] E. Gagliardo, *Proprietà di alcune classi di funzioni in più variabili*, Ric. Mat., **7** (1958), 102-137.

[Gag2] E. Gagliardo, *Ulteriori proprietà di alcune classi di funzioni in più variabili*, Ric. Mat., **8**:1 (1959), 24-51.

[GSh] I.M. Gelfand and G.E. Shilov, *Generalized Functions*, Vol. 1, *Operators on them*, Academic Press, NY, 1964.

[Ger] P. Germain, *Multipliers, paramultipliers, and weak-strong uniqueness for the Navier-Stokes equations*, J. Differential Equations, **226**:2 (2006), 373-428.

[GG] G. Geymonat and P. Grisvard, *Problemi ai limiti lineari ellittici negli spazi di Sobolev con peso*, Matematiche (Catania), **22** (1967), 212-249.

[GM] L. Grafakos and C. Morpurgo, *A Selberg integral formula and applications*, Pacific J. Math., **191**:1 (1999), 85-94.

[GR] V. Gol'dshtein, and Yu.G. Reshetnyak, *Quasiconformal Mappings and Sobolev Spaces*, Translated and revised from the 1983 Russian original. Mathematics and its Applications (Soviet Series), vol. 54. Kluwer Academic Publishers, Dordrecht, 1990.

[Gu1] A. Gulisashvili, *Multipliers in Besov spaces*, Zapiski Nauchn. Sem. LOMI, **135** (1984), 36-50.

[Gu2] A. Gulisashvili, *Multipliers in Besov spaces and traces of functions on subspaces of Euclidean spaces*, Dokl. Akad. Nauk SSSR, **281**:4 (1985), 777-781; English translation: Soviet Math. Dokl., **31**:2 (1985), 332-336.

[Gus] W. Gustin, *Boxing inequalities*, J. Math. Mech., **9** (1960), 229-239.

[Guz] M. de Guzman, *Covering lemma with applications to differentiability of measures and singular integral operators*, Studia Math., **34**:3 (1970), 299-317.

[Ha] B. Hanouzet, *Applications bilinéaires compatibles avec un système à coefficients variables. Continuité dans les éspaces de Besov*, Comm. Partial. Diff. Eq., **10**:4 (1985), 433-465.

[Hed1] L.-I. Hedberg, *On certain convolution inequalities*, Proc. AMS, **36** (1972), 505-510.

[Hed2] L.-I. Hedberg, *Nonlinear potentials and approximation in the mean by analytic functions*, Math. Zeitschr., **129** (1972), 299-319.

[Hed3] L.-I. Hedberg, *Approximation in the mean by solutions of elliptic equations*, Duke Math. J., **40** (1973):1, 9-16.

[Her] C.S. Herz, *Lipschitz spaces and Bernstein's theorem on absolutely convergent Fourier transforms*, J. Math. Mech., **18**:4 (1968), 283-323.

[Hi1] I.I. Hirschman, *On multiplier transformations, II*, Duke Math. J., **28** (1961), 45-56.

[Hi2] I.I. Hirschman, *On multiplier transformations, III*, Proc. AMS, **13** (1962), 851-857.

[H1] L. Hörmander, *Linear Partial Differential Operators*, Springer, 1963.

[H2] L.Hörmander, *The Analysis of Linear Partial Differential Operators*, vol.2, Springer, 1983.

[Ja1] S. Janson, *On functions with conditions on the mean oscillation*, Ark. Mat., **14**:2 (1976), 189-196.

[Ja2] S. Janson, *Mean oscillation and commutators of singular integral operators*, Ark. Mat. **16**:2 (1978), 263-270.

[JK1] D.S. Jerison and C.E. Kenig, *The Dirichlet problem in nonsmooth domains*, Ann. of Math., **113** (1981), 367-382.

[JK2] D.S. Jerison and C.E. Kenig, *The Neumann problem on Lipschitz domains*, Bull. AMS, **4** (1981), 203-207.

[JN] F. John and L. Nirenberg, *On functions of bounded mean oscillation*, Comm. Pure Appl. Math., **14** (1961), 415-426.

[Jo] J. Johnsen, *Pointwise multiplication of Besov and Triebel-Lizorkin spaces*, Math. Nachr., **175** (1995), 85-133.

[Kal] A. Kałamajska, *Pointwise interpolative inequalities and Nirenberg type estimates in weighted Sobolev spaces*, Studia Math., **108**:3 (1994), 275-290.

[K1] G.A. Kalyabin, *Conditions for multiplicative property of Besov and Lizorkin-Triebel function spaces*, Dokl. Akad. Nauk SSSR, **251**:1 (1980), 25-26.

[K2] G.A. Kalyabin, *Descriptions of functions in classes of Besov-Triebel-Lizorkin type*, Trudy Math. Inst. Steklov, **156** (1980), 82-109.

[K3] G.A. Kalyabin, *Criteria of the multiplication property and the embedding in C of spaces of Besov-Triebel-Lizorkin type*, Mat. Zametki, **30** (1981), 517-526.

[Ka1] T. Kato, *Schrödinger operators with singular potentials*, Israel J. Math., **13** (1972), 135-148.

[Ke1] C.E. Kenig, *Boundary value problems of linear elastostatics and hydrostatics on Lipschitz domains*, Semin. Goulaouic-Meyer-Schwartz, Equation Deriv. Partielles 1983-1984, Exp. N 21, 1-12.

[Ke2] C.E. Kenig, *Harmonic analysis techniques for second order elliptic boundary value problems*, CBMS Regional Conference Series in Mathematics, **83**, AMS, Providence, 1994.

[KeS] R. Kerman and E. Sawyer, *The trace inequality and eigenvalue estimates for Schrödinger operators*, Ann. Inst. Fourier (Grenoble), **36** (1986), 207-228.

[KoS] H. Koch and W. Sickel, *Pointwise multipliers of Besov spaces of smoothness zero and spaces of continuous functions*, Rev. Mat. Iberoamericana, **18** (2002), 587-626.

[KN] J.J. Kohn and L. Nirenberg, *An algebra of pseudo-differential operators*, Comm. Pure Appl. Math., **18**:1-2 (1965), 269-305.

[KZPS] M.A. Krasnoselskii, P.P. Zabreyko, E.I. Pustylnik, P.E. Sobolevskii, *Integral Operators in Spaces of Summable Functions*, Noordhoff, Leiden, 1976.

[KP] S.G. Krantz and H.R. Parks, *The Implicit Function Theorem. History, Theory, and Applications*, Birkhäuser, 2002.

[Kr] A.S. Kronrod, *On functions of two variables*, Usp. Mat. Nauk, **5**:1 (1950), 24-134.

[KWh] D.S. Kurtz and R.L. Wheeden, *Results on weighted norm inequalities for multipliers*, Trans. AMS, **255** (1979), 343-362.

[Lad] O.A. Ladyzhenskaya, *The Mathematical Theory of Viscous Incompressible Flow*, Gordon and Breach, 1969.

[Las] I. Lasiecka, *Finite-dimensional attractors of weak solutions to von Karman plate model*, J. Math. Systems, Estimation, and Control, **7**:3 (1997), 251-275.

[LR] P.G. Lemarié-Rieusset, *Recent Developments in the Navier-Stokes Problem*, Chapman and Hall, Research Notes in Math. 431 (2002).

[LRM] P.G. Lemarié-Rieusset and R. May, *Uniqueness for the Navier-Stokes equations and multipliers between Sobolev spaces*, Nonlinear Anal., **66**:4 (2007), 819-838.

[LeL] J. Leray and J.-L. Lions, *Quelques résultats de Višik sur les problèmes elliptiques non-lineaires par les méthodes de Minty-Browder*, Bull. Math. Soc. France, **93** (1965), 97-107.

[LeM] A.V. Levin and V. Maz'ya, *Asymptotics of densities of harmonic potentials near the vertex of a cone*, Z. Anal. Anwend., **8**:6 (1989), 501-514.

[Lew] J. Lewis, *Uniformly fat sets*, Trans. AMS, **308** (1988), 177-196.

[LL] E.H. Lieb and M. Loss, *Analysis*, Second Edition, AMS, Providence, RI, 2001.

[LiM1] J.-L. Lions and E. Magenes, *Problèmes aux limites non homogènes, IV*, Ann. Scuola Norm. Sup. Pisa, **15** (1961), 311-326.

[LiM2] J.-L. Lions and E. Magenes, *Non-homogeneous Boundary Value Problems and Applications*, Vol. I. Die Grundlehren der Mathematischen Wissenschaften, Band 181. Springer, 1972.

[Liz] P.I. Lizorkin, *On function characteristics of interpolation spaces* $(L_p(\Omega), W_p^1(\Omega))_{\theta,p}$, Trudy Mosk. Matem. Inst., **134** (1975), 180-203.

[Mal] J. Malý, *Sufficient conditions for change of variables in integral*, Proceedings on Analysis and Geometry. International conference in honor of the 70th birthday of Professor Yu. G. Reshetnyak, Novosibirsk, Russia, August 30-September 3, 1999. Novosibirsk: Izdatel'stvo Instituta Matematiki Im. S. L. Soboleva SO RAN. 370-386 (2000).

[MaMi] M. Marcus and V. Mizel, *Absolute continuity on tracks and mappings of Sobolev spaces*, Arch. Rat. Mech. Anal., **45**:4 (1972) 294-320.

[MMP] M. Marcus, V. Mizel, and Y. Pinchover, *On the best constant for Hardy's inequality in* $\mathbb{R}^n$, Trans. AMS, **350** (1998), 3237-3255.

[MaPa] F. Marchand and M. Paicu, *Remarques sur l'unicité pour le systéme de Navier-Stokes tridimensionnel*, C. R. Math. Acad. Sci. Paris, **344**:6 (2007), 363-366.

[Mar1] J. Marschall, *Some remarks on Triebel spaces*, Studia Math., **87** (1987), 79-92.

[Mar2] J. Marschall, *On the boundedness and compactness of nonregular pseudo-differential operators*, Math. Nachr., **175** (1995), 231-262.

[Mar3] J. Marschall, *Remarks on nonregular pseudo-differential operators*, Z. Anal. Anwendungen, **15** (1996), 109-148.

[MM] S. Mayboroda and M. Mitrea, *Sharp estimates for Green potentials on non-smooth domains*, Math. Res. Lett., **11**:4 (2004), 481-492.

[Maz1] V.G. Maz'ya, *Classes of domains and embedding theorems for functional spaces*, Dokl. Akad. Nauk SSSR, **133** (1960), 527-530.

[Maz2] V.G. Maz'ya, *On the theory of the n-dimensional Schrödinger operator*, Izv. Akad. Nauk SSSR, ser. Matem., **28** (1964), 1145-1172 (Russian).

[Maz3] V.G. Maz'ya, *On certain integral inequalities for functions of many variables*, Probl. Math. Anal., **3**, Leningrad Univ. (1972), 33-68. English translation: J. Soviet Math., **1** (1973), 205-234.

[Maz4] V.G. Maz'ya, *Weak solutions of the Dirichlet and Neumann problems*, Trudy Mosk. Matem. Obsh., **20** (1969), 137-172.

[Maz5] V.G. Maz'ya, *The degenerate problem with oblique derivative*, Mat. Sb., **87** (1972), 417-454.

[Maz6] V.G. Maz'ya, *The removable singularities of bounded solutions of quasi-linear elliptic equations of arbitrary order*, Zap. Nauchn. Sem. LOMI, **27** (1972), 116-130. English translation: J. Math. Sci., **3**:4 (1975), 480-492.

[Maz7] V.G. Maz'ya, *The (p, l)-capacity, embedding theorems, and the spectrum of a selfadjoint elliptic operator*, Izv. Akad. Nauk SSSR, ser. Matem., **37** (1973), 356-385.

[Maz8] V.G. Maz'ya, *On the local square summability of convolution*, Zap. Nauchn. Sem. LOMI, **73** (1977), 211-216.

[Maz9] V.G. Maz'ya, *On capacitary strong type estimates for fractional norms*, Zap. Nauchn. Sem. LOMI, **73** (1977), 161-168.

[Maz10] V.G. Maz'ya, *Multipliers in Sobolev spaces.* In the book: Application of function theory and functional analysis methods to problems of mathematical physics. Pjatoe Sovetso-Čehoslovackoe Soveščanie, 1976, Novosibirsk, 1978, 181-189.

[Maz11] V.G. Maz'ya, *On summability with respect to an arbitrary measure of functions in Sobolev-Slobodezkii spaces*, Zap. Nauch. Sem. LOMI, **92** (1979), 192-202.

[Maz12] V.G. Maz'ya, *An imbedding theorem and multipliers in pairs of Sobolev spaces*, Trudy Tbilis. Mat. Inst., **66** (1980), 59-69.

[Maz13] V.G. Maz'ya, *The integral equations of potential theory in domains with piecewise smooth boundary*, Usp. Mat. Nauk, **36**;4 (1981), 229-230.

[Maz14] V.G. Maz'ya, *Boundary integral equations of elasticity in domains with piecewise smooth boundaries*, Equadiff 6, Proc. Int. Conf., Brno/Czech., Lect. Notes Math., **1192**, (1985), 235-242.

[Maz15] V.G. Maz'ya, *Sobolev Spaces*, Springer, 1985.

[Maz16] V.G. Maz'ya, *Potential theory for the Lamé equations in domains with piecewise smooth boundary*, In: Proc. All-Union Symp., Tbilisi, April 21-23 (1982), Metsniereba: Tbilisi, 1986, 123-129. (Russian)

[Maz17] V.G. Maz'ya, *Boundary integral equations*, Encyclopaedia of Mathematical Sciences, **27**, Springer, 1991, 127-233.

[Maz18] V.G. Maz'ya, *Conductor and capacitary inequalities for functions on topological spaces and their applications*, J. Funct. Anal., **224** (2005), 408-430.

[MH1] V.G. Maz'ya and V. Havin, *Nonlinear analogue of Newton potential and metric properties of (p, l)-capacity*, Dokl. Akad. Nauk SSSR, **194**:4 (1970), 770-773.

[MH2] V.G. Maz'ya and V. Havin, *Nonlinear potential theory*, Usp. Mat. Nauk, **27**:6 (1972), 67-138.

[MN] V.G. Maz'ya and Y. Netrusov, *Some counterexamples for the theory of Sobolev spaces on bad domains*, Potential Analysis, **4** (1995), 47-65.

[MP] V.G. Maz'ya and S.P. Preobrazhenski, *Estimates for capacities and traces of potentials*, Internat. J. Math. Math. Sci., **7**:1 (1984), 41-63.

[MSh1] V.G. Maz'ya and T.O. Shaposhnikova, *Multipliers in function spaces with fractional derivatives*, Dokl. Akad. Nauk SSSR, **244**:5 (1979), 1065-1067.

[MSh2] V.G. Maz'ya and T.O. Shaposhnikova, *Multipliers in Sobolev spaces*, Vestnik Leningrad. Univ. Mat. Mekh. Astr., no. 2 (1979), 33-40.

[MSh3] V.G. Maz'ya and T.O. Shaposhnikova, *On traces and extensions of multipliers in the space W_p^l*, Usp. Mat. Nauk, **34**:2 (1979), 205-206.

[MSh4] V.G. Maz'ya and T.O. Shaposhnikova, *Multipliers in spaces of differentiable functions*, Trudy Sem. S.L. Soboleva, Novosibirsk, no. 1 (1979), 37-90.

[MSh5] V.G. Maz'ya and T.O. Shaposhnikova, *On conditions for the boundary in the L_p-theory of elliptic boundary value problems*, Dokl. Akad. Nauk SSSR, **251**:5 (1980), 1055-1059.

[MSh6] V.G. Maz'ya and T.O. Shaposhnikova, *Multipliers of Sobolev spaces in a domain*, Math. Nachr., **99** (1980), 165-183.

[MSh7] V.G. Maz'ya and T.O. Shaposhnikova, *A coercive estimate for solutions of elliptic equations in spaces of multipliers*, Vestnik Leningrad. Univ. ser. Mat. Mekh. Astr., no. 1 (1980), 41-51.

[MSh8] V.G. Maz'ya and T.O. Shaposhnikova, *Theory of multipliers in spaces of differentiable functions and their applications*, Theory of cubature formulas and numerical mathematics (Proc. Conf. , Novosibirsk, 1978), Nauka, Novosibirsk, 1980, 225-233.

[MSh9] V.G. Maz'ya and T.O. Shaposhnikova, *Multipliers in spaces of Bessel potentials*, Math. Nachr., **99** (1980), 363-379.

[MSh10] V.G. Maz'ya and T.O. Shaposhnikova, *On the regularity of the boundary in L_p-theory of elliptic boundary value problems*, Part I: Trudy Sem. S.L. Soboleva, Novosibirsk, no. 2 (1980), 39-56; Part II: Trudy Sem. S.L. Soboleva, Novosibirsk, no. 1 (1981), 57-102.

[MSh11] V.G. Maz'ya and T.O. Shaposhnikova, *Multipliers in pairs of spaces of differentiable functions*, Trudy Moskov. Mat. Obsh., **43** (1981), 37-80.

[MSh12] V.G. Maz'ya and T.O. Shaposhnikova, *Multipliers on the space $\mathring{W}_p^m$ and their applications*, Vestnik Leningrad. Univ., ser. Mat. Mekh. Astr. no. 1 (1981), 42-47.

[MSh13] V.G. Maz'ya and T.O. Shaposhnikova, *Sufficient conditions for belonging to classes of multipliers*, Math. Nachr., **100** (1981), 151-162.

[MSh14] V.G. Maz'ya and T.O. Shaposhnikova, *Change of variables as an operator on a pair of Sobolev spaces*, Vestnik Leningrad. Univ., ser. Mat. Mekh. Astr., no. 1 (1982), 43-48.

[MSh15] V.G. Maz'ya and T.O. Shaposhnikova, *Theory of multipliers in spaces of differentiable functions*, Uspekhi Mat. Nauk, **38**:3 (1983), 23-86.

[MSh16] V.G. Maz'ya and T.O. Shaposhnikova, *Theory of Multipliers in Spaces of Differentiable Functions*, Monographs and Studies in Mathematics, **23**, Pitman, Boston–London, 1985.

[MSh17] V.G. Maz'ya and T.O. Shaposhnikova, *On pointwise interpolation inequalities for derivatives*, Math. Bohemica, **124**:2-3 (1999), 131-148.

[MSh18] V.G. Maz'ya and T.O. Shaposhnikova, *Maximal algebra of multipliers between fractional Sobolev spaces*, Proceedings of Analysis and Geometry, S.K. Vodop'yanov (Ed.), Sobolev Institute Press, Novosibirsk, 2000, pp. 387-400.

[MSh19] V.G. Maz'ya and T.O. Shaposhnikova, *Pointwise interpolation inequalities for Riesz and Bessel potentials*, Analytical and Computational Methods in Scattering and Applied mathematics, Chapman and Hall, London, 2000, pp. 217-229.

[MSh20] V.G. Maz'ya and T.O. Shaposhnikova, *Maximal Banach algebra of multipliers between Bessel potential spaces*, Problems and Methods in Mathematical Physics, The Siegfried Prössdorf Memorial Volume, J.

Elschner, I. Gohberg, B. Silbermann (Eds.), Operator Theory: Advances and Application, Vol. 121, Birkhäuser, 2001, pp. 352-365.

[MSh21] V.G. Maz'ya and T.O. Shaposhnikova, *Characterization of multipliers in pairs of Besov spaces*, Operator Theory. Advances and Applications, Vol. 147 (2004), 365-386.

[MSh22] V.G. Maz'ya and T.O. Shaposhnikova, *Traces of multipliers in pairs of weighted Sobolev spaces*, J. Function Spaces Appl., **3** (2005), 91-115.

[MSh23] V.G. Maz'ya and T.O. Shaposhnikova, *Higher regularity in the classical layer potential theory for Lipschitz domains*, Indiana Univ. Math. J., **54**:1 (2005), 99-142.

[MV1] V.G. Maz'ya and I.E. Verbitsky, *Capacitary estimates for fractional integrals, with applications to partial differential equations and Sobolev multipliers*, Arkiv för Matem., **33** (1995), 81-115.

[MV2] V.G. Maz'ya and I.E. Verbitsky, *The Schrödinger operator on the energy space: boundedness and compactness criteria*, Acta Math., **188** (2002), 263-302.

[MV3] V.G. Maz'ya and I.E. Verbitsky, *The form boundedness criterion for the relativistic Schrödinger operator*, Ann. Inst. Fourier (Grenoble), **54** (2004), 317-339.

[MV4] V.G. Maz'ya and I.E. Verbitsky, *Form boundedness of the general second order differential operator*, Comm. Pure Appl. Math., **59**:9 (2006), 1286-1329.

[Me] N.G. Meyers, *A theory of capacities for potentials of functions in Lebesgue classes*, Math. Scand., **26** (1970), 255-292.

[MiP] S.G. Mikhlin and S. Prössdorf, *Singuläre Integraloperatoren*, Berlin, Akademie-Verlag, 1980.

[Mir] C. Miranda, *Partial Differential Equations of Elliptic Type*, Springer, 1970.

[MT1] M. Mitrea and M. Taylor, *Boundary layer methods for Lipschitz domains in Riemannian manifolds*, J. Funct. Anal., **163** (1999), 181-251.

[MT2] M. Mitrea and M. Taylor, *Potential theory on Lipschitz domains in Riemannian manifolds: L^p, Hardy, and Hölder space results*, Comm. Anal. Geom., **9** (2001), 369-421.

[MT3] M. Mitrea and M. Taylor, *Potential theory on Lipschitz domains in Riemannian manifolds: Sobolev-Besov space results and the Poisson problem*, J. Funct. Anal., **176** (2000), 1-79.

[MT4] M. Mitrea and M. Taylor, *Potential theory on Lipschitz domains in Riemannian manifolds: Hölder continuous metric tensors*, Comm. PDE, **25** (2000), 1487-1536.

[MT5] M. Mitrea and M. Taylor, *Potential theory on Lipschitz domains in Riemannian manifolds: the case of Dini metric tensors*, TAMS, **355**:5 (2002), 1961-1985.

[Mi] A. Miyachi, *Multiplication and factorization of functions in Sobolev spaces and in C_p^α spaces on general domains*, Math. Nachr., **176** (1995), 209-241.

[Mo] A.P. Morse, *The behavior of a function on its critical set*, Ann. Math., **40** (1939), 62-70.

[Na1] E. Nakai, *Pointwise multipliers for functions of weighted bounded mean oscillation*, Studia Math., **105** (1993), 105-119.

[Na2] E. Nakai, *Pointwise multipliers on weighted BMO spaces*, Studia Math., **125**:1 (1997), 35-56.

[NY1] E. Nakai and K. Yabuta, *Pointwise multipliers for functions of bounded mean oscillation*, J. Math. Soc. Japan, **37** (1985), 207-218.

[NY2] E. Nakai and K. Yabuta, *Pointwise multipliers for functions of weighted bounded mean oscillation on spaces of homogeneous type*, Math. Japon., **46**:1 (1997), 15-28.

[Ne] J. Nečas, *Les Méthodes Directes en Théorie des Équations Elliptiques*, Academia, Prague, 1967.

[Net] Yu. Netrusov, *Theorems on traces and multipliers for functions in Lizorkin-Triebel spaces*, Zap. Nauchn. Sem. St.-Petersburg. Otdel. Mat. Inst. Steklov. (POMI), **200**:24 (1992), 132-138. English translation: J. Math. Sci. **77**:3 (1995), 3221-3224.

[Nik] O. Nykodim, *Sur une classe de fonctions considérées dans le problème de Dirichlet*, Fundam. Mat., **21** (1933), 129-150.

[Nir] L. Nirenberg, *On elliptic partial differential equations: Lecture 2*, Ann. Sc. Norm. Sup. Pisa, Ser. 3, **13** (1959), 115-162.

[Pa] R. Palais, *Seminar on the Atiyah-Singer Index Theorem*, Princeton University Press, Princeton, 1965.

[Pe1] J. Peetre, *On the differentiability of the solutions of quasilinear partial differential equations*, Trans. Amer. Math. Soc., **104**:3 (1962), 476-482.

[Pe2] J. Peetre, *New Thoughts on Besov Spaces*, Duke Univ. Math. Ser., Durham, 1976.

[Poh1] S.I. Pohozhaev, *On eigenfunctions of the equation $\Delta u + \lambda f(u) = 0$*, Dokl. Akad. Nauk SSSR, **165**:1 (1965), 36-39.

[Poh2] S.I. Pohozhaev, *On higher order quasi-linear elliptic equations*, Diff. Uravneniya, **17**:1 (1981), 115-128.

[Pol1] J.C. Polking, *A Leibniz formula for some differential operators of fractional order*, Indiana Univ. Math. J., **27**:11 (1972), 1019-1029.

[Pol2] J.C. Polking, *Approximation in L^p by solutions of elliptic differential equations*, Amer. Math. J., **94** (1972), 1231-1244.

[RS1] M. Reed and B. Simon, *Methods of Modern Mathematical Physics. I: Functional Analysis*, Academic Press, New York–London, 1980.

[Re] Yu.G. Reshetnyak, *Spatial mappings with bounded distortion*, Sib. Mat. Ž., **8**:3 (1967), 629-658.

[Ru] T. Runst, *Mapping properties of non-linear operators in spaces of Triebel-Lizorkin and Besov type*, Anal. Math., **12** (1986), 313-346.

[RS] T. Runst and W. Sickel, *Sobolev Spaces of Fractional Order, Nemytskij Operators, and Nonlinear Partial Differential Equations*, Walter de Gruyter, Berlin-New York, 1996.

[RY] T. Runst and A. Youssfi, *The Jacobian-determinant equation on Besov and Triebel-Lizorkin spaces*, Nonlinear World, **4** (1997), 267-282.

[SW] E.T. Sawyer and R.L. Wheeden, *Weighted norm inequalities for fractional integrals on Euclidean and homogeneous spaces*, Amer. J. Math., **114** (1992), 813-874.

[Sch] M. Schechter, *Hamiltonians for singular potentials*, Indiana Univ. Math. J., **22** (1972), 483-503.

[Se] R.T. Seeley, *Complex powers of an elliptic operator*, Proc. Symp. AMS, Jan. 1967, Boston, 1967, 288-307.

[Sha] E. Shamir, *Une propriété des espaces $H^{s,p}$*, C.R. Acad. Sci. Paris, Ser. A-B, **255** (1962), A448-A449.

[Sh1] T. Shaposhnikova, *Equivalent norms in spaces with fractional or functional smoothness*, Sibir. Mat. Ž., **21** (1980), 184-196.

[Sh2] T. Shaposhnikova, *On the spectrum of multipliers in Bessel potential spaces*. Časopis Pěst. Mat., **110**:2 (1985), 197-206.

[Sh3] T. Shaposhnikova, *Bounded solutions of linear elliptic equations as multipliers in spaces of differentiable functions*, Zapiski Nauchn. Semin. LOMI, **149** (1986), 165-176.

[Sh4] T. Shaposhnikova, *An implicit mapping theorem for multipliers in spaces of Bessel potentials*, Izv. Akad. Nauk Azerbaĭdzhan. SSR Ser. Fiz.-Tekhn. Mat. Nauk, **8**:1 (1987), 14-18.

[Sh5] T. Shaposhnikova, *The superposition operator in classes of multipliers of S. L. Sobolev spaces*, Seminar Analysis (Berlin, 1986/87), 181-190, Akad. Wiss. DDR, Berlin, 1987.

[Sh6] T. Shaposhnikova, *Solvability of quasilinear elliptic equations in spaces of multipliers*, Izv. Vissh. Uchebn. Zaved. Math., no. 8 (1987), 74-81.

[Sh7] T. Shaposhnikova, *Applications of multipliers in S. L. Sobolev spaces to L_p-coercivity of the Neumann problem*, Dokl. Akad. Nauk SSSR **305**:4 (1989), 786-789; translation in Soviet Math. Dokl. **39**:2 (1989), 344-347.

[Sh8] T. Shaposhnikova, *Multipliers in the space of Bessel potentials as traces of multipliers in weighted classes*, Trudy Tbiliss. Mat. Inst. Razmadze Akad. Nauk Gruzin. SSR, **88** (1989), 59-63.

[Sh9] T. Shaposhnikova, *Traces of multipliers in the space of Bessel potentials*, Mat. Zametki, **46**:3 (1989), 100-109. English translation: Math. Notes, **46**:3-4 (1990), 743-749.

[Sh10] T. Shaposhnikova, *Applications of multipliers to the problem of coercivity in W_p^l of the Neumann problem.*, Translated in J. Soviet Math., **64**:6 (1993), 1381-1388. Probl. Mat. Anal., 11, Nonlinear equations and variational inequalities. Linear operators and spectral theory (Russian), 237-248, Leningrad. Univ., Leningrad, 1990.

[Sh11] T. Shaposhnikova, *On continuity of singular integral operators in Sobolev spaces*, Math. Scand., **76** (1995), 85-97.

[Sh12] T. Shaposhnikova, *Sobolev multipliers in the theory of integral convolution operators*, Mathematical aspects of boundary element methods (Palaiseau, 1998), 285-295, Chapman and Hall/CRC Res. Notes Math., **414**, Chapman and Hall/CRC, Boca Raton, FL, 2000.

[Sh13] T. Shaposhnikova, *Sobolev multipliers in the L_p theory of boundary integral equations of elasticity on non-smooth surfaces*, Problemi Attuali dell' Analisi e della Fisica Matematica, Gaetano Fichera memorial volume, Aracne, Rome, 2000, 161-166.

[Sh14] T. Shaposhnikova, *Description of pointwise multipliers in pairs of Besov spaces $B_1^k(\mathbb{R}^n)$*, Z. Anal. Anwend., **28**:1 (2009).

[Sh] M.A. Shubin, *Pseudodifferential Operators and Spectral Theory,* Second edition, Springer, 2001.

[Sic1] W. Sickel, *On pointwise multipliers in Besov-Triebel-Lizorkin spaces*, Seminar Analysis of the Karl-Weierstrass-Institute 1985/1986, Teubner-Texte Math. Vol. 96, Teubner, Leipzig, 1987.

[Sic2] W. Sickel, *Pointwise multiplication in Triebel-Lizorkin spaces*, Forum Math., **5** (1993), 73-91.

[Sic3] W. Sickel, *On pointwise multipliers for $F^s_{p,q}(\mathbb{R}^n)$ in case $\sigma_{p,q} < s < n/p$*, Ann. Mat. Pura Appl. **76** (1999), 209-250.

[SS] W. Sickel and I. Smirnow, *Localization properties of Besov spaces and its associated multiplier spaces*, Jenaer Schriften Math/Inf 21/99, Jena, 1999.

[ST] W. Sickel and H. Triebel, *Hölder inequalities and sharp embeddings in function spaces of $B^s_{p,q}$ and $F^s_{p,q}$ type*, J. Anal. Appl., **14**:1 (1995), 105-140.

[SY] W. Sickel and A. Youssfi, *The characterization of the regularity of the Jacobian determinant in the framework of potential spaces*, J. London Math. Soc., **59**:1 (1999), 287-310.

[Sj] T. Sjödin, *Capacities of compact sets in linear subspaces of $\mathbb{R}^n$*, Pacif. J. of Math., **78**:1 (1978), 261-266.

[Sob] S.L. Sobolev, *Some Applications of Functional Analysis to Mathematical Physics*, Translations of Mathematical Monographs, 90. AMS, Providence, RI, 1991.

[Ste1] D.A. Stegenga, *Bounded Toeplitz operators on H^1 and applications of duality between H^1 and the functions of bounded mean oscillations*, Amer. J. Math., **98** (1976), 573-589.

[Ste2] D.A. Stegenga, *Multipliers on the Dirichlet space*, Illinois J. of Math., **24** (1980), 113-139.

[St1] E.M. Stein, *The characterization of functions arising as potentials*, Bull. AMS, **67** (1961), 102-104.

[St2] E.M. Stein, *Singular Integrals and Differentiability properties of Functions*, Princeton University Press, Princeton, 1970.

[St3] E.M. Stein, *Harmonic Analysis: Real-Variable Methods, Orthogonality, and Oscillatory Integrals,* Princeton University Press, Princeton , New Jersey, 1983.

[Str] R.S. Strichartz, *Multipliers on fractional Sobolev spaces*, J. Math. and Mech., **16**:9 (1967), 1031-1060.

[Tr1] H. Triebel, *Multiplication properties of the spaces $B^s_{p,q}$ and $F^s_{p,q}$. Quasi-Banach algebras of functions*, Ann. Mat. Pura Appl., **113**:4 (1997), 33-42.

[Tr2] H. Triebel, *Multiplication properties of Besov spaces*, Ann. Mat. Pura Appl., **114**:4 (1997), 87-102.

[Tr3] H. Triebel, *Interpolation Theory. Function Spaces. Differential Operators*, Berlin, VEB Deutscher Verlag der Wissenschaften, 1978.

[Tr4] H. Triebel, *Theory of Function Spaces. II*, Monographs in Mathematics, **84**, Birkhäuser, 1992.

[Tru] N.S. Trudinger, *On imbeddings into Orlicz spaces and some applications*, J. Math. Mech., **17** (1967), 473-483.

[Yu] V.I.Yudovich, *On certain estimates connected with integral operators and solutions of elliptic equations*, Dokl. Akad. Nauk SSSR, **138**:4 (1961), 805-808.

[Usp] S.V. Uspenskii, *Imbedding theorems for classes with weights*, Tr. Mat. Inst. Steklova, **60** (1961), 282-303 (Russian), English translation: AMS Transl., **87** (1970), 121-145.

[Va] V. Valent, *A property of multiplication in Sobolev spaces. Some applications*, Rend. Sem. Mat. Univ. Padova, **74** (1985), 63-73.

[Ver1] I.E. Verbitsky, *Imbedding and multiplier theorems for discrete Littlewood–Paley spaces*, Pacific J. Math., **176** (1996), 529-556.

[Ver2] I.E. Verbitsky, *Superlinear equations, potential theory, and weighted norm inequalities*, Nonlinear Analysis, Function Spaces and Applications, Vol. 6 (Prague, 1998), Acad. Sci. Czech Repub., Prague, 1999, 223-269.

[Ver3] I.E. Verbitsky, *Nonlinear potentials and trace inequalities*, The Maz'ya Anniversary Collection, Eds. J. Rossmann, P. Takáč, G. Wildenhain, Operator Theory: Advances and Applications, Vol. 110, Birkhäuser, 1999, 323-343.

[Verc] G. Verchota, *Layer potentials and regularity for the Dirichlet problem for Laplace's equation in Lipschitz domains*, J. Funct. Anal., **59**:3 (1984), 572-611.

[VG] S.K. Vodop'yanov and V.M. Gol'dshtein, *Quasi-conformal mappings and spaces of functions with the first generalized derivatives*, Sib. Mat. Z., **16**:3 (1976), 515-531.

[VGR] S.K. Vodop'yanov, V.M. Gol'dshtein, and Yu.G. Reshetnyak, *The geometric properties of functions with generalized first derivatives*, Uspehi Matem. Nauk, **34**:1 (1979), 17-65.

[VP] L.R. Volevich and B.P. Paneyah, *Some spaces of generalized functions and embedding theorems*, Usp. Mat. Nauk, **20** (1965), 3-74.

[Wa] S.E. Warschawski, *On conformal mapping of infinite strips*, Trans. AMS, **51** (1942), 280-335.

[Wl] J. Wloka, *Partial Differential Equations*, Cambridge Univ. Press, 1987.

[Wu] Z. Wu, *Strong type estimate and Carleson measures for Lipschitz spaces*, Proc. Amer. Math. Soc., **127** (1991), 3243-3249.

[Ya] K. Yabuta, *Pointwise multipliers of weighted BMO spaces*, Proc. AMS, **117** (1993), 737-744.

[Yam] M. Yamazaki, *A quasi-homogeneous version of paradifferential operators I: Boundedness on spaces of Besov type*, J. Fac. Sci. Univ. Tokyo Sect. IA Math. **33** (1986), 131-174. *A quasi-homogeneous version of paradifferential operators II: A symbolic calculus. Ibidem* **33** (1986), 311-345.

[Ye] D. Ye, *Prescribing the Jacobian determinant in Sobolev spaces*, Ann. Inst. Henri Poincaré, **11** (1994), 275-296.

[Yo] A. Youssfi, *Commutators on Besov spaces and factorization of the paraproduct*, Bull. Sci. Math., **119** (1995), 157-186.

[Zo] J.L. Zolesio, *Multiplication dans les espaces de Besov*, Proc. Roy. Soc. Edinburgh, **78**:1-2 (1977), 113-117.

[Zy] A. Zygmund, *Trigonometric Series*, Cambridge, 1959.

List of Symbols

Classes of boundaries:

Operators:

Set functions:

Spaces:

Author and Subject Index

Grundlehren der mathematischen Wissenschaften

A Series of Comprehensive Studies in Mathematics

A Selection

248. Suzuki: Group Theory II
249. Chung: Lectures from Markov Processes to Brownian Motion
250. Arnold: Geometrical Methods in the Theory of Ordinary Differential Equations
251. Chow/Hale: Methods of Bifurcation Theory
252. Aubin: Nonlinear Analysis on Manifolds. Monge-Ampère Equations
253. Dwork: Lectures on ρ-adic Differential Equations
254. Freitag: Siegelsche Modulfunktionen
255. Lang: Complex Multiplication
256. Hörmander: The Analysis of Linear Partial Differential Operators I
257. Hörmander: The Analysis of Linear Partial Differential Operators II
258. Smoller: Shock Waves and Reaction-Diffusion Equations
259. Duren: Univalent Functions
260. Freidlin/Wentzell: Random Perturbations of Dynamical Systems
261. Bosch/Güntzer/Remmert: Non Archimedian Analysis – A System Approach to Rigid Analytic Geometry
262. Doob: Classical Potential Theory and Its Probabilistic Counterpart
263. Krasnosel'skiĭ/Zabreĭko: Geometrical Methods of Nonlinear Analysis
264. Aubin/Cellina: Differential Inclusions
265. Grauert/Remmert: Coherent Analytic Sheaves
266. de Rham: Differentiable Manifolds
267. Arbarello/Cornalba/Griffiths/Harris: Geometry of Algebraic Curves, Vol. I
268. Arbarello/Cornalba/Griffiths/Harris: Geometry of Algebraic Curves, Vol. II
269. Schapira: Microdifferential Systems in the Complex Domain
270. Scharlau: Quadratic and Hermitian Forms
271. Ellis: Entropy, Large Deviations, and Statistical Mechanics
272. Elliott: Arithmetic Functions and Integer Products
273. Nikol'skiĭ: Treatise on the shift Operator
274. Hörmander: The Analysis of Linear Partial Differential Operators III
275. Hörmander: The Analysis of Linear Partial Differential Operators IV
276. Liggett: Interacting Particle Systems
277. Fulton/Lang: Riemann-Roch Algebra
278. Barr/Wells: Toposes, Triples and Theories
279. Bishop/Bridges: Constructive Analysis
280. Neukirch: Class Field Theory
281. Chandrasekharan: Elliptic Functions
282. Lelong/Gruman: Entire Functions of Several Complex Variables
283. Kodaira: Complex Manifolds and Deformation of Complex Structures
284. Finn: Equilibrium Capillary Surfaces
285. Burago/Zalgaller: Geometric Inequalities
286. Andrianaov: Quadratic Forms and Hecke Operators
287. Maskit: Kleinian Groups
288. Jacod/Shiryaev: Limit Theorems for Stochastic Processes
289. Manin: Gauge Field Theory and Complex Geometry
290. Conway/Sloane: Sphere Packings, Lattices and Groups

291. Hahn/O'Meara: The Classical Groups and K-Theory
292. Kashiwara/Schapira: Sheaves on Manifolds
293. Revuz/Yor: Continuous Martingales and Brownian Motion
294. Knus: Quadratic and Hermitian Forms over Rings
295. Dierkes/Hildebrandt/Küster/Wohlrab: Minimal Surfaces I
296. Dierkes/Hildebrandt/Küster/Wohlrab: Minimal Surfaces II
297. Pastur/Figotin: Spectra of Random and Almost-Periodic Operators
298. Berline/Getzler/Vergne: Heat Kernels and Dirac Operators
299. Pommerenke: Boundary Behaviour of Conformal Maps
300. Orlik/Terao: Arrangements of Hyperplanes
301. Loday: Cyclic Homology
302. Lange/Birkenhake: Complex Abelian Varieties
303. DeVore/Lorentz: Constructive Approximation
304. Lorentz/v. Golitschek/Makovoz: Construcitve Approximation. Advanced Problems
305. Hiriart-Urruty/Lemaréchal: Convex Analysis and Minimization Algorithms I. Fundamentals
306. Hiriart-Urruty/Lemaréchal: Convex Analysis and Minimization Algorithms II. Advanced Theory and Bundle Methods
307. Schwarz: Quantum Field Theory and Topology
308. Schwarz: Topology for Physicists
309. Adem/Milgram: Cohomology of Finite Groups
310. Giaquinta/Hildebrandt: Calculus of Variations I: The Lagrangian Formalism
311. Giaquinta/Hildebrandt: Calculus of Variations II: The Hamiltonian Formalism
312. Chung/Zhao: From Brownian Motion to Schrödinger's Equation
313. Malliavin: Stochastic Analysis
314. Adams/Hedberg: Function spaces and Potential Theory
315. Bürgisser/Clausen/Shokrollahi: Algebraic Complexity Theory
316. Saff/Totik: Logarithmic Potentials with External Fields
317. Rockafellar/Wets: Variational Analysis
318. Kobayashi: Hyperbolic Complex Spaces
319. Bridson/Haefliger: Metric Spaces of Non-Positive Curvature
320. Kipnis/Landim: Scaling Limits of Interacting Particle Systems
321. Grimmett: Percolation
322. Neukirch: Algebraic Number Theory
323. Neukirch/Schmidt/Wingberg: Cohomology of Number Fields
324. Liggett: Stochastic Interacting Systems: Contact, Voter and Exclusion Processes
325. Dafermos: Hyperbolic Conservation Laws in Continuum Physics
326. Waldschmidt: Diophantine Approximation on Linear Algebraic Groups
327. Martinet: Perfect Lattices in Euclidean Spaces
328. Van der Put/Singer: Galois Theory of Linear Differential Equations
329. Korevaar: Tauberian Theory. A Century of Developments
330. Mordukhovich: Variational Analysis and Generalized Differentiation I: Basic Theory
331. Mordukhovich: Variational Analysis and Generalized Differentiation II: Applications
332. Kashiwara/Schapira: Categories and Sheaves. An Introduction to Ind-Objects and Derived Categories
333. Grimmett: The Random-Cluster Model
334. Sernesi: Deformations of Algebraic Schemes
335. Bushnell/Henniart: The Local Langlands Conjecture for GL(2)
336. Gruber: Convex and Discrete Geometry
337. Maz'ya/Shaposhnikova: Theory of Sobolev Multipliers. With Applications to Differential and Integral Operators
338. Villani: Optimal Transport: Old and New

Printing: Krips bv, Meppel, The Netherlands
Binding: Stürtz, Würzburg, Germany